DATE DUE

DEMCO 38-296

HANDBOOK OF FLUID DYNAMICS AND FLUID MACHINERY

HANDBOOK OF FLUID DYNAMICS AND FLUID MACHINERY

VOLUME II: EXPERIMENTAL AND COMPUTATIONAL FLUID DYNAMICS

Edited by

Joseph A. Schetz and Allen E. Fuhs

A WILEY-INTERSCIENCE PUBLICATION

JOHN WILEY & SONS, INC.

New York • Chichester • Brisbane • Toronto • Singapore

Library of Congress Cataloging in Publication Data:
Handbook of fluid dynamics and fluid machinery/editors, Joseph A.
   Schetz and Allen E. Fuhs
      p.   cm.
   Includes index.
   ISBN: 0-471-12597-0 (Volume II)
   ISBN: 0-471-87352-7 (set)
   1. Fluid mechanics. 2. Hydraulic machinery. I. Schetz, Joseph E.
II. Fuhs, Allen E.
   TA357.H286   1996
   620.1'06—dc20   95-5671
Printed in the United States of America

10 9 8 7 6 5 4 3 2 1

EDITORIAL REVIEW BOARD

LIST OF CONTRIBUTORS

DANIEL F. ANCONA,   U.S. Department of Energy, Washington, D.C.

JOHN D. ANDERSON, JR.,   University of Maryland, College Park, MD

W. KYLE ANDERSON,   NASA Langley Research Center, Hampton, VA

TAKAKAGE ARAI,   Muroran Institute of Technology, Muroran, Japan

GEORGE D. ASHTON,   U.S. Army Cold Regions Research and Engineering Laboratory, Hanover, NH

FRITZ H. BARK,   The Royal Institute of Technology, Stockholm, Sweden

ALAN BEERBOWER (Deceased),   University of California at San Diego, La Jolla, CA

MAREK BEHR,   University of Minnesota, Minneapolis, MN

R. BYRON BIRD,   University of Wisconsin–Madison, Madison, WI

RODNEY D. W. BOWERSOX,   Air Force Institute of Technology, Dayton, OH

BOBBIE CARR,   Naval Postgraduate School, Monterey, CA

CAMPBELL D. CARTER,   Systems Research Laboratories, Dayton, OH

MICHAEL V. CASEY,   Sulzer Innotec/Sulzer Hydro, Winterthur/Zurich, Switzerland

JACK E. CERMAK,   Colorado State University, Fort Collins, CO

CHAU-LYAN CHANG,   High Technology Corporation, Hampton, VA

HERBERT S. CHENG,   Northwestern University, Evanston, IL

PASQUALE CINNELLA,   Mississippi State University, Starkville, MS

FAYETTE S. COLLIER, JR.,   NASA Langley Research Center, Hampton, VA

J. ROBERT COOKE,   Cornell University, Ithaca, NY

MALCOLM J. CROCKER,   Auburn University, Auburn, AL

CLAYTON T. CROWE,   Washington State University, Pullman, WA

NICHOLAS A. CUMPSTY,   University of Cambridge, Cambridge, UK

WILLIAM G. DAY, JR.,   David Taylor Model Basin, Carderock, MD

REINER DECHER,   University of Washington, Seattle, WA

JAMES D. DE LAURIER,   University of Toronto, Toronto, Canada

Note: All Contributors to Volumes I, II, and III are mentioned in this Section of the handbook.

SERGE T. DEMETRIADES,   STD Research Incorporated, Arcadia, CA

AYODEJI O. DEMUREN,   Old Dominion University, Norfolk, VA

THOMAS E. DILLER,   Virginia Polytechnic Institute and State University, Blacksburg, VA

PANAYIOTIS DIPLAS,   Virginia Polytechnic Institute and State University, Blacksburg, VA

LOUIS V. DIVONE,   U.S. Department of Energy, Washington, D.C.

FRANKLIN T. DODGE,   Southwest Research Institute, San Antonio, TX

EARL H. DOWELL,   Duke University, Durham, NC

DONALD J. DUSA (Retired),   General Electric Aircraft Engines, Evandale, OH

HARRY A. DWYER,   University of California, Davis, CA

PETER R. EISEMAN,   Program Development Corporation, White Plains, NY

GEORGE EMANUEL,   University of Oklahoma, Norman, OK

JOHN F. FOSS,   Michigan State University, East Lansing, MI

ALLEN E. FUHS (Emeritus),   U.S. Naval Postgraduate School, Monterey, CA

SUSAN E. FUHS,   Rand Corporation, Santa Monica, CA

WALTER S. GEARHART (Retired),   Pennsylvania State University, State College, PA

WILLIAM K. GEORGE,   State University of New York at Buffalo, Buffalo, NY

ALFRED GESSOW,   University of Maryland, College Park, MD

K. N. GHIA,   University of Cincinnati, Cincinnati, OH

U. GHIA,   University of Cincinnati, Cincinnati, OH

CARL H. GIBSON,   University of California—San Diego, La Jolla, CA

RICHARD J. GOLDSTEIN,   University of Minnesota, Minneapolis, MI

SANFORD GORDON,   Sanford Gordon & Associates, Cleveland, OH

ROBERT A. GREENKORN,   Purdue University, West Lafayette, IN

EDWARD M. GREITZER,   Massachusetts Institute of Technology, Cambridge, MA

BERNARD GROSSMAN,   Virginia Polytechnic Institute and State University, Blacksburg, VA

FREDRICK G. HAMMITT (Deceased),   University of Michigan, Ann Arbor, MI

EVERETT J. HARDGRAVE, JR. (Retired),   Applied Physics Laboratory, The Johns Hopkins University, Laurel, MD

A. GEORGE HAVENER,   U.S. Air Force Academy, Colorado Springs, CO

STEPHEN HEISTER,   Purdue University, West Lafayette, IN

ROBERT E. HENDERSON,   Pennsylvania State University, State College, PA

JACKSON R. HERRING, National Center for Atmospheric Research, Boulder, CO

JOHN L. HESS (Retired), Douglas Aircraft Company, Long Beach, CA

RAYMOND M. HICKS, NASA Ames Research Center, Moffett Field, CA

GUSTAVE J. HOKENSON (Deceased), Air Force Institute of Technology, Dayton, OH

TERRY L. HOLST, NASA Ames Research Center, Moffett Field, CA

DAVID P. HOULT, Massachusetts Institute of Technology, Cambridge, MA

THOMAS J. R. HUGHES, Stanford University, Stanford, CA

EVANGELOS HYTOPOULOS, Automated Analysis Corp., Ann Arbor, MI

TAKAO INAMURA, Tohoku University, Sendai, Japan

DAVID JAPIKSE, Concepts ETI, Incorporated, Wilder, VT

SAM P. JONES, TCOM, L.P., Columbia, MD

HELMUT KECK, Sulzer Innotec/Sulzer Hydro, Winterthur/Zurich, Switzerland

JAMES L. KEIRSEY (Retired), Applied Physics Laboratory, The Johns Hopkins University, Laurel, MD

LAWRENCE A. KENNEDY, University of Illinois at Chicago, Chicago, IL

JOHN KIM, University of California Los Angeles, Los Angeles, CA

ANJANEYULU KROTHAPALLI, Florida State University, Tallahassee, FL

PAUL KUTLER, NASA Ames Research Center, Moffett Field, CA

K. KUWAHARA, Institute of Space and Astronautical Science, Sagamihara, Japan

E. EUGENE LARRABEE (Emeritus), Massachusetts Institute of Technology, Cambridge, MA

J. GORDON LEISHMAN, University of Maryland, College Park, MD

PETER E. LILEY (Retired), Purdue University, West Lafayette, IN

RAINALD LÖHNER, The George Mason University, Fairfax, VA

LUIZ M. LOURENCO, Florida State University, Tallahassee, FL

MUJEEB R. MALIK, High Technology Corporation, Hampton, VA

JAMES F. MARCHMAN, III, Virginia Polytechnic Institute and State University, Blacksburg, VA

CRAIG SAMUEL MARTIN, Georgia Institute of Technology, Atlanta, GA

HUGH R. MARTIN, University of Waterloo, Waterloo, Ontario, Canada

C. D. MAXWELL, STD Research Incorporated, Arcadia, CA

UNMEEL B. MEHTA, NASA Ames Research Center, Moffett Field, CA

JOHN E. MINARDI, University of Dayton, Dayton, OH

ALBERT M. MOMENTHY, Boeing Commercial Airplane Company, Seattle, WA

THOMAS B. MORROW, Southwest Research Institute, San Antonio, TX

HANY MOUSTAPHA, Pratt and Whitney Canada, Montreal, Canada

S. NAKAMURA, Ohio State University, Columbus, OH

RICHARD NEERKEN (Retired), The Ralph M. Parsons Company, Pasadena, CA

WAYNE L. NEU, Virginia Polytechnic Institute and State University, Blacksburg, VA

NEIL OLIEN, National Institute of Standards and Technology, Boulder, CO

BHARATAN R. PATEL, Fluent Incorporated, Lebanon, NH

VICTOR L. PETERSON (Retired), NASA Ames Research Center, Moffett Field, CA

J. LEITH POTTER (Retired), Vanderbilt University, Nashville, TN

THOMAS H. PULLIAM, NASA Ames Research Center, Moffett Field, CA

SAAD RAGAB, Virginia Polytechnic Institute and State University, Blacksburg, VA

RICHARD H. RAND, Cornell University, Ithaca, NY

EVERETT V. RICHARDSON (Emeritus), Colorado State University, Fort Collins, CO

DONALD O. ROCKWELL, Lehigh University, Bethlehem, PA

COLIN RODGERS, Los Angeles, CA

JOHN P. ROLLINS, Clarkson University, Potsdam, NY

PHILIP G. SAFFMAN, California Institute of Technology, Pasadena, CA

MANUEL D. SALAS, NASA Langley Research Center, Hampton, VA

P. SAMPATH, Pratt and Whitney Canada, Montreal, Canada

TURGUT SARPKAYA, Naval Postgraduate School, Monterey, CA

JOSEPH A. SCHETZ, Virginia Polytechnic Institute and State University, Blacksburg, VA

LEON H. SCHINDEL, Naval Surface Warfare Center, Silver Spring, MD

JOHN E. SCHMIDT (Retired), Boeing Commercial Airplane Company, Seattle, WA

WILLIAM B. SHIPPEN (Retired), Applied Physics Laboratory, The Johns Hopkins University, Laurel, MD

TERRY W. SIMON, University of Minnesota, Minneapolis, MI

HELMUT SOCKEL, Technical University of Vienna, Vienna, Austria

GEOFFREY R. SPEDDING, University of Southern California, Los Angeles, CA

PHILIP C. STEIN, JR., Stein Seal Company, Kulpsville, PA

WILLIAM G. STELTZ (Retired), Westinghouse Electric Company, Orlando, FL

KENNETH G. STEVENS, NASA Ames Research Center, Moffett Field, CA

PAUL N. SWARZTRAUBER, National Center for Atmospheric Research, Boulder, CO

ROLAND A. SWEET, University of Colorado at Denver, Denver, CO

JULIAN SZEKELY, Massachusetts Institute of Technology, Cambridge, MA

JIMMY TAN-ATICHAT, California State University, Chico, Chico, CA

RICHARD S. TANKIN, Northwestern University, Evanston, IL

TAYFUN E. TEZDUYAR, University of Minnesota, Minneapolis, MN

JAMES L. THOMAS, NASA Langley Research Center, Hampton, VA

CHANG LIN TIEN, University of California, Berkeley, CA

EUGENE D. TRAGANZA, Naval Postgraduate School, Monterey, CA

STEVENS P. TUCKER, Naval Postgraduate School, Monterey, CA

ERNEST W. UPTON, Bloomfield Hills, MI

MICHAEL W. VOLK, Oakland, CA

JAMES WALLACE, University of Maryland, College Park, MD

CANDACE WARK, Illinois Institute of Technology, Chicago, IL

FRANK M. WHITE, University of Rhode Island, Kingston, RI

JOHN M. WIEST, Purdue University, West Lafayette, IN

JAMES C. WILLIAMS, III, Auburn University, Auburn, AL

SCOTT WOODWARD, State University of New York at Buffalo, Buffalo, NY

TERRY WRIGHT, University of Alabama at Birmingham, Birmingham, AL

GEORGE T. YATES, Consultant, Boardman, OH

H. C. YEE, NASA Ames Research Center, Moffett Field, CA

HIDEO YOSHIHARA (Retired), Boeing Company, Seattle, WA

TSUKASA YOSHINAKA, Concepts ETI, Incorporated, Wilder, VT

VIRGINIA E. YOUNG, Virginia Polytechnic Institute and State University, Blacksburg, VA

JAMES L. YOUNGHANS, General Electric Aircraft Engines, Evandale, OH

HENRY C. YUEN, TRW Space and Technology Group, Redondo Beach, CA

EDWARD E. ZUKOSKI, California Institute of Technology, Pasadena, CA

CONTENTS

15 Instrumentation for Fluid Dynamics

TERRY W. SIMON
RICHARD J. GOLDSTEIN
University of Minnesota
Minneapolis, MI

GUSTAVE J. HOKENSON (Deceased)

RODNEY D. W. BOWERSOX
Air Force Institute of Technology
Dayton, OH

LUIZ M. LOURENCO
ANJANEYULU KROTHAPALLI
Florida State University
Tallahassee, FL

JAMES F. MARCHMAN III
JOSEPH A. SCHETZ
THOMAS E. DILLER
Virginia Polytechnic Institute and State University
Blacksburg, VA

CHANG LIN TIEN
University of California
Berkeley, CA

JOHN F. FOSS
Michigan State University
East Lansing, MI

JAMES WALLACE
University of Maryland
College Park, MD

CANDACE WARK
Illinois Institute of Technology
Chicago, IL

CAMPBELL D. CARTER
Systems Research Laboratories
Dayton, OH

Handbook of Fluid Dynamics and Fluid Machinery, Edited by Joseph A. Schetz and Allen E. Fuhs
ISBN 0-471-12598-9 Copyright © 1996 John Wiley & Sons, Inc.

LAWRENCE A. KENNEDY
University of Illinois at Chicago
Chicago, IL

JIMMY TAN-ATICHAT
California State University, Chico
Chico, CA

SCOTT WOODWARD
WILLIAM K. GEORGE
State University of New York at Buffalo
Buffalo, NY

15.1 RANDOM ERROR, SYSTEMATIC ERROR, AND CALIBRATION
Terry W. Simon and Richard J. Goldstein

In any real measurement, one cannot determine exactly the true value of the quantity being measured. This inability is an inevitable consequence of uncontrolled variables influencing the system studied, the instrument, or the manner in which the instrument and system interact. Statistical techniques can be used to draw the maximum amount of information from the measurement, within a predetermined risk of drawing false conclusions—no conclusions can be drawn from experimental data with zero risk.

The difference between the value indicated by an instrument and the true value, *error*, falls in one of two general categories—*random error* and *systematic error*. Consider measuring a quantity which is known and constant. If the instrument string (sensor, reading instrument, intermediate instrument such as a filter or A/D converter, and recorder) indicates differing magnitudes of that measured quantity from reading to reading there is a *random error* (also called *precision error* or *stochastic error*). The degree to which the measured values differ from reading to reading for a suitably large number of readings indicates the magnitude of the random error or the *imprecision* of the measurement. Generally, the most appropriate value to choose as the indicated output given by the instrument string is the mean of a large number of such readings. The difference between this mean indicated value and the true

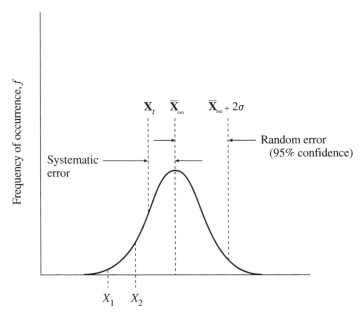

FIGURE 15.1 The frequency distribution of the indicated value of a measured quantity, X.

value is the *systematic error* (also called *accuracy error,** deterministic error*, or *bias error*). Thus, *accuracy* has to do with closeness to the truth or the degree of correctness, while *precision* indicates closeness together or reproducibility of different measurements.

Suppose that an infinite number of readings (measurements) were taken and that each can be taken with infinitesimally fine resolution of the indicated output, X. The plot of the relative frequency (number) of occurrence, f, versus the indicated value of the measured quantity, X, would ideally be similar to that of Fig. 15.1. Shown in this figure are the true value, X_t, the average value, $\overline{X}_\infty$ given as

$$\overline{X}_\infty = \lim_{N \to \infty} \left[\left\{ \sum_{n=1}^{N} X_n \right\} \Big/ N \right] \tag{15.1}$$

the systematic error, $\overline{X}_\infty - X_t$, and the random error. The shape of the curve, $f(X)$, is from a mathematical model that seems to fit many experimental observations. It is the *normal distribution* (also called the *normal curve of errors*), or *Gaussian distribution* and is expressed in functional form as

$$f = f_o \exp\left(-h^2(X - \overline{X}_\infty)^2\right) \tag{15.2}$$

*It should be noted that using accuracy error (or accuracy) synonymously with systematic or bias error is not accepted by all. Eisenhart (1963) and Murphy (1961), point out that a set of measurements with large random error (or low precision) but small bias error should not be called ''accurate.''

where $f_o(=h/\sqrt{\pi})$ is the relative frequency of occurrence for $X = \overline{X}_\infty$ and h is the modulus or index of precision. The normal distribution is symmetric about a central value, $\overline{X}_\infty$, and asymptotically approaches zero far from $\overline{X}_\infty$. This Gaussian distribution, $f(X)$, would be an appropriate model if the random error consisted of a large number of small, independent contributions, each of equal magnitude and each equally probable of being positive or negative (free of bias) about the mean. There is no proof that the normal distribution applies to experimental observations in general, but its use is widely accepted. Care should be taken, however, not to apply it in biased observational systems—e.g., filtered samples. [A discussion of skewed distribution methodology can be found in Lassahn (1983).]

Note that h is a parameter that indicates the *fullness* or width of the region below the curve or the degree to which the readings vary from the mean. A larger value of h would be associated with a smaller random error, making h a convenient indicator of *precision* but a rather awkward parameter to quantify imprecision or random error.

The *standard deviation*, σ, of a series of measurements can be calculated from

$$\sigma = \lim_{N \to \infty} \left[\frac{\sum\limits_{n=1}^{N} (X - \overline{X}_\infty)^2}{N} \right]^{1/2} \tag{15.3}$$

Assuming a Gaussian distribution, σ can be related to the *index of precision*, h, by $\sigma = 1/(h\sqrt{2})$. When the random error is large, the curve of Fig. 15.1 is full, σ is large, and h is small. The parameter σ which increases with the magnitude of the random error, is sometimes labeled the *precision index*.[*] The area under the curve of Fig. 15.1 represents all values of the measured quantity, X. The fraction of that total area occupied by the area under the curve from X_1 to X_2 represents the probability that the measured value will be between X_1 and X_2. Assuming a normal distribution, Eq. (15.2), X will be between $\overline{X}_\infty + 2\sigma$ and $\overline{X}_\infty - 2\sigma$ approximately 95 out of 100 times, or, with 95% probability (20:1 odds). Frequently, then, the magnitude of the random error is taken to be 2σ, recognizing that the measured reading will be found within this band with 95% confidence.[†] The random error need not be based on 95% confidence, but the 95% confidence interval is the most common confidence interval used. For instance, the random error could be presented as σ and the confidence interval, the percentage of the total area under the Gaussian distribution curve which is between $\overline{X}_\infty - \sigma$ and $\overline{X}_\infty + \sigma$, would be 68%. Table 15.1 shows the ranges which correspond to other confidence levels. It is important to note that a random error without the accompanying confidence level is meaningless.

In practice, the sample size is finite, and a mean value of the measurement sample, $\overline{X}$, and standard deviation of that sample, s, are used. These approximations are

$$\overline{X} = \frac{1}{N} \sum_{n-1}^{N} X_N \tag{15.4}$$

---

[*]The standard deviation, σ, might be better called the *index of imprecision*.
[†]A deviation of 2σ actually encloses about 95.5% of all the population, but this is usually taken as approximately 95% or 20:1 odds.

TABLE 15.1 Confidence Intervals and Associated Random Error Ranges

Confidence Interval	Odds	Random Error Range
0.500	2:1	$\overline{X}_\infty \pm 0.675\sigma$
0.683	3.15:1	$\overline{X}_\infty \pm 1.000\sigma$
0.800	5:1	$\overline{X}_\infty \pm 1.282\sigma$
0.900	10:1	$\overline{X}_\infty \pm 1.645\sigma$
0.950	20:1	$\overline{X}_\infty \pm 1.960\sigma$
0.990	100:1	$\overline{X}_\infty \pm 2.576\sigma$
0.997	333:1	$\overline{X}_\infty \pm 2.965\sigma$
0.999	1000:1	$\overline{X}_\infty \pm 3.290\sigma$

and

$$s = \left[\left\{ \sum_{n=1}^{N} (X_n - \overline{X})^2 \right\} \Big/ (N - 1) \right]^{1/2} \tag{15.5}$$

The denominator in Eq. (15.5) is $N - 1$, because each entry in the series is based on the difference from the sample mean, $\overline{X}$, not the total population mean, $\overline{X}_\infty$. The use of the sample mean, $\overline{X}$, in Eq. (15.5) reduces the number of degrees of freedom by one ($X_n - \overline{X}$ of a sample of one will always be zero), hence the denominator is reduced to $(N - 1)$.

Frequently, it is inconvenient to determine the mean, $\overline{X}$, and then determine the standard deviation using Eq. (15.5). An alternate form of the above equation for calculating s is

$$s = \left[\frac{\sum_{n=1}^{N} X_n^2 - \left(\sum_{n=1}^{N} X_n \right)^2 \Big/ N}{(N - 1)} \right]^{1/2} \tag{15.6}$$

This formula may be preferred where samples are being taken and processed in sequence, and it is inconvenient to store the magnitude of each sample. With Eq. (15.6) as each sample is processed, the series ΣX_n^2 and ΣX_n are updated, thus only their present value need be stored.

The t *distribution*, frequently called *Student's t*, can be used to estimate the random error versus confidence interval for small sample sizes. The random error is given as $t \cdot s$ where t, the t distribution parameter, is given in Table 15.2 based upon the sample size, N, and the confidence level. For instance, if the sample size were five, the random error based on 95% confidence would not be $\overline{X} \pm 2s$, the value for a large sample, but would be $\overline{X} \pm 2.776s$.

Suppose the population (of measured values) with a given mean $\overline{X}$ and a standard deviation, s, were divided into M samples of N readings each and the mean of each sample j, $\overline{X}_j$, were determined. The *standard deviation of the mean* (or *standard error of the mean*) defined as

$$s_{\overline{X}} = \left[\frac{\sum_{j=1}^{M} (\overline{X}_j - \overline{X})^2}{(M - 1)} \right]^{1/2} \tag{15.7}$$

TABLE 15.2 Values of the Student's t Parameter

Sample Size N	Degrees of Freedom	t Confidence Level			
		0.500	0.900	0.950	0.990
2	1	1.000	6.314	12.706	63.657
3	2	0.816	2.920	4.303	9.925
4	3	0.765	2.353	3.182	5.841
5	4	0.741	2.132	2.776	4.604
6	5	0.727	2.015	2.571	4.032
7	6	0.718	1.943	2.447	3.707
8	7	0.711	1.895	2.365	3.499
11	10	0.700	1.812	2.228	3.169
16	15	0.691	1.753	2.131	2.947
21	20	0.687	1.725	2.086	2.845
31	30	0.683	1.697	2.042	2.750
61	60	0.679	1.671	2.000	2.660
∞	∞	0.674	1.645	1.960	2.576

is

$$s_{\bar{X}} = s/\sqrt{N} \tag{15.8}$$

and the random error is $t \cdot s_{\bar{X}}$. The parameter t is taken from Table 15.2 and is consistent with the chosen confidence level and the number of samples, M. The mean of any one of the M samples, $\bar{X}_j$, is expected to be within the interval $\bar{X} \pm t \cdot s_{\bar{X}}$ with the chosen confidence level. This would indicate how representative $\bar{X}_j$ is of $\bar{\bar{X}}$.

We now have a means for evaluating random and systematic error when the true value is known. The systematic error is in the form $\bar{X}_\infty - X_t$ and the random error in terms of s, with a confidence level. When the random error is excessive, consideration should be given to replacement of the instruments which are significant contributors to this error with instruments that are more *precise* (have higher precision or smaller random error).

Every instrument experiences random error. It is often difficult to determine its magnitude, however. An appropriate method of determining random error comes from the above discussion: A number of independent readings are taken of a measured quantity known to be constant. The mean, $\bar{X}$, and standard deviation, s, are evaluated and the random error, based upon the chosen confidence level, is assigned.

Random error for an instrument can vary over a range. It could be found by taking a large number of readings for each of many representative states spanning its expected domain of operation. This is usually not practical, and an approximate random error found by pooling samples is substituted. A sample of size, n_1, and standard deviation, s_1, centered around $\bar{X}_1$ can be pooled with another sample of size n_2 and standard deviation s_2 centered around $\bar{X}_2$ (a value near $\bar{X}_1$) according to

$$s_p^2 = \frac{(n_1 - 1)s_1^2 + (n_2 - 1)s_2^2}{(n_1 - 1) + (n_2 - 1)} \tag{15.9}$$

The pooled standard deviation, s_p^2, is applied as an average for the interval from $\overline{X}_1$ to $\overline{X}_2$. Note that the individual sample variances are weighted according to each sample's number of degrees of freedom. In general, for k samples to be pooled

$$s_p^2 = \frac{\displaystyle\sum_{i=1}^{k} [(n_i - 1)s_i^2]}{\displaystyle\sum_{i=1}^{k} (n_i - 1)} \tag{15.10}$$

where n_i is the number of measurements about each central value, $\overline{X}_i$, and s_i is the standard deviation from the measurements about each central value $\overline{X}_i$. The pooled standard deviation could be applied over the range of $\overline{X}_i$'s.

The random error may have short-term and long-term components. To include the long-term component, the readings used to compute s should be taken over the full time interval over which the instrument string is in service.

Random errors are caused by a number of uncontrolled variables. Frequently, close study will show the existence of assignable causes which can be located and perhaps eliminated. As each major component of error is identified and eliminated, random error is reduced. For example, sensitivity of a meter to variations in room temperature would be perceived as random variations. If this cause of error were determined, the room temperature could be controlled and this *random* component might be eliminated, reducing the random error of the measurement. As major contributors are eliminated, only minor, background, effects remain and the variations become more random—more amenable to statistical analysis.

The second category of error, systematic error, can be subcategorized according to *illegitimate error*, *constant bias error*, or *variable bias error*. The first subcategory includes mistakes or blunders, such as transposing numbers, or equipment malfunction. Errors of this type are frequently obvious; if so, they can be immediately corrected or the measurement discarded. When such errors are not large enough to be obvious, the readings must be checked against a rejection criterion to determine if they are bad observations (*outliers*). One such criterion comes from *Thompson's tau technique* [Thompson (1935) and Beckman and Cook (1983)]. A reading, X, for which $|X - \overline{X}| > s \cdot \tau$ is an outlier and is to be discarded. The magnitude of τ, given in Table 15.3, depends on the level of significance (the fraction of the total readings that are to be rejected) and the sample size.

The second subcategory of systematic error is the constant bias error. When the source of the constant bias error is in the instrument string, it is called an *instrument error*. When it is excessive, the instruments which are the significant contributors should be removed from the string and *calibrated*. An instrument is calibrated by using it and a reference instrument to independently measure the magnitude of the quantity being determined. The reference instrument could be a basic or primary instrument, such as a manometer used to calibrate a pressure transducer or a triple-point-of-water cell used to calibrate a thermocouple, or it may be a calibrated instrument, such as a platinum resistance thermometer which has been calibrated against another standard known to be accurate. The reference instrument must have a random error which is much less than the error within which the calibration is to be made. The reference instrument is presumed to have no systematic error (after applying its own calibration). The calibration procedure will yield a curve of the necessary correction based upon a sufficient number of calibration runs, $X_c - \overline{X}$,

TABLE 15.3 Thompson's Tau[a]

Sample Size N	Level of Significance			
	0.1	0.05	0.02	0.01
3	1.3968	1.4099	1.41352	1.414039
4	1.559	1.6080	1.6974	1.7147
5	1.611	1.757	1.869	1.9175
6	1.631	1.814	1.973	2.0509
7	1.640	1.848	2.040	2.142
8	1.644	1.870	2.087	2.207
9	1.647	1.885	2.121	2.256
10	1.648	1.895	2.146	2.294
11	1.648	1.904	2.166	2.324
12	1.649	1.910	2.183	2.348
13	1.649	1.915	2.196	2.368
14	1.649	1.919	2.207	2.385
15	1.649	1.923	2.216	2.399
16	1.649	1.926	2.224	2.411
17	1.649	1.928	2.231	2.422
18	1.649	1.931	2.237	2.432
19	1.649	1.932	2.242	2.440
20	1.649	1.934	2.247	2.447
21	1.649	1.936	2.251	2.454
22	1.649	1.937	2.255	2.460
23	1.649	1.938	2.259	2.465
24	1.649	1.940	2.262	2.470
25	1.649	1.941	2.264	2.475
26	1.648	1.942	2.267	2.479
27	1.648	1.942	2.269	2.483
28	1.648	1.943	2.272	2.487
29	1.648	1.944	2.274	2.490
30	1.648	1.944	2.275	2.493
31	1.648	1.945	2.277	2.495
32	1.648	1.945	2.279	2.498
∞	1.64485	1.95996	2.32634	2.57582

[a]ANSI/ASME PTC 19.1 recommends multiplying each of these τ values by $\sqrt{(N-1)/N}$.

versus the mean indicated value, $\overline{X}$. Calibrations are based upon observed calibration data which are not precise or accurate (possess random and systematic error). The corrections, $X_c - \overline{X}$, are applied to each reading taken by each instrument in the instrument string. Of course, each instrument should be recalibrated from time to time. When the calibration correction changes significantly from calibration to calibration, the source of the *drift* should be determined. If the instrument is at fault, it should be either discarded, repaired, or recalibrated more frequently, depending on the severity of the drift. If the test procedure is at fault (e.g., poor control of the fluid temperature during hot-wire anemometer measurements), either the conditions should be improved or a means should be devised for continuous compensation of this drift, e.g., a temperature-compensated probe in the case of hot-wire anemometer measurements. Such errors due to poor control fall into the third subcategory of systematic error, variable bias error.

Calibration of each instrument of the instrument string may not correct all the systematic error of the measurement. For example, a thermocouple which has been calibrated in a separate fixture can be used free of systematic error in the calibration device environment, but, when that thermocouple is placed in its duty environment, it can suffer systematic error due, for example, to radiation from the thermocouple junction to a nearby cold wall. This type of systematic error is a *constant bias error* called a *sensing error*, which is the difference between the quantity being measured and the quantity at the sensing element. Once this error is recognized, appropriate corrections can be made. The instrument may be recalibrated under conditions more like its duty conditions, or calculated corrections might be made using an analytical model of the system, i.e., calculation of the radiation loss correction based upon estimated emissivities and estimated or measured surrounding temperatures. Note that such a calculated correction would leave residual systematic errors (they should be smaller than the corrected error) and would introduce new random errors, i.e., the random error associated with measuring the surrounding temperatures. If each instrument of the string has been calibrated and all other systematic errors have been removed from the measurement system, only random errors remain, and the measurement is *zero-centered* [Moffat (1982)].

The instrument string may be called upon to record the value of the measured quantity with only one reading, especially under transient conditions. If we want this reading to be within certain limits, we must be assured that random and systematic errors are acceptably small. For example, if the magnitude of the measured quantity is to be in the interval $\pm E$ about the true value, X_t, with a 95% confidence level, the systematic error (which remains after calibration and application of the calculated correction) combined with the random error based upon a 95% confidence level must be less than E; high precision is a necessary prerequisite to accuracy for single-sample or small-sample determinations.

Many standards recommend stating random and systematic error separately. Frequently, though, the combined result is to be given. There are two methods used to combine systematic and random errors—add the two or add their squared values then take the square root (*root-sum-square method*). The combining method which has been chosen must be specified when combined results are presented.

15.1.1   General Bibliography

Arnberg, B. T., "Practice and Procedures of Error Calculations," *Flow—Its Measurement and Control in Science and Industry*, Vol. 1, Part 3, pp. 1267–1284, Instrument Society of America, Pittsburgh, PA, 1974.

ANSI/ASME PTC 19.1, *Instruments and Apparatus, Part 1 Measurement Uncertainty*, Amer. Soc. Mech. Engr., New York, 1985.

ASTM Designation E-177-61T, American Society for Testing Materials, Philadelphia, PA, 1961.

Eisenhart, C., "Use of the Terms 'Precision' and 'Accuracy' as Applied to Measurement of a Property of a Material," *J. Research of the National Bureau of Standards-C. Engineering and Instrumentation*, Vol. 67C, No. 2, pp. 45–185, 1963.

Harris, F. K., "Measurement Errors," *Handbook of Applied Instrumentation*, Considine, D. M. (Ed.), McGraw-Hill, New York, 1964.

Holman, J. P., *Experimental Methods for Engineers*, McGraw-Hill, New York, 1978.

Moffat, R. J., "Contributions to the Theory of Single-Sample Uncertainty Analysis," *J. Fluids Engineering*, Vol. 104, pp. 250–260, 1982.

Schenck, H., *Theories of Engineering Experimentation*, Hemisphere Publishing, Washington, D.C., 1979.

15.2 METHODS OF DATA ANALYSIS
Terry W. Simon and Richard J. Goldstein

When the true value of a measured quantity is known, the systematic error can be evaluated as discussed in the previous section. In general, the true value is not known; if it were, there would be no need to make the measurement. Although the errors cannot be evaluated without knowing the true value, there are means of *estimating* the measurement error. Such an estimated error is called the *uncertainty*, and the evaluation and application of uncertainties comprises the *uncertainty analysis*. Uncertainty is an estimate of what the error would be if one were able to measure it.

The true value of a quantity, X_t, is expected to be within the range $\overline{X} \pm \delta X$ (*confidence interval*) with the stated confidence level (or odds), where $\overline{X}$ is the nominal value and δX is the uncertainty. The uncertainty is meaningless without knowing its associated confidence level.

Uncertainty analysis employs the statistical techniques introduced in Sec. 15.1. In doing so, it is assumed that elements of a large sample, X, would be distributed randomly and symmetrically about $\overline{X}$. It is also usually assumed that the shape of this distribution is Gaussian. Because the true value is unknown, systematic error cannot be directly evaluated. Two procedures have evolved for dealing with this problem.

In one procedure [Moffatt (1982)], an attempt is made to eliminate every known systematic error component. This is done, as discussed in the previous subsection, by calibration and by application of best-guess correction models. In so doing, the recognized systematic errors are reduced by calibration corrections and calculated corrections which add components to the random error. These new random components are presumed to be equally likely to be positive or negative with a distribution which is Gaussian. If all significant contributions to systematic error are processed in this fashion and reduced to corrections and random errors, the experiment is said to be *zero-centered*.

In the other procedure [Eisenhart (1968) and ISA (1980)], systematic error and random error are treated separately. From experience with the measurement process and knowledge of its sensitivity to uncontrolled factors, one must place reasonable bounds on the likely systematic error. Systematic error and random error are reported separately (with documentation of their evaluations). An overall uncertainty computed as either the sum or the root-sum-square of the estimated systematic error and the estimated random error (based upon a chosen confidence level) may also be presented.

The following discussion focuses on the random error components of the uncertainty analysis; note that this is the entire uncertainty analysis of the zero-centered system (the first procedure).

15.2.1   Propagation of Uncertainty

Typically, one is concerned with systems in which several instrument strings are each measuring different quantities. The uncertainty associated with each of these measurements must be combined to determine the uncertainty of the resultant measurement. For instance, evaluation of the uncertainty of the local skin friction coefficient, $c_f/2 = \tau_w/\rho U_\infty^2$, requires appropriately combining uncertainties associated with the measurement of the wall shear stress, τ_w, the fluid density, ρ, and the freestream velocity, U_∞. In general, the resultant, X_R, can be expressed as a function of all identifiable components, $X_R = X_R(X_1, X_2, \cdots X_N)$. Each of these component terms may be evaluated from basic measurements. When combining component uncertainties to give a resultant uncertainty, it is presumed that: 1) the component uncertainties are all based upon the same confidence level, 2) for each component value, the random fluctuations about the mean have the same distribution (the Gaussian distribution is usually assumed*), 3) each component is independent of the others, and 4) each measurement is taken independently and is a single sample. The sensitivity of the resultant of the measurement, X_R, to small changes of a particular component value, X_i, is the partial derivation $\partial X_R/\partial X_i$. The contribution of the component i uncertainty is $(\partial X_R/\partial X_i)\delta X_i$.

One method for combining these contributions that would give a maximum possible resultant uncertainty, δX_R, is to merely add their absolute values

$$\delta X_R = \sum_i \left| \frac{\partial X_R}{\partial X_i} \delta X_i \right| \tag{15.11}$$

This worst-case propagation of errors will yield a much larger confidence level (higher odds) for the resultant uncertainty than that of the component uncertainties. For this reason, the straight summation method is rarely used.

A widely-used propagation method which yields the same confidence level for the resultant uncertainty as for the component uncertainties is the root-sum-square method [Kline and McClintock (1953)].

$$\delta X_R = \left[\sum_i \left(\frac{\partial X_R}{\partial X_i} \delta X_i \right)^2 \right]^{1/2} \tag{15.12}$$

This propagation method is the recommended form except, possibly, where the penalty for failure is extreme and one may wish to use the worst-case form. Occasionally, a simple relationship can be written for the resultant in terms of the components. Then, the sensitivity coefficients can be easily evaluated. Thus for $c_f/2 = \tau_w/\rho U_\infty^2$, the uncertainty of the skin friction coefficient, $\delta(c_f/2)$, can be written

$$\delta(c_f/2) = \left[\left(\frac{1}{\rho U_\infty^2} \delta \tau_w \right)^2 + \left(\frac{\tau_w}{\rho^2 U_\infty^2} \delta \rho \right)^2 + \left(\frac{2\tau_w}{\rho U_\infty^3} \delta U_\infty \right)^2 \right]^{1/2} \tag{15.13}$$

*A discussion of skewed distribution methodology can be found in Lassahn (1983).

or

$$\frac{\delta(c_f/2)}{\overline{c_f}/2} = \left[\left(\frac{\delta\tau_w}{\overline{\tau}_w}\right)^2 + \left(\frac{\delta\rho}{\overline{\rho}}\right)^2 + \left(\frac{2\delta U_\infty}{\overline{U}_\infty}\right)^2 \right]^{1/2} \tag{15.13a}$$

In general, if the resultant depends upon the components in a power law form $X_R = X_1^a X_2^b X_3^c \cdots$, the fractional uncertainty of the resultant is

$$\frac{\delta X_R}{\overline{X}_R} = \left[\left(a\frac{\delta X_1}{\overline{X}_1}\right)^2 + \left(b\frac{\delta X_2}{\overline{X}_2}\right)^2 + \left(c\frac{\delta X_3}{\overline{X}_3}\right)^2 + \cdots \right]^{1/2} \tag{15.14}$$

Frequently, the resultant is a much more complicated function of the components, possibly one that is found in a data reduction computer program where mean or nominal values of the components are input values and the nominal value of the resultant is calculated in the program. This program could be used to calculate the sensitivity coefficients for the uncertainty analysis. After the nominal run is made, perturbation runs are made where each component value, X_i, is perturbed a small amount, ΔX_i, one at a time, and each corresponding change of the resultant, $\Delta X_{R,i}$, is noted. The sensitivity coefficients are then approximated for each component, i, as

$$\frac{\partial X_R}{\partial X_i} \cong \frac{\Delta X_{R,i}}{\Delta X_i} \tag{15.15}$$

The uncertainty of the resultant is next evaluated from Eq. (15.12). This resultant uncertainty is based upon the same confidence level as used to obtain the component uncertainties. If the component uncertainties, δX_i, were provided to the program, each component's contribution to the summation in Eq. (15.12), $\delta X_{R,i}$, could be evaluated.

$$\delta X_{R,i} = \frac{\Delta X_{R,i}}{\Delta X_i} \delta X_i \tag{15.16}$$

If the resultant uncertainty is too large, those components with larger contributions to the uncertainty should be scrutinized first in search of methods for reducing their contributions. A component with a contribution, $\delta X_{R,i}$, that is less than one-third the largest $\delta X_{R,i}$ usually need not be improved because it accounts for less than 10% of the resultant uncertainty. Components whose contributions are less than one-fifth the maximum component contribution represent less than 4% of the resultant uncertainty and probably could be ignored in the uncertainty analysis.

Many of the instruments in the string will be calibrated, and calibrations cannot be made without error. One may choose to treat the calibration uncertainty separately or to introduce it into the uncertainty analysis. In the following, the experiment has been zero-centered where the calibration is applied to correct the reading and only random error associated with the calibration remains. Because the calibration uncertainty is statistically independent of the instrument reading uncertainty, $\delta X_{i,r}$, it can be combined using the constant confidence interval, single-sample rules discussed above to give an uncertainty for that instrument, δX_i, of

$$\delta X_i = [\delta X_{i,\text{cal}}^2 + \delta X_{i,r}^2]^{1/2} \tag{15.17}$$

where $\delta X_{i,\text{cal}}$ is the uncertainty associated with the calibration of the ith instrument. The calibration uncertainty must come from evidence assembled during the calibration run using uncertainty techniques applied to the calibration. Ingredients include precision of the instruments used in the calibration, steadiness of the calibration run, and calibration uncertainties of the instruments used for the calibration run. The degree of care necessary in evaluating $\delta X_{i,\text{cal}}$ is indicated in the estimated contribution of this uncertainty to the resultant uncertainty. Because of the root-sum-square rule of combining uncertainties, a $(\partial X_R/\partial X_i)\delta X_{i,\text{cal}}$ value of one third of the maximum $(\partial X_R/\partial X_i)\delta X_i$ value contributes less than 10% to the overall uncertainty and can be rather coarsely estimated. Similarly, if the calibration uncertainty is less than one-fifth the maximum component uncertainty, the calibration can be regarded as certain—set $\delta X_{i,\text{cal}} \cong 0$.

The formulation of the random component of the uncertainty analysis is now complete. In the first procedure, where the systematic component had been eliminated and the system is zero-centered, the random component of the uncertainty is the only component. In the second procedure, where the systematic and random components are treated separately, they could be reported separately or combined as the sum or the root-sum-square to give an overall uncertainty. If the root-sum-square combination method is employed, the two procedures give essentially the same magnitude of the overall uncertainty. By the definition of uncertainty, one would expect that if the true value of the resultant were known, it would lie within the overall uncertainty band $\overline{X}_R \pm \delta X_R$ (confidence level) with the stated confidence.

Whenever possible, results of repeated trials could be compared, to check the uncertainty analysis. Examples of such comparisons are mass, momentum, or energy balances, e.g., verifying that: 1) all the flow rates in a system sum to the measured total flow rate, 2) the measured drag on a body is equal to the momentum deficit of its wake, or 3) the total heat transfer from a wall to a flowing fluid can be checked in an energy balance on the flowing fluid. If the uncertainty analysis is correct, one can expect *closure* (agreement) of these checks to within the overall uncertainty with the stated confidence. If closure is not attained within the predicted uncertainty interval, the predicted interval is in error. Either poor estimates of the inputs to the random error were made, or one or more important contributors to the systematic error were overlooked or poorly modeled. Other checks may also be measurements taken in systems that have already been studied and for which previous results are generally accepted as correct, e.g., measurement of a velocity or turbulent shear stress profile in a fully-developed turbulent channel flow. Satisfactory *baseline* checks should be achieved before the instrument string is put into service.

The value of the uncertainty is an important output of the measurement program— one that should be reported along with the nominal value of the resultant. It indicates the expected range about the nominal value within which the true value should lie (with the stated confidence level). But, the value of an uncertainty analysis may be greater. The analysis may also be used in the planning, development, and shake-down phases of the measurement program [Moffat (1985)]. Often, when developing an instrumentation program, there are many options in choice of instruments and configurations as well as a desire for simplicity and constraints on time and expenditures. The uncertainty analysis can provide guidance in these decisions, if incorporated appropriately and done sufficiently early in the program. It would be a poor use of resources to purchase a precise instrument and apply it to the measurement of a component which has a low sensitivity coefficient with the resultant, $\partial X_R / \partial X_i$.

The above formulation of the uncertainty analysis was based on the assumptions that the component uncertainties, δX_i, could be found by experience with the instruments. Moffatt (1982) discusses the evaluation of δX_i's further, tying this evaluation to the various end uses of the uncertainty analysis which include planning, developing, and qualifying the measurement program.

Uncertainty documentation is becoming a requirement for publication in numerous journals; examples of the specifications are given as Rood and Telionis (1991) and Kim *et al.* (1993).

15.2.2 General Bibliography

"American National Standard ASME Performance Test Codes Supplement on Instruments and Apparatus—Part I: Measurement Uncertainty," ANSI/ASME PTC 19.1, 1983.

Eisenhart, C., "Realistic Evaluation of the Precision and Accuracy of Instrument Calibration Systems," *J. Res. Nat. Bureau Standards—Eng. and Instrumentation*, Vol. 67C, No. 2, pp. 161–187, 1963.

Harris, F. K., "Measurement Errors," *Handbook of Applied Instrumentation*, Considine, D. M. (Ed.), McGraw-Hill, New York, 1964.

15.3 PRESSURE MEASUREMENT
Terry W. Simon and Richard J. Goldstein

5.3.1 Static, Dynamic, and Total Pressure

Pressure is defined as the normal force per unit area acting upon some real or imagined boundary. The *static pressure*, p, in a fluid stream is the normal force per unit area on a boundary moving with the flow. Assuming local thermodynamic equilibrium, this definition is consistent with the thermodynamic definition of pressure. Static pressure is a descriptor of the intrinsic thermodynamic state of the fluid.

Another pressure used in flow studies is the *stagnation* (also called *impact*, or *total*) *pressure*, p_0. This is the static pressure that would exist in a fluid stream that had been decelerated from its velocity, V, to zero velocity in a reversible and adiabatic fashion—isentropic stagnation. The stagnation pressure contains information about, and can be expressed in terms of, the local static pressure of the fluid and the magnitude of the local fluid velocity, V. The relationship between p_0, p, and V depends on the behavior of the fluid as it is isentropically decelerated. If the fluid is incompressible the relationship is

$$p_0 = p + 1/2\rho V^2 \qquad (15.18)$$

where ρ is the fluid density. In general, the total pressure is determined from the change in enthalpy with constant entropy as the fluid is brought to rest. This change in enthalpy is $\rho V^2/2$. If the fluid behaves like an ideal gas with constant specific heats during deceleration, and the Mach number, M, near the measurement point just upstream of the beginning of deceleration is less than one, the stagnation pressure can be calculated from

$$\frac{p_0}{p} = \left[1 + \left(\frac{\gamma - 1}{2} \right) M^2 \right]^{\gamma/\gamma - 1} \qquad (15.19)$$

where γ is the specific heat ratio. For flows in which the constant specific heat assumption cannot be made but the gas obeys the perfect gas law, tables are available for determining p_0/p [e.g., Keenan *et al.* (1980)]. For non-ideal gases, the stagnation state is found using real gas data [e.g., Reynolds (1979)]. A third pressure term frequently used in fluid measurements, the *dynamic pressure*, is the difference between the stagnation and static pressures. For incompressible flows

$$p_{\mathrm{dyn}} = p_0 - p = \rho V^2/2 \qquad (15.20)$$

Often pressures are measured not in absolute terms but with reference to another pressure. The most common reference pressure is the instrument ambient (generally atmospheric) pressure. The pressure difference $p - p_{\mathrm{amb}}$ is called the *gage pressure*. Knowing the gage pressure, an independent measurement of the ambient pressure is required to determine the absolute (static or stagnation) pressure.

The remaining sections discuss the measurement of the static, stagnation, and dynamic pressure.

15.3.2   Sensors

Sensing Static Pressure.   The static pressure can be difficult to measure accurately. It is generally measured using a wall tap or a static pressure probe. Errors that may arise when using a static pressure measuring probe (e.g., Fig. 15.2) include:

(i) *Misalignment of the probe with the flow:* The effect of misalignment is shown in Fig. 15.3. Sensitivity to misalignment can be reduced by proper choice of the static port positions around the tube as discussed in Chue (1975) and Capone (1961). When alignment with the flow can be assured, the static ports* should be equally spaced around the probe.

(ii) *Influence of the probe tip region on the downstream static ports:* If the static ports are in the region of curved streamlines or are in a separated flow zone in the vicinity of the tip, the measured static pressure will be in error. It is usually recommended that the static ports be approximately eight diameters downstream of the tip. Sensitivity of this error to the distance from the tip is shown on Fig. 15.2.

(iii) *Blockage of the flow path:* Increased local velocity due to blockage of the probe tip results in a decreased local static pressure.

---

*The terms *port* and *tap*, frequently used in this context, can be considered synonymous.

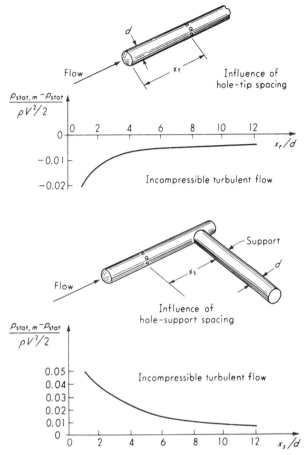

FIGURE 15.2 Static pressure tap errors. (Reprinted with permission of McGraw-Hill Book Co., Doebellin, E. O., *Measurement Systems: Application and Design*, 1990.)

(iv) *Influence of the downstream stagnation region:* If the sensing ports are too near the support, the apparent static pressure will be too high due to locally decelerated flow approaching the stagnation region on the probe support. Sensitivity of this error to the separation distance between the ports and the stem is shown in Fig. 15.2.

Static pressure readings may also be taken using taps located in duct or channel walls. Three important considerations when measuring static pressure through taps in channel and duct walls as well as probe walls are:

(i) *Hole size:* There is a difference in pressure between the surface and the bottom of a pressure port. The magnitude of this difference scales on the wall shear stress, τ_w, the hole diameter, d, and the hole depth, l [Shaw (1960)]. If the non-dimensional hole size parameter, $(d/\nu)\sqrt{\tau_w/\rho}$, is less than 150, and l/d is greater than 1.5, the ratio of static pressure error (a higher than actual pressure is measured) to the wall shear stress, $\Delta p/\tau_w$, will be less

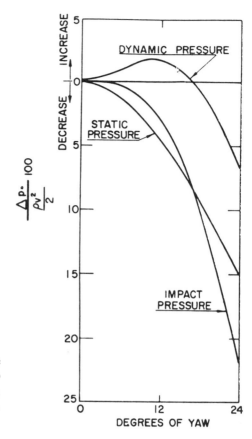

FIGURE 15.3 Variation of static pressure with yaw angles for a hemispherical-tip probe. (Reprinted with permission, Folsom, R. G., "Review of the Pitot Tube," *Trans. ASME*, pp. 1450–1460, 1956.)

than 0.5, which is probably acceptable. If not, $\Delta p / \tau_w$ can be further reduced by reducing the Reynolds number, $(d/\nu)\sqrt{\tau_w/\rho}$.

(ii) *Burrs:* The static pressure hole must be clean, have square edges, and be free of burrs. A burr protruding into the mainstream 1/100 the hole diameter can result in a static pressure error of seven times the hole size error [Shaw (1960)].

(iii) *Turbulence:* Fluctuating inflow and outflow through the static tap hole due to the compressibility of the fluid in the instrument lines may lead to errors in mean static pressure. This flow is in response to fluctuating surface pressure. Usually the error due to this effect is small, but it may become significant where surface pressure fluctuations are large (i.e., in highly turbulent flows) [Bradshaw and Goodman (1966)].

Sensing Total Pressure. Possible sources of error encountered when measuring total (or impact) pressure with a *total pressure tube*, e.g., Fig. 15.4, inserted into the flow are:

(i) *Misalignment with the flow:* This displaces the stagnation point at the sensing port and prevents reading the true stagnation pressure. This error is shown on Fig. 15.5. Special probes have been designed to tolerate large misalign-

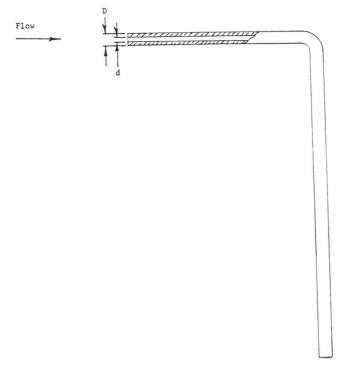

FIGURE 15.4 Total pressure tube.

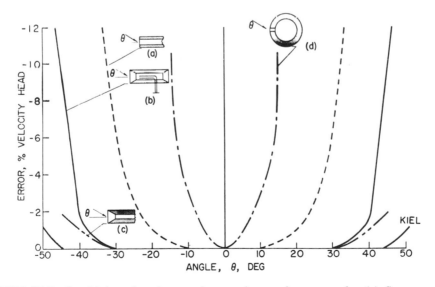

FIGURE 15.5 Sensitivity of various probes to changes in yaw angle. (a) Square-ended tube, (b) shielded tube, (c) chamfer tube, and (d) transverse cylinder. (Reprinted with permission of ISA Proceedings, Vol. 7, Gettelman, C. C. and Krause, L. N., ''Considerations Entering into the Selection of Probes for Pressure Measurement in Jet Engines,'' ISA, 1952.)

ment [Chue (1975), Gracey *et al.* (1951a), (1951b), Gracey (1957), Holman (1978), and Gettelman and Krause (1952)]. To assure alignment with the flow, a total pressure tube installed in a duct should be in a straight section with 8–10 diameters of development length between the probe and upstream valves, fittings, bends, etc. that may be installed no less than 1.5 diameters upstream of the probe.

(ii) *Mean Shear:* When the velocity field is not uniform, as in a shear flow, the measured pressure corresponds to an average stagnation pressure. Because of the non-linear relationship between velocity and pressure, this averaging tends to disproportionately weight the higher velocities. Data showing the effective displacement of the measurement point from the tube center, δ, for boundary layer flow due to this non-linear behavior are given in Folsom (1956) and MacMillan (1956). The displacement, δ, is approximately 0.1 of the tube external diameter. The effective center is displaced in the direction of increasing velocity.

(iii) *Turbulence:* In turbulent flow, the indicated mean pressure read from a total pressure tube is high by the amount $\rho \overline{v'^2}/2$. The quantity $\overline{v'^2}$ is the mean square value of fluctuations of the component of velocity which is in line with the probe axis. This contribution is usually small. For example if the turbulence intensity, defined as $\sqrt{\overline{v'^2}}/V$, is 0.2, the error in the mean velocity measurement is 2% [Chue (1975)].

(iv) *Proximity to a Surface:* When a total pressure probe is used very near a wall, the presence of the wall distorts the flow preventing the probe from measuring the true stagnation pressure. The magnitude of this error for a round, plane-faced tube (e.g., Fig. 15.4) is shown in Fig. 15.6 as an additive velocity, v, to the indicated velocity V [MacMillan (1956)].

(v) *Low Reynolds Number Effects:* Viscous shear may exert additional forces on the fluid near the stagnation hole preventing the total pressure tube from measuring the isentropic stagnation pressure. The result is an indicated stagnation pressure, $p_{0,m}$, which is usually larger than the true isentropic stagnation pressure. The difference between the measured stagnation pressure and the static pressure is given as

$$p_{0,m} - p = C_p(\rho V^2/2) \qquad (15.21)$$

The coefficient, C_p, which is equal to 1.0 for inviscid deceleration of an incompressible flow, has been the subject of considerable analytical and experimental investigation over the last sixty years. The results of this work are discussed in Chue (1975), where he notes and discusses the significant discrepancies. Figure 15.7 shows a low Reynolds number correction and its supporting data [Ng (1986), Hurd *et al.* (1953)] for a plane-faced tube (shown on Fig. 15.4). This correction depends on the Reynolds number of the probe, given as

$$\text{Re}_r = Vr/\nu \qquad (15.22)$$

The radius, r, is the probe internal radius (radius of the hole). Figure 15.7 shows that the C_p is within 2% of 1.0 (correction on velocity is less than 1%) for $\text{Re}_r > 30$ when using a plane-faced tube (shown in Fig. 15.4).

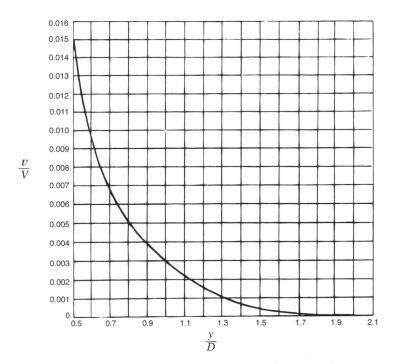

FIGURE 15.6 The wall effect expressed as a function of distance of the probe center from the wall, y. (From Macmillan, 1956. With permission. British Crown Copyright/MOD.) The correction, v, is to be added to the measured velocity V. The parameter D is the probe outside diameter.

(vi) *Shock Waves:* When the oncoming flow is supersonic, a compression wave will be established ahead of the total pressure tube. Downstream of the wave the flow will be subsonic. If the wave is sufficiently weak, the phenomenon can be analyzed as a normal shock [Sherman (1953) and Shapiro (1953)] giving the following expression for the measured stagnation pressure (downstream of the shock) as a function of the static pressure (upstream of the shock) and the Mach number of the undisturbed flow, M

$$\frac{p_0}{p} = \left(M^2 \left(\frac{\gamma + 1}{2} \right) \right)^{\gamma/(\gamma - 1)} \bigg/ \left(\frac{2\gamma M^2 - \gamma - 1}{\gamma + 1} \right)^{1/(\gamma - 1)} \tag{15.23}$$

Stagnation pressure and temperature have been measured with an aspirating probe as discussed in Ng (1986). The technique employs two hot-wires operated at different temperatures. Through calibration, the two voltage levels supplied to the two wires provide sufficient information to deduce the stagnation thermodynamic state.

Sensing Dynamic Pressure. The dynamic pressure, which is the difference between the stagnation or total pressure and the static pressure, can be directly measured in incompressible flow with a *Pitot-static tube*, Fig. 15.8. This probe is a combined total pressure probe and a static pressure probe with separate tubing allowing each pressure to be sensed separately. Errors that may be encountered are those listed above for static pressure and total pressure measurements. Dynamic pressure is mea-

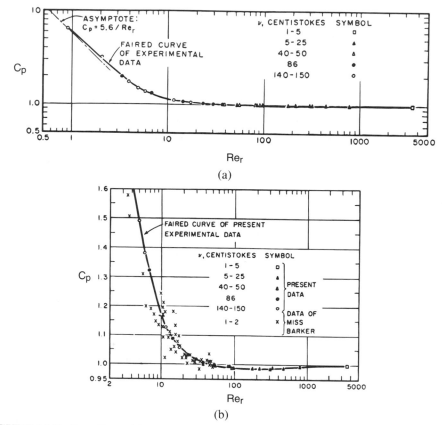

FIGURE 15.7   Low Reynolds number correction; (a) over a large Reynolds number range and (b) expansion of curve in (a). (Reprinted with permission, Hurd, C. W., Chesky, K. P., and Shapiro, A. H., "Influence of Viscous Effects on Impact Tubes," *J. Appl. Mech.*, pp. 253–256, 1953.)

sured by attaching a differential pressure element across the two channels of the probe. If the flow is incompressible, the pressure difference is $\rho V^2/2$, [cf. Eq. (15.20)]. When the flow is subsonic but compressible, the total and static pressures must be measured independently and Eq. (15.19) is employed to evaluate the velocity or Mach number. In supersonic flows, static taps on the side of Pitot-static tubes will give a close approximation to the static pressure upstream of the shock, provided that they are placed at least ten tube diameters downstream of the tip [Shapiro (1953)]. The preferred method is to use a slender probe with a sharp tip [Liepmann and Roshko (1957)]. Static taps on the side wall of a duct located where they do not experience shock effects would be preferred over taps on probes inserted into the stream.

15.3.3   Pressure Measuring Elements

Primary standards are the most accurate, repeatable, and fundamental of the pressure measuring instruments. Others must be calibrated against a primary standard.

In choosing among gages, one must be concerned with accuracy. Manufacturers present accuracy information based on testing done in their laboratories. Care must

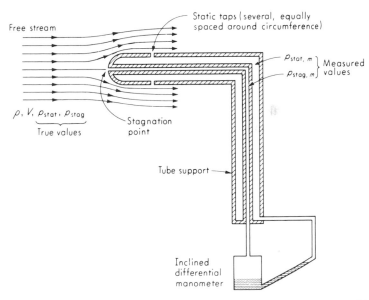

FIGURE 15.8 Pitot-static tube. (Reprinted with permission of McGraw-Hill Book Co., Doebelin, E. O., *Measurement Systems: Application and Design*, 1990.)

be taken that the actual environment does not degrade this stated accuracy to an unacceptable degree by, for example, temperature effects, uncleanliness, acceleration, etc. Accuracy of a primary pressure standard is typically presented as *percent of reading*, whereas, for other pressure gages, accuracy is frequently presented as *percent of full scale* or *percent of range*. In the lower portions of a gage's range, the difference between these two accuracies can be large.

The units listed in Table 15.4 are frequently used in describing pressure. In the following sections, mm Hg will be used for lower pressures and *bars* will be used for higher pressure.

Standards: Pressure is a force per unit area and can be derived from the fundamental quantities of mass, length, and time. Pressure standards, which use as force the weight of a measured mass of solid or liquid in a measured gravitational field, depend on fundamental relationships among these quantities. For pressures ranging from a medium vacuum (10^{-1} mm Hg) to high pressures (14,000 bars), the basic pressure standards are precision manometers and deadweight piston gages. The

TABLE 15.4 Pressure Units

Pascal (N/m^2)	mm Hg (torr)	in H_2O	kPa	psi (lbf/in^2)	Bar	Standard Atmosphere
1 N/m^2 (Pa) = 1	7.50×10^{-3}	4.02×10^{-3}	10^{-3}	1.450×10^{-4}	10^{-5}	0.987×10^{-6}
1 mm Hg = 133.3	1	0.536	0.133	0.01933	1.333×10^{-3}	1.316×10^{-3}
1 in H_2O = 249.1	1.868	1	0.2491	0.0361	2.490×10^{-3}	2.458×10^{-3}
1 kPa = 10^3	7.50	4.02	1	0.1450	10^{-2}	9.869×10^{-3}
1 psi = 6895	51.71	27.68	6.895	1	0.0689	0.0680
1 bar = 10^{+5}	750	401.5	100	14.50	1	0.9869
1 atm = 1.013×10^5	760	406.8	101.3	14.69	1.013	1

choice between these two standards is not usually made by accuracy considerations but is a matter of convenience. At low pressures, a manometer gage may be more convenient because of its simplicity, but, at high pressures, the manometer column height would be excessive.

The McLeod vacuum gage is considered the standard for low pressures down to 10^{-3} mm Hg. Measurement of pressures below 10^{-3} mm Hg can be made using a technique of pressure-dividing with precision orifices to relate a low downstream pressure with the higher upstream pressure which is measured by the McLeod gage [Roehrig and Simons (1963)]. For lower pressures, a hot-cathode or an ionization vacuum gage can be used, as discussed in Simons (1963).

The following section discusses these standards and other pressure measuring elements which span the pressure range from 10^{-14} mm Hg to 10^6 bars. Figure 15.9 shows the approximate range over which each type of element is typically used.

Mechanical Elements for Moderate Pressure Measurements: Pressure measurement techniques can be categorized into the following three:

   (i) Techniques which balance the pressure force with weights or a column of fluid.
  (ii) Techniques where the deformation of a solid barrier is measured.
 (iii) Techniques which correlate pressure with electrical properties.

The first two are generally used for moderate pressures, whereas the last is most common for low pressure (rarefied gases) or very high pressures. In the following, pressure measuring devices are discussed according to pressure range and gage configuration.

DEADWEIGHT TESTER GAGES: A deadweight tester gage is shown schematically in Fig. 15.10. The gage, which is used as a calibration standard, consists of a piston-cylinder arrangement upon which standard weights are added. Readings are taken when the piston is floating and the pressure below the piston is equal to the sum of the

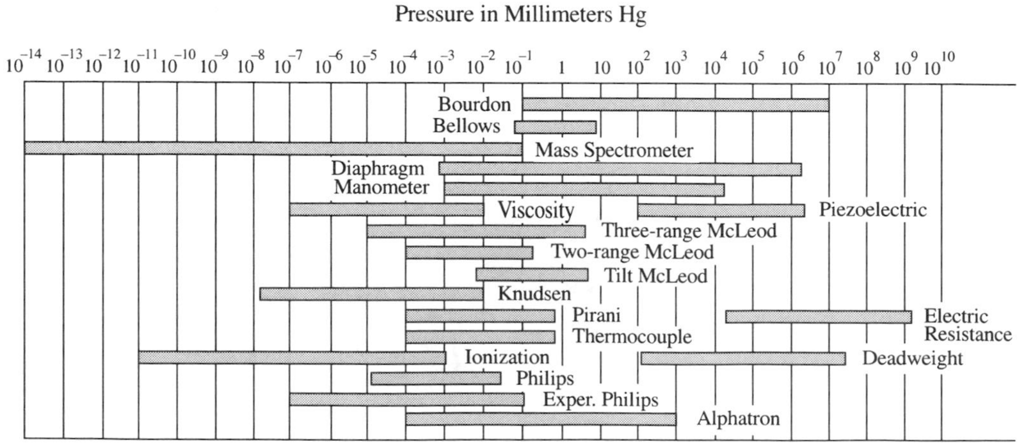

FIGURE 15.9 Ranges of commercially available gages.

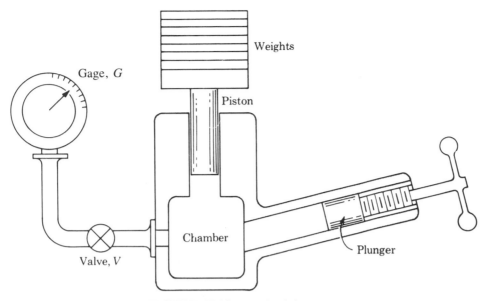

FIGURE 15.10 Deadweight tester.

weight of the piston and added standard weights, divided by the piston area. Accuracy is improved by making minor corrections for: 1) clearance between the cylinder and the piston, 2) thermal expansion, 3) variation in the local acceleration due to gravity, and 4) surface tension effects. Static friction effects are eliminated by rotating the piston. Corrections for many small effects including those listed above are discussed in Sweeney (1977). The accuracy of deadweight testers is usually limited by the friction between the cylinder and piston and the uncertainty of the effective area of the piston. As an example, one gage (Ruska Instr. Co., Cat. No. PS82) is designed for pressure measurements in two ranges: 0.2–86.3 MPa in increments of 0.005 MPa and 0.04–17.26 MPa in increments of 0.001 MPa. It has a stated inaccuracy of 0.01% of reading or 9.5 kPa on the high-pressure range and 0.01% of reading or 0.07 kPa (whichever is larger) on the low-pressure range. The imprecision (two standard deviations of measurement fluctuations) is stated as 14 parts in a million parts.

When the piston is vertical, the minimum pressure which can be measured (the *tare* pressure) is the weight of the piston divided by the area. Lower pressures can be reached by inclining the piston. The correction for the force on the top of the piston assembly due to the piston's ambient pressure becomes more important at lower pressures. Therefore, low pressure measurements require accurate measurement of the pressure inside the evacuated space.

MANOMETERS: A manometer is similar to a deadweight gage in that both are steady-state instruments which compare the pressure force with the weight of a known mass. In the manometer, however, the weight is due to a column of liquid, which is usually deflected upon loading, whereas a null measurement is taken with a deadweight gage. The manometer position is continuous, whereas the deadweight gage pressures are incremental according to the selection of the calibration masses.

A *U-tube manometer* is shown in Fig. 15.11. It is used to measure pressure differences by balancing the forces on a manometer liquid column in the U-shaped tube. A static force balance gives

$$h = \frac{p_1 - p_2}{(\rho_m - \rho_f)g} \tag{15.24}$$

where $p_1 - p_2$ is the difference in the static pressures on the manometer fluid column, ρ_m is the density of the manometer fluid, ρ_f is the density of the transmitting fluid (often air) above the manometer fluid, and g is the local acceleration due to gravity. If leg 2 is open to the ambient, the column deflection indicates the gage pressure of leg 1 (positive as shown, or a vacuum when the surface in leg 1 is below the surface in leg 2). The most common manometer fluids are mercury, water, alcohol, and oil. Mercury is attractive when the pressures are large because of its high density. Oil is attractive because the evaporation loss and corrosion, or other deterioration of manometer materials, are minimized. Alcohol is attractive because of its low surface tension and hence flatter *meniscus*.

Meniscus is the name given to the interfacial surface between the manometer fluid and the line (or transmitting) fluid. Because the meniscus shape is important in the reading of precision manometers, the factors which influence it must be discussed. The meniscus is usually not flat. This is due to surface tension, or capillary, effects. Capillary action is due to the cohesion among the liquid molecules and the adhesion of the liquid molecules to the molecules of the surface of the tube. The shape of the meniscus is, therefore, dependent upon the fluids and the nature of the tube wall (including films which may be coating the wall). A discussion of meniscus shapes for several fluids in manometers is given in Daugherty and Franzini (1977). For clean glass tubes and standard manometer fluids, capillary effects are generally avoided if the tube internal diameter is 1 cm (3/8 in.) or larger. When capillary effects are significant, the same point on the meniscus must be used each time the column height is read. Generally, the extremum of the meniscus is used for all readings, whether the tube is vertical or slanted.

Corrections which must be made to realize an accurate manometer reading include: 1) corrections for the thermal expansion of the scale due to a difference between the recording temperature and the temperature at which the scale was calibrated, 2) corrections for the change of manometer and transmitting fluid densities with pressure and temperature, 3) corrections for the local acceleration due to gravity, g, and 4) corrections for the angle of inclination by using the component of the acceleration of gravity parallel to the tube. Care must be taken to read the elevation of each leg using the same position on the meniscus. The tube walls must remain clean so that the meniscus shape does not change. Special sightglasses and vernier scales may be used to give a more accurate location of the manometer fluid height than could be obtained with the naked eye. A float may be used with a sharp-edged or graduated indicator so that the readings are not subject to the arbitrariness of viewing a meniscus. One optical device which is useful in reading manometer levels is a cathetometer. It is a telescope which is horizontally mounted on a carriage that can move on an accurately calibrated vertical guide. The height of the vertical axis, the meniscus level, is read on the scale when the reticle is positioned on the extre-

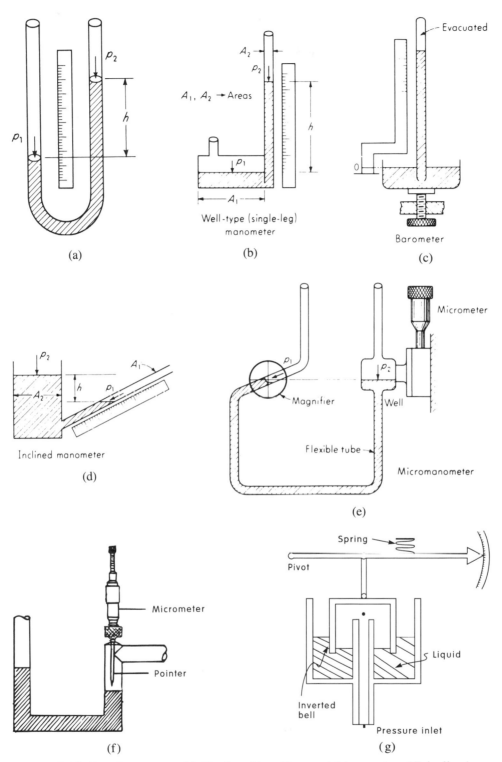

FIGURE 15.11 Manometers: (a) U-tube, (b) well-type, (c) barometer, (d) inclined manometer, (e) nulled micromanometer, (f) displaced micromanometer, and (g) bell-type. (Reprinted with permission of Instrument Society of America, from ''Basic Instrumentation Lecture Notes and Study Guide,'' Moore, R. L., 1982.)

mum of the meniscus. As an example, with a precision cathetometer, small level changes in a manometer approximately 1 meter away can generally be measured to a precision of at least 0.025 mm.

There are many variations on the U-tube manometer. One is the *well-type manometer* [Fig. 15.11(b)] which is similar to the U-tube manometer except that one leg has a very large cross-sectional area; the fluid level in this leg changes little from one reading to the next. Then, only one liquid level needs to be read. Because the actual reservoir size is finite, the manometer fluid level in the reservoir moves somewhat from reading to reading. This can be compensated for by sliding the zero of the scale until it aligns with the reservoir level before each reading is taken. Experience has shown that the well-type manometer is less accurate than a conventional U-tube manometer.

If the column of a well-type manometer is evacuated, and the reservoir is at the ambient pressure, the column height can be used to determine the absolute ambient pressure; this is a standard design for a *barometer* [Fig. 15.11(c)]. The evacuated space cannot be precisely at absolute zero pressure, but is at the vapor pressure of the manometer fluid at the ambient temperature. Corrections are made for this effect, and these are combined with other corrections discussed above for U-tube and well-type manometers. Increased sensitivity of well-type and U-tube manometers is achieved by tilting the columns, e.g., the *inclined manometer* of Fig. 15.11(d).

Systems that allow readings of very small pressure differences are called *micromanometers*. In one scheme [Fig. 15.11(e)], zero is found by connecting the legs ($p_1 = p_2$) then accurately positioning the reservoir until the opposite leg is at the null position. Sensitivity is enhanced by making the tube section nearly horizontal at the null position [Fig. 15.11(e)]. Small pressure differences are measured by moving the well a precisely measured distance until the opposite column level is again on the null position. In another scheme [Fig. 15.11(f)], the fluid level is precisely located by use of an electric circuit. A micrometer-mounted pointer is lowered until it contacts the manometer fluid level. Contact is detected when completion of an a.c. circuit activates a bridge rectifier which produces a signal for indication on a sensitive d.c. microammeter. For example, fluid level changes of no more than 5 cm can be measured to a precision of 0.01 mm. A micromanometer, which is nulled by moving a reservoir, would be less susceptible to tube non-uniformities and to fluid contamination which may change the shape of the meniscus. Micromanometers can be used to read pressures with a precision of typically 0.005–0.025 mm of the manometer fluid. They are usually limited in range to about 5 cm of the working fluid differential pressure, though some can be used with pressure differences as high as 50 cm of working fluid. One precision manometer is capable of reading pressure as high as 280 cm of manometer fluid (mercury) to a stated measurement accuracy of 8 μmHg + 0.003% of reading. It consists of a U-tube manometer with a cistern on each end. The instrument is nulled by traversing one cistern until capacitors which use the mercury interface as lower conducting surfaces indicate nulled capacitance. The position of the traversing cistern is then read. The device can be computer automated.

Other micromanometers use large floats to reduce surface tension effects, with a mounted mirror. An optical beam arrangement amplifies the float displacement making the device very sensitive.

The manometer displacements discussed above are read directly by the experimenter. Other techniques used to automate or amplify the measurement include sonar reflection, light interference, floating mirrors, and light beam balance.

The *Bell-Type Manometer* [Fig. 15.11(g)] has an inverted bell which traps a gas volume at the pressure to be measured. The bell is immersed in and sealed by a reservoir of manometer fluid. Increased line pressure results in a change in the difference in levels between the inside and outside liquid surfaces giving increased buoyancy to the bell. This buoyancy is balanced by the forces exerted by a resisting spring and the weight of the bell and support assembly. Movement of the bell is sensed by a linkage and dial arrangement or by a displacement transducer. The force applied to the bell, with a given pressure difference, increases as the bell size increases. Thus, low pressure differences can be accurately measured with gages having large bells. Typically, inverted bell designs operate in the 0–2.5 kPa range [Miller (1983)].

MANOMETER DYNAMICS: Though manometers are instruments for measuring differences of steady pressure, the column moves from one reading to the next, and one must be concerned with its dynamics. Detailed discussions of manometer dynamics are given in Doebelin (1990), Ury (1962), and Richardson (1963). If channel curvature, surface tension, and line fluid inertial effects are neglected and the flow remains laminar, the level of the manometer fluid surface in a constant-diameter leg responds to variations in pressure according to the linear differential equation [Doebelin (1990)]

$$\frac{1}{\omega_n^2}\frac{d^2x}{dt^2} + \frac{2\zeta}{\omega_n}\frac{dx}{dt} + x = k(p_1 - p_2) \tag{15.25}$$

The dependent variable, x, is the meniscus position relative to the equilibrium position when $p_1 - p_2$ is zero. The natural frequency is $\omega_n = \sqrt{3g/2L}$ where L is the entire length, both columns and the bend, of the liquid column. The damping factor, ζ, is given by

$$\zeta = 2.45\nu\,\frac{\sqrt{L/g}}{R^2} \tag{15.26}$$

where R is the liquid column radius and ν is the liquid kinematic viscosity. The restoring constant, k, is given by $k = 1/(2\rho g)$ where ρ is the liquid density. For a highly-damped manometer tube, the response to a step change in pressure is $k(p_1 - p_2)(1 - e^{-\tau/t})$. The time constant, τ, is $2\zeta/\omega_n = 4\nu L/R^2 g$.

If readings are taken visually, transient behavior is obvious. If the readings are automatically acquired, checks must be made to verify that the column is stationary. The above analysis can provide guidance in estimating the settling time.

BOURDON TUBES: In the *Bourdon tube*, a pressure difference causes a deflection of an elastic element. It is a curved tube of noncircular cross section [Fig. 15.12(a)]. In response to a positive pressure difference, $p_1 - p_2$, where p_1 is the pressure inside the tube and p_2 is the external pressure, the Bourdon tube attempts to assume a circular cross-section causing the free end to deflect, rotating a pointer or driving a transducer. *C-type* Bourdon tubes [Fig. 15.12(a)] with nearly circular cross sections

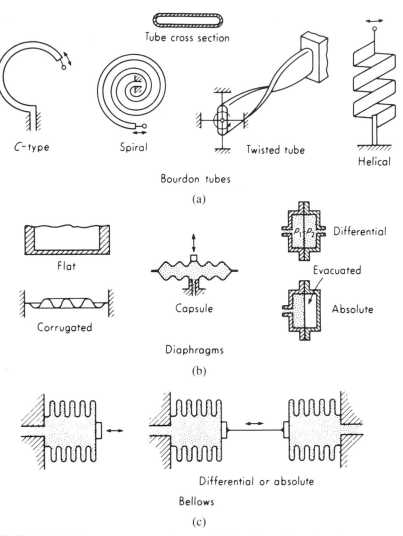

FIGURE 15.12 Elastic pressure transducers: (a) Bourdon tubes, (b) diaphragms, and (c) bellows. (Reprinted with permission of Instrument Society of America, from "Basic Instrumentation Lecture Notes and Study Guide," Moore, R. L., 1982.)

have been used to measure gage pressures as large as 10,000 bars. *Spiral-* and *helical-types*, which are sensitive to small pressure differences, have been used to about 3000 bars, but are typically used below about 70 bars—some as slow as 1–2 bars. Industrial grade Bourdon tubes have accuracies typically from ±0.5% to 5% full scale, while more precise C-type Bourdon test gages have accuracies of 0.1% of full scale. These gages have bimetal components in their mechanical linkages that compensate for thermal effects. The displacement of the free end of the tube is read on a dial indicator or processed as an electrical signal from a displacement transducer. Usual materials for construction are brass, bronze, phosphor–bronze, beryllium–copper, monel, inconel, steel, and 304 or 316 stainless steel.

DIAPHRAGM AND BELLOWS GAGES: Other pressure gage configuations which use the elastic deflection to measure pressure are diaphragm-type and bellows-type gages. Bellows-type gages are typically used to measure small pressure differences (Fig. 15.9). The geometry of the diaphragm and bellows differs from gage to gage, as indicated on Fig. 15.12(b) and (c). Bellows are usually made of phosphor–bronze, beryllium, copper, stainless steel, or brass. The deflection of the diaphragm, or the free end of the bellows, is read on a dial driven by linkage and gears, as in the all mechanical Bourdon gage, or as an electric signal from a displacement transducer. Diaphragm gages of increased sensitivity can be attained by corrugating the diaphragm [see Fig. 15.12(b)]. Properties of *corrugated diaphragm* gages are discussed in Wildhack *et al.* (1957).

A special type of diaphragm gage used to measure small pressure differences is a *slack diaphragm* gage. The slack diaphragm offers minimal resistance, so the pressure force balances that of a linear return spring. The advantage is that the diaphragm deflection is linear with pressure. The deflection is sensed with a linkage and dial arrangement or by a displacement transducer. In some arrangements, the spring is replaced by a force transducer. *Flat diaphragm* pressure elements may have *strain gages* attached to the diaphragm, or slack diaphragm elements may have strain gages attached to a bending beam which restrains the pressure force. Proper choice of the strain gage configuration on the flat diaphragm maximizes the output signal for a given deflection [Measurements Group (1974)]. Strain gages, typically used in flat diaphragm and slack diaphragm gages, include bonded metal foils, vacuum- or sputter-deposited thin-metal films or bonded or diffused semiconductor surfaces. Vacuum-deposited, sputter-deposited, or diffused gages usually suffer less from creep or hysteresis, and therefore require less frequent calibration [Farmer (1981)]. Creep in bonded gages is usually attributed to creep of the bonding agent. Semiconductor strain gages provide larger output voltages and have higher sensitivity, but they may suffer from temperature instability and non-linearity, and they typically have low natural frequencies. Construction details of some of these gages are presented in Doebelin (1990), Whittier (1980), and Sensym (1983). Strain gage elements are typically fast-responding (d.c. to ~5 kHz), have low source impedance, have minimal mechanical motion, and are small in size and weight. They occasionally suffer a loss of repeatability due to hysteresis and usually have rather costly output measuring devices.

Figure 15.13 [Winteler and Gantschi (1979)], shows a flat-diaphragm semiconductor *piezoresistive pressure transducer*. The deformation of certain materials such as silicon or quartz is accompanied by the generation of an electric charge on the surface. This is the *piezoelectric effect*. This effect is used by constructing the diaphragm of piezoelectric material and arranging for electrical amplification of the weak electric signal. Because the deflections are very small, piezoelectric transducers can resolve high-frequency pressure fluctuations. Natural resonance frequencies typically lie from a few tens of kilohertz to two-hundred kilohertz. Sensors designed for high-frequency resolutions can be applied to 10–20 kHz [Smol'yakov and Tkachenko (1983)]. The measuring cell of Fig. 15.13 consists of a sealed chamber with single-crystal silicon chip walls. Mounted on the wall, which deflects under pressure as a diaphragm, is a piezoresistive element which senses both thermal and pressure loading strains. On the opposite, rigid wall are the remaining elements of a bridge circuit—piezoelectric elements which are sensitive to temperature changes.

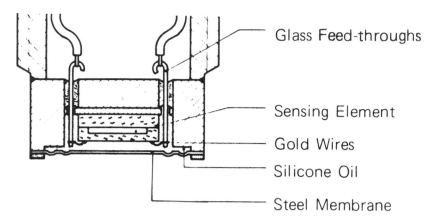

FIGURE 15.13 Diffused semiconductor strain-gage transducer. (Reprinted with permission of Kistler Instruments, Winteler, H. R. and Gautschi, G. H., "Piezoresistive Pressure Transducers," 1979.)

Bridge imbalance indicates pressure loading across the silicon chip wall. The steel diaphragm does not support a pressure difference; it retains silicone oil inside the sensor which protects the chip from the line fluid. When the transducer is constructed for vacuum measurements, care is taken to rid the oil space of air, including that which may be dissolved in the silicone oil. This gage is constructed with various degrees of deflection wall rigidity for measuring pressure differences of 0–2 to 0–200 bars. Its natural frequency is greater than 70 kHz. Attractive features of silicon-base piezoresistive transducers include [Winteler and Gantschi (1979)]: 1) long cycle life, 2) excellent stability, 3) high reproducibility, 4) small size, 5) low acceleration (vibration and shock) sensitivity, 6) high output voltage, and 7) high natural frequency (due to potential small size).

An example of a low-cost strain-gage pressure transducer from Sensym Inc. is shown in Fig. 15.14. The strain-sensitive element is diffused into a wall which bears

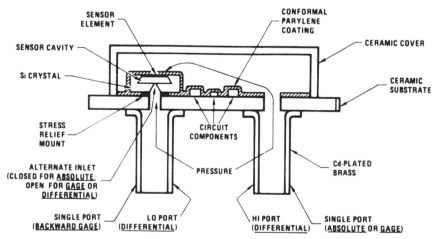

FIGURE 15.14 Diffused sensor transducer. (Reprinted with permission of Sensym Inc., Sunnyvale, CA, "Pressure Transducer Handbook," 1983.)

the pressure difference. The remaining elements of the bridge are diffused into an unloaded wall. They respond to, and compensate for, temperature changes. The transient response of this gage is fast, with a diaphragm natural frequency of 50–100 kHz. Differential pressures as large as 20 bars can be measured.

A *carbon-type* pressure sensitive element has a chamber filled with loosely compacted carbon granules or discs [Lawford (1974)]. The carbon is compressed by the pressure force acting through a slack-type diaphragm. The carbon bed acts as both a spring, returning the diaphragm upon removal of the load, and a transducer. The resistance change of the bed is sensed as a Wheatstone bridge imbalance.

Inductive transducers [e.g., Fig. 15.15(a)] use a change in magnetic flux linkage due to the changing position of a ferromagnetic core of a coil to measure the diaphragm deflection. Inductive transducers have fast response and low sensitivity to acceleration and vibration, but display sensitivity to temperature and external magnetic fields. Commercial models permit measurement of pressure differences to as low as 2×10^{-3} mm Hg.

Capacitive transducers [e.g., Fig. 15.15(b)] record movements of plates separated by a dielectric [Slomiana (1979a), (1979b), Behar (1951), and Lee (1970)]. Changes in capacitance can be seen as a change in the resonance frequency of an

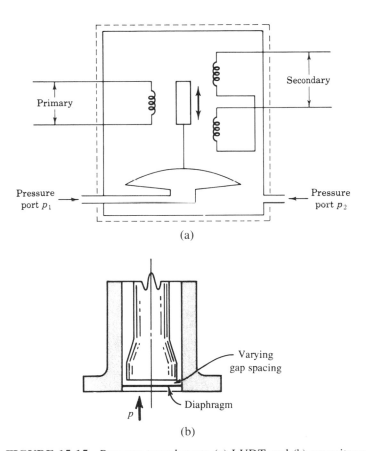

(a)

(b)

FIGURE 15.15 Pressure transducers: (a) LVDT and (b) capacitance.

RC oscillator. Advantages in their use include high sensitivity to diaphragm movement, minimum hystersis due to friction, fast response, and minimal self-heating. The high sensitivity to diaphragm deflection makes the capacitive-type gage well suited to high-frequency pressure measurements. They compete in this function with piezoelectric transducers showing greater sensitivity at lower pressures as with turbulent air flows, but they have diaphragms which are usually too fragile for turbulent pressure fluctuation measurements in liquids [Lawford (1974)]. Disadvantages may include errors by charge leakage, nonlinearity of output, and cost of output sensing equipment. In capacitive-type gages, the diaphragm movement is small, and changes in gage temperature can result in significant zero drift. Some precise gages have built-in control units which maintain constant gage temperature. Gages which do not have such temperature control must be frequently compensated for zero-drift when the ambient temperature is not constant. Highly fluctuating pressures, such as would be experienced in turbulent flows, are usually measured by flush-mounted diaphragm pressure sensing elements with either piezoelectric or capacitive type transducers.

Discussions of the applications of the above differential pressure transducers can be found in Lee and Pfeifer (1981), Behr and Giachino (1981), and Lee and Wise (1982).

Careful specification of the electrical performance of the transducers discussed above should be made. One example is a clear representation of the thermal sensitivity. Unfortunately, complete documentation of transducer characteristics is not universally done. A collection of transducer terms and their definitions for transmitting complete transducer documentation can be found in Lee and Wise (1982). An example of a complete specification sheet is given in Whittier (1980).

A selection guide for gages to measure pressures from atmospheric pressure to 10,000 bars is presented in Hall (1981).

Very High Pressure Measurements: Pressures as high as 7000 atmospheres can be measured with diaphragm cells and Bourdon tubes. Bourdon tubes at high pressure can be accurate to about 1% of full scale with an additional temperature error of about 4%/100 C. Strain gage diaphragm cells with temperature compensation can be accurate to 0.25% of full scale over a large temperature range.

Bridgeman-type gages based on the change of resistance of manganin or gold-chrome wire with hydrostatic pressure are used for pressures above 7000 atmospheres [Behar (1959) and Howe (1955)]. An example is shown in Fig. 15.16; the sensitive wire is wound in a loose coil and encased in a non-conducting, liquid-filled region which is pressurized to the system pressure. The resistance change is sensed in a Wheatstone bride. The transient response is excellent. Gages of this type are commercially available to pressures of 14,000 atmospheres with accuracies of 0.1% to 0.5% of full scale. Since these gages are subject to aging, frequent recalibration may be necessary.

Vacuum Measurements: Pressure units commonly used in vacuum measurements are the *torr* (mm of Hg) and the *micrometer* (μm, or *micron*) of Hg. Vacuum pressures are given as absolute, not gage, pressures. Bourdon gages are used down to about 10 torr, manometers and bellows gages to 0.1 torr, and diaphragm gages to 10^{-3} torr. At lower pressures, gages built specifically for vacuum measurements are

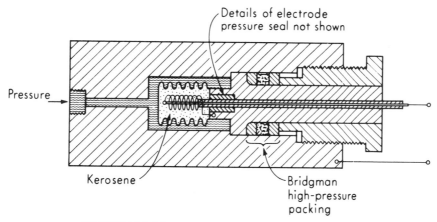

FIGURE 15.16 Wire resistance high-pressure gage.

needed. Some of these are described below; more specialized information can be found in Dushman and Lafferty (1962) and Morrison (1964).

The *McLeod gage* is a standard for vacuum measurement, because the pressure can be computed from the gage dimensions. It can be used to measure pressures directly to about 10^{-4} torr and can be used with a pressure dividing technique to lower pressures [Kreisman (1964)]. Its inaccuracy can be as small as 1% of reading when used in the direct mode, though it can be much larger for high vacuum measurements. In the McLeod gage, a low pressure gas sample is compressed to a pressure which is sufficiently high that it can be read with a manometer [Fig. 15.17(a)]. First, the reservoir is lowered, or a plunger is withdrawn, allowing the sample gas of unknown pressure, p_i, to enter the bulb and capillary region of the gage. The reservoir is next raised, or the plunger is pushed in, and the working fluid (mercury or butyl phthalate) level rises (without compression), sealing the sample into a known volume, V, of the capillary tube. Continued reservoir or plunger motion compresses the gas until the working fluid column level is at the zero mark. It is presumed that no vapors are condensed from the sample upon compression. The sample pressure, after compression, is read from the manometer scale as a height, h, above the reference pressure. From Boyle's law

$$p_i V = p A_t h \tag{15.27}$$

where $p = p_i + \rho h g$ and ρ is the working fluid density. The area, A_t, is the tube cross-sectional area.

Eliminating p gives an expression for the unknown pressure,

$$p_i = \frac{\rho g A_t h^2}{V - A_t h} \tag{15.28}$$

The McLeod gage requires a discrete sample of gas and, therefore, cannot give continuous readings. It also has low-pressure limitations due to the working fluid vapor pressure.

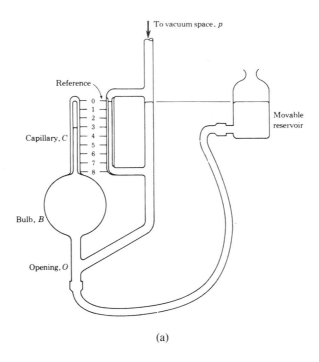

(a)

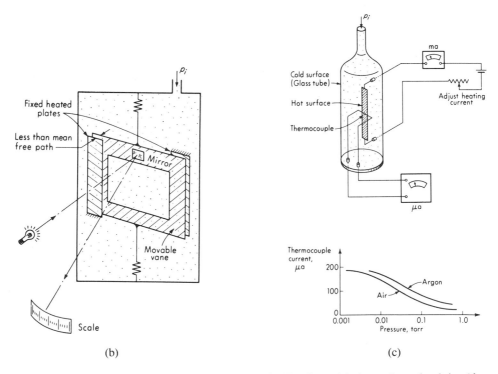

(b)

(c)

FIGURE 15.17 Vacuum gages: (a) McLeod, (b) Knudsen, (c) thermal conductivity (thermocouple-type), (d) thermal conductivity (Pirani type), (e) ionization, and (f) alphatron.

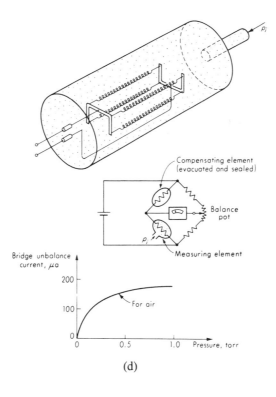

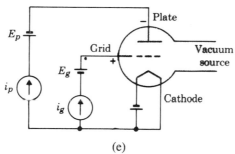

(d)

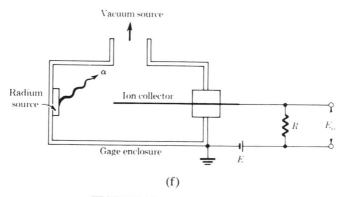

(e)

(f)

FIGURE 15.17 (*Continued*)

In the *Knudsen gage* [Fig. 15.17(b)], a low pressure sample enters a chamber in which two fixed plates at temperature, T_f, and a spring-restrained movable plate at temperature, T_v, are located. The spacing between the fixed and movable plates is less than the mean free path of the gas. Kinetic theory of gases shows that gas molecules rebound from the hotter plate with greater momentum than from the cooler plate. The total momentum exchange between the two plates is a function of molecular density which is dependent on the temperature and pressure of the gas. The momentum exchange causes a net twisting motion on the movable plate supports. The deflection is sensed and from the geometry and spring torsional constant, the net turning force is computed. Kinetic theory gives

$$p_i = \frac{2F}{\sqrt{T_f/T_v - 1}} \tag{15.29}$$

where F is the measured twisting force. The gage is insensitive to gas composition except for the effect on the *accommodation coefficient*, the extent to which rebounding molecules reach equilibrium with the plate. Knudsen gages presently can be used for pressures ranging from 10^{-8}–10^{-2} torr.

Viscosity gages are used for measuring pressures below 10^{-2} torr, where kinetic theory of gases predicts that the viscosity of a gas is proportional to the pressure. Unfortunately, this viscosity is species dependent. The viscosity is measured as an applied torque necessary to rotate one cylinder inside another. For a particular species, the gage can be calibrated as viscosity versus pressure. Such gages have been used to 10^{-7} torr.

Thermal Conductivity gages utilize the finding that a linear relationship exists between thermal conductivity and pressure for pressures below 10^{-2} torr. Conductivity gages have an element which is electrically heated with a constant energy input [cf. Benson (1963)]. This element will seek a temperature at which the energy input balances heat transfer away by conduction and radiation. For a given gas composition, this equilibrium temperature is a measure of the gas pressure. To enhance sensitivity, care is taken to minimize the emissivity of the surface. Figure 15.17(c) shows such a device where the temperature of the element is sensed with a thermocouple. The cold surface is the glass wall of the container which is near ambient temperature. Generally, sensors must be calibrated against a standard, possibly a McLeod gage. Thermocouple gages are available for measuring over the range of 10^{-4} to 1 torr. In the *Pirani-type* thermal conductivity gage, the heating and temperature sensing functions are in the same element. The element consists of tungsten wires inside a glass tube which is at the unknown internal pressure. An identical tube, which has been evacuated, holds an identical heating element. This element, placed in a Wheatstone bridge circuit with the first element, serves as a compensating element correcting for changes in ambient temperature [Fig. 15.17(d)]. When voltage is applied to the bridge, the wires in each tube are heated. The wires in the tube with the gas sample will be cooled by gas conduction and radiation, whereas the wires in the evacuated tube will be cooled by radiation only. The wires in both tubes will conduct the same current, but, because of the resistance change with temperature, the wire in the evacuated tube will have a higher voltage drop. The Pirani gage displays a bridge imbalance which varies with sample gas pressure. The gage should be calibrated against a standard. Pirani gages are used in the range 10^{-4}

to 1 torr. The lower limit is where heat transfer by radiation and axial conduction in the heated wire overshadows heat transfer by conduction through the sample gas. The transient response of the Pirani gage is poor with several minutes required to achieve thermal equilibrium at low pressures.

The *ionization gage* is one of the oldest and most widely used devices for measuring gas pressure under high-vacuum conditions. In the ionization gage [Fig. 15.17(e)], an electric potential accelerates electrons from a hot cathode to an anode in a fashion similar to an ordinary vacuum tube. If the electron strikes a sample gas molecule, a positive ion may be created, freeing a secondary electron. The rate of creation of positive ions is proportional to the sample gas pressure. The ions are attracted to the negatively charged electrode where they combine with electrons. Recombination produces a current which is proportional to the sample gas pressure. Conventional ionization gases are suitable for measurements between 10^{-8} torr and 10^{-3} torr. The ionization gage may cause decomposition of some gases at the hot cathode filament or may cause contamination of the sample gas by vaporization of the hot filament. Hot cathode problems are overcome in the *Philips cold-cathode* gage [Lafferty and Vandestlice (1963)] where a high (~ 2000 V) potential is substituted for cathode heating to accelerate the electrons from a cold zirconium or thorium cathode. The electrons released by bombardment of the positive ions are accelerated by the potential. Philips gages are generally used for measuring pressures in the range 10^{-5}–10^{-2} torr.

Magnetron gages, ionization gages with applied mangetic fields, are available for pressures down to 2×10^{-14} torr [Lafferty and Vanderslice (1963)]. A magnetic field causes the traversing electrons to travel around the cathode many times before being collected, thus increasing the path length and increasing the probability of collision and ionization. Mass spectrometer techniques can be used to determine partial and total pressures below 10^{-14} torr [Lafferty and Vanderslice (1963)].

The *Alphatron gage* is a radioactive ionization gage [shown schematically in Fig. 15.17(f)]. A radium source emits α particles which ionize the sample gas. The degree of ionization is determined by the measured voltage, E_0. The linear range of the gage is typically 10^{-4}–10^{3} torr. The output characteristics are species-dependent.

15.3.4 Transient Pressure Measurements

Dynamic Effects of Sensors and Connecting Tubing. One must be concerned about the lags associated with the volumes and resistances of the sensor and connecting tubing even if steady pressures are to be measured. The question frequently arises concerning how long, after a change of pressure, one must wait before an accurate reading can be taken. The transient response of pressure measuring instruments is dependent upon two factors—the response of the sensing element and the response of the pressure transmitting fluid and connecting tubing. The second factor usually dominates the overall frequency response of the system. Often a direct calibration of the system must be relied upon for determining the system response. Several configurations are simple enough to be amenable to analysis, however.

If the sensing element of a pressure gage could be mounted at the wall of the system containing the test flow or on the tip of pressure probe, the only lag between system pressure and gage response would be due to the sensor inertia. This lag would be small. More commonly, a pressure tap or port and the measuring element are separated by a connecting line. The volume of the pressure measuring element and the volume of the connecting line represent fluid volumes that change with pressure. The connecting line between the system and the sensor represents resistance to the flow of mass, which is required for the system to come into pressure equilibrium after a change in system pressure.

The dynamics of this re-equalization process are discussed in Doebelin (1990), White (1949), Iberall (1950), Taback (1949), and Black (1983). If the connecting line is rigid and of constant diameter, the fluid is incompressible, the flow is laminar and quasi-steady and the inertia of the diaphragm and fluid are negligible, the difference between the system pressure and measured pressure (error), $(p - p_m)$, following a step change in $(p - p_m)$, $(p - p_m)_0$, is given as

$$\frac{p - p_m}{(p - p_m)_0} = e^{-t/\tau} \tag{15.30}$$

In this expression, t is the time after the step change of pressure, $(p - p_m)_0$, and τ is $128\mu LC/\pi d^4$, where μ is the dynamic viscosity, L is the connecting tube length, C is the *sensor compliance* which is the sensor volume change per change in pressure, and d is the tube diameter. The responsiveness of the sensor is increased by reducing the tube length, L, or the compliance of the sensor, C, or increasing the tube diameter. If the fluid were compressible, increasing the tube diameter may reduce the system response because of the increased fluid volume in the connecting line.

Equation (15.30) presumes that the delay associated with the pressure wave traveling through the connecting line is small. This delay time is given by

$$T_{\text{delay}} = L/a \tag{15.31}$$

where the speed of sound, a, in a fluid of bulk modulus, E_L, and density, ρ_L, in an elastic tube which has a wall bulk modulus, E_t, a wall thickness, t, and an inner diameter, d, is

$$a = \left(\frac{E_L}{\rho_L}\right)\frac{1}{1 + dE_L/(tE_t)} \tag{15.32}$$

It is clear that many variations on this are possible due to different geometry, fluid, tube conditions, etc. The above serves as a reminder that these dynamics may be important. The specific dynamics of each user's configuration should be considered. Because of the degraded frequency response of differential pressure sensors due to connecting line dynamics, one may wish to consider measuring the two pressures separately with flush-mounted gages, then subtracting the two signals electrically. This technique introduces larger error in the pressure difference reading due to bias errors like thermal drift, however these can be overcome with care in the design and operation. For example, checks for drift could be frequently made during the test if the facility were designed so that the two gages could be introduced to the same static pressure and then adjusted so that the net electrical signal indicated zero pressure difference.

Dynamic Testing. Methods for testing the dynamic response of pressure measuring systems include impulse, step, and frequency-response type tests. A comprehensive review of this topic is given in Schweppe (1963), a few points follow here.

Step function tests are usually made by bursting a thin diaphragm in a tube separating a high pressure reservoir from the pressure sensor. A pressure wave moves down the tube to the sensor causing a step in sensor loading upon impact. If the step test is to excite the natural frequencies of the sensor, the rise time of the pressure increase should be less than one-fourth the natural period of the system.

An impulse test could be as simple as dropping a steel ball on the sensor diaphragm. The impact excites the sensor, and the response waveform is recorded and analyzed to determine the natural frequency and damping ratio. This test is discussed further in Bentley and Walter (1963).

A frequency-response test could be made with an electrodynamic vibrator shaker that applies a sinusoidal force of adjustable frequency and amplitude to a piston creating a sinusoidal pressure variation in a liquid-filled chamber. The test would involve comparing a reference pressure sensor, which is capable of following the pressure fluctuations up to the maximum frequency, with the pressure sensor under test. Test results would be the phase shifts and attenuations versus pressure signal frequencies. Pressure sensors which can, via on-board computational capability, determine a shift of response time, are documented in Hashemian *et al.* (1986).

15.3.5 General Bibliography

"Accuracy in Measurements and Calibrations," NBS Tech. Note 262, National Bureau of Standards, Washington, D.C., 1965.

Beckwith, T. G., Maragoni, R. D., and Leinhard, J. H. (V), *Mechanical Measurements*, 5th ed., Addison–Wesley, Reading, MA, 1993.

Blake, W. K., "Differential Pressure Measurements," *Fluid Mechanics Measurements*, Goldstein, R. J. (Ed.), Hemisphere Publishing, Washington, D.C., 1983.

Chue, S. H., "Pressure Probes for Fluid Measurement," *Prog. Aerospace Sci.*, Vol. 16, No. 2, pp. 147–223, 1975.

Doebelin, E. O., *Measurement Systems: Application and Design*, 4th ed., McGraw-Hill, New York, 1990.

Lawrence, R. B., "A Survey of Gauges for Measurements of Low Absolute Gas Pressures," *Chem. Eng. Prog.*, Vol. 50, No. 3, pp. 155–160, 1954.

Moore, R. L. (Ed.), *Basic Instrumentation Lecture Notes and Study Guide—Measurement Fundamentals*, Instrument Society of America, Pittsburgh, PA, 1976.

15.4 VELOCITY MEASUREMENT

15.4.1 Introduction
Gustave J. Hokenson

The use of pressure probes, along with some knowledge of the physics of the flow and probe calibrations, allows the local velocity and/or Mach number to be measured, as discussed in Sec. 15.3. Therefore, in this section, other commonly-used devices will be discussed. Due to the complexity of some of the systems, only a general understanding of the instrument capabilities and design can be provided. Some useful references are provided for the reader seeking those details needed to construct and operate a particular apparatus.

The emphasis on calibration of the instruments will be intense. Many of the devices will be useful only after their response is measured in a flow whose properties are determined by some accepted standard measurement instrument or technique. Whether the standard to which the instrument is compared is primary or secondary will not be as important as the relevance of the calibration flowfield to the experimentally-tested flowfield, as determined by the comparability of the appropriate dimensionless groups. Of special interest, therefore, are instruments which provide a measurement of velocity from primary length and time measurements and require no calibration. In addition, the use of any of the instruments for the measurement of complex situations, e.g., nonsteady, multiphase, nonuniform flows, etc., will depend on the need for an adequacy of any calibration.

15.4.2 Propeller, Cups, and Vane Anemometers
Gustave J. Hokenson

These mechanical devices have been used by meteorologists to measure wind speed and direction for many decades. Basically, each device moves in response to the forces applied to it, which are generated by the flow moving around it. The output may be either instantaneous angular velocity or position or a (time-integrated) length of the column of air which passes over the device.

Although these instruments have their origins in atmospheric applications, similar devices are available to measure velocity fields in the ocean. Additionally, and particularly for shrouded propeller anemometers, the recent needs to meet environmental standards in interior spaces, e.g., mines, chemical plants, hospitals, etc., have forced the use of these instruments at very low velocities. In all such applications, and particularly at low velocities, the instrument calibration is critical. At the low end of the velocity range it is often appropriate to move the instrument a given distance through a stagnant fluid over a given period of time. At higher speeds, calibration wind tunnels such as those available at NIST are useful.

In the traditional *cup anemometer* shown in Fig. 15.18, various numbers of (generally) hemispherical, opposed cups are mounted on arms leading away from a bearing on a shaft orthogonal to the plane of the rotor. The angular velocity of the rotor responds to the wind speed (primarily) in the plane of the rotor and produces a cup tangential velocity approximately in the range 0.27–0.33 times the (steady) wind speed. The precise value of the proportionality *constant* and its constancy over the wind speed range of interest for a given device is determined from calibration. The directional response characteristics, also derived from calibrations, typically indicate an insensitivity to wind direction within 10° of ideal alignment where the shaft is orthogonal to the flow.

As discussed by Wyngaard (1981) cup anemometers may be used in non-steady (turbulent) flows as well, providing a good representation of the instrument dynamics is available. For small wind speed fluctuations, the cup anemometer responds as a linear (first-order lag) system, with a time constant, τ, which is proportional to the scale of the device, L, and inversely proportional to wind speed, U. Most cases of

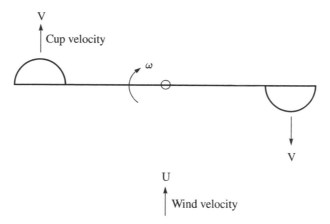

FIGURE 15.18 Opposed, hemispherical cup anemometer.

interest, however, require nonlinear forcing terms in the rotor dynamics equation to model cup overspeeding, for example. The so-called *WBL equation* in Wyngaard (1981) not only includes nonlinear terms in the wind speed and cup speed fluctuations, but also in the velocity, w, normal to the plane of the rotor

$$\tau \frac{dv}{dt} + v = u - 0.23v^2 + 0.96u^2 + 0.67w^2 - 0.73uv \qquad (15.33)$$

where the lower case u and v denote the time-varying U and V of Fig. 15.18.

Facilities exist which can impose a known, large amplitude periodic oscillation on the mean flow to which the anemometer is exposed. In this manner, the instrument transfer function can, in theory, be determined as a function of the amplitude and frequency of flow field fluctuations. The ease with which the nonlinear response may be used to infer the input is in question, and other instruments may be more appropriate for large fluctuations.

Propeller anemometers in most general use today have helocoid rotors and date, therefore, from the 1880's and the work of W. H. Dimes. In the 1960's, the propeller anemometer was simplified and improved by Gill (1973) thereby leading to the commonly-denoted Gill-type anemometers, depicted schematically in Fig. 15.19.

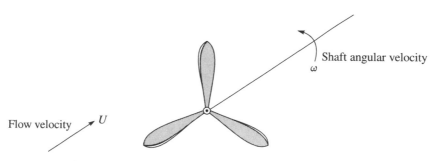

FIGURE 15.19 Gill-type helicoid propeller anemometer.

Whether the propeller is flat or helical, has two, three or four blades, is shrouded or open, it is generally assumed to respond to the flow velocity as a linear, first-order system. The steady rotor tip speed relationship to a steady oncoming flow velocity normal to the rotor plane is a constant over a range of speeds and is derived from calibration.

Dynamically, the linear time constant of the device is related to the mean wind speed and the characteristic diameter of each given design as with the cup anemometer. For large fluctuations, the propeller anemometer responds nonlinearly, however, detailed dynamic analysis of its operational characteristics apparently has not been accomplished. In order to calibrate the dynamic response of a propeller anemometer, it may be rotated in a steady mean flow, as shown in Fig. 15.20, so that the steady mean flow velocity plus the ω_s support arm rotation provides the periodic pulsation. Alternatively, an oscillating flow wind tunnel may be used with a fixed position propeller anemometer.

Vanes are devices which rotate with changes in the flow direction to provide directional information, as in Fig. 15.21, or to orient cup and propeller anemometers into the wind. Most familiar of all is the antiquated *wind sock* which can still be observed at small airfields.

In slowly varying flows, the vane will align itself with the flow, except for a narrow hysteresis region around the null point due to bearing friction. Nonrotating vanes which provide an output force signal as a function of flow angle are also used primarily to indicate flow direction and not to orient flow speed indicators.

For flows with spectral components up to 30 Hz and relatively large amplitude fluctuations, the vane response β (the angle of the vane relative to the wind) may be understood by means of the second order response equation [see Wyngaard (1981)]

$$\frac{d^2\beta}{dt^2} + 2\xi \left(\frac{U}{L}\right) \frac{d\beta}{dt} + \left(\frac{U}{L}\right)^2 (\beta - \phi) = 0 \tag{15.34}$$

If U is constant, U/L is the natural frequency where L is related to the instrument size. ξ is damping ratio, and ϕ is the wind direction. Both ξ and L may be determined from the mechanical properties of the vane and its aerodynamic derivative $\partial C_L/\partial\beta$.

A particularly novel application of vanes is due to Chimonas (1980). He demonstrated the application of a bi-vane to measure the Reynolds stress in a turbulent flow.

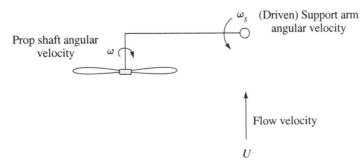

FIGURE 15.20 Experimental arrangement to measure propeller response to fluctuating flow.

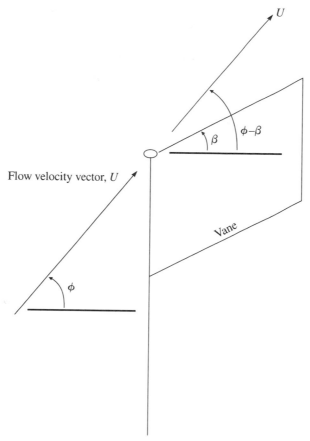

FIGURE 15.21 Vane configuration with flow and vane angles, ϕ and β, relative to fixed axes.

15.4.3 Thermal Anemometry
Rodney D. W. Bowersox

Thermal hot-wire, or hot-film, anemometry is a technique in which fluid properties are measured via a very small heated wire or thin film. The thermal balance between the small heated element and the fluid stream is the fundamental concept at the heart of thermal anemometry. The fluid mass flux, or Reynolds number, can be determined via a heat transfer law (calibration curve) by equating the power across the element to the heat transfer between the element and the fluid. *Constant temperature* and *constant current* anemometry are the two common modes of operation. The constant temperature method monitors the voltage required to maintain the element at a constant temperature (or resistance), where the constant current technique fixes the current across the element.

Hot-wire and hot-film anemometry have proven to be one of the most effective and popular instrument for measuring turbulence [Hinze (1975)]. The bibliography of researchers using hot-wire/film anemometry is extensive [Freymuth (1980)]. The use of hot-wire anemometry to measure gas motion can be first traced to King (1914), where most of the early work involved incompressible flow. The application of hot-wire anemometry to supersonic flow became prevalent in the 1950's [Kovasznay (1950), Spangenberg (1955), Morkovin (1956), and Laufer and McLellan (1956)]. Researchers began to apply hot-wire techniques to hypersonic flows in the 1970's [Fischer *et al.* (1971) and Demetriades and Laderman (1973)]. With the advancements in modern electronics, hot-wire and hot-film anemometry continue to be very powerful research tools.

Hot-wire probes can be operated at many different orientations relative to the fluid flow. Many popular probe arrangements [e.g., normal-wire or cross-wire probes (see Fig. 15.22)] are commercially available. However, just about any angle can be custom made.

First, this section briefly describes hot-wire and hot-film probes and anemometers. Secondly, a detailed discussion of data analysis techniques is presented. The reduction methods are derived for the general case of supersonic or hypersonic flow, however it is shown that the incompressible methods are easily obtained from the compressible equations provided that the density is held fixed. Finally, the errors associated with linearizing the hot-wire response are quantified.

Probes and Anemometers. Representative hot-wire probes are shown schematically in Fig. 15.22. The hot-wire probe consists of a sensor(s), metal prongs or connections, and a probe body. The hot-wire sensor diameters can range from about 0.25–13 μm, with the sensor lengths ranging from approximately 0.4–2.5 mm. Table 15.5 summarizes properties for some typical hot-wire sensors. For aerodynamic testing, the materials listed in Table 15.5 have proven to very popular. This is due to their

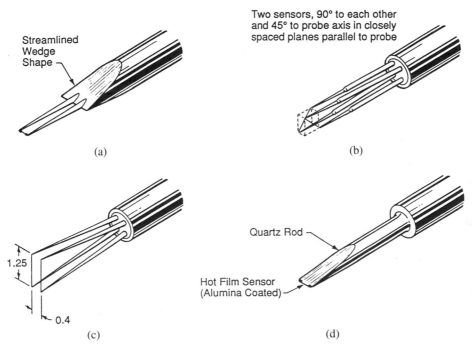

FIGURE 15.22 (a) Normal-wire, (b) cross-wire, (c) parallel-wire, and (d) hot-film probes. (Courtesy of Dantec Measurement Technology A/S and TSI Incorporated.)

TABLE 15.5 Typical Hot-Wire Sensor Materials[1]

Material	Typical Diameter [μm]	Temperature Coefficient of Resistance at 20 C [%/C]	Maximum Sensor Temperature [C]
Platinum Plated Tungsten Wire	4.0	0.42	300
Platinum Wire	5.0[2]	0.30	800
Platinum/Iridium Alloy Wire	6.3	0.09	800
Platinum/Rhodium Alloy Wire	10.0	0.16	800
Alumina Coated Platinum Film	25.0	0.24	350
Quartz Coated Platinum Film	25.0	0.24	80

[1]Data was compiled from DANTEC and TSI probe catalogs. (Courtesy of Dantec Measurement Technology A/S.)

[2]Also available with 1 μm diameter probe suitable for temperature measurement, however maximum current is 1 mA.

relatively low cost, small size, and ease of manufacture. Manufacturers typically offer probe designs suitable for subsonic, supersonic, and hypersonic applications.

The length-to-diameter-ratio (L/d) of the hot-wire sensor is an important parameter in optimizing performance. Considerations include spatial resolution, durability, and nonuniform temperatures along the wire. Obviously, small hot-wire elements are desirable to minimize measurement volume effects. Also, smaller aspect ratio (L/d) wires increase the durability and stiffness of the sensor element. From an application and data analysis perspective, it is desirable that the wire temperature be adequately represented by a single value. This corresponds to a wire of infinite length, uniform cross section, and homogeneous material property distribution. However, real wire sensors are finite in length and are supported by relatively massive prongs (see Fig. 15.22). In fact, in the theoretical analysis of the sensor temperature distribution, the prongs are generally assumed to be heat sinks at the ambient fluid temperature. Corrsin (1963) presents a detailed analysis for the case of a wire cylinder in a uniform flow. Champagne *et al.* (1966) measured the wire temperature distribution for various values of aspect ratio. In general, they found that as L/d was increased, the central region of the wire with a uniform temperature also increased. Thus, large aspect ratio sensors are desirable. Hence, based on the above described desirable attributes, commonly used values of L/d range from about 200–600. Even though the wire temperature is not completely uniform at these aspect ratios, it will be assumed for the remaining discussion that the wire temperature can be adequately represented by a single value.

Hot-film probes are also very popular due to increased strength and durability, and they are quite often preferred for liquid applications. These probes consist of a sensor element (e.g., platinum film) mounted onto a substrate support (e.g., a quartz rod machined to a desired shape). Figure 15.22(d) presents a schematic of a wedge hot-film sensor. Typically, the sensing film will be coated with quartz for electrical insulation from the possibly conducting liquid. Recently, hot-film probes have also been used in hypersonic flows [Stetson and Kimmel (1992)]. These probes do not require electrical insulation, so they are usually coated with alumina, which has high abrasion resistance. Hot-film probes are available in a wide variety of shapes and sizes; for example, cylinders, wedges, cones, and flush mounting.

The same issues that govern the choice of hot-wire selection (spatial resolution, durability, and temperature distribution) apply to hot-films as well. However, since hot-films are available in a variety of shapes and sizes, there are many more variables in the selection process. In fact, in many situations, probe ruggedness becomes the deciding factor in probe selection. For cylindrical probes, the sensor diameters typically range from about 25–150 μm. The sensor lengths range from 0.25 to 2.0 mm. Table 15.5 also includes some typical hot-film properties.

Constant Current Anemometry (CCA) and *Constant Temperature Anemometry* (CTA) are the two common modes of hot-wire operation. A constant current Wheatstone bridge anemometer circuit is shown schematically in Fig. 15.23(a). Kovasznay (1954) discusses the operating principle of this system. The thermal lag of the sensor (wire or film) limits the frequency response to the order of 1000 Hz. However, through electronic compensation, the frequency response can be increased to as high as 500 kHz. The constant temperature Wheatstone bridge circuit is shown in Fig. 15.23(b). The CTA system automatically maintains a balanced bridge and corrects for the thermal lag of the sensor. CTA systems can also deliver a frequency response on the order of 500 kHz.

Bestion *et al.* (1983) present a detailed comparison of both methods as applied to supersonic flow. In summary, both systems were found to perform satisfactorily, however there are advantages and disadvantages associated with each. For example, the frequency response of the CTA system was found to decrease with decreasing wire temperature. Hence, there exists a lower limit on the wire temperature. On the other hand, CTA systems are readily available and are usually simpler to use than CCA systems.

Data Reduction. In this section, the general thermal anemometry response equations will be discussed. First, the general theory of multiple overheat anemometry will be reviewed, where the *effective Reynolds number* concept is utilized to generalize the response to include subsonic (incompressible), supersonic and hypersonic flow fields, as well as cross-wires, normal-wire/film probes or any other wire orientation. Second, a simpler single overheat method will be discussed. Third, the most common modes of operation, normal-wire/film and cross-wire, will be presented. Last, methods of measuring turbulence structure, e.g., length scales, structure angles, and convective velocities will be described.

For turbulent compressible flow, the Nusselt number, the dimensionless heat transfer for a cylinder in crossflow has the following functional form [Kovasznay (1950)]

$$\text{Nu} = f(L/d, M, \text{Pr}, \text{Re}_e, \tau) \qquad (15.35)$$

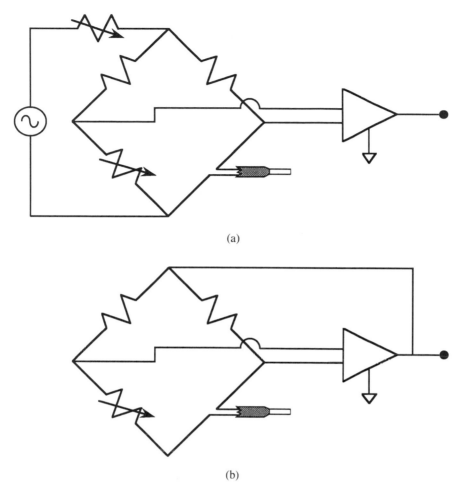

(a)

(b)

FIGURE 15.23 Constant current (a) and constant temperature (b) anemometry Wheatstone bridge circuit diagrams.

L/d is the sensor aspect ratio, M is the Mach number, Pr is the Prandtl number, Re_e is the effective Reynolds number (based on wire diameter), and τ is the temperature loading factor, $\tau = (T_w - T_e)/T_t$, where T_w is the sensor temperature and T_t is the total temperature. T_e, the equilibrium temperature, is the temperature the unheated sensor would attain if placed in the flow. For Reynolds numbers greater than about 20, T_e is about 97% of T_t. For flows where $M > 1.2$ or $M \sin \phi \geq 1$ [Spangenberg (1955)], constant Pr, and $L/d \gg 1$, then Eq. (15.35) reduces to [Kovasznay (1950)]

$$\mathrm{Nu} = f(\mathrm{Re}_e, \tau) \tag{15.36}$$

In general, the convective heat transfer from a hot-wire or film can be expressed as

$$\mathrm{Nu} = [a(\tau)\mathrm{Re}_e^n + b(\tau)] (1 + c\tau)^m \tag{15.37a}$$

where a, b, n, and m are to be determined from calibration. For cylindrical probes, n is usually near 1/2. Aside from fixing n to 1/2, Eq. (15.37a) can be factored, and rewritten without any loss of generality as

$$\text{Nu}_i = a_i \sqrt{\text{Re}_e} + b_i \tag{15.37b}$$

where the subscript i denotes a particular wire temperature, which implies that the hot-wire must be calibrated at each wire temperature. Note, this is the same functional form that King (1914) was able to derive for incompressible flow, and Laufer and McClellan (1956) have experimentally verified this relation for supersonic flow for $M \in (1.3, 4.5)$. Figure 15.24 presents the classical incompressible calibration curve of King (1914), as well as supersonic calibrations, for various wire temperatures over a Mach number range of 1.15–4.54. For subsonic, compressible flows, the hot-wire response has been found to depend on, in addition to the Reynolds number and wire temperature, the local Mach number. It is commonly believed that a family of curves that depend on Mach number spans the region between the incompressible subsonic curve and the supersonic calibrations. Even though the square root dependence of Eq. (15.37b) has proven to be adequate for a large range of applications, other calibration curves have been used. The following analysis is readily extendable to practically any calibration law.

The Nusselt number is defined here as

$$\text{Nu} = \frac{q_w}{\pi k_t L (T_w - T_e)} \tag{15.38}$$

where $q_w = i_w^2 R_w$ (wire or film heat transfer = wire power). From anemometer circuit analysis, $i_w = V_w / (R_w + R_s + R_L)$. Here, R_w is the sensor resistance, R_L is

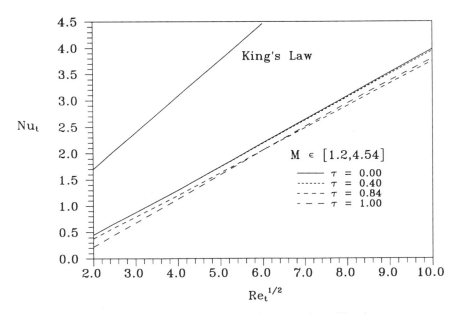

FIGURE 15.24 Typical hot-wire heat transfer calibrations.

the sensor lead resistance, and R_s is the resistance of a resistor in series with the probe; it is common practice for anemometers to include a resistor in series with the probe (R_s is typically around 20–50 Ω).

Usually, the equilibrium temperature is assumed to be the same as the total temperature, which is not too restrictive, since T_t is just slightly larger than T_e [Kovasznay (1950)]. Minimal error will result if the assumption is consistently applied to both the calibration and data reduction.

The thermal conductivity and viscosity are based on T_t in calculating Nu and Re (as indicated by the subscripts t in Fig. 15.24). This is reasonable, since for supersonic flow a bow shock is in front of the wire ($T_2/T_t \approx 1$), and most of the heat transfer takes place in the stagnation region.

To evaluate the mean flow response, the viscosity can be computed using the Sutherland formula, and the thermal conductivity can be computed from a curve fit of tabulated data. However, to evaluate the turbulent fluctuation equations, the simpler power law formulations for μ_t and k_t are used, and are given by

$$k_t = k_o \left(\frac{T_t}{T_o}\right)^{n_k}; \qquad \mu_t = \mu_o \left(\frac{T_t}{T_o}\right)^{n_\mu} \tag{15.39}$$

where the subscript o denotes a reference condition. For air in a temperature range of 200–1000 K, $n_\mu = 0.677$, $n_k = 0.785$, $\mu_o = 30.3 \times 10^{-6}$ Pa-s, $k_o = 45.6 \times 10^{-3}$ W/(m-K), where $T_o = 600$ K. Combining Eqs. (15.37)–(15.39), the *thermal anemometry response equation* can be obtained, and is given by

$$\frac{i_w^2 R_w}{\pi k_o L} = \left(\frac{T_t}{T_o}\right)^{n_k} \left[a\sqrt{\text{Re}o_e} \left(\frac{T_t}{T_o}\right)^{-n_\mu/2} + b\right] (T_w - T_t) \tag{15.40}$$

where $\text{Re}o_e$ is the effective Reynolds number with $\mu = \mu_o$. For the CTA system, R_w and T_w are constant, and for the CCA system, i_w is constant, but T_w and R_w both contain a mean and fluctuating part. The relationship between the sensor resistance and temperature is usually assumed linear and given by

$$T_w = T_{\text{ref}} + \frac{a_{\text{ref}}}{\gamma_{\text{ref}}} \tag{15.41}$$

where a_{ref}, the reference overheat ratio, is defined as $(R_w - R_{\text{ref}})/R_{\text{ref}}$. The reference temperature is typically 20 C, and γ_{ref} is the temperature coefficient of resistance which depends on the sensor material (see Table 15.5 for typical values).

Replacing i_w (or V_w), $\text{Re}o_e$, and T_t by their mean plus a fluctuating component, and applying the *Binomial Theorem*, retaining only the first order terms, noting that

$$\frac{\overline{i_w^2} R_w}{\pi k_o L} = \left(\frac{\overline{T_t}}{T_o}\right)^{n_k} [a\sqrt{\overline{\text{Re}}_e} + b] (T_w - \overline{T_t}) \tag{15.42}$$

and then solving for $v_w'/\overline{V}_w$ leads to the *thermal anemometry fluctuation equation* which is given by

$$\frac{v'_w}{V_w} = f\left(\frac{Reo'_e}{Reo_e}\right) + g\left(\frac{T'_t}{T_t}\right) \tag{15.43}$$

where the *CTA sensitivities* are given by

$$f_{CTA} = \frac{1}{4}\left(1 + \frac{b}{a\sqrt{\overline{Re}_e}}\right)^{-1}; \quad g_{CTA} = \frac{-\overline{T}_t}{2(T_w - \overline{T}_t)} + \frac{n_k}{2} - f_{CTA}n_\mu \tag{15.44}$$

The *CCA sensitivities* are given by

$$\frac{f_{CCA}}{f_{CTA}} = \frac{g_{CCA}}{g_{CTA}} = -2\bar{a}_t \tag{15.45}$$

where $\bar{a}_t = (\overline{R}_w - \overline{R}_t)/\overline{R}_t$. It is important to notice here that $Reo'/\overline{Reo}$ is identically $(\rho u)'/\overline{\rho u}$ for supersonic or hypersonic flow fields. For incompressible flows, $Reo'/\overline{Reo}$ reduces to $u'/\bar{u}$. However, the Reynolds number nomenclature will be consistently applied throughout the discussion.

Instead of using the analytical formulations for the sensitivities, some researchers determine f and g from a direct calibration. However, if Eqs. (15.44) or (15.45) are to be used, then the mean quantities $\sqrt{Re_e}$ and $\overline{T}_t$ must be measured. This can be accomplished with the thermal anemometry probe and Eq. (15.42), which is rewritten here as

$$\sqrt{\overline{Re}_e} + x_i\overline{T}_t\sqrt{\overline{Re}_e} + y_i\overline{T}_t = z_i \tag{15.46}$$

where

$$x_i = -\frac{1}{T_{w_i}}$$

$$y_i = -\frac{b_i}{(a_iT_{w_i})}$$

$$z_i = \frac{\overline{V}^2_{w_i}}{(C_ia_iT_{wi})} - \frac{b_i}{a_i}$$

$$C_i = \frac{(R_w + R_s + R_L)^2}{R_w}\pi Lk_t \tag{15.46a}$$

The subscript i indexes the overheat ratio. At a minimum, two sensor temperatures are required to resolve the unknowns $\sqrt{Re_e}$ and $\overline{T}_t$. If more than two overheat ratios are used, then a *least squares analysis* yields

$$N\sqrt{\overline{Re}_e} + \overline{T}_t\left(\sum y_i - \sum x_iz_i\right) + 2\overline{T}_t\sqrt{\overline{Re}_e}\sum x_i$$

$$+ \overline{T}^2_t\sum x_iy_i + \overline{T}^2_t\sqrt{\overline{Re}_e}\sum x^2_i = \sum z_i \tag{15.47a}$$

$$\sqrt{\overline{Re_e}}\left(\sum y_i - \sum x_i z_i\right) + \overline{T_t} \sum y_i^2 + 2\sqrt{\overline{Re_e}}\,\overline{T_t} \sum x_i y_i$$

$$+ \sqrt{\overline{Re_e}}^2 \sum x_i + \sqrt{\overline{Re_e}}^2 \overline{T_t} \sum x_i^2 = \sum y_i z_i \qquad (15.47b)$$

where N is the number of overheat ratios, and the summations are over the i overheat ratios. The solution of the above two nonlinear equations is iterative, where the *Secant method* can be incorporated to drive the residual to zero. Figure 15.25 demonstrates the excellent agreement that can be achieved with the CTA cross and normal-wire mean flow methods as compared to those obtained by conventional mean flow probes for a supersonic turbulent mixing layer.

To obtain the turbulence results, Eq. (15.43) is squared and time-averaged, yielding

$$f_i^2 \overline{\left(\frac{Reo_e'}{Reo_e}\right)^2} + 2f_i g_i \overline{\left(\frac{Reo_e'}{Reo_e}\frac{T_t'}{T_t}\right)} + g_i^2 \overline{\left(\frac{T_t'}{T_t}\right)^2} = \overline{\left(\frac{v_w'}{V_w}\right)^2}_i \qquad (15.48)$$

Three overheat ratios are required to resolve the three turbulence terms. If more than three wire temperatures are used, then a least squares analysis or the classical *fluctuation diagram method* of Kovasznay (1953) can be applied. Generally speaking, a minimum of 6–10 overheat ratios is desired. Since computers are commonplace, the present discussion will concentrate on the readily automated least squares method. The *General Least Squares Method* (GLS-Method) can be derived if a least squares analysis is applied directly to Eq. (15.48), yielding the following 3 × 3 system

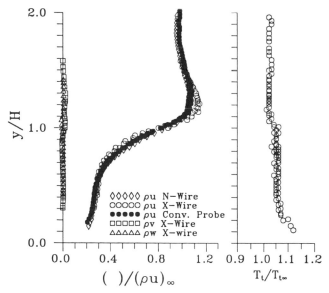

FIGURE 15.25 Comparison of multiple overheat (8 wire temperatures) hot-wire results to conventional mean flow methods.

$$
\begin{bmatrix}
\sum f_i^4 & 2\sum f_i^3 g_i & \sum f_i^2 g_i^2 \\
\sum f_i^3 g_i & 2\sum f_i^2 g_i^2 & \sum f_i g_i^3 \\
\sum f_i^2 g_i^2 & 2\sum f_i g_i^3 & \sum g_i^4
\end{bmatrix}
\begin{bmatrix}
\overline{\left(\dfrac{\mathrm{Reo}_e'}{\overline{\overline{\mathrm{Reo}_e}}}\right)^2} \\[2ex]
\overline{\left(\dfrac{\mathrm{Reo}_e'}{\overline{\overline{\mathrm{Reo}_e}}}\dfrac{T_t'}{\overline{T_t}}\right)} \\[2ex]
\overline{\left(\dfrac{T_t'}{\overline{T_t}}\right)^2}
\end{bmatrix}
=
\begin{bmatrix}
\overline{\sum f_i^2 \left(\dfrac{v_w'}{V_w}\right)^2} \\[2ex]
\overline{\sum f_i g_i \left(\dfrac{v_w'}{V_w}\right)_i^2} \\[2ex]
\overline{\sum g_i^2 \left(\dfrac{v_w'}{V_w}\right)_i^2}
\end{bmatrix}
$$

$$(15.49)$$

where the summations are over the N overheat ratios. This least squares method is general in the sense that the forms of f and g are not required. Hence, the above analysis is independent of the anemometer system (i.e., CTA or CCA), or the sensitivities f and g can be evaluated by direct calibration.

The thermal anemometry equations derived in this section are also general in the sense that the form of Re_e has not been prescribed. The two special cases of Re_e to be discussed below correspond to the popular normal-wire/film and cross-wire probes. However, the analysis can be applied to practically any probe orientation. Further, the hot-wire response equations derived here are general in the sense that they are valid for supersonic, hypersonic, and incompressible flows. Recall, for subsonic compressible and transonic flows, the response has the additional complication of depending on the local Mach number.

Normal-wire/film anemometry has been well established in both the incompressible and supersonic flow regimes. For this application, Re_e is simply the x-component of the Reynolds number in the wind axes, i.e., Re_{wx}. Thus, the *normal-probe response equation* can be obtained by substituting Re_{wx} into Eqs. (15.40)–(15.49).

Incompressible cross-wire anemometry has also been well established [Freymuth (1980)]. However, the cross-wire (or the swept-wire), in supersonic flow, has not received a great deal of attention. Early pioneering work has been published by Spangenberg (1955) concerning the response of a swept wire at supersonic speeds. Spangenberg (1955) found that if $M \sin \phi \geq 1$, then the cross-wire response was independent of Mach number. Therefore, Eq. (15.36) is valid.

Many workers have developed techniques to correct for tangential wire cooling in incompressible flow fields [see Blackwelder (1981) for a complete listing]. These cooling laws have led to effective Reynolds number formulations (or for incompressible flow, effective velocity) that have been used routinely in Eq. (15.37). The most popular method is that of Champagne *et al.* (1966). However, a tangential cooling law has not been developed for inclined wires in supersonic flows. In the present analysis, the methods of Champagne *et al.* (1966) will be generalized to include supersonic and hypersonic flows through an inclusion of oblique shock theory. In addition, this method has the advantage that if the tangential cooling constant is set to zero, then the methods are consistent with the early results of Spangenberg (1955), where to within measurement uncertainty, the cross-wire response was found to be adequately described by the cosine law, i.e., depending only on the normal component of Reynolds number.

The equations are derived in Cartesian (or *tunnel*) coordinates, where the angle, ϕ, is defined as the angle between the x-axis and the normal to the wire (positive counter-clockwise), see Fig. 15.26. The relationships between the normal and tangential Reynolds number components to the x and y components are given by the following matrix transformation

$$\begin{pmatrix} \text{Re}_n \\ \text{Re}_t \end{pmatrix} = \begin{bmatrix} \cos(\phi) & \sin(\phi) \\ -\sin(\phi) & \cos(\phi) \end{bmatrix} \begin{pmatrix} \text{Re}_x \\ \text{Re}_y \end{pmatrix} \tag{15.50}$$

For the x-z orientation (Fig. 15.26), the signs on the sin terms are switched. The functional form of Re_e to be used in Eq. (15.37) is given by

$$\begin{aligned} \text{Re}_e^2 &= \text{Re}_n^2 + k_c^2 \text{Re}_t^2 \\ &= A_1 \text{Re}_x^2 + 2 A_2 \text{Re}_x \text{Re}_y + A_3 \text{Re}_y^2 \end{aligned} \tag{15.51}$$

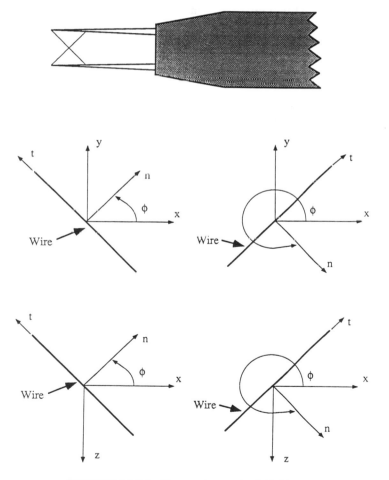

FIGURE 15.26 Cross-wire angle definitions.

where A_i are given by

$$A_1 = \cos^2(\phi) + k_c^2 \sin^2(\phi)$$

$$A_2 = (1 - k_c^2) \cos(\phi) \sin(\phi) \qquad (15.52)$$

$$A_3 = k_c^2 \cos^2(\phi) + \sin^2(\phi)$$

A possible determination of k_c stems from the analysis of Champagne *et al.* (1966) with the inclusion oblique shock theory. The normal component of mass flux is conserved across the oblique shock just upstream of the wire. However, the tangential velocity is conserved, as opposed to the mass flux. Hence, the tangential Reynolds number behind the shock can be written as

$$\mathrm{Re}_{t2} = \frac{\rho_2}{\rho_1} \mathrm{Re}_{t1} \qquad (15.53)$$

where 1 refers to the conditions just prior to the shock and 2 represents the conditions between the shock and the wire. After making first-order approximations

$$k_c = \frac{\overline{\rho_1}}{\overline{\rho_2}} k \qquad (15.54)$$

where $k = f(L/d)$, and must be determined experimentally (see Table 15.6). Notice, that if $\rho_2 = \rho_1$, then this method reduces back to the incompressible results of Champagne *et al.* (1966). Also, if k is assumed zero, then the method reduces to the cosine law.

Replacing Re_{o_e}, Re_{o_x}, and Re_{o_y} by their mean plus the fluctuating components, applying the Binomial Theorem, and using the following definitions

$$R_o = \frac{\overline{\mathrm{Re}_{o_y}}}{\overline{\mathrm{Re}_{o_x}}}$$

$$B_1 = \frac{A_1}{B_3} \qquad (15.55)$$

$$B_2 = \frac{A_2}{B_3}$$

$$B_3 = A_1 + 2A_2 R_0$$

TABLE 15.6 Dependence of k on L/d[1]

L/d	k^2
600	0.00
450	0.09
375	0.12
250	0.18
200	0.21

[1]Data from Champagne *et al.* (1966).
[2]Standard deviation for k is 0.04.

it can be shown that mean flow cross-wire response is given by

$$\overline{Reo_{ej}} = \overline{Reo_x}\,\sqrt{B_{3j}} \tag{15.56}$$

The turbulence response equation can be written as

$$\left(\frac{Reo'_e}{\overline{Reo_e}}\right)_j = B_{1j}\left(\frac{Reo'_x}{\overline{Reo_x}}\right) + B_{2j}\left(\frac{Reo'_y}{\overline{Reo_x}}\right) \tag{15.57}$$

where j indexes one of the two wires on the cross-wire probe. In the derivation of Eqs. (15.56) and (15.57), it was assumed that $R_o^2 \ll 1$.

The mean x and y components of the Reynolds number can be found from Eq. (15.56) as

$$\overline{Reo_x}^2 = \frac{\overline{Reo_{e1}}^2/A_{21} - \overline{Reo_{e2}}^2/A_{22}}{A_{11}/A_{21} - A_{12}/A_{22}}$$

$$\overline{Reo_y} = \frac{1}{2\overline{Reo_x}}\,\frac{\overline{Reo_{e1}}^2/A_{11} - \overline{Reo_{e2}}^2/A_{12}}{A_{21}/A_{11} - A_{22}/A_{12}} \tag{15.58}$$

The turbulent x and y components of the Reynolds number can be found via the crosswire fluctuation response equation, Eq. (15.57). Thus,

$$\overline{\left(\frac{Reo'_x}{\overline{Reo_x}}\right)^2} = \frac{1}{D_2^2}\left[\frac{1}{B_{21}^2}\overline{\left(\frac{Reo'_e}{\overline{Reo_e}}\right)^2}_1 - \frac{2}{B_{21}B_{22}}\overline{\left(\frac{Reo'_e}{\overline{Reo_e}}\right)_1\left(\frac{Reo'_e}{\overline{Reo_e}}\right)_2} + \frac{1}{B_{22}^2}\overline{\left(\frac{Reo'_e}{\overline{Reo_e}}\right)^2}_2\right]$$

$$\overline{\left(\frac{Reo'_y}{\overline{Reo_x}}\right)^2} = \frac{1}{D_1^2}\left[\frac{1}{B_{11}^2}\overline{\left(\frac{Reo'_e}{\overline{Reo_e}}\right)^2}_1 - \frac{2}{B_{11}B_{12}}\overline{\left(\frac{Reo'_e}{\overline{Reo_e}}\right)_1\left(\frac{Reo'_e}{\overline{Reo_e}}\right)_2} + \frac{1}{B_{12}^2}\overline{\left(\frac{Reo'_e}{\overline{Reo_e}}\right)^2}_2\right]$$

$$\overline{\left(\frac{Reo'_x}{\overline{Reo_x}}\frac{Reo'_y}{\overline{Reo_x}}\right)} = \frac{1}{2B_{11}B_{21}}\left[\overline{\left(\frac{Reo'_e}{\overline{Reo_e}}\right)^2}_1 - B_{11}^2\overline{\left(\frac{Reo'_x}{\overline{Reo_x}}\right)^2} - B_{21}^2\overline{\left(\frac{Reo'_y}{\overline{Reo_x}}\right)^2}\right] \tag{15.59}$$

where $D_1 = (B_{21}/B_{11} - B_{22}/B_{12})$ and $D_2 = (B_{11}/B_{21} - B_{12}/B_{22})$. The GLS-Method or the Kovasznay (1953) fluctuation diagram can be used to solve for all the terms on the right-hand side of Eq. (15.59) except the middle correlation between wire 1 and wire 2. The correlation term can be found from the covariance between the two wires on the cross-wire probe. The relationship for the covariance can be expressed as

$$\overline{\left(\frac{v'_w}{\overline{V_w}}\right)_1\left(\frac{v'_w}{\overline{V_w}}\right)_2} = f_1 f_2 \overline{\left(\frac{Reo'_e}{\overline{Reo_e}}\right)_1\left(\frac{Reo'_e}{\overline{Reo_e}}\right)_2} + g_1 g_2 \overline{\left(\frac{T'_t}{\overline{T_t}}\right)}$$

$$+ f_1 g_2 \overline{\left(\frac{Reo'_e}{\overline{Reo_e}}\frac{T'_t}{\overline{T_t}}\right)_1} + f_2 g_1 \overline{\left(\frac{Reo'_e}{\overline{Reo_e}}\frac{T'_t}{\overline{T_t}}\right)_2} \tag{15.60}$$

With Eq. (15.60), the fluctuation x and y components of the Reynolds number can now be computed.

The remaining turbulence variables to decompose into x and y components are the Reynolds number-total temperature correlations. Therefore, it can be shown that

$$\frac{\overline{Reo'_x\, T'_t}}{\overline{Reo_x}\,\overline{T_t}} = \frac{1}{D_2}\left[\frac{1}{B_{21}}\overline{\left(\frac{Reo'_e\, T'_t}{\overline{Reo_e}\,\overline{T_t}}\right)}_1 - \frac{1}{B_{22}}\overline{\left(\frac{Reo'_e\, T'_t}{\overline{Reo_e}\,\overline{T_t}}\right)}_2\right]$$

$$\frac{\overline{Reo'_y\, T'_t}}{\overline{Reo_x}\,\overline{T_t}} = \frac{1}{D_1}\left[\frac{1}{B_{11}}\overline{\left(\frac{Reo'_e\, T'_t}{\overline{Reo_e}\,\overline{T_t}}\right)}_1 - \frac{1}{B_{12}}\overline{\left(\frac{Reo'_e\, T'_t}{\overline{Reo_e}\,\overline{T_t}}\right)}_2\right]$$

(15.61)

Many workers operate the CTA at the largest possible overheat (limited by the wire material), to minimize the total temperature sensitivity g [Eq. (15.44)]. Figure 15.27 illustrates the sensitivities, for a hot-wire probe, as a function of the wire temperature, for the nominal conditions of $T_t = 300$ K and Re $= 100$. The error analysis of Bowersox, Ng, and Schetz (1991) indicated that the errors associated with this methodology can be as high as 25%, and the correction techniques they developed are generalized here to include cross-wire as well as normal-wire probes.

Many flow fields exhibit only small mean total temperature gradients, implying that the total temperature fluctuations should be small, and mainly due to coupling with the mass flux fluctuations. If $d(T_t) = 0$ (adiabatic flow field) and Pr $= 1$, then the *Crocco Integral* implies [Schetz (1993)] that, for a calorically perfect gas, $T = T(u)$. Therefore, it was postulated that $T' = T'(\bar{u}, u')$, and with this, the energy equation can be used to obtain a correlation between the temperature and velocity fluctuations. First, if $T'_t = 0$, then

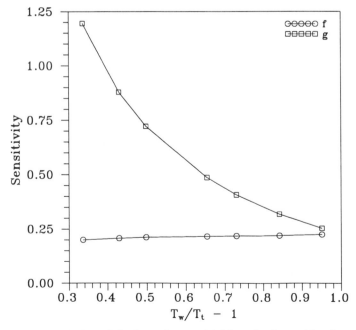

FIGURE 15.27 Variation of the hot-wire sensitivities, f and g, with wire temperature.

$$\frac{T'}{\overline{T}} = -(\gamma - 1) M^2 \frac{u'}{\overline{u}} \tag{15.62}$$

which is the familiar *Strong Reynolds Analogy*, presented by Morkovin (1956). However, Pr $\neq$ 1, and even for flow fields that are adiabatic in the mean, $T'_t \neq 0$. Thus, it was postulated that

$$\frac{T'}{\overline{T}} = -\kappa(\gamma - 1) M^2 \frac{u'}{\overline{u}} \tag{15.63}$$

where κ is a correction factor, which was determined empirically using multiple overheat hot-wire data from Kistler (1959), Bowersox, Ng, and Schetz (1991), and Bowersox (1992). For supersonic wall boundary layers and wake-type flows that are nominally adiabatic in the mean, the value of κ was found to be about 0.5 (see Fig. 15.28). Defining $\theta \equiv -k(\gamma - 1)M^2$, using Eq. (15.63), and assuming $R_o^2 \ll 1$, then

$$\frac{u'}{\overline{u}} = \frac{1}{1 - \theta} \frac{(\rho u)'}{\overline{\rho u}} + \overline{\rho u} \tag{15.64}$$

and

$$\frac{v'}{\overline{u}} = \frac{R_0 \theta}{1 - \theta} \frac{(\rho u)'}{\overline{\rho u}} + \frac{(\rho v)'}{\overline{\rho u}} \tag{15.65}$$

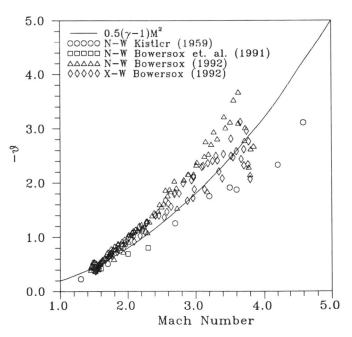

FIGURE 15.28 Comparison of single overheat correction parameter (θ) with experimental velocity-temperature correlation data.

Therefore, the total temperature fluctuation equation becomes

$$\frac{T_t'}{\overline{T_t}} = \frac{\beta + \alpha\theta}{1 - \theta} \frac{(\rho u)'}{\overline{\rho u}} + \overline{\rho u} \tag{15.66}$$

where, again, $\alpha = [1 + \frac{1}{2}(\gamma - 1)M^2]^{-1}$ and $\beta = (\gamma - 1)\alpha M^2$. With Eqs. (15.63)–(15.66), the crosswire response can be approximated as

$$[S_{ij}] \begin{bmatrix} \overline{\left(\dfrac{\mathrm{Re}o_x'}{\overline{\mathrm{Re}o_x}}\right)^2} \\[2em] \dfrac{\overline{\mathrm{Re}o_x'}}{\overline{\mathrm{Re}o_x}} \dfrac{\mathrm{Re}o_y'}{\overline{\mathrm{Re}o_x}} \\[2em] \overline{\left(\dfrac{\mathrm{Re}o_y'}{\overline{\mathrm{Re}o_x}}\right)^2} \end{bmatrix} = \begin{bmatrix} \overline{\left(\dfrac{v_w'}{\overline{V_w}}\right)^2_1} \\[2em] \overline{\left(\dfrac{v_w'}{\overline{V_w}}\right)^2_2} \\[2em] \overline{\left(\dfrac{v_w'}{\overline{V_w}}\right)_1 \left(\dfrac{v_w'}{\overline{V_w}}\right)_2} \end{bmatrix} \tag{15.67}$$

where the coefficients of the single overheat matrix, S, are given by

$$S_{11} = f_1^2 B_{11}^2 + 2f_1 g_1 R_{mT} R_{uT} B_{11} + g_1^2 R_{mT}^2$$

$$S_{12} = 2[f_1^2 B_{11} B_{21} + f_1 g_1 R_{mT}(R_{uT} R_0 B_{11} + R_{vT} B_{21}) + g_1^2 R_{mT}^2 R_0]$$

$$S_{13} = f_1^2 B_{21}^2 + 2f_1 g_1 R_{mT} R_{vT} R_0 B_{21}$$

$$S_{21} = f_2^2 B_{12}^2 + 2f_2 g_2 R_{mT} R_{uT} B_{12} + g_2^2 R_{mT}^2$$

$$S_{22} = 2[f_2^2 B_{12} B_{22} + f_2 g_2 R_{mT}(R_{uT} R_0 B_{12} + R_{vT} B_{22}) + g_2^2 R_{mT}^2 R_0]$$

$$S_{23} = f_2^2 B_{22}^2 + 2f_2 g_2 R_{mT} R_{vT} R_0 B_{22}$$

$$S_{31} = f_1 f_2 B_{11} B_{12} + f_1 g_2 R_{mT} R_{uT} B_{11} + f_2 g_1 R_{mT} R_{uT} B_{12} + g_1 g_2 R_{mT}^2$$

$$S_{32} = f_1 f_2(B_{11} B_{22} + B_{12} B_{21}) + f_1 g_2 R_{mT}(R_{uT} R_0 B_{11} + R_{vT} B_{21})$$
$$\quad + f_2 g_1 R_{mT}(R_{uT} R_0 B_{12} + R_{vT} B_{22}) + 2g_1 g_2 R_{mT}^2 R_0$$

$$S_{33} = f_1 f_2 B_{21} B_{22} + f_1 g_2 R_{mT} R_{vT} R_0 B_{21} + f_2 g_1 R_{mT} R_{vT} R_0 B_{22} \tag{15.68}$$

where

$$R_{mT} = \frac{\beta + \alpha\theta}{1 - \theta}$$

$$R_{uT} = \frac{\overline{(\rho u)' T_t'}}{\sqrt{\overline{(\rho u)'^2}} \sqrt{\overline{T_t'^2}}} \tag{15.69}$$

$$R_{vT} = \frac{\overline{(\rho v)' T_t'}}{\sqrt{\overline{(\rho v)'^2}} \sqrt{\overline{T_t'^2}}}$$

Equation (15.68) reduces to the normal-wire results of Bowersox, Ng, and Schetz (1991), since $B'_1 = 1$, $B'_2 = 0$, and $R_o = 0$. Also, setting $R_{mT}(k = 1)$ equal to zero reduces Eq. (15.68) to the typical method of neglecting the total temperature terms, or setting $R_{mT} = 1$ corresponds to the *Strong Reynolds Analogy* of Morkovin (1956).

Figure 15.29 compares the various single overheat reduction methods for a cross-wire probe to the more rigorous GLS-method multiple overheat technique. Interest-

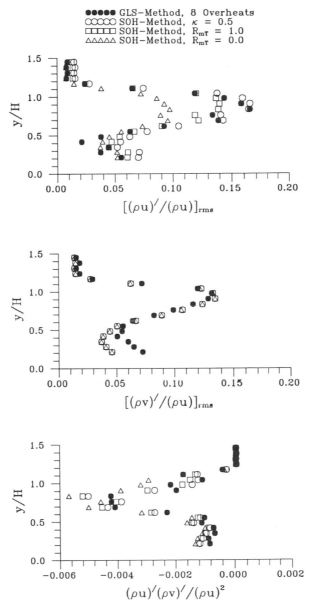

FIGURE 15.29 Comparison of various single overheat techniques (overheat ratio was set to 0.95) to GLS-method multiple overheat results (8 wire temperatures).

ingly, the largest uncertainties associated with the single overheat methods were associated with the $(\rho u)'$ turbulence intensity. The errors noticeable in the uncorrected data are consistent with the analysis of Bowersox, Ng, and Schetz (1991).

The procedure for the separation of the mass flux and total temperature terms into velocity and density fluctuation results for the normal-wire probe has been well established. The underlining assumption in the uncoupling of the turbulence data is that for a thermally perfect gas $(p = \rho R T)$, the effects of the pressure fluctuations on the hotwire response equations are small compared to the density and temperature fluctuations. After making first-order assumptions, the following results, where the p' terms will be included for completeness, can be obtained

$$\frac{u'}{\bar{u}} = \frac{(\rho u)'}{\overline{\rho u}} - \frac{\rho'}{\bar{\rho}}$$

$$\frac{v'}{\bar{u}} = \frac{(\rho v)'}{\overline{\rho u}} - R_0 \frac{\rho'}{\bar{\rho}} \qquad (15.70)$$

$$\frac{\rho'}{\bar{\rho}} = \frac{1}{\alpha + \beta} \left[\beta \frac{(\rho u)'}{\overline{\rho u}} - \frac{T_t'}{T_t} + \alpha \frac{p'}{p} \right]$$

With Eq. (15.70) and assuming $\alpha p' \approx 0$, one can arrive at all the turbulent shear terms in the Reynolds (time) averaged Navier–Stokes equations [Bowersox and Schetz (1994)].

To obtain the terms in the Favre (mass-weighted time) averaged Navier–Stokes equations, the following relationships can be used

$$\frac{\overline{u_j''}}{U} = - \frac{\overline{\rho' u_j'}}{\overline{\rho u_i}}$$

$$\overline{u_i'' u_j''} - \overline{u_i''}\,\overline{u_j''} = \overline{u_i' u_j'} \qquad (15.71)$$

$$\overline{\rho h_o'' u_j''} = \overline{h_o'(\rho u_j)'} + U_i \overline{\rho' h_o'}$$

where $(\ )''$ correspond to Favre (mass) averaged fluctuating components, $U = \overline{\rho u}/\bar{\rho}$, and i and j index coordinate directions, and no repeated indicies are implied.

The assumption that $p' \approx 0$ is problematical. Kistler (1959), suggested that $p' \sim u'^2$ and demonstrated that this assumption was valid in boundary layers for $M_e \in (1.7, 4.8)$. Referring to the density fluctuation equation in Eq. (15.70), the effects of p' can be seen. It is important to notice that the p' term is multiplied by a factor α, which is always less than 1, and as $M \to \infty$, $\alpha \to 0$. Thus, even if p' is not equal to zero, the effects on separating the hot-wire variables may still be small, especially for hypersonic flow fields. One can also notice that if R_o is small (i.e., $\bar{v}$ is on the order of v'), then $(\rho v)'$ can be separated without the p' assumption.

When a parallel-wire [Fig. 15.22(c)] is subjected to a flow where there is a significant amount of organized structure, the angle of this structure can be estimated from a crosscorrelation between the two wires. The structure angle associated with the average large-scale motion is given by [Bowersox and Schetz (1994)]

$$\theta = \tan^{-1} \left(\frac{w}{\bar{u} \tau_{\max}} \right) \qquad (15.72)$$

where w is the wire separation distance, $\bar{u}$ is the mean stream wise velocity or the convective velocity (discussed below), and τ_{max} is the time delay between the signals (which corresponds to the time of the peak on the cross correlation).

The integral length scale (Λ) and the micro length scale (λ) of the turbulence [Cebeci and Smith (1974)] can be estimated from the normalized autocorrelation function (normalized by the variance), where the time axis, in correlation space, has been transformed into length by multiplying by the local mean velocity (i.e., an application of *Taylor's Hypothesis*). The integral scale is found by integrating the autocorrelation function from $\bar{u}\tau = 0$ to ∞. The microscale is the value of $\bar{u}\tau$ found by setting the *osculating* parabola, which is determined by fitting a parabola through the first few points of the autocorrelation, equal to zero.

The convective velocity can be estimated from the cross correlation between two parallel-wires [Spina *et al.* (1991)]. The convective velocity is given by

$$U_c = \frac{w}{\tau_{max}} \tag{15.73}$$

where w is the wire separation distance, and τ_{max} is the time delay between the signals which corresponds to the time at the peak on the cross correlation).

Linearization Error Analysis. To arrive at the thermal anemometry response equations, the response was linearized via the Binomial theorem. Here, we discuss the errors associated with this approximation. In order to estimate the magnitude of the neglected terms, the second order terms in the binomial expansion are included, thus

$$\frac{V'_w}{\overline{V}_w}(1 + e_V) = f\frac{Reo'}{\overline{Reo}}(1 + e_R) + g\frac{T'_t}{\overline{T}_t}(1 + e_T) \tag{15.74}$$

where the errors are given by

$$e_V = \frac{1}{2}\frac{V'_w}{\overline{V}_w}$$

$$e_R = -\frac{1}{4}\frac{Reo'}{\overline{Reo}} + \left(\frac{n_\mu}{2} + n_x\right)\frac{T'_t}{\overline{T}_t}$$

$$e_T = H\frac{T'_t}{\overline{T}_t} \tag{15.75}$$

where H is defined for convenience as

$$H = -\frac{f}{g}n_\mu\left[\frac{1}{2}\left(\frac{n_\mu}{2} - 1\right) + n_R - 2g\right] \tag{15.76}$$

Equation (15.75) implies that the linearization errors are first order.

15.4.4 Time-of-Flight Devices
Gustave J. Hokenson

The flow-invasive devices discussed in Secs. 15.3 (pressure probes) and 15.4.2 and 15.4.3 (mechanical devices and heated-element devices) rely on calibration, flow physics, and modeling the instrument-flow interaction for relating the instrument output to the flow velocity. In this and the subsequent parts of Sec. 15.4, various (essentially) non-invasive instruments will be described whose flow velocity measuring capabilities are grounded in primary measurements of length and time.

Due to its simplicity, the (low density flow) ion-tracer technique of Lillienfeld *et al.* (1967), shown in Fig. 15.30, later studied by Brown and Good (1982) will be discussed first. Two electrically-conducting probes are located in the flow separated by a fixed, known distance. A high voltage pulse is applied to the upstream probe which ionizes some of the gas which is subsequently advected downstream and collected by the second probe. The time delay between the applied pulse and collected current divided into the known probe spatial separation provides a measure of the flow speed. This technique is applicable to low density flows, and the calibrations of Brown and Good (1982) indicate that, with suitable precautions, the instrument may be used to infer flow velocities without calibration. Note that it is assumed that the devices which measure length and time are well-characterized.

In subsequent sections, the spirit of this device in relating the flow velocity to primary measurements of time interval and fluid particle displacement distance will be carried forward.

15.4.5 Laser Velocimetry
Gustave J. Hokenson

The technology of laser velocimetry is rooted in the interaction between an illuminating monochromatic light source and scattering centers (particles) distributed throughout the flow [see Penner and Jerskey (1973), Durrani and Greated (1977), Durst *et al.* (1981), and Adrian (1993)]. For sufficiently small, neutrally-buoyant particles, it may be assumed that the scattering center motions are in equilibrium with (identical to) the Lagrangian fluid *point* at that spatial and temporal location. In this case, the measured particle scattering characteristics are used to infer the local flowfield properties. If the particle concentration is sufficiently dilute, it is

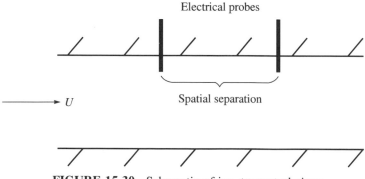

FIGURE 15.30 Schematic of ion–tracer technique.

assumed that the inferred flowfield characteristics are unaffected by the presence of the particles. In fact, many laboratory flows naturally contain an adequate supply of scatterers. For truly two-phase flows with large flow-interacting particles of significant concentration, laser velocimetry may be used to provide information concerning the particles. However, inference of the flowfield properties is then experimentally difficult.

One of the earliest forms of laser velocimetry is depicted in Fig. 15.31 and conveys one essence of the technique which is computing velocities by measuring the primary quantities of length and time (time of flight). In this configuration the light scattered by a particle as it passes two spatially-separated beams is composed of two temporally-staggered bursts. If these spatial and temporal scales are measured accurately, the particle velocity may be computed directly, *without calibration*. This *non-Doppler* technique may be expanded to include the extraction of information on particle size and two velocity components. In addition, by analyzing the scattered light of (small) particles, their velocity normal to the mean may be inferred from a single-beam system [see Hirleman (1982)] in which the scattered intensity, as a function of time, is related to the particle passage velocity through the beam. Clearly, both of these techniques are related to the detection of events of varying intensity at varying times.

More typical of most laser velocimetry systems is the fact that, independent of the scattered light intensity, its frequency is Doppler shifted from the incident illumination by the particle motion. The amount of the frequency shift is $\bar{\mathbf{k}} \cdot \bar{\mathbf{U}}$ where $\bar{\mathbf{k}}$ is the relative wavenumber vector between the incident and scattered light, $\bar{\mathbf{k}} = \bar{\mathbf{k}}_i - \bar{\mathbf{k}}_s$. A direct measurement of this small shift in a high frequency signal is generally beyond the scope of available instruments. However, by heterodyning the scattered light with the incident light, the beat frequency may be readily measured. A typical configuration is depicted in Fig. 15.32.

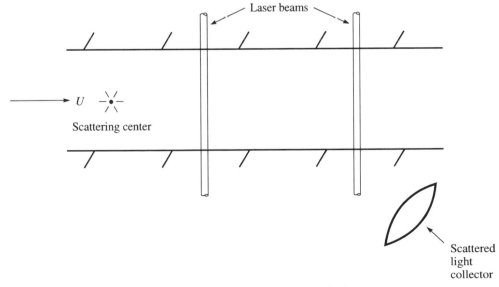

FIGURE 15.31 A simplified laser velocimeter.

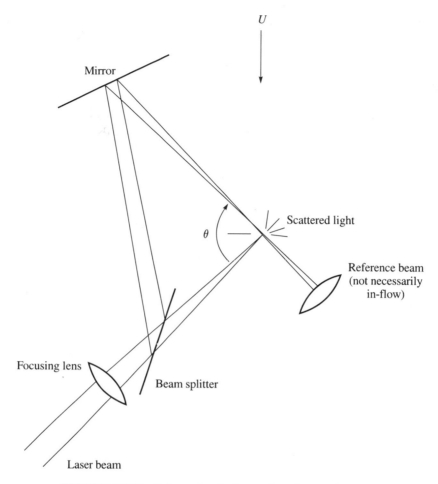

FIGURE 15.32 Schematic of a heterodyne laser velocimeter.

The relationships between the scattering center velocity and the mixed (beating) signal between these beams is

$$\omega = \frac{2U}{\lambda} \sin \frac{\theta}{2} \qquad (15.77)$$

which is linear and requires no calibration. Only the measurement of physical properties is necessary—namely the illumination wavelength, λ, and the angle, θ, between the reference and illumination beams. Note that these beams are symmetrically disposed about the normal to U.

An extension of the heterodyne configuration is to mix and focus two laser beams at a given point in the flowfield, as in Fig. 15.33. As a result, interference fringes are formed within the focal volume, and, as a particle traverses the focal volume, the scattered light intermittently exhibits peaks. The fringe spacing and particle velocity determine the frequency or temporal spacing of the peaks in scattered light intensity. For this *real fringe* system, the relationship between particle velocity and

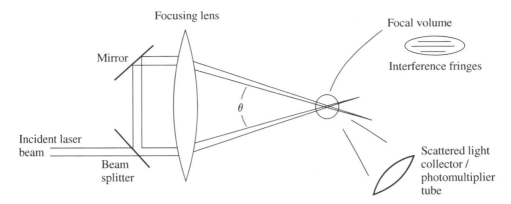

FIGURE 15.33   Real fringe laser Doppler velocimeter schematic.

scattered light characteristics is identical to the previously discussed heterodyne system. For this reason, the heterodyne system is sometimes referred to as the *apparent fringe* system.

Either of these systems may be extended to measure multiple components of velocity. Additionally, depending on the scattering particle concentration, various signal analysis techniques including frequency trackers (for nearly continuous signals) and frequency burst counters (for dilute particle concentrations) are utilized [see Buchhave *et al.* (1979)]. Frequency domain processors are now available to extract low uncertainty velocity information from noisy signals [Shinpaugh *et al.* (1992)].

An inherent problem of the basic systems discussed so far is the ambiguity relative to flow direction. This problem may be resolved by putting some directional bias into the configuration. For example, the original two beam (time of flight) arrangement may be modified so that one beam is more intense than the other. Therefore, one *blip* of detected scattered light will be more intense and the direction of particle motion may be inferred.

With regard to the real fringe LDV system, the frequency of one of the beams may be modulated (e.g., by a Bragg cell) so as to produce moving fringes in the focal volume. With this *biasing*, a particle which is stationary in space within the focal volume will scatter light at a given frequency. Depending on the particle velocity and direction relative to these moving fringes, the scattered light will be frequency shifted relative to the known stationary particle value.

Various configurations of the LDV allow multiple velocity components to be measured. In addition, the scattered light may be collected in the back-scatter mode utilizing the same optical components which form the focal volume. System alignment is thereby facilitated. When combined with spectroscopy techniques, flow temperature, pressure, and species may also be measured. Particle size, velocity, and concentrations in multiple-phase flows are clearly accessible by this technique.

Prior to use of this superficially simple system, some knowledge of the subtleties of data interpretation in the various data collection modes is required [see Buchhave *et al.* (1979) and Adrian (1993)]. Included in the various phenomena which have been encountered is the so-called *biasing* of particle measurements toward the higher velocities discussed by McLaughlin and Tiederman (1973) and others [Adrian (1993)].

15.4.6 Particle Image Velocimetry
Luiz M. Lourenço and Anjaneyulu Krothapalli

One of the most challenging problems in fluid mechanics is the measurement of the overall flow field properties, such as the velocity, vorticity, and pressure fields. Local measurements of the velocity field (i.e., at individual points) are now done using Hot-Wire (HW) and Laser Doppler Velocimetry (LDV). However, many flow fields such as coherent structures in shear flows or wake flows, are highly unsteady. HW or LDV data of such flows are difficult to interpret, as both spatial and temporal information of the entire flow field are required, and these methods are commonly limited to simultaneous measurement at only a few spatial locations. Hence, a whole field velocity measurement technique is valuable in the characterization of these flow fields. One such technique is Particle Image Velocimetry (PIV).

The principal behind PIV is that instantaneous fluid velocities can be evaluated by recording the position of images produced by small tracers suspended in the fluid at successive time instants. The underlying assumption is that these tracers follow closely, and with minimal lag, the fluid motion. This assumption holds true for a wide variety of flows of interest provided that the tracers are small enough and/or that their density approaches that of the fluid.

The terms *Laser Speckle Velocimetry* and *Particle Image Velocimetry* are often used interchangeably. However, these terms reflect an important distinction related to the particle density in the flow. To understand these differences, it is first necessary to describe how the application of this technique to measurement of fluid flows developed. The term *speckle*, refers to the granular appearance that diffusely reflecting or transmitting surfaces take on when illuminated by a laser beam. This is caused by constructive and destructive interference of coherent light scattered from a surface whose roughness is large compared with the wavelength of the laser.

The speckle effect is used for the measurement of displacements in solid mechanics as follows. A surface is first illuminated by a laser beam, as shown in Fig. 15.34. When this surface is imaged through a lens onto a photographic plate,

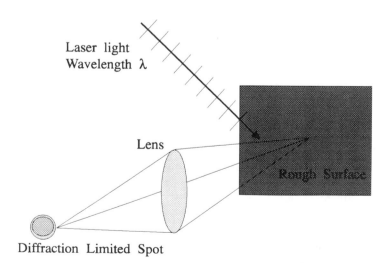

FIGURE 15.34 The diffraction limited image.

the interference of the scattered light wavelets gives rise to a speckle pattern. The speckle size is a statistical average of the distance between adjacent regions of maximum and minimum brightness and can be estimated [Erf (1980)] by the Rayleigh resolution criterion

$$d_s = 2.44\lambda f\#(1 + M) \tag{15.78}$$

where d_s is the size of a speckle grain, λ is the wavelength of the illuminating laser, $f\#$ is the f-number of the recording optics, and M is the magnification.

A doubly exposed photograph of the speckle pattern is then taken, once before and once after a lateral displacement is introduced. This specklegram, contains two correlated grids which can be analyzed as a nonuniform diffraction grating. The technique can only be used when the displacement between exposures is greater than the speckle size, d_s, but not so great as to destroy correlation. Thus, the individual speckle size sets the lower measurable limit. Analysis of this specklegram can be performed using the Young's fringe analysis method as shown by Burch and Tokarski (1968).

The application of this method to the measurement of velocity in a fluid involves the generation of a surface within the flow field. The orientation of this plane is such that it contains the dominant flow direction. The plane itself is created by seeding the flow with small tracers, such as those used in LDV applications, and illuminating them with a sheet of very intense pulsed light, as depicted in Fig. 15.35. The light scattered by the tracer particles in the illuminated plane provides the moving pattern.

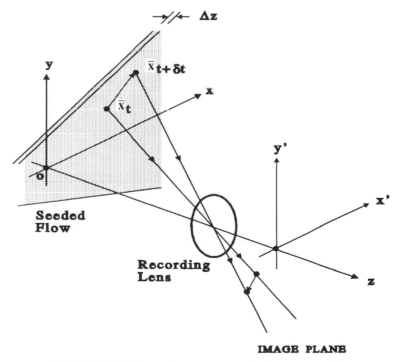

FIGURE 15.35 Basic arrangement for image recording.

Although speckle photographs obtained in solid mechanics and fluid mechanics are similar, there are two fundamental differences. The first is that the fluid is illuminated by a laser sheet whose thickness is Δz. Therefore, scattering occurs from a volume distribution rather than a surface distribution. Secondly, the number density of particles per unit volume (seeding concentration) can vary over a wide range of values. Strictly speaking, for a true speckle pattern to exist, the number of scattering particles must be so high that the images overlap and interfere. When the particle concentration is lower than this level, discrete images of the particles will be photographed instead. Operation of the technique with a low particle concentration is referred to as *Particle Image Velocimetry*. A more detailed description of the technique can be found in recent review articles by Lourenco *et al.* (1989) and Adrian (1990).

Principle of Operation. The measurement of the velocity field using PIV is based on the ability to accurately record and measure the positions of small tracers suspended in the flow as a function of time. The velocity is then deduced from the relationship

$$\overline{U} = \frac{\overline{x}_{t+\delta t} - \overline{x}_t}{\delta t} \tag{15.79}$$

where $\overline{x}$ is the position vector of the fluid tracers, measured at closely spaced time intervals and $\overline{U}$ the approximation to the local fluid velocity.

The technique can operate in two distinct modes. In the *autocorrelation mode*, a recording of two or more instantaneous image patterns created by the seeds are recorded in the same frame. In the *crosscorrelation mode*, the individual instantaneous patterns are kept in separate frames. The time interval between exposures is chosen such that the tracer particles will have moved only a few diameters, far enough to resolve their motion, but less than the smallest fluid macroscale.

The Autocorrelation Approach: Let us consider a small region of a doubly exposed digital frame with uniform particle displacements. The ensemble of the particle image doublets is represented by the function $D(x, y)$

$$D(x, y) = I(x, y) + I(x + \Delta x, y + \Delta y) \tag{15.80}$$

where $I(x, y)$ and $I(x + \Delta x, y + \Delta y)$ represent the images of the particles recorded during the first and the second exposures, and $\overline{s}(\Delta x, \Delta y)$ is the average particle displacement. For the sake of simplicity, it is assumed that all the particle images have a corresponding pair, i.e., purely two-dimensional motion, and that the size of the region being interrogated is large when compared with the particle image displacement. Equation (15.80) is written in the following form

$$D(x, y) = I(x, y) \otimes [\delta(x, y) + \delta(x + \Delta x, y + \Delta y)] \tag{15.81}$$

where $\delta(x, y)$ is the Dirac delta function centered at $\overline{x}(x, y)$ and $\otimes$ is the convolution operator.

Fourier analysis provides us with a convenient means to obtain the autocorrelation function. Two Fourier transformations are required. The Fourier transform of $D(x,$

y) is represented by $F\{D(x, y)\} = D^*(\omega_x, \omega_y)$, where F is the Fourier transform operator, $j = \sqrt{-1}$ and ω_x, ω_y are the spatial frequencies along the x and y directions respectively,

$$D^*(\omega_x, \omega_y) = I^*(\omega_x, \omega_y) [1 + e^{-2\pi j(\Delta x \omega_x + \Delta y \omega_y)}] \qquad (15.82)$$

where $I^*(\omega_x, \omega_y)$ represents the Fourier transform of the collection of particle images. The Spectrum is obtained by multiplication of Eq. (15.82) with its complex conjugate, resulting in

$$|D^*(\omega_x, \omega_y)|^2 = 2|I^*(\omega_x, \omega_y)|^2 (1 + \cos [2\pi(\Delta x \omega_x + \Delta y \omega_y)]) \qquad (15.83)$$

where the term $|I^*(\omega_x, \omega_y)|^2$ represents the spectrum of the collection of particle images at random locations. Visually this is seen as a *diffraction halo* modulated by a fine speckle grain due to multiple interferences. This halo is also modulated by the cosine term in Eq. (15.83) resulting in a straight fringe pattern, known as Young's fringes. The envelope of the *halo* and the diffraction limited particle images are related through the Fourier transform and are both closely approximated by a Gaussian distribution. The visual Young's fringe pattern can also be generated by an optical transformation method as shown by Burch and Tokarski (1968). The autocorrelation is found by means of a second Fourier transformation as follows,

$$F\{|D^*(\omega_x, \omega_y)|^2\} = \iint |D^*|^2 \, e^{-2\pi j(\Delta x \omega_x + \Delta y \omega_y)} \, d\omega_x \, d\omega_y \qquad (15.84)$$

Equation (15.84) is rewritten as

$$\iint |D^*|^2 \, e^{-2\pi j(\Delta x \omega_x + \Delta y \omega_y)} \, d\omega_x \, d\omega_y = 2H(0, 0) + H(\Delta x, \Delta y) + H(-\Delta x, -\Delta y)$$

$$(15.85)$$

where $F\{|D^*(\omega_x, \omega_y)|^2\}$ is the Fourier transform of the intensity *halo*. The envelope of $H(0, 0)$ is also closely approximated by a Gaussian. Equation (15.85) clearly shows the three main terms of the autocorrelation function. These are $H(0, 0)$ centered at the origin and shifted by $(\Delta x, \Delta y)$ and $(-\Delta x, -\Delta y)$. Thus, the topology of the autocorrelation function allows for the determination of the average particle displacement in the interrogated region simply by finding the location of the peak of either $H(\Delta x, \Delta y)$ or $H(-\Delta x, -\Delta y)$.

The Cross-Correlation Approach: In this mode of operation, the technique requires that each image produced by the two illumination pulses be stored in separate frames. The distance and direction traveled by the tracer particles is then unambiguously determined by performing the cross correlation between the two small and corresponding regions of the image frame acquired in quick succession. Consider the two small regions corresponding to the first and second exposure, respectively

$$D_1(x, y) = I(x, y) \qquad (15.86)$$

and

$$D_2(x, y) = I(x + \Delta x, y + \Delta y) = I(x, y) \otimes [\delta(x + \Delta x, y + \Delta y)] \quad (15.87)$$

The cross correlation may also be obtained using the Fourier transform operator. However, three Fourier transforms are required. First, the transforms of $D_1(x, y)$ and $D_2(x, y)$ are evaluated

$$F\{D_1(x, y)\} = I*(x, y) \quad (15.88)$$

and

$$F\{D_2(x, y)\} = I*(x, y) \, (e^{-2\pi j(\Delta x \omega_x + \Delta y \omega_y)}) \quad (15.89)$$

The cross-correlation function, $G(x, y)$, is obtained by evaluating the Fourier transform of the product $F\{D_1(x, y)\} \cdot F\{D_2(x, y)\}$,

$$F\{D_1(x, y)\} \cdot F\{D_2(x, y)\} = I*(x, y)^2 \cdot (e^{-2\pi j(\Delta x \omega_x + \Delta y \omega_y)}) \quad (15.90)$$

and

$$G(x, y) = \int\!\!\int I*^2 \, e^{-2\pi j(\Delta x \omega x + \Delta y \omega y)} \, e^{-2\pi j(\Delta x \omega x + \Delta y \omega y)} \, d\omega_x \, d\omega_y \quad (15.91)$$

resulting in

$$G(x, y) = F\{I*^2\} \otimes [\delta(\Delta x, \Delta y)] \quad (15.92)$$

The cross correlation $G(x, y)$ is thus a function centered at $(\Delta x, \Delta y)$, the average displacement coordinates. In contrast with the auto correlation, there is only one correlation peak, with much larger signal amplitude. The main advantages of the cross correlation approach over the autocorrelation are: 1) the displacement is obtained without any directional ambiguity, and 2) the correlation peak signal is carries more signal strength, and thus is more immune to noise. The main disadvantages of the cross-correlation approach are that: 1) the computation is more *expensive* in time, as three two-dimensional Fourier transforms are required instead of two for the autocorrelation, and 2) the image acquisition system (camera) must cope with the necessity of acquiring two image frames in quick succession in synchronization with the laser illumination pulses and register the frame position with respect to the flow with absolute precision. (Presently, this requirement can only be met by video cameras in low speed flows—less than 1 m/sec.)

Therefore, the autocorrelation approach is currently the method of choice for whole field velocity measurements. However, the cross correlation is an effective means if the flow velocity does not exceed a few centimeters/sec. When the autocorrelation method is employed, additional means are required to obtain information about the velocity direction.

Detailed Considerations. The technique relies on the ability to detect and record on a photographic plate, or solid state sensor, the images of the tracer particles. This image is a function of the scattering power of the particles, the amount of light, the exposure time, magnification, and the sensitivity of the recording media. Additional details on the parameters playing a role in photographic recording of PIV images are contained in Lourenco and Krothapalli (1987).

The autocorrelation approach is the most widely used method. However, one disadvantage of using a doubly or multiply exposed recording is the 180 degree ambiguity in determining the direction of the velocity vector. Measuring the separation of the tracer image pairs provides the magnitude of the velocity at that point, but is insufficient to give the direction of the velocity vector field.

A method to resolve both this ambiguity, as well as to improve the dynamic range of the measurements, was first introduced by Ewan (1979). The velocity vector direction ambiguity was resolved by translating the recording photograph by a known, uniform amount, between the first and second exposure. The same method was reintroduced by Adrian (1986) and Lourenco *et al.* (1986). This method is termed *spatial image shifting* or *velocity bias technique*. The method consists conceptually of recording the flow field in a moving reference frame, thus superposing a known velocity bias to the actual flow velocity. The image can be shifted using a moving camera, or by optical means (e.g., using rotating or scanning mirrors). Other image shifting methods use light polarization effects. Landreth and Adrian (1988) proposed the use of a calcite crystal as a *filter* to separate and shift images of particles illuminated with either *s* or *p* polarized light. An alternate method using the light polarization effect was proposed by Lourenco (1989), (1993) which replaces the calcite crystal with a beam splitter cube and two isolators (1/4 wave plate) arrangement.

The basic assumption in all techniques that use tracer particles added to the flow, is that the motion of the particle accurately follows the motion of the fluid. The particle dynamics must be especially considered in flows with large velocity gradients, vortical structures and flows containing shocks. The effect of particle velocity lag is discussed by Ross *et al.* (1994) and Crisler *et al.* (1994). The choice of the type of particle and the mechanism used for their introduction in the fluid stream, must be carefully considered to minimize particle agglomeration and provide uniform distribution in the fluid stream.

The minimum detectable particle diameter is a function of the recording optics and the laser input energy. Particles ranging from 0.5–10 μ are fairly typical. It should be noted that, for particles of this size, the diameter of the recorded image is relatively insensitive to the actual particle diameter, because the image size is dominated by diffraction effects from the lens.

In applications of PIV to liquid flows, particles must be relatively small to follow the fluid motion and closely match the density of the fluid to avoid buoyancy effects. Metallic coated microballons are probably the best choice because of their reduced diameter, which ranges from 10–50 μm, high efficiency as light scatterers, and specific gravity of about 1.03.

Extremely bright light sources are usually required because of the low efficiency of the side scattering process used in PIV recordings. The amount of laser light energy required is a function of tracer size and concentration, recording lens aperture and magnification, and sensor sensitivity. For a successful detection and recording of a particle image, the mean exposure of an individual particle image must be greater than the sensor sensitivity at the wavelength of the illuminating laser. In the case of photographic film this minimum sensitivity is sometimes referred to as the *gross fog* level [Adrian and Yao (1985)]. The fog level is defined as the exposure level below which the transmissivity of the film is independent of the incident intensity. In analytical terms, this is expressed as

$$\epsilon = \Gamma \delta t > CE_0 \tag{15.93}$$

where ϵ is the mean exposure of an individual particle image, Γ is the average intensity of light scattered by a particle, E_0 is the film fog level, δt is the duration of the illumination pulse and C is a constant between 1 and 10.

The mean intensity of the light, Γ, per unit surface area, of the light scattered by a single particle can be expressed as

$$\Gamma = I_0 \frac{\lambda^2 \int \sigma^2 \, d\Omega}{\pi^3 d_i^2} \tag{15.94}$$

where λ is the wavenumber of the illuminating laser light, d_i is the nominal diameter of the particle including diffraction, I_0 is the intensity of the illuminating sheet, σ is the Mie parameter, and Ω is the solid angle subtended by the camera lens. The effective dimension of the particle image, d_i, can in turn be expressed as

$$d_i = (M^2 d_p^2 + d_s^2)^{1/2} \tag{15.95}$$

where d_p is the actual particle diameter, and d_s is the diffraction-limited spot diameter of the particle image, given in Eq. (15.78). When a pulsed laser is used, and assuming that the particle is stationary during exposure, the laser power required is determined from Eq. (15.93) where the recommended value of the constant C is three to five to account for the film reciprocity effects caused by very short exposures.

The Mie parameter is a function of the refractive index ratio between the particle and the fluid. Because the refractive index of water is 1.3 times that of air, the scattering efficiency of particles in air is much more effective than in water. Therefore, in experiments in water, the particles need to be larger than those in air. However, because the much smaller velocities involved in liquid flows, the particle size, of the order of 10 μm, is still able to track the flow.

When using conventional photographic film to record the particle images, it is essential to maximize the film resolution while maintaining a good level of sensitivity. Table 15.7 presents some commonly used photographic films and their characteristics.

TABLE 15.7 Resolution and Sensitivity of Current Films

Film Type	Exposure for Density = 1 (ergs/cm²)	Resolving Power (Line Pairs/mm)
Kodak Holographic Plate		
Type 120 − 20 @ 694.3 mm	330	1250
Kodak Technical Pan		
@ 694.3 nm	1.5	320
Kodak TMAX-100		
@ 532 nm	3.0	100

To compensate for the many cases when limited laser power is available, the film used should have good sensitivity, without sacrificing resolution. Unfortunately, for commercially available photographic films, speed and resolving power are inversely related. Hence, choices range between high speed, low resolving power and low speed, high resolving power. For most practical applications good precision is necessary, so the advantage lies with the high resolution films (with about 100–300 line pairs/mm).

There are practical bounds to the particle concentration in PIV applications. The upper boundary is set simply at that value above which a speckle pattern is formed. If C_p is the particle concentration and Δz the width of the laser sheet, the maximum concentration is given by

$$\{1/(\Delta z C_p)\}^{1/2} \gg d_i/M \qquad (15.96)$$

where d_i is the image diameter, given in Eq. (15.95). In addition to contributing to unwanted attenuation of the illumination sheet and interactions with the flow, large concentrations of particles are not easy to generate and may contribute to loss in signal correlation in flows with out-of-plane velocity components [Lourenco and Krothapalli (1987)].

The lower end in the PIV mode can be determined by the criterion that a minimum number of particle image pairs must be present in the area being scanned by the interrogation beam. The case of a single particle image pair is ideal, because it yields fringes with optimum signal-to-noise ratio. However, this situation can only be achieved by lightly seeding the flow, which gives rise to signal drop-out. An interesting case occurs when two particle pairs are present in the interrogation area. The resulting diffraction pattern includes multiple, equally intense fringe patterns due to cross interference of noncorresponding image pairs. In this situation, the local displacement cannot be resolved. As the number of particle image pairs in the interrogation area increases, the cross interference fringes become weaker in comparison with the main fringe pattern, which reflects the local displacement. These cross interference fringes are sometimes called *background speckle noise*. Experience shows that at least four particle image pairs should be present in the interrogation area.

The time between exposures determines the temporal resolution and is chosen in accordance with the maximum expected velocity in the flow field and the required spatial resolution. The spatial resolution, which in turn is proportional to the area of the interrogated region, is dictated by the scales associated with the fluid motion. The spatial resolution should be less than the smallest scale in the flow being studied.

The time between exposures, Δt, is determined by the maximum permissible displacement of a particle such that a valid correlation is obtained. A necessary condition to obtain a valid correlation is that the distance between adjacent particle images be less than a fraction of the linear dimension of the region being interrogated. In practice, the maximum permissible displacement that can be detected corresponds to 1/3 to 1/4 of the dimension of the analysis region, d_{int}. This is also valid in the case of the optical fringe processor, where d_{int} would represent the interrogating laser beam diameter. Thus, the time between exposures to obtain displacements less than 1/4 the analysis region size, everywhere in the field is set according to the maximum expected velocity, U_{max}

$$\Delta t = \frac{d_{\text{int}}}{4MU_{\text{max}}} \tag{15.97}$$

This points up an advantage of this technique, in that velocity sensitivity range can easily be shifted by altering the pulse separation, Δt.

For very short exposures, the recorded particle images are identical to the diffraction limited particle images, as the particles appear to be stationary during the exposure. When the exposure time is increased, the recorded images becomes streaks whose length is directly proportional to the exposure time. The diffracted light in the spectrum is concentrated in a band whose width is inversely proportional to the streak length. For optimum exposure, the exposure time must be less than $\Delta t/10$.

The *dynamic range* of the technique is defined as the largest velocity difference that can be detected. The low end of the dynamic range is determined by the requirement that the spacing between successive particle images be well resolved. That is, corresponding particle doublets should not overlap. This requirement can be expressed as

$$l = d_i + U_{\text{min}}\delta t < \Delta t U_{\text{min}} \tag{15.98}$$

where l is the spacing between successive particle images, and U_{min} is the minimum measurable flow velocity. It was shown above that Δt, the time between exposures, is related to the maximum velocity U_{max} thus, assuming that $\delta t \approx 0$, we obtain

$$U_{\text{min}} = \frac{4Md_iU_{\text{max}}}{d_{\text{int}}} \tag{15.99}$$

The velocity dynamic range is then defined as the normalized velocity difference

$$\frac{(U_{\text{max}} - U_{\text{min}})}{U_{\text{min}}} = \frac{d_{\text{int}}}{4Md_i} - 1 \tag{15.100}$$

Uncertainties in the velocity measurement include errors introduced during the recording of the multiple exposure photograph, such as the ones introduced by the distortion of the scene being recorded by the camera lens, limited lens and film resolution, three-dimensional effects [Lourenco and Krothapalli and Smith (1989)], bias introduced by large velocity gradients [Adrian (1988) and Keane and Adrian (1990)], and the inaccuracies due to the processing algorithms. The latter is essentially determined by how close the displacement $\bar{s}(\Delta x, \Delta y)$ is approximated. Because the correlation is represented digitally, the maximum displacement is initially obtained by the peak value of the discrete correlation array. For improved accuracy, i.e., in order to resolve the peak coordinates with sub-grid accuracy, this estimate needs to be further refined. The accuracy of the measurement is thus determined, in part, by the choice of sub-grid interpolation algorithm. Lourenco and Krothapalli (1995) have shown that the Gaussian interpolator is the strategy that combines both accuracy and efficiency.

Examples of Application. The turbulent near-wake of a circular cylinder placed in a uniform stream is experimentally investigated in a water towing tank using the technique by Shankar *et al.* (1995). The Reynolds number, based on the cylinder diameter, D, and the freestream velocity, U_∞, is 3900, and the corresponding Strouhal number defined as $f_s D/U_\infty$, where f_s is the measured shedding frequency, is found to be 0.195.

A recording camera with a scanning mirror system is mounted on a platform that extends from the cylinder carriage. The scanning mirror is used to introduce a velocity bias in order to resolve flow reversals and stagnant flow regions. A CW laser with a rotating polygon provided the laser sheet illumination. The time separation between the laser pulses can be varied with the rotation speed of the polygon. Metallic coated particles, with an average diameter of 11 μm are used as the flow tracers. A phase-triggered 35 mm camera (Nikon F-3) with a 105 mm Macro lens is used in the recording of the flow field, at a rate of 6 frames/sec. A total of 93 recordings corresponding to 30 cycles of the vortex shedding were obtained.

The multiply exposed photographic film is processed to produce the velocity field using a Young's fringe analysis system. The velocity field is obtained by interrogation of the photographic image on a regular grid (60 $\times$ 40).

Figure 15.36 shows four consecutive velocity and the corresponding vorticity fields of the near wake. These figures demonstrate the turbulent character of the wake as well as that vortices are shed alternatively from the top and bottom of the cylinder. Because the time dependent velocity field is acquired with more than twice the vortex shedding frequency, it is possible to reconstruct the velocity at any other time instant.

In spite of its remarkable capabilities, the PIV technique based on photographic recording can be difficult to implement. This difficulty stems from the fact that before the analysis of the doubly or multiply exposed photographic recordings, there are one or two dark room development steps involved, such as the wet processing of the film negative and eventual contact printing of the film negative to improve the data quality. It is, therefore, desirable to develop an alternative approach that facilitates this set-up stage by making the image acquisition step less laborious and time consuming.

A new, fully digital and operator interactive PIV system has been introduced by Lourenco *et al.* (1994). This instrument uses high resolution CCD area sensors and state of the art microcomputer hardware and software, and unlike earlier digital approaches it is capable of recording flows within a wide range of velocities (from a few millimeters per second up to several hundreds of meters per second).

The *On-Line* PIV system described was also used to map the supersonic flow issuing from a convergent-divergent nozzle by Krothapalli *et al.* (1994). The nozzle pressure ratio (stagnation pressure/ambient pressure) was 7.8, and the Mach number is 2. The jet was seeded with 0.3 μm Al_2O_3 particles, while the ambient air was seeded with 1 μm smoke particles. The central plane of the jet over the first six diameters was illuminated by a light sheet created by a Lumonics double pulse (pulse width = 20 nsec) Ruby laser operating at 1 Hz. The time between exposures was about 1 μsec. The image was acquired using a 1.4 MB (1340 $\times$ 1035) Kodak digital camera in conjunction with FFD MkIII[TM] PIV system. To capture the details of the shear layers a velocity bias was applied using a high speed (6000 rpm) spinning mirror. For details of the experiment reference can be made to Wishart (1994).

A typical double exposure image is shown in Fig. 15.37(a). The image extends from the nozzle exit to about 4 jet exit diameters. The maximum velocity in front of the normal shock is about 515 m/sec. The corresponding velocity field is shown

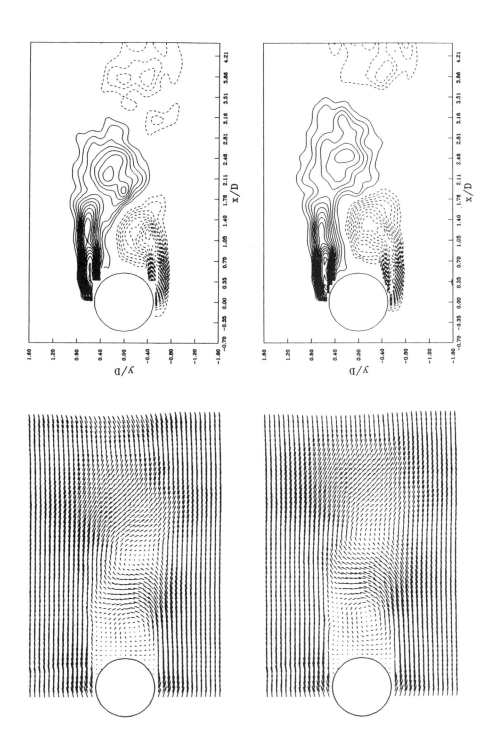

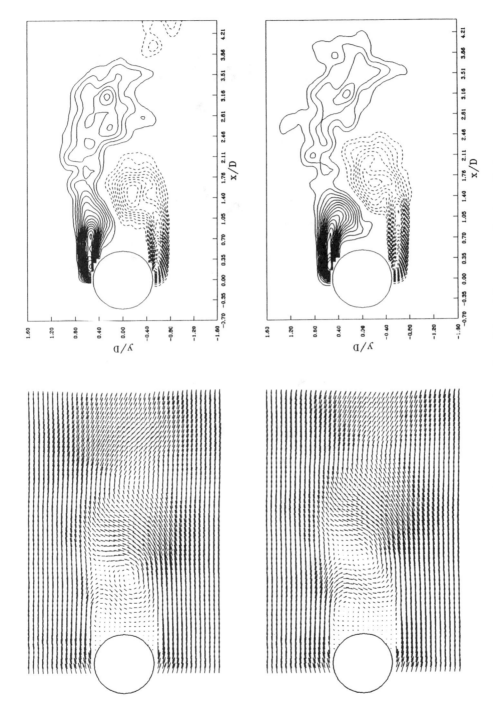

FIGURE 15.36 Synoptic velocity and vorticity of flow past a circular cylinder.

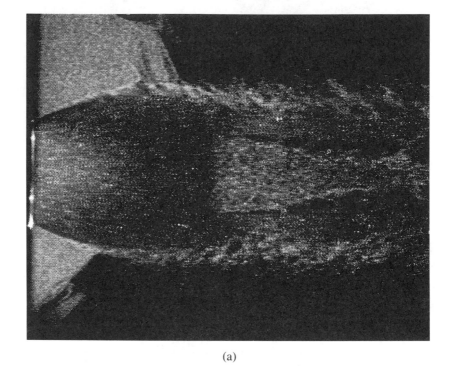

(a)

(b)

FIGURE 15.37 PIV studies of an underexpanded supersonic jet. (a) Doubly exposed photograph of under expanded jet and (b) velocity field obtained from the images depicted in (a).

in Fig. 15.37(b). This picture shows the detailed velocity field in a shock cell containing the following features: a normal shock, slipstream shear layers inside the jet, and three-dimensional shear layers at the jet boundary.

Three-Dimensional Measurements. Until recently, applications of PIV have been limited to two-dimensional flows. The seeding particles must be in the illuminated sheet during both exposures. Any out-of-plane motion of particles into or out of this sheet reduces the particle correlation and can result in a loss of signal. However, many flows of interest are three-dimensional, and ways of extending PIV to the study of such flows are of obvious interest. One such method was implemented by Gauthier and Riethmuller (1988). The method is a stereoscopic scheme in which the flow in the illuminated plane is viewed from two different directions simultaneously. Thus, measurements over a single plane are still made, but all three components of velocity are obtained. The constraint is that the time delay between exposures must be chosen such that the maximum particle motion normal to the illuminated plane is much less than the thickness of the sheet. With this scheme, it is often necessary to add a velocity bias to the particle images to keep them in the range of the analysis system.

In the stereoscopic method, the flow is viewed from two directions, and the three velocity components are obtained from their projections and the geometrical characteristics of the optical system. Two different stereoscopic systems have been investigated. The first is called an *angular displacement* method where the optical axes are not perpendicular to the illuminated sheet but are inclined at an angle in relation to the normal to the sheet. The second stereoscopic scheme is referred to as the *translation method*. In this technique, the optical axis of each camera is perpendicular to the illuminated sheet, and the distance between the two axes provides the stereoscopic effects.

Gauthier and Riehtmuller (1988) applied each of these methods to a simulated 3D flow by measuring the uniform flow in a rectangular duct (25 × 40 mm) with a laser sheet at an angle of 20 degrees to the duct axis. These basic experiments demonstrated the applicability of PIV for the instantaneous measurements of the three components of velocity in a plane of a fluid flow. Prasad and Adrian (1992) applied essentially the same technique to map the velocity distribution on the liquid flow of a rotating disk. They also concluded that the lateral stereoscopic technique performed measurements of both the in-plane and out-of-plane velocity components with a high degree of accuracy. Errors of 0.2% full scale and 0.8% of full scale for the in-plane and out-of-plane components, respectively, were reported.

Current research by the authors and co-workers [Van Eck (1994)] is aimed at developing a holographic recording method by means of which the three velocity components can be recorded in three-dimensional regions of the flow. The particulate tracer images in the flow are recorded holographically, i.e., both their amplitude and phase, thus making it possible, in principle to reconstruct their position in the three-dimensional space, in contrast with traditional photographic methods which only allow for two-dimensional positioning.

The dual reference beam, *off-axis*, holographic apparatus for image recording is presented in Fig. 15.38. The object beam consists of multiple laser sheets, and the laser sheets have a sizable thickness, Δz. The purpose of the multiple laser sheets is twofold. First, the small particles are made more visible, and the intensity of the

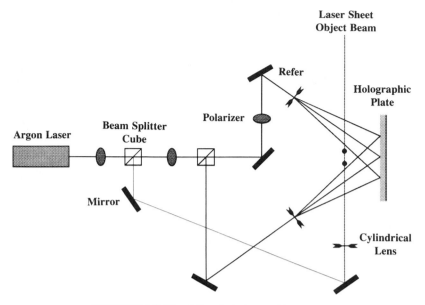

FIGURE 15.38 Off-axis holographic set up.

scattered beam is controlled independently of the reference beam. Second, spurious speckle formation is greatly reduced by limiting the illumination to a few selected planes. With this apparatus, the quality of the particle images improves very significantly, and images of very small (sub micron) particles are recorded.

Similarly to the conventional PIV technique, the holographic plate is doubly exposed. In order to resolve the problem of directional ambiguity, two holograms, though recorded onto the same holographic plate, are generated. One corresponding to the first light pulse is formed with the reference beam positioned at 45 degrees incidence with the plate and the other corresponding to the second light pulse with the reference beam positioned at −45 degrees incidence with the plate. These two holograms are reconstructed to produce, independently, the images corresponding to the first and the second pulses.

The velocity of the particulates is reduced from the holographic recording using conventional means. A high resolution video camera images the reconstructed images corresponding to each illumination pulse, one plane at a time. These images can be combined electronically and processed using the auto correlation approach to obtain the in-plane velocity field. A cross correlation algorithm can also be used. The out-of-plane velocity components can be obtained by imaging the same slices with a lateral decentering of the camera to produce stereoscopic views. The in-plane velocity field obtained from these views are then combined to yield the third component.

15.5 FLOW RATE MEASUREMENT
Gustave J. Hokensen

Flow rate characterizations include both volumetric and mass flow rate measurements. For many applications, the fluid density is known to reasonable accuracy, and it suffices to measure the volumetric flow rate to obtain mass flow rate. In those cases where the density is not known, the mass flow rate may be obtained directly by various instruments to be discussed or from separate but simultaneous measurements of density and volume flow rate.

Calibration procedures include the so-called *catch and weigh* technique which applies a known total flow to the device over a known time thus establishing the flow rate through the instrument being calibrated. By differentiating the collected flow which passes through the device, the instantaneous flow rate and instrument response may be obtained. Calibration over a range of flow rates is thereby facilitated. In operation, not only the flow rate but also the total flow over a period of time during which the flow rate has been varying is of interest. Therefore, mechanical and electrical integrators are often applied to the instrument output signal.

Calibration of devices or operational measurement of flow rates from the spatial integration of detailed flowfield probing data is also a valuable technique.

15.5.1 Pressure Drop Devices

In this category of flow rate measurement instruments, an element is inserted into the fluid stream which either changes the local streamwise distribution of flow area or applies frictional resistance to the flow. A few typical examples of the former approach are: 1) Venturi/Dall tube/flow nozzle, 2) orifice plates, and 3) choked nozzle, whereas the frictional resistance pressure drop may be generated by inserting large wetted area material such as a honeycomb or a pipe with large L/d into the flow.

The three types of *constriction* meters above are generically related to the simple Venturi shown in Fig. 15.39. The *Venturi tube* is a smooth contraction in the flow with static pressure taps located at the minimum area (2) and at the upstream maximum area (1). If the equation of momentum for steady, inviscid, incompressible flow is applied to a control volume between (1) and (2), it may be shown that the ideal volumetric flow rate, Q_i, is related to the measured pressures and the geometry by

$$Q_i = A_1 \times \frac{AR}{(1 - AR^2)^{1/2}} \left(\frac{2}{\rho}\right) (p_1 - p_2)^{1/2} \qquad (15.101)$$

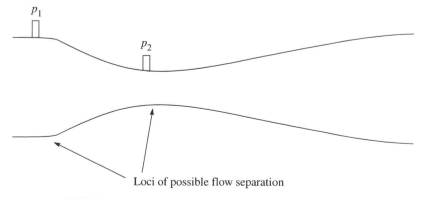

FIGURE 15.39 Schematic of a typical Venturi tube.

where AR is the ratio A_2/A_1. Due to losses in the device, however, the actual flow rate, Q_a, is computed from

$$Q_a = C_d Q_i \qquad (15.102)$$

where the *discharge coefficient*, C_d, is obtained from calibrations and is Reynolds number dependent. At values of $\mathrm{Re} = 10^6$, $C_d \simeq 0.99$, whereas at $\mathrm{Re} = 10^4$, C_d decreases to approximately 0.94.

Unfortunately, the streamlined Venturi has not been standardized, and for accurate measurements, calibration is generally required. This is particularly true if there are nonideal effects in the upstream flow unique to the particular application such as: 1) swirl, 2) high turbulence intensity, and/or 3) unsteadiness. Even in the absence of such phenomena, Venturis have been constructed which appear similar to that in Fig. 15.39 but have produced much lower values of C_d. This has been attributed to flow separation near the points of maximum wall curvature. In general, however, the overall pressure loss of a good Venturi is in the range of 10% of the measured pressure drop for AR between 0.05 and 0.65.

Variations of the Venturi include the so-called Dall flow tube and the flow nozzle. The *Dall flow tube* has the appearance of a Venturi with sharp corners, as shown in Fig. 15.40.

The Dall tube has some surprising idiosyncracies, however. In spite of its unstreamlined shape, the Dall flow tube has a pressure loss at least half that of a Venturi whereas a typical discharge coefficient (for $AR = 0.05$) *drops* from 0.68 at $\mathrm{Re} = 10^5$ to 0.66 at $\mathrm{Re} = 10^6$. The reduced pressure loss across the overall device may be due to the fact that the aforementioned separations are localized by the sharp corners. A beneficial consequence of the low discharge coefficient is the elevated sensitivity of the pressure measurement to flow rate. It is, however, relatively complex to fabricate and, due to its lack of standardization, must be calibrated for each particular design and site-unique application.

Another member of the Venturi family is the *flow nozzle*. This device is essentially half of a venturi followed by an abrupt expansion, shown in Fig. 15.41. This device has relatively high discharge coefficients (similar to the aforementioned Venturi trends) with a high pressure loss due to the downstream flow separations. From the standpoint of sensitivity to measure low flow rates and high fluid pumping power requirements, this device apparently has no redeeming features. It may, however, be advantageous (relative to the subsequently discussed orifice plate) in, e.g., slurries which might clog a device with an unstreamlined entrance.

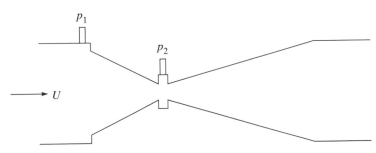

FIGURE 15.40   Schematic of a Dall flow tube.

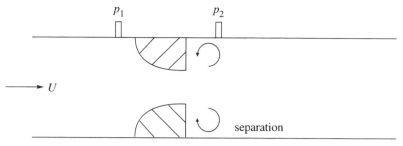

FIGURE 15.41   Schematic of flow rate measurement nozzle.

The flow nozzle leads naturally to the most commonly used flow rate device, the *orifice plate*. The orifice plate is basically an abrupt contraction and abrupt expansion back-to-back, shown in Fig. 15.42. The dividing streamline between the separated flow regions and the main flow is reminiscent of the Venturi. The orifice plate, however, has generally much lower discharge coefficients, shown in Fig. 15.43, with a pressure loss which is the largest of all the constriction devices being approximately $\Delta p \times (1 - AR)$ where Δp is the maximum pressure drop.

The pressure difference across an orifice plate which is used to infer flow rate may be taken [see Doebelin (1983)]

1. Near the plate (flange taps), typically 2.5 cm upstream and downstream,
2. Slightly away from the plate (vena-contract taps), typically one diameter upstream and one-half diameter downstream to get the maximum pressure drop,
3. Far upstream and far downstream (pipe taps), typically 2-1/2 diameters upstream and 8 diameters downstream to measure the pressure loss.

For compressible flow situations, Eq. (15.101) may be modified (if the pressure ratio is nearly one, say, 0.98, so that the pressure drop is relatively small) with Q_a replaced by the weight flowrate, and ρ replaced by v_1/g where v_1 is the (measured, known, or computed) upstream specific volume. This is the limiting (small pressure drop) form of the general isentropic compressible formula derived in Eckman (1950).

Location of various pressure taps

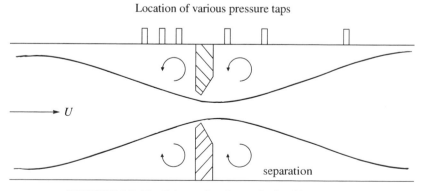

FIGURE 15.42   Schematic of a typical orifice plate.

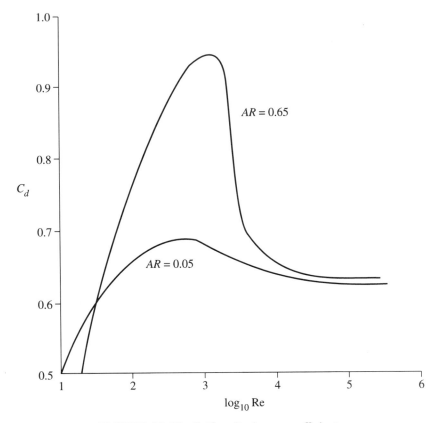

FIGURE 15.43 Orifice discharge coefficients.

Due to the more *lossy* nature of the orifice plate, however, a compressibility factor, Γ, is typically multiplied times the discharge coefficient with a form dependent on the pressure tap location [see ASME (1927)].

$$Y = 1 - (0.41 + 0.35\beta^4)(1 - p_2/p_1)^{\gamma - 1}$$

$$Y = 1 - [0.333 + 1.145(\beta^2 + 0.7\beta^5 + 12\beta^{13})](1 - p_2/p_1)^{\gamma - 1} \quad (15.103)$$

for vena contracta and flange taps or pipe taps, respectively, where β is the diameter ratio (≤ 1), and γ is the ratio of specific heats.

In some compressible flow situations, the pressure drop across the constriction is large, and the flow chokes. That is, the Mach number equals one at the throat (minimum area) and, for a given size throat and upstream conditions, mass flow rate is maximized and is known. This eliminates the need for measuring the downstream pressure.

In contrast to devices which constrict the flow by changing the cross-sectional area distribution and affecting pressure drops by both inviscid and viscous (lossy) mechanisms, it is often beneficial or required to maintain the cross-sectional area of the flow passage intact. In this case, the flow-wetted area distribution may be altered to induce a purely viscous pressure drop. Most common is the introduction of a *honeycomb* section which, if the entrance flow is laminar, results in a linear relationship between volumetric flow rate and pressure drop

$$Q \sim \frac{D^4}{\mu L} \Delta p. \tag{15.104}$$

Note that, for sufficiently large L/D, this method is useful for measuring very small flow rates.

Inasmuch as, for a given D, this (fully-developed pipe flow) formula shows these devices to be analogous to electrical resistors ($\Delta p \sim QL$), control and measurement of small flow rates may be achieved. One such device is essentially a screw through which the helical flow is established of a desired length with the length depending on the setting of the screw. With appropriate calibrations and theoretical models, these instruments are applicable to turbulent flow as well.

The honeycomb device is significantly more tolerant of upstream flow aberrations than the aforementioned area change devices. However, in flows with significant nonsteady components, devices with linear steady flow Q–Δp transfer functions will be suspect as well as the more obvious nonlinear devices. This is because the transfer function itself will change with significant unsteadiness and almost nothing is learned about evaluating unsteady flows with the transfer function derived from steady flow— except that the unsteadiness will confuse the meaning of the measurement.

15.5.2 Variable Area Devices

In contrast to the fixed geometry instruments in Sec. 15.5.1 which establish a pressure drop (and somewhat smaller overall pressure loss) in response to a given flow rate, there are variable geometry devices which indicate the flow rate by adjusting the cross sectional area with a fixed pressure drop. One such device is the *rotameter* which consists of a flotation device located inside a tapered tube, as in Fig. 15.44. The net downward force on the float (its weight) is essentially balanced by the pressure drop across the float times its cross-sectional area. Therefore, for a given float and fluid, the pressure drop across the float must be independent of flow rate. Since

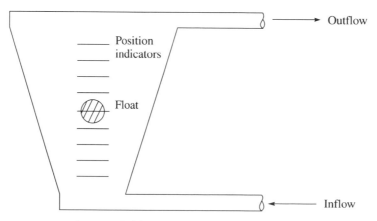

FIGURE 15.44 Schematic of typical rotameter.

$\Delta p \sim \rho U^2/2$, as Q changes, the float position must vary in order to seek a location with an average velocity, $U = Q/A_f$, and hence Δp which is constant. Therefore, for small flow areas, A_f, relative to the tube area, A_t

$$Q \sim A_f = A_t - A_F \tag{15.105}$$

where A_F is float area. If the tube area varies linearly with height along the tapered tube, y

$$A_t = a + by \tag{15.106a}$$

then the float position will be a linear reflection of the flowrate

$$Q = a' + by \tag{15.106b}$$

The linearity allows this device to be useful over a much wider range of flow rates than the square root devices in Sec. 15.5.1.

There are a variety of float and tapered tube designs for various fluids and applications. All, however, work on the same basic principle. The *area meter*, for example, is a device in which a piston is located in a straight cylindrical tube with openings cut into the wall. The piston position is displaced according to the flowrate in order to provide the flow area required. The heritage of the area meter is linked to the antiquated passive weighted-flapper valves which are in infrequent use due to their propensity for oscillation.

Similar in spirit, but for different applications, are various types of *weirs*. A weir is a free surface (open) channel device which forms a variable area flow constriction. The simplest form of the weir is a sharp-edge plate which spans the entire flow, shown in Fig. 15.45. The theoretical flowrate over such a (rectangular-suppressed) weir of width, w, is

$$Q = \tfrac{2}{3} C_d wh\sqrt{2gh} \tag{15.107}$$

The classical *Francis formula*, in units of meters and seconds, is

$$Q = 1.29 wh^{3/2} \tag{15.108}$$

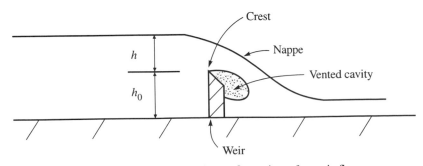

FIGURE 15.45   Schematic configuration of a weir flow.

which reflects a discharge coefficient of 0.62. The Francis formula is accurate to within 1 to 3% for heads above 10 cm when the approach velocity is small. For larger approach velocities, the *Rehbock formula**

$$Q = \left(3.234 + \frac{5.347}{320h\text{-}3} + 0.428\,\frac{h}{h_0}\right) h^{3/2} \qquad (15.109)$$

provides values within 1 percent of experiment in weirs from 6–120 cm high and heads from 3–60 cm, as long as $h/h_0 \leq 4.0$.

The (finite width) rectangular-notch weir in Fig. 15.46 contracts the flow at each side as well as the crest. The flow rate is reduced as though the weir width were decreased by 0.1 h per side. The resulting Francis formula is

$$Q = 1.29(w - 0.2h)h^{3/2} \qquad (15.110)$$

which is valid for $w \geq 2h$.

According to Cipolletti [see, e.g., Avallone and Baumeister (1987)], the trapezoidal weir in Fig. 15.47 compensates for this end effect allowing the Francis formula to be employed. Again $w \geq 2h$ and the channel walls and bottom should be more than $3h$ away from the weir to avoid corrections for velocity of approach.

Triangular notch weirs provide a convenient method of measuring high and low flow rate discharges with the same degree of accuracy. The theoretical flow rate is

$$Q = \frac{8}{15}\,C_d h^2 \sqrt{2gh}\,\tan\frac{\alpha}{2} \qquad (15.111)$$

where α is the vertex angle. Based on experiments at Cornell University quoted by Baumeister and Marks (1987), for 60 and 90° weirs, $C_d \simeq 0.58$ whereas for 28° and 120° weirs, $C_d \simeq 0.59$.

Finally, the parabolic-shaped *Sutro weir* provides a linear relationship between flow rate and h.

*Due to the formula's dimensional inhomogeneity, the constants are appropriate only for h measured in feet.

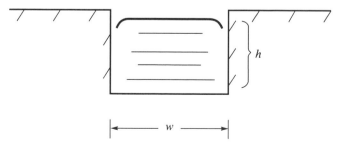

FIGURE 15.46 Rectangular notch weir (the flow direction is into the page).

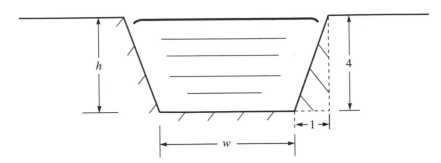

FIGURE 15.47   Trapezoidal notch weir.

15.5.3   Turbine Flow Meters

By immersing a turbine rotor in a flow channel, the interaction between the oncoming flow and the turbine blades will cause the wheel to rotate. The measured angular velocity of rotation, ω, becomes a function of the flow rate past the device. According to dimensional analysis [see Hochreiter (1958)]

$$\frac{Q}{\omega D^3} = f\left(\frac{\omega D^2}{\nu}\right) \tag{15.112}$$

if the effects of bearing loss and shaft output are ignored.

A typical calibration in Fig. 15.48 shows that, at high Reynolds number

$$\frac{Q}{\omega D^3} = \text{constant} \tag{15.113}$$

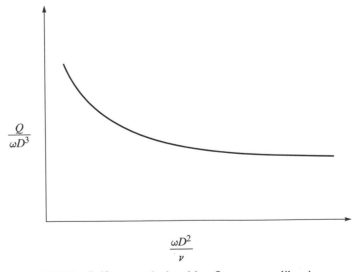

FIGURE 15.48   A typical turbine flowmeter calibration.

and, for a given turbine wheel,

$$Q \sim \omega \qquad (15.114)$$

with a constant of proportionality dependent only on the wheel geometry. Although calibrations are available to allow the use of turbine flow meters in the low $\omega D^2/\nu$ range, generally the flow meter will be changed for one with a range such that the linear Q–ω relationship is maintained.

15.5.4 Positive Displacement Meters and Pumps

In certain applications, the device which provides the energy to establish the flow also may serve to measure the flow rate induced. Such devices are positive displacement pumps. For each rotation of the pump, chambers of known volume are first filled with a supply fluid and then evacuated to the system of interest. In this manner, the fluid quantity is known from the monitored mechanical displacement which sets the fluid in motion.

Similarly, but operating only on the pressure that is available from other driving forces, positive displacement meters provide a measure of the flow rate which displaces a known volume inside the instrument and causes a shaft rotation in proportion to the displacement. Generally, even with care taken to minimize the bearing friction and inertia, a significant pressure drop is inherent across these devices.

15.5.5 Drag Instruments

The concept behind these devices is simply to measure the force exerted on a bluff body immersed in the flow, as in Fig. 15.49. In the example shown, a sphere is mounted on an elastic sting whose deflection is monitored, e.g., by strain gages isolated from the flow. As the flow rate is increased, the average velocity in the tube increases. This increases the force on the sphere according to

$$F = \tfrac{1}{2}\rho U^2 A C_D \qquad (15.115)$$

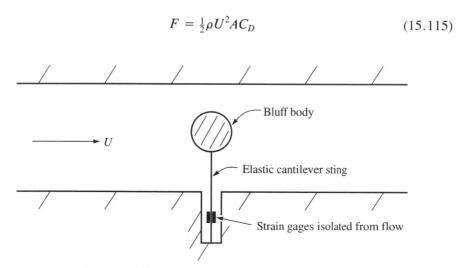

FIGURE 15.49 Schematic of drag flowmeter.

and increases the deflection of the sting mount. If desired, the nonlinear response is easily linearized with available square-root electronic modules.

Depending on the dynamic characteristics of the specific design, these devices can be made to have very good dynamic response. There is one difficulty in this virtue, however. C_D is very sensitive to the geometry and flow details. Therefore, unless one knows the value of C_D in the unsteady flow being measured, the steady flow C_D must be assumed to be invariant.

15.5.6 Vortex Shedding Devices

Depending on the Reynolds number of the bluff body immersed in the flow of Sec. 15.5.5, a second method is available to infer the flow rate. It is known that the flow separates from bluff bodies and, depending on the Re, sheds vortices at frequencies determined by the Strouhal number, $Su = \omega L/U$, where ω is the frequency, and L is the characteristic body diameter. A typical plot of Su vs. Re for a circular cylinder is shown in Fig. 15.50 [see Roshko (1954)]. By monitoring the vortex shedding-

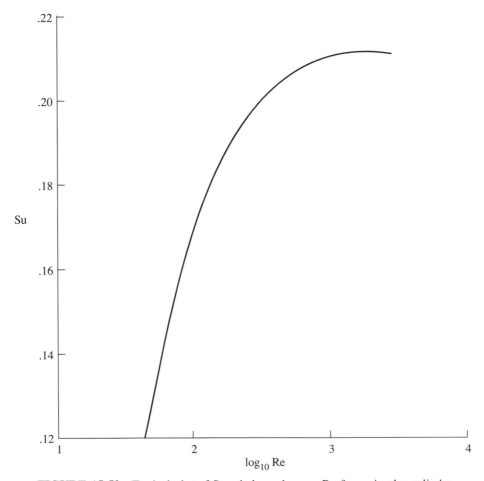

FIGURE 15.50   Typical plot of Strouhal number vs. Re for a circular cylinder.

induced flowfield fluctuations, for example by microphone, the average velocity and, therefore, flow rate past the bluff body, may be computed.

15.5.7 Ultrasonic and Electromagnetic Flow Meters

Sonic, or more typically, *ultrasonic flow meters* operate on the principle that small-amplitude pressure disturbances propagate at the speed of sound, a, through the fluid relative to the flow. That is, if the fluid has some velocity, U, relative to laboratory frame coordinates, sound will propagate at $U + a$ in the direction of flow and at $-U + a$ against the flow. Suppose now that an acoustic transmitter (e.g., an electrically-excited piezoelectric crystal) and receiver (pressure transducer or microphone) are separated by a distance L in the flow direction as in Fig. 15.51.

Transmitted pulses will be received with different time lags depending on the average flow velocity (directly related to the bulk flow rate). The difference between the time lag for a given average velocity and that for zero flow is

$$\Delta t = \frac{L}{a} - \frac{L}{U + a} = \frac{LU}{a(U + a)} \simeq \frac{LU}{a^2} \tag{15.116}$$

If two transmitter–receiver pairs are located within the flow with opposing orientations as in Fig. 15.52, the difference between the time lags in the upstream and downstream directions is

$$\Delta t = \frac{L}{-U + a} - \frac{L}{U + a} = \frac{2LU}{a^2 - U^2} \simeq \frac{2LU}{a^2} \tag{15.117}$$

which provides twice the sensitivity of the single set arrangement. The essential difficulty with both of these configurations lies in the sensitivity to a^2 and variations therein due, particularly, to temperature. A clever way to circumvent this difficulty is to close both circuits in the dual transmitter-receiver system with a feedback arrangement such that the pulses are self-excited. If the two oscillating signals are

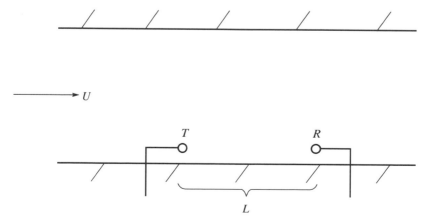

FIGURE 15.51 Ultrasonic anemometer schematic with single R and T. T is transmitter, and R is receiver.

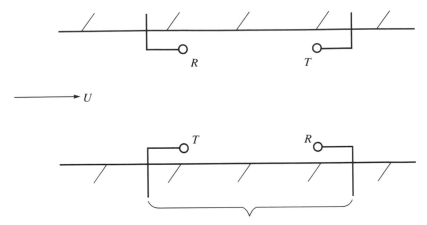

FIGURE 15.52 Schematic of ultrasonic anemometer with dual R and T. T is transmitter, and R is receiver.

multiplied together, their beat frequency (difference) may be measured. This frequency difference will be

$$\Delta\omega = \frac{U + a}{L} - \frac{-U + a}{L} = \frac{2U}{L} \qquad (15.118)$$

which is independent of a and, therefore, insensitive to variations in the physical properties which affect it.

   Various other configurations, including skewed (to the flow) versions of Fig. 15.52 [Doebelin (1990)] and cross-flow transmission in Fig. 15.53, in which the difference between the two received signals may be related to the flow rate, are in use.

   These devices all operate on the same, fundamental principle of pressure wave propagation relative to the flow. The flow meters respond to the spatial average velocity profile and, thus, provide a direct measure of volumetric flow rate. Com-

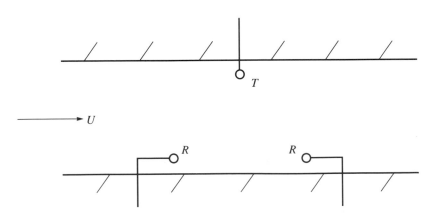

FIGURE 15.53 Schematic of ultrasonic anemometer with dual R and single T. T is transmitter and R is receiver.

mercially available units are expensive, yet reasonably accurate (a few percent of full scale) and provide a relatively wide operating range with upper end capabilities to 10 m/s and a span of up to 25 : 1.

Electromagnetic flow meters are essentially electrical generators which operate on the same magnetic flux cutting principle as rotating mechanical generators. According to Faraday, a conductor of length, L, which cuts magnetic *lines* of flux with a velocity, U, induces a potential across the conductor of magnitude

$$E = BLU \qquad (15.119)$$

Therefore, if a conductive fluid, contained in a nonconductive flow passage, is immersed in a magnetic field orthogonal to the flow with a flux density, B, electrodes oriented orthogonal to both the flow and B directions will register a voltage proportional to the average velocity, channel width, and magnitude of $\overline{B}$ as in Fig. 15.54.

Typically, the applied B field is generated by a–c (rather than d–c) currents flowing through fixed coils. Primarily, this is to minimize the direct (*magnetohydrodynamic, MHD*) effect of the applied magnetic field on the flow. A d–c field of sufficient magnitude could distort the mean velocity profile resulting in some asymmetry. As long as the profile is symmetric, however nonuniform, the induced voltage in the flow meter will respond to the average velocity in the channel. An a–c magnetic field of sufficiently high frequency on the average cancels the *MHD* forces, and symmetry is maintained because the fluid inertia prevents the flow from responding to high frequency forces. In addition to the *MHD* consideration, for some conductive fluids, a d–c field establishes an ion migration to the electrodes which interferes with the current passage. Finally, an a–c signal is simpler to deal with in terms of bandpass filtering and amplification.

The electromagnetic systems are limited to highly conductive fluids with liquid metals probably being optimum. The electromagnetic meters are noninvasive, but they do interact with the flow (via *MHD* forces). They provide the capability of

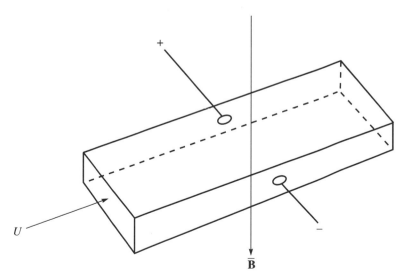

FIGURE 15.54 Electromagnetic flowmeter configuration.

monitoring (bulk) flow speed and directionality of symmetric, highly nonuniform flows. These meters may be used in unsteady flow, and they are insensitive to the fluid properties.

15.5.8   Mass Flow Rate Measurements

In the preceding sections, devices which measure the volumetric flow rate have been discussed. Many of these instruments provided an output signal independent of fluid properties (i.e., density, viscosity, etc.). However, in some cases, the fluid density must also be known to compute the volumetric flow rate (see, e.g., Sec. 15.5.1). Once the fluid density and the volumetric flow rate are measured, the mass flow rate is then known also from

$$\dot{m} = \rho Q(\rho) \tag{15.120}$$

For devices which do not require knowledge of ρ to compute Q, $\dot{m}$ may be computed by multiplying the known Q times the subsequently measured ρ

$$\dot{m} = \rho Q \tag{15.121}$$

It is apparent that, to measure mass flow rate, density measurements are an important topic to be discussed. In addition, however, some so-called direct mass flow rate measurement devices will be described in which $\dot{m}$ is inferred directly.

Fluid density measurements are differentiated between discrete sampling measurements and continuous, in-flow measurements. The *discrete sampling* type are generally either weight or buoyant force measurements. If a known volume of liquid is withdrawn from the flow system and weighed, its density may be computed. Alternatively, if a material of known weight and volume is immersed in the (liquid) sample and weighed, the buoyant force will provide the sample density. Utilizing the ideal gas law, a fluid sample may be collected (in equilibrium with the local flow pressure and temperature) and the density computed from $\rho = p/RT$ after the pressure and temperature of the sample are measured.

With regard to continuous, in-flow measurements both weighing a known volume of the fluid, isolated from the rest of the piping, and also using the buoyant force approach are conceivable, as is measuring the flow pressure and temperature *in situ* and computing density for gases. An interesting possibility, utilizing the relationships obtained from volume flow measurement, is to positively displace a (bypassed) fraction of the flow at the measurement point of interest and measure the pressure drop across any device which has a transfer function

$$\Delta p \sim \rho Q^2 \tag{15.122}$$

Since Δp is measured and Q is controlled, ρ may be computed as long as the device calibration (discharge) coefficient is well characterized. Finally, average fluid density through a flow passage may be measured by illuminating the flow with acoustic radiation or radiation from a radioisotope source [Doebelin (1990)].

When the complications of making separate density and volumetric flow rate measurements to infer mass flow rate become excessive, various instruments may be employed to measure mass flow rate directly. The majority of these instruments rely on the application of the conservation laws of fluid mechanics for their operating principle. For steady flow, the conservation of linear momentum states the sum of forces on a control volume equals the change in momentum across it

$$\Sigma\overline{\mathbf{F}} = \int\!\!\int_s \overline{\mathbf{U}}(\rho\overline{\mathbf{U}} \cdot d\overline{\mathbf{A}}) \tag{15.123}$$

If appropriately averaged quantities and the conservation of mass are used, this may be written

$$\Sigma\overline{\mathbf{F}} = \dot{m}\Delta\overline{\mathbf{U}} \tag{15.124}$$

where $\Delta\overline{\mathbf{U}}$ is the change in velocity across the control volume experienced by the flow. Clearly, if the force, $\overline{\mathbf{F}}$, and $\Delta\overline{\mathbf{U}}$ are measured or known, $\dot{m}$ may be computed.

In application, it has been more convenient to apply this principle to steady rotating systems wherein the governing expression relates applied torques, $\overline{\mathbf{T}}$, to angular momenta

$$\Sigma\overline{\mathbf{T}} = \dot{m}\Delta(r\overline{\mathbf{U}}_t) \tag{15.125}$$

This equation very simply guides the design of a typical device as in Fig. 15.55. This is a rotating device with flow passages near the periphery, which is inserted into an initially nonrotating flow. The relative length of the device ensures that the exit flow is in equilibrium with the mechanical rotation velocity. Since the exit angular momentum is $\dot{m}rV_t = \dot{m}r^2\omega$, the applied torque, T, is

$$T = \dot{m}r^2\omega. \tag{15.126}$$

The symbol V_t denotes tangential component of velocity. If the device is operated at constant speed, the measured torque is a direct linear measure of the mass flow rate. Particularly in the low mass flow rate range, viscosity and changes to viscosity within the rotor introduce offset and hysteresis problems. Therefore, it is equally possible to operate at constant torque and measure the angular velocity, ω, such that ω^{-1} is the reflection of $\dot{m}$. If ω^{-1} is obtained by measuring the time between pulses, e.g., by a proximity sensor or gear wheel arrangement, then ω^{-1} and, therefore, $\dot{m}$ are measured directly.

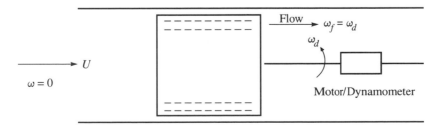

FIGURE 15.55 Schematic of angular momentum mass flow rate meter. ω_d is the angular velocity of the driven motor and ω_f is angular velocity of the fluid.

In order to operate this type of device in the torque measurement mode and avoid problems with viscosity at low mass flow rates, the following variation is employed. The rotating device is used only to impart a known angular momentum to the fluid. Close downstream, a sufficiently long, nonrotating bladed wheel absorbs all of the flow angular momentum, and the shaft torque is measured. Since the $r^2\omega$ approaching this wheel is known, the measured torque provides the mass flow rate indication. The spacing and interior passage details, along with the nonrotating T measurement, minimize the viscous problems at low flow rates [see Doebelin (1990)].

Various other instruments are described in the literature but two of particular interest will be discussed here, one for its curiosity and both for their similarity to certain aspects of contant temperature hot-wire anemometers.

The first, shown in Fig. 15.56, is an arrangement of four identical orifice plates operated at high Re (for which C_d is constant) in a bridge circuit. The pump makes the flowrate in the upstream legs asymmetric which causes a pressure drop across the pump to be

$$\Delta p \sim \rho \left[\frac{Q - \Delta Q}{2} \right]^2 - \rho \left[\frac{Q + \Delta Q}{2} \right]^2 + \rho Q \Delta Q \qquad (15.127)$$

and therefore

$$\Delta p \sim \rho Q \qquad (15.128)$$

As Q varies, ΔQ could be changed to maintain Δp constant in a feedback control loop.

The final example is that of *a simple heating element* applied to a flow passage. The conservation of energy shows that for essentially constant density flow,* the input electrical power, P, is converted into fluid thermal energy such that

$$\dot{m}\Delta h = P \qquad (15.129)$$

The mechanism by which Δh is affected by P is by heat transfer through the wall boundary layer. This is empirically represented by the so-called *law of cooling*.

*The kinetic energy remains constant along a constant area duct.

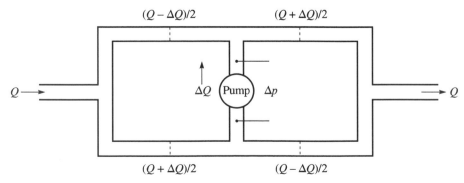

FIGURE 15.56 An orifice-plate bridge mass flow rate meter.

$$\dot{q} = \rho U A_w \, \text{St} \, (h_w - h_\infty) \tag{15.130}$$

where St is the Stanton number. This may be rewritten

$$\dot{q} = \dot{m} \, \frac{A_w}{A} \, \text{St} \, (h_w - h_\infty) \tag{15.131}$$

Two options exist with regard to developing a mass flow rate meter. First, the electrical power ($=\dot{q}$ for a system well insulated to the surroundings), wall and freestream fluid enthalpies may be measured such that

$$\dot{m} \, \text{St}(\dot{m}) = \frac{P}{\dfrac{A_w}{A} \, (h_w - h_\infty)} \tag{15.132}$$

where St ($\simeq \text{Pr}^{2/3} \, C_f /2$) is dependent upon the Reynolds number and, therefore, for a given fluid and channel geometry on $\dot{m}$. If P is varied to maintain $h_w - h_\infty$, it becomes the measure of $\dot{m}$.

A much less sensitive technique to obtain $\dot{m}$ is to measure the downstream flow enthalpy after the flow has become uniform, in lieu of P, and equate input heat to overall energy increase

$$\text{St} = \frac{\Delta h}{h_w - h_\infty} \, (A/A_w) \tag{15.133}$$

and use the variation of St with Re alone to infer the mass flowrate. Again, if P is varied to maintain $h_w - h_\infty$ constant, Δh becomes a function of $\dot{m}$ through the (weak) dependence of St on Re. This may be useful, however, when the heater cannot be insulated to the surroundings. The essential problem with this technique is the lack of precise knowledge of the dependence of St on Re.

15.6 FLOW VISUALIZATION

15.6.1 Introduction
Gustave J. Hokensen

Optical techniques of various types are available to study qualitatively different aspects of flow phenomena. In some cases, quantitative information is available if the experiment is done with care. In this section, two general types of flow visualization approaches will be analyzed—indicators (invasive devices placed within the flow) and noninvasive devices which are usually fluid index of refraction measurement instruments. The second type of device responds to variations in certain aspects of the density field and its interaction with an illumination source. Flow visualization is discussed in detail in Werle (1973), and beautiful examples may be found in Van Dyke (1982).

The flow indicators are typically *mechanical* devices inserted to interact with some aspect of the flowfield, typically, but not exclusively, the velocity. Due to this method of operation, some care must be exercised in visual data interpretation. There are three primary types of lines or surfaces conceivable in a flow relative to visualization. Other surfaces of tangential relevance to the subject will be mentioned when appropriate. The three lines are: 1) pathlines, 2) streaklines, and 3) streamlines. A *pathline* is the line which traces out the motion of a fluid point. A *streakline* is the line connecting all fluid points which, at some time in their history, passed through a given spatial location. A *streamline* is the line which is everywhere tangent to the local velocity vector. In steady flow, these lines coincide. However, in time-varying flows, each has its own characteristic and is generated by a particular type of indicator.

15.6.2   Flow Line Indicators
Gustave J. Hokenson

In many cases, this general approach, which involves the addition of elements to the flow for the purpose of visualization (ideally at such a dilution so as not to disrupt the flow due to the observation), falls into the category of *seeding* [see Merzkirch (1974)]. Meteorological balloons are an example of fluid particle followers used to indicate wind speed and direction profiles. In the laboratory flows of air or water, small *bubbles* are used to indicate the local velocity or displacement field. In water, an electrical current can be passed through the fluid (doped with a material which dissociates into ions and makes the fluid conductive), and the water molecules dissociate forming hydrogen bubbles which advect with the flow. In air, small helium-filled soap bubbles are employed. In both cases, it is assumed that the bubble acts approximately as a fluid mass point. Therefore, it is important that the bubbles be neutrally-buoyant and smaller than any scales of fluid motions which are to be resolved.

Similarly, in water, a *dye* may be added locally to the flow or, in air, a stream of *smoke*, vapor or particles may be injected. As discussed by Maltby (1962) the application of smoke to a gaseous flow is done via wire, tube, or screen apparatus. To form the smoke wire, a thin resistive wire is coated with oil and electrically pulse heated to form an oil vapor which is advected away by the flow. The smoke tube injects a continuous stream of smoke at a given location selectable during the experiment. The smoke (or vapor) screen is a combination of invasive and noninvasive techniques wherein smoke (or vapor) is added to the main flow over, say, a wing in a wind tunnel. As the sheet of smoke (or vapor) is distorted by the wing, a light beam transverse to the flow is used to visualize variations in the smoke and cause a unique shadow to form, as in Fig. 15.57.

The use of *dyes* in liquids also can be quite instructive. Naturally visible additives can be used to form a discrete or continuous trace by injecting the material into the flow at a given location. Alternatively, and for differing purposes, an invisible dyed additive may be made visible by application of an electric current or a pulse of, e.g., UV illumination. A unique application of the second technique to measure velocity profiles via the so-called time line is accomplished with UV laser illumination orthogonal to the flow as depicted in Fig. 15.58. This method is apparently due to R. E. Falco.

Note in the smoke or vapor screen and the photo-chemically-induced dye method the use of an external illumination source (other than the required ambient light) to

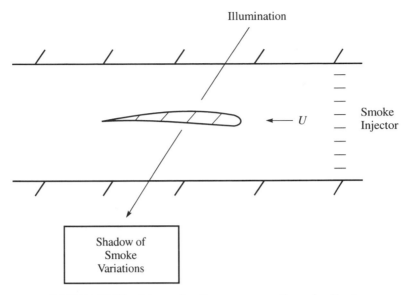

FIGURE 15.57 Schematic of vapor screen flow visualization.

interact with certain aspects of the flow. Similar in spirit to the time line, is the use of a *spark* or electrical arc (either intermittent or continuous) to indicate the relative velocity via the spark distortion. This method is generally used in high speed flows.

If the flow in question has additives, e.g., particulates, similar to the LDV application discussed earlier, then fluid motions may be viewed optically. Perhaps holographic interferometry may be used.

Note that this brief survey of flow visualization techniques relies on an additive injected at a given point. Therefore, each fluid particle passing that point is *tagged*. If the additive is continuous, or nearly so as in a rapidly pulsed mode, a single snapshot of the flow provides a streakline. Successive snapshots of each tagged point provide the family of pathlines. If a sequence of snapshots of pathlines are differenced, an instantaneous streamline may be formed with sufficient number of tracers.

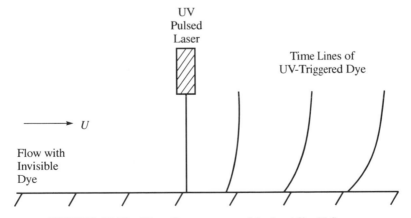

FIGURE 15.58 Time lines generated in dyed liquid flows.

In contrast to this in-flow type of visualization, various surface-mounted systems exist including: tufts, oil films, temperature sensitive paints, and ablatives (e.g., china clay) [refer to Maltby (1962)]. The *tufts* are essentially miniature wind vanes located along the surface to indicate wind direction and, in particular, possible flow separations. Alternatively, wake flow tufts may be mounted on screens in the wakes of bodies to indicate vortical flow patterns transverse to the oncoming flow. *Oil films* also indicate flow separation and surface streamline directions, whereas the ablative surfaces typically indicate transition due to more rapid erosion in turbulent flow with enhanced (measured at different selected times) surface transport rates (mass-heat-momentum). *Temperature sensitive coatings* reflect the varying surface temperature due to the heat transfer established by the flow. *Ablative* materials also are nonuniformly scoured from the body surface by the varying surface transport.

A method of flow visualization on surfaces that also gives quantitative information is the use of pressure-sensitive paints [McLachlan (1992)]. Such materials respond to the local density of oxygen atoms which can be related to pressure (and temperature).

For unsteady liquid flows or particulate laden gaseous flows, the use of a strobed illumination source is invaluable—particularly in periodic devices such as turbomachines.

15.6.3   Refractive Optical Flow Visualization Techniques
Rodney D. W. Bowersox

Many problems in fluid mechanics involve mediums that are colorless, transparent, and nonluminous. In many cases, the flow phenomena of interest will cause changes in the refractive index, which can be detected using optical methods that depend on the effects of these changes on the transmission of light. These methods are often termed *schlieren methods*, because they were originally used in Germany for the detection of inhomogeneous regions, usually streaks (*schliere*) in optical glass. Schlieren methods are used extensively in aerodynamic research associated with high speed flow. The three common modes of refractive optical flow visualization are *shadowgraph*, *schlieren*, and *interferometry*. The photographs in Fig. 15.59 are examples of each mode. It is important to note that these methods are not duplicative, but are in fact supplementary. The shadowgraph is sensitive to the second derivative of the index of refraction, the schlieren method responds to the first derivative, and the interferometer gives the index directly. Hence, when possible, all three methods should be used.

This section describes all three methods including basic optical configurations and data analysis procedures. But first, the relationship between the index of refraction and gas density and how changes in the index field effect a traversing light beam are briefly discussed.

Index of Refraction. Merzkirch (1981) shows that the Clausius–Mosotti relation, which was derived from the radiation interaction theory of Lorentz, reduces to

$$n - 1 = K\rho \tag{15.134}$$

for dilute gases, where the index of refraction is very nearly unity (e.g., for air at STP and $\lambda = 5893$ Å, $n = 1.000292$). K is the Gladstone–Dale constant or more

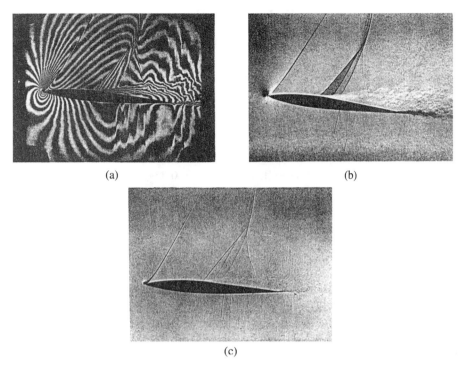

FIGURE 15.59 Examples of the three modes of refractive visualization of the flow past a two-dimensional aerofoil. (a) Interferometer photograph, (b) Toepler schlieren photograph, and (c) direct-shadow photograph. (From Holder and North, 1963, with permission from Her Majesty's Stationery Office.)

formally called the *specific refractivity*, and the form for K can be found in Merzkirch (1981). Table 15.8 summarizes K for a variety of fluids. Since the electric field vector and dipole moment are linearly related, the specific refractivity for a gas mixture of i-species is given by [Merzkirch (1981)]

$$K = \sum K_i \frac{\rho_i}{\rho} \qquad (15.135)$$

Analysis of Index of Refraction Field. The basic phenomenon underlying many flow visualization techniques for compressible flow is that variations in the index of refraction field affect a traversing light beam. The changes in the index of refraction across the field are recorded via photography or computer imaging. Here, we address the interpretation of the recorded index of refraction field using concepts from geometrical optics.

The propagation of a light ray through an index field is determined by Fermat's principle, which states that the first variation of the integral along the light path of the local index of refraction $n(x, y, z)$ must vanish, i.e. [Weyl (1954)]

$$\delta \int n(x, y, z) \, ds = 0 \qquad (15.136)$$

where s denotes the ray arc length.

TABLE 15.8 Gladstone–Dale Constants for Various Fluids (5893 Å, 0 C, and 76 cm Hg)

Gas	K (cm^3/g)[1]
Acetone	0.421
Air	0.226
Ammonia	0.495
Argon	0.158
Benzene	0.506
Carbon dioxide	0.230
Carbon disulfate	0.437
Carbon monoxide	0.268
Chlorine	0.488
Chloroform	0.508
Ethanol	0.424
Helium	0.202
Hydrochloric acid	0.275
Hydrogen	1.478
Hydrogen sulfide	0.411
Methane	0.622
Methanol	0.410
Nitrogen	0.238
Nitrous oxide	0.263
Oxygen	0.190
Pentane	0.532
Sulfur dioxide	0.240

[1]Generated from data given in CRC Handbook (1955). (Reprinted with permission from Smithsonian *Physical Tables, 9th ed.*, Smithsonian Institution Press, 533, Washington, D.C., 1969.)

Figure 15.60 presents a comparison of a light path through an index field to that of the undisturbed ray. In the absence of index changes, the light ray would have arrived at the screen, S, at the location Q, with deflection angle, θ, and time, t. With the flow or index field changes present, the light ray will arrive at S at (Q^*, θ^*, t^*). Various optical arrangements have been devised to record on the screen the *phase lag*, $\tau = t^* - t$, *deflection*, $\epsilon = \theta^* - \theta$, or the *displacement*, $Q \rightarrow Q^*$ [Weyl (1954)].

Shadowgraph Photography. Shadowgraph systems are the least complex of the refractive flow visualization techniques. Figure 15.61 illustrates a simple shadowgraph arrangement, comprised of a light source (usually a short, 10^{-8} s, duration point source), a parabolic mirror (or lens) to produce a parallel or collimated light beam through the wind tunnel test section (not required if magnification is acceptable) and a photographic plate.

The name *shadowgraph* stems from the notion that the light from point Q is displaced to Q^*. Therefore, Q^* receives more light, and no light remains at Q, hence forming a shadow of that respective object point. Shadowgraphs are interpreted as measuring the displacement. Under the assumption that the slopes of the light rays

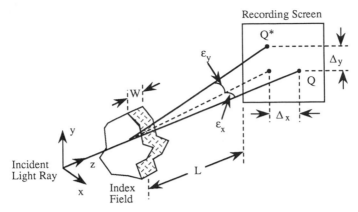

FIGURE 15.60 Distortion of a light ray through an inhomogeneous index of refraction (density) field.

are small (i.e., dx/dz, $dy/dz \ll 1$, where z is direction of the light ray, see Fig. 15.60), then Eq. (15.136) reduces to [Merzkirch (1981)]

$$\overline{(QQ^*)}_x = L \int_w \frac{1}{n} \frac{\partial n}{\partial x} \, dz$$

$$\overline{(QQ^*)}_y = L \int_w \frac{1}{n} \frac{\partial n}{\partial y} \, dz$$

(15.137)

where L is the distance from the center of the flow field to the screen, and w is the width of the index field.

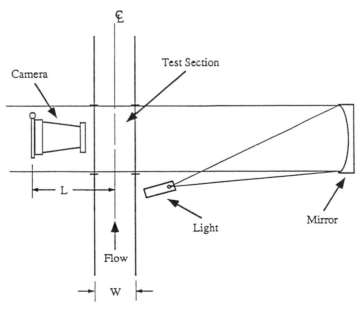

FIGURE 15.61 Arrangement of optics for a typical shadowgraph.

Let $I(x, y)$ represent the light intensity at the screen of the undisturbed field, while $I^*(x^*, y^*)$ depicts the intensity with the flow present. The intensity I^* results from the intensities of all $I_i(x, y)$ which are mapped to (x^*, y^*). Since the area illuminated by a particular light beam is deformed due to the mapping, the intensity per unit area changes. Hence in determining I^*, each I_i should be divided by the Jacobian of the mapping [Weyl (1954)]. The resulting intensity at (x^*, y^*) is

$$I^*(x^*, y^*) = \sum_i \frac{I_i(x_i, y_i)}{\left| \dfrac{\partial(x^*, y^*)}{\partial(x, y)} \right|} \tag{15.138}$$

where the term in the denominator is the Jacobian. If the mapping is assumed to be one-to-one, and the displacements (Δ_x, Δ_y) and the displacement derivatives are small, then the Jacobian can be linearized as

$$\left| \frac{\partial(x^*, y^*)}{\partial(x, y)} \right| \simeq 1 + \frac{\partial \Delta_x}{\partial x} + \frac{\partial \Delta_y}{\partial y} \tag{15.139}$$

Also, if the mapping is one-to-one, then the summation in Eq. (15.138) can be dropped. Since photographic techniques are sensitive to relative changes of light intensity, we will use the following approximations

$$\Delta_x \simeq L \int_W \frac{\partial}{\partial x} \ln n \, dz \simeq L \int_W \frac{\partial n}{\partial x}$$

$$\Delta_y \simeq L \int_W \frac{\partial}{\partial y} \ln n \, dz \simeq L \int_W \frac{\partial n}{\partial y} \tag{15.140}$$

to reduce Eq. (15.138) to

$$\frac{\Delta I}{I^*} = \frac{I(x, y) - I^*(x, y)}{I^*(x, y)} = L \int_W \left(\frac{\partial^2 n}{\partial x^2} + \frac{\partial^2 n}{\partial y^2} \right) dz \tag{15.141}$$

Hence, the shadowgraph is sensitive to the *second derivative* of the index of refraction or density for the case of constant K.

Shadowgraph Processing for Density Fluctuations. Shadowgraphs can provide a great wealth of qualitative flow field information and limited quantitative data. For example, turbulent regions show up clearly on shadowgraphs, and statistical turbulence information about a turbulent flow can be deduced from shadowgraphs [Bowersox and Schetz (1994)]. The contrast is defined as $h = \Delta I / I^*$ as given in Eq. (15.141). It is important to note that the distance to the screen from the test section, L, (see Fig. 15.61) should be made as small as possible in order to insure one-to-one correspondence between the shadowgraph and the actual index of refraction field [Kovasznay (1954)].

Replacing h and n by their mean plus fluctuating components, it can be shown that

$$h' = L \int_W \left(\frac{\partial^2 n'}{\partial x^2} + \frac{\partial^2 n'}{\partial y^2} \right) dz \qquad (15.142)$$

In order to analyze the shadowgraph, a relation between the two-dimensional (2D) film contrast and the three-dimensional (3D) index of refraction field must be established. Taking the autocorrelation of both sides of Eq. (15.142), one can show that

$$R_{hh} = \alpha^* \int_{-\infty}^{\infty} \int_{-\infty}^{\infty} (f_x^4 + f_y^4) \, e^{2\pi i(f_x \zeta + f_y \eta)} \int_{-\infty}^{\infty} S_{nn} \frac{\sin^2 (\pi W f_z)}{(\pi W f_z)^2} \, df_z \, df_x \, df_y \qquad (15.143)$$

where $R_{hh} = R_{hh}(\zeta, \eta)$, ζ and η are correlation space coordinates, $\alpha^* = (LW)^2 (2\pi)^4$, $S_{nn} = S_{nn}(f_x, f_y, f_z)$, which is the 3D, double-sided, autospectra function of the index of refraction fluctuations, and f_i are frequencies in m^{-1}.

Equation (15.143) can be further simplified by taking advantage of the inherent averaging along the axis of the light path. Thus, the 3D autospectra can be replaced by a 2D, x–y planar, spectrum that is averaged along z. Therefore, denoting the 2D, z-averaged, autospectra, by $S_{nn}(f_x, f_y)$ and noting that

$$\int_{-\infty}^{\infty} \frac{\sin^2 (\pi W f_z)}{(\pi W f_z)^2} \, df_z \approx 4.37 \qquad (15.144)$$

Equation (15.143) becomes

$$R_{hh}(\zeta, \eta) = \alpha^{**} \int_{-\infty}^{\infty} \int_{-\infty}^{\infty} (f_x^4 + f_y^4) \, e^{2\pi i(f_x \zeta + f_y \eta)} \, S_{nn}(f_x, f_y) \, df_x \, df_y \qquad (15.145)$$

where $\alpha^{**} = 4.37 \, \alpha^*$.

Thus, assuming K is constant and defining $\alpha = 4.37 \, (KLW)^2 \, (2\pi)^4 \, (= 3.86 \times 10^{-4} \, (LW)^2 \, m^{10}/kg^2$ for air), then from Eq. (15.145) it becomes obvious that the film contrast autospectra and the density autospectra are related by the following

$$S_{\rho\rho}(f_x, f_y) = \frac{1}{\alpha} \frac{S_{hh}(f_x, f_y)}{f_x^4 + f_y^4} \qquad (15.146)$$

Hence, the density autocorrelation function can be written as

$$R_{\rho\rho}(\zeta, \eta) \approx 2 \int_0^{\infty} \int_0^{\infty} S_{\rho\rho}(f_x, f_y) \cos (2\pi(f_x \zeta + f_y \eta)) \, df_x \, df_y \qquad (15.147)$$

where the approximation is included to emphasize the averaging over z. With this, Eqs. (15.146) and (15.147) can be combined to give an estimate of the variance of the density fluctuations as

$$\sigma_\rho^2 = \frac{1}{\alpha} \int_0^{\infty} \int_0^{\infty} \frac{G_{hh}(f_x, f_y)}{f_x^4 + f_y^4} \, df_x \, df_y \qquad (15.148)$$

where G_{hh} is the single-sided autospectra function ($= 2S_{hh}$), and is valid for f_x, f_y ≥ 0. The frequency resolution is limited to approximately $1/W$.

Bowersox and Schetz (1994) compared the root-mean-square density fluctuation levels obtained from the above analysis to those measured with hot-wire anemometry in a supersonic free mixing layer, and excellent agreement was found between the two techniques.

Schlieren Photography. Schlieren techniques utilize optical components in order to record the deflection, ϵ, of the light ray on the screen. The best known method is that of Toepler (1866). Figure 15.62 illustrates the basic components of various

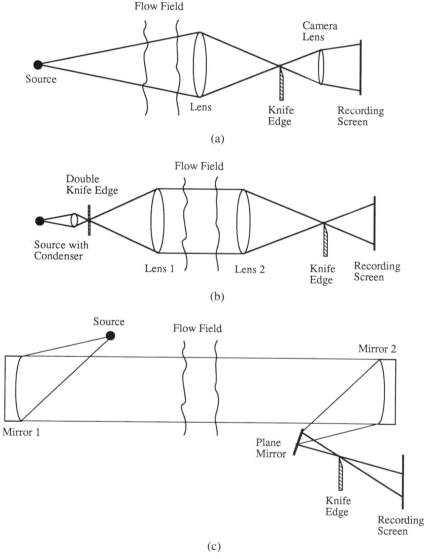

FIGURE 15.62 Various Toepler schlieren arrangements. (a) Single lens (or schlieren head) system, (b) two lens parallel light system, and (c) two mirror parallel light system.

typical Toepler schlieren arrangements, and both mirrors or lenses can be used. The arrangement consists of a light source (typically a rectangular source with a nominal duration of 10^{-6} s), a collimating mirror or lens to produce parallel light through the test section (not a necessity if magnification is acceptable), a second mirror or lens to produce an image of the source in its focal plane where a knife edge is located to block a portion of the undisturbed light, and a focusing lens which may be used to give an image of the flow field on the screen. The double knife edge arrangement (also known as a diaphragm couple) shown in Fig. 15.62 has proven to be very popular, whether lenses or mirrors are used. One of the main reasons for this popularity is that a variety of knife edge geometries can be utilized to yield different schlieren effects. For a more complete inventory of commonly used diaphragm couples, see Philbert, Surget, and Veret (1985).

Under the assumption that the slopes of the light rays are small (i.e., dx/dz, $dy/dz \ll 1$, where z is direction of the light ray, see Fig. 15.60), then Eq. (15.136) reduces to [Merzkirch (1981)]

$$\tan \epsilon_x = \int_W \frac{1}{n} \frac{\partial n}{\partial x} \, dz$$

$$\tan \epsilon_y = \int_W \frac{1}{n} \frac{\partial n}{\partial y} \, dz \qquad (15.149)$$

where W is the width of the flow field. Figure 15.63 illustrates the basic principle behind the Toepler method. Essentially, the knife edge is adjusted so that part of the source image is blocked. It is important that the screen illumination decrease uniformly. If it does not, then the optics or knife edge may have to be realigned, providing that the lenses are free from chromatic or spherical aberration and astigmatism [Beams (1954)]. If, when the optical disturbance from the flow is introduced,

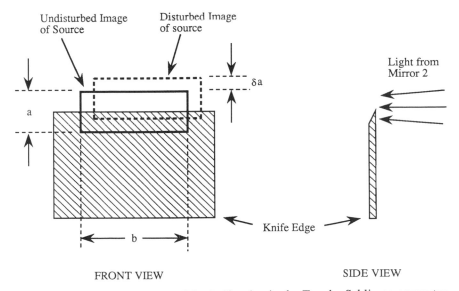

FIGURE 15.63 Arrangement of the knife edge in the Toepler Schlieren apparatus.

part of the image of the source is displaced as shown in Fig. 15.63, then the intensity on the screen will either decrease or increase by an amount proportional to the deflection angle.

If light losses are neglected, one can show that when the knife edge is absent, the intensity on the screen is given by

$$I_o = \frac{Bhb}{m^2 f_1^2} \tag{15.150}$$

where h and b are the height and width of the light source, respectively (see Fig. 15.63). B is the light source luminance, m is the image magnification, and f_1 is the focal length of the first mirror or lens. The dimensions of the source at the knife edge are $(f_2/f_1)b \times (f_2/f_1)h$. If all but a portion, a, parallel to the width of the source is cut off by the knife edge (see Fig. 15.63), then, for a homogeneous index of refraction field, the illumination on the screen decreases uniformly to

$$I = \frac{Bab}{m^2 f_1 f_2} \tag{15.151}$$

where f_2 is the focal length of the second mirror or lens. For a horizontal knife edge, it is only necessary to consider the vertical displacements. An optical disturbance will produce a vertical shift, δa, of the source image, and δa is related to the deflection angle by

$$\delta a = f_2 \tan \epsilon_y \simeq f_2 \epsilon_y \tag{15.152}$$

where the approximation is valid for small ϵ_y. Hence, the change in intensity is

$$\delta I = \frac{Bb\delta\epsilon}{m^2 f_1} \tag{15.153}$$

The contrast of this part of the image with respect to the background is given by

$$C = \frac{\delta I}{I} = \frac{\delta a}{a} = \frac{f_2 \delta \epsilon}{a} \tag{15.154}$$

Now with Eq. (15.149), Eq. (15.154) can be rewritten as

$$\frac{\delta I}{I} = \frac{f_2}{a} \int_w \frac{1}{n} \frac{\partial n}{\partial y} \, dz \tag{15.155}$$

Thus, we have shown that in this case, the schlieren method is sensitive to the *first derivative of the index of refraction* in the y direction. If the knife edge is rotated by 90°, then the analysis would indicate a sensitivity to the x gradient. For many gases $(1/n) \approx 1$, thus Eq. (15.155) reduces to

$$\frac{\delta I}{I} = \frac{Kf_2}{a} \int_w \frac{\partial \rho}{\partial y} \, dz \tag{15.156}$$

which demonstrates that the schlieren method is sensitive to changes in the density gradient for compressible flows.

Determination of the amount of light to cut off at the knife edge is of great importance. Obviously, the above analysis ceases to apply if the source is completely off or completely on the knife edge. Hence, the maximum range of source displacement is $(f_2/f_1)h$ of the image, and the corresponding range of angular deflection is $(\delta e)_{max} = h/f1$ [Holder and North (1963)]. Thus, the range of the apparatus can be increased by replacing the first mirror or lens with one of shorter focal length or by increasing the height of the source. The height of the source can be increased physically or by means of an optical system consisting of a condenser lens. Usually, the schlieren system is set up so that the range is the same for deflections toward or away from the knife edge, i.e.,

$$a_R = \frac{f_2(\delta\epsilon)_{max}}{2} \qquad (15.157)$$

Thus, the corresponding intensity is reduced to $I_o/2$.

The schlieren sensitivity, S, can be determined from Eq. (15.155) as

$$S = \frac{dC}{d\epsilon} = \frac{f_2}{a} \qquad (15.158)$$

Hence, the sensitivity can be increased by increasing the focal length or decreasing the height a. As mentioned above, increasing the focal length would involve changing the optical components, thus this may be an undesirable or expensive modification. In addition, there are practical limits to the minimum height a. For example, the film emulsion speed limits the minimum light intensity for a given exposure time. A second limitation is the requirement to measure both positive and negative deflections [i.e., $a_{min} \geq f_2(\delta\epsilon)_{max}$]. As discussed above, this is usually satisfied by Eq. (15.157). Diffraction is a third limiting factor. It has been noted [Holder and North (1963)] that the effects of diffraction usually result in increased illumination on the screen near the boundaries of the aperture (e.g., the wind tunnel walls). The magnitude of these effects is inversely proportional to the height a. For typical systems, these effects are usually confined to very narrow bands of bright illumination near the image of the boundaries.

In most high subsonic, transonic, and supersonic wind tunnels, the density gradients are easily detected by typical schlieren systems. However, for wind tunnels operating at very high supersonic or hypersonic velocities, the density may become too low for the gradients to be detected by schlieren methods. Holder and North (1963) report that typical schlieren methods are adequate for supersonic flows if the free stream density exceeds about 0.01 kg/m^3.

The *sharp-focusing schlieren systems* were developed to obtain an image of a given plane in the flow field. The sharp focusing technique of Kantrowitz and Trimpi (1950) employs multiple sources and slits, with each source-slit pair a conventional schlieren system in itself. The beam is oblique to the optical axis as it traverses the tunnel. Consequently, the light intensity changes that are caused by the density gradients in the focal plane conjugate to the screen superimpose exactly, thereby forming an image. The light intensity due to density gradients in other planes superimpose offset from one another, and are thus blurred out. With an appropriate experimental procedure, one can change the z-location of the desired plane. *Holography* can be an important aid for this method, since for a phase hologram the instantaneous index of refraction field can be *frozen*. Afterwards, the focusing schlieren method can then be applied to the reconstructed density field scanning various z-locations [Trolinger (1974) and Merzkirch (1981)]. *Combined Schlieren–Laser Doppler* techniques have been used to measure the velocity of flame fronts [Schwar and Weinberg (1969)] and unsteady shock waves in shock tubes [Merzkirch and Erdmann (1974)].

The *graded-filter methods* [North (1952)] were developed to improve sharpness and to allow for the use of a point light source. Instead of a knife edge in the plane of the second mirror, a graded-filter is used. The improvement in sharpness is due to the different effects that the filter and knife edge have on the defracted light from the fine details of the flow. The arguments for this method over the knife edge decrease in importance as the sensitivity of the system is increased.

Color schlierens can be obtained from black-and-white arrangements by including a prism [Holder and North (1952)] or a diffraction grating [Maddox and Binder (1971)] in the focal plane of the first mirror (or lens). A color spectrum is formed at the focal plane of the second mirror, and the color of the image on the screen can be adjusted by moving the slit (or double knife edge), which is parallel to the bands of the light spectrum [Holder and North (1952)]. With the flow present, part of the spectrum is displaced, thus the image of the density gradient is a different color. Color methods can also be used with a circular double knife edge arrangement. A combination of a multicolor source filter in front of the light source and an appropriate aperture in the focal plane of the second mirror (or lens) allows the discrimination of the light deflections [Settles (1970) and Merzkirch (1981)].

A simpler method to obtain color schlierens is to use the graded filter apparatus, but replace the usual graded filter with one containing different bands of different colors in the focal plane of the second mirror (or lens) [North (1954)]. Or, the knife edge of the conventional Toepler method can be replaced by a tricolor filter consisting of three colored strips parallel to the rectangular light source [Kessler and Hill (1966)]. The width of the middle strip is equal to the width of the light source. The colors above and below should be chosen to give the maximum contrast.

Many other schlieren method variations have been developed over the years. For a more complete list of techniques see Merzkirch (1981), Settles (1980), or Volluz (1961).

Schlieren Processing for Density Fluctuations. Schlieren photographs can provide great amount of qualitative and limited amounts of quantitative mean flow data. However, as was the case for shadowgraphs, schlieren photographs can also be processed for statistical turbulence properties [Clay *et al.* (1965)]. The contrast for both the horizontal and vertical knife edge schlieren photographs is given by

$$C_h(x, y) = SK \int_W \frac{\partial \rho(x, y, z)}{\partial y} \, dz$$
$$C_v(x, y) = SK \int_W \frac{\partial \rho(x, y, z)}{\partial x} \, dz$$

$$(15.159)$$

S is the sensitivity, where for the double pass system of Clay *et al.* (1965), $S = 4f/a$. Taking the autocorrelation of both sides of the horizontal contrast equation in

Eq. (15.159), and simplifying yields

$$R_{ch}(\zeta, \eta) = 2\pi S^2 K^2 W \int_{-\infty}^{\infty} \int_{-\infty}^{\infty} k_y^2 S_\rho(k_x, k_y, 0) e^{i(k_x\zeta + k_y\eta)} dk_x dz_y \quad (15.160)$$

where k_i are wave numbers. The derivation can be found in Clay *et al.* (1965). The correlation function can also be expressed in terms of the density correlation function as

$$R_{ch}(\zeta, \eta) = S^2 K^2 W \int_{-\infty}^{\infty} \frac{\partial^2 R_\rho(\zeta, \eta, \xi)}{\partial \eta^2} d\xi \quad (15.161)$$

where ζ, ξ, and η correspond to the correlation space coordinates. From this equation, and its vertical contrast counterpart, one gets four correlation functions and four spectra relations. To simplify the analysis, the density field will be assumed statistically isotropic [Clay *et al.* (1965)]. However, the 2D spectra analysis of Bowersox and Schetz (1994) can be applied here as well as to the shadowgraph as discussed above, which eliminates the necessity of the isotropic assumption. In any event, the following autocorrelation relations can be derived for isotopic turbulence

$$R_{hx}(\tau) = R_{vy}(\tau) = -\frac{\beta}{2\pi\tau} \frac{\partial}{\partial \tau} \int_0^{\infty} R_\rho[(\xi^2 + \tau^2)^{1/2}] d\xi$$

$$R_{vx}(\tau) = R_{hy}(\tau) = -\frac{\beta}{2\pi} \frac{\partial^2}{\partial \tau^2} \int_0^{\infty} R_\rho[(\xi^2 + \tau^2)^{1/2}] d\xi \quad (15.162)$$

where $\beta = 4\pi S^2 K^2 W$ [Clay *et al.* (1965)]. For the autospectra, the equations are given by

$$S_{hx}(\kappa) = S_{vy}(\kappa) = \beta \int_0^{\infty} k_y^2 S_\rho[(k_y^2 + \kappa^2)^{1/2}] dk_y$$

$$S_{vx}(\kappa) = S_{hy}(\kappa) = \beta\kappa^2 \int_0^{\infty} S_\rho[(k_y^2 + \kappa^2)^{1/2}] dk_y \quad (15.163)$$

where κ is a wave number. The variance can be found from any of the expressions given in Eqs. (15.162) and (15.163) as

$$\sigma_\rho = \frac{4}{\beta} \int_0^{\infty} R_{hx}(\tau) d\tau = \frac{4}{\beta} \int_0^{\infty} \frac{1}{\tau} \left[\int_0^{\tau} R_{vx}(\tau') d\tau' \right] d\tau$$

$$= \frac{4\pi}{\beta} S_{hx}(0) = \frac{4\pi}{\beta} \int_0^{\infty} \frac{1}{\kappa} S_{hy}(\kappa) d\kappa \quad (15.164)$$

The above expressions describe the relations between the measured statistical properties of the film contrast and the gas density. The frequency resolution is limited to $1/W$. Clay *et al.* (1965) successfully applied these techniques to estimate the turbulent density variance in the wake of a Mach 8 projectile.

Interferometry. Interferometry techniques utilize optical components to record the light ray phase or time shift, τ, on the recording screen. The first to be applied to fluid mechanics is the now famous Mach–Zehnder Interferometer [Zehnder (1891) and Mach (1892)]. The development and interpretation of interferograms relies on the electromagnetic wave theory of light. For most applications, it is sufficient to consider a linearly polarized propagation plane wave, which can be represented as

$$E = a \cos \left[2\pi f(t - z/v) \right] \qquad (15.165)$$

where a is the amplitude, f is the frequency, and v $(= c/n)$ is the velocity of the propagation of the wave [Hariharan (1985)]. It is usually convenient to define the phase difference, ϕ, as $2\pi fz/v$ or $2\pi nz/\lambda$, where λ is the wavelength of the light in a vacuum $(=c/f)$. Also, a very important parameter in interferometry is the optical path, P, where between point z and the origin, P is given by nz.

If two monochromatic waves propagating in the same direction and having the same polarization plane are superimposed at a point, the total light intensity at that point is

$$I = I_1 + I_2 + 2\sqrt{I_1 I_2} \cos \Delta\phi \qquad (15.166)$$

where $I_{1,2}$ are the intensities of the two waves acting separately, and $\Delta\phi = \phi_2 - \phi_1$. Also, the change in optical path length ΔP is given by $\lambda \Delta\phi / 2\pi$. The intensity in Eq. (15.166) has a maximum when

$$N = m, \quad \Delta P = m\lambda, \quad \Delta\phi = 2m\pi \qquad (15.167)$$

where $N = \Delta P/\lambda$ is the *order of interference*, and m is an integer. Equation (15.167) has a minimum when

$$N = \frac{(2m + 1)}{2}, \quad \Delta P = \frac{(2m + 1)}{2} \lambda, \quad \Delta\phi = (2m + 1)\pi \qquad (15.168)$$

These interference or fringe patterns are at the heart of all interferometry techniques. Another way of understanding the formation of fringe patterns is to invision imaginary wave fronts [Philpert, Surget, and Veret (1985)], which are perpendicular to the direction of wave propagation. The fringe pattern is a result of the difference between the recombined disturbed and undisturbed wave fronts being separated by integer multiples of the light wavelength.

Under the assumption that the slopes of the light rays are small (i.e., dx/dz, $dy/dz \ll 1$, where z is direction of the light ray, see Fig. 15.60), then Eq. (15.136) reduces to [Merzkirch (1981)]

$$c\Delta t = \Delta P = \int_W [n(x, y, z) - n_o]dz \qquad (15.169)$$

where W is the width, and n_o is the index of refraction of the undisturbed test field. Thus, here we have demonstrated that interferometry is directly *sensitive to the index of refraction*. For gases with a constant Gladstone–Dale constant, the interferometer is sensitive to the density field directly.

Fringe patterns can be obtained by the superposition of two monochromatic light rays with the same polarization propagating in the same direction. The reference beam method creates the change in optical path ΔP by passing one beam through the test section, while a second conjugate beam traverses exterior to the test section, thus remaining undisturbed. The Mach–Zehnder interferometer operates on this principle.

Figure 15.64 shows the basic components of the Mach–Zehnder interferometer. The dual light beams are created by splitting the collimated light with the beam splitter M_1' (semireflecting mirror). This method of obtaining two beams from a single source falls under the category of amplitude division. One beam traverses the flow field, while the second reference or conjugate beam does not. The two beams are recombined to create the interference pattern with a second beam splitter M_2'. Usually, the traversing distances of the two beams are designed to be equal. In addition, if the test beam passes through glass windows, the reference beam is usually passed through identical windows. The recombined beam is then focused onto the recording screen through a schlieren head (mirror or lens) and a camera lens (if necessary). For 2D flows, the equation for bright fringes can be given as

$$KW\,[\rho(x,\,y)\,-\,\rho_o]\,=\,m\lambda \qquad (m\,=\,\pm 1,\,\pm 2,\,\ldots) \qquad (15.170)$$

Thus, the curves are lines of constant density. With this arrangement, the undisturbed field is uniform in color, thus this method is termed the *infinite fringe with alignment interferometer* [Merzkirch (1981)]. Figures 15.59(a) and 15.65(a) present examples of 2D infinite fringe interferograms.

By inclining one or several mirrors, one may produce, for an undisturbed flow,

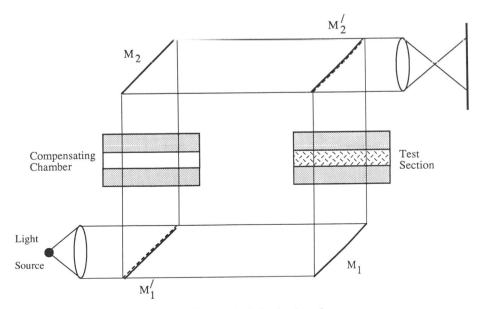

FIGURE 15.64 Mach–Zehnder interferometer.

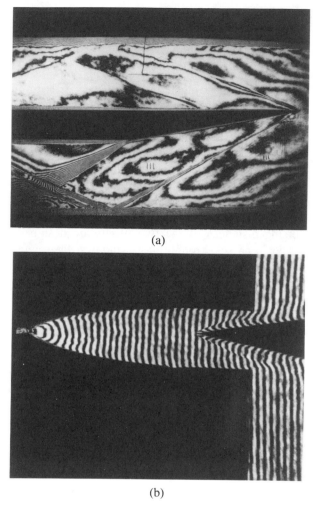

(a)

(b)

FIGURE 15.65 Comparison of infinite and finite fringe interferometry techniques: (a) Infinite fringe interferogram of supersonic flow. The shock wave, expansion wave, and wall boundary layers are clearly depicted; (b) finite fringe interferogram of supersonic flow. The shock wave and boundary layers are clearly noticeable.

a system of parallel fringes in any preferred direction. This method is called the *finite fringe width method*. Kinder (1946) developed the *single-plate controlled interferometer*, which produces the parallel finite fringe pattern by rotating a single mirror. The effect of rotating one or more mirrors is the same as placing a transparent wedge into one of the beams [Ladenburg and Bershader (1954)]. Further, the flow field and the fringe pattern must be in focus at the recording plane; s is the width of the undisturbed fringes, and the relative change in s is a measure of the density. The relationship between the change in fringe width and the flow density for a 2D flow is given by [Merzkirch (1981)]

$$\frac{\Delta s}{s} = \frac{KW}{\lambda} [\rho(x, y) - \rho_o] \qquad (15.171)$$

By superimposing the positive of the undisturbed fringes with the negative of the disturbed fringes, one gets a *Moire pattern* which is equivalent to the infinite fringe pattern of the disturbed field. Figure 15.65 presents examples of both the finite and infinite fringe interferograms.

Laser light sources have led to the emergence of *holographic interferometry* [Heflinger and Wuerker (1966), Philbert, Surget, and Veret (1985)], which is a different class of reference beam interferometers. With these systems, the spatial separation between the reference and test beam is replaced by temporal separation. Holographic interferometers are usually obtained by means of a double exposure onto the same holographic plate [Merzkirch (1981)]. One exposure is made in the absence of flow in the test section (i.e., the reference beam), while the second is made with the flow field present (i.e., the test beam). Upon illumination of the double exposed hologram, the superposition of the two wave pattern is formed. If the interferogram is obtained in a 2D flow, the observed fringe pattern will depend on the viewing direction [Witte and Wuerker (1970)]. Hence, the hologram contains 3D flow field information [Merzkirch (1981)]. Thus, a very attractive feature of holographic interferometry is this unique ability to resolve 3D flow fields. In addition, holographic interferometers are less expensive than conventional interferometers, hence they are now widely used in aerodynamic research.

Schlieren or *shearing interferometers* are a class of dual beam interferometers, where both of the beams tranverse the test section separated by a small distance d. If the shearing distance, d, is very small, then the change in the optical path is given by [Merzkirch (1974)]

$$\Delta P = \int_W [n(x, y + d/2, z) - n(x, y - d/2, z)] \, dz \qquad (15.172)$$

Expanding n in a Taylor series and retaining only the first term, then Eq. (15.172) reduces to

$$\Delta P = Kd \int_W \frac{\partial \rho(x, y, z)}{\partial y} \, dz \qquad (15.173)$$

where the Gladstone–Dale relation [Eq. (15.134)] has been incorporated. Referring to Eq. (15.173), one can immediately notice that unlike the reference beam interferometer, the shearing interferometer is sensitive to the density gradient. Because of this, shearing interferometry is also referred to as schlieren interferometry (recall, schlieren techniques are sensitive to the density gradient). For a 2D flow, the equations for bright fringes are given by

$$KWd \frac{\partial \rho(x, y)}{\partial y} = m\lambda \qquad (m = \pm 1, \pm 2, \ldots) \qquad (15.174)$$

Turning the interferometer 90° allows the measurement of the density gradient with respect to the x-axis. The shearing interferometer is also capable of operating in either the infinite or finite fringe mode.

The *single-plate schlieren interferometer* is shown schematically in Fig. 15.66(a) [Kiss, Schetz, and Moses (1993)]. The collimated beam after traversing the test

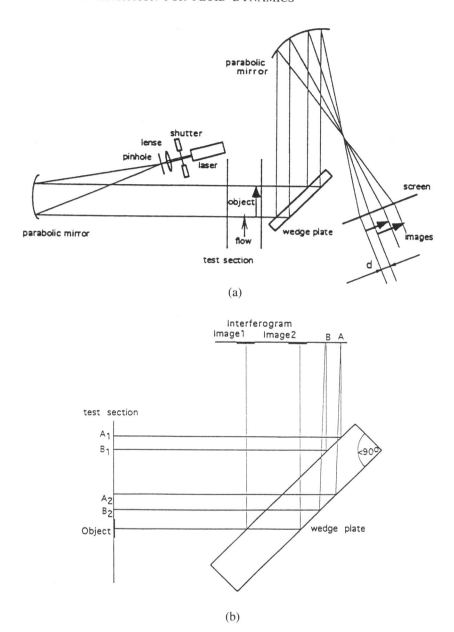

FIGURE 15.66 Single plate interferometer. (a) Schematic of the single-plate interferometer and (b) the applied orientation of the wedge plate.

section hits a glass wedge plate, which has a very small wedge angle (10–30 seconds of an arc). Plane plates can also be used to create fringes [Hariharan (1985)]. If the wedge plate orientation and thickness creates a small displacement of the beams, then this system will behave as a schlieren interferometer. However, if the displacement, d, is relatively large, then retaining only the first term in the Taylor series in the derivation of Eq. (15.173) becomes inaccurate. Kiss, Schetz, and Moses (1993) addressed this issue from a different perspective. They traced the rays of their *double image* as shown in Fig. 15.66(b) back to the test field. In particular, they were able to show that

$$(\rho_{A_1} - \rho_{B_1}) - (\rho_{A_2} - \rho_{B_2}) = -\frac{\lambda}{K}(m - m_{nf}) \qquad (15.175)$$

where m and m_{nf} are the number fringes between points A and B with and with out flow, respectively. Thus, with a relatively large separation distance, d, the Kiss–Schetz–Moses technique is sensitive to the absolute density (i.e., their method is a *single-plate absolute density interferometer*). It is important to note here that a portion of each ray of the beam undergoes multiple internal reflections and transmissions from the upper and lower glass surfaces. If the intensity of these effects is significant, the system is not a true interferometer [Merzkirch (1981)]. This is due to the fact that there will be interference at the same point on the screen from beams separated by d, $2d$, ... However, by a proper choice of the reflection coefficients and the orientation of the plate, the intensity of the additional rays can be made insignificant.

A second type of shearing interferometer is the *simple Wollaston schlieren interferometer* described by Merzkirch (1965), (1974), (1981). This method has the attractive feature in that the knife edge in a typical schlieren system is replaced by an optical system consisting of two crossed polarizers with a Wollaston prism in between. The shearing distance, d, is determined by the focal length of the second focusing lens or mirror of the schlieren system (i.e., the schlieren head) and the separation angle of the Wollaston prims. A second very attractive feature of this method is that it can be converted from infinite fringe to finite fringe mode by moving the Wollaston prism along the optical or z-axis.

Interferograms provide both qualitative and quantitative flow field information. For 2D flows, the interpretation of interferograms is straightforward. However, for 3D flows, the situation is much more complicated.

Optical tomography is a technique that allows quantitative analysis of 3D flow fields from a set of 2D line-of-sight images [Hesselink (1985)]. Any of the classical refractive flow visualization techniques discussed above can be used. However, the technique is almost exclusively applied to holographic interferograms. By image reconstruction, a 3D image can be obtained from a series of planar images. The reconstructions are considered approximate in the sense that they are created from a finite number of planar images. Currently, there are several schemes to perform the reconstruction [see Hesselink (1985) for scheme descriptions]. Tomographic techniques can be applied to stationary (steady), as well as unsteady or turbulent flows.

Optical Component Quality. The optical component quality requirement is a very important consideration in the design of refractive flow visualization systems. These requirements depend not only on the visualization method (shadowgraph, schlieren, or interferometry) but also on the nature of the flow field under investigation. The reader is referred for detailed information to Beams (1954), Holder, North, and Wood (1956), Volluz (1961), Sedvey *et al.* (1966), Holder and North (1963), Buzzard (1968), Ladenburg and Bershader (1956), Merzkiran (1981), and Harihan (1985).

15.7 FLOW ANGLE MEASUREMENT
Gustave J. Hokensen

In the measurement of local flow angles, two basic techniques using a given instrument exist—nulling and calibration. In addition, the available devices generally fall into two categories: 1) flow angle deduced from velocity components, and 2) flow angle directly (often from force or pressure measurements). Although, depending on how the usage, an instrument may fall into either category.

The most common operational technique for a given device is *nulling* wherein the response is minimized or maximized by varying the orientation of the probe such that the device becomes aligned with the local flow. The instrument angle relative to the laboratory frame of reference is then measured, and the flow angle is known. This approach has the advantage of being insensitive to probe calibration. Alternatively, the *calibration* technique relies on a fixed orientation probe immersed in the flow. The flow angle is subsequently determined from angular response calibration curves obtained by measuring the instrument output over a range of known angles. This technique is more useful when many readings, or a continuous profile (in time or space) of the flow angle is required. The accuracy and usefulness of this approach depends on the characteristics of the calibrations for each probe.

Discussions in subsequent sections emphasize two-dimensional flows for simplicity. Extensions to 3D cases are noted, where applicable.

15.7.1 Hot-Wire Anemometer

Operated in the nulling mode, either a single wire [see Champagne *et al.* (1967)] (normal or slant) or an X-wire probe (measuring two velocity components, U and V) [see Morrison *et al.* (1972)] may be used to determine the local flow angle. With an uncalibrated, balanced X-wire, the minimum-v maximum-u angle may be sensed which indicates flow alignment and, therefore, flow angle relative to the flow facility. Alternatively, the X-wire probe may be calibrated (see Sec. 15.4.2) to provide U and V, and the flow angle may be computed from $\alpha \equiv \tan^{-1} (V/U)$. With a calibrated hot-wire and the associated electronic hardware, the instantaneous flow angle may be measured well into the 10^4 Hz region. Multi-wire probes may be used to obtain flow angles in fully 3D flows. Some *hot-film probes* may also be used for flow angle measurement.

The laser Doppler velocimeter may be used for flow angle measurement in a manner identical to the hot-wire.

15.7.2 Differential Pressure and Force Gauges

In this approach, probes of various configurations, which are shown in Fig. 15.67, are inserted into the flow. The differential pressure across the device is either nulled or measured and converted to flow angle via calibration curves. Alternatively, the lift on the probe is nulled or converted to flow angle via calibration at the same Mach and Reynolds numbers.

Force-type devices which are self-nulling such as a wind vane, also exist.

The use of a static pressure probe to measure small normal velocities (and, therefore, angles by either primary technique) has been reviewed by Nenni *et al.* (1982).

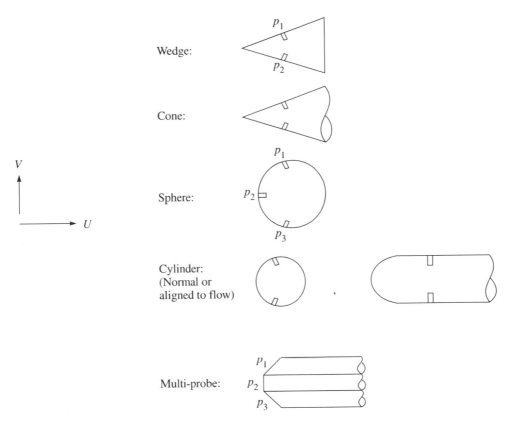

FIGURE 15.67 Differential pressure gages to measure flow angles.

Also, an effective calibration technique has been obtained by Dudzinski and Krause (1969) to allow for two- or three-dimensional flow angle measurements with fixed orientation multi-probes.

15.7.3 Flow Visualization

In subsonic flows, many of the tracer techniques discussed in Sec. 15.6, combined with sequential or strobe photography, may be used to optically recover local flow angles.

In supersonic flow, Mach waves or shock waves will spread from the point or leading edge of a sharp probe at a given angle relative to the flow; the wave angle depends on the local Mach number.

15.7.4 Electrical and Electromagnetic Probes

Devices mentioned earlier, including particularly the corona-discharge probe, exhibit a strong sensitivity to the local flow angle which is useful for its measurement. Fuhs (1965) discusses an electromagnetic indicator which senses the local angle of a hypersonic plasma flow; such a flow occurs during reentry into the earth's atmosphere. Figure 15.68 indicates the output signal from a pair of electrical conductiv-

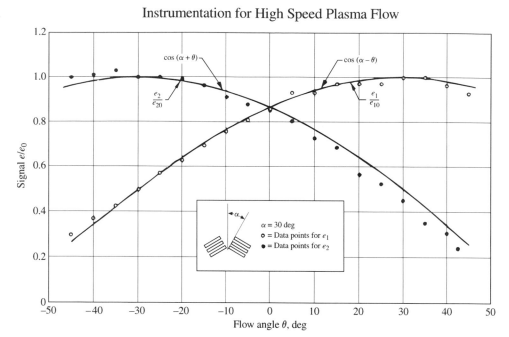

FIGURE 15.68 Plots of observed signals e_1 and e_2 as a function of θ and of $\cos(\alpha \pm \theta)$ for $\alpha = 30$ degrees for an electromagnetic plasma flow angle indicator. (From Fuhs, 1965.)

ity–velocity, σ-U, transducers oriented at angle α relative to a reference axis. The angle θ in Fig. 15.68 is the angle between the plasma flow and the reference axis. From the signal ratio, e_2/e_1, the angle θ can be determined. Note that the algebraic sign of θ is measured also.

15.8 FORCES AND MOMENTS
James F. Marchman III

The primary purpose of most wind (or water) tunnel and towing tank tests is to measure aero/hydrodynamic forces and moments on a scale model. Over the years such measurements have been made in many different ways, using wires, springs, moving weight beam balances, struts, stings and internally or externally mounted strain gage balances. Measurement of forces and moments is never simple. Most good wind tunnel facilities are equipped with several different balance systems in order to accommodate a wide variety of testing. In aircraft aerodynamic testing, lift is usually the predominant force and pitch is the primary moment with both being at least an order of magnitude greater than other forces or moments. In tests of *non-aerodynamic objects* such as other vehicles, buildings, signs, outdoor lights, etc., drag is usually by far the predominant force. The system designed for aircraft tests may be totally inappropriate for testing automobiles.

In all wind (or water) tunnel force or moment measurement schemes there must be some mechanism for mounting the test model in the center of the tunnel, neces-

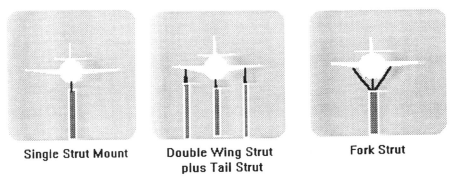

Single Strut Mount   Double Wing Strut   Fork Strut
                          plus Tail Strut

FIGURE 15.69 Types of strut mounts.

sitating an accounting for the effects of the mounting system itself on the flow about the test model and for any forces transmitted to the balance system due to aerodynamic forces on the mounting device. Most tunnels employ strut or sting mounts (see Fig. 15.69). With strut systems, one or more struts go from a rigid base or external balance system outside of the tunnel test section, passing through the tunnel floor or ceiling to the model. Multiple strut systems are often used with external mechanical or strain gage which employ different struts for measurement of separate forces or moments. A common multiple strut system shown in Fig. 15.70 uses two struts (one mounted to each wing of an aircraft model) to measure lift and drag plus a third, tail-mounted strut used to measure pitching moment. Most modern systems use a single strut which is either instrumented with strain gages to directly measure the forces and moments transmitted to the strut or is connected to an external balance system (see Fig. 15.71).

In a strut mount system, the struts are normally shrouded with a streamlined fairing that extends almost to the test model and which ensures that the loads measured by the strut are primarily those from the model. It is likely that small portions

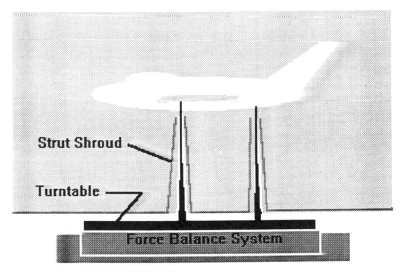

FIGURE 15.70 Wing/tail strut mount arrangement.

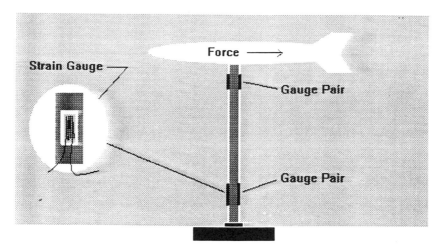

FIGURE 15.71   Simple strain gage arrangement.

of the strut immediately adjacent to the model will remain unshrouded, necessitating an evaluation of the *tare* forces on this exposed strut section.

A major concern with strut mount systems is the extent to which the model forces are altered by the presence of the strut(s). There are fairly reliable methods for the evaluation of such interference effects. These usually involve the use of *image* struts which are noninstrumented duplicates of the balance struts placed on the opposite side of the model from the active struts. Rae and Pope (1984) provides a thorough overview of various types of strut balance systems and the evaluation of their interference and *tare* effects.

While strut mounting systems are adequate for many aerodynamic tests, the flow wake from the strut may interfere significantly with the flow over parts of the model for tests of aircraft at high angles of attack. The use of wing mounted dual struts can alleviate this problem as far as fuselage/strut wake interactions are concerned, but this still permits some wake interaction with wing flap systems. The mounting of choice in such cases is usually a *sting* or tail mount with a round sting extending from the tail of the model rearward to a pylon or *quadrant* system which employs a linkage or screw drive mechanism to translate and/or rotate the model in the tunnel test section as shown in Fig. 15.72. This system is normally configured to allow rotation of the model about its center of gravity or other selected point without vertical translation in the test section. More complex systems also allow model rotation in yaw.

Although sting systems eliminate strut wake/model interactions, they also introduce their own sources of error in their effect on the flow over the rear of the model and in the model wake. Keeping the sting diameter small and allowing a reasonable distance between the model tail and the sting support system can minimize this effect, however the magnitude of this error must be assessed for any test in which a sting is used. The ultimate choice between a sting or a strut mounting system must include an assessment of the relative interference errors and may lead to some use of both types of mounts in order to gain an understanding of the true aerodynamic forces on the model itself.

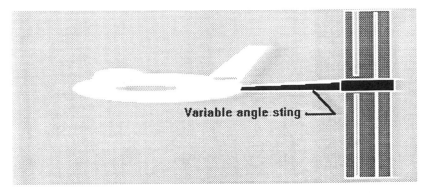

FIGURE 15.72 Tail mounted sting and pylon balance system.

While sting mounts can be used with external force and moment measurement systems, they usually use an electronic strain gage balance located at the front end of the sting, inside the model. The use of such a balance eliminates concerns about any aerodynamic loads on the balance itself, since the forces and moments measured are only those between the model and sting caused by loads on the model itself.

Another type of model mounting system is used in testing semi-span models (see Fig. 15.73) or two dimensional models. The use of semi-span models allows larger scale tests by assuming model symmetry at the tunnel wall or floor and that the forces and moments measured are half of the total. It is assumed that the presence of the wall or floor at the model plane of symmetry results in a mirror image effect in accurately simulating the influences of the flow (vortices, etc.) from one side of the plane of symmetry on the other side of the model. In such mounting, the balance can be internal to the model or outside of the tunnel.

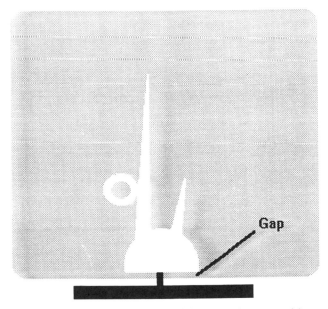

FIGURE 15.73 Semi-span model mounted on turntable.

Two dimensional testing of airfoils usually involves extending a constant chord wing section between the tunnel walls or between two *end-plates* placed in the tunnel. The walls or end plates are supposed to ensure that there is a truly two-dimensional flow about the airfoil section.

With both two-dimensional and semi-span testing arrangements, there are corrections which must be made for the effects of the wall or end-plate boundary layers and for gap effects. Gap effects occur because it is necessary to leave a small space between the model and the wall or end plates to prevent false force and moment indications due to friction between the model and wall or plate. This type of gap effect is discussed in Rae and Pope (1984) and is examined in detail in Marchman and Kuppa (1987) for low Reynolds number aerodynamic testing.

It is common now to make most wind tunnel force and moment measurements with electronic strain gage balances, although some facilities still have excellent mechanical balances. The mechanical balances are commonly external systems with a strut or sting system attached to a massive floating frame located below the test section or surrounding the section in a yoke arrangement. Such systems employ very sensitive linkages and flexures which are used to isolate the various force and moment components and transmit them to automatically nulling scales which mechanically or electronically display their measurements. Rae and Pope (1984) provides an excellent review of this type of balance, their construction and operation.

Strain gage balances offer significant advantages in force and moment measurement in that they can be made small enough to fit entirely inside the test model or be made part of the strut or sting support system. The small size strain gage balance also makes it possible to separately instrument various parts of a test model to examine individual component contributions to the overall forces on a model. In principle, strain gage balances are easy to construct by gluing a small electrical resistance element to a beam and measuring the change in resistance of the gage as it is stretched or compressed under beam bending due to an applied force. The gage output can be calibrated against the moment caused by the force on the beam. Thermal expansion or axial loads on the beam can also cause resistance changes but these can be canceled out by placing a second, identical gage on the opposite side of the beam and connecting the gages in a Wheatstone bridge circuit such that their outputs subtract, resulting in a zero net output when purely axial loads or thermal expansion is present.

The two-gage beam described above actually measures a bending moment rather than a force; however a second pair of gages can be added at another axial location on the beam measuring a second moment due to the same applied force. This extra pair of gages are used to complete the Wheatstone bridge circuit such that the difference in the two measured moments gives an output proportional to the applied force multiplied by the fixed distance between the two pairs of gages regardless of the actual force location. Proper use and calibration of such a set of four gages can give accurate measurement of the force normal to the beam as well as the moment caused by the force.

There are many other physical arrangements of strain gages which can be used to measure desired forces and moments, and proper combinations of such systems can be used to measure multiple forces and moments. The culmination of this is the six-component strain gauge balance capable of measuring lift, drag and side force, and the pitch, roll and yawing moments simultaneously. Rae and Pope (1984) shows excellent examples of such gage and balance arrangements.

Strain gage balances normally have a designated center about which moments are measured, and, ideally, the balance would be designed to allow mounting of the wind tunnel test model so its center of gravity, aerodynamic center or other relevant point is coincident with the balance center.

A major concern with strain gage and other balances is the degree to which the various force and moment measurements are independent of one another. Modern balances can be designed and constructed to be relatively free of such interactions as long as the prescribed load and moment range limits for the balance are observed, however, balance calibration should carefully document all possible interactions. Also, while temperature effects can be compensated for to a large degree, the user of such balances in a situation where temperatures may vary widely during testing should ascertain that thermal effects do not limit the repeatability of test data.

Two other methods sometimes used to measure forces on a model in a wind tunnel involve measurement of pressures. Normal and chord wise forces are often found by integrating pressure distributions measured about wings and other shapes, particularly in two-dimensional model tests. Such measurements can be taken at multiple locations, such as at different stations along a wing span, and are valuable for comparison with calculations of local pressures and forces found from panel or other computational methods. The forces found from these pressure distributions are due to pressures only and do not include friction forces or such things as induced drag.

Drag measurements are also sometimes made by examining the momentum change in a flow as seen in a momentum deficit in the wake behind a test model. This is essentially a two-dimensional test technique that assumes that all momentum change is linear. As such, it is usually employed in tests of 2D airfoil sections. Where significant rotational momentum changes are present, such as in separated flows and flows producing vortical motions, the value of such measurements is questionable. While this technique is very limited in value, it does provide a means of getting some indication of forces where external or internal balance systems and internal pressure measurement systems cannot be used.

A final note related to force measurement in wind tunnels concerns the definition of the axis system used. External balance systems normally read forces and moments with respect to a *wind axis* system where drag is identified as the force parallel to the wind with lift and side force normal to drag and moments defined about these axes. If the model is pitched or yawed using a tail strut and turntable or similar device attached to a large balance frame or yoke, the measured forces and moments are still defined relative to the wind.

In cases where a strain gaged strut is used, the balance rotates in yaw with the model. This results in a *local horizontal* axis system with lift still perpendicular to the wind but with the *drag* force measured in a horizontal plane normal to the lift vector, pointed toward the tail of the model. The *side force* is normal to the other two forces, pointed out the right wing. Pitch, roll, and yaw moments are measured about these force vectors. It may be necessary to resolve these measured values into a wind axis or other system for comparison with other data.

When internally mounted sting type strain gage balances are used, forces and moments are defined relative to a *body axis* system and the two main forces are more appropriately thought of as *normal* and *axial forces* rather than as lift and drag. The axial force runs through the centerline of the model fuselage and the balance, and other forces are normal to this, with appropriately defined moments.

15.9 SKIN FRICTION MEASUREMENT
Joseph A. Schetz

Values for the skin friction (wall shear) on the wall, τ_w, are usually given in terms of the dimensionless *skin friction coefficient*, $C_f = \tau_w/(\rho V^2/2)$, where ρ is the density, and V is the velocity at the edge of the boundary layer. Measuring frictional drag is important for both practical and scientific reasons, such as performance assessment and validation of computer codes which predict performance. Skin friction is also the key component in the so-called *friction velocity*, $v*$, that is used as the scaling velocity in correlating turbulent boundary layer velocity profiles (see Chap. 4). These correlations are central to all turbulent transport model development. Thus, the results of experimental tests can provide accurate skin friction values that will permit calibration and refinement of available calculation methods and turbulence models.

Experimental skin friction data can be obtained by using both direct and indirect measurement techniques. The direct method implies an actual measurement of the surface shear force on a small area without requiring the use of any assumed correlation. Indirect methods are based on measurement of other flow quantities that are then related back to skin friction through an associating principle. For example, one could experimentally measure heat flux and attempt to infer skin friction from that. This will work for simple flowfields. But the relationship between skin friction and heat transfer, called the *Reynolds Analogy*, does not accurately extend to complex compressible or strongly three-dimensional flows or flows with injection. However, an important compressible analysis does exist for simple flows. This variable density, variable property Reynolds Analogy was derived by van Driest (1956). Various indirect methods are shown in Fig. 15.74.

One of the most common indirect shear measurement techniques available is the *Preston tube*. This utilizes a standard Pitot pressure measurement on the wall and makes use of the similarity laws for the velocity profile of the boundary layer to relate to a corresponding wall shear stress [Preston (1954)]. One of the reasons this method has been so successful is that the relationship between the total pressure measurement and the skin friction is independent of pressure gradient. A thorough review of the Preston tube and other indirect methods can be found in Nitsche *et al.* (1984). Another indirect technique which is described in that paper is a sublayer fence which, like the Preston tube, also uses a pressure measurement, but only within the laminar sublayer. This simplifies the relation to shear stress. The sublayer fence also has some advantage over the other techniques in that it is capable of making measurements in transitional flow, separated flow, and three-dimensional flow, although it is still limited to flows without a high total temperature.

The surface hot film technique measures the heat transfer and employs Reynolds Analogy to relate the skin friction to the heat transfer. A wall-fixed doublet hot wire method is also relatively flexible. This method has an advantage in that two velocities above the wall are measured directly, rather than indirectly, as in the case of a Computational Preston Tube, which uses the same law of the wall to relate the flow properties to shear.

The major differences between the two methods of measuring surface shear force is that the indirect methods require the use of analytical correlations, while the direct method requires relying on a floating element of the wall to measure the shear in the

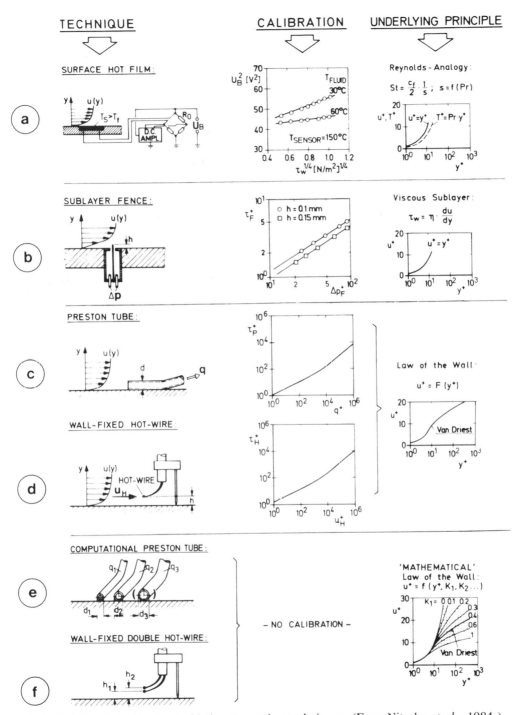

FIGURE 15.74 Indirect skin friction measuring techniques. (From Nitsche *et al.*, 1984.)

flow without being intrusive. It is obvious that a direct measurement is preferable to an indirect one because of the uncertainties that are introduced by the required assumptions, although, making a direct measurement can be more difficult.

A history of early direct measurement skin friction gages is given by Winter (1977). This summary includes floating element gages which were first introduced by Dhawan (1953). Dhawan's floating element gage configuration took into consideration such important issues as the effects of gaps, external vibrations and temperature changes, as well as static and dynamic calibration. A sketch of the four-bar linkage arrangement used is shown in Fig. 15.75. Dhawan tested and verified his gage in both laminar and turbulent incompressible flow, and also in turbulent compressible flow. Since then, other skin friction balances have been developed for measurements in supersonic flow, including work by Bruno and Risher (1968), Allen (1976), (1980), and Voisinet (1979). For the most part, these gages were mechanically complex, using some parallel linkage arrangement to sense the movement of the floating element. A schematic of this configuration from Allen is included as Fig. 15.76. Also, most gages are of the *nulling* design, where some mechanism is used to move the floating element against the shear force back to its position with no force applied. Another direct measurement device has been suggested by Vakili and Wu (1988), using a moving belt technique. Most of these have been used only in subsonic or moderate total temperature supersonic flows.

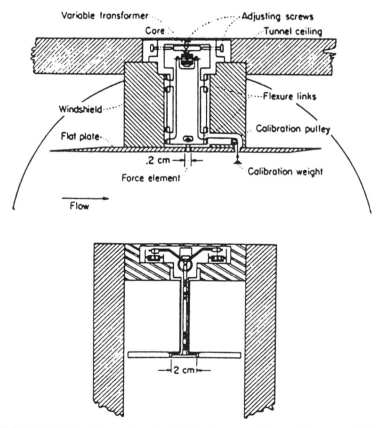

FIGURE 15.75 Dhawan's skin friction balance. (From Dhawan, 1953.)

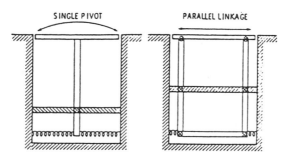

FIGURE 15.76 Parallel linkage arrangement of Allen (1976).

A unique deflection sensing device was first used in a cantilever arrangement by Schetz and Nerney (1977), where the key advance was a non-nulling design made possible by the use of very sensitive crystal strain gages. The difference between a nulling device and a non-nulling device is the mechanical complexity necessary to measure the shear. The non-nulling device allows the simple cantilevered floating element to move (albeit slightly), thus the actual bending of the base of the beam is being measured. This arrangement is much simpler than a nulling device, and consequently less susceptible to error. A nulling design also has a much slower time response. The concept of a non-nulling instrument using crystal strain gages allows measurement of the small surface shear forces with very small deflections of the sensing head. For very small head displacements, the tilting of the head is small enough so that errors due to misalignment are minimal. Successful experiments were conducted using this sensor for gages measuring skin friction in low temperature, low speed flows and in unheated supersonic flows, including cases with injection through the floating head.

The specific geometry of the floating element is an important design consideration, because it can introduce various potential error sources at the surface. As always, when using a floating element device to measure deflection, misalignment effects are a concern. These errors can be introduced because of the interaction between the floating element and the surrounding wall. The important parameters, such as misalignment with the surface, gap size, lip size, and local pressure gradients between the flow and the underside of the element are defined in Fig. 15.77 from Allen (1976). These concepts are discussed at length in the study by Allen (1976), (1980) for cold flow conditions. It is advantageous to fill the inside of the gage with oil. This serves to minimize the effects of axial pressure gradients on the surface and facility vibrations. It also aids in thermal protection in hot flows.

The non-nulling design has been successfully extended to very high enthalpy flows by the addition of cooling systems to protect the thermally sensitive crystal strain gages used [DeTurris *et al.* (1990), Schetz *et al.* (1993), and Chadwick *et al.* (1993)]. It is also important to ensure that the floating head of the gage is at the same temperature as the surrounding wall, or the fluid flow field can be disturbed. In some high heat flux cases, it was found necessary to actively cool the floating head.

Finally, this gage design concept has been extended for use in short-duration facilities such as shock tunnels where the test times are only 1–10 msec [Bowersox and Schetz (1993) and Bowersox *et al.* (1995)]. The basic idea is to use a beam and floating head made of a low density material to minimize acceleration loads in comparison to the shear loads to be measured as shown in Fig. 15.78.

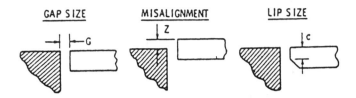

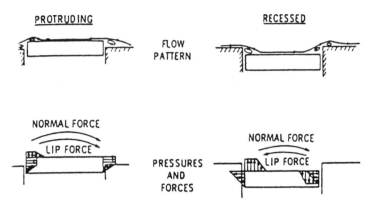

FIGURE 15.77 Definitions and error sources for floating element gages. (From Allen, 1976.)

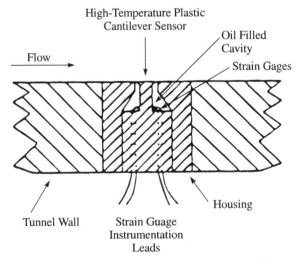

FIGURE 15.78 Skin friction gage for use in short-duration facilities. (From Bowersox *et al.*, 1994.)

15.10 TEMPERATURE AND HEAT FLUX MEASUREMENT
T. E. Diller and C. L. Tien

15.10.1 Introduction

Measurements of fluid and body surface temperatures and heat flux are often required to monitor or control the thermal behavior of industrial or aerodynamic processes. They can provide information about the fluid interaction with the surface and the performance and structural requirements of the material. One critical parameter for structural survival often is the temperature, while the required cooling system is dependent upon the surface heat flux that must be removed to maintain the needed temperature control. The heat transfer at the material surface is usually composed of a combination of convection with a fluid and radiation. Energy conservation at the surface requires a balance of these fluxes with the conduction in the solid material.

The relationship between temperature and heat flux is common in thermal problems and can be expressed by the energy conservation throughout the solid material

$$\rho c_p \frac{\partial T}{\partial t} + \nabla \cdot \bar{\mathbf{q}} = 0 \tag{15.176}$$

where $\bar{\mathbf{q}}$ is the heat flux vector, which is proportional to the gradient of the temperature, T, in the material, $\bar{\mathbf{q}} = -k\nabla T$. The time rate of change of the temperature is, therefore, related to the spatial gradient of the heat flux. Sufficient knowledge or assumptions about either the temperature or heat flux may then be used to calculate the other. The most common form uses the assumption of a locally one-dimensional, semi-infinite material. Cook and Felderman (1966) developed a numerical approximation for heat flux from discrete temperature steps using a piecewise linear signal of the surface temperature trace, which can be simplified to

$$q(t_n) = \frac{2\sqrt{k\rho c_p}}{\sqrt{\pi}} \sum_{j=1}^{n} \frac{T_j - T_{j-1}}{\sqrt{t_n - t_j} + \sqrt{t_n - t_{j-1}}} \tag{15.177}$$

where $k\rho c_p$ is the product of thermal conductivity, density and specific heat of the material and $q(t_n)$ is the heat flux at the specified time. This uses a time-derivative operation on the surface temperature to calculate the corresponding surface heat flux. Unfortunately, this tends to increase the effects of random noise in the signal. The opposite can also be used as demonstrated by Baker and Diller (1993) to calculate the change in surface temperature from measurements of the surface heat flux as a function of time

$$T_s(t_n) - T_o = \frac{2}{\sqrt{\pi}\sqrt{k\rho c_p}} \sum_{j=0}^{n-1} q_j \left[\sqrt{t_n - t_j} - \sqrt{t_n - t_{j+1}} \right] \tag{15.178}$$

Because this is a time-integration operation on the heat flux signal, the noise of the signal is diminished through the signal processing.

A number of different measurement methods are briefly discussed in the following sections on temperature and heat flux. More detailed treatments are available in the references listed for specific techniques. Some newer methods which make use of microsensors and high-speed digital data processing are emphasized. These have made possible time resolution of high frequency components and detailed spatial resolution on surfaces.

15.10.2 Temperature Measurement

Because temperature affects many properties of materials, there are numerous methods for measuring temperature. The simplest methods employ thermal expansion. The most common electrical-based sensors are thermocouples and resistance elements. A number of textbooks [Benedict (1984) and Figliola and Beasley (1991)] cover operation of these point measurement devices. Optical methods such as liquid crystals, infrared detectors and thermographic phosphors have the advantage of non-contact measurement. They can provide a two-dimensional temperature field over a surface, but it is more difficult to obtain absolute temperatures and they require more sophisticated data acquisition equipment. Other optical methods are discussed in Secs. 15.12 through 15.14. The approximate useful range of the various devices is illustrated in Fig. 15.79.

Thermal Expansion Methods. In the medium temperature range, many convenient and useful temperature sensing devices are based on the phenomenon of thermal expansion of materials. Most common and well known are the liquid-in-glass thermometers, particularly the mercury-in-glass thermometers, which have an operating temperature range from the freezing point of 235.2 K to the boiling point, which can be increased by pressurizing the inert gas in the capillary space above the liquid. In the mercury case, it can go up to about 800 K. For lower temperatures, alcohol is applicable to 210 K, toluol to 180 K, pentane to 70 K, and a mixture of propane and propylene to 60 K.

Pressure thermometers, which are widely used for industrial applications, consist of three basic elements—a sensitive bulb containing a liquid, gas or vapor, an interconnecting capillary tube, and a pressure-sensing device such as a Bourdon tube, bellows or diaphragm. A variation in temperature of the bulb results in a volume change of the fluid, thus causing a pressure change which can be calibrated to record

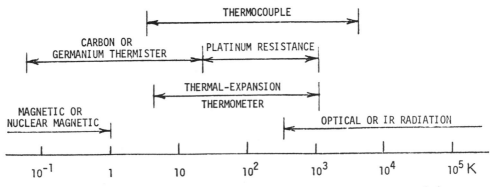

FIGURE 15.79 Operating temperature range of various measurement techniques.

the temperature variation. Capillary tubes of 70 m length have been used, but proper corrections must be introduced for temperature variations along the capillary. Liquid filled sensors operate from 170–670 K with xylene and 230–860 K with mercury, while gas-filled sensors range from 30–920 K.

The phenomenon of thermal expansion of metals is employed in *bimetallic thermometers*. When two strips of metals with different thermal expansion coefficients are bonded together, a change in temperature will result in a deflection of the strip due to differential expansion. The common bimetallic sensors, which are made of Invar, a nickel steel with a nearly zero expansion coefficient, and another high-expansion alloy, have an operating temperature range of about 200–810 K. They are widely used in on–off temperature control systems.

Thermocouples. The most important of all temperature sensors is the thermoelectric sensor, commonly known as the *thermocouple*. The operating principle can be easily described through Fig. 15.80. When two wires of dissimilar metals, *A* and *B*, are joined to form a circuit with the two junctions maintained at two different temperatures T_1 and T_2, there results an electromotive force, *E*, that can be measured by a voltmeter or potentiometer. This phenomenon is known as the *Seeback effect*, one of the many effects in complex thermoelectric phenomena. The *Peltier effect* and the *Thomson effect* are also closely related to the thermocouple operation but will not be elaborated here. If the thermoelectric properties of the two metals are known and one of the junctions is maintained at a known reference temperature (frequently by employing an ice bath), the voltage can be accurately related to the temperature of the other junction.

Many materials can be used for thermocouples [Kinzie (1993)], but in most industrial and scientific applications, thermocouples are made of standardized materials. They are now identified by a type letter rather than by proprietary names. Table 15.9 shows some common thermocouple combinations and their respective operating temperature ranges. Thermocouples are usually made from wires with diameters of 0.002–0.32 cm, with thicker wires being used for long life in hostile environments, while the thinner ones are used for fast response, smaller conduction and radiation errors and precision of junction location.

For temperature measurements in flows, protective shields may encase the thermocouple to eliminate large radiation errors (in high temperature environment), particle erosion (in gas–solid flows), or droplet attachment (in gas–liquid flows). Care must also be exercised about the temperature correction due to conduction loss through lead wires. In high-speed flows, the stationary sensor gives a reading of the *recovery temperature* instead of the conventional (static) temperature or the stag-

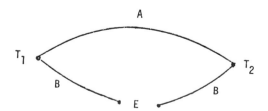

FIGURE 15.80 Thermocouple phenomenon.

TABLE 15.9   Characteristics of Common Thermocouples

Type	Materials[1]	Temperature Range K
B	platinum-30% rhodium/platinum-6% rhodium	273–2000
E	chromel/constantan	3–1150
J	iron/constantan	63–1030
K	chromel/alumel	3–1530
R	platinum-13% rhodium/platinum	223–1750
S	platinum-10% rhodium/platinum	223–1750
T	copper/constantan	3–640

[1]Note: constantan is copper–nickel alloy ($Cu_{55}Ni_{45}$), chromel is nickel–chromium ($Ni_{90}Cr_{10}$), and alumel is nickel–aluminum ($Ni_{94}Mn_3Al_2Si_1$).

nation (total) temperature [Eckert and Drake (1972)]. While correlations have been developed to relate the recovery temperature to various flow parameters, careful calibration of the probe which normally consists of a thermocouple and a protective shield must still be conducted at different flow conditions.

Two books are dedicated to the details of the theory and practical application of thermoelectric circuits [Pollock (1991), ASTM (1993)]. One of the more intriguing aspects is that the voltage generated across a given temperature difference can be multiplied by adding pairs of thermocouple junctions in series to form a *thermopile*.

One of the drawbacks of thermocouples is the small voltage output, which is typically a fraction of a millivolt per °C. This makes signal noise and grounding issues of concern. In addition, connection methods are important to limit measurement errors. Whenever thermocouple wires are introduced, there is the likelihood of thermal disruption of the system. The best thermal and physical contact between the thermocouple and a surface is afforded by a direct press fit into the surface or welding the wires to the surface. In either case, the measurement side of the surface should be left physically unperturbed. The wires may also conduct thermal energy to or from the surface and alter the local measured temperature of the surface. Wires should be run parallel with the surface for some distance from the junction to limit this error. The use of thin films for the thermocouple elements in place of wires is an easy way to limit the effects of thermal disruption. However, because the thermoelectric voltage of sputtered thin films is typically different from that of the bulk materials [Kreider (1993)], junctions between thin film thermocouples and lead wires of the same materials can cause additional voltages to occur in the circuit.

Resistance Elements.   The electrical resistance of most materials changes with temperature. Because electrical circuits for measuring resistance are common, this is an easy method for measuring temperature. The resulting sensors have a variety of names. RTD's generally refer to metal wires wrapped around an insulator to minimize the effects of strain. An RTS usually refers to a metal thin film deposited on an insulator to give fast time response. Metals typically have a resistance change of less than 1% per °C. Common metal resistive elements are made of platinum, nickel, copper, tungsten or nickel/iron alloys. The platinum resistance thermometer, in particular, exhibits great precision and stability and is used to define the international temperature scale. *Thermistors* use semiconductor materials with one to two orders of magnitude larger resistance change over a small temperature range. Because they are highly nonlinear over extended temperature ranges, they are generally applicable only over limited specified temperatures. Thermistors are commonly made of metal oxides of manganese, nickel cobalt, copper, iron, or titanium.

The resistance measurement is usually performed by passing a small known current through the resistor and monitoring the corresponding voltage drop. The current must be tailored to the resistor to provide sufficient voltage output while maintaining small self-heating effects. A current less than 1 mA is usually desired. Effects of lead wire resistance can also be significant and may require a four-wire circuit to subtract this effect. For surface measurements, thin-film sensors are desired. They are typically deposited by sputtering or painting and firing on a ceramic substrate.

Radiation Methods. All the previous temperature measuring techniques require that the sensor be placed at the location under consideration and in contact with the body whose temperature is to be determined. The radiation sensor, however, can be remotely situated, thus offering special convenience and advantages in many applications. It measures from a source, the object of interest, a certain amount of spectral or total radiation which is then interpreted in terms of the source temperature. If the source or the interstitial medium is a distributed one and radiation-participating, radiation measurements can be used to map or scan the whole temperature field.

All material objects emit and absorb radiation at different wavelengths as a function of temperature. For a blackbody, which has an emissivity and absorptivity of unity, the radiation-temperature relation is prescribed by the Planck distribution law at each wavelength or the Stefan–Boltzmann law on a total integrated basis. The spectrum of wavelengths containing the most radiating energy ranges is from ultraviolet (about 0.3 μm) to visible (about 0.4–0.7 μm) and into the infrared (up to 100 μm). The primary energy containing region moves to shorter wavelengths with increasing temperatures as indicated by *Wien's displacement law* [Eckert and Drake (1972)].

There exist many different kinds of radiation sensors which are also given with a large variety of names. Radiation sensors are often called detectors or *radiometers*, as well as others like optical (or radiation) *pyrometer* and *radiation thermometers*. Basically, there are two broad classes of radiation sensors—the thermal detectors and the photon counters. Thermal detectors measure the temperature rise due to incident radiation using a resistance thermometer–thermistor, thermocouple (thermopile), or even a thermal-expansion device (e.g., Golay cell). Photon counters detect the photoelectric effect caused by the incident flux of photons which interact with the electronics in the detector structure. They can operate in photoconductive, photovoltaic or photoelectromagnetic modes.

A standard device widely used in industry is the optical pyrometer, because it is commercially available, relatively inexpensive, portable and accurate. To avoid large errors due to inadequate knowledge of surface emissivities especially at high temperatures, the two-color pyrometer has been made available for two brightness measurements at two different wavelengths. High-speed scanning radiometers have opened the way for useful two-dimensional surface temperature measurements.

Liquid Crystals. Liquid crystals are composed of long chain molecules that reversibly change color as a function of mechanical stresses and temperature. They are often microencapsulated to increase their stability and to decrease their sensitivity to factors other than temperature. For transient temperature measurements, the chiral-nematic form is preferred, because the time response is on the order of milliseconds [Ireland and Jones (1987)]. A commercially available paint containing the liquid crystals in microcapsules can be sprayed or screen printed onto a surface in layers ~ 10 μm thick. Although the usual temperature range for color sensitivity is 25°C to 40°C, liquid crystals that operate to 100°C have been reported [Kenning (1992)]. Interpretation and calibration of the results remains the most difficult aspect of obtaining quantitative temperature information [Moffat (1990)]. Both the lighting system and angle of viewing are important aspects of their proper use. Image processing systems are being developed that aid in evaluating two-dimensional surface temperature fields.

Thermographic Phosphors. A recent development in the measurement of surface temperature is the use of thermographic phosphor luminescence [Noel *et al.* (1991) and Chyu and Bizzak (1994)]. This is a non-contact method requiring the illumination of the surface with an appropriate ultraviolet light source. The phosphors are deposited as small grains of ceramic material doped with rare-earth elements. When illuminated with ultraviolet, they emit a spectrum of narrowband visible wavelengths. The amplitude and decay time of these emission lines are characteristic of the temperature. Although the ratio of amplitudes at two of the wavelengths has been demonstrated for temperature measurement, the preferred method is to use the rate of luminescence decay. A pulsed laser is used to provide the ultraviolet illumination in short bursts. The source and measurement can be accomplished through fiber optics or an image processing system for two-dimensional distributions.

15.10.3   Heat Flux Measurement

For research in low-speed wind tunnels, the easiest method of measuring heat transfer is to actively heat the surface electrically and directly measure the required power. For other situations, particularly high heat flux, the methods that can be used are limited to essentially two categories. The first is to infer the heat transfer from the transient change of surface temperature, while the second uses the temperature difference across a spatial distance. Some of the standard methods are reviewed in the following sections along with promising new methods. A comprehensive review of the field is given by Diller (1993).

Transient Methods. All of the transient techniques use some form of solution of Eq. (15.176) to relate the change in stored thermal energy of the wall material (first term) to the distribution of heat flux (second term) during a change in thermal conditions. The solutions necessarily must reflect the details of the substrate thermal properties and geometry. For accurate measurements, this can entail considerable data processing to properly reconstruct the heat flux from the temperature signal. The solution often assumes one-dimensional heat transfer in the substrate material. Because such solutions are generally valid only for short periods of time, the measurements must be performed quickly. This usually requires a model to be rapidly inserted into a flow or the flow to be quickly started, such as in a shock tunnel. Although the applications are necessarily limited, the instrumentation is simple. The disruption caused by the presence of the sensor is also of concern.

One of the simplest methods is the *slug calorimeter*. A piece of high-conductivity copper is placed in the wall and is assumed to remain at a uniform, but unsteady temperature. A *lumped capacitance* solution of Eq. (15.176) can then be used, which is a simple exponential for convection with a constant heat transfer coefficient and a step change in flow temperature from T_i to T_∞ at time $t = 0$

$$\frac{T - T_\infty}{T_i - T_\infty} = e^{-t/\tau} \qquad (15.179)$$

where the *time constant* is

$$\tau = \frac{mc_p}{hA} \qquad (15.180)$$

The mass and specific heat of the slug are represented by m and c_p, while h is the heat transfer coefficient acting over the surface area of the sensor, A. Although details of the use of these sensors are given by an ASTM Standard (1988a), they are rarely used today.

One improvement of the slug calorimeter concept has been developed by Liebert (1994), termed the *plug-type heat flux gage*. Instead of using a single temperature measurement to represent the entire slug, his sensor uses a number of thermocouples positioned along a central shaft. This greatly improves the time response and accuracy of the method. Practical sensors have been produced using electrical discharge machining from the backside of the wall. Therefore, the sensor does not physically disrupt the flow surface and is composed of the same material as the rest of the wall. It has typically been used under conditions of high heat flux and high temperature.

Another approach to improving the response of a slug calorimeter is to move the thermocouple closer to the surface of the sensor by carefully machining a cavity from the back of the sensor. The resulting calorimeter is termed a *null calorimeter* and is designed with geometric proportions so that the temperature response corresponds closely to that of a semi-infinite solid with the thermocouple at the surface. Therefore, the heat flux can be determined from the temperature measurement using Eq. (15.177). Kidd (1990) has performed detailed computations to identify the optimum parameters for these sensors. In addition, an ASTM Standard (1988b) details the proper use of null calorimeters. They are typically made of high-conductivity copper and used under conditions of extremely high heat flux.

Coaxial thermocouples operate with a similar principle, but they are easier to fabricate. Consequently, they are more widely used in practice. Instead of a slug of material with a thermocouple attached, the thermocouple and sheathing is directly press fit into the wall to be tested. One thermocouple material forms the center wire, which is surrounded by an electrical insulator followed by a sheath of the second thermocouple material. When installed in a metal wall, electrical isolation with the wall should be insured to avoid the effect of a second thermoelectric effect with the wall material, which can affect the output of the sensor [Kidd *et al.* (1994)]. The thermal properties of the coaxial sandwich (product of $k\rho c_p$) are matched with those of the wall material to minimize the temperature disruption caused by the sensor [Neumann (1989)]. The actual thermocouple junction is formed by bridging the two thermocouple materials with a thin film ($\sim 10\ \mu$m) at the tip of the sensor. The resulting fast time response allows resolution of heat flux in less than 1 ms. Equation

(15.177) can again be used to reduce the temperature data to the corresponding heat flux, although smoothing of the temperature measurements is usually required. Tests can only last as long as the wall can still be assumed to be semi-infinite,

$$t < \frac{L^2}{16\alpha} \tag{15.181}$$

where L is the wall thickness and α is its thermal diffusivity.

A more direct approach to determining the heat flux is to simply measure the temperature at the surface of the wall material itself. The most common method is to use a thin-film resistance element deposited on the surface to measure the surface temperature. Because these elements can be fabricated a fraction of a micrometer (μm) thick, the initial response time is very fast ($< 1 \mu$s). As can be noted from Eq. (15.178), the surface temperature change resulting from a given heat flux for a semi-infinite material is inversely proportional to the square root of the product of $k\rho c_p$. Consequently, a low conductivity material for the wall gives a higher measurement sensitivity. Moreover, the sensor must be electrically insulated from the wall. Therefore, ceramic models are often used with the semi-infinite, transient technique, although metal covered with ceramic layers can also be used for the substrate [Doorly and Oldfield (1986)]. Temperature is determined by measuring the sensor resistance using a small current source (~ 1 mA) and a suitable bridge or operational amplifier circuit. The sensor resistance is typically 100 ohms or greater. This combination usually keeps the self-heating effects negligible. A calibration of the temperature-resistance relationship is required along with the value of $k\rho c_p$ of the substrate to determine the heat transfer. The actual data analysis can use Eq. (15.177) although increased noise inherent in this differentiation process may mask the details of time-resolved heat flux. Consequently, analog circuits have been developed to minimize the noise problem [Schultz and Jones (1973)].

Optical techniques of measuring the surface temperature have the advantages of no electrical connections with the measurement surface and the possibility of complete two-dimensional profiles of the surface. The basic methods of optical temperature measurement have been discussed in Sec. 15.3.2. All of these methods have been used to determine heat flux using the transient, semi-infinite technique.

Infrared imaging was made practical with the advent of high-speed scanning radiometers. The emissivity of the surface and the background radiation must be known to convert the absorbed radiation into absolute temperatures for heat flux measurement. Resolution is limited at low temperatures because of the corresponding small energy flux. Infrared measurements were used by de Luca *et al.* (1992) to measure heating after injection of a model into hypersonic flow.

The use of liquid crystals for temperature measurement was discussed in Sec. 15.3.2. They can also be used to deduce heat flux as above. The biggest limitations are the limited operating temperature range (typically 25–40°C) and the temperature resolution that can be achieved. The former limits the applications and the latter limits the accuracy. New methods for intepretation of the color changes are being developed [Camci *et al.* (1993)].

The use of thermographic phosphors for heat flux measurement has been demonstrated by Horvath (1990) and Buck (1991) in hypersonic flow. Digital image processing was used.

Spatial Temperature Difference Methods. Sensors that measure heat flux by means of a spatial temperature difference have the advantage of giving a continuous output that is proportional to the heat flux. Although there are many potential geometries for the location of the temperature measurements relative to the heat flux, only a few have had substantial impact. One important criterion is that the sensor should not substantially distort the local temperature field and the associated heat flux. As discussed in later sections, if this criterion is not followed, large errors in the measurements can result.

One of the simplest geometries is a layered sensor, the concept of which is shown in Fig. 15.81. It is placed on the surface with the temperature difference across the thermal resistance layer proportional to the heat flux perpendicular to the surface. The temperature difference is measured with thermocouples, resistance sensors, or thermographic phosphors. The advantages of thermocouples are that they generate the voltage output themselves, and the signal can be amplified on the sensor by linking multiple thermocouple pairs together in series to form a differential thermopile. A number of these different sensors are currently available commercially. Ortalano and Hines (1983) describe a thermopile sensor using metal foils around a plastic thermal resistance layer. Hager *et al.* (1993) have successfully sputter coated the entire thermopile sensor using microfabrication techniques. The resulting sensor is less than 2 μm thick, with a corresponding response time of 10 μs. The temperature limits are also high ($>1000°C$) because of the high-temperature materials which can be used throughout the sensor. The time response of the sensor was tested in a shock tunnel by Holmberg *et al.* (1994) during the passage of an incident shock. An example of the resulting curve is shown in Fig. 15.82. Because a thin-film temperature sensor is deposited alongside of the layered heat flux sensor, simultaneous comparisons between the spatial and transient temperature methods are possible. A good match of time-resolved results for the two methods was reported by Baker and Diller (1993).

Resistance-temperature sensors have also been used across a thermal resistance layer to measure heat flux by Hayashi *et al.* (1989) and Epstein *et al.* (1986). The latter group places nickel layers on either side of a polyamide sheet, which limits the temperature capability. They used numerical data reduction to increase the frequency response, modeling the transient response of the upper temperature measurement for short times as described in the previous section.

An optical method has been developed by Noel *et al.* (1994) using layers of thermographic phosphors above and below a transparent thermal resistance layer. Because the wavelength characteristics of the two layers are different, the tempera-

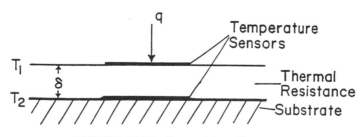

FIGURE 15.81 Layered heat flux sensor.

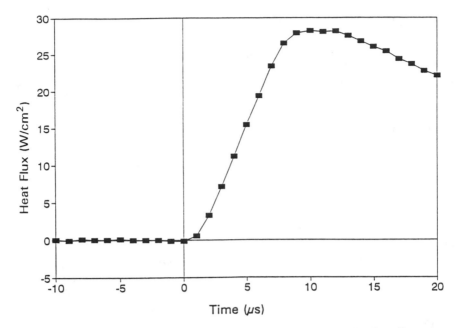

FIGURE 15.82   Shock tunnel measurement of an incident shock with a heat flux microsensor. [From Holmberg *et al.* (1994).]

ture difference can be determined at the same points on the surface. Operating temperature is currently limited by the thermal resistance layer and accuracy by the optical analysis.

An additional method of creating multiple thermocouple junctions in a sensor has been to wrap wire around an insulated material and then plate one half of the wrapped wire. The use of these sensors, commonly called *Schmidt–Boelter gages*, for aerospace measurements is reviewed by Neumann (1989). The same concept is used to produce transducers for low heat flux measurements as described by Hauser (1985).

A circular foil sensor, commonly called a *Gardon gage*, measures the temperature difference parallel to the surface. It consists of a hollow metal cylinder inserted into the surface with a foil of a different metal attached over one end to form the surface and a wire of the same metal as the cylinder attached to the center of the foil from below. This creates a differential thermocouple from the center to edge of the foil. The thermal energy that flows into the foil surface, must exit through the cylinder, which creates a corresponding temperature difference from the center to the edge of the foil. If the heat flux is uniform, the temperature difference can be expressed as

$$\Delta T = \frac{qR^2}{4k\delta} \qquad (15.182)$$

where k is the thermal conductivity of the foil and δ and R are the thickness and radius. This is a reasonable assumption for radiation to the surface, however, as explained by Kuo and Kulkarni (1991), this is often not a good assumption when convection is present. Therefore, when measuring convection, corrections are generally required for the measured heat flux, and special care must be taken to keep the temperature difference across the foil small compared to the total temperature difference between the surface and the fluid.

Heat Flux Sensor Calibration and Potential Measurement Errors. One goal of any sensor is that its presence does not affect the measured quantity. This is particularly difficult for heat flux sensors based on the spatial temperature difference, because quite often the temperature disruption is the basis for the measurement. Consequently, the goal must be to obtain the best acceptable measurement sensitivity with the least temperature change. This means that thermopile-type sensors have a decided advantage over single temperature measurements. The temperature disruption of the surface is particularly a problem for convection measurements where the thermal boundary layer is locally disturbed. Diller (1993) gives a graphic example to emphasize the extent of error that can occur when a sensor disrupts the local thermal condition. This is a particular problem with high-temperature testing, if the heat flux gage is water cooled to keep its temperature from exceeding specifications. A local cold spot is created precisely at the measurement location.

A more subtle effect results from different types of surface boundary conditions on heat flux measurements. Common test conditions are specified as either constant surface temperature or constant heat flux over the surface of the material. A number of investigators have demonstrated that different distributions of heat transfer coefficient are measured for constant temperature versus constant heat flux conditions on the same surface under otherwise identical situations.

The absolute measurement of heat flux is no better than the calibration of the sensor. Although investigators have reported a wide variety of calibration methods, the most common is radiation, probably because it is easier to establish a known value of heat flux. Unfortunately, different calibration methods often do not give the same results, because different boundary conditions and temperature distributions are created at the surface. Consequently, a sensor calibrated in radiation may not give correct results when used for convection measurements or a mixture of convection and radiation. The most difficult aspect of convection calibrations is to establish a known and repeatable heat flux. In addition, the overall sensitivity of sensors is usually a function of the sensor temperature. Therefore, sensors used at elevated temperatures should be calibrated over the appropriate temperature range.

Data Reduction of Convection Heat Flux Measurements. The most common method of characterizing convection heat flux is with a heat transfer coefficient based on the temperature difference between the fluid, T_∞, and face, T_w

$$h = \frac{q_{\text{conv}}}{T_\infty - T_w} \tag{15.183}$$

In many situations, however, there is more than one fluid temperature to consider. One example is high-speed flows where the *recovery temperature* (or *adiabatic wall temperature*) is usually the preferred fluid temperature used to define the heat transfer coefficient. In film cooling, the adiabatic wall temperature, T_{aw}, is usually used to define the heat transfer coefficient.

$$h = \frac{q_{\text{conv}}}{T_{aw} - T_w} \tag{15.184}$$

Moffat (1990) suggests that Eq. (15.184) also be used in situations where the surface temperature of the wall changes. Moreover, the reference temperature used to evaluate fluid properties can have a large impact on the reported value of heat transfer coefficient.

15.11   VORTICITY MEASUREMENTS
John Foss, James Wallace, and Candace Wark

15.11.1   Basic Considerations

The *vorticity*, $\overline{\omega}$, of a fluid particle is given in terms of the velocity field, $\mathbf{V}$

$$\overline{\omega} = \nabla \times \overline{\mathbf{V}} \tag{15.185a}$$

or, in tensor notation

$$\omega_K = \epsilon_{ijk} \frac{\partial u_i}{\partial x_j} \tag{15.185b}$$

Its companion variable is the *circulation*

$$\Gamma = \oint_C \overline{\mathbf{V}} \cdot d\overline{\mathbf{s}} = \int_A \overline{\omega} \cdot \overline{\mathbf{n}} \, dA \tag{15.186}$$

where A is the area bounded by the contour C.

An important attribute of the vorticity is the solenoidal condition

$$\nabla \cdot \overline{\omega} = \nabla \cdot \nabla \times \overline{\mathbf{V}} \equiv 0 \tag{15.187}$$

with the consequence that a vorticity filament (similar in concept to a streamline) will, in general, form closed loops in a flow field. Foss and Wallace (1989) identify some of the properties of such fields and provide commentary and references for interpretive analyses of vorticity fields.

The challenge to the experimentalist is well stated by Eqs. (15.185a,b). If this definition is to be used to extract components of the vorticity vector, then the procedure must provide velocity components in sufficiently close proximity that their spatial gradients can be evaluated. A limitation is also present if the spacing is too small. Specifically, as the spacing approaches the turbulent Kolmogoroff scale, the velocity difference evaluation may be overwhelmed by the noise level of the measurement [Wallace and Foss (1995)].

The measurement methods discussed in this section are those which (the authors suggest) can be readily adapted by other investigators. Sufficient details are given to permit such usage. It is important to emphasize that the following is *not* a comprehensive list of the available techniques. Wallace and Foss (1995) have sought to provide such a list, and the interested reader is encouraged to consult that source for laser Doppler and other optically based techniques.

15.11.2 Vorticity Measurements with Thermal Anemometry

Streamwise Vorticity. Different strategies are employed for the interpretation of one or more components of $\overline{\omega}$ using multiple wire probes. The historical context is discussed in Foss and Wallace (1989) and Wallace and Foss (1995).

Multiple wire probes (up to 12) have been used to infer the three velocity components (u, v, w) and the velocity gradients required to evaluate $(\omega_x, \omega_y, \omega_z)$. Less complicated five- and six-wire probes have also been used to infer the three velocity and two vorticity components. A six-wire probe was used to evaluate (u, v, w) and either (ω_x, ω_y) or (ω_x, ω_z) as time series. A five-wire probe can provide similar results, albeit with a greater need for invoking approximations. The least complicated anemometry technique is a four-wire array.

Kovasznay (1950), (1954) conceived the first vorticity probe. This four-wire array, in a pattern of orthogonal x-arrays on common prongs was used in the study of the viscous superlayer by Corrsin and Kistler (1954). Vukoslavčević and Wallace (1981) and Park and Wallace (1993) provide a discussion of such probes to the effect that it is not fruitful to attempt their use in trubulent flows.

If, on the contrary, a probe is sought to identify the magnitude of ω_x in a flow field with an *ordered* streamwise vorticity distribution, then the Kovasznay style probe can be recommended. Note that an even simpler, and correspondingly less quantitative, technique can be utilized in such a situation. Namely, a cross vane vorticity meter as used, for example, by Holdeman and Foss (1975) in a bounded jet and by Wigeland *et al.* (1978) can be used to indicate the qualitative or quantitative presence of ω_x in a flow. The cross vane meter, when viewed from the front, appears as $+$; its length is typically twice its diameter and a diameter of 8–10 mm is typically required to overcome bearing friction and to permit its rotation.

An eight-wire array, four x-arrays in a square pattern, is a logical extension to the four-wire probe (see Fig. 15.83). This eight-wire probe can be interpreted as providing $[u, w]$ $(x_o, y_o \pm \delta y, z_o)$ and $[u, v]$ $(x_o, y_o, z_o \pm \delta z)$. These perimeter values can be averaged to provide $[u, v, w]$ (x_o, y_o, z_o) as well as processed to provide

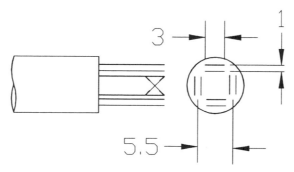

FIGURE 15.83 An eight-wire array to evaluate $\omega_x(t)$ (dimensions in mm). Notes: The active length of the 5 μm wire is 1 mm. There are 50 μm diameter plated ends on the 3 mm long wires.

$\omega_x(x_o, y_o, z_o)$. For the latter quantity,

$$\delta\Gamma = \oint \overline{\mathbf{V}} \cdot d\overline{\mathbf{s}} \tag{15.188a}$$

where $\overline{\mathbf{V}}$ is known at $(x_o, y_o \pm \delta y, z_o \pm \delta z)$. Hence,

$$\delta\Gamma = [+w(\delta y) - v(\delta z) - w(-\delta y) + v(-\delta z)]\pi D \tag{15.188b}$$

and

$$\langle \omega_x \rangle = 4\delta\Gamma/\pi D^2 \tag{15.188c}$$

Kock am Brink and Foss (1993) have used such a probe for $\omega_x(t)$ measurements in the shear layer of a lobed splitter plate. This probe, using conventional 5 μm wire with 1 mm length active sections and plated extensions has a contour diameter (see Eq. 15.188) of 5.5 mm.

This study utilized the *speed-wire/angle-wire* calibration/processing technique [Foss (1981)], a quite accurate method (given relatively large angles) for the inference of the velocity magnitude and the incidence angle.* The calibration data are used to define the coefficients in the expression

$$E^2 = A(\gamma) + \beta(\gamma)Q^{n(\gamma)} \tag{15.189}$$

where Q is the velocity magnitude in the plane of the slant wires and γ is the pitch angle of the velocity with respect to the probe shaft. Note that Eq. (15.189) is a generalization of the form recommended by Collis and Williams (1959). For a probe aligned in the x-direction, $\tan \gamma = v/u$. Consider that calibration data provide discrete (e.g., each 6 degrees) γ values in, for example, the range $-36 \leq \gamma \leq 36$ degrees. Such data permit the coefficients A, B and n to be defined for each wire and, as shown below, for these coefficients to recover Q and γ for each data sample. Specifically, the computation of Q and γ can proceed as (for wires 1 and 2):

i) For the first approximation, use the cosine effective cooling relationships and the technique of Bradshaw (1975) to infer (Q, γ) from (E_1, E_2). For $\beta_i = $ angle of the normal to wire i from the probe axis, one can define an effective cooling relationship as

$$E_1^2 = A_1 + B_1'[Q \cos(\beta_1 - \gamma)]^{n_1} \tag{15.190a}$$

and, for wire 2,

$$E_2^2 = A_2 + B_2'[Q \cos(\beta_2 - \gamma)]^{n_2}. \tag{15.190b}$$

Using Eqs. (15.190a) and (15.190b), one can use two functions of the voltage to permit a single computation of (u, v). The functions f_1, f_2 are defined as

---

*An alternative data processing style is in widespread use for smaller incidence angles; specifically, the Jorgenson technique is described below in the context of three component vorticity measurements.

$$f_1 = \left[\frac{E_1^2 - A_1}{B'} \right]^{1/n_1} = Q(\cos \beta_1 \cos \gamma + \sin \beta_1 \sin \gamma)$$

$$= u \cos \beta_1 + v \sin \beta_1$$

(15.191)

and, for wire 2,

$$f_2 = u \cos \beta_2 + v \sin \beta_2 \qquad (15.192)$$

The matrix defined by Eqs. (15.191) and (15.192) can be inverted to solve for (u, v). The second and continuing approximations follow from the steps:

ii) Knowing γ from step (i), designate the wire most normal to the velocity vector as the *speed wire* and the wire most tangent as the *angle wire*.

iii) Use Q from step (i) as the first estimate of the flow speed. Compute γ using the expression

$$\gamma = \gamma(\eta) \text{ where } \eta = \frac{E_a(\gamma, Q)}{E_a(0, Q)} \qquad (15.193)$$

Note that $E_a(\gamma, Q)$ is the measured voltage of the angle wire and $E_a(0, Q)$ is computed, for the angle wire, from $A_a(0)$, $B_a(0)$, $n_a(0)$ and the current estimate for Q. The $\gamma = \gamma(\eta)$ curves are polynomial functions at discrete Q values.

iv) Compute a new estimate for Q using the speed wire voltage and the discrete γ values that are closest to the step (iii) γ value. Namely,

$$Q_{\text{Low}} = \left[\frac{E_s^2 - A(\gamma_+)}{B(\gamma_+)} \right]^{1/n(\gamma_+)} \qquad (15.194)$$

and

$$Q_{\text{Hi}} = \left[\frac{E_s^2 - A(\gamma_-)}{B(\gamma_-)} \right]^{1/n(\gamma_-)} \qquad (15.195)$$

Note that these two estimates for Q are derived from the calibration γ values that are nearest to but lesser than (i.e., γ_-) and greater than (i.e., γ_+) the pitch angle inferred in step (iii). Q_{Low} and Q_{Hi} are shown in Fig. 15.84. The required interpolation between these two velocity estimates will make use of a local (i.e., between γ_- and γ_+) *cosine law* approximation. If γ from step (iii) were closest to γ_+, then, since

$$Q_{\text{eff}} = Q \cos (\beta - \gamma) = Q_{\text{Low}} \cos (\beta - \gamma_+) \qquad (15.196a)$$

$$Q \text{ (step iv)} = Q_{\text{Low}} \frac{\cos (\beta - \gamma_+)}{\cos (\beta - \gamma)} \qquad (15.196b)$$

v) The above computations can continue until an arbitrary convergence criterion is achieved.

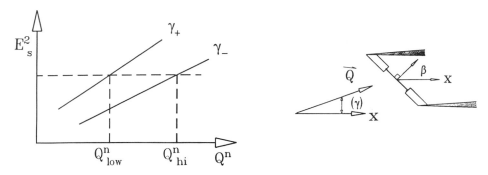

FIGURE 15.84 Definitions of the Q_{Hi} and Q_{Low} quantities used in the speed-wire voltage $\rightarrow$ velocity interpolation scheme; see Eq. (15.190a,b).

Binormal Vorticity. For x as the streamwise direction, the ω_z or ω_y values can be obtained with a four-wire array of probes as shown in Fig. 15.85. These *transverse* directions will be referred to as *binormal* in this discussion given the *streamwise* and *normal* usage identified below. This probe, and its extensive algorithms for calibration and data processing, have been developed by Foss (1976), (1981), Haw *et al.* (1989), and Foss and Haw (1990). (This probe is referred to as the *Mitchell probe* since R. M. Mitchell first developed the technique to *thread* the fourth wire, a parallel wire, onto the prongs given the prior mounting of the other three wires.) A similar probe configuration has also been developed independently by Antonia and Rajagopalan (1990).

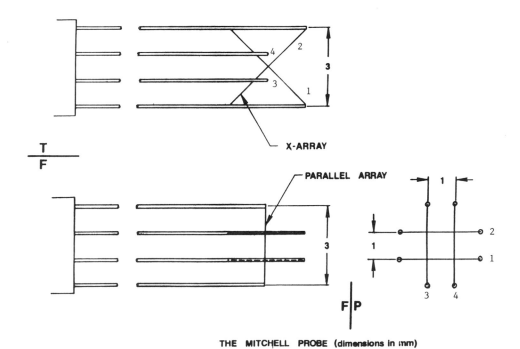

THE MITCHELL PROBE (dimensions in mm)

FIGURE 15.85 The MSU compact (Mitchell) four-wire transverse vorticity probe. Notes: Dimensions in mm. The wire numbers are shown in the top and profile views.

The presence of the four wires permits a more accurate assessment of the speed and angle than that which would be provided by an x-array. Specifically, the two parallel wires (3, 4) provide values of the speed as

$$Q_3 = \left[\frac{E_3^2 - A_3(\gamma)}{B_3(\gamma)} \right]^{1/n_3(\gamma)} \tag{15.197}$$

(These Q values are interpreted as the projection of the velocity vector onto the plane perpendicular to the No. 3 and 4 wires.) The computation of Q_3 and Q_4 can proceed with $\gamma = 0$ as the angle for the first iteration to be followed by the use of the γ dependent coefficients once the γ value is determined using the slant wire voltages.

The two slant wires can be used to provide two independent estimates for the γ values. Specifically, for E_1 and E_2 as their respective voltages, and for $\gamma(\eta)$ curves as introduced in Eq. (15.193), but now extended to all angles ($\pm 36°$), the known Q value, i.e.,

$$Q = (Q_3 + Q_4)/2 \tag{15.198}$$

permits γ_1, γ_2 to be evaluated [see Eq. (15.193)].

It is appropriate to assume that γ in the center of the probe is that obtained as the average of these two γ values; that is,

$$\gamma = (\gamma_1 + \gamma_2)/2 \tag{15.199}$$

Hence, at each time step, estimates for Q_3, Q_4 and γ are provided by the processing algorithm.

Equation (15.186) provides the basis for this measurement technique. Namely, the velocity field will be considered to *sweep out* a micro-circulation domain in the $\bar{s}(\tau)$ direction where $\bar{s}$ is the average streamwise direction over the time period of the required A/D conversion steps. The τ value is the elapsed time associated with the formation of the micro-domain. The reader is referred to Foss and Haw (1990) for a complete description.

Two-Component Probes. A five-wire (hot-film) array developed and utilized by Eckelmann and co-workers at the Max Planck Institute in Göttingen [see Eckelmann *et al.* (1977)] provides streamwise and one transverse vorticity component as well as the three velocity components. This probe is best suited for flows with a strong transverse vorticity given the geometry and the required assumptions about the velocity gradients.

A six-wire array (see Fig. 15.86), developed by Fiedler and co-workers [Kim (1989)] provides the three velocity components, the streamwise vorticity and a transverse component. Simultaneous voltage measurements provide six experimental quantities for the three velocity components and the four derivatives. The developers of this method utilize the Taylor hypothesis for the x-derivative—$(\partial v/\partial x)$ for ω_z or $\partial w/\partial x$ for ω_y. Their computing algorithm for ω_x, that is fully described by Kim (1989) and that is too complex to be described here, relies upon the *normal component cooling* hypothesis and combinations of these cooling velocities to create the cross stream derivative (i.e., $\partial w/\partial y$ of Fig. 15.86). Representative experimental data are presented in Kim and Fiedler (1989).

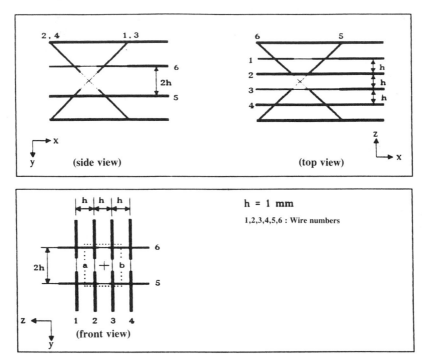

FIGURE 15.86   The T. U. Berlin six-wire probe. Note: This is Fig. 4.3.2.a of the Kim (1989) thesis.

Three-Component Probes. Wallace and co-workers [see Balint *et al.* (1987), Vukoslavčević *et al.* (1991), and Balint *et al.* (1991)] pioneered in the development of nine-wire probes constructed in a remarkably small (about 1.7 mm diameter) sensing volume. They introduced the idea of expressing the velocity vector cooling each wire in a Taylor series expansion to first order. These cooling velocity vectors are thus given in terms of the three velocity components at the centroid of the projection of the probe's sensing volume on the transverse plane and the six velocity gradients in this plane. Nine nonlinear algebraic equations describing the response of the nine wires to the instantaneous flow thereby can be written in terms of these nine unknowns. They calculated the streamwise gradients from the temporal gradients using Taylor's hypothesis [Piomelli *et al.* (1989)], thus, they were able to obtain estimates of all the components of the instantaneous velocity vector and the velocity gradient tensor. All three components of the vorticity vector therefore can also be obtained instantaneously. Tsinober *et al.* (1992) have improved this probe by adding an additional wire to each array of three wires. These additional wires serve the purpose of relaxing the otherwise restrictive cone of angles of attack of the velocity vector for which unique solutions of the wire response equations can be obtained. A twelve-wire probe of this type, arranged so that each of the three four-sensor arrays are placed at one of the vertices of an equilateral triangle as sketched in Fig. 15.87, has been constructed by Vukoslavčić. A generalized calibration and data reduction procedure for it and other multi-wire probes, described below, has been developed by Marasli *et al.* (1993).

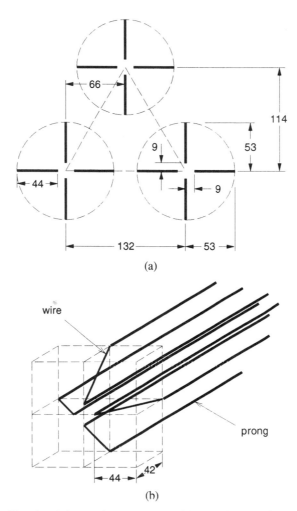

FIGURE 15.87 Sketch of the twelve-sensor vorticity probe: (a) front view and (b) perspective view of one four-sensor array. All dimensions are in 1/100 mm. [From Marasli *et al.* (1993). Courtesy of *Exp. Fluids*, Springer-Verlag.]

The procedure begins by assuming the Jorgensen (1971) directional response equation for a single wire,

$$U_{\text{eff}}^2 = u_N^2 + k^2 u_T^2 + h^2 u_B^2 \qquad (15.200)$$

where u_N, u_T, with u_B are the normal, tangential and binormal components of the velocity vector with respect to the wire, and k and h are the tangential and binormal cooling coefficients, respectively. When only relatively small pitch, α, and yaw, β, angles ($-20° \leq \alpha, \beta \leq 20°$) are considered, k and h can be assumed constant, as made clear by Vagt (1979), Adrian *et al.* (1984), and Lekakis *et al.* (1989). For larger angles, the angular dependence of these coefficients can be incorporated into the procedure. In order to express the effective cooling velocity, U_{eff}, in terms of the velocity vector components $(u, v, w) \equiv \bar{\mathbf{u}}$ with respect to a rectangular coordinate system fixed to the laboratory, u_N, u_T and u_B can be written as

$$u_N = N_1 u + N_2 v + N_3 w \tag{15.201a}$$

$$u_T = T_1 u + T_2 v + T_3 w \tag{15.201b}$$

$$u_B = B_1 u + B_2 v + B_3 w \tag{15.201c}$$

where N_i, T_i and B_i ($i = 1, 2, 3$) are coefficients of the coordinate transformation from the laboratory coordinates to the coordinates along the wire. In principle, these coefficients can be determined by careful measurement of the orientation of the wires with respect to the probe axis. However, for a multi-sensor miniature probe such measurements are difficult to achieve with sufficient accuracy. Rather, N_i, T_i, and B_i are better treated as unknowns and determined through the calibration process. Substitution of Eqs. (15.201a–c) into Eq. (15.202) yields

$$U_{\text{eff}}^2 = b_0 u^2 + b_1 v^2 + b_2 w^2 + b_3 uv + b_4 uw + b_5 vw \tag{15.202}$$

where b_m ($m = 0$–5) are functions of N_i, T_i, B_i, k, and h. Thus, the geometrical and thermal coefficients are lumped together to be determined by direction calibration. In the present scheme, U_{eff} is related to the anemometer bridge voltage by a fourth order polynominal rather than by King's (1914) law. The response equation of the wires thus becomes

$$u^2 - K_1 v^2 - K_2 w^2 - K_3 uv - K_4 uw - K_5 vw = P(e) \tag{15.203}$$

where

$$P(e) = A_0 + A_1 e + A_2 e^2 + A_3 e^3 + A_4 e^4 \tag{15.204}$$

Here e represents the anemometer bridge voltage, and Eq. (15.202) has been divided throughout with b_0 to make the coefficient of u^2 unity. The coefficients K_m ($m = 1$–5) and A_n ($n = 0$–4) can be obtained via calibration by subjecting the wires to a uniform flow of variable but known velocity magnitude, Q, at various known pitch, α, and yaw, β, angles. The velocity components are then given by

$$u = Q \cos \alpha \cos \beta \tag{15.205a}$$

$$v = Q \sin \alpha \cos \beta \tag{15.205b}$$

$$w = Q \sin \beta \tag{15.205c}$$

as illustrated in Fig. 15.88. A minimum of ten calibration measurements are required to solve for the unknowns K_m and A_n, but many more calibration values should be used to obtain a good set of coefficients. Given a set of calibration velocities and the corresponding values of the anemometer bridge voltages e, the unknown coefficients can be determined by the method of least-squares [Marasli et al. (1993)].

When a multi-wire probe is used in a turbulent flow, the velocity vector seen by each wire will be different, because the flow is not uniform across the sensing volume. The effects of the velocity gradients on multiple-wire probe performances have been demonstrated by Vukoslavčević and Wallace (1981) and Park and Wallace

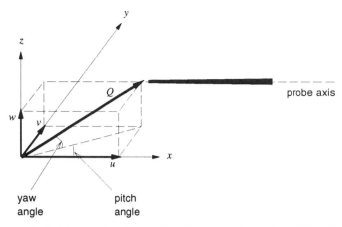

FIGURE 15.88 A sketch defining the pitch and yaw angles. [From Marasli *et al.* (1993). Courtesy of *Exp. Fluids*, Springer-Verlag.]

(1992). The data reduction for the twelve-sensor probe follows Vukoslavčević *et al.* (1991) by expanding the velocity vector, u_{ij}, sensed by wire $j(=1$–$3)$ or array $i(=1$–$3)$ in a Taylor series to first order around the centroid of the projection of the probe on the transverse plane. This expansion gives

$$u_{ij} = u + C_{ij1} \frac{\partial u}{\partial y} + C_{ij2} \frac{\partial u}{\partial z} \qquad (15.206)$$

where the right-hand-side is evaluated at the centroid where the velocity is u, and the constants C_{ij1} and C_{ij2} are the vertical and horizontal distances from the center of each wire to the centroid. Thus, in this procedure, instead of the velocity, the velocity *gradient* is assumed to be uniform across the sensing volume of the three arrays of the probe. Substitution of Eq. (15.206) into Eq. (15.203) yields twelve nonlinear algebraic equations given by

$$
\begin{aligned}
f_{ij} \equiv & -P_{ij} + u^2 + 2C_{ij1}u \frac{\partial u}{\partial y} + 2C_{ij2}u \frac{\partial u}{\partial z} \\
& -K_{1ij}\left[v^2 + 2C_{ij1}v \frac{\partial v}{\partial y} + 2C_{ij2}v \frac{\partial v}{\partial z} \right] \\
& -K_{2ij}\left[w^2 + 2C_{ij1}w \frac{\partial w}{\partial y} + 2C_{ij2}w \frac{\partial w}{\partial z} \right] \\
& -K_{3ij}\left[uv + C_{ij1}\left(u \frac{\partial v}{\partial y} + v \frac{\partial u}{\partial y} \right) + C_{ij2}\left(u \frac{\partial v}{\partial z} + v \frac{\partial u}{\partial z} \right) \right] \\
& -K_{4ij}\left[uw + C_{ij1}\left(u \frac{\partial w}{\partial y} + w \frac{\partial u}{\partial y} \right) + C_{ij2}\left(u \frac{\partial w}{\partial z} + w \frac{\partial u}{\partial z} \right) \right] \\
& -K_{5ij}\left[vw + C_{ij1}\left(v \frac{\partial w}{\partial y} + w \frac{\partial v}{\partial y} \right) + C_{ij2}\left(v \frac{\partial w}{\partial z} + w \frac{\partial v}{\partial z} \right) \right] = 0
\end{aligned}
$$

$$(15.207)$$

in terms of the 9 unknowns

$$U \equiv U_k \equiv \left(u,\ v,\ w,\ \frac{\partial u}{\partial y},\ \frac{\partial u}{\partial z},\ \frac{\partial v}{\partial y},\ \frac{\partial v}{\partial z},\ \frac{\partial w}{\partial y},\ \frac{\partial w}{\partial z} \right) \tag{15.208}$$

The nonlinear system can be solved in a least-square sense using Newton's method, where the error defined by

$$F = \sum_{i=1-3} \sum_{j=1-4} f_{ij}^2 \tag{15.209}$$

is minimized. An initial guess for the velocity components can be obtained by treating each four-wire array as two X-array probes. The initial gradients are computed by differencing the velocities from the three four-wire arrays. Convergence, within a specified tolerance, is accomplished within five iterations. The pitch and yaw angles encountered by each sensor should then be checked *a posteriori* in order to verify the integrity of the solution.

The uniqueness of the response equations solutions for the twelve-sensor probe can be determined to leading order (ignoring the gradients) by treating each four-wire array as a four-wire probe. The uniqueness of solutions for a given four-wire array is demonstrated by examining the solution convergence process making use of the measured calibration voltages. For this purpose the velocity vector with the highest pitch, α, and yaw, β, angles [$\alpha = +20°$ and $\beta = -20°$] used in a calibration will be considered. This is the most severe case for convergence within the calibration range.

Fig. 15.89(a) shows the contours of the error, plotted in the u–v plane as defined by Eq. (15.209). The w component of the velocity was eliminated by substitution using one of the equations represented by Eq. (15.207) in which w was expressed in terms of the u and v components and the measured voltage, P_{ij}. The unique minimum in the error corresponds to the physical solution, and convergence is achieved within a few iterations. With these u and v solution values at this minimum, the corresponding w component value can be found from Eq. (15.207).

Similar contours are depicted in Fig. 15.89(b) for the same calibration velocity

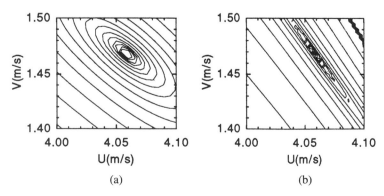

(a)   (b)

FIGURE 15.89   Contours of the error defined by Eq. (15.209) using: (a) four sensors and (b) three sensors. [From Marasli *et al.* (1993). Courtesy of *Exp. Fluids*, Springer-Verlag.]

data by disregarding one of the wires in the four-sensor probe, thereby yielding a triple-wire configuration. In this case, there are only three response equations in three unknowns for each array. The resulting error contours show a valley of possible solutions with several minima. In this case the algorithm iterated numerous times jumping from one solution *well* to another thus demonstrating the uniqueness problem which occurs with only three sensors. These observations are consistent with the conclusions of Döbbeling *et al.* (1990).

15.11.3 Vorticity Measurements with Particle Image Velocimetry

Particle Image Velocimetry (PIV) is based upon the fundamental definition of velocity; i.e.,

$$\overline{V} = \frac{\lim}{\Delta t \rightarrow 0} \frac{\Delta \overline{x}(\overline{x}, t)}{\Delta t} \tag{15.210}$$

In PIV, a marker is introduced into the flow field and illuminated at two instances in time, separated by Δt. Typically, Δt is precisely known (within nanoseconds) and the displacement vector of the marker, $\Delta \overline{x}$, is then determined to yield the velocity, $\overline{V}$, of the fluid particle located at $\overline{x}$ at time t. A more detailed description of PIV can be found in Sec. 15.4.6 of this handbook.

PIV simultaneously provides the investigator with multiple components of the velocity at several discrete locations in space, and this can be processed to give vorticity. Having the spatial information regarding the velocity and vorticity field represents a significant advantage over thermal anemometry techniques. However, the temporal resolution of the vorticity obtainable using PIV is severely limited when compared to vorticity measurements with thermal anemometry. In addition, the accuracy of PIV vorticity measurements in turbulent flows has not been adequately investigated. If PIV proves to be a useful tool to study the vorticity field in turbulent flows, then thermal anemometry and PIV techniques would best be utilized in a complementary fashion to extract the temporal and spatial characteristics of the vorticity field respectively.

The majority of PIV investigations to date have measured two components of the velocity in a two-dimensional grid; i.e., planar PIV. Stereo PIV, which utilizes two cameras that nominally image the same planar region in the flow to provide three components of the velocity, is gaining some popularity [Prasad and Adrian (1993)]. However, both planar and stereo PIV provide only one component of the vorticity. The component of the vorticity that can be obtained from PIV is the out-of-plane component (with respect to the illuminated 2-D plane in the flow field). For example, the velocity components $u(x, y)$ and $v(x, y)$ would be obtained if the light sheet is oriented in the x-y plane; thereby yielding the necessary velocity derivatives [see Eq. (15.185b)] or velocity contour integral [see Eq. (15.186)] to determine ω_z.

Accuracy of Vorticity Measurements. The accuracy of the vorticity calculated using PIV will be dependent on three factors: 1) the spatial resolution of the measurement technique, 2) the numerical scheme used to approximate the velocity derivatives in Eq. (15.185) or the contour integral in Eq. (15.186), and 3) the accuracy of the velocity measurements.

Spatial Resolution Errors: The spatial resolution of the calculated vorticity is dependent upon the size of the interrogation region used to determine the velocity and the separation between adjacent interrogation spots. Since the measured velocity is an average of the velocity of all particle pairs within the interrogation spot, the accurate determination of the vorticity relies on minimizing the velocity differences occurring *within* the interrogation spot. By decreasing the interrogation spot size, the particle seeding density must increase to maintain a sufficient number of particle pairs within the spot.

The application of Eq. (15.185b) or (15.186) relies on using measured velocity data from adjacent interrogation spots. Since there is typically overlap between adjacent interrogation spots (see Fig. 15.90), the measured velocity is not statistically independent, however this does minimize the spatial resolution of the calculated vorticity.

Numerical Scheme Errors: Using a second-order accurate, central difference technique to approximate the derivatives in Eq. (15.185b) in the calculation of the out-of-plane component of vorticity [e.g., ω_z from $u(x, y)$ and $v(x, y)$] results in the following.

$$\omega_{z_{ij}} = \frac{\partial v}{\partial x} - \frac{\partial u}{\partial y} \approx \left[\frac{v_{i+1,j} - v_{i-1,j}}{2\Delta x} \right] - \left[\frac{u_{i,j+1} - u_{i,j-1}}{2\Delta y} \right] \qquad (15.211)$$

where i and j correspond to the respective x and y locations of the discrete velocity measurements and Δx and Δy represent the spatial separation between the discrete velocity measurements (see Fig. 15.90). Note that Δx and Δy are equivalent to the interrogation spot size in the x and y directions respectively. This approximation neglects third and higher order terms resulting in an error on the order of Δx^3 (or Δy^3). This implies that as the size of the interrogation region decreases, the error associated with the numerical scheme given by Eq. (15.211) also decreases. Using this method, the effective spatial resolution of the vorticity measurement becomes $(2\Delta x, 2\Delta y)$.

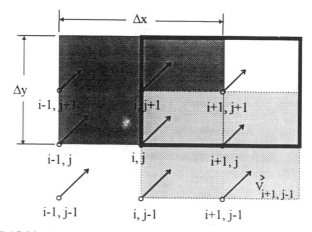

FIGURE 15.90 Schematic depicting discrete measurement locations in PIV.

Alternatively, one could choose Eq. (15.186) to estimate the vorticity. Choosing the contour around point i, j to include the eight neighboring points denoted as open symbols in Fig. 15.90 will give [Reuss *et al.* (1989)]

$$
\omega_{z_{ij}} \approx \frac{1}{4\Delta x \Delta y}
$$

$$
\cdot \begin{bmatrix} (v_{i+1,j} - v_{i-1,j})\Delta y + (v_{i+1,j-1} + v_{i+1,j+1} - v_{i-1,j-1} - v_{i-1,j+1})\frac{1}{2}\Delta y \\[2mm] + (u_{i,j-1} - u_{i,j+1})\Delta x + (u_{i-1,j-1} + u_{i+1,j=1} - u_{i-1,j+1} + u_{i+1,j+1})\frac{1}{2}\Delta x \end{bmatrix}
$$

$$
(15.212)
$$

It is appropriate to *smooth* the discrete velocity measurements provided by the PIV technique. If one defines a *filtered velocity* in the following manner

$$
\hat{u}_{i,j} = \tfrac{1}{2}u_{i,j} + \tfrac{1}{4}(u_{i-1,j} + u_{i+1,j}) \text{ and } \hat{v}_{i,j} = \tfrac{1}{2}v_{i,j} + \tfrac{1}{4}(v_{i-1,j} + v_{i+1,j})
$$

$$
(15.213)
$$

and substitutes Eq. (15.213) into Eq. (15.212), ω_z can be approximated as

$$
\omega_{z_{ij}} \approx \left[\frac{\hat{v}_{i+1,j} - \hat{v}_{i-1,j}}{2\Delta x} \right] - \left[\frac{\hat{u}_{i,j+1} - \hat{u}_{i,j-1}}{2\Delta y} \right] \qquad (15.214)
$$

This filtering acts to smooth the velocity field and reduce the noise in the calculated vorticity data. That is, the filtering operation attenuates the noise in the velocity field in the same sense that the differentiation inherent in Eq. (15.211) amplifies it. Comparing Eqs. (15.214) and (15.211) one can see that the vorticity calculated from PIV data via contour integration is equivalent to that determined from a straightforward second-order accurate central difference technique with a local filtering [Eq. (15.213)] of the velocity. However, it should be noted that this smoothing implies a further loss in spatial resolution.

Velocity Measurement Errors: The accuracy of the velocity measurement is determined by the precision with which one can determine the correct displacement peak, $\Delta \bar{x}$, within the correlation plane. As discussed by Adrian (1991) and Sec. 15.4.6 of this handbook, the errors in the measurement of $\Delta \bar{x}$ can originate from a variety of sources.

Comparison of PIV Vorticity Measurements with Prior Investigations. A relatively small number of PIV investigations present results regarding the vorticity field compared with either numerical results or more conventional methods of measuring vorticity. In addition, the present authors could only find two investigations [Urushihara *et al.* (1993) and Eggels *et al.* (1993)] that compare second-moment statistics regarding the *fluctuating* vorticity field as calculated from PIV data. Comparisons of the *mean* or *first moment* vorticity distributions are given by Westerweel (1992) in a turbulent flow and Robinson and Rockwell (1993) and Lourenco and Krothapalli (1994) in laminar flows. Additional studies which investigate the spatial character of the instantaneous vorticity field can be found in Reuss *et al.* (1989), Willert and Gharib (1991), Liu *et al.* (1991), Rockwell (1992), and Towfighi and Rockwell (1994), among others, however statistics regarding the vorticity field are either not given or not compared with other methods in these additional studies.

Urushihara *et al.* (1993) employed very high resolution PIV to study the logarithmic layer of fully developed turbulent pipe flow at moderate Reynolds numbers ($Re_D \approx 50,000$). The high resolution was achieved by imaging the flow field onto large format photographic film with a magnification of $6\times$. With this, they were able to obtain a measurement volume significantly smaller than that reported with thermal anemometry probes. They compare their r.m.s. azimuthal vorticity results with two turbulent boundary layer studies—Direct Numerical Simulation (DNS) of a low Reynolds number turbulent boundary at $Re_\theta = 670$ and a moderate Reynolds number turbulent boundary layer at $Re_\theta \approx 2900$ [Balint *et al.* (1991)] utilizing thermal anemometry. The PIV results are larger in magnitude than both the DNS and the thermal anemometry results for all y (distance from the wall) locations investigated except for locations close to the wall where the agreement with DNS is fairly good.

Aside from the fact that all three studies employed wall-bounded turbulent shear flows at different Reynolds numbers, the authors point out that the high rms values for the PIV investigation could result from the amplification of noise in the velocity field due to the differentiation inherent in Eq. (15.214).

Comparisons of vorticity measurements from PIV data with those resulting from a DNS of turbulent pipe flow are given by Eggels *et al.* (1993). Excellent agreement is obtained when comparing the r.m.s. value of the fluctuating azimuthal vorticity throughout most of the pipe, except near the wall. This is most probably due to the relatively large velocity gradients that exist near the wall which, result in an increased uncertainty in the velocity measurement.

Westerweel (1993) investigated turbulent pipe flow using PIV and presents the mean azimuthal vorticity profile compared with Reichardt's formula for fully developed turbulent pipe flow. The measurements and the Reichardt formula show good agreement for $y^+ > 44$ with a maximum deviation of approximately 15% at $y^+ \approx 66$. However, for smaller y^+ values the data trend is qualitatively, incorrect since the mean vorticity magnitude is decreasing with decreasing y^+ values.

Lourenco and Krothapalli (1994) calculate the spanwise vorticity distribution in a laminar wall jet experiment and compare the results with those from a numerical simulation of the same flow. The former exhibit variations of order 10% (nondimensional values) with respect to the mean.

15.12 LASER-BASED RAYLEIGH AND MIE SCATTERING METHODS
Campbell D. Carter

15.12.1 Introduction

Laser-based measurements of gas properties such as temperature, velocity, and species concentration have given researchers new and valuable insight into the chemistry and physics of reacting and nonreacting flows. Laser-based techniques have significant advantages over physical probes (e.g., hot wires and thermocouples), because they provide measurements that are spatially and temporally resolved (resolutions of <0.1 mm^3 and <10 ns) and nonintrusive. In addition, instantaneous two- or even three-dimensional measurements are feasible with these methods. Thus, one can obtain instantaneous spatial and temporal correlations. In this section, we focus on Rayleigh and Mie scattering techniques which, because of the modest requirements for equipment (principally, a cw or pulsed laser), are among the simplest of the laser-based methods.

The purpose of this section is to acquaint the reader with the Rayleigh and Mie scattering techniques as applied to the investigation of gaseous flows. First, we review the fundamental aspects of the scattering techniques, presenting the descriptive equations. Second, we describe some (but certainly not all) of the applications that have appeared in the literature. These applications have been divided into those involving the following: 1) species-concentration or mixture-fraction measurements when one gas mixes with another, 2) temperature and density measurements, particularly in reacting flows, and 3) velocity measurements in subsonic and supersonic flows.

15.12.2 Background

Both Rayleigh and Mie scattering (sometimes referred to as Lorentz–Mie scattering) processes are *elastic* in nature, i.e., the scattered radiation experiences no change in frequency or energy in the scattering process. Throughout, we use the term *Mie scattering* to refer to scattering from particles too large to be covered by Rayleigh theory, which is appropriate for particles much smaller than the wavelength of the incident radiation. Strictly speaking, however, this terminology is misleading, since Mie theory describes scattering from a spherical, homogeneous particle of *arbitrary* size when illuminated by plane waves [van de Hulst (1957) and Kerker (1969)]. The problem of scattering from a spherical particle is, in fact, one of the few cases where an exact analytical solution of the Maxwell equations has been found; in general, one must use approximate or numerical methods to describe the scattering process. Since generally the particles in question are not spherical (agglomerated soot, for example), an *equivalent sphere* often must be assumed if the scattering is to be quantified [Jones (1979)].

Another scattering technique, not treated in this section but reviewed elsewhere [Long (1977), Lederman (1977), and Eckbreth (1988)], is spontaneous Raman scattering. As with Rayleigh scattering, the scattered radiation is dipole-like. However, the scattering process is inelastic; that is, the scattered photons are shifted in energy by the separation of 1) the vibrational states with vibrational Raman scattering, 2) the rotational states with rotational Raman scattering, and 3) the electronic states with electronic Raman scattering. Since each molecule has a distinct energy spacing, molecular species can be identified by detecting the scattered radiation with a spectrometer or with optical filters (i.e., colored glass or interference filters). A disadvantage of the Raman technique, however, is that the scattering cross sections are small—about three orders of magnitude less than those for Rayleigh scattering [Eckbreth (1988)].

Rayleigh Scattering. Rayleigh theory is appropriate for a spherical particle when the following condition is satisifed for the particle size

$$\alpha \equiv \frac{2\pi a}{\lambda_i} \ll 1 \tag{15.215}$$

where a is the particle radius, λ_i is the wavelength of the incident radiation, and α is the dimensionless size parameter. Equation (15.215) indicates that the particle should be sufficiently small that the incident electric field is uniform over the extent of the particle. With the Rayleigh process, the scattering may be thought of as emission from radiating dipoles induced by the incident electric field [Born and Wolf (1980), Fabelinskii (1968), and van de Hulst (1957)]. Consequently, the scattered intensity depends on the polarization of the incident beam and the point of observation. In general, the scattered irradiance I_s (W/cm^2) from N_p identical independent particles in a gas can be written as

$$I_s = \frac{I_i}{r^2} N_p \sigma_p \tag{15.216}$$

where I_i (W/cm^2) is the incident irradiance and r is the distance to the point of observation ($r \gg$ the dimensions of volume V containing the N_p particles). The particle scattering *cross section*, σ_p (cm^2), is dependent on the characteristics of the particles and on the angle between the incident-beam polarization and the scattered radiation.

Consider a collimated beam traveling in the y direction, linearly polarized in the z (vertical) direction, and impinging on a spherical, homogeneous dielectric particle. The cross section describing the vertically polarized scattering in the x–y (horizontal) plane is given by [Kerker (1969)]

$$\sigma_{Vp} = \frac{\alpha^6 \lambda_i^2}{4\pi^2} \left(\frac{n_p^2 - 1}{n_p^2 + 2} \right)^2 = \frac{9\pi^2 V_p^2}{\lambda_i^4} \left(\frac{n_p^2 - 1}{n_p^2 + 2} \right)^2 \tag{15.217}$$

where n_p is the index of refraction of the particle and V_p is the volume of the individual particles. For horizontally polarized incident radiation, the scattering is in the horizontal plane, but the magnitude depends on the observation angle θ (see Fig. 15.91),

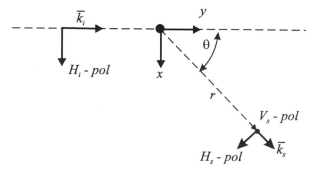

FIGURE 15.91 Plan-view schematic of scattering from a particle for horizontally polarized incident radiation. H-pol and V-pol denote horizontal and vertical polarization, respectively. Vertical polarization is normal to the plane of the page (the z-direction).

$$\sigma_{Hp} = \frac{\alpha^6 \lambda_i^2}{4\pi^2} \left(\frac{n_p^2 - 1}{n_p^2 + 2} \right)^2 \cos^2 \theta \tag{15.218}$$

These equations can be generalized for arbitrary polarization by considering the incident beam to be linearly polarized in the x–z plane, with components $I_{i, \text{vert}} = I_{i,z}$ and $I_{i, \text{horz}} = I_{i,x}$. If the beam were unpolarized, or if the polarization were at 45° with respect to the x-axis, then $I_{i, \text{horz}} = I_{i, \text{vert}} = I_i/2$, and the scattering would be proportional to $(I_i/2)(\cos^2 + 1)$. Note that these equations are appropriate for a nonabsorbing particle. When the particle is not *highly* absorptive or reflective [Kerker (1969)], n_p can be replaced by the complex refractive index $m = n_p - ik$, where the imaginary part describes absorption by the particle.

To determine the scattered power, $P_s(W)$, within some solid angle, $\Delta\Omega$, (defined by the scattering collection optics), one must integrate over the range of detected angles (see Fig. 15.92),

$$P_s = \int_{\Delta\Omega} I_s r^2 \, d\Omega \tag{15.219}$$

The collection optics and beam diameter define a *probe* volume $V = lA_c$, where l is the probe length (defined by the detector aperture) and A_c is the cross-sectional area defined by the beam (Fig. 15.92). The scattered power impinging on a detector is

$$P_s = \epsilon \int_{\Delta\Omega} I_i N V \sigma \, d\Omega = \epsilon \int_{\Delta\Omega} P_i N l \sigma \, d\Omega \tag{15.220}$$

where N (cm^{-3}) is the number density of particles, $P_i(W)$ is the incident power, and ϵ is the efficiency of the collection optics. In the event that the variation in θ is small over the collected solid angle, one can use the approximation $P_s \cong \epsilon I_i N V \sigma \Delta\Omega$.

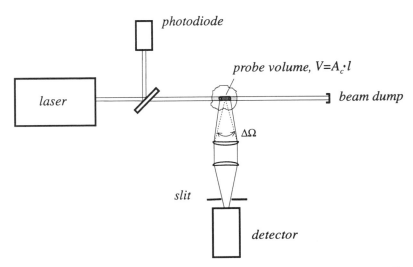

FIGURE 15.92 Plan-view schematic of a point-wise scattering measurement. The photodiode is used to record the beam energy or power.

As written, Eqs. (15.219) and (15.220) are valid whether the medium is considered to be a gas or a group of particles (or even a single particle); however, for a gas, the cross section is normally written as a function of n, the *gas* index of refraction. One can relate the refractive index of the gas to that for the individual particles/molecules using the Lorenz–Lorentz formula,

$$\left(\frac{n_p^2 - 1}{n_p^2 + 2}\right) N_p V_p = \left(\frac{n^2 - 1}{n^2 + 2}\right) V \cong \frac{2}{3} (n - 1)V \qquad (15.221)$$

where V is the volume occupied by N_p particles, the gas number density is $N = N_p/V$, and the approximation is valid for $n \cong 1$ (appropriate for a gas). In addition, this equation leads to the important conclusion that $(n - 1)/N$ is independent of temperature and pressure.

Although scattering from the ideal spherical, homogeneous particle (and from atoms) is completely polarized (e.g., scattering from a vertically polarized beam will be vertically polarized), scattering from molecules, in general, is not. This *depolarization*, ρ_v, results from the anisotropy of the scatterer's polarizability and the random orientation of the molecules in a gas; typically ρ_v is small for gases (Table 15.10). For vertically polarized incident radiation, the vertically polarized scattering cross section for a gas is

$$\sigma_{VV} = \sigma_o = \left[\frac{4\pi^2(n_o - 1)^2}{N_o^2 \lambda_i^4} \cdot \left(\frac{3}{3 - 4\rho_v}\right)\right] \qquad (15.222)$$

Here N_o and n_o are the respective number density and index of refraction at STP ($N_o = 2.687 \times 10^{19}$ cm^{-3} for an ideal gas at $T = 273$ K and $P = 760$ Torr). This equation describes the scattering cross section for a wide range of temperatures and pressures, though not including the neighborhood of the critical point [Fabelinskii (1968)]; implicit in this derivation is that scattered radiation is collected from the complete Doppler–Brillouin structure in addition to any rotational Raman structure [Rowell *et al.* (1971)]. The transverse scattering cross sections resulting from the depolarization σ_{VH} (referring to vertically polarized incident radiation and horizontally polarized scattered radiation) and σ_{HV} (referring to horizontally polarized incident radiation and vertically polarized scattered radiation) have the same magnitude,

$$\sigma_{VH} = \sigma_{HV} = \rho_v \sigma_o \qquad (15.223)$$

Of course, the cross section for horizontally polarized incident and scattered radiation is a function of the angle θ

$$\sigma_{HH} = [(1 - \rho_v) \cos^2 \theta + \rho_v]\sigma_o \qquad (15.224)$$

When one is using a vertically polarized laser beam and detecting both polarization states, the total scattering cross section is simply $\sigma_o(1 + \rho_v)$, independent of angle in the horizontal plane. Likewise, for a horizontally polarized incident beam, the total scattering cross section (i.e., collecting both polarization states) is $\sigma_{HH} + \sigma_{HV}$. Note that the scattering cross section is often described as a *differential* cross section

TABLE 15.10 Refractivity Data

[Values are for STP conditions ($T = 273$ K, and $P = 760$ Torr), and the cross sections are given for $\lambda_i = 488$ nm. The indices of refraction, except for those for Freon 12 and 22, are derived from the data of Gardiner et al. (1981), and the cross sections were calculated using Eq. (15.222) and assuming ideal gas behavior (i.e., $N_o = 2.687 \times 10^{19}$ cm^{-3}). The cross sections for Freon 12 and 22 are derived from the data of Shardanand and Rao (1977), and the corresponding n_o were derived from Eq. (15.222), assuming ideal gas behavior and $\rho_v = 0$.]

Species	$n_o - 1$	$100 \cdot \rho_v$	σ_o $(10^{-28}$ cm$^2)$
Ar	0.000284	0	7.75
CH_4	0.000446	0	19.2
CO	0.000339	0.52[a]	11.1
CO_2	0.000450	4.12[a]	20.6
C_2H_4	0.000727	1.27[a]	51.9
He	0.000035	0	0.118
H_2	0.000140	0.95[b]	1.92
H_2O	0.000255	1.0[c]	6.33
NO	0.000297	1.54[d]	8.71
N_2	0.000301	1.08[b]	8.87
O_2	0.000273	2.91[b]	7.50
$CHClF_2$ (Freon 22)	0.000990[e]	—	94.5
CCl_2F_2 (Freon 12)	0.00127[e]	—	155

[a]Bogaard et al. (1978).
[b]Rowell et al. (1971).
[c]D'Alessio (1981).
[d]$\lambda_i = 632.8$ nm; Bridge and Buckingham (1966).
[e]Shardanand and Rao (1977).

and typically denoted by $d\sigma/d\Omega$ (cm^{-2}/steradian); this differential cross section and the one described above have the same magnitude.

For a mixture of gases, σ_o is simply the sum of the mole-fraction-weighted cross sections

$$\sigma_{o,\text{mix}} = \sum_i X_i \sigma_{o,i} \qquad (15.225)$$

where X_i and $\sigma_{o,i}$ are the mole fraction and cross section of species i, respectively. The most comprehensive tabulations of the refractive indices of gases are those given by Landolt–Börnstein (1962). Based, in part, on these data, Gardiner et al. (1981) compiled refractivity data for a large number of species. In general, the index of refraction *and* depolarization depend on wavelength [Alms et al. (1975)], although this dependence is generally small throughout the visible spectrum [Landolt–Börnstein (1962) and Shardanand and Rao (1977)]. Nonetheless, for each species considered, Gardiner and co-workers have fit the wavelength dependence of the refractive index with the *dispersion relation*, $n_o - 1 = a/(b - \lambda_i^{-2})$. We have used the data of Gardiner *et al.* (1981) along with published depolarization ratios to calculate the scattering cross sections for $\lambda_i = 488$ nm shown in Table 15.10.

Mie Scattering. In this subsection, we will treat only the basic aspects of Mie scattering theory. For a more in-depth description of the Mie scattering process, the reader is referred to the review article and the chapter by Jones (1979), (1993) and books by Kerker (1969), Bohren and Huffman (1983), van de Hulst (1957), and Born and Wolf (1980). Whereas we stated above that Rayleigh theory is valid when the particle is much smaller than the incident wavelength, we now modify that criterion by adding that the particle should *also* be much smaller than the wavelength *within* the particle [van de Hulst (1957)], i.e., $\alpha|m| \ll 1$. A more specific guideline was given by Kerker *et al.* (1978) who found that for $\alpha|m| < 0.2$, the agreement between the scattering behavior and that predicted by Rayleigh theory was generally within $\sim 1\%$. Furthermore, Jones (1979) suggests that the error incurred from using Rayleigh theory to predict the total scattering and/or absorption is within $\sim 10\%$ for $\alpha|m| < 0.6$. As the particle size increases, the departure from the angular dependence observed in Rayleigh scattering increases. The ratio of forward to backward scattering increases, and the angular mode structure becomes increasingly complex; that is, lobed structures (the number of which is on the order of $\alpha|m|$) appear in the polar diagram [Jones (1979)]. In addition, the scattering intensity does not depend on the sixth power of the particle diameter (as is true for a Rayleigh particle). With nonabsorbing dielectric spheres, for example, the scattering efficiency, $Q_{sca} \equiv \sigma_p/\pi a^2$, increases monotonically with a to values of $Q_{sca} \geq 3$ (depending on the exact value of m) until $\alpha(m - 1) \approx 2$ [Kerker (1969)]. With increasing $\alpha(m - 1)$, the scattering efficiency undergoes a damped oscillation about $Q_{sca} = 2$ [Kerker (1969)]. Thus, when Mie scattering is used as a diagnostic tool, the collected scattering may depend on 1) the exact angle of detection, 2) the shapes and size distribution of the particles, and 3) the homogeneity of the particles and their surface characteristics. In general, one must use caution in relating the collected scattering to quantitative parameters (mixture fraction, temperature, etc.).

Ideally, scattering by a group of particles in a gas is an extension of scattering by a single particle. This criterion is satisfied under the following conditions [Jones (1979)]: 1) the particles are separated by greater than three particle radii and, thus, no electrical interaction takes place between them [Kerker (1969)]; 2) the *scattered* radiation does not interact with the particles; and 3) the number of particles is sufficiently large that no optical interference occurs between radiation scattered by different particles. The first condition leads to a maximum particle number density; with 1-μm particles, the maximum seeding density would, therefore, be on the order of 10^{10} particles/cm^3. The second condition, however, places a greater restriction on the maximum desirable seeding density. Consider the transmission through a nonuniform medium, which is described by a Beer's Law relation,

$$I = I_i \exp\left(-\int_0^L K_{ext} \cdot dx\right) = I_i \exp(-\tau) \qquad (15.226)$$

where K_{ext} is the extinction coefficient, L is the optical path length, and τ is the turbidity. When the particles are identical, the extinction coefficient is the product of the particle number density and the extinction cross section, $K_{ext} = N\sigma_{ext}$; otherwise the extinction coefficient is equal to the integral over the particle size distribution. Kerker (1969) points out that multiple scattering can be important for $\tau >$ 0.1 (a violation of the second condition above). For soot particles with $a = 0.1$ μm and an optical path length of 1 m, Jones (1979) reports that the maximum number density consistent with $\tau = 0.1$ is $N = 2 \times 10^7$ cm^{-3}. From the standpoint of flow diagnostics, in particular the use of scattering as a measure of the gas density or temperature, high turbidity across the region of interest can lead to two sources of measurement bias. First, with a large τ, the laser beam irradiance decreases as the beam traverses the region of interest, and thus the scattering intensity at a particular region depends on the particle number density encountered by the beam *prior* to reaching the measurement point. Second, the scattering can be attenuated as it traverses the particle-laden gas between the measurement region and the detector. This source of measurement bias, referred to as *radiation trapping*, can often be minimized through the choice of viewing position or angle.

In view of the restriction on the number density, a principal limitation of Mie scattering for flow diagnostics is *marker shot noise* or *ambiguity noise*, which results from a finite number of particles within the probe volume [Becker *et al.* (1967)]. That is, even in the absence of gas-number-density variations, the scattering varies from one spatial location to the next (or from one time to the next), depending on the exact number of particles within the probe volume. (A similar problem known as *photon shot noise* is often encountered when the number of collected photons is small.) Thus, reduction of this source of measurement noise requires the use of more small particles rather than fewer large particles. The problem of marker shot noise is exacerbated with a polydisperse group of particles, since scattering is dominated by a few large particles. Because the seeding density is limited, this noise source imposes a bound on the minimum spatial resolution; as the dimensions of the probe volume become smaller, the number of particles decreases and the marker shot noise increases.

Smaller particles also have a higher velocity frequency response, and thus are more capable of tracking turbulent fluctuations of the gas. Becker *et al.* (1967) have calculated the velocity frequency response—the frequency for which the particle velocity amplitude is 10% below the value of the gas—for a range of particle sizes in air near STP. For a particle 1 μm in diameter having a 1 gm/cm^{-3} density, the response is ~22 kHz; the frequency response of a 3-μm-diameter particle, however, is ~2.7 kHz. As a *rule of thumb*, the mass fraction of the seed should be limited to 1% to minimize the effect on the gas properties (e.g., density and viscosity). In addition, the influence on the turbulent kinetic energy should be small for a particle mass fraction of 1% or less [Squires and Eaton (1990)]. Of course, the diffusivity of the particles is much lower than that of the gas. Finally, caution must be exercised in interpreting the scattering intensity in reacting flows. Moss (1980), for example, found that as TiO_2 seed particles passed through the flamefront of a premixed flame, the scattering signal decreased by a factor of 25 rather than 8, as expected from the density variation across the flame. Stepowski (1992) reports that for combusting flows, the more stable Al_2O_3 and ZrO_2 seed is preferable.

Spectral Distribution of Rayleigh/Mie Scattering. Scattered radiation can be both *Doppler broadened* by the thermal motion of the scatterers and *Doppler shifted* by the bulk motion of the scatterers. The spectral profile resulting from this thermal motion is described by a Gaussian function [Fabelinskii (1968)] having a characteristic full width Δv_D (Hz), measured at half the maximum height of the profile, of

$$\Delta v_D = 2v_i \left(\frac{8 k_B T \ln 2}{mc^2} \right)^{1/2} \sin \frac{\theta}{2} = 1.4325 \times 10^8 \, v_i \left(\frac{T}{M} \right)^{1/2} \sin \frac{\theta}{2}$$

(15.227)

where T (K) is the gas temperature, m (gm) and M (a.m.u.) refer to the molecular/atomic weight, v_i (Hz) is the incident frequency, k_B is the Boltzmann constant, c is the speed of light, and θ is the angle between detection and beam-propagation wave vectors $\overline{\mathbf{k}}_s$ and $\overline{\mathbf{k}}_i$, respectively (see Fig. 15.93). In addition to this thermal broadening, the scattering can be broadened by the *Brillouin effect*, which is an interaction of the incident radiation and *thermal* acoustic waves [Fabelinskii (1968)]. The relative importance of Brillouin scattering is deduced from the product of the mean free path, l, between scatterers and the magnitude of the wave vector associated with momentum transfer in the scattering process [Pitz *et al.* (1976) and Miles and Lempert (1990b)]

$$\frac{1}{|\overline{\mathbf{k}}_s - \overline{\mathbf{k}}_i| l} = \frac{\lambda_i}{4\pi \sin (\theta/2)}$$

(15.228)

As before, θ is the angle between $\overline{\mathbf{k}}_s$ and $\overline{\mathbf{k}}_i$ (Fig. 15.93). When this ratio is on the order of one or greater, Brillouin scattering is significant, and the spectral profile is not described by a simple Gaussian function [Sandoval and Armstrong (1976)].

With both Rayleigh and Mie scattering, the radiation scattered by a moving particle undergoes a change in frequency. This Doppler frequency shift (Hz) between the incident and scattered radiation, as seen by the stationary observer, is described by the relation [Jones (1993)]

$$\Delta v = \frac{1}{2\pi} (\overline{\mathbf{k}}_s - \overline{\mathbf{k}}_i) \cdot \overline{\mathbf{u}} = v_i \frac{2 |\overline{\mathbf{u}}|}{c} \sin \frac{\theta}{2} \sin \left(\alpha - \frac{\theta}{2} \right)$$

(15.229)

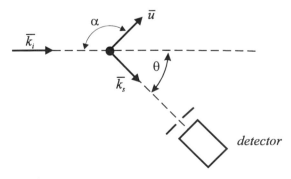

FIGURE 15.93 Plan-view schematic of the interaction between the incident radiation and a particle with velocity vector $\overline{\mathbf{u}}$ in the plane of observation (plane of the page).

where $\bar{\mathbf{u}}$ is the velocity vector within the plane of observation (defined by the vectors $\bar{\mathbf{k}}_s$ and $\bar{\mathbf{k}}_i$) and α is the angle between $\bar{\mathbf{k}}_i$ and $\bar{\mathbf{u}}$ (Fig. 15.93). In reality, of course, the detector will *see* a range of θ.

15.12.3 Applications

One requirement for the use of Rayleigh scattering is that the flow be essentially free of particles. Thus, the technique is best suited to laboratory flows, where one can effectively remove particles. Of course, occasional particles do not necessarily compromise an experiment. Because the scattering intensity from the particles is so much greater than that from atoms or molecules, one can often distinguish atomic/molecular scattering from particle scattering, whether point or field measurements are used. Furthermore, Rayleigh and Mie scattering measurements near surfaces can be difficult because of scattering from these surfaces. Optical arrangements for Rayleigh/Mie-scattering (as well as spontaneous-Raman-scattering) measurements often resemble those shown schematically in Figs. 15.92 and 15.94. Vertically polarized laser radiation is directed to the probe volume, and scattering is collected at 90° to the beam-propagation axis. For field (i.e., 2D) measurements, one can use a cylindrical lens or a combination of cylindrical lenses to expand the laser beam into a *sheet*. Two less common methods of forming a laser sheet involve use of a rotating mirror to sweep the laser beam through the region of interest and use of a multi-pass cell, formed by two cylindrical concave mirrors.

Because the scattering intensity is proportional to the laser intensity, monitoring the laser intensity is generally important when relating the strength of the scattering signal to some physical property of the gas (e.g., concentration of fuel). Of course, for field measurements, monitoring the laser intensity is more difficult, since a 1D detector is often necessary (though a linear detector is appropriate only for a collimated laser sheet). Another approach is to infer the irradiance distribution of the laser sheet from the Rayleigh/Mie scattering in a region of flow that is known to be uniform in density. If the sheet irradiance does not vary with time, then one can record the irradiance distribution once and use it to correct subsequent images. Two other factors can affect the quantitative nature of field measurements: 1) the 2D detector may have a nonuniform response (to a uniformly illuminated field), and 2) the solid angle for collection of the scattered light generally varies across the detector

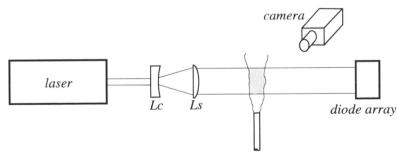

FIGURE 15.94 Side-view schematic of a typical 2D scattering measurement. Here, *Lc*, a plano-concave cylindrical lens, and *Ls*, a plano-convex spherical lens, are the sheet-forming optics. The diode array is used to record the beam energy/power in the vertical direction.

(e.g., the edge of the detector will generally subtend a smaller solid angle and, thus, *see* less scattering).

Species Concentration. Perhaps the best application of Rayleigh scattering for a quantitative measurement is in the investigation of the mixing of two fluids at constant temperature and pressure. Dowling and Dimotakis (1990) employed Rayleigh scattering with an argon-ion (Ar^+) laser to study the mixing of a free, turbulent axisymmetric jet with the surrounding quiescent fluid. For one set of measurements with a jet-exit Reynolds number of 5000, they used C_2H_4 (ethylene) as the jet fluid and N_2 as the ambient fluid. Due to the significant difference in scattering cross section (Table 15.10), these two fluids can easily be distinguished, and the scattering signal is linearly related to the jet mole fraction,

$$P_s \propto X_{jet}(\sigma_{o,jet} - \sigma_{o,amb}) + \sigma_{o,amb} \qquad (15.230)$$

With this arrangement, they were able to make point measurements having high signal-to-noise ratios (which, of course, limits the precision of the *mixedness* measurement) at frequencies as high as 100 kHz [Dowling *et al.* (1989)].

In a nonpremixed, turbulent jet flame, Muss *et al.* (1994) applied Rayleigh scattering to the measurement of a fuel mixture fraction (defined as the ratio of the mass originating from the fuel jet to the total value) using a fuel mixture of 80% He and 20% H_2. The small σ_o of He—in addition to its large concentration in the jet fluid— ensures that the scattering signal will decrease monotonically as a function of mixture fraction. Thus, they measured the mixture fraction in a field and were able to determine the scalar dissipation.

To investigate the mixing of a turbulent jet into a coflowing stream, Long *et al.* (1979), (1981) measured the concentration of jet fluid in a field using Mie scattering. The fluid was doped with an aerosol of sugar particles (diameter ≤ 1 μm). The large scattering cross section of the particles enabled them to use a 2.5 W cw Ar^+ laser. The scattering from a 2×2 cm^2 area was recorded using a TV camera that was gated on for a period of 10 μs, effectively *freezing* the fluid motion. The images were digitized in a 100×100 pixel format, yielding a spatial resolution of $0.2 \times 0.2 \times 0.15$ mm, with the third dimension (0.15 mm) being set by the thickness of the laser sheet. They investigated the spatial structures and length scales of turbulent flows for different jet-exit Reynolds numbers. In addition, Long and co-workers (1981) demonstrated the feasibility of measuring a 1D scattering image (along the line of the laser beam) as a function of time. They removed the cylindrical lens from the laser focusing optics and added a rotating mirror to the scattering collection optics, which enabled them to sweep the 1D scattering along the second dimension of the camera.

To circumvent the problem of limited spatial resolution with Mie scattering (due to marker shot noise), Escoda and Long (1983) applied molecular Rayleigh scattering to a turbulent jet flow. To increase the scattering signal, they used the second harmonic of a Q-switched Nd:YAG laser ($\lambda_i = 532$ nm), delivering 20 mJ per 15 ns pulse (whereas a 2.5-W cw Ar^+ laser delivers only 25 μJ during a 10 μs exposure); in addition, they used Freon 12 as the jet fluid since its scattering cross section is much larger than that of the coflowing N_2 (see Table 15.10). They were able to resolve fluid structures measuring 100 to 200 μm with an improved spatial resolution of ~ 50 μm. Furthermore, they investigated the effect of marker diffusivity (or lack thereof) on the jet-concentration images. The images derived from molecular Rayleigh scattering clearly showed a more diffuse boundary than those derived from Mie scattering because of molecular diffusion.

More recently, Long and co-workers have proved the feasibility of both 2D concentration measurements at high framing rates and 3D concentration measurements. Winter *et al.* (1987) compared three approaches to obtaining 2D high-framing-rate measurements. The highest framing rate reported (10 kHz) was achieved for a system including a rotating mirror and a slow vidicon camera; the drawback of this arrangement was that it allowed only a limited number of exposures. For measuring gas concentrations in 3D, Yip and Long (1986) employed two lasers and two cameras to record molecular Rayleigh scattering in two adjacent planes (separated by 300 μm) simultaneously within a turbulent jet, again using Freon as the jet fluid. To distinguish the scattering planes, they used different laser wavelengths, 532 nm from a Nd:YAG and 563 nm from an Nd:YAG-pumped dye laser, in conjunction with wavelength filters in front of each camera. They were able to determine the joint probability density function (PDF) of the mixture fraction and the magnitude of the mixture–fraction gradient.

Subsequent to this experiment, Yip *et al.* (1988) used a somewhat different approach to record the scattering from multiple planes *swept* through a turbulent jet flow of Freon. A pulsed laser is required to obtain sufficient scattering signal; however the pulse must be of sufficient duration to be swept through the gas jet. Consequently, they employed a flashlamp-pumped dye laser having a pulse duration of ~1 μs and a pulse energy of 550 mJ. An electronic framing camera recorded as many as twelve 10 ns exposures, each separated by 50 ns. The sweep rate of the laser sheet coupled with the 50 ns delay between exposures resulted in consecutive images being separated by 250 μm. They were able to record surfaces of constant jet-fuel concentration (equivalent to mixture fraction in this case). Unfortunately, the signal-to-noise ratio (where the noise component was probably photon shot noise) was poor; to solve this problem they used laser-induced fluorescence of biacetyl, which was mixed with jet fluid.

To study turbulent mixing, Roquemore *et al.* (1986) seeded $TiCl_4$ vapor into the fuel side of a bluff-body combustor. Upon mixing with a moist gas, the $TiCl_4$ quickly forms micron-sized TiO_2 particles through a reaction with H_2O; as a consequence, the *mixed* fluid can be visualized with the Mie-scattering technique. Through photographs and high-speed movies of the scattering from a laser sheet from an Ar^+ laser, Roquemore and co-workers were able to characterize regimes for the operation of the bluff-body combustor.

Smith *et al.* (1989) employed an argon-fluoride excimer laser ($\lambda_i = 193$ nm) to obtain 2D molecular Rayleigh scattering images within a Mach-2.5 air boundary layer. Use of a deep UV wavelength has two significant advantages: 1) the scattering cross section is much larger than in the visible region (for example, they report that for air, σ_o at $\lambda_i = 193$ nm is 80 times higher than at $\lambda_i = 532$ nm) as a result of the explicit λ^{-4} dependence and the smaller wavelength dependence of the index of refraction, and 2) scattering from surfaces and particles is suppressed due to stronger absorption by most materials in the UV. The latter point is particularly important for measurements near surfaces. They determined the density PDFs, where the density decreases in the boundary layer as a result of frictional heating. Note that the flow was effectively frozen, even at the high velocities, with the 10 ns pulse.

Several investigators have made use of the so-called *vapor–screen technique*

[McGregor (1961)] to study supersonic flows. Vapor, either naturally occurring (usually water) or added (for example, ethanol), condenses in a homogeneous nucleation process to form small ice clusters. Clemens and Mungal (1991), (1992) used this approach to study a supersonic mixing layer. In one experiment, they seeded the ethanol vapor in the subsonic stream; mixing with the supersonic stream resulted in the formation of the ethanol ice particles that effectively scattered the incident radiation. Unfortunately, as Clemens and Mungal (1991) point out, the condensed vapor is not a conserved marker, since sources and sinks for the markers exist throughout the flowfield. Furthermore, assuming that the particles are sufficiently small to be Rayleigh scatterers [for example, Clemens and Mungal (1991) estimated that $a = 0.02$ μm under their conditions], the scattering signal scales as the volume squared; consequently, as the ice particles coagulate, the signal can increase significantly without an increase in the total mass of particles. Others have used the vapor–screen method in conjunction with an I_2 filter, either to record images near surfaces [Forkey *et al.* (1994)] or to collect 2D images of velocity [Miles and Lempert (1990a), (1990b), Miles *et al.* (1991), and Elliott *et al.* (1994)]. The use of an atomic/molecular filter, referred to as the *Filtered Rayleigh–Mie Scattering technique*, is described in more detail below in connection with temperature and velocity measurements.

Temperature/Density Measurements. In constant pressure flows where the gas index of refraction is constant (e.g., a heated stream of air flowing into air) or can be determined, one can easily relate the variation in signal to temperature, mass density, or number density; the temperature is proportional to $\sigma_{o,\text{mix}}/P_s$, and the mass density is proportional to $M_{\text{mix}}P_s/\sigma_{o,\text{mix}}$, where M_{mix} is the molecular weight of the mixture. Of course, when σ_o varies throughout the flowfield (due to mixing of gases or chemical reactions), deriving the temperature or mass density from the scattering signal can be difficult. In flames with premixed fuel and air, the problem is ameliorated by the large concentration of N_2 in the reactants and products. Nevertheless, Namer and Schafer (1985) found that the variation in the net scattering cross section could be large, leading to a large error in the inferred temperature, especially in rich flames because of the large concentrations of H_2 and CO. Interpretation of the signal in terms of mass density was somewhat better, since the variation in $\sigma_{o,\text{mix}}/M_{\text{mix}}$ was less than that in $\sigma_{o,\text{mix}}$ alone. They concluded that for premixed H_2 flames and lean premixed CH_4 and C_2H_4 flames, the errors may be within acceptable limits ($< 10\%$ error in the inferred temperature).

In the special cases where the species concentrations can be measured, calculated, or estimated, this error can be reduced or eliminated, insofar as the scattering cross sections are known. Inbody (1992) measured temperatues in premixed $C_2H_4/O_2/N_2$ laminar flames using an iterative approach. Knowledge of the species concentrations across the flame, which he calculated from an estimated temperature profile, enabled him to obtain a more accurate temperature; this, in turn, was used to calculate more accurate species concentrations. Barat *et al.* (1991) used a similar approach to derive temperatures from Rayleigh scattering employing the second harmonic of a Q-switched Nd:YAG within a premixed, jet-stirred combustor. The $\sigma_{o,\text{mix}}$ was calculated based on the assumption that the combustion was adiabatic and that the mixture composition consisted of reactants and/or products (no intermediate species). The temperature indicated the relative concentrations of products and reactants and, thus, the scattering cross section. Because the collected scattering consisted of both the Rayleigh signal and a large background interference (caused by scattering from surfaces within the combustor), they developed a background–subtraction scheme that included splitting and recording both the horizontally polarized scattering component (consisting of background scattering and the small Rayleigh depolarization) and the vertically polarized component (consisting of Rayleigh scattering plus background scattering). In a variety of nonpremixed flames, Dibble *et al.* (1987) have used a combination of Rayleigh and vibrational Raman scattering for determining the temperature and the concentration of the major species; thus, $\sigma_{o,\text{mix}}$ is easily calculated from Eq. (15.225).

In general, the variation in the scattering cross section can be much larger across a nonpremixed flame than across a premixed one (e.g., consider the variation in $\sigma_{o,\text{mix}}$ across H_2/air and CH_4/air nonpremixed flames), which makes the measurement of temperatures more difficult. Dibble and Hollenbach (1981) circumvented this problem by choosing a fuel compoistion of H_2 and CH_4 having the same $\sigma_{o,\text{mix}}$ as that of air. They assumed, of course, that the scattering from the intermediate species was not substantially different from that of the fuel and air. Using an Ar^+ laser, they were able to record temperatures at a data rate of 5 kHz. In a later experiment employing this fuel mixture, Long *et al.* (1985) recorded time-resolved 2D images of temperature (along with 2D images of fuel concentration using vibrational Raman scattering) in a turbulent jet flame.

A different approach was taken by Pitz *et al.* (1976) who inferred the temperature from the measured Doppler width of the scattered radiation. Using a single-mode Ar^+ laser and a Fabry–Perot interferometer for measuring the spectral width of the scattered light, they measured the temperature in an H_2-air flame and found good agreement with the calculated adiabatic flame temperature ($T \cong 2300$ K). Under their conditions, Brillouin scattering had only a minor effect on the shape of the spectral profile; however even in cases where Brillouin scattering is significant (e.g., with N_2 near STP) [Sandoval and Armstrong (1976)], one can still deduce the temperature from the spectral profile. For example, Shimizu *et al.* (1986) and Voss *et al.* (1994) have proposed the use of Rayleigh scattering in conjunction with an atomic filter (using Cs or Pb vapor, for instance) for making atmospheric temperature measurements. When the laser wavelength is tuned to the center of the atomic filter transition, the atmospheric aerosol scattering is strongly absorbed by the filter. The fraction of Rayleigh scattering that is transmitted by the filter depends on the spectral width of the scattering (relative to the spectral width of the filter) and, therefore, on the gas temperature.

Velocity Measurements. The most familiar velocimetry technique employing Mie scattering is, of course, Laser Doppler Velocimetry (LDV); because of the maturity of LDV and the wealth of available books and review articles, we will not discuss this technique here. Like LDV, Particle Imaging Velocimetry (PIV; see Sec. 15.4.5 for a more detailed discussion of the PIV technique) can be applied to a wide range of flow velocities. Unlike the LDV method, however, the PIV technique can be applied to the measurement of the time-resolved velocity field; consequently, one can obtain velocity gradients and vorticity at an instant in time. Seed particles are illuminated by two collinear laser pulses which are temporally separated; the resulting Mie scattering from both laser pulses is recorded by a photographic or CCD (charge-coupled device) camera. The velocity field within the laser sheet is then deduced from the known time delay between the two pulses and from a measurement of the displacement of the particles from the first to the second laser pulse.

If the flow velocity is sufficiently large, one can determine the velocity of the gas in the probe volume by measuring the frequency (or Doppler) shift. One simple approach to determining the Doppler shift involves combining the scattered radiation with a portion of the laser beam. When the frequency shift is less than the bandwidth of the detection system, one can illuminate the photodetector with the combined beams; the frequency shift is then given by the beat frequency and is detectable with a spectrum analyzer [Yeh and Cummins (1964)]. When measuring velocities in high-speed flows, where the $\Delta \nu$ is beyond the bandwidth of the detector, a Fabry–Perot interferometer can be employed [Self (1974)]. Unfortunately, these techniques are difficult to extend to time-resolved, 2D velocity measurements. However, the Filtered Rayleigh–Mie Scattering approach does allow measurements of velocity in the plane of a laser sheet. Here, one takes advantage of the spectral coincidence of a tunable narrow-linewidth laser and the absorption lines of gas-phase molecules or atoms. In supersonic flows with ice particles for scatterers, Miles et al. (1991) and Elliott et al. (1994) have employed the second harmonic of an injection-seeded (where seeding results in a narrow linewidth) Q-switched Nd : YAG laser and iodine vapor for the molecular filter. When the laser is tuned to the edge of an I_2 absorption transition, the scattered light which has not been Doppler shifted is absorbed by the iodine filter that has been placed in front of the detector. However, light that has been frequency shifted is not absorbed or is only partially absorbed by the iodine transition and can thus be detected. Of course, the Filtered Raleigh–Mie Scattering approach is not limited to supersonic flows. Komine et al. (1991) compared results using an Ar^+ laser ($\lambda_i = 514$ nm) and an injection seeded Nd : YAG laser (for time-resolved measurements) with an aerosol of oil droplets in a small table-top wind tunnel. Rather than setting the laser wavelength so that the static scattering was strong, Komine and co-workers chose λ_i such that the transmission on the unshifted light was $\sim 50\%$, thus they were able to record both positive and negative $\Delta \nu$. By using three detection systems—where each system was composed of a *Doppler-image* camera (i.e., a camera in conjunction with an I_2 cell), a reference camera, and electronics for analog division of the Doppler image by the reference image—they were able to record the three velocity components. Using a simliar approach, Lee et al. (1993) [also see Meyers and Cavone (1991)] have recorded velocities in a plane above the wing of an F/A-18 aircraft using propylene–glycol seed particles and an Ar^+ laser.

By controlling the gas mix (using N_2 along with the I_2, for example), the temperature, and the pressure within the filter, one can control the dependence of attenuation (through absorption) on frequency shift. Note that the desired characteristics of a filter used for temperature measurements are somewhat different from those used for velocity measurements; for recording temperatures the spectral width for absorption should be less than that for scattering, while for velocity measurements the linewidth for scattering should be much smaller than that for absorption. This criterion for velocity measurements is more easily met when the scatterers are ice clusters (or even larger particles) since the particle thermal velocity (due to the mass) and number density (due to the restriction on turbidity) are small. Of course, the frequency-shifted scattering will still depend on the local number density and laser irradiance, consequently for a valid measurement of velocity, one must normalize frequency-shifted scattering with a second *unfiltered* reference scattering measurement (i.e., one that is independent of the Doppler shift).

15.12.4 Summary

In this section, we have reviewed Rayleigh and Mie scattering theory and diagnostic applications. Although the intensity from Mie scattering is much larger than that from Rayleigh scattering (and thus is more easily detected), the application of Mie scattering for quantitative measurements is somewhat more difficult, since one must use seed material to visualize the flow. In general, the seed mass density should be limited to ~ 1% (to avoid altering the turbulent kinetic energy of the flow or the bulk properties of the fluid), and the seed number density should be limited such that the scattered radiation is not scattered (which generally requies that the beam extinction be < 10%). In addition, when applying Mie scattering for concentration measurements, one should use more small particles to reduce the marker shot noise. Nonetheless, because of their relative simplicity and modest requirements for equipment, Rayleigh and Mie scattering are attractive optical diagnostic techniques; as with other optical methods, they offer excellent spatial and temporal resolution and, when used properly, do not perturb the flow. Despite their simplicity, the laser-based Rayleigh and Mie scattering methods are powerful tools for the investigation of the physics and chemistry of gaseous flows, since they allow the measurement of gas properties such as species concentration, temperature, and velocity at a point, along a line, in a plane, or even in a volume.

15.13 LASER RAMAN GAS DIAGNOSTICS
Lawrence A. Kennedy

Experimental techniques based upon laser spectroscopy offer the potential for remote, *in-situ*, nonperturbing point flow measurements [Lederman (1977) and Eckbreth (1988)]. In principle, they have the capability of simultaneously achieving high spatial ($< 10^{-3}$ cm^3) and temporal (10^{-8}–10^{-6} sec) resolution [Lapp and Penny (1979) and Long (1977)].

These techniques [Bloembergen (1965)] arise from interactions between the E field of an electromagnetic light wave and the electric polarization, P. The latter can be expressed as a power series of $E(\omega_1)$ which is characterized by the linear and nonlinear susceptibilities of the medium. Raman scattering arises due to the oscillating polarization induced through the linear susceptibility. The higher order polarizations are weaker, and the ratio of succeeding polarizations is small unless very large laser intensities are employed. Due to inversion symmetry in isotropic media such as gases, there are no second-order effects. Thus, in a gas, the lowest order nonlinearities arise through the third order nonlinear susceptibility. Examples of third-order processes are Coherent Anti Stokes Scattering (CARS) and stimulated Raman gain/loss spectroscopy.

15.13.1 Spontaneous Raman Scattering

Raman scattering arises due to the inelastic collision processes between photons and the molecules of the medium [see Long (1977) for a detailed discussion]. The application of this phenomenon offers a number of diagnostic advantages [Kennedy (1978) and Eckbreth *et al.* (1978)]. With proper detectors, many species can be simultaneously measured, and only a single laser is required. The laser does not have to be tuned to the resonances of the molecules being probed, and the scattered intensities are unaffected by collisional quenching. Absolute calibration is readily accomplished. The principal drawback is that spontaneous Raman scattering is very weak, which limits its applicability.

When the incident photons strike a molecular, the photons can either gain or lose energy. The latter is termed *Stokes Scattering* and the former is termed *Anti-Stokes*. For Stokes scattering, the molecule becomes excited, whereas for Anti-Stokes scattering, the molecule is de-excited. Since few molecules participate in the scattering, the molecular energy distribution is negligibly affected. The distribution of molecular energy states is characteristic of a given species, hence the measured spectral distribution of the Raman scattering is determined solely by the incident photon wavelength and the molecule from which it scatters. If a change in the vibrational quantum number occurs, the scattering is termed *Vibrational Raman*; if no change occurs, it is termed *Rotational Raman*. Rotational Raman scattering has not been widely used since the overlap of the rotational spectra smears the various signals [Aeschliman *et al.* (1979)]. In principle, the major species can be measured simultaneously using a multichannel optical detector [Socket *et al.* (1979)]. It should be noted that Raman scattering is not dependent on intensity. Figure 15.95 is a schematic illustration of a typical spontaneous Raman facility. A laser beam of energy, Q_i, is passed through the fluid to be studied. A spectral analyzer is placed at an angle, θ, to this beam and collects the Raman scattering from the sample volume over a solid angle, Ω. While not required in principle, the incident beam is normally focused to enhance the spatial resolution. The strength of the scattered radiation, Q_r, is

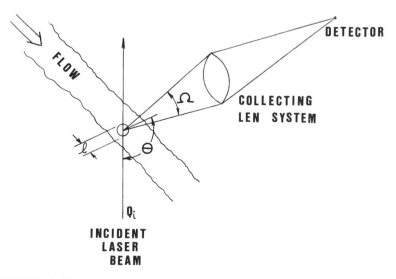

FIGURE 15.95 Schematic illustration of typical spontaneous Raman facility.

$$Q_r = f\left(Q_i,\, n,\, \Omega,\, l,\, e,\, \frac{\partial \sigma}{\partial \Omega}\right) \tag{15.231}$$

where n is the molecular number density in the appropriate energy state, e is the collection efficiency of the receiving optics, and $\partial\sigma/\partial\Omega$ is the Raman scattering cross section.

Density measurements are obtained from Q_r and are typically accurate to 5–10% depending upon the concentration level [Tolles *et al.* (1977)]. Temperature measurements can be obtained from the fact that the shape of the Raman scattering spectral distribution is proportional to the quantum energy state populations. Thus, spectal band contours [Hill *et al.* (1977)], line intensity [Drake and Rosenblatt (1976)] or band peak height ratio, [Stricker (1976)] all yield temperature information. However, the fitting of the contours is the preferable approach [Nibler *et al.* (1977)]. Another common technique is to ratio the Anti-Stokes to Stokes vibrational scattering spectra produced by a single laser pulse [Lederman *et al.* (1979)]. This approach has the advantage that it is independent of the laser power and number density. Temperature can typically be measured to within 2 to 5%.

The choice of a laser for these techniques depends mainly on the temporal resolutions desired and the background luminosity. For steady flows having little luminosity, continuous wave (CW) lasers are often employed using photon counting instrumentation [Schoenung and Mitchell (1979)]. As in conventional spectroscopy, the beam is *chopped* which allows background radiation to be subtracted. The photons are typically accumulated by employing a gated detector. A number of schemes are possible to enhance the Raman signal, such as use of double pass schemes [Lapp and Hartley (1976)], a laser cavity [Bailey *et al.* (1976)], or multipass cells [Hill *et al.* (1977)].

In turbulent flows, time average properties will be influenced by bias errors [Satchell (1977)]. If the detection bandwidth is sufficiently broad, average density measurements can be obtained. However, temperature measurements are open to errors. Quantitatively, these effects are a strong function of the magnitude of the fluctuations. For small fluctuations, the effects probably will not be serious. Species probability distribution functions (pdf) and power spectra can also be obtained [Birch *et al.* (1978)].

Pulsed lasers are typically used to make instantaneous measurements, and, in this case, the laser power is of primary importance. In turbulent flows, single-pulse measurements allow determination of the pdf which, in turn, permits average properties and the magnitude of fluctuations to be determined. When time-averaging pulsed Raman data, erroneous averages may be obtained, particularly with respect to temperature [Eckbreth (1978)]. If the cycle to cycle fluctuations are small in repetitive pulses, the averaging difficulties are not likely to be serious.

With sufficiently large laser pulse energies, Raman measurements employing a single pulse become possible. Examples in the literature include Q-switched ruby lasers [Lederman *et al.* (1979)], dye lasers [Tolles *et al.* (1977)], and $2 \times$ Nd:YAG lasers [Smith (1980)]. If the fluid density is sufficiently high, CW lasers have been employed in time resolved experiments. Real time turbulent flow may be studied using a free running, spiking ruby laser [Pealat *et al.* (1977)]. For any of these approaches to work, a statistically significant number of photons within the pulse or

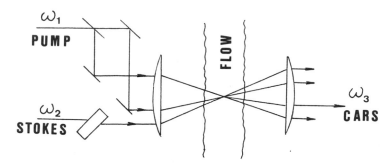

FIGURE 15.96   Schematic of a CARS facility.

temporal resolution period is required. Applications to combustion and high temperature flows are illustrated in Lapp (1980) and Drake *et al.* (1982a).

15.13.2   Coherence Anti-Stokes Raman Spectroscopy (CARS)

CARS has been the subject of a number of comprehensive reviews [e.g., Tolles *et al.* (1977)]. Figure 15.96 is a schematic of a CARS facility. Incidents laser beams at frequencies ω_1 and ω_2 (termed the *Pump* and *Stroke Beams*) are mixed and interact through the third-order, nonlinear susceptibility of the fluid. This generates a polarization field which produces coherent radiation at the frequency $\omega_3 = 2\omega_1 - \omega_2$. For efficient generation of the CARS signal, the incident beams must be aligned such that the mixing of the three waves is properly phased. This can be achieved by aligning the incident beams either parallel or colinear to each other. Colinear phase matching often has poor spatial resolution, since the CARS radiation is integrated along the optical path. Crossed beam phase matching as in Fig. 15.96, which results in CARS being generated only at the beam intersection, removes this problem [Eckbreth (1978) and Prior (1980)]. Measurements of the fluid properties are obtained from either the intensity of the CARS radiation or the shape of the spectral signature. For more details the reader is referred to Druet and Taran (1979) and Eckbreth (1978).

15.14   LASER INDUCED GAS FLUORESCENCE
Lawrence A. Kennedy

Laser Induced Fluorescence (LIF) has received much attention, since it is capable of measuring concentrations at the parts per million levels or smaller [Eckbreth (1988)]. Fluorescence measurements are similar to Raman scattering (Fig. 15.95) with the additional requirement that the laser must be *tunable*. Fluorescence is the spontaneous emission of radiation from an upper electronic state which is excited in a variety of ways. This discussion is restricted to excitation by absorption of laser radiation tuned to coincide with a molecular resonance. Two types of fluorescence are commonly experienced. The fluorescence may be at the same wavelength as the exciting wavelength (*Resonance Fluorescence*) or shifted in wavelength (generally to longer wavelengths). The former can suffer from interference arising from elastic scattering processes (Mie scattering), thus it is preferable to view the *wavelength-shifted radiation*.

In order to make fluorescence measurements, the emission spectrum of the molecule must be known and the absorption wavelength of the molecule must be able to be reached by a tunable laser. Since the fluorescence is proportional to the rate of decay, the Einstein transition probability and Einstein coefficient of the excited state must be known for quantitative measurements. Additionally, the fluorescence efficiency needs to be evaluated. In these types of measurements, the electronically excited state may also be de-excited due to collisions. This is termed *quenching* and is a process which reduces the fluorescence efficiency and for which corrections must be applied.

The basic fluorescence equation based upon a two level model which is used to relate the measured fluorescence signal to the total number density for weak excitations can be written,

$$S_f = \eta \, \frac{\Omega}{4\pi} \, V f_B \, \frac{A_{21}}{Q_{21}} \, B_{12} E_\nu n_t \tag{15.232}$$

Here, S_f is the fluorescence signal, η is the collection optics efficiency, Ω is the collection solid angle, V is the collection volume defined by the laser sheet thickness and the dimension along the beam sheet imaged onto the detector element, f_B is the Boltzmann fraction, n_t is the total number density, E_ν is the laser spectral fluence, A and B are, respectively, the Einstein transition probability and Einstein coefficient for stimulated emission and Q is the quenching rate.

A number of approaches have been used to circumvent the need to correct for quenching. Much attention has been directed towards saturated laser fluorescence [Stepowski and Cottereau (1979)]. In this approach, the intensity of the incident laser radiation is sufficiently large so that the collision quenching rate is negligible compared to the absorption and stimulated emission rates. This also maximizes the fluorescence signal. Experimental difficulties connected with saturated fluorescence include spectral intensity restrictions in the laser sources. Other restrictions are discussed in Berg and Shackleford (1979) and Blackburn *et al.* (1979). Applications of these techniques have been directed towards atmospheric, aerodynamic and combustion flows [Hartley (1977) and Drake *et al.* (1982b)].

Laser induced fluorescence permits detecting the population densities of molecules or atomic species, and this information can be used in combustion applications to obtain such quantities as mole fractions, density and temperature [Hanson (1988)]. Such point measurements can be extended to multipoint measurement for fluorescence imaging, e.g., planar laser induced fluorescence, (PLIF) [Hanson (1986)]. The basic arrangement for PLIF imaging is shown in Fig. 15.97. A two-dimensional laser sheet passes through the flame zone. The light is partially absorbed by certain species and reemitted as fluorescence which is collected by an intensified camera. The two-dimensional image is interpreted to yield the properties of interest. Planar distribution of species and temperature can be obtained. It is possible to image motion in real turbulent flows and to obtain time resolved distribution of species. A review of this area is given in McManus *et al.* (1993).

In recent years, considerable progress has been accomplished in laser induced fluorescence. Continued development of laser sources and major improvements in solid state detector technology promise additional advances.

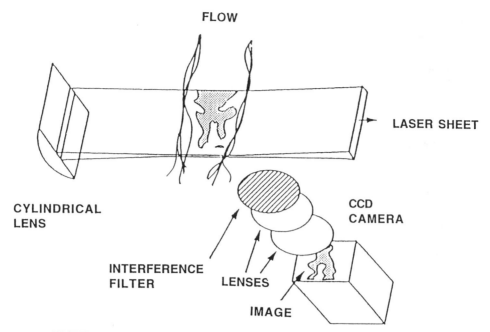

FIGURE 15.97 Schematic of planar laser induced fluorescence imaging.

15.15 USE OF COMPUTERS FOR DATA ACQUISITION AND PROCESSING
Jimmy Tan-atichat, Scott Woodward, and William K. George

15.15.1 Digital Versus Analog: Instrumentation and Processing

The design of any data acquisition and processing system involves a decision as to whether the data collected will ultimately be analyzed by analog or digital techniques. Instrumentation for measurement systems can be broadly classified as analog, digital or hybrid systems. Hybrid systems incorporate both analog and digital devices and/or signal processing techniques in an effort to combine the more desirable features inherent in each. Signals that vary in a continuous manner and can assume an infinity of values in any given range are called *analog signals*. Devices which produce or directly process such signals are called analog devices. In contrast, signals or data which vary in discrete steps and can therefore take on only a finite number of different values in a given range are known as *digital signals* (or digital data). Thus, devices which produce such signals or are used to perform desired calculations by numerical operations are called digital devices or digital instruments. Some devices are inherently digital (e.g., switches and most types of mechanical or electronic counters), while others such as ordinary mercury-in-glass thermometers, strain gages, D'Arsonval voltmeters are analog in nature. However, in many applications the instrumentation for a particular measurement system can be implemented using devices which are analog, digital or both. In general, digital instrumentation and data processing techniques can offer higher accuracy and analysis speed than is available through analog processing techniques alone. It is also very difficult to design and build analog electronic equipment which is immune to the large number of interference factors caused by ambient temperature and other environmental condition variations (e.g., electrical and electromagnetic noise, component-aging, etc.) that tend to degrade the performance and limit the accuracy of such systems to typically 0.5–1.0% (equivalent to approximately 7 bits). While the early digital instruments and measurement systems offered higher accuracy than their analog counterparts, they were also much more expensive and difficult to operate, thus limiting their use to specialists in large research laboratories. Advances in integrated circuit technology in the 1960's and early 1970's paved the way for the extensive use of digital instrumentation, signal processing, and digital control because of increased functionality, ease of use, and frequent price reductions due to competition and high manufacturing yields. Undoubtedly the two most important contributing factors for the widespread use of digital data acquisition and processing systems were the availability of algorithms for the fast computation of discrete Fourier transformations on a digital computer (*FFT* routines) and low-cost general purpose digital computers, first brought about by the mini-computer in the mid and late 1960's and then by the invention and rapid development of the MOS microprocessor/microcomputer in the early 1970's. With the low cost of microprocessor-based systems and the advent of dedicated digital signal processing (DSP) chips, analog signal processors and hard-wired digital logic instrumentation systems of the late 1960's and early 1970's genre are all but gone. The newer systems are programmed to perform more functions, give results with higher accuracy and sell at lower prices. Although logic functions implemented with software or firmwave and executed by a microprocessor are substantially slower than the hard-wired counterpart, partly due to the different semiconductor technologies involved, the speed is more than adequate in all but the most time-critical high-speed control applications. Digital systems designed with microprocessors are more versatile than hard-wired logic systems, since the logic functions can be easily altered by changing the software being executed. Such changes can be accomplished by firmwave (Read-Only-Memory or *ROM*) replacement for microprocessor-based systems, whereas new circuits would have to be designed and built to modify hard-wired digital processing equipment. As more powerful and faster microprocessors are being developed, new applications in high speed digital data acquisition and processing systems are being found. Despite these advantages of digital instrumentation and data analysis techniques, data presentation to human observers using analog displays are far superior to digital (numerical) displays. For example, it is much easier to visually comprehend the pertinent characteristics or to extract salient features of a time history of some signal (i.e., its waveform) on an oscilloscope or a graphic display than it is to comprehend the equivalent numerical time history of the same signal on a numerical readout device. Therefore, if the data has been processed digitally, a means of converting the results to an analog form for display purposes is very desirable. In control applications, it is often necessary to convert the digital information to an analog form for use as inputs. This is accomplished by digital-to-analog (D/A) converters.

15.15.2 Data Acquisition

The applications of data acquisition systems can be roughly organized into three categories:

i) Measurement—In a measurement application the process model is not known. In fact, the quantities being measured may not be well understood. Thus, measurement is the gathering of sufficient data in an attempt to construct a model of the unknown.

ii) Testing and Calibration—A testing or calibration approach to acquisition represents a situation where a device is being checked against its design standards or measurements of certain parameters are made to establish the values of other calibration parameters. Variables to be measured and requirements for accuracy and precision have been established earlier and are known prior to the tests and/or calibrations.

iii) Control—In a control application, the system initiates a series of actions, measures them, and takes corrective action if the desired results are not achieved.

General Considerations in Data Acquisition System Selection. In selecting an optimum data acquisition system, perhaps the most relevant question that needs to be answered is: "What sort of information is ultimately going to be extracted from the data acquired?" or equivalently, "What processing functions are going to be applied to the data?" These questions will be addressed in detail later. Depending on the application, data acquisition systems can range from a simple low-speed data logger to systems which include simultaneous multichannel high-speed digitizing, real-time signal analysis and graphic displays that are controlled by dedicated fast computers. Some important points that need to be considered are:

i) Overall Features and Capabilities—The selection process can be aided by providing answers to the following series of questions:
   a. How many channels are required for the proposed work?
   b. Can the number of channels be easily increased? If so, how will such expansions affect other system performance specifications such as acquisition speed, power supply capability, portability, etc.?
   c. What types of transducers will be used and what are their output ranges, output impedances, signal conditioning requirements, etc.?
   d. What peripherals or options (if any) will be required to implement a complete, operational data acquisition system with the desired capabilities?
   e. What performance boosting options such as enhanced graphics or floating-point processors might be desirable?

ii) Accuracy—An analysis of the application and the desired accuracy of the final result will reveal whether transducers, signal conditioners, A/D and D/A converters which may only be accurate to within a few percent are acceptable or whether state-of-the-art instrumentation modules must be used. Most requirements typically fall somewhere between the two extremes mentioned above. In others, a compromise between speed and accuracy must be made due to limitations in the present state-of-the-art. The compromise that must be made for high accuracy is the extra cost of the system components, and very often it comes at a reduction in the maximum conversion speed of the available A/D converters. High absolute accuracy (with calibrations traceable to NIST, formerly the National Bureau of Standards) may be a prime concern for some applications, while in others systems exhibiting only high relative accuracy (i.e., high precision) may be needed. In either case, temperature and long-term stability of the instrumentation system could be an important consideration.

iii) Acquisition Speed—The key to optimal acquisition of data which can be defined as technically correct for the data processing anticipated and the most economic in terms of data storage and processing time often lies in the proper selection of the acquisition or the sample rate. The optimal sampling rate will be discussed in detail later.

iv) Data Storage Compatibility—Although many acquisition and processing systems can operate in a stand-alone mode, the complexity of certain data analysis procedures such as computation of discrete Fourier transforms with a large number of data points sometimes requires that the data be transported and analyzed on a more powerful computer system. On these occasions, it is important to ensure that the method (usually tape, disks, or network) used for transferring the data from one system to another be compatible, both with respect to the physical medium and to the format of the data created by the user or the operating system software. With the increasing use of computer networking, the problem of data transfer media compatibility has been partially alleviated because collected data can be transferred between computer systems through the network without using a physical transfer medium. One must still ascertain that the internal or binary representation of the data is compatible on both computers.

v) System Portability—When data acquisition and control systems are required in the field, electrical generators or battery powered inverters can provide an AC power source. However, in many applications, lightweight, battery-powered computers and data acquisition systems can be used.

vi) Total System and Operating Costs—Depending on the technical requirements and system specifications, data acquisition and control systems can cost from a few hundred to several hundred thousand dollars or more to purchase. Besides the initial capital equipment investment, other costs, direct and indirect, involved in the operation and maintenance of the system and especially in programming it should be considered as well. For highly specialized applications, system reliability and the level of technical support available could also play an important role in the selection of a data acquisition system.

General Considerations in Analog to Digital Conversion. An Analog-to-Digital (A/D) converter provides a series of numbers, each of which corresponds to the weighted integral of the analog signal over some time period, Δ_s, called the *aperture* which is less than or equal to the sampling period Δ. Several examples are shown in Fig. 15.98. Thus if $u(t)$ is the signal and u_n is the digital sample,

$$u_n = (1/\Delta) \int_{t_n - \Delta}^{t_n} u(t)\, dt + \mu \qquad (15.233)$$

μ is the *quantization noise* discussed below.

If Δ_s/Δ is much less than unity, as is often the case, a convenient continuous

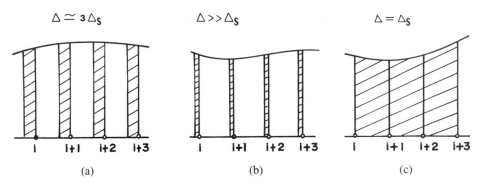

FIGURE 15.98   A/D converter output corresponds to shaded areas.

approximation to the sampled signal is

$$u_o(t) = \Delta\delta(t - n\Delta)u(t); \quad n = 1, 2, 3, \ldots \qquad (15.234)$$

where the δ represents the impulse function. It is easy to see that the time average of $u_o(t)$ is exactly the average of the numbers generated by the A/D, i.e.,

$$U_{oT} = (1/T) \int_0^T u_0(t) \, dt = (1/N) \sum_{n=1}^N u_n \qquad (15.235)$$

since the number of samples N is equal to T/Δ.

Although there are literally dozens of A/D architectures known and in use (see Table 15.11 for some representative types), two types represent over 90% of all A/D in the market. They are: 1) the shift-programmed successive approximation A/D for high-speed applications, and 2) the dual-slope integrating A/D for low speed, low cost applications. The only inherent advantage of successive-approximation A/D is speed. In all other respects, including cost and reliability, the integrating A/D is most likely to be superior. In recent years, the desirable characteristics of the integrating A/D have led to the development of multi-slope versions that extend its applications into the moderate speed range without major compromises.

Quantization Errors. Quantization noise arises from the approximation of a continuous signal by a finite set of numbers. In an ideal A/D coverter, the quantizing error

TABLE 15.11   Representative Types of Analog-to-Digital Converters

Type of A/D Converter	Application
Shift-Programmed Successive Approximation	General Purpose, High Speed
Dual-Slope Integrating	Low Cost, Low Speed
Multi-Slope Integrating	Moderate Speed
Parallel-Threshold (Flash)	Highest Possible Speed
Iterative ADC	Extremely High Speed, Less Expensive than Parallel

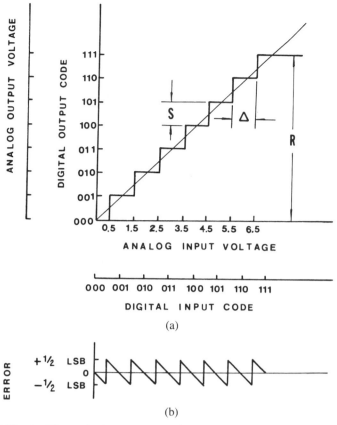

FIGURE 15.99 (a) Theoretical transfer function of 3-bit A/D and 3-bit D/A (outer set of axes) converters and (b) A/D converter quantizing error.

is the difference between the value of the analog input and the digital output due to the finite resolution of the A/D. Figure 15.99 shows the approximation of a simple ramp signal by a simple three-bit A/D converter. The quantization error (denoted as μ above) acts as a noise and is introduced by the sampling process.

There are a number of different models which can be chosen to compute the effect of the quantization noise on the statistics of the signal. In some cases, as for the ramp shown above and for periodic signals, the quantization noise is periodic. For many random signals, a better approximation to the actual noise is obtained by assuming all values of μ between two adjacent levels to be equally probable. For this case, the mean square noise is $s^2/12$, where s is the resolution of the A/D, and the noise spectrum is flat with respect to the sampling frequency.

The experimenter must provide enough gain to insure that the mean square value of his signal and its spectral levels are sufficiently above the quantization noise. Sometimes it is necessary to boost the spectral level of the analog signal before digitization by some form of *prewhitening* (e.g., differentiation) to insure that interesting portions of the signal spectrum are not buried in the quantization noise. Alternately, a higher resolution A/D converter can be used to reduce the quantization noise to acceptable levels.

Clipping Errors. There exist a wide variety of ways in which an A/D converter can be configured. Important considerations include the input voltage range and polarity (unipolar/bipolar) required to generate the full scale output code and whether the input lines are to be differential or single-ended. Regardless of the configuration, *clipping errors* can arise from the fact that the A/D converter has a finite range. In Fig. 15.99 the range is $[0, R]$ and the sampled signal is shown to be clipped at those values.

Clipping can substantially affect the measurement of averaged moments. The higher the moment, the greater the possibility of error. This can be illustrated as follows. If $B(u)$ is the probability density function of the signal $u(t)$, then the nth moment of u, say $\overline{u^n}$, is given by

$$\overline{u^n} = \int_{-\infty}^{\infty} u^n B(u) \, du \tag{15.236}$$

But, the value obtained from the sampled signal is only

$$\overline{u^n_{\text{meas}}} = \int_{A_2}^{A_1} u^n B(u) \, du \tag{15.237}$$

where A_1 and A_2 represent the upper and lower clipping levels. The higher the value of n, the more the tails of $B(u)$ are weighted. Since it is the tails of the probability density which are clipped, the more the tails are weighted, the greater the error. Figure 15.100(a) illustrates this problem. Since the areas under the curves shown correspond to the first and second moments, it is clear that the latter will be seriously in error while the same clipping will have only a relatively small effect on the mean. Higher moments will be even more seriously affected than the second.

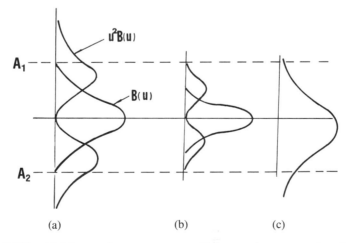

(a)   (b)   (c)

FIGURE 15.100 (a) Measured mean square will be too low, even though most of the probability density is in the data window, (b) input signal is properly scaled for measurement of mean square, and (c) gain and D.C. offset are large enough that clipping will substantially affect all moments, including the mean.

Since the clipping levels of the A/D converter are often fixed, it is usually desirable to precondition the analog signal before digitization to minimize the errors for the particular moment being measured. This can be accomplished by a judicious application of D.C. offset and gain. Figure 15.100(b) illustrates a situation where the gain has been reduced to provide an improved estimate of the second moment at the expense of increased quantization noise. Note however, that the estimate of the lower moments could be contaminated unnecessarily by the quantization noise. Figure 15.100(c) illustrates the necessity for offsetting the mean to avoid the bias which can be introduced by clipping.

It is often not possible to optimize the A/D system for measuring all statistical quantities to a sufficient degree of accuracy from a single digitized record. Since it is impossible to correct for errors unless the answers are known in advance (in which case measurement would be unnecessary), the experimenter must often digitize the signal on several channels, each channel having the appropriate preconditioning for the statistical quantity desired.

The Sampling Rate. Table 15.12 shows typical values of A/D conversion rates and how these rates depend on the accuracy required. Since the requirements for high speed, high accuracy, and low cost are mutually exclusive, the sampling rate/accuracy trade-off can be very important in the selection of an A/D converter. Like the choices outlined above for quantization and clipping problems, the correct or optimal sampling rate is very much a function of the particular experiment and of the type of processing that will be carried out on the sampled data. If only probability analysis (i.e., mean values, moments, or histograms) are desired, the optimal sampling rate is determined by the need to have a sufficient number of statistically independent data points over a sufficiently long time interval to insure statistical convergence. Thus, the problem is primarily *how slowly* (as opposed to *how rapidly*) the data must be taken for a specified period (set by statistical convergence criteria) to insure that adequate storage is available and to minimize computational time. On the other hand, if spectral analysis or time series reconstruction is to be carried out with the data, both minimum sampling rate and minimum record length criteria must be met. These conflicting criteria are usually most efficiently resolved by digitizing at two or more different rates when both time series and statistical information are desired.

Probability Analysis. The most important factor determining the optimal sampling rate for probability analysis is the requirement to have enough statistically indepen-

TABLE 15.12 Resolution versus Typical Conversion Speed A/D Converter

Resolution (Bits)	Typical Conversion Speed (μs)
8	0.03
10	0.07
12	0.10
14	0.50
16	2.0

dent samples to insure statistial convergence. Convergence (or accuracy) can be measured by the *variability* which is defined as the ratio of the *standard deviation* of the estimate for the statistical quantity to the true value of that quantity. For the *mean value*, an estimate based on N statistically independent samples has variability

$$\epsilon^2 = \frac{\text{var }\{U_N\}}{\bar{u}^2} = \frac{1}{N}\frac{\text{var }\{u\}}{\bar{u}^2} \tag{15.238}$$

where U_N is the estimate for the mean value, $\bar{u}$ is the true mean value, and var $\bar{u}$ is the true variance of the fluctuating signal. Thus, the percent error of the measurement depends on the variance of the phenomenon and the inverse square root of the number of independent samples.

The N samples in Eq. (15.238) must be statistically independent. In general, samples taken in an arbitrary manner from a continuous process will not be statistically independent, but will be correlated. The time over which a signal is correlated with itself can be measured by an appropriately defined integral scale. It can be shown that samples separated by two or more integral scales affect convergence as though they were statistically independent. Therefore, if Υ is the integral scale and Δ is the time between samples, the effective number of independent samples is

$$N_i = \begin{cases} N\Delta/(2\Upsilon), & \Delta < 2 \\ N, & \Delta \geq 2 \end{cases} \tag{15.239}$$

Thus, if the sampling rate is faster than one sample every two integral scales ($\Delta < 2\Upsilon$), the additional samples contribute nothing to convergence and only increase the memory and data storage requirements. On the other hand, if the sampling rate is slower than one sample every two integral scales, all samples are effectively independent, but the total length of time required to achieve a given accuracy is longer. The optimal sampling data is clearly equal to one sample every two integral scales. This rate will always be lower than the rate required for time series analysis (see below), although the total length of record required will be similar. Thus, if only probability analysis is required, not only can there be a considerable savings in data storage requirements, but the A/D hardware complexity is reduced, since the A/D conversion rate requirements are much less.

Time Series Analysis. An entirely different set of sampling constraints applies if spectral analysis or time reconstruction of the signals is to be carried out. The basic problem is that of spectral *folding* or *aliasing*, which arises from the fact that the Fourier transform of a digitally sampled signal is a periodic version of the Fourier transform of the original signal. In brief, the transform of the original signal is repeated at intervals corresponding to the rate at which the data are sampled. Thus if $u(f)$ is the Fourier transform of $u(t)$, the Fourier transform of the digitized signal is

$$u_o(f) = \sum_{-\infty}^{\infty} u(f - n/\Delta) \tag{15.240}$$

This is illustrated in Fig. 15.101.

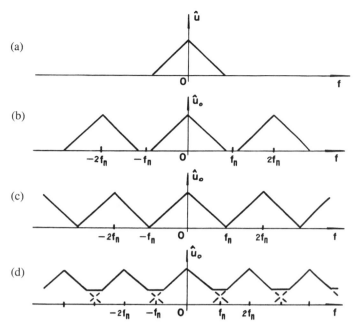

FIGURE 15.101 (a) Transform of input analog signal, (b) sample rate greater than twice the highest frequency, (c) sample rate equal to twice the highest frequency, and (d) sample rate less than twice the highest frequency. Original transform no longer recoverable.

If all signals were bandlimited [as in Fig. 15.101(a) and (b)] to frequencies less than half the sampling rate, problems would never arise. That serious problems can arise is illustrated in Fig. 15.101(d) where the highest frequency of the signal is greater than half the sampling rate. The spectra are folded into each other so that it is impossible to determine the original spectrum. If folding occurs, the data are said to be *aliased*. It is impossible to recover time series information once it is lost due to aliasing.

The folding frequency, which is half the sampling rate, is referred to as the *Nyquist frequency*, thus $f_N = 1/2\Delta$. Aliasing can be avoided only if the Nyquist frequency is greater than any frequency present in the signal, the so-called *Nyquist criterion*. Figure 15.102 illustrates what happens when a sine wave is sampled at

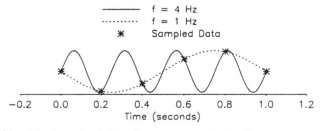

FIGURE 15.102 Aliasing of a 4 Hz sine wave to a 1 Hz sine wave by a 5 Hz sampling rate.

something less than twice its frequency. Instead of appearing at the proper frequency, say f_o, the peak is aliased to the frequency $2f_N - f_o$, which corresponds to the negative side of the first repeated version of the Fourier Transform of the original signal.

The phenomenon of aliasing does not occur if the rate at which the signal is sampled is at least twice the highest frequency. In fact, all of the information present in the original signal is also present in the digitized signal (but only if the record length is infinite!). This fact is the essence of the *Shannon Sampling Theorem* which states that the original time series can be reconstructed if the Nyquist criterion is satisified. A corollary of this result is the fact that once a signal has been sampled in a manner which introduces aliasing, the original time series cannot be recovered and the spectral information is aliased forever, regardless of how much digital processing is carried out.

It is not always convenient to sample at rates higher than twice the highest frequency, because of the need to have sufficiently long records to insure that the lowest frequencies are captured. Also, in some situations the highest frequencies are difficult to determine in advance, or may be uninteresting. In such situations, it is necessary to precondition the analog signal by analog low-pass filtering prior to sampling. Since no analog filter completely removes the high frequencies, the signal must be digitally sampled at a rate higher than twice the cutoff frequency of the filter, the exact value being determined by the roll-off of the analog filter and the signal itself.

As a final note it should be pointed out that aliasing does not affect the probability measures of a signal, only its spectral character. Therefore, the Nyquist criterion should not be applied if only probability measures are desired, since the only effect will be to greatly increase the amount of data which must be stored with no improvement in statistical accuracy.

Simultaneous Sample and Hold. Almost all A/D converters employ a sample and hold circuit to latch the voltage which is to be converted. The sample-and-hold amplifier is a circuit that can be digitally commanded to operate in two sequential modes: 1) the sample mode, and 2) the hold mode. Ideally, the output is an exact reproduction of the input in the same mode. In the hold mode, the output is *frozen* and maintained without decay or drift, regardless of the length of time or the amount of external influences. To reduce hardware cost and complexity, the various analog inputs are often multiplexed into a single sample-and-hold amplifier at a rate determined by the user and/or the device clock. [A *multiplexer* (MUX) is a circuit designed to connect one of a number of input signal paths (*channels*) to an output load, switching from channel to channel in an arbitrary, consecutive, or changing sequence, in accordance with a digitally coded *channel address* instruction from the system program control source. Most MUX modules contain internal decoding circuitry to convert the binary address input to one-in-eight, one-in-sixteen, or whatever else is required to switch on one channel at a time.] The rate at which each channel of data is digitized is the rate at which the A/D converter is operating divided by the number of channels multiplexed into it. Thus, if the number of channels is high, the required conversion rate can be quite high to achieve even modest Nyquist frequencies on the individual channels.

A consequence of the multiplexing of analog data is that all channels of data are not digitized simultaneously, but rather in sequence. These time delays (or *phase lags*) between channels can have a serious effect on both probability and time series analyses. For example, if the sample interval is Δ, the time between the conversion of channel m and channel n is $(m - n)\Delta$. Thus, the cross-covariance between the two channels is not $\overline{u^2}$, but rather $\overline{u(t)u(t + (m - n)\Delta)}$ which can be substantially different depending on the nature of the cross-correlation $\overline{u(t)u(t + \tau)}$ and the lag time $(m - n)\Delta$. As is shown below, the correct cross-moments can be found by first carrying out a spectral analysis of the data.

This phase lag also has an effect on spectral analysis. The cross spectrum can be shown to be given by

$$S_{\text{meas}}^{mn}(f) = S_{\text{true}}^{mn}(f) \exp\left(-i(m - n)2\pi f \Delta\right) \tag{15.241}$$

From Eq. 15.241, it is clear that the correct cross-spectrum can be obtained from the measured cross-spectrum by simply multiplying the latter by the complex conjugate of the frequency dependent exponential. Since the cross-correlation is the inverse Fourier transform of the cross-spectrum, the true cross-correlation can be obtained by making the phase correction before transforming. If this is done, then the true cross moments can be evaluated, either from the cross-correlation at zero-lag or from the integral of the cross-spectrum.

An alternative to these correction procedures is to avoid the problem altogether. One method is to introduce multiple sample-and-hold amplifiers before the multiplexer so that all channels are latched simultaneously before being read sequentially by the A/D. Another is to simply provide each channel with its own A/D converter and synchronize their operation. When used in conjunction with dedicated data buffers, extremely high acquisition rates can be achieved in this manner since no multiplexing is required.

Digital-to-Analog Conversion. Sometimes it is desirable to reconstruct an analog signal from digital samples using a D/A converter. Digital-to-analog converters are devices that accept digital signals (coded sets of bits) at their input terminals, and generate an analog current or voltage output that is exactly representative of the input code, if each bit were perfectly weighted. Figure 15.99 shows the theoretical transfer function for the three least-significant bits of a binary input code. Note that these three bits yield all the output values for the eight possible binary input codes from 000 to 111. The transfer function for an 8-bit D/A converter would have 256 steps and a 12-bit converter would have 4,096.

An analog signal is best generated by using an approximation to the *Whittaker interpolation formula* on the digital data before D/A conversion. This formula is given by

$$w(t) = \sin\left(\pi t/\Delta\right)/(\pi t/\Delta) \tag{15.242}$$

Other digital low-pass filters can also be used if their cutoff frequency is sufficiently below the sampling frequency so that a reasonably smooth output is produced. Often the digital samples are fed directly to a D/A whose output is analog low-pass filtered to smooth it. If this is done, the number of samples per unit time fed to the D/A should be 5–10 times higher than the cutoff of the filter to insure a reasonable reproduction of the original analog signal.

15.15.3 Processing of Digital Data

Data Validation. There are a variety of errors which can be introduced into digitized data by random hardware and transmission errors. These may involve only a single bit or may represent a more serious loss of data. It is important before investing a substantial effort in data processing that the data be subjected to some preliminary screening. If possible, the data should be plotted out as a time series so the user can inspect it visually, since few computer algorithms are as effective at spotting data problems as the human eye-mind combination of a skilled investigator.

Probability analysis can also provide useful clues as to the validity of data. In particular, the probability density function can be especially useful in spotting *wild points* and clipping. Statistical criteria can often be utilized to remove or replace bad data. The exact treatment depends on the nature of the errors, the signal itself, and the type of analysis to be carried out.

Computation of Moments. Assuming that a sufficient number of data are available to insure statistical convergence, the moments can be computed in a straightforward manner. For example, the mean and variance are computed from

$$u_N = \frac{1}{N} \sum_{i=1}^{N} u_i \tag{15.243}$$

and

$$\text{var } \{\bar{u}_N\} = \frac{1}{N-1} \sum_{i=1}^{N} (u_i - u_N)^2 \tag{15.244}$$

The factor of $(N-1)$ in the last expression is preferable to N, since it can be shown to have less variability. The primary problem which can occur is that the summation becomes so large that succeeding data make no contribution, since their scaled magnitude is less than the lowest significant bit carried by the machine in the summation. This problem is called *underflow* and can be very important when computing statistical quantities from large blocks of data. As a preventive measure, it is a good practice to use at least double precision variables for summation variables.

Computation of Probability Density Functions. The estimation of marginal or joint *probability density functions* from digital data requires only an algorithm to sort the data into appropriately defined boxes. The boxes must be sufficiently small that the shape of the measured density is not distorted by the curvature of $B(u)$ across the box. The variability of such estimators is inversely proportional to the expected number of independent occurrences in each box. Samples may be assumed independent if separated by at least two integral scales in time.

For example, if $B(u)$ is the marginal probability density and Δu is the box width at u, the expected number of occurrences in this box is $N \cdot B(u) \cdot \Delta u$ where N is the total number of samples in the ensemble. Similarly for the joint probability density function $B(u, v)$, the expected number of occurrences in a particular box is $N \cdot B(u, v) \cdot \Delta u \cdot \Delta v$. (If the sampling rate is faster than one sample every two integral scales, this number must be reduced by a factor $2\Upsilon/\Delta t$ so that it corresponds to the number of independent samples in the box.) Thus, extremely long records are required to accurately estimate values of the probability density function in the far tails which can be very small. Obviously, the situation can be relieved somewhat by increasing the size of the boxes in the tails, but at the risk of increased bias due to curvature effects.

The Autocorrelation and Cross-Correlation. Direct computation of the auto- and cross-correlations from digitized signals can easily be implemented on a digital computer by simply cross-multiplying the data strings with the appropriate time lag and then averaging. Thus

$$C_N^{uv}(\tau) = \frac{1}{N} \sum_{n=1}^{N-m} u_n v_{n+m} \tag{15.245}$$

where $\tau = m\Delta t$. Note that the sum is divided by N instead of $N - m$, since this provides the least variability in the estimator. As always, care must be taken to avoid underflow.

The statistical error in the estimate for the correlation is

$$\epsilon^2 = \frac{\text{var } \{C_N(\tau)\}}{C^2(\tau)} \tag{15.246}$$

$$= \frac{2\Upsilon}{T} \frac{C(0)^2}{C(\tau)} \tag{15.247}$$

where $T = N\Delta t$ is the record length and Υ is the integral scale for the process. Note that since $C(\tau)/C(0) \to 0$ as $\tau \to 0$ because of stationarity, the variability becomes unbounded. Thus, the smaller the correlation, the larger the record length must be relative to the integral scale in order to achieve a specified accuracy. Note that the variability is not dependent on the A/D conversion rate, $1/\Delta t$, therefore Δt can be chosen for a convenient estimation of the correlation at the desired lags, as long as the spectrum is not to be computed from the result. (If the spectrum is to be computed, the Nyquist criteria must be satisfied.) Alternately, computations can be minimized by computing only those lags desired (as opposed to all those possible).

A considerable savings in computational time can be achieved if only pairs of points separated by two integral scales are utilized. This is because pairs closer together are not statistically independent and, therefore, do not contribute to reduce the variability.

An alternate means by which the correlation can be estimated is to first compute the spectrum by means of a Fast Fourier Transform and then inverse transform it to get the correlation. This method is sometimes referred to as the *direct method* and will be discussed below under *Spectral Analysis*.

Fourier Analysis. The digital Fourier transform is defined as

$$\tilde{u}_k = \sum_{n=1}^{N} u_n e^{-i2\pi nk/N} \tag{15.248}$$

This can readily be shown to be the counterpart of the conventional continuous transform

$$\bar{u}(f) = \int_{-\infty}^{\infty} u(t)e^{-i2\pi ft} \, dt \qquad (15.249)$$

if the frequency is taken equal to the product of the index, k, and the bandwidth, $B = 1/T$, where T is the record length. Note that only $N/2$ frequencies are generated, since both real and imaginary parts must be determined from the N data points.

In many situations, there is a considerable advantage to implementing Eq. (15.248) by the so-called *Fast Fourier Transform* (FFT). The FFT can be implemented in a variety of ways, all of which consist of shuffling the data into bit-reversed order and introducing the appropriate phase factors. The savings in computational time can be substantial, since only $N \cdot \ln(N)$ operations are required for an N-point transform as opposed to N^2 operations for the conventional implementation. Which particular FFT algorithm is chosen is very much a function of the particular machine on which it is implemented and the speed and accuracy desired.

Since the Fourier transform of even a real signal is in general complex, it would appear that complex computations must be carried out. This is not the case, since a number of algorithms exist which treat the real and imaginary parts separately using only real variables. In fact, it is possible to utilize the complex nature of the transform to simultaneously transform two signals which have been interwoven; i.e., u_1, v_1, u_2, v_2, etc.

Spectral Analysis. There are two methods by which an estimator for the spectrum can be generated. The first of these requires computing the correlation function first, then Fourier transforming it to obtain the spectrum. This method follows directly from the fact that the correlation and spectrum are a Fourier transform pair; that is,

$$S(f) = \int_{-\infty}^{\infty} C(\tau)e^{-i2\pi f\tau} \, d\tau \qquad (15.250)$$

and

$$C(\tau) = \int_{-\infty}^{\infty} S(f)e^{+i2\pi f\tau} \, df \qquad (15.251)$$

It is usually more efficient on a digital computer to utilize an *FFT* algorithm directly on the time series data, and then compute the spectrum from the resulting Fourier coefficients. A *direct estimator* for the spectrum is given by

$$S_T(f) = \frac{|\bar{u}(f)|^2}{T} \qquad (15.252)$$

where T is the record length. The frequencies are given in multiples of the bandwidth, $B = 1/T$ (i.e., $f = B, 2B, \ldots, NB/2$), the index corresponding to the index in the transform vector. Note that even though the spectrum is real, it is determined by both the real and imaginary parts of the Fourier transform.

If the signal which is being analyzed is a realization of a random signal, its Fourier transform and the resulting spectrum will also be random. It can be shown that the variability of a spectral estimator is given by

$$\epsilon^2 = \frac{\text{var } \{S_T(f)\}}{S^2(f)} = \frac{1}{BT} \tag{15.253}$$

where $S(f)$ is the true spectrum, B is the bandwidth and T is the length of the record from which the spectral estimate is made. Since the bandwidth is $B = 1/T$, the r.m.s. fluctuation of the spectral estimate is equal to the spectrum itself. This is not acceptable, since every realization of the signal can be expected to yield spectra which are completely different. Therefore, some means of averaging or smoothing the spectral data must be utilized.

The simplest means of reducing the fluctuations in the spectrum is to simply average the spectral estimates obtained from independent data records, the variability being reduced by the inverse of the number of estimates utilized. If these blocks of data are obtained successively by subdividing a record into m-subrecords, the reduction in the variability of the estimator can be shown to result from the increased bandwidth, since $B = 1/T_m = m/T$. This procedure is usually referred to as *block averaging*.

An alternate method of reducing the spectral variability to acceptable levels is by averaging adjacent frequency estimates, thereby increasing the effective bandwidth of the estimate and reducing the number of independent frequencies for which estimates are made. This operation corresponds to convolving the original spectral estimate with a filter window, the window being determined by the weighting used when summing adjacent coefficients. In continuous variables

$$S_{fil}(f) = \int_{-\infty}^{+\infty} S(f_1)W(f - f_1) \, df_1 \tag{15.254}$$

where $W(f)$ is the frequency space representation of the window.

Several commonly used filter windows and their Fourier transforms are shown in Figs. 15.103 and 15.104, respectively. It is easy to show that the block averaging method employed above is equivalent to convolving the spectrum with a *Bartlett (or triangular) window*. Note that several of the windows have negative side lobes which can yield physically unrealistic negative spectral estimates. These side lobes can lead to a problem called *leakage* which occurs when the estimate at a given frequency is determined primarily by the spectrum at different frequencies which contain more energy and have been sampled by the side lobes. Leakage can be avoided by using windows which are sufficiently narrow or have side lobes which roll off more rapidly than the spectrum itself.

Convolution in frequency space corresponds to multiplication in time space and *vice versa*. This can be used to facilitate the smoothing of spectra. First, the spectrum is inverse-transformed (by an *FFT* for example) to yield the correlation function. This correlation function is multiplied point-by-point by the appropriate time window, then the result is transformed to yield the smoothed spectra.

It should be noted that even though low frequency data may not be desired, the record length must always be much greater than the integral scale for the process,

Lag Window Name	Lag Window Formula
Rectangular	$D_R(\tau) = \begin{cases} 1 & ,|\tau| \le M \\ 0 & ,|\tau| > M \end{cases}$
Bartlett	$D_B(\tau) = \begin{cases} 1 - \frac{|\tau|}{M} & ,|\tau| \le M \\ 0 & ,|\tau| > M \end{cases}$
Tukey	$D_T(\tau) = \begin{cases} 1/2\left\{1 + \cos\frac{\pi\tau}{M}\right\} & ,|\tau| \le M \\ 0 & ,|\tau| > M \end{cases}$
Parzen	$D_P(\tau) = \begin{cases} 1 - 6\left\{\frac{\tau}{M}\right\}^2 + 6\left\{\frac{|\tau|}{M}\right\}^3 & ,|\tau| \le \frac{M}{2} \\ 2\left\{1 - \frac{|\tau|}{M}\right\}^3 & ,\frac{M}{2} \le |\tau| \le M \\ 0 & ,|\tau| > M \end{cases}$

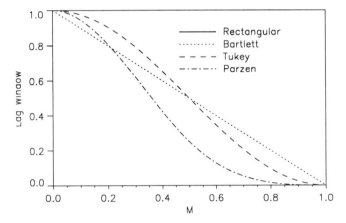

FIGURE 15.103 Lag windows commonly in use.

unless the time the process is correlated with itself is reduced by high-pass filtering. This is because the expected value of the spectral estimate (for the direct case) is really

$$S_T(f) = \int_{-T/2}^{T/2} C(\tau)e^{-i2\pi f\tau}\left\{1 - \frac{|\tau|}{T}\right\} d\tau \tag{15.255}$$

whereas what we desired is given by Eq. (15.250). The finite limits in Eq. (15.255) are responsible for the $1/T$ bandwidth discussed earlier. It is clear that the estimate converges to the desired result only if the factor $(1 - |\tau|/T)$ is approximately unity over the range for which $C(\tau)$ is non-zero. If not, spectral leakage will be introduced by the *default* window, the record length. Similar considerations can be shown to apply for the indirect estimator as well.

Cross Spectra. There are fundamentally no differences between computing spectra and cross-spectra by either method outlined above. If the direct method is used, the transform of each channel is first computed, then the cross-spectrum is obtained as

$$S_T^{uv}(f) = \frac{\bar{u}(f)\bar{v}^*(f)}{T} \tag{15.256}$$

where * denotes the complex conjugate.

Lag Window Name	Spectral Window Formula	Effective Bandwidth ($\Delta\omega$)
Rectangular	$W_R(\omega) = \frac{M}{\pi}\left\{\frac{\sin\omega M}{\omega M}\right\}$	$\frac{\pi}{M}$
Bartlett	$W_B(\omega) = \frac{M}{2\pi}\left\{\frac{\sin\omega M/2}{\omega M/2}\right\}^2$	$\frac{3\pi}{M}$
Tukey	$W_T(\omega) = \frac{M}{2\pi}\left\{\frac{\sin\omega M}{\omega M}\right\}\left[\frac{1}{1-(\omega M/\pi)^2}\right]$	$\frac{8\pi}{3M}$
Parzen	$W_P(\omega) = \frac{3M}{8\pi}\left\{\frac{\sin\omega M/4}{\omega M/4}\right\}^4$	$\frac{3.77\pi}{M}$

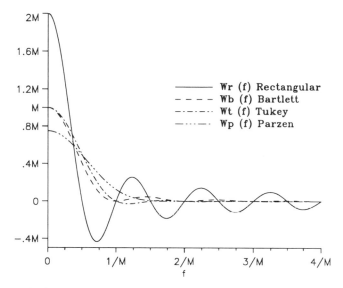

FIGURE 15.104 Spectral windows commonly in use.

As before, these cross-spectra must be either block-averaged or smoothed to reduce the variability. The procedure differs only in that the negative frequencies must be accounted for separately, since the cross-spectra are in general neither symmetric nor real.

The constraints on minimum record length can be more severe than for the spectrum, and a significantly longer record may be required for cross-spectra than for spectra. This is easily seen from the estimator which can be shown to reduce (for the direct case) to

$$S_T^{uv}(f) = \int_{T/2}^{T/2} C^{uv}(\tau)e^{-i2\pi f\tau}\left[1 - \frac{|\tau|}{T}\right]d\tau \qquad (15.257)$$

Unlike the autocorrelation, the cross-correlation does not in general peak at $\tau = 0$ and in fact may have its peak far from the origin. The record must still be long enough to insure that the factor $(1 - |\tau|/T)$ makes a negligible contribution over any region in which $C^{uv}(\tau)$ differs from zero.

To illustrate the kind of problems which can arise, consider two probes responding to the same signal, but with one delayed with respect to the other. The record must be sufficiently long to insure that there is enough overlapping information (i.e., the same detail sensed by both probes although at different times) to permit averaging long enough to reduce the variability to acceptable levels. This problem often arises with hard-wired analyzers and in small computers where the record length is limited by the available memory. In such situations, an external delay must be introduced into one input to move the peak of the cross-correlation to the origin, thereby reducing the record length requirement to approximately that for the spectrum alone. The phase introduced into the cross-spectrum by the external delay can be removed by complex multiplication, since the spectrum computed in this manner is readily shown to be given by

$$S_{\text{lag}}^{uv}(f) = e^{-i2\pi f \tau_{\text{lag}}} S^{uv}(f) \tag{15.258}$$

where τ_{lag} is the externally introduced delay.

REFERENCES

Adrian, R. J. (Ed.), *Selected Papers on Laser-Doppler Anemometry*, SPIE Milestone Series, Vol. MS 78, SPIE Optical Engineering Press, Bellingham, WA, 1993.

Adrian, R. J. and Yao, C. S., "Pulsed Laser Technique Application to Liquid and Gaseous Flows and the Scattering Power of Seed Materials," *App. Opt.*, Vol. 24, pp. 44–52, 1985.

Adrian, R. J., Johnson, R. E., Jones, B. G., Merati, P., and Tung, T. C., "Aerodynamic Disturbances of Hot-Wire Probes and Directional Sensitivity," *J. Phys. E: Sci. Instrum.*, Vol. 17, pp. 62–71, 1984.

Adrian, R. J., "Image Shifting Technique to Resolve Directional Ambiguity in Double Pulsed Velocimetry," *Appl. Opt.*, Vol. 25, pp. 3855–3858, 1986.

Adrian, R. J., "Particle Imaging Techniques for Experimental Fluid Mechanics," *Ann. Rev. Fluid Mech.*, Vol. 23, pp. 261–304, 1991.

Aeschliman, D. P., Cummings, J. C., and Hill, R. A., "Raman Spectroscopy Study of a Laminar Hydrogen Diffusion Flame in Air," *J. Quant. Spect. Rad. Trans.*, Vol. 21, p. 293, 1979.

Allen, J. M., "Improved Sensing Element for Skin-Friction Balance Measurement," *AIAA J.*, Vol. 18, No. 11, pp. 1342–1345, 1980.

Allen, J. M., "Systematic Study of Error Sources in Supersonic Skin-Friction Balance Measurement," NASA TN D-8291, 1976.

Alms, G. R., Burnham, A. K., and Flygare, W. H., "Measurement of the Dispersion in Polarizability Anisotropies," *J. Chem. Phys.*, Vol. 63, pp. 3321–3326, 1975.

Antonia, R. A. and Rajagopalan, S., "Performance of Lateral Vorticity Probe in a Turbulent Wake," *Exp. Fluids*, Vol. 9, pp. 118–20, 1990.

ANSI/ASME PTC 19.1, *Instruments and Apparatus, Part 1 Measurement Uncertainty*, Amer. Soc. Mech. Engr., New York, 1985.

ASME, *Fluid Meters: Their Theory and Application*, ASME, New York, 1927.

ASTM Standard E457-72, "Standard Method for Measuring Heat-Transfer Rate Using a Thermal Capacitance (Slug) Calorimeter," *Annual Book of ASTM Standards*, Vol. 15.03, pp. 299–303, 1988a.

ASTM Standard E598-77, "Standard Method for Measuring Extreme Heat Transfer Rates From High-Energy Environments Using a Transient Null-Point Calorimeter," *Annual Book of ASTM Standards*, Vol. 15.03, pp. 381–387, 1988b.

ASTM, *Manual on the Use of Thermocouples in Temperature Measurement*, 4th ed., MNL 12, Philadelphia, 1993.

Avallone, E. A. and Baumeister, T., III, *Marks' Standard Handbook for Mechanical Engineers*, 9th ed., McGraw-Hill, New York, 1987.

Bailly, R., Pealat, M., and Taran, J. P. E., "Raman Investigation of a Subsonic Jet," *Opt. Comm.*, Vol. 17, p. 68, 1976.

Baker, K. I. and Diller, T. E., "Unsteady Surface Heat Flux and Temperature Measurements," ASME Paper No. 93-HT-33, 1993.

Balint, J-L., Vukoslavčević, P., and Wallace, J. M., "A Study of the Vortical Structure of the Turbulent Boundary Layer," *Advances in Turbulence, First European Turbulence Conference 1986*, Compte-Bellot, G. and Mathieu, J. (Eds.), pp. 456–64, Springer, Berlin/Heidelberg/New York, 1987.

Balint, J-L., Wallace, J. M., and Vukoslavčević, P., "The Velocity and Vorticity Vector Fields of a Turbulent Boundary Layer. Part 2. Statistical Properties," *J. Fluid Mech.*, Vol. 228, pp. 53–86, 1991.

Barat, R. B., Longwell, J. P., Sarofim, A. F., Smith, S. P., and Bar-Ziv, E., "Laser Rayleigh Scattering for Flame Thermometry in a Toroidal Jet Stirred Combustor," *Appl. Opt.*, Vol. 30, pp. 3003–3010, 1991.

Beams, J. W., "Shadow and Schlieren Methods," *Physical Measurements in Gas Dynamics and Combustion*, Princeton University Press, Princeton, NJ, pp. 26–47, 1954.

Becker, H. A., Hottel, H. C., and Williams, G. C., "On the Light-Scatter Technique for the Study of Turbulence and Mixing," *J. Fluid Mech.*, Vol. 30, pp. 259–284, 1967.

Beckman, R. J. and Cook, R. D., "Outlier," *Techometrics*, Vol. 25, No. 2, pp. 119–163, 1983.

Behar, M. F., "Pressure and Vacuum," *Handbook of Measurement and Control*, Behar, M. F. (Ed.), Instruments Publishing, Pittsburgh, PA, 1951.

Behr, M. and Giachino, J., "A Miniature Pressure Sensor for Automotive Applications," IEEE Third Int. Conf. on Automotive Electronics, London, 1981.

Benedict, T. P., *Fundamentals of Temperature, Pressure, and Flow Measurement*, 3rd ed., John Wiley & Sons, New York, 1984.

Benson, J. M., "Thermal Conductivity Vacuum Gages," *Instr. and Cont. Sys.*, Vol. 36, pp. 98–101, 1963.

Bentley, C. C. and Walter, J. J., "Transient Pressure Methods Research," Princeton University, Aero Engineering Department, Rept. 595g, 1963.

Berg, J. O. and Schackleford, W. L., "Rotational Redistribution Effect on Saturated Laser Induced Fluorescence," *Appl. Opt.*, Vol. 18, p. 2093, 1979.

Bestion, D., Gaviglio, J., and Bonnet, J., "Comparison Between Constant-Current and Constant-Temperature Hot-Wire Anemometers in High Speed Flows," *Rev. Sci. Instr.*, Vol. 54, 1983.

Birch, A. D., Brown, D. R., Dodson, M. S., and Thomas, J. R., "The Turbulent Concentration Field of a Methane Jet," *J. Fluid Mech.*, Vol. 88, p. 431, 1978.

Blackburn, M. B., Mermet, J. M., Bautilier, J. D., and Winefordner, J. D., "Saturation in Laser Excited Atomic Fluorescence Spectrometry: Experimental Verification," *Appl. Opt.*, Vol. 18, p. 1804, 1979.

Blackwelder, R. F., "Hot-Wire and Hot-Film Anemometry," *Methods of Experimental Physics*, Vol. 18, Part A, Academic Press, New York, p. 259.

Blake, W. K., "Differential Pressure Measurements," *Fluid Mechanics Measurements*, Goldstein, R. J. (Ed.), Hemisphere Publishing, Washington, D.C., 1983.

Bloembergen, N., *Nonlinear Optics*, Benjamin, 1965.

Bogaard, M. P., Buckingham, A. D., Pierens, R. K., and White, A. H., "Rayleigh Scattering Depolarization Ratio and Molecular Polarizability Anisotropy for Gases," *J. Chem. Soc. Faraday Trans. I*, Vol. 74, pp. 3008–3015, 1978.

Bohren, C. F. and Huffman, D. R., *Absorption and Scattering of Light by Small Particles*, John Wiley & Sons, New York, 1983.

Born, M. and Wolf, E., *Principles of Optics*, 6th ed., Pergamon Press, Oxford, 1980.

Bowersox, R. and Schetz, J., "Measurements of Compressible Turbulence in a High-Speed High Reynolds Number Mixing Layer," *AIAA J.*, Vol. 32, No. 5, 1994.

Bowersox, R. D. W. and Schetz, J. A., "Compressible Turbulence Measurements in a High-Speed High Reynolds Number Mixing Layer," *AIAA J.*, Vol. 32, No. 4, 1994a.

Bowersox, R. D. W. and Schetz, J. A., "Measurements of Compressible Turbulent Flow Structure in a High-Speed High Reynolds Number Mixing Layer," AIAA-94-0819, 1994b.

Bowersox, R. D. W. and Schetz, J. A., "Skin Friction Gages for High Enthalpy Impulsive Flows," AIAA 93-5079, 1993.

Bowersox, R. D. W., Ng, W. F., and Schetz, J. A., "Hot-Wire Techniques Evaluated in the Wake of 2-D Supersonic Compressor Cascade," Yokohama International Gas Turbine Conference, IGTC-84, 1991, pp. 303–309.

Bowersox, R. D. W., Schetz, J. A., Chadwick, K., and Deiwert, S., "Technique for Direct Measurement of Skin Friction in High Enthalpy Impulsive Scramjet Experiments," *AIAA J.*, Vol. 33, No. 7, pp. 1286–1291, 1995.

Bowersox, R. D. W., *Compressible Turbulent in a High-Speed High Reynolds Number Mixing Layer*, Ph.D. dissertation, Virginia Polytechnic Institute and State University, Blacksburg, VA, 1992.

Bradshaw, P. and Goodman, D. G., "The Effect of Turbulence on Static-Pressure Tubes," A.R.C. R. and M. 3527, 1966.

Bradshaw, P., *An Introduction to Turbulence and its Measurement*, Pergamon, Oxford/New York/Toronto/Sydney/Braunschweig, 1975.

Bridge, N. J. and Buckingham, A. D., "The Polarization of Laser Light Scattered by Gases," *Proc. Roy. Soc. (Lond.) Ser. A.*, Vol. 295, pp. 334–349, 1966.

Brown, J. H. and Good, R. E., "A Probe for Low-Speed Measurements in Low-Density Flows," *AIAA J.*, Vol. 20, No. 1, pp. 156–158, 1982.

Bruno, J. R. and Risher, D. B., "Balance for Measuring Skin Friction in the Presence of Ablation," NOLTR 68-163, U.S. Naval Ordnance Lab, White Oak, MD, 1968.

Buchhave, P., George, W. K., Jr., and Lumley, J. L., "The Measurement of Turbulence with the Laser-Doppler Anemometer," *Ann. Rev. Fluid Mech.*, Vol. 11, pp. 443–503, 1979.

Buck, G. M., "Surface Temperature/Heat Transfer Measurement Using a Quantitative Phosphor Thermography System," AIAA 91-0064, 1991.

Burch, J. M. and Tokarski, J. M., "Production of Multiple Beam Fringes from Photographic Scatters," *Optica Acta*, Vol. 15, pp. 101–111, 1968.

Buzzard, R. D., "Description of Three-Dimensional Schlieren System," *High-Speed Photography*, Proc. 8th Int. Cong. High Speed Photography, Stockholm, June 23–29, 1968.

Camci, C., Kim, K., Hippensteele, S. A., and Poinsatte, P. E., "Evaluation of a Hue Capturing Based Transient Liquid Crystal Method for High-Resolution Mapping of Convective Heat Transfer on Curved Surfaces," *J. Heat Transfer*, Vol. 115, pp. 311–318, 1993.

Capone, F. J., "Wind Tunnel Tests of Seven Static-Pressure Probes at Transonic Speeds," NASA TN D-947, 1961.

Cebeci, T. and Smith, A. M. O., *Analysis of Turbulent Boundary Layers*, Academic Press, New York, 1974.

Chadwick, K. M., DeTurris, D. J., and Schetz, J. A., "Direct Measurements of Skin Friction in Supersonic Combustion Flowfields," *J. Eng. Gas Turbines and Power*, Vol. 115, No. 3, pp. 507–514, 1993.

Champagne, F. H., Sleicher, C. A., and Wehrmann, O. H., "Turbulence Measurements with Inclined Hot-Wires, Part 1. Heat Transfer Experiments with Inclined Hot-Wire Response Equations,'' *J. Fluid Mech.*, Vol. 28, Part 1, pp. 153–182, 1967.

Chimonas, G., ''Reynolds Stress Deflections of the Bi-Vane Anemometer,'' *J. Appl. Meteorol.*, Vol. 19, pp. 329–333, 1980.

Chue, S. H., ''Pressure Probes for Fluid Measurements,'' *Prog. Aerospace Sci.*, Vol. 16, No. 2, pp. 147–223, 1975.

Chyu, M. K. and Bizzak, D. J., ''Surface Temperature Measurement Using a Laser-Induced Fluorescence Thermal Imaging System,'' *J. Heat Transfer*, Vol. 116, pp. 263–266, 1994.

Clark, R., Ng, W., Rettew, A., Walker, D., and Schetz, J., ''Turbulence Measurements in a High-Speed Shear Flow Using a Dual-Wire Probe,'' AIAA-88-3055A.

Clay, W. G., Herrmann, J., and Slattery, R. E., ''Statistial Properties of the Turbulent Wake Behind Hypervelocity Spheres,'' *Phy. Fluids*, Vol. 8, No. 10, pp. 1792–1801, 1965.

Clemens, N. T. and Mungal, M. G., ''A Planar Mie Scattering Technique for Visualizing Supersonic Mixing Flows,'' *Exp. Fluids*, Vol. 11, pp. 175–185, 1991.

Clemens, N. T. and Mungal, M. G., ''Two- and Three-Dimensional Effects in the Supersonic Mixing Layer,'' *AIAA J.*, Vol. 30, pp. 973–981, 1992.

Collis, D. C. and Williams, M. J., ''Two-Dimensional Convection from Heated Wires at Low Reynolds Numbers,'' *J. Fluid Mech.*, Vol. 51, Part 3, p. 487, 1959.

Cook, W. J. and Felderman, E. M., ''Reduction of Data From Thin Film Heat-Transfer Gages: A Concise Numerical Technique,'' *AIAA J.*, Vol. 4, pp. 561–562, 1966.

Corrsin, S. and Kistler, A. L., ''The Free-Stream Boundaries of Turbulent Flows,'' NACA TN 3133, 1954.

Corrsin, S., *Encyclopedia of Physics*, Vol. 8, Part 2, Springer-Verlag, Berlin, p. 555.

Crisler, W., Krothapalli, A., and Lourenco, L., ''PIV Investigation of High Speed Flow Over a Pitching Airfoil,'' AIAA Paper 94-0533, 1994.

D'Alessio, A., ''Laser Light Scattering and Fluorescence Diagnostics of Rich Flames Produced by Gaseous and Liquid Fuels,'' *Particulate Carbon Formation During Combustion*, Siegla, D. C. and Smith, G. W. (Eds.), Plenum, New York, 1981.

Dally, J. W., Riley, W. F., and McConnel, K. G., *Instruments for Engineering Measurements*, John Wiley & Sons, 1984.

Dantec, *Probe Catalog*, Dantec Measurement Technology, Mahwah, NJ, No. 193-108-01.

Daugherty, R. L. and Franzini, J. B., *Fluid Mechanics with Engineering Applications*, McGraw-Hill, New York, 1977.

de Luca, L., Cardone, G., Carlomagno, G. M., Aymer de la Chevalerie, D., and Alziary de Roquefort, T., ''Flow Visualization and Heat Transfer Measurement in a Hypersonic Wind Tunnel,'' *Experimental Heat Transfer*, Vol. 5, pp. 65–78, 1992.

Demetriades, A. and Laderman, A., ''Reynolds Stress Measurements in a Hypersonic Boundary Layer,'' *AIAA J.*, Vol. 11, No. 11, 1973.

DeTurris, D., Schetz, J. A., and Hellbaum, R., ''Direct Measurements of Skin Friction in a Scramjet Combustor,'' AIAA 90-2342, 1990.

Dhawan, S., ''Direct Measurements of Skin Friction,'' NACA Report 1121, 1953.

Dibble, R. W. and Hollenbach, R. E., ''Laser Rayleigh Thermometry in Turbulent Flames,'' *Eighteenth Symposium (International) on Combustion*, The Combustion Institute, Pittsburgh, PA, pp. 1489–1499, 1981.

Dibble, R. W., Masri, A. R., and Bilger, R. W., ''The Spontaneous Raman Scattering Technique Applied to Nonpremixed Flames of Methane,'' *Combust. Flame*, Vol. 67, pp. 189–206, 1987.

Diller, T. E., ''Advances in Heat Flux Measurements,'' *Advances in Heat Transfer*, Vol. 23, pp. 279–368, 1993.

Döbbeling, K., Lenze, B., and Leuckel, W., "Basic Considerations Concerning the Construction and Usage of Multiple Hot-Wire Probes for Highly Turbulent Three-Dimensional Flows," *Meas. Sci. Technol.*, Vol. 1, pp. 924–933, 1990.

Doebelin, E. O., *Measurement Systems: Application and Design*, 4th ed., McGraw-Hill, New York, 1990.

Doorly, J. E. and Oldfield, M. L. G., "New Heat Transfer Gages for Use on Multilayered Substrates," *J. Turbomachinery*, Vol. 108, pp. 153–160, 1986.

Dowling, D. R. and Dimotakis, P. E., "Similarity of the Concentration Field of Gas-Phase Turbulent Jets," *J. Fluid Mech.*, Vol. 218, pp. 109–141, 1990.

Dowling, D. R., Lang, D. B., and Dimotakis, P. E., "An Improved Laser-Rayleigh Scattering Photodetection System," *Exp. Fluids*, Vol. 7, pp. 435–440, 1989.

Drake, M. C. and Rosenblatt, G. M., "Flame Temperatures for Raman Scattering," *Chem. Phys. Letters*, Vol. 94, p. 313, 1976.

Drake, M. C., Bilger, R. W., and Starner, S. H., "Raman Measurements and Conserved Scalar Modeling in Turbulent Diffusion Flames," *19th Symp. (Int.) Combust.*, The Combustion Institute, 1982b.

Drake, M. C., Lapp, M., and Penney, C. M., "Use of the Raman Effect for Gas Temperature Measurements," *Temperature, Its Measurement and Control in Science and Industry*, Vol. 5, Schooley, J. F. (Ed.), American Inst. Physics, 1982a.

Druet, S. and Taran, J. P., *Chemical and Biological Applications of Lasers*, Moore, C. E. (Ed.), Academic Press, New York, 1979.

Ducruet, C., "A Method for Correcting Wall Pressure Measurements in Subsonic Compressible Flow," *J. Fluids Eng.*, Vol. 113, pp. 256–260, 1991.

Dudzinski, T. J. and Krause, L N., "Flow-Direction Measurements with Fixed-Position Probes," NASA TM X-1904, 1969.

Durrani, T. S. and Greated, C. A., *Laser Systems in Flow Measurement*, Plenum Press, New York, 1977.

Durst, F., Melling, A., and Whitelaw, J. H., *Principles and Practice of Laser-Doppler Anemometry*, 2nd ed., Academic Press, NY, 1981.

Dushman, S., *Scientific Foundations of Vacuum Technique* 2nd ed., Lafferty, J. M. (Ed.), John Wiley & Sons, New York, 1962.

Eckbreth, A. C., *Laser Diagnostics for Combustion Temperature and Species*, Abacus Press, Cambridge, MA, 1988.

Eckbreth, A. C., "Averaging Considerations for Pulsed, Laser Raman Signals from Turbulent Combustion Media," *Combust. Flame*, Vol. 31, p. 231, 1978.

Eckelmann, H., Nychas, S. G., Brodkey, R. S., and Wallace, J. M., "Vorticity and Turbulence Production in Pattern Recognized Turbulent Flow Structures," *Phys. Fluids*, Vol. 20, pp. S225–S231, 1977.

Eckert, E. R. G. and Drake, R. M., Jr., *Analysis of Heat and Mass Transfer*, McGraw-Hill, New York, 1972.

Eckman, D. P., *Industrial Instrumentation*, John Wiley & Sons, New York, 1950.

Eggels, J. G. M., Westerweel, J., Nieuwstadt, F. T. M., and Adrian, R. J., "Comparison of Vortical Flow Structures in DNS and PIV Studies of Turbulent Pipe Flow," *Near Wall Turbulent Flows*, Elsevier Science Publishers, Amsterdam/London/New York/Tokyo, 1993.

Eisenhart, C., "Expression of the Uncertainties of Final Results," *Science*, Vol. 160, pp. 1201–1204, 1968.

Eisenhart, C., "Realistic Evaluation of the Precision and Accuracy of Instrument Calibration Systems," *J. Research of the National Bureau of Standards—Engineering and Instrumentation*, Vol. 67C, No. 2, pp. 161–187, 1963.

Elliott, G. S., Samimy, M., and Arnette, S. A., "Details of a Molecular Filter-Based Velocimetry Technique," AIAA 94-0490, 1994.

Epstein, A. H., Guenette, G. R., Norton, R. J. G., and Cao, Y., "High-Frequency Response Heat-Flux Gauge," *Rev. Sci. Instrum.*, Vol. 57, pp. 639–649, 1986.

Erf, R. K., *Speckle Metrology*, Academic Press, New York, 1978.

Escoda, M. C. and Long, M. B., "Rayleigh Scattering Measurements of the Gas Concentration Field in Turbulent Jets," *AIAA J.*, Vol. 21, pp. 81–84, 1983.

Ewan, B. C. R., "Particle Velocity Distribution Measurement by Holography," *Appl. Opt.*, Vol. 18, pp. 3156–3160, 1979.

Fabelinskii, I. L., *Molecular Scattering of Light*, Plenum Press, New York, 1968.

Farmer, E., "Making Pressure Measurements," *Instr. Cont. Sys.*, Vol. 54, No. 5, pp. 75–80, 1981.

Figliola, R. S. and Beasley, D. E., *Theory and Design for Mechanical Measurements*, John Wiley & Sons, New York, 1991.

Fischer, M., Maddalon, L., and Wagner, R., "Boundary-Layer Pitot and Hot-Wire Surveys at $M_\infty \approx 20$," *AIAA Journal*, Vol. 9, No. 5, pp. 826–834, 1971.

Folsom, R. G., "Review of the Pitot Tube," *Trans. ASME*, pp. 1447–1460, 1956.

Forkey, J. N., Lempert, W. R., Bogdonoff, S. M., Miles, R. B., and Russell, G. R., "Volumetric Imaging of Supersonic Boundary Layers Using Filtered Rayleigh Scattering Background Suppression," AIAA 94-0491, 1994.

Foss, J. F. and Haw, R. C., "Transverse Vorticity Measurements Using a Compact Array of Four Sensors," *The Heuristics of Thermal Anemometry*, ASME-FED, Vol. 97, Stock, D. E., Sherif, S. A., and Smits, A. J. (Eds.), pp. 71–71, 1990.

Foss, J. F. and Wallace, J. M., "The Measurement of Vorticity in Transitional and Fully Developed Turbulent Flows," *Advances in Fluid Mechanics Measurements*, Lecture Notes in Engineering, No. 45, Springer-Verlag, Berlin, 1989.

Foss, J. F., "Accuracy and Uncertainty of Transverse Vorticity Measurements," *Bull. Am. Phys. Soc.*, Vol. 21, p. 1237 (Abstr.), 1976.

Foss, J. F., "Advanced Techniques for Transverse Vorticity Measurements," *Proc. of Biennial Symp. on Turbulence*, 7th Symp., Rolla, MO, pp. 208–218, 1981.

Freymuth, P., "A Bibliography of Thermal Anemometry," TSI Incorporated, St. Paul, MN 55164, 1982.

Fuhs, A. E., *Instrumentation for High Speed Plasma Flow*, Gordon and Breach Science Publishers, New York, 1965.

Gardiner, W. C., Jr., Hidaka, Y., and Tanzawa, T., "Refractivity of Combustion Gases," *Combust. Flame*, Vol. 40, pp. 213–219, 1981.

Gauthier, V. and Riethmuller, M. L., "Application of PIDV to Complex Flows: Measurement of the Third Component," von Karman Institute Lecture Series 1988-06, 1988.

Gill, G. C., "The Helicoid Anemometer," *Atmosphere*, Vol. 11, pp. 145–155, 1973.

Gettleman, C. C. and Krause, L. N., "Considerations Entering into Selection of Probes for Pressure Measurement in Jet Engines," *ISA Proc.*, vol. 7, pp. 134–137, 1952.

Gracey, W., Coletti, D. E., and Russell, W. R., "Wind Tunnel Investigation of a Number of Total-Pressure Tubes at High Angles of Attack, Supersonic Speeds," NACA TN 2261, 1951a.

Gracey, W., "Wind Tunnel Investigation of a Number of Total-Pressure Tubes at High Angles of Attack, Subsonic, Transonic and Supersonic Speeds," NACA Rept. 1303, 1957.

Gracey, W., Letko, W., and Russel, W. R., "Wind Tunnel Investigation of a Number of Total-Pressure Tubes at High Angles of Attack, Subsonic Speeds," NACA TN 2331, 1951b.

Hager, J. M., Onishi, S., Langley, L. W., and Diller, T. E., "High Temperature Heat Flux Measurements," *J. Thermophysics*, Vol. 7, pp. 531–534, 1993.

Hall, J., "Demand Grows for Solid State Pressure Transducers," *Instr. Contr. Sys.*, Vol. 54, No. 4, pp. 59–62, 1981.

Hanson, R. K., "Combustion Diagnostics: Planar Imaging Techniques," *21st Symp. (Int.) Combust.*, The Combustion Institute, Pittsburgh, PA, 1986.

Hanson, R. K., "Planar Laser Induced Fluorescence," *J. Quant. Spectrosc. Radiative Trans.*, Vol. 40, pp. 343–362, 1988.

Hariharan, P., *Optical Interferometry*, Academic Press, Australia, 1985.

Hartley, D. L., "Laser Scattering Diagnostics for Temperature and Concentration Measurements," Vol. 53, *Prog. Astronautics and Aeronautics*, Finn, R. (Ed.), AIAA, 1977.

Hashemian, H. M., Kerlin, T. W., and Petersen, K. M., "New Methods for Response Time Testing of Industrial Temperature and Pressure Sensors," *Pressure and Temperature Measurement*, Kim, J. H. and Moffat, R. J. (Eds.), FED Vol. 44 and HTD Vol. 58, ASME, New York, 1986.

Hauser, R. L., "Construction and Performance of *In Situ* Heat Flux Transducers," in *Building Applications of Heat Flux Transducers*, ASTM STP 885, Bales, E. *et al.* (Eds.), ASTM, Philadelphia, PA, pp. 172–183, 1985.

Haw, R. C., Foss, J. K., and Foss, J. F., "Vorticity Based Intermittency Measurements in a Single Stream Shear Layer," *Proc. Second European Turb. Conf. Advances in Turbulence 2*, Fernholz, H. H. and Fiedler, H. E. (Eds.), Springer-Verlag, Berlin, 1989.

Hayashi, M., Aso, S., and Tau, A., "Fluctuation of Heat Transfer in Shock Wave/Turbulent Boundary-Layer Interaction," *AIAA J.*, Vol. 27, pp. 399–404, 1989.

Heflinger, L. and Wueker, R., "Holographic Interferometry," *J. Appl. Phys.*, Vol. 37, p. 642, 1966.

Hill, R. A., Mulac, A. J., and Hackett, C. E., "Retroreflecting Multipass Cell for Raman Scattering," *Appl. Optics*, Vol. 16, p. 2004, 1977.

Hinze, J. O., *Turbulence*, 2nd ed., McGraw-Hill, New York, 1975 (reissued 1987).

Hirleman, E. D., "Non-Doppler Laser Velocimetry: Single Beam Transit-Time LDV," *AIAA J.*, Vol. 20, No. 1, pp. 86–87, 1982.

Hochreiter, H. M., "Dimensionless Correlation Coefficients of Turbine-Type Flowmeters," *Trans. ASME*, p. 1363, 1958.

Holdeman, J. D. and Foss, J. F., "On the Initiation, Development and Decay of Secondary Flow in a Bounded Jet," *J. of Fluids Engr.*, Vol. 97, No. 3, Series 1, 1975.

Holder, D. W. and North, R. J., "Schlieren Methods for Observing High-Speed Airflow," *The Aeroplane*, Vol. 82, No. 2111, p. 16, Jan. 1952.

Holder, D. W. and North, R. J., "Schlieren Methods," *NPL, Notes on Applied Sciences*, No. 31, 1963.

Holder, D. W., North, R. J., and Wood, G. P., "Optical Methods for Examining the Flow in High-Speed Wind Tunnels," AGARDograph 23, 1956.

Holman, J. P., *Experimental Methods for Engineering*, McGraw-Hill, New York, pp. 191 and 253, 1978.

Holmberg, D. G., Mukkamala, Y. S., and Diller, T. E., "Shock Tunnel Evaluation of Heat Flux Sensors," AIAA 94-0730, 1994.

Horvath, T. J., "Aerothermodynamic Measurement on a Proposed Assured Crew Return Vehicle (ACRV) Lifting Body Configuration at Mach 6 and 10 in Air," AIAA 90-1774, 1990.

Howe, W. H., "The Present Status of High Pressure Measurement and Control," *ISA J.*, pp. 77–113, Mar. 1955.

Hurd, C. W., Chesky, K. P., and Shapiro, A. H., "Influence of Viscous Effects on Impact Tubes," *J. Appl. Mech.*, pp. 253–256, 1953.

Iberall, A. S., "Attenuation of Oscillatory Pressure in Instrument Lines," *Trans. ASME*, Vol. 72, pp. 689–695, 1950.

Inbody, M. A., *Measurements of Soot Formation and Hydroxyl Concentration in Near-Critical Equivalence Ratio Premixed Ethylene Flames*, Ph.D. dissertation, School of Mechanical Engineering, Purdue University, West Lafayette, IN, 1992.

Ireland, P. T. and Jones, T. V., "The Response Time of a Surface Thermometer Employing Encapsulated Thermochromic Liquid Crystals," *J. Phys. E. Sci. Instrum.*, Vol. 20, pp. 1195–1199, 1987.

ISA, *Measurement Uncertainty Handbook*, IS&N: 87664-483-3, Instrument Society of America, Pittsburgh, PA, 1980.

ISA, "Electrical Transducer Nomenclature and Terminology," Standard ISA-S 37.1, Instrument Society of America, Research Triangle Park, NC, 1969.

Jones, A. R., "Light Scattering for Particle Characterization," *Instrumentation for Flows with Combustion*, Taylor, A. M. K. P. (Ed.), Academic Press, London, 1993.

Jones, A. R., "Scattering of Electromagnetic Radiation in Particulate Laden Fluids," *Prog. Energy Combust. Sci.*, Vol. 5, pp. 73–96, 1979.

Jorgensen, F. E., "Directional Sensitivity of Wire and Fiber Film Probes," *DISA Inform.*, Vol. 11, pp. 31–37, 1971.

Kantrowitz, A. and Trimpi, R. L., "A Sharp Focusing Schlieren System," *J. Aeronaut. Sci.*, Vol. 17, pp. 311–319, 1950.

Keenan, J. H., Chao, J., and Kay, J., *Gas Tables*, 2nd ed., John Wiley & Sons, New York, 1980.

Kennedy, L. A. (Ed.), "Turbulent Combustion," Vol. 58, *Prog. Astronautics and Aeronautics*, AIAA, 1978.

Kenning, D. B. R., "Wall Temperature Patterns in Nucleate Boiling," *Int. J. Heat and Mass Transfer*, Vol. 35, pp. 73–86, 1992.

Kerker, M., Scheiner, P., and Cooke, D. D., "The Range of Validity of the Rayleigh and Thomson Limits for Lorenz-Mie Scattering," *J. Opt. Soc. Am.*, Vol. 68, pp. 135–137, 1978.

Kerker, M., *The Scattering of Light and Other Electromagnetic Radiation*, Academic Press, New York, 1969.

Kessler, T. J. and Hill, W. G., *Aeronaut. Astronaut.*, Vol. 4, No. 38, 1966.

Kidd, C. T., "Extraneous Thermoelectric EMF Effects Resulting From the Press-Fit Installation of Coaxial Thermocouples in Metal Models," in *Proc. 40th Int. Instr. Symp.*, ISA, Research Triangle Park, NC, pp. 317–335, 1994.

Kidd, C. T., "Recent Developments in High Heat-Flux Measurement Techniques at the AEDC," in *Proc. 36th Int. Instr. Sym.*, ISA, Research Triangle Park, NC, pp. 477–492, 1990.

Kim, J. H. and Fiedler, H. E., "Vorticity Measurements in a Turbulent Mixing Layer," *Advances in Turbulence 2*, 2nd European Conference, Fernholz, H. H. and Fiedler, H. E. (Eds.), Springer, Berlin/Heidelberg/New York, 1989.

Kim, J. H., Simon, T. W., and Viskanta, R., "Journal of Heat Transfer Policy on Reporting Uncertainties in Experimental Measurements and Results," *J. Heat Transfer*, Vol. 115, pp. 5–6, 1993.

Kim, J. H., *Wirbelstärkemessungen in einer turbulenten scherschict*, Doktor-Ing. thesis, Technischen Universität Berlin, Germany, 1989.

Kinder, W., "Theorie des Mach-Zehnder-Interferometer und Beschreibung eines Getates mit Einspielgeleinstellung," *Optik*, Vol. 1, pp. 413–448, 1946.

King, L., "On the Convection of Heat from Small Cylinders in a Stream of Fluid: Determination of the Convection Constants of Small Platinum Wire, with Applications to Hot-Wire Anemometry," *Proc. Roy. Soc. (London)*, England, Vol. 90, pp. 563–570, 1914.

Kinzie, P. A., *Thermocouple Temperature Measurement*, John Wiley & Sons, New York, 1973.

Kiss, T., Schetz, J., and Moses, H., "Experimental and Numerical Study of Transonic Turbine Cascade Flow," AIAA-93-3064, July 1993.

Kistler, A., "Fluctuation Measurements in a Supersonic Turbulent Boundary Layer," *Phys. Fluids*, Vol. 2, p. 220, 1959.

Kline, S. J. and McClintock, F. A., "Describing Uncertainties in Single-Sample Experiments," *Mech. Eng.*, Vol. 75, pp. 3–8, 1953.

Knöös, S., *Proc. Int. Cong. High-Speed Photogr., 8th, 1968*, p. 346, 1969.

Kock am Brink, B. and Foss, J. F., "Enhanced Mixing via Geometric Manipulation of a Splitter Plate," AIAA 3rd Shear Flow Control Conference, 1993.

Komine, H., Brosnan, S. J., Litton, A. B., and Stappaerts, E. A., "Real Time, Doppler Global Velocimetry," AIAA 91-0337, 1991.

Kovasznay, L. S. G., *Q. Prog. Rep. Aero. Dept. Contract NORD-8036-JHB-39*, The Johns Hopkins Univ., 1950.

Kovasznay, L. S. G., "Hot Wire Method," *Physical Measurements in Gas Dynamics and Combustion*, Princeton University Press, Princeton, NJ, pp. 219–276, 1954.

Kovasznay, L. S. G., "The Hot-Wire Anemometer in Supersonic Flow," *J. Aeronaut. Sci.*, Vol. 17, pp. 565–584, 1950.

Kovasznay, L. S. G., "Turbulence in Supersonic Flow," *J. Aeronaut. Sci.*, Vol. 20, No. 10, 1953.

Kovasznay, L. S. G., "Turbulence Measurements—Optical Methods," *Physical Measurements in Gas Dynamics and Combustion*, Princeton University Press, Princeton, NJ, pp. 277–285, 1954.

Kovasznay, L. S. G., "Turbulent Measurements," *High Speed Aerodynamics and Jet Propulsion*, Landenbuerg, R. W., Lewis, B., Pease, R. N., and Taylor, H. S. (Eds.), Princeton, NJ, 1954.

Kreider, K. G., "Sputtered High Temperature Thin Film Thermocouples," *J. Vac. Sci. Tech. A*, Vol. 11, pp. 1404–1405, 1993.

Kreisman, W., "Extension of the Low Pressure Limit of McLeod Gages," NASA CR-52877, 1964.

Krothapalli, A., Wishart, D., and Lourenco, L., "Near Field Structure of a Supersonic Jet: "On-Line" PIV Study," *Proc. 7th Symp. Appl. Laser Tech. Fluid Mech.*, Ladoan-Inst. Super. Tech., 1994.

Kuo, C. H. and Kulkarni, A. K., "Analysis of Heat Flux Measurement by Circular Foil Gages in a Mixed Convection/Radiation Environment," *J. Heat Transfer*, Vol. 113, pp. 1037–1040, 1991.

Ladenburg, R. and Bershader, D., "Interferometry," *Physical Measurements in Gas Dynamics and Combustion*, Princeton University Press, Princeton, NJ, pp. 277–285, 1954.

Laderman, A. and Demetriades, A., "Turbulent Shear Stresses in a Compressible Boundary Layer," *AIAA J.*, Vol. 17, No. 7, 1979.

Lafferty, J. M. and Vanderslice, T. A., "Vacuum Measurement by Ionization," *Instr. Contr. Sys.*, Vol. 36, No. 3, pp. 90–96, 1963.

Landolt–Börnstein Tablellen, Vol. II, *Eigenschaften der Materie in ihren Aggergatzustän-den, Part 8, Optische Konstanten*, Springer-Verlag, Berlin, 1962.

Landreth, C. C. and Adrian, R. J., "Electro-Optical Image Shifting for Particle Image Velocimetry," *Appl. Opt.*, Vol. 27, pp. 4216–4220, 1988.

Lapp, M. and Hartley, D. L., "Raman Scattering Studies of Combustion," *Combust. Sci. Tech.*, Vol. 13, p. 199, 1976.

Lapp, M. and Penney, C. M. (Eds.), *Laser Raman Gas Diagnostics*, Plenum Press, New York, 1974.

Lapp, M., "Raman-Scattering Measurements of Combustion Properties," *Laser Probes for Combustion Chemistry*, Crosley, D. R. (Ed.), American Chem. Soc. Symp., Vol. 134, 1980.

Lassahn, G. D., "LOFT Uncertainty Analysis Methodology," ASME paper 83-HT-107, 1983.

Laufer, J. and McClellan, R., "Measurements of Heat Transfer from Fine Wires in Supersonic Flows," *J. Fluid Mech.*, Vol. 1, p. 276, 1956.

Lawford, V. N., "Differential-Pressure Instruments: The Universal Measurement Tools," *Instr. Tech.*, Vol. 21, No. 12, pp. 30–40, 1974.

Lederman, S., Celentano, A., and Glaser, J., "Temperature, Concentration and Velocity in Jets, Flames and Shock Tubes," *Phys. Fluids*, Vol. 22, p. 1065, 1979.

Lederman, S., "The Use of Laser Raman Diagnostics in Flow Fields and Combustion," *Prog. Energy Combust. Sci.*, Vol. 3, pp. 1–34, 1977.

Lee, C. Y. and Pfeifer, J. L., "Quartz Capsule Pressure Transducer for the Automotive Industry," Technical Paper 810374, Society of Automotive Engineers, 1981.

Lee, J. W., Meyers, J. F., and Cavone, A. A., "Doppler Global Velocimetry Measurements of the Vortical Flow Above an F/A-18," AIAA 93-0414, 1993.

Lee, S. Y., "Variable Capacitance Signal Transducers and the Comparison with Other Transducer Schemes," *Fund. Aero. Instrum.*, Vol. 3, p. 1, 1970.

Lee, Y. S. and Wise, K. D., "A Batch-Fabricated Silicon Capacitive Pressure Transducer with Low Temperature Sensitivity," *IEEE Trans. Electronic Devices*, Vol. 29, No. 1, pp. 42–48, 1982.

Lekakis, I. C., Adrian, R. J., and Jones, B. G., "Measurement of Velocity Vectors With Orthogonal and Non-Orthogonal Triple-Sensor Probes," *Exp. Fluids*, Vol. 7, pp. 228–240, 1989.

Liebert, C. H., "Miniature Convection Cooled Plug-Type Heat Flux Gauges," in *Proc. 40th Int. Instr. Symp.*, ISA, Research Triangle Park, NC, pp. 289–302, 1994.

Liepmann, H. W. and Roshko, A., *Elements of Gas Dynamics*, John Wiley & Sons, New York, 1957.

Lillienfeld, P., Solon, L., and DiGiovanni, H., "Ion Tracer Anemometer for the Measurement of Low Density Flows," *Rev. Sci. Instr.*, Vol. 38, pp. 405–409, 1967.

Liu, Z. C., Landreth, C. C., Adrian, R. J., and Hanratty, T. J., "High Resolution Measurement of Turbulent Structure in a Channel With Particle Image Velocimetry," *Exp. Fluids*, Vol. 10, pp. 301–312, 1991.

Long, D. A., *Raman Spectroscopy*, McGraw-Hill Int., New York, 1977.

Long, M. B., Chu, B. T., and Chang, R. K., "Instantaneous Two-Dimensional Gas Concentration Measurements by Light Scattering," *AIAA J.*, Vol. 19, pp. 1151–1157, 1981.

Long, M. B., Levin, P. S., and Fourguette, D. C., "Simultaneous Two-Dimensional Mapping of Species Concentration and Temperature in Turbulent Flames," *Opt. Lett.*, Vol. 10, pp. 267–269, 1985.

Long, M. B., Webber, B. F., and Chang, R. K., "Instantaneous Two-Dimensional Concentration Measurements in a Flow by Mie Scattering," *Appl. Phys. Lett.*, Vol. 34, pp. 22–24, 1979.

Lourenco, L. and Krothapalli, A., "The Role of Photographic Parameters in Laser Speckle or Particle Image Displacement Velocimetry," *Experiments in Fluids*, Vol. 5, pp. 29–32, 1987.

Lourenco, L. and Krothapalli, A., "On the Accuracy of Velocity and Vorticity Measurements with PIV," *Exp. Fluids*, Vol. 18, pp. 421–428, 1995.

Lourenco, L., Gogenini, S., and LaSalle, R., "On-Line PIV: An Integrated Approach," *Appl. Opt.*, Vol. 33, pp. 2465–2470, 1994.

Lourenco, L., Krothapalli, A., and Smith, C. A., "Particle Image Velocimetry," *Advances in Fluid Mechanics Measurements*, Springer-Verlag, Berlin, 1989.

Lourenco, L., Krothapalli, A., Buchlin, J. M., and Riethmuller, M. L., "A Non-Invasive Experimental Technique for the Measurement of Unsteady Velocity and Vorticity Fields," *Aerodynamic and Related Hydrodynamic Studies Using Water Facilities*, AGARD CP-413, 1986.

Lourenco, L., "A Passive Velocity Bias Technique for PIDV," *Bull. Amer. Phys. Soc.*, Vol. 34, p. 2267, 1989.

Lourenco, L., "Velocity Bias Technique for Particle Image Velocimetry Measurements of High Speed Flows," *Appl. Opt.*, Vol. 32, pp. 2159–2162, 1993.

Mach, L., "Uber Ein Neuer Interferenzfrakor. Z.," *Instrumentenk*, Vol. 12, pp. 89–93, 1892.

MacMillan, F. A., "Experiments on Pitot Tubes in Shear Flow," A.R.C. R. and M. 3028, 1956.

Maddox, A. R. and Binder, R. C., "New Dimension in the Schlieren Technique: Flow Field Analysis Using Color," *Appl. Opt.*, Vol. 10, p. 474, 1971.

Maltby, R. L. (Ed.), "Flow Visualization in Wind Tunnels Using Indicators," AGARDograph 70, 1962.

Marasli, B., Nguyen, P., and Wallace, J. M., "A Calibration Technique for Multiple-Sensor Hot-Wire Probes and Its Application to Vorticity Measurements in the Wake of a Circular Cylinder," *Exp. Fluids*, Vol. 15, pp. 209–218, 1993.

Marchman, J. F. II and Kuppa, S., "Semi-Span Model/End Plate Gap Effect on Low Reynolds Number Aerodynamic Test Results," AIAA 87-2350, 1987.

Mardin, H. R., "The Color Schlieren Systems," Report AL-2052, North American Aviation, Sept. 1954.

McGregor, I., "The Vapour-Screen Method of Flow Visualization," *J. Fluid Mech.*, Vol. 11, pp. 418–511, 1961.

McLachlan, B. G., "Flight Testing of a Luminescent Surface Pressure Sensor," NASA TM 103970, 1992.

McLaughlin, D. K. and Tiederman, W. G., "Biasing Correction for Individual Realization of Laser Anemometer Measurements in Turbulent Flow," *The Physics of Fluids*, Vol. 16, No. 12, pp. 2082–2088, 1973.

McManus, K., Yip, B., and Candel, S., "Emission and Laser Induced Fluorescence Imaging Methods in Experimental Combustion," *Experimental Heat Transfer, Fluid Mechanics, and Thermodynamics*, Kelleher, M. D. (Ed.), Elsevier Science, New York, 1993.

Measurements Group, Inc., "Design Considerations for Diaphragm Pressure Transducers," TN-510, Raleigh, NC, 1974.

Merzkirch, W. and Erdmann, W., "Measurements of Shock Wave Velocity Using the Doppler Principle," *Appl. Phys.*, Vol. 4, p. 363, 1974.

Merzkirch, W., *Flow Visualization*, Academic Press, New York, 1974.

Merzkirch, W., "A Simple Schlieren Interferometer System," *AIAA J.*, Vol. 3, No. 10, pp. 1974–1976, 1965.

Merzkirch, W., "Density Sensitive Flow Visualization," *Method of Experimental Physics*, Vol. 18, Part A, Academic Press, New York, pp. 345–403, 1981.

Merzkirch, W., "Generalized Analysis of Shearing Interferometers as Applied for Gas Dynamics," *Appl. Opt.*, Vol. 13, No. 2, pp. 409–413, 1974.

Meyers, J. F. and Cavone, A. A., "Signal Processing Schemes for Doppler Global Velocimetry," Presented at the 14th International Congress on Instrumentation in Aerospace Simulation Facilities, Rockville, MD, 1991.

Miles, R. B. and Lempert W. R., "Two-Dimensional Measurement of Density, Velocity, and Temperature in Turbulent High-Speed Air Flows by UV Rayleigh Scattering," *Appl. Phys.*, Vol. B 51, pp. 1–7, 1990b.

Miles, R. B. and Lempert, W. R., "Flow Diagnostics in Unseeded Air," AIAA 90-0624, 1990a.

Miles, R. B., Lempert, W. R., and Forkey, J., "Instantaneous Velocity Fields and Background Suppression by Filtered Rayleigh Scattering," AIAA 91-0357, 1991.

Miller, R. W., *Flow Measurement Engineering Handbook*, McGraw-Hill, New York, 1983.

Moffat, R. J., "Contributions to the Theory of Single-Sample Uncertainty Analysis," *J. Fluids Eng.*, Vol. 104, pp. 250–260, 1982.

Moffat, R. J., "Using Uncertainty Analysis in the Planning of an Experiment," *J. Fluids Eng.*, Vol. 107, pp. 173–178, 1985.

Moffat, R. J., "Experimental Heat Transfer," *Heat Transfer 1990*, Vol. 1, Hestroni, G. (Ed.), Hemisphere, Washington, D.C., pp. 187–205, 1990.

Moore, R. L., *Basic Instrumentation Notes and Study Guide*, 3rd ed., ISA, Research Triangle Park, NC, 1982.

Morkovin, M., "Fluctuations and Hot-Wire Anemometry in Compressible Flow," AGARDograph 24, 1956.

Morrison, G. L., Perry, A. E., and Samuel, A. E., "Dynamic Calibration of Inclined and Crossed Hot-Wires," *J. Fluid Mech.*, Vol. 52, Part 3, pp. 465–474, 1972.

Morrison, J. M., "Pressure and Vacuum Measurement," *Handbook of Applied Instrumentation*, McGraw-Hill, New York, 1964.

Moss, J. B ., "Simultaneous Measurements of Concentration and Velocity in an Open Premixed Turbulent Flame," *Combust. Sci. Technol.*, Vol. 22, pp. 119–129, 1980.

Murphy, R. B., "On the Meaning of Precision and Accuracy," *Materials Research and Standards*, Vol. 4, pp. 264–267, 1961.

Muss, J. A., Dibble, R. W., and Talbot, L., "A Helium-Hydrogen Mixture for the Measurement of Mixture Fraction and Scalar Gradient in Non-Premixed Reacting Flows," AIAA 94-0612, 1994.

Namer, I. and Schefer, R. W., "Error Estimates for Rayleigh Scattering Density and Temperature Measurements in Premixed Flames," *Exp. Fluids*, Vol. 3, pp. 1–9, 1985.

Nenni, J. P., Erickson, J. C., and Wittliff, C. E., "Measurements of Small Normal Velocity Components in Subsonic Flows by Use of a Static Pipe," *AIAA J.*, Vol. 20, No. 8, pp. 1077–1083, 1982.

Neumann, R. D., "Aerothermodynamic Instrumentation," AGARD Report No. 761, 1989.

Ng, W. F., "Review—Simultaneous Measurements of Stagnation Temperature and Pressure Using an Aspirating Probe," *Pressure and Temperature Measurement*, Kim, J. H. and Moffat, R. J. (Eds.), FED Vol. 44 and HTD Vol. 58, ASME, New York, 1986.

Nibler, J. W., Shaub, W. M., McDonald, J. R., and Harvey, A. B., "Coherent Anti-Stokes Raman Spectroscopy," *Vibrational Spectra and Structure*, Vol. 6, Durig, J. R. (Ed.), Elsevier, New York, 1977.

Nitsche, W., Haberland, C., and Thunker, R., "Comparative Investigations of Friction Drag

Measuring Techniques in Experimental Aerodynamics," *14th ICAS Cong.*, ICAS-84-2.4.1, 1984.

Noel, B. W., Borella, H. M., Lewis, W., Turley, W. D., Beshears, D. L., Capps, G. J., Cates, M. R., Muhs, J. D., and Tobin, K. W., "Evaluating Thermographic Phosphors in an Operating Turbine Engine," *J. Eng. Gas Turbines and Power*, Vol. 113, pp. 242–245, 1991.

Noel, B. W., Turley, W. D., and Allison, S. W., "Thermographic-Phosphor Temperature Measurements: Commercial and Defense-Related Applications," *Proc. 40th Int. Instr. Symp.*, ISA, Research Triangle Park, NC, pp. 271–288, 1994.

North, R. J., "A Colour Schlieren System Using Multicolour Filters of Simple Construction," NPL, Aero. 266, 1954.

North, R. J., "Schlieren Systems Using Graded Filters," A.R.C. No. 15099, 1952.

O'Hanlon, J. F., *A User's Guide to Vacuum Technology*, John Wiley & Sons, New York, 1980.

Ortolano, D. J. and Hines, F. F., "A Simplified Approach to Heat Flow Measurement," *Adv. Instr.*, Vol. 38, Part II, ISA, Research Triangle Park, pp. 1449–1456, 1983.

Park, S. R. and Wallace, J. M., "The Influence of Instantaneous Velocity Gradients on Turbulence Properties Measured With Multi-Sensor Hot-Wire Probes," *Exp. Fluids*, Vol. 16, pp. 17–26, 1993.

Pealat, M., Bailly, R., and Taran, J. P. E., "Real Time Study of Turbulence in Flames by Raman Scattering," *Opt. Comm.*, Vol. 22, p. 91, 1977.

Penner, S. S. and Jerskey, T., "Use of Lasers for Local Measurements of Velocity Components, Species Densities, and Temperature," *Ann. Rev. Fluid Mech.*, Vol. 5, pp. 9–30, 1973.

Perry, A. E. and Morrison, G. L., "A Static and Dynamic Calibration of Constant-Temperature Hot-Wire Systems," *J. Fluid Mech.*, Vol. 47, Part 4, pp. 765–777, 1971.

Perry, A. E. and Morrison, G. L., "A Study of the Constant-Temperature Hot-Wire Anemometer," *J. Fluid Mech.*, Vol. 47, Part 3, pp. 577–599, 1971.

Philbert, M., Surget, J., and Veret, C., "Interferometry," *Handbook of Flow Visualization*, Yang, W. J. (Ed.), Hemisphere Publishing, pp. 203–210, 1989.

Philbert, M., Surget, J., and Veret, C., "Shadowgraph and Schlieren," *Handbook of Flow Visualization*, Yang, W. J. (Ed.), Hemisphere Publishing, pp. 189–201, 1989.

Piomelli, U., Balint, J.-L., and Wallace, J. M., "On the Validity of Taylor's Hypothesis for Wall-Bounded Turbulent Flows," *Phys. Fluids A*, Vol. 1, pp. 609–611, 1989.

Pitz, R. W., Cattolica, R., Robben, F., and Talbot, L., "Temperature and Density in a Hydrogen-Air Flame from Rayleigh Scattering," *Combust. Flame*, Vol. 27, pp. 313–320, 1976.

Pollock, D. D., *Thermocouples: Theory and Practice*, CRC Press, Boca Raton, FL, 1991.

Prasad, A. K. and Adrian, R. J., "Stereoscopic Particle Image Velocimetry Applied to Liquid Flows," *Exp. Fluids*, Vol. 15, pp. 49–60, 1993.

Prasad, A. K. and Adrian, R. J., "Stereoscopic Particle Image Velocimetry Applied to Liquid Flows," *Proc. 6th Int. Symp. Appl. Laser Tech. Fluid Mech.*, Ladoan-Inst. Super Tech., 1992.

Preston, J. H., "The Determination of Turbulent Skin Friction by Means of Pitot Tubes," *J. Roy. Aeronaut. Soc.*, Vol. 58, pp. 109–121, 1954.

Prior, Y., "Three Dimensional Phase Matching in Four Wave Mixing," *Appl. Opt.*, Vol. 19, p. 1741, 1980.

Rae, W. T., Jr and Pope, A., *Low Speed Wind Tunel Testing*, Wiley–Interscience, New York, 1984.

Reuss, D. L., Adrian, R. J., Landreth, C. C., French, D. T., and Fansler, T. D., "Instantaneous Planar Measurements of Velocity and Large-Scale Vorticity and Strain Rate in an Engine Using Particle-Image Velocimetry," SAE Technical Paper Series 890616, 1989.

Reynolds, W. C., *Thermodynamic Properties in SI*, Stanford University, Department of Mechanical Engineering, 1979.

Richardson, P. D., "Comments on Viscous Damping in Oscillating Liquid Columns," *Int. J. Mech. Sci.*, Vol. 5, pp. 415–418, 1963.

Robinson, O. and Rockwell, D., "Construction of Three-Dimensional Images of Flow Structure via Particle Tracking Techniques," *Exp. Fluids*, Vol. 14, pp. 257–270, 1993.

Rockwell, D., "Quantitative Visualization of Bluff-Body Wakes via Particle Image Velocimetry," *Bluff-Body Wakes, Dynamics and Instabilities*, Springer-Verlag, Göttingen, 1992.

Roehrig, J. R. and Simons, J. C., "Calibrating Vacuum Gages to 10^{-9} Torr," *Instr. Contr. Sys.*, pp. 107–111, 1963.

Rood, E. P. and Telionis, D. P., "Journal of Fluids Engineering Policy on Reporting Uncertainties in Experimental Measurements and Results," *J. Fluids Eng.*, Vol. 113, pp. 313–314, 1991.

Roquemore, W. M., Tankin, R. S., Chiu, H. H., and Lottes, S. A., "A Study of a Bluff-Body Combustor Using Laser Sheet Lighting," *Exp. Fluids*, Vol. 4, pp. 205–213, 1986.

Roshko, A., "On The Drag and Shedding Frequency of Two-Dimensional Bluff Bodies," NACA TN 3169, 1954.

Ross, C. B., Lourenco, L., and Krothapalli, A., "Particle Image Velocimetry Measurements in a Shock Containing Supersonic Flow," AIAA Paper 94-0047, 1994.

Rowell, R. L. and Aval, G. M., "Rayleigh-Raman Depolarization of Laser Light Scattered by Gases," *J. Chem. Phys.*, Vol. 54, pp. 1960–1964, 1971.

Rudder, R. R. and Bach, D. R., "Rayleigh Scattering of Ruby-Laser Light by Neutral Gases," *J. Opt. Soc. Am.*, Vol. 58, pp. 1260–1266, 1968.

Sandborn, V. and Laurence, J., "Heat Loss from Yawed Hot Wires at Subsonic Mach Numbers," NACA TN 3563, 1955.

Sandoval, R. P. and Armstrong, R. L., "Rayleigh-Brillouin Spectra in Molecular Nitrogen," *Phys. Rev. A*, Vol. 13, pp. 752–757, 1976.

Satchell, R. E., "Time Averaged Measurements in Turbulent Flames Using Raman Spectroscopy," Vol. 53, *Prog. in Astronautics and Aeronautics*, AIAA, 1977.

Schetz, J. A. and Nerney, B., "Turbulent Boundary Layer with Injection and Surface Roughness," *AIAA J.*, Vol. 15, No. 9, pp. 1288–1294, 1977.

Schetz, J. A., Vinogradov, V. A., Marshakov, A., and Petrov, V., "Direct Measurements of Skin Friction in a Scramjet Combustor," AIAA-93-2443, 1993.

Schetz, J. A., *Boundary Layer Analysis*, Prentice Hall, Englewood Cliffs, NJ, 1993.

Schoenung, S. M. and Mitchell, R. E., "Comparison of Raman and Thermocouple Temperature Measurements in Flames," *Combust. Flame*, Vol. 35, pp. 207, 1979.

Schultz, D. L. and Jones, T. V., "Heat Transfer Measurements in Short Duration Hypersonic Facilities," AGARDograph 165, 1973.

Schwar and Weinberg, *Proc. Roy. Soc. (London), Ser. A*, Vol. 311, p. 469, 1969.

Schweppe, J. L., "Methods for the Dynamic Calibration of Pressure Transducers," NBS Monograph 67, National Bureau of Standards, Washington, D.C., 1963.

Self, S. A., "Laser-Doppler Anemometer for Boundary Layer Measurements in High Velocity, High Temperature MHD Channel Flows," *Proc. 2nd Int. Workshop on Laser Velocimetry*, Purdue University, Vol. II, pp. 44–67, 1974.

Sensym Inc., *Pressure Transducer Handbook*, Sunnyvale, CA, 1983.

Settles, G. S., "A Direction-Indicating Color Schlieren System," *AIAA J.*, Vol. 8, p. 2282, 1970.

Settles, G. S., "Color Schlieren Optics—A Review of Techniques and Applications," Merzkirch, W. (Ed.), *Flow Visualization II*, Hemisphere Publishing, New York, 1980.

Shapiro, A. H., *The Dynamics and Thermodynamics of Compressible Fluid Flow*, Vol. I, Ronald Press, New York, 1953.

Shardanand and Prasad Rao, A. D., "Absolute Rayleigh Scattering Cross Sections of Gases and Freons of Stratospheric Interest in the Visible and Ultraviolet Regions," NASA, TN D-8442, 1986.

Shaw, R., "The Influence of Hole Dimensions on Static Pressure Measurements," *J. Fluid Mech.*, Vol. 7, Part 4, pp. 550–564, 1960.

Sherman, F. S., "New Experiments on Impact-Pressure Interpretation in Supersonic and Subsonic Rarefied Air Streams," NACA TN 2995, 1953.

Shimizu, H., Noguchi, K., and She, C.-Y., "Atmospheric Temperature Measurements by a High-Spectral-Resolution Lidar," *Appl. Opt.*, Vol. 25, pp. 1460–1466, 1986.

Shinpaugh, K. A., Simpson, R. L., Wicks, A. L., Ha, S.-M., and Fleming, J. L., "Signal-Processing Techniques for Low Signal-to-Noise Ratio Laser-Doppler Velocimetry Signals," *Experiments in Fluids*, Vol. 12, pp. 319–328, 1992.

Simons, J. C., "On Uncertainties in Calibration of Vacuum Gages and the Problem of Traceability," *Trans. 10th Nat. Vacuum Symp.*, Macmillan, New York, 1963.

Sirohi, R. S. and Krishna, H. C., *Mechanical Measurements*, 2nd ed., John Wiley & Sons, New York, 1983.

Slomiana, M., "Using Differential Pressure Sensors for Level, Density, Interface, and Viscosity Measurements," *Instr. Tech.*, Vol. 26, No. 9, pp. 63–68, 1979a.

Slomiana, M., "Selecting Differential Pressure Instrumentation," *Instr. Tech.*, Vol. 26, No. 8, pp. 32–40, 1979b.

Smith, J. R., "Temperature and Density Measurements in an Engine by Pulsed Raman Spectroscopy," SAE Paper No. 80-0137, 1980.

Smith, M., Smits, A., and Miles, R. B., "Compressible Boundary-Layer Density Cross Sections by UV Rayleigh Scattering," *Opt. Lett.*, Vol. 14, pp. 916–918, 1989.

Smol'yakov, A. V. and Tkachenko, V. M., *The Measurement of Turbulent Fluctuations— An Introduction to Hot-Wire Anemometry and Related Transducer*, p. 92, trans. by Chomet, S. and Bradshaw, P. (Eds.), Springer-Verlag, Berlin, 1983.

Socket, L. R., Lucquin, M., Bridoux, M., Crunell-Cras, M., Grase, F., and Delhaye, M., "Use of Multichanneled Pulsed Raman Spectroscopy as a Diagnostic Technique in Flames," *Combust. Flame*, Vol. 36, p. 109, 1979.

Spangenberg, W. G., "Heat-Loss Characteristics of Hot-Wire Anemometers at Various Densities in Transonic and Supersonic Flow," NACA TN 3381, 1955.

Spina, E., Donovan, J., and Smits, A., "Convection Velocity in Supersonic Turbulent Boundary Layers," *Physics of Fluids A*, Vol. 3, No. 12, 1991.

Squires, K. and Eaton, J., "Particle Response and Turbulence Modification in Isotropic Turbulence," *Phys. Fluids A*, Vol. 2, pp. 1191–1203, 1990.

Stepowski, D. and Cotterau, M. J., "Direct Measurement of OH Local Concentration in a Flame from the Fluorescence Induced by a Single Laser Probe," *Appl. Opt.*, Vol. 18, p. 354, 1979.

Stepowski, D., "Laser Measurements of Scalars in Turbulent Diffusion Flames," *Prog. Energy Combust. Sci.*, Vol. 18, pp. 463–491, 1992.

Stetson, K. and Kimmel, R., "On Hypersonic Boundary-Layer Stability," AIAA-92-0737.

Strickler, W., "Local Temperature Measurements in Flames by Laser Raman Spectroscopy," *Combust. Flame*, Vol. 27, p. 133, 1976.

Sweeney, R. J., *Measurement Techniques in Mechanical Engineering*, John Wiley & Sons, New York, 1977.

Taback, I., "The Response of Pressure Measuring Systems to Oscillating Pressure," NACA TN 1819, 1949.

Thompson, W. R., "On a Criterion for the Rejection of Observations and the Distribution of the Ratio of Deviation to Sample Standard Deviation," *Ann. Math. Stat.*, Vol. 6, pp. 214–219, 1935.

Toepler, A., "Beobachtungen Nach Einer Neuen Optischen Methode," *Poggendorf's Ann. d. Phys. u. Chem.*, Vol. 9, 556, 1866; Vols. 128, 126, 1866; Vols. 133, 33, and 180, 1868.

Tolles, W. M., Nibler, J. W., McDonald, R., and Harvey, B., "Review of the Theory and Application of Coherent Anti-Stokes Raman Spectroscopy," *Appl. Spect.*, Vol. 31, p. 253, 1977.

Towfighi, J. and Rockwell, D., "Flow Structure from an Oscillating Nonuniform Cylinder: Generation of Patterned Vorticity Concentrations," *Physics of Fluids*, Vol. 6, No. 2, pp. 531–536, 1994.

Tritton, D. J., "The Use of a Fiber Anemometer in Turbulent Flows," *J. Fluid Mech.*, Vol. 16, pp. 269–281, 1963.

Trolinger, J. D., "Laser Instruments for Flowfield Diagnostics," *AGARDograph*, No. 186, 1974.

TSI, *Hot WIRE/Hot Film Anemometry Probes & Accessories*, TSI Inc., Glastonbury, CT, 1988.

Tsinober, A., Kit, E., and Dracos, T., "Experimental Investigation of the Field of Velocity Gradients in Turbulent Flows," *J. Fluid Mech.*, Vol. 242, pp. 169–192, 1992.

Urushihara, T., Meinhart, C. D., and Adrian, R. J., "Investigation of the Logarithmic Layer in Pipe Flow Using Particle Image Velocimetry," *Near Wall Turbulent Flows*, Elsevier Science Publishers, Amsterdam/London/New York/Tokyo, 1993.

Ury, J. F., "Viscous Damping in Oscillating Liquid Columns, Its Magnitude and Limits," *Int. J. Mech. Sci.*, Vol. 4, pp. 349–370, 1962.

Vagt, J.-D., "Hot-Wire Probes in Low Speed Flow," *Prog. Aerospace Sci.*, Vol. 18, pp. 271–323, 1979.

Vakili, A. D. and Wu, J. M., "Direct Measurement of Skin Friction with a New Instrument," *Fluid Control and Measurement*, 1988 (organized by the Society of Instrument and Control Engineers, Japan).

van de Hulst, H. C., *Light Scattering by Small Particles*, John Wiley & Sons, New York (reprinted in 1981 by Dover Publications, New York), 1957.

van Driest, E. R., "The Problem of Aerodynamic Heating," *Aeronaut. Eng. Rev.*, Vol. 15, pp. 26–41, 1956.

Van Dyke, M. H., *An Album of Fluid Motion*, Parabolic Press, Stanford, CA, 1982.

Van Eck, A., "Holographic Particle Image Velocimetry Measurements in a Rectangular Jet," M.S thesis, Florida State University, Tallahassee, Apr. 1994.

Voisinet, R. L. P., "Combined Influence of Roughness and Mass Transfer on Turbulent Skin Friction at Mach 2.9," AIAA-79-0003, 1979.

Volluz, R. J., "Handbook of Supersonic Aerodynamics, Section 20, Wind Tunnel Instrument and Operation," NAVORD Report 1488, Vol. 6, 1961.

Voss, E., Weitkamp. C., and Michaelis, W., "Lead-Vapor Filters for High-Spectral-Resolution Temperature Lidar," *Appl. Opt.*, Vol. 33, pp. 3250–3260, 1994.

Vukoslavčević, P. and Wallace, J. M., "Influence of Velocity Gradients on Measurements of Velocity and Streamwise Vorticity With Hot-Wire X-Array Probes," *Rev. Sci. Instrum.*, Vol. 52, No. 6, pp. 869–879, 1981.

Vukoslavčević, P., Wallace, J. M., and Balint, J.-L., "The Velocity and Vorticity Vector Fields of a Turbulent Boundary Layer. Part 1, Simultaneous Measurement by Hot-Wire Anemometry," *J. Fluid Mech.*, Vol. 228, pp. 25–51, 1991.

Wallace, J. M. and Foss, J. F., "The Measurement of Vorticity in Turbulent Flows," *Ann. Rev. Fluid Mech.*, 1995.

Werlé, H., "Hydrodynamic Flow Visualization," *Ann. Rev. Fluid Mech.*, Vol. 5, pp. 361–382, 1973.

Westerweel, J., *Digital Particle Image Velocimetry*, Delft University Press, Delft, The Netherlands, 1993.

Weyl, F. J., "Analysis of Optical Methods," *Physical Measurements in Gas Dynamics and Combustion*, Princeton University Press, Princeton, NJ, pp. 3–25, 1954.

White, G., "Liquid Filled Pressure Gages Systems," *Statham Instr. Notes*, Statham Instruments, Inc., Los Angeles, CA, 1949.

Whittier, R. M., "Basic Advantages of the Anisotropic Etched, Transverse Gage Pressure Transducer," *Prod. Dev. News*, Vol. 16, No. 3, Endevco Corp., San Juan Capistrano, CA, 1980.

Wigeland, R. A., Ahmed, M., and Nagib, H. M., "Vorticity Measurements Using Calibrated Vane-Vorticity Indicators and Cross-Wires," *AIAA J.*, Vol. 16, No. 12, pp. 1229–1234, 1978.

Wildhack, W. A., Dressler, R. F., and Lloyd, E. C., "Investigation of the Properties of Corrugated Diaphragms," *Trans. ASME*, Vol. 79, pp. 65–82, 1957.

Willert, C. E. and Gharib, M., "Digital Particle Image Velocimetry," *Exp. Fluids*, Vol. 10, pp. 181–193, 1991.

Winteler, H. R. and Gautschi, G. H., "Piezoresistive Pressure Transducers," *Transducer '79 Conference*, available through Kistler Instrument Corp., Amherst, NY, 1979.

Winter, K. G., "An Outline of the Techniques Available for the Measurement of Skin Friction in Turbulent Boundary Layers," *Prog. Aerospace Sci.*, Vol. 18, pp. 1–57, 1977.

Winter, M., Lam, J. K., and Long, M. B., "Techniques for High-Speed Digital Imaging of Gas Concentration in Turbulent Flows," *Exp. Fluids*, Vol. 5, pp. 177–183, 1987.

Wishart, D., "On the Structure of an Axisymmetric Supersonic Jet," Ph.D. dissertation, Florida State University, Tallahassee, Dec. 1994.

Witte, A. and Wuerker, R., "Laser Holographic Interferometry Study of High-Speed Flowfields," *AIAA J.*, Vol. 8, p. 581, 1970.

Wyngaard, J. C., "Cup, Propeller, Vane and Sonic Anemometers in Turbulence Research," *Ann. Rev. Fluid Mech.*, Vol. 13, pp. 399–423, 1981.

Yeh, Y. and Cummins, H. Z., "Localized Fluid Flow Measurements with an He-Ne Laser Spectrometer," *Appl. Phys. Lett.*, Vol. 4, pp. 176–178, 1964.

Yip, B. and Long, M. B., "Instantaneous Planar Measurements of the Complete Three-Dimensional Scalar Gradient in a Turbulent Jet," *Opt. Lett.*, Vol. 11, pp. 64–66, 1986.

Yip, B., Schmitt, R. L., and Long, M. B., "Instantaneous Three-Dimensional Concentration Measurements in Turbulent Jets and Flames," *Opt. Lett.*, Vol. 13, pp. 96–98, 1988.

Zehnder, L., "Ein Neuer Interferenzfrakor. Z.," *Instrumentenk*, Vol. 11, pp. 275–285, 1891.

16 Fluid Dynamics Ground Test Facilities

LEON H. SCHINDEL
Naval Surface Warfare Center
Silver Spring, MD

CONTENTS

Handbook of Fluid Dynamics and Fluid Machinery, Edited by Joseph A. Schetz and Allen E. Fuhs
ISBN 0-471-12598-9 Copyright © 1996 John Wiley & Sons, Inc.

16.1   INTRODUCTION

A fluid mechanics facility provides a test medium in which phenomena of interest can be observed and measured under known and controlled conditions. The various types of facilities, therefore, are designed to reproduce or simulate the appropriate environmental parameters and provide for the measurement of the effects produced on a test specimen (for example, a scale model of a vehicle) by these flow conditions.

In this chapter, descriptions are given of a variety of facilities in which fluid mechanics phenomena can be observed and measured. The objective here is to acquaint the reader with what types of facilities are available and how they function. Details on facility design and operation may be found in the references. In addition, the references include information on particular existing facilities where more data may be available.

The next four sections give brief descriptions of the different types of facilities. The list is not all-inclusive. Special facilities with very narrow application are omitted, and others are covered elsewhere in this handbook.

In general, aerodynamic and hydrodynamic tests deal with measurements of phenomena associated with the flow over the external surfaces of a model. Even when propulsive effects, such as propeller slipstreams or rocket plumes, are simulated, tests are concerned with their external influence rather than with the performance of the propulsion system itself. In the propulsion tests, on the other hand, the function of the prime mover is of greatest interest, and measurements are made of internal flow phenomena and their effects on engine performance. In some cases, such as inlets for air-breathing engines, although internal flows are to be measured, the shape of the entire upstream configuration influences the internal flow, and therefore the testing is done in so-called "aerodynamic" facilities.

The tables in Sec. 16.2 summarize the kinds of tests that might be made in major types of fluid mechanics facilities. The entries in the table indicate parameters to be simulated. In many cases, it is not feasible to duplicate free flight values, so that analytical extrapolations are required. For example, the effects of boundary layer thickness on aerodynamic forces may be estimated to account for discrepancies between free flight Reynolds number and test conditions. On the other hand, to study properties of boundary layer turbulence, it is necessary to make measurements in a turbulent boundary layer.

The following parameters are shown as entries in the tables: Mach number, $M \equiv$

U/a; Reynolds number, $Re \equiv \rho UL/\mu$; velocity, U; stagnation temperature, T_0; altitude, h; Froude number, $Fr = U/\sqrt{gL}$; and cavitation number, $\sigma \equiv 2(p_\infty - p_v)/\rho U^2$. Engine test conditions are frequently expressed in terms of total temperature and pressure, p_0, rather than Mach number and static pressure (or altitude). Consequently, these parameters are included in the propulsion facility tables. These facilities are also characterized by the mass flow of air, $\dot{m}$, which they can provide, thus indicating the maximum size of the engine that can be tested.

16.2 TABLES OF FACILITIES

TABLE 16.1 Aerodynamic Facilities with Scaling Parameters

Quantities to be measured \ Facility	Subsonic Wind Tunnel	Transonic Wind Tunnel	Supersonic Hypersonic Wind Tunnel	Ballistics Ranges	Arc Tunnels Shock Tubes
Static/dynamic forces	Re	M, Re	M, Re	M, Re	
Pressure distribution on vehicle	Re	M, Re	M, Re		
Aeroelastic phenomena	Re, U	U, M, Re	M, Re		
Skin Friction	Re	M, Re	M, Re	M, Re	
Heat transfer		M, Re	M, Re, T_0	M, Re, T_0	Re, T_0
Flow separation	Re	M, Re	M, Re		
Boundary-layer transition	Re	M, Re	M, Re	M, Re, T_0	Re, T_0
Store separation	U, Re	U, M, Re	U, M		
Environmental effects	U, Re				
Flow field properties	Re	M, Re	M, Re, T_0	M, Re, T_0	Re, T_0
Ablation			M, Re, T_0	M, Re, T_0	Re, T_0
Chemical reactions				M, T_0	T_0

TABLE 16.2 Hydrodynamic Facilities with Scaling Parameters

Quantities to be measured \ Facility	Water Table	Open Channel	Water Tunnel	Towing Tank	Hydro Tank
Static/dynamic forces			Re, Fr, σ	Re, Fr	Re, Fr, σ
Pressure distribution	M (analog)		Re, Fr, σ	Re, Fr	Re, Fr, σ
Skin friction		Re	Re	Re	Re
Flow separation		Re	Re	Re	Re
Boundary-layer transition		Re	Re	Re	Re
Flow field		Re	Re, Fr, σ	Re, Fr, σ	Re, Fr, σ

TABLE 16.3 Propulsion Facilities with Scaling Parameters

Quantities to be measured	Test Stands	Inlet, Diffuser, Nozzle Facilities	Combustion Facilities	Engine Test Facilities
Thrust, Impulse Pressure recovery, mass flow	$\dot{m}, h, T_0, p_0, w$	$\dot{m}, M, \mathrm{Re}, T_0, p_0$ $\dot{m}, M, \mathrm{Re}, T_0, p_0$	$\dot{m}, M, \mathrm{Re}, T_0, p_0$	$\dot{m}, M, \mathrm{Re}, T_0, p_0$ M, Re
Flow field properties		M, Re, T_0, p_0	M, Re, T_0, p_0	$\dot{m}, M, \mathrm{Re}, T_0, p_0$
Combustion process	$\dot{m}, h, T_0, p_0, w$		$\dot{m}, M, \mathrm{Re}, T_0, p_0$	$\dot{m}, M, \mathrm{Re}, T_0, p_0$

16.3 AERODYNAMIC FACILITIES

16.3.1 Introduction

Aerodynamic facilities provide for the measurement of the effects of the relative motion of air or other gases. Usually, the desired quantity is the aerodynamic force on some vehicle such as an airplane or missile. But, forces on automobiles, buildings, windmills, and many other devices are also of interest. Forces are deduced from measurements made on scaled models subjected to known and controlled flow conditions. Besides forces, other aerodynamic influences include heat transfer, load distribution, and the distribution of flow field properties such as velocity, pressure, density, radiation, species concentration, etc.

The main parameters of interest are Mach number, M, and Reynolds number, Re. The Mach number is a measure of the effects of compressibility, while the Reynolds number is a measure of the effects of viscosity. At lower subsonic speeds ($M < 0.6$), any effects of compressibility can usually be inferred from scaling laws. Hence, simulation of Reynolds number is of primary interest in subsonic facilities. Most aerodynamic phenomena are rather insensitive to Reynolds number, and the effects of changes in Reynolds number from that of a given test can often be reliably estimated provided the boundary layer (or other viscous phenomenon) remains laminar or turbulent. Predictions of the value of an aerodynamic quantity under turbulent boundary layer conditions from measurements in which the boundary layer is laminar can be very unreliable. Therefore, considerable effort had been expended in the design and construction of facilities which can reproduce the high Reynolds numbers associated with turbulent boundary layers. Sometimes, however, it is necessary to artificially induce turbulence on a test specimen by a boundary layer trip or similar device. Furthermore, noise and other factors cause perturbations in wind tunnels which are absent in free flight. Hence, quantities related to boundary layer transition are seldom inferred from wind tunnel measurements.

At high supersonic speeds, the effects of aerodynamic heating may dominate vehicle design considerations. It then becomes necessary to test in facilities which will reproduce high temperature environments. This will be especially important when real gas effects alter the properties of the test medium. Measurements must then be made in a facility which simulates such high temperature environments. The chemistry of the medium is of interest also in simulations of high speed flight in planetary media, in conditions in which the fluid medium interacts with the test specimen, and conditions where electrical properties can be of importance, as in electromagnetic transmission, for example.

At very high altitudes, the density is so low that air can no longer be regarded as a continuum. Then, the Knudsen number (ratio of mean free path to characteristic length) becomes a significant parameter. Facilities in which high Knudsen number may be simulated are indicated in the subsection entitled Hypervelocity Wind Tunnels in Sec. 16.3.6. Rarefied gas dynamics is discussed in Chap. 11.

16.3.2 Subsonic Wind Tunnels

The simplest form of wind tunnel is illustrated schematically in Fig. 16.1. It consists of an inlet open to the atmosphere and an exhaust fan which pulls air through the test section. The inlet is followed by flow-straightening vanes and screens to produce a smooth flow. The test section is usually of reduced cross section to give a region of high flow velocity and higher Reynolds number and reduced relative flow variations. It may have a window for viewing the test specimen.

The facility provides some means of supporting test models and instrumentation; it may include an external balance for measuring forces and moments on the model, or the forces may be measured by a balance mounted inside the model or its support structure.

For boundary layer research, stream turbulence can be minimized by specially-designed screens accompanied by a large contraction ratio between test section area and inlet area. Furthermore, the fan may be isolated from the test section by screens or ducting to minimize upstream propagation of noise and vibrations.

Reynolds numbers associated with subsonic aircraft flight are typically of the order of 10^7 or 10^8; boundary layers will tend to be thin and turbulent. To simulate such high Reynolds numbers in an open wind tunnel requires very large size and high power. The situation can be relieved somewhat by closing the wind tunnel with a return leg as shown in Fig. 16.2. In a closed cycle facility, the power supplied by the fan need only be sufficient to overcome the losses in the circuit, associated primarily with friction at the walls and at the flow straightening screens, and with

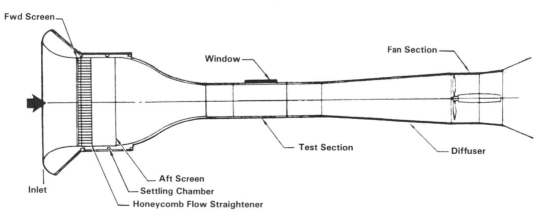

FIGURE 16.1 Open subsonic wind tunnel. (From Pirrello *et al.*, 1971a.)

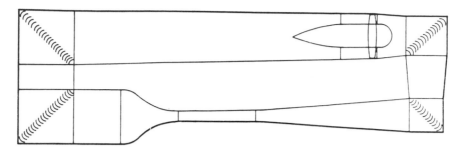

FIGURE 16.2   Closed return tunnel. (From Pirrello *et al.*, 1971a.)

dissipation at the corners. These losses can be minimized by making the return flow piping of large cross-sectional area and by smoothing the corner flow by means of turning vanes as indicated in Fig. 16.2.

It is necessary, however, to extract the heat added to the medium by the fan. Thus, some heat exchanger is required, and it will also introduce an obstruction in the flow. The drag of the heat exchanger can be eliminated by cooling the walls of the facility or the turning vanes instead of adding a separate cooler.

High Reynolds number is achieved in such facilities by a combination of large size, high velocity, and high density. For a given amount of power, the Reynolds number will be highest if the stream is accelerated to high velocity with a large contraction ratio. The allowable stream velocity is limited, however, in a subsonic wind tunnel. Consequently, the tunnel size and air density are increased to produce high Reynolds numbers. More detailed information on subsonic wind tunnel design and operation may be found in Pope and Harper (1966) and *Problems of Wind Tunnel Design and Testing* [AGARD (1973)]. Pirrello *et al.* (1971a) contains descriptions of most major wind tunnels in the United States covering all speed ranges.

Since subsonic phenomena and characteristics can often be inferred from incompressible flow data, hydrodynamic facilities, such as water tunnels, can also be used. Such facilities are described in Sec. 16.4. Allowance must be made, however, for the effects of added mass, free surface, cavitation and other phenomena peculiar to liquid media.

Environmental effects, such as wind loads on buildings and other facets of wind engineering, can also be measured in subsonic wind tunnels. Such facilities, which may incorporate special simulation and measurement techniques, are described in Sec. 24.3.

In most tests involving flight vehicles, some means must be provided for supporting the model. The support structure then interferes with the flow over the model. Various techniques, described in Sec. 16.7.2, are employed to subtract out, minimize, or eliminate these undesirable effects. In addition, vertical wind tunnels have been constructed [Vayssaire (1973)] in which the model weight is counterbalanced by its drag such that the model floats in the vertical stream of air. Spin recovery, drag, and certain dynamic characteristics can be measured in such facilities.

16.3.3   Transonic Facilities

Special Test Sections.   In the range of Mach numbers near unity ($0.6 < M < 1.2$), compressibility effects produce large changes in aerodynamic characteristics, and shock-wave/boundary-layer interactions can cause violent dynamic phenomena, such as buffeting. Testing in this flight regime presents special difficulties. If the empty wind tunnel is operating at $M = 1$ in the test section, then when a model and support system are inserted into the flow the tunnel will choke (reach sonic speed locally) in the vicinity of the model. The undisturbed free stream ahead of the model will then be subsonic. Thus, in such a facility, it is possible to simulate sonic flight only in the limit of an infinitesimally small model. At low supersonic speeds, shock waves will tend to reflect from the wind tunnel walls onto the model thus distorting the flow simulation.

Both of these problems can be relieved by equipping the test section with slotted or perforated walls which allow the flow to expand into a plenum surrounding the test section. The walls can be designed to allow some of the air to escape into the plenum chamber, thus relieving the tendency to choke the flow at the model. In addition, shock waves reflecting from solid parts of the tunnel wall will be cancelled by expansion waves reflected from the holes. The precise flow patterns depend on the model being tested. Hence, in some facilities slot configurations are adjustable; and the adjustment can even be made automatically in accordance with some measured flow parameter. Design features of these porous test sections are described in Vayssaire (1973) and Murman (1972). Slotted or perforated test sections may be incorporated in continuous flow facilities or blowdown types.

Blowdown Wind Tunnels. Since shock-wave/boundary-layer interactions strongly influence transonic flow phenomena, it is especially important to correctly simulate both Mach number and Reynolds number in transonic facilities. High power requirements make continuous flow wind tunnels costly to build and operate. Therefore, *blowdown wind tunnels* are often employed. In fact, the cost savings make such facilities popular in all high-speed test regimes.

A blowdown wind tunnel is shown schematically in Fig. 16.3. In this facility, air or other test gas is stored under pressure. To make a test run, a valve or diaphragm is opened, initiating flow in the test section. The Mach number is determined by the shape of the nozzle, which may be slotted or perforated for transonic flows. The pressure is regulated to maintain constant conditions in the test section. It may

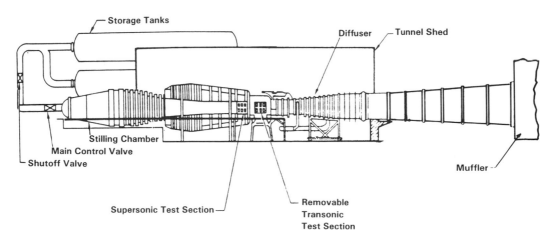

FIGURE 16.3 Blowdown wind tunnel. (From Pirrello *et al.*, 1971a.)

be necessary to heat the air also in order to maintain a constant flow temperature and to avoid air condensation at high Mach numbers.

At high Mach numbers, the pressure ratio required to drive the air through the tunnel may be larger than available with an atmospheric discharge. In that case, the system includes ejectors or an evacuated outlet tank and associated vacuum pumps and possibly a cooler. More complete descriptions of blowdown wind tunnels may be found in Pope and Goin (1965).

Tube Facilities. High Reynolds number test conditions can also be achieved in a *tube wind tunnel* [Ludwieg (1957) and Enkenhus and Merritt (1971)] which provides uniform flow conditions for a period of time determined by the length of the tube.

The principle of operation of the Ludweig tube, for example, is illustrated in Fig. 16.4. When a diaphragm is ruptured, a shock wave travels along the tube at approximately sonic velocity. Air stored behind the diaphragm then flows into the test section at a Mach number determined by the nozzle geometry. The properties of the air in the reservoir and throughout the test region remain constant until the opening shock wave travels to the end of the storage tube and its reflection comes back. The run time in a large facility can be sufficiently long to permit reasonable measurement quantity and quality. By storing air in a large diameter pipe at high pressure, it is feasible to achieve high Reynolds number transonic flow at relatively low cost.

The shock tube, as a device for producing a high temperature flow, is discussed in Sec. 16.3.6.

Cryogenic Wind Tunnels. Another means of producing high Reynolds number transonic flow is to run at low temperature. Since viscosity is reduced at low temperature, a blowdown wind tunnel using nitrogen gas that has been boiled off a liquid nitrogen source will have 4–5 times the Reynolds number of an equivalent room-temperature facility. A different type of cryogenic wind tunnel circuit, described in Sec. 16.3.6, is used to simulate high altitude conditions.

Track Facilities. To avoid wind tunnel wall interference, it is sometimes desirable to mount a model on the front of a sled-like vehicle that is pushed along a straight track by rockets. Examples of such facilities are described in Pirrello *et al.* (1971c).

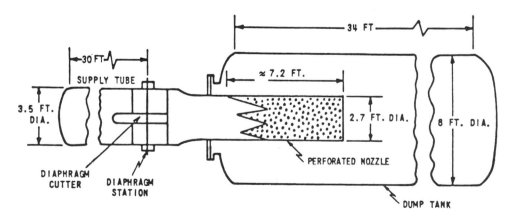

FIGURE 16.4   The Cornell Aeronautical Laboratory Ludwieg tube.

Aerodynamic forces can be measured by mounting the model on a balance, but it is not generally feasible to vary angle of attack or other model parameters during the short time of the test.

Although much effort goes into the construction of a smooth track, the sled vibrates somewhat making the data noisy. Nevertheless, valid and accurate data can be obtained in such facilities. Ground proximity also influences the flow, but not seriously. Since only atmospheric conditions are available, Reynolds number and Mach number (which can include the supersonic regime) can not be adjusted independently except by model scaling. The technique works better for missiles than for aircraft.

A useful feature of this type of facility is the ability to test the effects of rain and dust on missile survivability by controlling the environment of the test. It is also possible to simulate some features of store ejection from aircraft without endangering airplanes or pilots. The high acceleration provided by the rocket booster is also useful for tests of the structural integrity of vehicle components.

16.3.4 Supersonic Wind Tunnels

A closed-return supersonic wind tunnel facility is illustrated in Fig. 16.5. The Mach number is controlled by the shape of the supersonic nozzle, while the Reynolds number can be established by adjusting the pressure level of the test medium by means of auxiliary compressors and vacuum pumps. Blowdown wind tunnels are also employed in supersonic testing. Generally, long test time is sacrificed for high Reynolds number.

As in subsonic wind tunnels, static, and dynamic aerodynamic forces and pressures can be measured on models of aircraft, missiles, or other vehicles. At supersonic speeds, aerodynamic heating can be important, hence temperature and heat transfer rate distributions are often measured. Flow field characteristics can be measured by means of probes or other techniques so that engine inlet conditions or shock wave patterns, for example, can be examined.

Rocket exhaust effects and air-breathing engine internal flows can be simulated

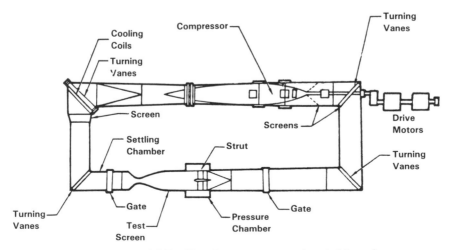

FIGURE 16.5 Closed return supersonic wind tunnel.

[Pirrello *et al.* (1971a)] in supersonic wind tunnels; and in some facilities, combustion can be accommodated [Ward *et al.* (1977)].

Supersonic wind tunnels are frequently used for research into flow field and boundary layer phenomena, but transition to turbulence is strongly influenced by inherent flow field noise. Transition location is, therefore, often studied in ballistics ranges (see Sec. 16.3.5), which also are useful for certain dynamic stability measurements, and, when pressure can be adjusted, for studies of effects of Reynolds number variation.

The Mach number, in some wind tunnels, is changed by replacing removable nozzles. Some installations incorporate flexible or sliding nozzles for changing the Mach number [Pirrello *et al.* (1971a), Kenny and Webb (1954), Rosen (1955), and Jackson *et al.* (1981)]. These facilities make possible the study of Mach-number dependent phenomena, such as the onset of buzz in inlets.

16.3.5 Hypersonic Facilities

Hypersonic Wind Tunnels. The range of *hypersonic* Mach numbers will be assumed here to include, very approximately, $5 < M < 10$ (we will denote $M > 10$ as the *hypervelocity* regime). In this regime, thermal effects are important as well as aeromechanical phenomena. Hence, it is necessary to be able to simulate Mach number, Reynolds number, and ambient temperature. Since, especially at the higher Mach numbers, it is difficult to simulate all these parameters simultaneously, various combinations are investigated in specialized facilities. Fortunately, aerothermal effects can usually be decoupled from aeromechanical functions.

Hypersonic wind tunnels are similar to supersonic facilities, operating in either a continuous or blow-down mode. A schematic drawing of a transonic/supersonic blowdown installation is shown in Fig. 16.3. However, for hypersonic flow, a heater must be installed ahead of the nozzle in order to prevent condensation of the test gas (usually nitrogen or air). A high Reynolds number hypersonic facility is illustrated in Fig. 16.6. At high Mach numbers, then, the nozzle must be round in order to avoid excessive erosion at the throat. Pirrello *et al.* (1971a), Geineder *et al.* (1967), and Schaefer (1965) describe various hypersonic wind tunnels.

Aerodynamic forces, pressures, and often even heat transfer rates can be scaled from measurements made in wind tunnels which simulate appropriate Mach number and Reynolds number regimes even though flow field temperatures are much lower than those that would be encountered in flight. The stagnation temperature of the flight or test medium can be calculated from the following formula

$$T_0 = T \left(1 + \frac{\gamma - 1}{2} M^2 \right) \tag{16.1}$$

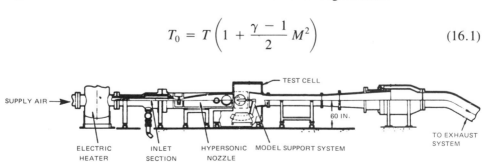

FIGURE 16.6 Hypersonic wind tunnel.

where T is the ambient temperature, M is the flow Mach number, and γ is the ratio of specific heat capacities of the test gas ($\gamma = 1.4$ for air).

Aerodynamic heating effects can also be simulated by mounting a test model in the exhaust of a rocket motor or other combustion generated flow. In such a facility, aerodynamic heating can be simulated at Reynolds numbers appropriate to laminar or turbulent boundary layers. However, since the test medium is chemically different than air, the actual aerodynamic environment is not represented correctly. Therefore, engine exhaust facilities are primarily used to test the survivability of structural components. It is also possible to burn other fuels in air to generate high temperatures and then add oxygen to the combustion products to approximate a high temperature air flow.

Arc-heated wind tunnels, described in Sec. 16.3.6, can also simulate high heat-transfer rate conditions. Since air may be used as a test medium, it is possible to correctly simulate hypervelocity temperature conditions. Maintaining the desired uniform and constant test conditions is very difficult in an arc-heated wind tunnel because local heating tends to wear out components of the facility. Successful testing has been achieved, however, in several large-scale arc wind tunnels.

Ballistic Ranges. Hypersonic flight conditions can be achieved by launching a test model at high speeds in a ballistics range [Edney *et al.* (1977)]. The facility consists of a launcher and an instrumented range. The launcher is a gun barrel. For very high speeds (which can exceed Mach number 20) light gas guns are used in which high pressures and temperatures are achieved by a gunpowder-driven free piston [Charters and Curtis (1964)]. The high pressure gas then accelerates the test specimen in a smooth or rifled barrel. The model is usually protected by a *sabot* that carries it through the launcher and then breaks apart to let the test article fly freely down the range.

Data is generally obtained optically at fixed camera stations. Drag can be deduced from the measured position of the model as a function of time. Static and dynamic coefficients can also be measured from observed model motions. Flow field characteristics are made visible by schlieren or shadowgraph techniques. At high Mach numbers, the model is often obscured by radiation from the heated air in its vicinity. To identify the model geometry precisely, it may be photographed under laser illumination or even X-rays (which are also useful in exposing internal model-sabot components). Onboard instrumentation can measure quantities like temperature or pressure and then transmit the data via a telemetry signal. The difficulty of building an instrument package that will survive the launch, only to be destroyed on impact tends to limit such measurements. However, temperatures can be inferred from pyrometric data obtained by optical means [Reda *et al.* (1975)]. Test time is determined by the length of the instrumented range which is on the order of hundreds of meters in large facilities.

It is possible to create regions of rain drops or other environmental conditions in the path of the projectile. Models may also be preheated or precooled to adjust the simulated conditions.

Since the test medium in the range is at rest, measurements of boundary layer transition are not distorted by extraneous flow disturbances. However, as the Mach number is pushed higher, the model size that can be accelerated gets smaller; hence Reynolds number simulation suffers. By pressurizing the range, the Reynolds number of the test can be increased. Similarly, high altitude phenomena can be studied in an evacuated range.

The model and the shock-heated air radiate in high speed tests. Spectrographic measurements are sometimes employed to infer chemical kinetics.

The ballistics range, with its high-speed launcher is also useful for studying impact phenomena. The effect of projectiles on targets and of targets on projectiles, for example, can be investigated.

16.3.6 HYPERVELOCITY FACILITIES

Hypervelocity Wind Tunnels. The term *hypervelocity* is often used to describe the regime of flight velocities greater than Mach 10. Reentry vehicles and the trans-atmospheric vehicle are primary subjects of study in this regime. The many types of phenomena encountered in this regime are not all addressable in any one facility, consequently several different types of equipment will be described.

Ideally, hypervelocity wind tunnels would simulate Mach number, Reynolds number, and air temperature for long periods of time under constant, known conditions. Unfortunately, the prospect of forcing an extremely hot gas through a small nozzle throat at very high pressure is very unattractive, so that continuous, simultaneous simulation of all hypervelocity flow conditions is not available.

In general, however, phenomena associated with pressure are Mach number-dependent, while thermal phenomena are temperature-dependent. Hence, the test philosophy is to measure forces in a facility that simulates Mach number and Reynolds number and to measure heat transfer, radiation, chemical kinetics, etc. in a facility that simulates appropriate temperatures.

In one type of hypervelocity wind tunnel (Fig. 16.7) the test medium is heated sufficiently to avoid condensation in the test section [Pirello *et al.* (1971a) and Hill (1976)]. Nitrogen is a convenient test gas, because it does not oxidize heater elements as hot air would. Nitrogen has the same γ as air. After heating, the test gas is fed into the wind tunnel as in a blowdown facility. To simulate high Reynolds

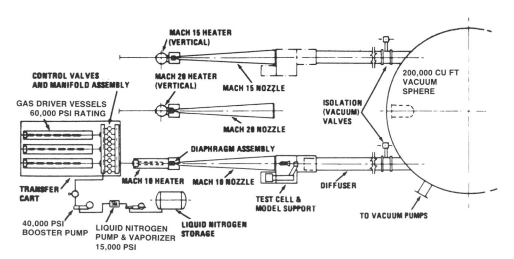

FIGURE 16.7 Naval Surface Warfare Center hypervelocity wind tunnel.

number, the supply pressure must also be very high. It is so difficult to heat this enormous flow of air (or nitrogen) that instead, a fixed mass of the gas is heated and forced through the nozzle, and the test run ends (on the order of seconds later) when the hot gas has been expended. At Mach numbers near 15, the pressure ratio required to start the flow through the nozzle is so large that it is necessary to evacuate the tunnel downstream of the nozzle.

Forces, pressures, temperatures, and other aerodynamic properties can be measured. Instrumentation and data reduction are complicated by the short run times, but it is possible to pitch or roll the model through the desired test range in one second, while taking data continuously.

So-called *hot-shot* wind tunnels [Pirrello *et al.* (1971a) and Lukaciewicz *et al.* (1961)] operate on the same principles but the test gas (nitrogen or air) is heated by an electric arc instead of a resistance heater. The mass of gas that can be heated is limited by the available power supply. Run times in hot shot wind tunnels are measured in hundredths of seconds (10–300 milliseconds). A hot-shot facility is illustrated schematically in Fig. 16.8.

Helium tunnels are also useful in the study of certain high Mach number fluid mechanics phenomena [Pirrello *et al.* (1971a) and Schaefer (1965)]. Helium has the advantage of a very low boiling point. Therefore, relatively little energy is required to accelerate cold helium to high Mach numbers, and heating to avoid condensation in the test nozzle is unnecessary. Thus, a helium wind tunnel is relatively inexpensive to build and operate. However, helium has a different ratio of specific heats, γ, than air, and γ is important for high-speed phenomena. A theoretical hypothesis expressible in terms of general gas properties can be tested in helium just as well as in air. In addition, many phenomena can be observed qualitatively in a helium atmosphere.

The hypervelocity wind tunnels described above do not provide aerodynamic heating levels commensurate with a free flight environment. Heating and related phenomena are very important in the hypervelocity flight regime, but they are even more difficult to simulate than in the hypersonic case. It is possible, however, to simulate the temperatures associated with hypervelocity flight in arc-heated wind tunnels [Pirrello *et al.* (1971a) and Davis and Carver (1992)]. High stagnation temperatures are obtained by heating air with an electric arc and then expanding the flow through a nozzle to supersonic speeds. Since the facility runs at modest supersonic Mach numbers ($2 < M < 4$, for example), the stagnation pressure is not high;

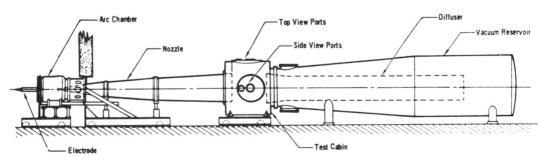

FIGURE 16.8 Hotshot wind tunnel. (From Pirrello *et al.*, 1971a.)

thus nozzle erosion problems and high power requirements are avoided. Of course, the flow can be expanded to higher Mach numbers, but then the density and Reynolds number are correspondingly reduced. Arc facilities may be of continuous or blowdown type, heat being supplied by an electric arc.

At the model, however, the local flow can experience high temperature effects, including dissociation, radiation, and other chemical kinetic phenomena. Heat and mass transfer conditions can be simulated for material screening and for determining cooling requirements, for example.

Since hypervelocity flight is often associated with reentry vehicles, high altitudes may accompany the high Mach numbers. At sufficiently high altitudes, the Knudsen number (ratio of mean free path to reference length) becomes large, and the laws of continuum fluid mechanics are invalid. For example, the fluid does not come to rest at the vehicle surface, and collisions among molecules are rare compared with collisions with the body (see Chap. 11). Achievement of a suitable test environment to simulate such conditions requires very low densities and related high vacuum technology.

Expanding a perfect gas to a Mach number of 15, for example, results in a density ratio of about 7×10^{-5} compared with conditions in the settling chamber. To begin to simulate rarefied flow regimes, the density in the settling chamber would have to be about 1/25 of atmospheric density. Then, in order to start the flow, the density downstream of the nozzle would have to be much lower. These conditions are achievable in a cryogenically-pumped wind tunnel [Schaef and Chambre (1958)] in which the working fluid (air, for example) is condensed downstream of the nozzle by cooling it, thereby reducing the local pressure to that associated with the vapor pressure of the liquid. *Slip flow*, at least, can be studied in cryogenically-pumped wind tunnels.

Shock Tube Wind Tunnels. In order to characterize chemical kinetics of air or other test media, it is necessary to determine rate constants for the various chemical reactions (dissociation, for example) that occur at high temperatures. Experiments in this regime can be performed in a shock tube which heats the test gas by passing a strong shock wave through it [Pirrello *et al.* (1971a) and Gaydon and Hurle (1963)]. The strong shock wave is produced by a driver gas stored at high pressure and separated from the test gas by a diaphragm. When the diaphragm ruptures, a shock wave from the driver compresses and heats the test gas. In this way, a supply of gas at uniform high temperature and high pressure conditions is produced. This gas may be expanded through a nozzle to supply a test environment for a few milliseconds, or the compressed gas can be studied by optical instruments to extract rate constants for chemical kinetic processes. Test times depend on the length of the shock tube. The shock tube may be operated at relatively low pressures as a cheap supersonic wind tunnel, but test times are rather short for such applications.

Shock tubes can be exploited in various configurations. In an expansion tube [Beach and Bushnell (1990)] the shock-heated driver gas (which is moving at a modest supersonic speed) can be expanded in a nozzle to the desired test Mach number, or it can be accelerated by expansion into a lower pressure extension of the shock tube. In either case, test time is limited to the few milliseconds that it takes for the supersonic shock-heated driver gas to exit from the shock tube.

In the reflected shock configuration [Holden (1986)], the initial shock wave reflects off the end wall of the shock tube, further heating the driver gas and turning it into a reservoir of test gas at stagnation conditions which can (after breaking a diaphragm) then be expanded through a nozzle into a test section. Meanwhile, the reflected shock wave travels upstream until it encounters the contact surface separating the driver gas from the driven gas, then the disturbance generated in this collision travels down the shock tube into the test section where it causes an abrupt change in the free stream conditions, ending the test. The initial conditions of the driver and driven gases can be selected such that the pressure behind the reflected shock matches the pressure at the contact surface, and no disturbance propagates from the collision. In this *tailored interface* condition [Witliff *et al.* (1959)], essentially the entire mass of the driver gas becomes the supply reservoir, and longer run times (tens of milliseconds) are feasible.

Another class of shock tube wind tunnel heats the driven gas with multiple shock waves bouncing back and forth between the end wall of the driven tube and the advancing contact surface [Schindel (1993)]. In this arrangement, the driven gas will be heated by several passes of the shock wave until the pressure in the driven gas essentially matches the driver gas pressure. Then, the entire reservoir of driven gas may be expanded through a nozzle to desired test conditions. This concept has an advantage over the tailored interface device in having a wider range of operating conditions, but it requires some external means of breaking a diaphragm (or opening a valve) to start the flow in the test section. By contrast, the high pressure jump across the reflected shock can be used to start the flow (break a diaphragm) automatically in that type of facility. In these devices (reflected shock and multiple shock), a high pressure, high temperature gas must be squeezed though a nozzle throat to achieve supersonic (hypersonic) test conditions. In the expansion tube, the shock-heated driven gas is initially supersonic, hence no throat is required.

Many other configurations are possible employing multiple diaphragms, valves, pistons, and various geometrical arrangements [Bogdanoff (1990)]. In general, performance calculations can be carried out with some suitable variation of a one-dimensional computer code [Seigel (1965)] originally developed for analysis of light-gas guns.

Analysis shows that performance is enhanced by packing as much energy as possible into the driver gas. Thus, heated light gases (hydrogen or helium) are often used as drivers. Since the run time is short, it is not necessary to heat the driver for a long period of time. Compressing the gas with a free piston is, therefore, a commonly used method of storing extra energy in the driver [Stalker (1966)].

Shock tubes have been combined with wind tunnels to study the effects of strong blast waves impinging on a vehicle in flight [Merritt and Aronson (1965)].

Hypervelocity Ranges. Ballistics ranges, described in Sec. 16.3.5, are capable of simulating hypervelocity environments and are useful for measurements of ablation and erosion as well as other aerodynamic properties. A special type of facility combines a range and a wind tunnel.* Here, the model is launched at high speeds through the test section of a wind tunnel giving a velocity equal to the sum of the model plus the flow velocity in the wind tunnel. Since the test occurs only during the transit of

*See *Research Facilities Handbook*, Arnold Engineering Development Center, Tullahoma, TN.

the wind tunnel test section, data from such a facility is limited. Radiation from shock waves is measured, for example.

If the launcher velocity is adjusted to cause the model to just reach the wind tunnel test section before slowing to a halt, then a so-called wind-tunnel free-flight test can be performed [Prislin (1966)]. Since the model can be made to undergo several oscillations during its short residence in the wind tunnel test section, it is possible to measure dynamic characteristics without the encumbrance of a model support.

Another special facility of this type combines a shock tube and a ballistics range in which effects of blast waves (produced by the shock tube) can be observed as the blast waves pass over the model [Baltakis *et al.* (1967)].

Norfleet *et al.* (1977) and Roper (1981) describe a ballistics range that is equipped with a track that restrains the model, which is propelled to high speed by a rocket launcher. Measurements of ablation, for example, can be made on an asymmetric model nose without causing the model to fly out of the field of view of the instrumentation.

Free Flight. Hypervelocity flight phenomena can be studied by launching a vehicle downward at high speed. A rocket first carries the model to a high altitude, then turns around and accelerates it downward. During its high-speed run, sensors may measure temperature, pressure, and other desired parameters. The data is either telemetered during the flight or recorded on board for analysis after recovery. Parachutes or other retarding devices can be used to decelerate the vehicle sufficiently to allow recovery of the instrumentation. Such a procedure can be used at lower speeds also, but its use is restricted by cost to situations which cannot be simulated by other means.

A number of free flight ranges are available for testing of projectiles [Edney *et al.* (1977)]. These ranges allow investigations of warhead penetration and effectiveness as well as some projectile flight characteristics.

Aircraft and missile flight test facilities will not be discussed here because their primary function is to provide real estate at which product-dependent testing can be conducted.

16.4 HYDRODYNAMIC FACILITIES

16.4.1 Introduction

Hydrodynamic facilities, such as water tunnels, can be used to study underwater flows (or flows in other incompressible media) where the primary simulation parameters are Reynolds number (Re $= \rho UL/\mu$), and cavitation number ($\sigma = 2(p_\infty - p_v)/\rho U^2$). Cavitation is discussed in Sec. 23.8. Also, because of the high density of water, it is sometimes convenient to simulate high Reynolds number internal or external air flows in water tunnels.

Water surface phenomena can be investigated in open channels or towing tanks. Here, besides the appropriate scaling parameters, it may also be necessary to simulate wave motions and other surface-related phenomena. The Froude number (Fr $= U/\sqrt{gL}$) is identified with scaling of free surface effects.

Effects associated with water entry, as with air-dropped mines, for example, can be studied in open ponds or other bodies of water. Better instrumentation and control can be obtained in special hydrotank facilities. The same similarity parameters apply, including surface waves. Water exit phenomena are also of interest in application to submarine-launched missiles, for example. Water exit phenomena may also be studied in water tank facilities.

Another interesting type of hydrodynamic facility is the water table which operates on analogy between local variations in water depth and pressure variations in a supersonic flow.

The following sections describe types of testing appropriate to these various facilities.

16.4.2 Water Tunnels

A water tunnel [Lewis (1947)] is defined here as a closed test section which is filled with a uniform flow of water. The water may flow from one reservoir to another under the action of gravity or may be circulated by a pump. Thus, the test section of a water tunnel is similar to that of a wind tunnel with water replacing air as the test medium. In fact, many experiments in incompressible fluid mechanics can be performed in either type of facility. Boundary layer phenomena, for example, may be studied in either medium.

However, forces on bodies are accompanied by an *added mass* effect [Milne–Thompson (1960)] which, like buoyancy, depends on the mass of the displaced fluid. The added mass is thus negligible in gases but not in liquids. Added mass is discussed in Chap. 1.

Furthermore, flow in water may be accompanied by *cavitation* if the local pressure in the medium is less than its vapor pressure [Birchoff and Zarantonello (1957)]. In this case, the flow is accompanied by bubbles or cavities containing the vaporized form of the liquid. Besides strongly affecting the force distribution on the moving body, the cavities tend to form and collapse with destructively high local pressures.

Water tunnel scaling may require simulation of two parameters: Reynolds number, Re, where viscous effects are important; and cavitation number, σ, when cavitation is present. Froude number, Fr, does not apply since free surface does not occur in a water tunnel.

Water tunnels are, therefore, used to measure forces on submerged bodies, such as torpedoes or submarines, and also to simulate cavitation effects associated with high local flow velocities (and therefore low pressures) as on propellers and high speed vehicles. Water tunnels may also be convenient facilities for studies of incompressible viscous flow phenomena when the only significant flow parameter is the Reynolds number.

16.4.3 Towing Tanks and Water Channels

Forces on surface ship models are often measured in towing tanks [Lehman and Summey (1984)]. If the Froude number of the test matches that of the full scale vehicle, then surface phenomena, such as bow wave patterns, will be correctly simulated. On the other hand, Reynolds number and cavitation number will not be duplicated. It is, therefore, necessary to account for viscous effects by calculated corrections.

Assuming that the model is smaller than full scale, the model velocity required to match the Froude number will also be reduced. At the slower speeds, cavitation is less likely to occur than would be the case at full scale. Therefore, if there is any likelihood of cavitation, it must be investigated in some facility where cavitation number can be correctly simulated.

Effects of surface waves on ship performance can be studied in a towing tank by generating appropriate wave conditions. Waves are produced by a two-dimensional plunging vane that spans the tank. Wave height and spacing are varied by adjusting the amplitude and frequency of the wave-maker. The orientation of the waves with respect to the model can be varied to some extent, but reflections from the walls of the towing tank can severely limit the availability of regions of simple wave behavior.

A basic towing tank experiment would be a measurement of forces on a model which is towed through the water at uniform speed by a moving platform on which the instrumentation is mounted. The test time is limited by the length of the tank. The layout of a towing tank facility is shown in Fig. 16.9.

It would seem to be more convenient to hold the model fixed and move the water in an open channel [Preston (1966)]. However, it is then necessary to pump or supply large quantities of water in order to maintain the circulatory flow. Another limitation on open channels is the ability to maintain quiet surface flow conditions. At low flow speeds, a carefully designed facility will have a quiet flow, but, as the flow speed increases, surface perturbations become unstable and smooth flow conditions cannot be maintained. For these reasons, open channels may be more desirable than towing tanks only for low speed experiments or for qualitative studies. Fig. 16.10 shows a schematic drawing of a water channel.

16.4.4 Water Entry Facilities

A water entry facility is defined here as a large body of water (natural or man-made) instrumented primarily for studies of water-entry and water-exit phenomena [*Morris Dam Test Range*, NUSC (1973)]. The problems that are typically addressed deal with the underwater trajectory of vehicles subsequent to water entry. It may be em-

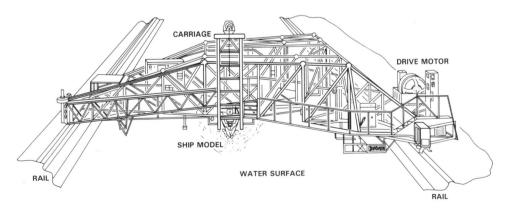

FIGURE 16.9 Towing tank section showing carriage equipped for photography. (From Lehman and Summers, 1984.)

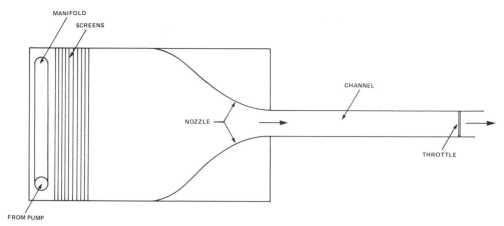

FIGURE 16.10 Sketch of the plan view of a water channel.

barrassing, for example, if an air-delivered mine broaches after water entry and comes to rest on dry land or in some other undesirable location.

Stresses on structural elements and shedding of disposable components (air inlets, for example) can also be studied in water entry facilities. Such investigations usually require the use of full-scale production hardware.

A water entry facility requires a launcher which can propel the test vehicle to the desired speed and direct it at the water surface along a prescribed path and orientation. The test object then enters a body of water where its underwater trajectory can be recorded photographically or located approximately by having it pierce underwater screens.

The vehicle may enter a pond or other natural body of nonflowing water. However, outdoor facilities have disadvantages of exposure to uncontrolled effects of winds and waves. Also, underwater photography in natural ponds is severely degraded by the poor transmission of the medium. For these reasons, and for convenience of operation and maintenance, the experiments are often conducted in indoor hydrotank facilities [Seigel (1966)]. The hydrotank is equipped with launchers and instrumentation for measuring velocities, pressures, stresses, and trajectories. The flight path of the model, as it enters and traverses the water, can be photographed through windows in the sides of the tank. The water in the facility is filtered to maintain good transmission for photography.

When it enters the water, the wake of the vehicle contains a pocket of air extending to the surface. As the vehicle continues into the water, the rear of this trailing cavity closes off. Depending on the underwater trajectory, the cavity may collapse entirely leaving the vehicle surrounded only by water. During the open- and closed-cavity stages of its flight, the vehicle motion is influenced strongly by the nature of the cavity in which it is encased. To correctly simulate the formation and development of the cavity, it is necessary to match both the Froude number and the cavitation number of the full scale vehicle.

Froude scaling requires that the velocity be proportional to the square root of a physical dimensional. The dynamic pressure under water, $\rho U^2/2$, will then scale with the model length. In order to maintain cavitation number simulation, the stream pressure must scale with the dynamic pressure. This relation will be maintained under water if the depth at which it travels is measured in the model scale. However, the pressure of the air at the surface must also be adjusted to correctly simulate the open-cavity phenomenon. Therefore, the space above the water must be pressurized or evacuated (for subscale models) to match a full scale cavitation number.

Water-entry facilities can be used to study the trajectories and other flow phenomena associated with mines, depth charges, torpedoes, and other vehicles that are designed to penetrate the water surface. It is also possible to study the flow around submerged vehicles, such as submarines and torpedoes, by towing models under water; although other facilities, such as water tunnels, may be better suited for investigations of this type. Water-entry-type facilities generally offer an opportunity for larger scale testing than is available in water tunnels.

Reynolds number cannot be simulated in a subscale test in which Froude number has been matched. In principle, a test medium with the proper viscosity coefficient would permit independent adjustment of Reynolds number. Low viscosity liquids are not, however, commonly available, hence viscous effects must be isolated and scaled analytically. The scaling process is more accurate if transition to turbulence is similar in both full scale and model test. Boundary layer trips may be useful for this purpose.

Waves may also be generated artificially by a plunging vane. The amplitude and wavelength of the waves are scaled to the model size for correct simulation of full scale conditions. Water entry phenomena can be significantly influenced by large surface waves.

Water-exit effects can also be studied in these facilities. In principle, the basic modification required is an underwater launcher. To simulate the proper trajectory conditions at the surface, it may be desirable to launch from a moving platform. Trajectories of projectiles, rockets, and other missiles fired from underwater platforms can be simulated. If the launch platform is moving when the missile is ejected, it is necessary to simulate this motion in the water exit experiment.

Structural loads at the air–water interface are not likely to be of great concern during water exit.

Capture and intact recovery of the model may not be feasible in the case of large vehicles that are not designed to withstand impact loads. However, many vehicles are traveling slowly enough at water exit that they can be stopped by nets right after they have traversed the water surface. In this manner, indoor facilities can also be conveniently used for water-exit investigations.

16.4.5 Water Tables

The equations of motion governing the height of waves over a shallow body of water are analogous to the equations for pressure distribution in two-dimensional compressible flow [Bryant (1956)]. The ratio of specific heats ($\gamma = 1.4$ for a diatomic gas such as air) which appears in the compressible flow equations is replaced by the factor 2 in the equations for wave height of a liquid. Therefore, many phenomena can be studied qualitatively by towing a model through a shallow body of water, or by directing a shallow flow over a two-dimensional model in a so-called water table facility. The model is two-dimensional rising from the surface of the table through the surface of the water.

Bow waves about complicated two-dimensional configurations can be observed conveniently in a water table facility. The wave patterns will be similar to shock waves around the same shape in supersonic flow. Pressure distributions in channels simulating supersonic nozzles and diffusers can be studied in a water table by relating the height of the water to the pressure distribution in a gas. Some three-dimensional phenomena, such as blast waves, can be visualized in a water table experiment, although only a two-dimensional representation is observed. The advantage of the water table over a wind tunnel is in its convenience and low cost. It is better suited for phenomenological investigations than for quantitative measurements.

16.5 PROPULSION TEST FACILITIES

16.5.1 Introduction

Engines used to drive moving vehicles, from lawnmowers to aircraft, operate in a flowing stream of air. At high speeds, the flow of air has a significant effect on the performance of the engine, consequently it is necessary to simulate the external flow in order to measure the engine characteristics properly.

In some cases, a propulsion test facility is simply a wind tunnel in which the model is an engine instead of a vehicle (or it may be an engine mounted on a vehicle). However, the quantities that are measured in an engine test are not those associated with vehicle aerodynamics, and propulsion phenomena may not scale in the same way. Therefore, wind tunnels and other facilities that are used for testing propulsion systems or components require special capabilities.

Rocket engine testing will be included in the following subsection; the remainder of this section will concentrate on air-breathing propulsion.

An air-breathing propulsion system consists of an air inlet and associated ducting and perhaps a diffuser or compressor to bring the air into the combustion chamber. Air is mixed with fuel and burned in the combustion chamber, where the heat from the chemical reaction raises the enthalpy of the working medium. This added chemical energy is then used to drive pistons or turbines and may be accelerated through a nozzle to provide jet thrust. Facilities must simulate these processes. In addition, it may be necessary to provide for air which by-passes the combustor and for simulation of afterburners which may follow the combustion chamber.

Therefore, in testing propulsion systems, measurements are required of the performance of the inlet system leading to the combustion chamber, of the combustion process, and of the progress of the products of combustion as power is produced. Before combustion, the fluid processes are related to those accompanying any configuration and may be scaled accordingly. Mach number and Reynolds number are the primary aerodynamic parameters that must be simulated in a test. The combustion process is subject to different scaling laws depending on the reaction rates of the chemical processes. The products of combustion again are subject to appropriate gas dynamic scaling providing that remaining chemical activity can be neglected. Propulsion test facilities described in the remainder of this section are designed to provide the necessary environments to permit measurements of each phase of the propulsion system process.

In engine test stands, many of the components of the complete engine can be operated together in full scale. This capability makes possible the optimization and adjustment of individual components and the determination of regimes of stable operation. Special tests can be devised for measurements of such quantities as exhaust emissions, acoustics, or effects of ingestion of foreign material (including rain and ice). In addition, mechanical integrity, durability and the adequacy of material properties under realistic thermal and vibrational conditions can also be evaluated, so that flight clearance can be established. Test stands operating at sea-level conditions are especially convenient for durability testing.

The performance of pumps, valves, and other mechanical components must also be measured and evaluated, but separate testing of mechanical components will not be treated in this section.

16.5.2   Static Test Cells

A test cell [Pirrello (1971b)] is a facility for measuring engine performance in a stationary environment. Usually, full-scale equipment is employed. This type of test is applicable to reciprocating engines, gas turbines, turbojets, and rockets.

At a minimum, the facility must provide a support for the engine, a means of measuring the output (shaft horsepower, torque, or thrust) and the input in terms of fuel rate. Other quantities, such as air flow rate, internal pressure and temperature distributions and the velocities of moving parts may also be determined.

For air-breathing engines, the measured quantity is static thrust or static power, and the measurements are corrected to standard sea-level conditions. The effects of altitude are often simulated by partial evacuation of the test cell. An engine test cell installation is shown in Fig. 16.11.

Rocket thrust, F, is given by the equation

$$F = \dot{m}V_e + A_e(p_e - p_\infty) \qquad (16.2)$$

where $\dot{m}$ is mass flow rate, kg/s; V_e is exhaust velocity at the nozzle exit plane, m/s, A_e is the area nozzle exit plane, m^2, p_e is the static pressure at the nozzle exit plane, Pa, and p_∞ is the ambient pressure, Pa. The term $\dot{m}V_e$ in the rocket thrust equation is unaffected by the ambient environment as long as the exhaust flow fills the exit nozzle and is entirely supersonic. Many rockets are designed for high-altitude operation, however, and the flow may separate inside the nozzle when the ambient pressure is much higher than the design condition. Therefore, a complete measurement of rocket performance requires tests at various environmental pressure levels (see Fig. 16.12). Of course, the pressure on the outside of the rocket will decrease when the pressure in the test cell is reduced. Then, the measured thrust of the rocket will increase even if the internal flow is unseparated and supersonic. This

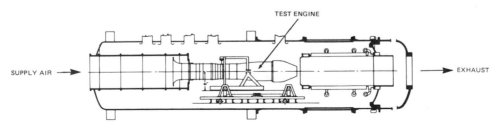

FIGURE 16.11   Jet engine test cell. (From AEDC.)

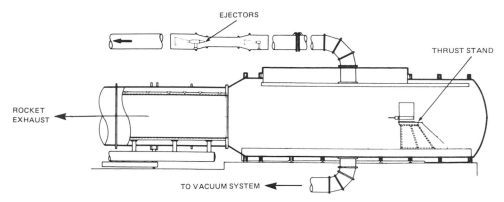

FIGURE 16.12   Rocket engine test cell. (From AEDC.)

effect, due to the ambient pressure acting on the nozzle exit cross-sectional area, can be easily taken into account in interpreting the performance data.

16.5.3   Inlet–Diffuser Test Facilities

An airbreathing propulsion system requires a combustion process or its equivalent to provide an energy source. The internal flow of air through the inlet plumbing is only indirectly affected by combustion and hence can be investigated independently.

The air intake system will begin at an inlet, which includes all components of a vehicle forebody that influence inlet performance. The inlet is followed by ducting and frequently by a diffuser where high pressure is recovered from the momentum of the inlet air. A compressor (fan) may also be incorporated in simulating turbojet inlet systems. For a discussion of air inlets, see Secs. 25.7 and 25.8.

Wind tunnels (described in Sec. 16.3) are applicable to the study of these air-handling devices. The fact that the flow is internal, rather than over the external surface of a vehicle, does not fundamentally alter the test requirements or conditions. At low subsonic speeds, the Reynolds number is the most important simulation parameter. At high speeds, the Mach number must also be simulated. Shock-boundary-layer interactions and associated flow separation can strongly influence inlet performance, therefore it is important to test in the proper Mach number and Reynolds number regimes.

Three types of test set-up are employed: *direct-connect, semi-free jet*, and *free jet*. In direct-connect testing, all of the air flow is internal. A nozzle feeds air directly into the inlet–diffuser model. The Mach number and mass flow of air are adjusted to the expected inlet conditions.

In semi-free jet tests, the model includes sections of the forebody ahead of the inlet so that their effect on the inlet flow may be investigated. As in the direct-connect type of experiment, the flow is entirely enclosed within the model–nozzle combination.

In free-jet tests, the entire forebody of the vehicle including the inlet is immersed in a wind tunnel with both internal and external flow being simulated. Free-jet tests can introduce special problems in supersonic wind tunnels. If the model is too large, or if the internal flow is not properly initiated, then the flow in the test section of the wind tunnel may be choked and never attain the design flow conditions. In any case, the internal flow must go somewhere. It may be exhausted into the tunnel downstream of the model, or it may be necessary to provide a separate exhaust system to remove the inlet air from the tunnel flow system. Boundary layer removal may also be required.

The quantities of major interest relating to air intake systems are the pressure and mass flow of the air introduced into the combustion section. The distributions of flow quantities are also of interest, so flow field properties are often surveyed with probes or other local measurement techniques. In addition, possible causes of poor performance, such as flow separation, vorticity, strong shock waves, or viscous effects may be investigated. Models with transparent walls with application of flow visualization are often of great value in this regard.

At high Mach numbers, heating effects can influence the design of inlet components, therefore heat transfer and temperature distribution may be measured. At hypersonic speeds, the air may be partially dissociated when it reaches the combustion chamber. Since the chemical kinetics influences the combustion process, it is then necessary to determine the actual state of the air at that point. True temperature simulation is very difficult to obtain. Arc-heated wind tunnels and other high-temperature facilities are described in the sections on hypersonic and hypervelocity facilities (Secs. 16.3.5 and 16.3.6).

16.5.4 Facilities for Testing Turbomachinery

Two-Dimensional Cascades. In the development of propulsion systems, the various components are optimized separately and are constrained to operate efficiently when combined. Thus, studies of the performance of isolated turbines and compressors contribute significantly to the design of the system. (See Chap. 27 for discussions of turbines and compressors.) Similarly, the flow in these components is investigated in various facilities in order to establish the data for design trade-offs. Turbines and compressors, for example, may be tested in facilities of three basic types. Two-dimensional cascade facilities are the simplest and, therefore, are widely exploited. Another type of facility tests nonrotating annular configurations in a swirling inlet flow, and, finally, rotating components are tested.

A cascade is a stationary array of two-dimensional airfoils which are arranged to simulate the flow conditions experienced by successive blades of a single stage of a rotating turbine or compressor. A cross section of a cascade array is shown schematically in Fig. 16.13. Since the flow is normally steady in the cascade, it is easy to measure forces and pressures on the airfoils and to probe the flow field. By changing airfoil sections and other geometric and flow parameters, the compressor or turbine can be designed for efficient operation, and off-design performance can also be inferred.

Air or simulated combustion products can be blown over the cascade at the desired pressure, temperature, and velocity, in a manner similar to other wind tunnel applications. To simulate the effects of all the blades on a rotating component, a large number of airfoils (5 to 15) may be used. In some supersonic installations, however, only the nearest neighbors influence each other, so that two blades may be sufficient.

Cascade facilities can also provide valuable flow field information needed for the development or verification of two-dimensional analyses. To reduce the influence

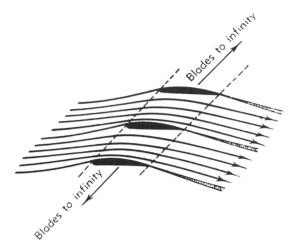

FIGURE 16.13 Schematic of cascade array. (Reproduced with permission, Erwin, J. R., "Experimental Techniques," Vol. X, Section D, *High Speed Aerodynamics and Jet Propulsion*, Princeton University Press, Princeton, NJ, 1964.)

of side walls on the two-dimensionality of the flow, blades of high aspect ratio may be required, or boundary-layer suction may be applied.

Annular Cascades. The three-dimensional characteristics of a turbine or compressor can be more appropriately investigated by blowing the test gas over the geometrically simulated, but nonrotating, component. The rotational velocity of the machine is simulated by using vanes to impart a swirling motion to the incoming gas. Thus, the model is stationary while the incoming gas rotates.

This type of facility permits investigations of the effects of secondary flows around blade edges and similar realistic geometrical details. Also, since the configuration is stationary, probes, pressure orifices, and other flow-measuring instrumentation can be conveniently installed. In addition, unsteady phenomena, such as *rotating stall*, can be studied in this type of facility.

While flow conditions can be readily modified, a change in geometry usually requires an expensive new model or even rebuilding the entire test region since the inlet annulus must mate with the compressor or turbine stage. Annular cascades are described more fully by Erwin (1964).

Rotating Rotor Test Facilities. Finally, the most realistic simulation of compressors and turbines requires a facility in which the components actually rotate. Rotors are driven by the test gas (or sometimes water) supplied by the facility (turbines) or by a separate power source (compressors). Either single, or multistage configurations can be investigated. The advantages of realistic simulation are counterbalanced somewhat by the difficulty of obtaining good flow field measurements.

In some installations [Erwin (1964)], space is provided for probes between stages or between rotating and stationary components. The performance of multistage equipment would be influenced by such an arrangement, but at least flow field measurements could be made, in principle. Actually, to resolve flow variations in the space between blades of a rapidly rotating device requires instrumentation with extremely rapid response times. Average or periodic data could probably be extracted with difficulty. Probes which rotate with the moving blades are also effective, but then there is the problem of transferring a signal from a moving probe to a fixed reference frame. Some kind of seal or slip ring arrangement must be employed, unless the data is transmitted by a form of electromagnetic radiation.

Besides the aerodynamic information, the rotating model type of facility can apply the centrifugal forces required for structural evaluation. Also, the effects of ingestion of foreign material or of nonuniform entrance flow can be investigated.

As in the case of the annular cascade, the effects of varying inlet and exhaust flows can be readily examined, but variations of geometrical parameters are expensive. Hence, the facilities are generally operated to provide a large amount of information on each particular configuration. Advances in measuring techniques are quickly exploited. [Lakshminarayana (1981)], for example, describes the application of fast response hot wire probes and associated signal processing to the measurement of blade wake velocity distribution including turbulence intensity.

Entire full scale turbine or compressor components are also tested to determine overall performance, stable operating regimes, structural integrity, and environmental effects.

16.5.5   Tests Involving Air-Breathing Combustion

Since chemical processes do not scale in the same ways as other aerodynamic quantities, it is desirable to test combustion chambers in full scale, although combustion-related phenomena can be studied independently in small facilities. In direct-connect tests [Pirrello *et al.* (1971b) and Dunsworth and Reed (1979)], air is fed into the combustion chamber in a manner that closely simulates the expected conditions there. In fact, all or part of the inlet may be modeled as well as the combustion chamber, as long as the model is not too large for the test facility. Thus, proper simulation requires that the air enter the combustion chamber with the pressure, temperature, and velocity that would prevail in free flight at the same location. The velocity distribution may be distorted by screens or other devices to simulate free flight. In a turbojet, the full engine, including turbine and compressor, can be tested. Then, the air supply simulates the free-flight conditions upstream of the compressor. If only the combustion process is being investigated, the air supply simulates conditions downstream of the compressor, and all of the rotating machinery is omitted. A combustion facility is illustrated in Fig. 16.14.

To examine ramjet (see Sec. 25.9) combustion, the supply conditions would correspond to the subsonic high-pressure, high-temperature air that leaves the diffuser of the free-flight engine. In a scramjet (see Sec. 25.9), the air will be supersonic at the combustor entrance. True simulation of scramjet combustion requires high stagnation temperature air corresponding to hypersonic flight conditions. The combination of high pressure, high temperature, and high mass flow that may be required for full-scale tests of scramjet combustors results in high cost at best, and limits the Mach number that can be simulated. The cost is related to how much air must be compressed, stored, and heated. Therefore, the available run time is likely to be limited as well.

Combustion tests fundamentally permit measurements of the properties and mass flow of air and fuel entering and leaving the combustion chamber. These properties

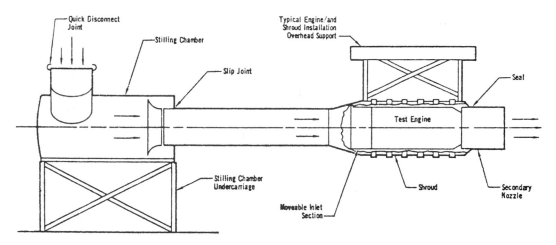

FIGURE 16.14   Direct-connect combustion facility. (From Pirrello *et al.*, 1971b.)

are measured with probes and other instruments. The combustion process itself may be of primary concern. The details of ignition and burning of the fuel can be observed, if the combustion chamber can be made of transparent materials. That may not be possible because of material limitations. It is usually feasible to look inside the combustion chamber from the ends, but then the optical equipment interferes with the air flow. Usually, the desired information is derived from flow field probing.

Shock tube wind tunnels can also be used for direct-connect testing of scramjet combustors [Parker *et al.* (1992)]. They are able to simulate combustor conditions corresponding to hypervelocity flight speed, but a small-scale short duration test may be difficult to scale to free-flight conditions.

Large facilities are available* in which entire engines (including exhaust nozzles) can be tested. In direct-connect, or semi-free jet tests, only the internal flow is simulated. The thrust can be derived from the change of momentum imparted to the gas or from surface pressure and shear force distributions, or from measurement of forces which the engine exerts on its mounting equipment. With curved inlets and vectored nozzles, the forces and moments in other directions besides the thrust axis are also of interest.

In free-jet testing, the engine is mounted in an appropriate flow field, and internal and external flow conditions are simulated. In such tests, the engine inlet and exhaust conditions are correctly represented, but the data is applicable strictly only to the particular geometry that is tested. Free-jet tests of complete missile models are, therefore, of direct application to the measurement of missile performance. Complete aircraft can be tested, also, with power on, but full-sized tests are usually impractical. Therefore, it is more advantageous to test an engine in the vicinity of those parts of the aircraft which determine the flow through and around the engine.

*See *Research Facilities Handbook*, Arnold Engineering Development Center, Tullahoma, TN and Dunsworth and Reed (1979).

16.5.6   Jet Nozzle and Exhaust Plume Tests

The exhaust nozzle on a jet engine (see Secs. 25.7 and 25.8) increases the momentum of the internal flow by expanding it to a lower pressure. The increased momentum of the stream results in thrust which appears as a pressure acting on the nozzle. The nozzle may be made asymmetrical in some manner so that the direction of the thrust vector can be controlled. The properties of interest concerning the exhaust nozzle include the magnitude and direction of the thrust, the distribution of the flow properties at the exit of the nozzle, and the heating in the vicinity of the nozzle throat.

Usually, nozzle performance is not greatly affected by chemical reactions, hence the hot flow in the nozzle can be simulated by that of a cold gas having similar properties. The most significant parameters are the ratio of specific heats of the gas, γ, and the Mach number. If thermal effects are to be measured, then the Reynolds number, Re, should be in the proper range to correctly simulate the nature of the boundary layer. The boundary layer on the nozzle started somewhere upstream in the engine, so the Reynolds number is high, and the free-flight boundary layer is almost always turbulent. Nozzle flow can be studied in a wind tunnel or engine test facility either separately or as part of a complete engine installation.

The flow in a well-designed nozzle is isentropic and can be predicted accurately. Although the performance of the nozzle is not sensitive to minor variations in flow property distributions, the net thrust of the engine can be significantly affected by nozzle efficiency. The heat transfer at the throat, and the effect of injected coolants, for example, are of great interest, and these properties can be measured with thermocouples and heat-transfer gauges.

More difficult to predict are the effects of the nozzle exhaust on neighboring surfaces, noise fields generated by the exit flow, and the radiation signature of hot nozzle plumes. All of these effects can be measured in propulsion test facilities where the nozzle is mounted on a full-scale operating engine. Noise distribution is primarily a pressure effect, and it can be simulated in a facility that provides a direct-connect internal flow. Tunnel wall interference may strongly affect nozzle plume characteristics (noise in particular); hence open-jet facilities may be preferable.

Plume pressure effects can be modeled in a cold-flow facility in which the exhaust from the nozzle impinges on scaled surfaces. Another type of problem arises, however, in connection with the effect of an exhaust plume on the aerodynamics of a missile. Simulation of the internal flow, even cold, greatly complicates and increases the cost of measurement of missile aerodynamics. Therefore, various techniques have been developed to simulate these effects without requiring internal flow in the model. A solid expanding support is sometimes used to shape the external flow to conform to the expected outside of the exhaust plume. The plume is thus simulated by a solid body. Another technique is to squirt a screen of gas laterally from the model support system in order to deflect the external flow properly. The difficulty in both cases is the design of the flow separation mechanism so that the external flow field is correctly simulated. Rules for these selections are provided in Nyberg and Argell (1981) and Henderson (1981).

16.5.7   Propeller Tests

Propellers driven by reciprocating engines or gas turbines (see Sec. 27.2) can enter into two distinct types of test. In one case, the engine and propeller are mounted on a vehicle, and the desired result is a measurement of the aerodynamic characteristics of the vehicle with power on. In the other type of test, the aerodynamic and structural performance of the propeller is the subject of investigation.

In power-on tests of propeller-driven vehicles, the internal working of the engine does not significantly affect the external flow field. Therefore, the engine outer shape is simulated, and scaled propellers are driven by electric motors or any other suitable power source. The engine power is scaled by matching the model thrust coefficient to the free-flight value. In this type of test, the flow induced by the propellers is represented in the measurement of the aerodynamic characteristics of the vehicle.

Propeller test facilities are designed to measure the aerohydrodynamic and structural properties of full-scale propellers. Such a facility must supply a controlled flow of free-stream air or water (partly driven by the propeller being tested), a drive motor and support, and a shield to catch flying propeller blades. Of course, aerohydrodynamic measurements apply to the propeller-engine-support system.

Aerohydrodynamic measurements would include thrust and efficiency of the propeller, and perhaps some flow field quantities, such as momentum distribution. Structural properties, both static and dynamic, including flutter, may also be measured. Propeller-generated noise is also of interest.

16.5.8 Helicopter and V/STOL Testing

Helicopters and other V/STOL aircraft (see Sec. 27.3) present certain problems. Basically, a helicopter is just another aerodynamic configuration, but, especially in proximity to the ground, the flow field created by the rotor system can drastically influence the engine performance. Other V/STOL aircraft encounter similar problems in ground proximity operations. Tests in which full-scale engine operation can be studied under these conditions require large test sections with provisions for horizontal air motion and removal of engine exhaust products. In some cases [Pirrello *et al.* (1971a)], a horizontal belt simulates ground motion relative to the aircraft.

Engine operation and performance characteristics can be measured as well as the aerodynamic forces on the entire vehicle. For some tests, it is also necessary to measure flow field quantities, particularly velocity distributions, in order to investigate the effects of engine and rotor-induced air currents.

16.6 SPECIAL FACILITIES

16.6.1 Introduction

The term *special facilities* might be applied to several categories of equipment. One category usually would be a facility that tests one particular line of equipment; an example is a wind tunnel arranged to blow air over a fuse assembly to test its arming mechanism. Such facilities are really parts of the development and production of particular products and will not be further described here.

Another category of special facility is a general-purpose facility in which special instrumentation has been installed to address a particular class of problems. One example is the installation of water-spraying equipment in sled track facilities for the investigation of rain erosion effects. This application is mentioned in Sec. 16.3.3. However, in some cases, the special instrumentation effectively creates a new facility which has general application to a limited type of testing. Store separation facilities, for example, fall into this class and are described in this section.

A third category of special facilities, which are also included in this section, covers some installations that are entirely devoted to certain types of testing. Vertical wind tunnels for investigating the dynamics of descending or hovering vehicles fall into this category.

16.6.2 Store Separation

The separation of stores, such as bombs, from aircraft is a process of vital concern to pilots and designers who may be more than embarrassed if the store collides with the launch airplane. Several techniques have been developed for deducing free-flight trajectories from wind tunnel measurements [Schindel (1975)]. In the *grid data bank* method, for example, the force on the store is measured in the wind tunnel as a function of store angle of attack and position relative to the airplane. Then, for each initial condition (store position and motion at the end of the ejection process) the motion of the store through the flow field of the aircraft is calculated, making use of the measured forces corresponding to the store location at the end of each time step. Although the wind tunnel measurements may require data at a large number of locations and for each of various aircraft flight attitudes and Mach numbers, once the data has been assembled, calculations can be made for arbitrary initial (end of stroke) conditions.

In a free drop simulation, a model store is ejected from an airplane model mounted in a wind tunnel. High-speed motion pictures then show the actual scaled trajectory. This process can be expensive if it is necessary to simulate several initial conditions, various aircraft attitudes, different Mach numbers, and different store carriage arrangements. It is also necessary to make some compromise in scaling if supersonic tests are required, since supersonic wind tunnels do not normally match both velocity and Mach number.

The captive trajectory technique makes use of a store model mounted on a movable support. The store is initially positioned at its carriage location next to an aircraft model. The forces on the store are measured in that location, and the position of the store at the end of a short time increment is calculated. Then, the store is moved to the new location, and the process is repeated until the store is safely (or unsafely) out of the aircraft flow field. By using the same on-line computer to calculate store position and drive the support mechanism, a complete trajectory can be traversed in seconds. Several large wind tunnels are equipped with captive trajectory systems [Carleton and Christopher (1968)].

16.6.3 Vertical Wind Tunnels

Vertical wind tunnels, in which a test model is supported by an upward flow of air, are useful for studying the dynamic characteristics of vehicles whose motion is mostly vertical descent. Spin recovery of aircraft, dynamic behavior of high-drag (retarded) bombs and similar items can be investigated.

Since the motion of an airplane in a flat spin is mostly vertical, it can be simulated by a dynamically scaled model supported by a vertical air flow. Spin recovery techniques can be investigated by remotely activating control and recovery devices. A schematic drawing of a vertical wind tunnel is shown in Fig. 16.15.

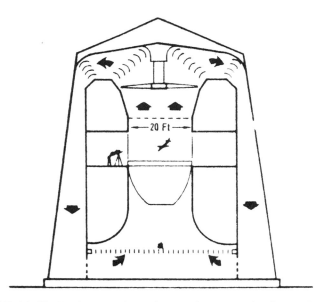

FIGURE 16.15 NASA Langley spin tunnel. (From Pirrello *et al.*, 1971a.)

Dynamic behavior of a high drag bomb is investigated by observing the response to perturbations of the vertical flight path. Full scale testing may be employed if the normal descent rate does not exceed the velocity of the vertical stream. Otherwise, dynamically scaled models can be tested. It may not be necessary to scale moment of inertia if only damping characteristics are of interest.

16.6.4 Icing Tests

Build-up of ice on wings, control surfaces, or engine inlets can cause serious operational problems on aircraft. Therefore a wind tunnel equipped to simulate flow conditions which cause icing [Pirrello *et al.* (1971a)] can be used to measure the rate of ice build-up under controlled conditions and the effectiveness of de-icing equipment. The test medium is a subsonic flow of cold air with high humidity controlled by injection of water.

16.6.5 Air Gun Facilities for Shock Tests

The object of a shock test is to subject a specimen (such as electronic components or systems) to a controlled acceleration pulse. In an air gun facility [Harris and Crede (1961) and *Shock Testing Facilities*, NOL (1967)], the acceleration is achieved by mounting the test object to a piston which is accelerated by high-pressure air along a tube (Fig. 16.16). In one technique, the piston is restrained until the air pressure behind it reaches the desired level, then the piston is suddenly released. An alternate approach is to accelerate a piston gradually with air at modest pressure. The piston acquires considerable momentum as it accelerates down the tube. Then, it is suddenly brought to rest when it compresses a measured quantity of gas at the closed far end of the tube. The shock test, in this case, occurs at the end of the piston travel rather than at the beginning.

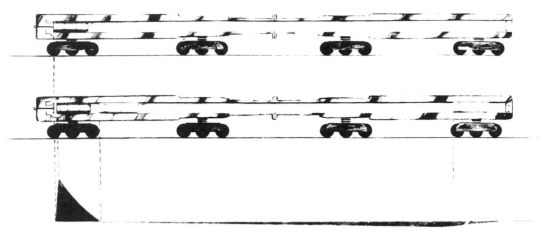

FIGURE 16.16   Schematic of air gun operation showing acceleration versus time. (Reproduced with permission from *Shock Testing Facilities*, NOL, 1967.)

Measurements are made of the acceleration as a function of time and damage to the test specimen. It is possible to extract electrical signals during the test in order to monitor the operation of the specimen. The peak acceleration depends on the weight of the specimen. Large masses can be accelerated to values approaching 5000 g; very small masses can be tested at more than 200,000 g.

16.6.6   Aeroelastic Testing

Aeroelastic (or hydroelastic) phenomena (see Sec. 23.4) can be studied in any suitable wind (or water) tunnel, but specialized techniques and equipment are required. The object of the test is to determine the mutual interactions between aerohydrodynamic loading and structural deformation. Of particular concern are cases of static instability (divergence) and dynamic instability (flutter) that can result in catastrophic structural failure.

The first problem is to construct an aerohydroelastically-scaled model. For correct scaling, the model must deflect to the same nondimensional shape as the full-scale vehicle under the scaled loading. This result can be achieved in a full-scale test which also duplicates full-scale pressure. A scaled model will also deflect proportionately if it is made of the same material and subjected to full-scale pressures. Otherwise, the material modulus must scale with the pressure level.

Dynamic aerohydroelastic models are difficult to construct on a small scale, since the mass must be distributed in such a way as to give the properly scaled dynamic response. Model construction techniques are discussed in Maget (1968).

Aerohydroelastic phenomena are load dependent. In a fixed-density subsonic wind tunnel, for example, the dynamic pressure will increase with air speed, therefore a quantity of interest might be the speed at which divergence appears. The facility must therefore provide a convenient means of slowly changing and monitoring the free stream velocity. Flutter is both speed and pressure dependent so that the wind tunnel density must be adjusted to the model scale, or vice versa.

At transonic and supersonic speeds, Mach number must also be simulated, and, especially in the transonic regime, the pressure distribution can also be dependent on the Reynolds number.

Aerohydroelastic data provides the load distribution and aerohydrodynamic characteristics of deformable bodies, conditions for divergence, and flutter speeds and frequencies.

16.6.7 Gas Dynamic and Chemical Lasers

Laser radiation is produced during the spontaneous relaxation toward equilibrium of an inverted distribution of molecular energy states. In the population inversion, the number of molecules at some particular high energy level exceeds the number at the lower level of the radiation-producing transition. The inversion may be produced by various pumping mechanisms for increasing the population of the high-energy state. Electric discharge, which pumps by collisions with high-energy electrons, is one of the most common gas laser devices. The gas dynamic laser and the chemical laser can be appropriately designated as fluid mechanics facilities in that properties of flowing gases are involved in the process.

The principle of operation might be regarded as a conversion of a time-dependent process into a position-dependent one by means of the velocity of the chemically active medium. Thus, chemical rate constants can be inferred from observations of the changes with position of the state of the medium.

In a nitrogen–carbon dioxide gas dynamic laser [Anderson (1975) and Monsler and Greenberg (1971)], for example, nitrogen is heated sufficiently, usually by combustion of carbon monoxide, to activate high energy levels of the gas. Nitrogen has a relatively slow relaxation rate, so that, if it is expanded rapidly through a nozzle, the number and rate of collisions among molecules are insufficient to bring the gas to equilibrium. The high-energy molecules then collide with molecules of carbon dioxide, inverting its energy state population and resulting in laser radiation at a wavelength of 10.6 μ (infrared). The gas dynamic laser is generally operated as an amplifier of power initially supplied by a low-power electric discharge laser. The amplification is achieved by passing a beam of radiation back and forth through the laser cavity which corresponds to the test section of the facility. A gas dynamic laser (Fig. 16.17) thus resembles a wind tunnel with mirrors. The gas dynamic laser is less efficient than an electric discharge laser, but it has the advantage of carrying away waste heat with the flowing stream so that it can deliver high power continuously. The high power output can be propagated over large distances because atmospheric absorption is small at this wavelength. Thus, energy or information can be transmitted.

A chemical laser [Gross and Bott (1976)] derives its population inversion from the inherent relative rates of the chemical reactions occurring in the medium. The nonequilibrium inverted energy states persist over the length of the laser cavity where energy is extracted as in the gas dynamic laser. The chemical laser usually operates as a combined oscillator and amplifier. Like the gas dynamic laser, it resembles a wind tunnel with mirrors.

The excited DF molecule which results from chemical reaction of deuterium with fluorine radiates at 3.8 μ. This wavelength propagates easily in the atmosphere. Needless to say, however, a vital part of such a facility is the equipment used to neutralize and dilute the exhaust products.

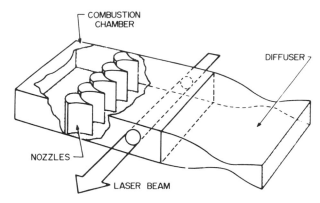

FIGURE 16.17   Gas dynamic laser. (Reproduced with permission, Anderson, J. D., Jr., "Gasdynamic Lasers: Review and Extension," *Acta Astronautica*, Vol. 2, pp. 911–927, 1975.)

16.6.8 Magnetohydrodynamic Facilities

In its broad sense, magnetohydrodynamics (see Sec. 13.8) (or magnetoplasmadynamics) deals with the properties and applications of conducting fluids. The discussion here will be limited to the facilities employing flowing ionized gases (plasmas) for the generation of electrical power [Sutton and Sherman (1965) and Branover (1978)] or using magnetic fields to accelerate the flow [Stratton (1965) and Cowling (1957)].

An ionized gas conducts electricity; its conductivity can be enhanced by seeding with readily-ionized particles. If the flowing plasma cuts magnetic lines of force, electrical power is generated that can be extracted via electrodes embedded in the gas. It is also possible to extract power inductively, thus avoiding the problem of electrode survival. If the magnetic field moves with the fluid, the efficiency of the generator is enhanced. Maximum efficiency occurs at zero slip between magnetic field velocity and fluid velocity. However, power is not extracted at this condition. Therefore, the flow velocity must exceed the magnetic field velocity in order to generate power, in which case some energy is lost in heating the gas and other components. It may be necessary, in fact, to set some level of plasma heating in

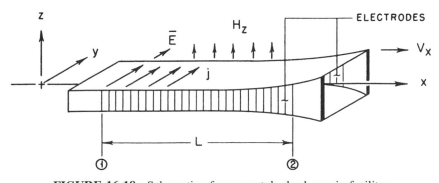

FIGURE 16.18   Schematic of a magnetohydrodynamic facility.

order to maintain the conductivity of the medium, thereby compensating for heat losses by radiation and conduction.

If the velocity of the traveling magnetic field exceeds that of the plasma, the system will operate as a motor rather than a generator, and the fluid is accelerated. Again, the condition of no slip results in maximum efficiency and no acceleration. The accelerated fluid may be used to provide a high Mach number, high temperature test medium or may be expelled through a nozzle to provide thrust. In the latter application, pulsed operation can be effective if discharges of electricity are used to ionize slugs of plasma and couple them to traveling magnetic pulses.

Figure 16.18 shows, schematically, a section of a magnetohydrodynamic facility. A conducting medium, moving with velocity U_x, crossing a magnetic field H_z, generates an electric field E, or is accelerated by applied field E.

REFERENCES

Anderson, J. D., Jr., "Gasdynamic Lasers: Review and Extension," *Acta Astronautica*, Vol. 2, pp. 911–927, 1975.

Baltakis, F. P., Merritt, D. L., and Aronson, P. M., "Two Techniques for Simulating the Interaction of a Supersonic Vehicle with a Blast Wave," Naval Ordnance Laboratory Report NOLTR 67-154, Oct. 1967.

Beach, H. L., Jr. and Bushnell, D. M., "Aeronautical Facility Requirements into the 2000's," AIAA Paper 90-1375, 1990.

Birchoff, G. and Zarantonello, E. H., *Jets, Wakes, and Cavities*, Academic Press, New York, 1957.

Bogdanoff, D. W., "Improvement of Pump Tubes for Gas Guns and Shock Tube Drivers," *AIAA Journal*, Vol. 28, No. 3, pp. 483–491, 1990.

Branover, H., *Magnetohydrodynamic Flow in Ducts*, Israel University Press, Jerusalem (John Wiley & Sons, New York), 1978.

Bryant, R. A. A., "The Hydraulic Analogy as a Distorted Dissimilar Model," *J. Aeron. Sci.*, Vol. 23, No. 3, pp. 282–283, Mar. 1956.

Carleton, W. E. and Christopher, J. F., "Captive-Trajectory System of the AEDC PWT 4-Foot Transonic Wind Tunnel," Arnold Engineering Development Center Report AEDC-TR-68-200, 1968.

Charters, A. C. and Curtis, J. S., "High Velocity Guns for Free Flight Ranges," *The High Temperature Aspects of Hypersonic Flow*, AGARDograph 68, Macmillan, New York, 1964.

Cowling, T. G., *Magnetohydrodynamics*, Interscience Publ., New York, 1957.

Davis, L. M. and Carver, D. B., "Initial Calibration of the HEAT H2 Arc-Heated Wind Tunnel," Arnold Engineering Development Center Report AEDC-TR- 91-16, Jan. 1992.

Dunsworth, L. C. and Reed, G. J., "Ramjet Engine Testing and Simulation Techniques," *J. Spacecraft and Rockets*, Vol. 16, No. 6, pp. 382–388, Nov./Dec. 1979.

Edney, B. E. *et al.*, "Ballistic Test Facility Guide," Joint Technical Coordinating Group on Aircraft Survivability Report JTCG/AS-76-D-001, May 1977.

Enkenhus, K. R. and Merritt, D. L., "Evaluation of Two Types of Facilities to Fulfill the Need for High Reynolds Number Transonic Testing," Naval Ordnance Laboratory Report NOLTR 71-147, July 1971.

Erwin, J. R., "Experimental Techniques," Vol. X, Section D, *High Speed Aerodynamics and Jet Propulsion*, Princeton University Press, Princeton, NJ, 1964.

Gaydon, A. G. and Hurle, I. R., *The Shock Tube in High-Temperature Chemical Physics*, Van Nostrand Reinhold, New York, 1963.

Geineder, F., Schlesinger, M. I., Baum, G., and Cornett, R., "The U.S. Naval Ordnance Laboratory Hypersonic Wind Tunnel," Naval Ordnance Laboratory Report NOLTR 67-27, Apr. 1967.

Gross, R. W. F. and Bott, J. F. (Eds.), *Handbook of Chemical Lasers*, John Wiley & Sons, New York, 1976.

Harris, C. M. and Crede, C. E., *Shock and Vibration Handbook, Vol. 2, Data Analysis, Testing, and Methods of Control*, McGraw-Hill, New York, 1961.

Henderson, J., "An Investigation for Modelling Jet Plume Effects on Missile Aerodynamics," Army Missile Command Tech. Report RD-CR-82-25, Jan. 1981.

Hill, J., "Hypervelocity Wind Tunnel: Test Planning Guide," Naval Surface Weapons Center Report WOL MP 76-2, Jan. 1976.

Holden, M. S., "A Review of Aeronautical Problems Associated with Hypersonic Flight," AIAA Paper 86-0267, Jan. 1986.

Jackson, C. M., Jr., Corlett, W. A., and Monta, W. J., "Description and Calibration of the Langley Unitary Plan Wind Tunnel," NASA Tech. Paper 1905, Nov. 1981.

Kenney, J. T. and Webb, L. M., "A Summary of the Techniques of Variable Mach Number Supersonic Wind Tunnel Nozzle Design," AGARDograph 3, NATO, Oct. 1954.

Lakshminarayana, B., "Techniques for Aerodynamic and Turbulence Measurements in Turbomachinery Rotors," *J. Eng. Power*, Vol. 103, pp. 374–392, 1981.

Lehman, A. F. and Summey, D. C., "Compendium of U.S. Incompressible Flow Facilities," Naval Coastal Systems Center, Technical Memorandum NCSC TM 394-84, May 1984.

Lewis, F. M., "Propeller Tunnel Notes," *Transactions, Society of Naval Architects and Marine Engineers*, pp. 1–12, 1947.

Ludwieg, H., "Tube Wind Tunnel—A Special Type of Blowdown Tunnel," AGARD Report No. 143, NATO, July 1957.

Lukaciewicz, J., Harris, W. G., Jackson, R., Van der Bliek, J. A., and Miller, R. M., "Development of Capacitance and Induction Driven Hotshot Tunnels," Arnold Engineering Development Center Report TN-60-222, Jan. 1961.

Maget, R. (Ed.), *Manual on Aeroelasticity, Part IV, Experimental Methods*, AGARD, NATO, 1968.

Merritt, D. L. and Aronson, P. M., "Study of Blast-Bow Wave Interactions in a Wind Tunnel," AIAA Paper 65-5, Jan. 1965.

Milne–Thompson, *Theoretical Hydrodynamics*, 4th ed., Macmillan, New York, 1960.

Monsler, M. J. and Greenberg, R. A., "The Effects of Boundary Layers on the Gain of a Gasdynamic Laser," AIAA Paper No. 71-24, Jan. 1971.

Morris Dam Test Range, Naval Ocean Systems Center Report NUC TN-1109, July 1973.

Murman, E. M., "Computations of Wall Effects in Ventilated Transonic Wind Tunnels," AIAA Paper 72-1007, Sept. 1972.

Norfleet, G. D., Hendrix, R. E., and Jackson, D., "Development of a Hypervelocity Track Facility at AEDC," AIAA Paper No. 77-151, Jan. 1977.

Nyberg, S. E. and Argell, J., "Investigation of Modelling Concepts for Plume-Afterbody Flow Interaction," FFA Final Technical Report on Grant No. DA-ERO-78-G-028, Nov. 1981.

Parker, T. E., Allen, M. G., Reinecke, W. A., Legner, H. H., Foutter, R. R., and Rawlins, W. T., "Supersonic Combustor Testing Using Optical Diagnostics and a High Enthalpy Shock Tunnel," AIAA Paper 72-0761, Jan. 1992.

Pirrello, C. J., Hardin, R. D., Heckart, M. V., and Brown, K. R., "An Inventory of Aeronautical Ground Research Facilities. Volume I: Wind Tunnels," NASA Report No. CR-1874, Nov. 1971a.

Pirrello, C. J., Hardin, R. D., Heckart, M. V., and Brown, K. R., "An Inventory of Aeronautical Ground Research Facilities, Volume II, Air-Breathing Engine Test Facilities," NASA Report CR-1875, Nov. 1971b.

Pirrello, C. J., Hardin, R. D., Heckart, M. V., and Brown, K. R., "An Inventory of Aeronautical Ground Research Facilities. Volume III: Structural and Environmental Facilities," NASA Report No. CR-1876, Nov. 1971c.

Pope, A. and Goin, K. L., *High Speed Wind Tunnel Testing*, John Wiley & Sons, New York, 1965.

Pope, A. and Harper, J. J., *Low Speed Wind Tunnel Testing*, John Wiley & Sons, New York, 1966.

Preston, J. H., "The Design of High Speed Free Surface Water Channels," *Proceedings, NATO Advanced Study Institute on Surface Hydrodynamics*, Bressanone, Italy, pp. 1–82, 1966.

Prislin, R. H., "Free-Flight and Free-Oscillation Techniques for Wind-Tunnel Dynamic-Stability Testing," NASA Jet Propulsion Laboratory Technical Report No. 32-878, Mar. 1966.

Problems of Wind Tunnel Design and Testing, AGARD Report No. 600, NATO, Dec. 1973.

Reda, D. C., Leverance, R. A., and Dorsey, W. G., Jr., "Application of Electro-Optical Pyrometry to Reentry Vehicle Nosetip Testing in a Hyperballistic Range," ICIASF, pp. 150–160, 1975.

Roper, R. M., "Boundary-Layer Transition on Large-Scale CMT Graphite Nosetips at Reentry Conditions," Arnold Engineering Development Center Report AEDC-TR-79-45, Jan. 1981.

Rosen, J., "The Design and Calibration of a Variable Mach Number Nozzle," *J. Aeron. Sci.*, July 1955.

Schaaf, S. A. and Chambre, P. L., "Flow of Rarefied Gases," Vol. III, *High Speed Aerodynamics and Jet Propulsion Series*, Princeton University Press, Princeton, NJ, 1958.

Schaefer, W. T., Jr., "Characteristics of Major Active Wind Tunnels at the Langley Research Center," NASA Tech. Memo X-1130, July 1965.

Schindel, L. H., "Characterization of the Performance of Shock-Tube Wind Tunnels," AIAA Paper 93-351, Jan. 1993.

Schindel, L. H., *Store Separation*, AGARDograph No. 202, NATO, June 1975.

Seigel, A. E., "The Hydroballistics Facility at NOL," Naval Ordnance Laboratory Report NOLTR 66-125, Aug. 1966.

Seigel, A. E., "The Theory of High Speed Guns," *Agardograph 31*, May 1965.

Shock Testing Facilities, Naval Ordnance Laboratory Report NOLR 1056 (3rd rev.), Nov. 1967.

Stalker, R. J., "The Free Piston Shock Tube," *Aeronautical Quarterly*, Vol. 17, Part 4, pp. 351–370, 1966.

Stratton, T. F., "High Current Steady-State Coaxial Plasma Accelerators," *AIAA Journal*, Vol. 3, pp. 1961–1963, Oct. 1965.

Sutton, G. W. and Sherman, A., *Engineering Magnetohydrodynamics*, McGraw-Hill, New York, 1965.

Vayssaire, J. C., ''Wall Interference Effects,'' Lecture Series 52: Large Transonic Wind Tunnels, Von Karman Institute, Brussels, 1973.

Ward, J. R., Baltakis, F. P., and Mancinelli, D. J., ''Wind Tunnel Experiments on the Effect of Combustion in the Wake Region of Supersonic Projectiles—Test Series 3,'' Ballistics Research Laboratory Report MR-2737, Apr. 1977.

Witliff, C. E., Wilson, M. R., and Hertzberg, A., ''The Tailored-Interface Hypersonic Shock Tunnel,'' *J. Aero. Sci.*, Vol. 20, No. 4, pp. 219–228, 1959.

17 Videotapes and Movies on Fluid Dynamics and Fluid Machines

BOBBIE CARR
Naval Postgraduate School
Monterey, CA

VIRGINIA E. YOUNG
Virginia Polytechnic Institute and State University
Blacksburg, VA

VIDEOTAPES AND MOVIES ON FLUID DYNAMICS AND FLUID MACHINES

The items listed below are videorecordings unless otherwise specified. The motion pictures are 16 mm. Addresses for ordering are listed at the end of this chapter.

Accumulators and Cylinders. Tel-A-Train, Chattanooga, TN, 1994. (Industrial Hydraulic Technology)

Operation of hydraulic accumulators and hydraulic cylinders, including discussion of adiabatic and isothermal charging and discharging.

Aerodynamic Generation of Sound, produced by Education Development Center under the direction of the NCFMF, Chicago, IL, Encyclopaedia Britannica, 1988.

Describes the mechanism of sound generation by turbulence resulting from an instability of steady flow, discusses quadrupoles in detail, and illustrates the fundamentals of jet noise generation.

Originally issued as a motion picture in 1969.

Annubar . . . The Natural Way [motion picture], Burt Munk Productions, Chicago, IL, Ellison Instruments, 1972.

Presents information for consulting and specifying engineers on the conduct of water under various conditions. Shows how water acts in its natural, free-flowing state, and how it flows when confined in pipes. Explains how fluid flow may be measured without interfering with nature's processes.

Applied Computer Fluid Dynamics [motion picture], Motion Picture/TV Production Group, Los Alamos Scientific Laboratory, Los Alamos, NM, 1979.

Examines sophisticated computer modeling which the Los Alamos Scientific Laboratory is using to examine complex hydrodynamic problems such as tornado

Handbook of Fluid Dynamics and Fluid Machinery, Edited by Joseph A. Schetz and Allen E. Fuhs
ISBN 0-471-12598-9 Copyright © 1996 John Wiley & Sons, Inc.

dynamics, tidal waves, reactor core behavior, and atomic nuclei collision dynamics. Computer generated motion pictures are used to illustrate computer applications to these problems.

Basic Aircraft Hydraulic Systems [motion picture], U.S. Navy, Washington, D.C., National Audiovisual Center, 1970.

Shows the basic operating principles of hydraulic systems.

Basic Turbine Theory [motion picture], U.S. Navy, Washington, D.C., released by National Audiovisual Center, 1969.

Traces the development of the turbine engine from steam power to the jet turbine.

Boundary Layer Control, David C. Hazen, produced by Educational Services Incorporated, Chicago, IL, Encyclopaedia Britannica, 1988. (Fluid mechanics films)

Illustrates means by which airfoil design can be modified to control transitions from laminar to turbulent flow and from laminar flow to flow separation.

Originally issued as a motion picture in 1965.

Caution, Wake Turbulence, United States Federal Aviation Administration, Washington, D.C., 1970. Distributed by National Audiovisual Center.

Graphically illustrates the phenomenon of wing tip vortices, telling how they are generated, what generates them, and their effects on light aircraft. Suggests pilot actions in order to avoid them.

Also issued as a motion picture.

Cavitation, Phillip Eisenberg, produced by Education Development Center, Chicago, IL, Encyclopedia Britannica, 1988.

Introduces and demonstrates cavitation, ''the process of formation of the vapor phase of a liquid when it is subjected to reduced pressures at constant ambient temperature.'' Addresses the role of nuclei and the effects of cavitation on machinery performance.

Originally issued as a motion picture in 1969.

Centrifugal Pumps, Tel-A-Train, Chattanooga, TN, 1979 and 1987, three cassettes. (Pumps)

Part 1 presents a comprehensive description of the theory, operation, and terminology of centrifugal pumps. Part 2 deals with design characteristics of centrifugal pumps, as well as layout and design requirements for a system incorporating centrifugal pumps. Part 3 covers maintenance.

Also issued as a motion picture.

Channel Flow of a Compressible Fluid, produced by Education Development Center, Chicago, IL, Encyclopaedia Britannica, 1988.

Presents a Schlieren flow visualization and a simultaneous display of the pressure distribution along a channel of varying area. Describes supersonic flow and compression, Mach waves and normal waves, and the phenomena of choking, blocking, and starting.

Originally issued as a motion picture in 1966.

Characteristics of Laminar and Turbulent Flow [motion picture], Iowa City, IA, University of Iowa, 1964. (Mechanics of fluids)

Couette flow in relation to viscosity measurement and various combinations of

Couette and plane Poiseuille flow are demonstrated. Fall of spheres singly and in groups illustrates deformation drag. Axisymmetric Poiseuille flow at the inlet of a tube and development of velocity distribution around an elliptic cylinder are related to variation in the Reynolds number.

Check Valves, Cylinders, and Motors, Tel-A-Train, Chattanooga, TN, 1976. (Industrial Pneumatic Technology)

Part I: Includes sizing of motors and cylinders, cylinder selection, and replacement. Part II: Tube selection, rod and cylinder sizing, and flow rate.

Compressors, Tel-A-Train, Chattanooga, TN, 1994.

Shows the various types of compressors and their applications. It covers positive displacement, vane, and oil flooded compressors, as well as piston compressors, and a variety of centrifugal compressors.

Conservation of Mass, M.I.T. Center for Advanced Engineering Study, Cambridge, MA, © 1985. (Fluid dynamics, lecture 5) (M.I.T. video course)

Includes a general equation of continuity for three-dimensional, time-varying flow and a continuity equation for a general one-dimensional, unsteady flow in a compliant tube.

Control of Hydraulic Energy, Tel-A-Train, Chattanooga, TN, 1994. (Industrial hydraulic technology)

Discusses the use, construction, and operation of pressure, directional, and flow control valves in a simple hydraulic system.

Also issued as a motion picture.

Control of Pneumatic Energy, Tel-A-Train, Chattanooga, TN, 1994.

Concentrates on controlling the pressure in a pneumatic system by explaining why pressure controls are needed, where they should be placed in a system, and how they operate.

Control Volume and Reynolds' Transport Theorem, M.I.T. Center for Advanced Engineering Study, Cambridge, MA, © 1985. (Fluid dynamics, lecture 8) (M.I.T. video course)

Includes definitions of control volume and control surface, Reynolds' theorem, particular forms of the control-volume theorem for conservation of mass, Newton's laws of motion and the first and second laws of thermodynamics, and guidelines for control-volume analysis.

Deformation of Continuous Media, John L. Lumley, produced by Educational Services Incorporated, Chicago, IL, Encyclopedia Britannica, 1989. (Fluid mechanics films)

Demonstrates deformation [local rotation and distortion (strain)] in a fluid medium (glycerine). The discussion is limited to flows at constant density and, generally, to two-dimensional flow.

Originally issued as a motion picture in 1969.

Dimensional Analysis, M.I.T. Center for Advanced Engineering Study, Cambridge, MA, © 1985. (Fluid dynamics, lecture 23) (M.I.T. video course)

Includes the principle of dimensional homogeneity, Buckingham's Pi Theorem, illustrative applications to drag on a body in a streaming flow, and important dimensionless groups: Reynolds, Froude, Rossby, Mach, and Weber numbers.

Directional Control Valves, Tel-A-Train, Chattanooga, TN, 1976 and 1994. (Industrial hydraulic technology), two cassettes.

Discusses various types of directional valves and their uses in hydraulic systems.

Also issued as a motion picture.

Double Input Mixers [motion picture], U.S. Dept. of the Air Force, Washington, D.C., released by National Audiovisual Center.

Identifies the components of a pentagrid tube, explains the construction used in the mixer, and discusses the two inputs and connections to the elements of the tube. Also treats the operation of the circuit, its adaptability to a receiver, and symptoms of component failure.

Drag, M.I.T. Center for Advanced Engineering Study, Cambridge, MA, © 1985. (Fluid dynamics, lecture 38) (M.I.T. video course)

Includes energy dissipation due to drag, comparison of curves of drag vs. Reynolds number for streamlined shapes and for bluff bodies, compilations of data for various shapes, base drag, drag in transonic flow, drag due to gravity waves, and boundary-layer control.

Dynamic Similitude and Model Testing, M.I.T. Center for Advanced Engineering Study, Cambridge, MA, © 1985. (Fluid dynamics, lecture 22) (M.I.T. video course)

Includes the concept and uses of similitude, geometric similarity, the conditions of dynamic similarity obtained by normalization of the continuity and Navier–Stokes equations and the boundary conditions, dimensionless parameters, and model testing and scaling laws.

Dynamics of Fluid Exchange, Directions for Education in Nursing via Technology, Center for Instructional Technology, released by Wayne State University, Detroit, MI, 1974.

This training film for nurses demonstrates the concepts of hydrostatic and osmotic pressure. Describes the fluid dynamics involved in the development of edema, and the major functions of the kidney tubules in relation to fluid and electrolyte balance.

Also issued as a motion picture.

Effects of Fluid Compressibility [motion picture], Motion Picture Unit, University of Iowa, Iowa City, IA, 1972. (Mechanics of fluids)

Uses sound-wave analogy to present concepts of fluid mechanics, including wave celerity, shock waves and surges, and wave reflection. Patterns of waves illustrate the effect of a changing Mach number.

Energy Transmission using a Pneumatic System, Tel-A-Train, Chattanooga, TN, 1994.

Covers air compression, air expansion, and transmission of energy. Also addresses heat, friction, changing fluid direction, and flow rate.

Eulerian and Lagrangian Descriptions in Fluid Mechanics, John L. Lumley, produced by Education Development Center, Chicago, IL, Encyclopaedia Britannica, 1988. (Fluid mechanics films)

Deals with the mathematical description of fluid motion.

Originally issued as a motion picture in 1969.

Examples of Flow Instability [motion picture], NCFMF, Chicago, IL, Encyclopaedia Britannica, 1968. (Fluid mechanics films)

Illustrates a demonstration experiment on a phase of fluid mechanics.

Flight Without Wings, National Aeronautics and Space Administration, made by John J. Hennessy Motion Pictures, Washington, D.C., distributed by National Audiovisual Center, 1980.

Discusses the aerodynamic principle of the lifting body and the progress that has been made in its development. Points out that the technology being developed will enable NASA to eventually build a craft that will carry men and materials to and from an Earth-orbiting space station.

Originally issued as a motion picture in 1969.

Flow at High Reynolds Number, M.I.T. Center for Advanced Engineering Study, Cambridge, MA, © 1985. (Fluid dynamics lecture, 25) (M.I.T. video course)

Includes Prandtl's viscous boundary layer, order-of-magnitude estimates for a flat-plate boundary layer, approximate boundary layer equations, and why flow at infinite Reynolds number is not entirely inviscid.

Flow Control Devices, Sterling Educational Films, New York, NY, 1970.

Demonstrates the properties of fluids: density, pressure, viscosity, and velocity. On-site applications of elements for measuring flow are illustrated, as are radio control applications and the blending of fluids.

Also issued as a motion picture.

Flow Control Valves, Silencers, and Quick Exhausts, Tel-A-Train, Chattanooga, TN, 1994. (Industrial Hydraulic Technology)

Operation of several types of control valves and circuit applications, including lunge control, meter-in and meter-out circuits, quick exhaust systems, and mufflers.

Flow Instabilities, Erik Mollo–Christensen, produced by Education Development Center, Incorporated, Chicago, IL, Encyclopaedia Britannica, 1989. (Fluid mechanics films)

Shows surface waves generated by wind in a water channel to illustrate the general features of flow instability.

Originally issued as a motion picture in 1969.

Flow Visualization, produced by Education Development Center, Chicago, IL, Encyclopaedia Britannica, 1988. (Fluid mechanics films)

Defines pathline, timeline, streakline, and streamline as concepts necessary to understanding visual images of flow patterns, and illustrates some techniques of flow visualization: marker, optical, wall trace, birefringence, and self-visible phenomena methods.

Fluid and Electrolyte Balance, National Naval Medical Center, Bethesda, MD, 1972.

Part 1 discusses the normal distribution and measurement of fluid and electrolytes in the body. Part 2 discusses the clinical signs and therapy of the following conditions: dehydration and overdehydration, hyponatremia and hypernatremia, hypokalemia and hyperkalemia, hypocalcemia and hypercalcemia, and hypomagnesia and hypermagnesia.

Fluid Dynamic Instabilities, M.I.T. Center for Advanced Engineering Study, Cambridge, MA, © 1985. (Fluid dynamics lecture, 39) (M.I.T. video course)

Includes Reynolds' classical experiment, Kelvin–Helmholtz instability, theoretical strategy for investigation of linear stability, the Orr–Sommerfield equation, Tollmien–Schlichting waves, boundary–layer transition, instability, and Taylor–Goertler waves.

The Fluid Dynamics of Drag, produced by Educational Services Incorporated, Film Division, Chicago, IL, Encyclopaedia Britannica, 1988.

Certain concepts and laws of drag are explained. Includes experiments, fundamental concepts, the laws of drag in fluids of high and low viscosity, and how to reduce drag.

Originally issued as a motion picture in 1960.

Fluid Mechanics, University of South Carolina, Columbia, SC.

Thirty-eight 50-minute lectures on the basic principles of fluid statics and dynamics. Conservation laws of mass, momentum, and energy are developed in the context of the control volume formulation. Topics include application of dimensional analyses, dynamic similitude, and steady-state laminar viscous flow. Applications of turbulent flow to pipe problems are introduced.

Fluid Motion in a Gravitational Fluid [motion picture], University of Iowa, Iowa City, IA, 1963. (Mechanics of fluids)

Action of gravity in modifying the pattern of accelerated flow once a free surface exists is demonstrated. Fluid motion in outflow, overflow, and underflow structures is illustrated.

Fluid Power Technology—Actuators, Meridian Education Corp., Bloomington, IL, 1993.

Construction and operation of actuators, with actual cutaways, industry examples, and computer animation.

Fluid Power Technology—Control Mechanics, Meridian Education Corp., Bloomington, IL, 1993.

Control of fluid power systems, from hand valves to programmable controllers, with live footage and computer animation.

Fluid Power Technology—Pumps, Lines, Filters, Meridian Education Corp., Bloomington, IL, 1993.

Operation of various pumps in the fluid power industry, with computer animation, and live action.

Fluid Power Technology at Work, Meridian Education Corp., Bloomington, IL, 1993.

Overview of hydraulics and pneumatics, from the steam engine to the space shuttle.

Fluids and Electrolytes, Robert J. Brady Co., Bowie, MD, 1981.

Identifies the principles related to fluid and electrolytes within the human body and applies these principles to the clinical situation. Module 1, The basics. Module 2, Extracellular fluid. Module 3, Intracellular fluid. Module 4, Potassium. Module 5, Calcium. Module 6, Ph. Module 7, Metabolic acidosis and alkalosis. Module 8, Respiratory acidosis and alkalosis.

Fluids in Weightlessness, U.S. National Aeronautics and Space Administration, Washington, D.C., distributed by National Audiovisual Center, 1980.

Illustrates basic principles of fluid dynamics by utilizing film footage from the Skylab in-flight science demonstrations.

Originally issued as a motion picture in 1975.

Fluids, Reservoirs, Coolers, and Filters [motion picture], Tel-A-Train, Chattanooga, TN, 1976. (Industrial hydraulic technology)

Describes the fundamental characteristics of a wide variety of hydraulic fluids, examines different types of reservoirs and air and water coolers used in hydraulic systems, and discusses various types of filters used in hydraulic circuits.

Force Transmission through a Fluid, Tel-A-Train, Chattanooga, TN, 1994.

Introduces the physics of compressed air. It addresses the workings of fluid power cylinders and vacuums. Pressure scales and intensifiers are also covered.

Form, Drag, Lift, and Propulsion [motion picture], University of Iowa, Iowa City, IA, 1966. (Mechanics of fluids)

The role of boundary–layer separation in producing longitudinal and lateral components of force on a moving body is explained. Also demonstrated are distribution of pressure around typical body profiles and its relation to the resulting drag, the variation of drag with body form, and pertinent aspects of the flow pattern.

Free-Surface Channel Flow, M.I.T. Center for Advanced Engineering Study, Cambridge, MA, © 1985. (Fluid dynamics, lecture 17) (M.I.T. video course)

Includes steady flow in an open channel under the action of gravity and friction; governing equations for a rectangular channel; Froude number, subcritical and supercritical flow; effects of friction, bottom slope, and change of channel width; critical depth; depth contours for various bottom slopes; frictionless, constant-width flow; and hydraulic jump.

Fundamental Principles of Flow [motion picture], University of Iowa, Iowa City, IA, 1962. (Mechanics of fluids)

Illustrates, through experiments and animation, basic concepts and physical relationships involved in the analysis of fluid motion.

Fundamentals of Boundary Layers, produced by Education Development Center, Chicago, IL, Encyclopaedia Britannica, 1988.

Illustrates the causes and behaviors of boundary layers, how they respond to pressure gradient, and the differences in laminar and turbulent layers.

Originally issued as a motion picture in 1969.

Gravitational Distribution [motion picture], John T. Fitch, Concord, MA, Kalmia Co., 1974. (Kinetic theory by computer animation)

Illustrates the gravitational distribution of gases.

Head Losses in Piping Systems, M.I.T. Center for Advanced Engineering Study, Cambridge, MA, © 1985. (Fluid dynamics, lecture 29) (M.I.T. video course)

Includes thermodynamics and the head loss; wall friction stress; friction coefficient; pumping power; incompressible flow in long pipes; Poiseuille and Prandtl–Nikuradse formulas; roughness effects; the Moody diagram; adiabatic flow of a compressible, perfect gas in long pipes; head losses for pipeline components; and diffusers.

Helmholtz's Vorticity Equation, M.I.T. Center for Advanced Engineering Study, Cambridge, MA, © 1985. (Fluid dynamics, lecture 32) (M.I.T. video course)

Includes derivation from Navier–Stokes equation for viscous but incompressible flow; generation or destruction of vorticity by rotational body forces and by viscosity, effects on vorticity due to rotation and stretching of vortex lines, conditions for vorticity to be locked in the fluid, derivation of vorticity equation for compressible but inviscid flow, how nonbarotropicity changes vorticity, and the vorticity equation in a rotating reference frame.

Helmholtz's Vortex Laws, M.I.T. Center for Advanced Engineering Study, Cambridge, MA, © 1985. (Fluid dynamics, lecture 33) (M.I.T. video course)

Includes geometric vortex laws for vortex tubes and vortex lines; dynamical vortex laws for inviscid, barotropic fluid, free of rotational body force; induced velocity field of a vortex, rotating reference frames; and the bathtub vortex.

High-Speed Gas Flow, M.I.T. Center for Advanced Engineering Study, Cambridge, MA, © 1985. (Fluid dynamics, lecture 37) (M.I.T. video course)

Includes the role of Mach number; pressure field of a moving disturbance, Mach lines, Mach angle, linearized theory for flow with small perturbations, Prandtl–Glauert rule for subsonic flow, thin airfoils at supersonic speed, transonic flows, supersonic simple-wave flow, and oblique shock waves.

How Airplanes Fly, U.S. Federal Aviation Administration, Washington, D.C., 1969. Distributed by National Audiovisual Center.

Shows what makes an airplane get off the ground and stay aloft. Combines animation and live sequences to explain basic aerodynamics. Describes forces of lift, weight, thrust, and drag in relation to flight.

Also issued as a motion picture.

Hydraulic Actuators, Tel-A-Train, Chattanooga, TN, 1994. (Industrial hydraulic technology)

An introduction to the fundamental characteristics and operation of hydraulic cylinders and motors.

Also issued as a motion picture.

The Hydraulic Jump [motion picture], NCFMF, Chicago, IL, Encyclopedia Britannica, 1969.

Demonstrates the phenomenon of the hydraulic jump which occurs when water flowing rapidly in an open channel is retarded by an obstruction or change in the channel slope.

Hydraulic Motors, Tel-A-Train, Chattanooga, TN, 1994. (Industrial hydraulic technology)

Demonstrates the construction, application, and operational characteristics of a wide variety of vane, gear, and piston motors.

Also issued as a motion picture.

Hydraulic Pumps, Tel-A-Train, Chattanooga, TN, 1994. (Industrial hydraulic technology)

Operation of vane pumps, internal and external gear pumps and piston pumps, and how to determine pump efficiency.

Hydraulic Transmission of Force and Energy, Tel-A-Train, Chattanooga, TN, 1994. (Industrial hydraulic technology)

Discusses the application of liquids in the transmission of force and the conversion of hydraulic pressure to mechanical force. Examines the effects of heat, friction, and changing direction of fluid flow in force transmission. Illustrates the use of hydraulic cylinders, intensifiers, accumulators, and pumps.

Also issued as a motion picture.

Hydrostatics, M.I.T. Center for Advanced Engineering Study, Cambridge, MA, © 1985. (Fluid dynamics, lecture 2) (M.I.T. video course)

Includes the scalar pressure field required for hydrostatic equilibrium with a body force, pressure distributions for incompressible and compressible fluids in the earth's gravity field and in the centrifugal field of a rotating fluid mass, and shapes of isobaric and isochoric surfaces.

Inertia-Free Flows, M.I.T. Center for Advanced Engineering Study, Cambridge, MA, © 1985. (Fluid dynamics, lecture 24) (M.I.T. video course)

Includes small Reynolds number as the inertia-free criterion; differential equations for the velocity, pressure, and vorticity fields; kinematic reversibility; similarity conditions and scaling laws; contrast between flows respectively dominated by inertia and by viscosity; Pitot tube at low Reynolds number; flow through porous media; and streaming flows past bodies.

Introduction to Fluid Power, Meridian Education Corp., Bloomington, IL, 1994.

Live action coverage of fluid power in industry. Questions are inserted for student discussion.

Introduction to Inviscid Flow I: Bernoulli's Equation, M.I.T. Center for Advanced Engineering Study, Cambridge, MA, © 1985. (Fluid dynamics, lecture 6) (M.I.T. video course)

Includes conservative and nonconservative body force fields, Euler's differential equations of inviscid motion, Bernoulli's integral, stagnation pressure and total head, and steady-flow examples and applications.

Introduction to Inviscid Flow II: Bernoulli's Equation, M.I.T. Center for Advanced Engineering Study, Cambridge, MA, © 1985. (Fluid dynamics, lecture 7) (M.I.T. video course)

Includes further examples and applications of Bernoulli's integral for steady flow, Euler's equations of inviscid motion in streamline coordinates, and effects of streamline curvature.

Introduction to the Study of Fluid Motion [motion picture]. University of Iowa, Iowa City, IA, 1962. (Mechanics of fluids)

Orients engineering students (at the beginning of a course) in the mechanics of fluids. Examples are drawn from a host of everyday experiences and laboratory and field demonstrations are used to illustrate the variety and range of flow phenomena.

Introduction to the Subject of Fluid Dynamics, M.I.T. Center for Advanced Engineering Study, Cambridge, MA, © 1985. (Fluid dynamics, lecture 1) (M.I.T. video course)

Introduces the scope and applications of fluid dynamics; the underlying physical principles; concepts of the continuous media, continuum properties, and continuum fields; categories of forces; stress at a point; the stress tensor; and the hydrostatic state of stress.

Kelvin's Circulation Theorem, M.I.T. Center for Advanced Engineering Study, Cambridge, MA, © 1985. (Fluid dynamics, lecture 31) (M.I.T. video course)

Includes derivation from Navier–Stokes equation; illustrations of the changes of circulation due to nonbarotropicity, rotational body forces, and viscosity; conditions for the permanency of circulation; streaming motion of a real fluid at high Reynolds number; the irrotational sink-vortex with vortical core; drainage of a rotating fluid mass; flow in a rotating reference frame; motion in a rotating flow at low Rossby number; and the Taylor–Proudman theorem.

Kinematics of Fluid Motion, M.I.T. Center for Advanced Engineering Study, Cambridge, MA, © 1985. (Fluid dynamics, lecture 4) (M.I.T. video course)

Includes definitions of kinematic concepts, Lagrangian and Eulerian description; the material, or substantial, derivative; the control volume; and fluxes of mass, momentum, energy, and entropy.

The Laminar Boundary Layer, M.I.T. Center for Advanced Engineering Study, Cambridge, MA, © 1985. (Fluid dynamics, lecture 26) (M.I.T. video course)

Includes the Blasius flat-plate boundary layer, self-similarity, Blasius' solution, displacement thickness, momentum thickness, laminar boundary layers in pressure gradients, the momentum integral method, application to a flat plate, and comparison with Blasius' solution.

Lectures, Adaptive Mesh Workshop, sponsored by the Center for Nonlinear Studies, Los Alamos National Laboratory and DOE Office of Basic Energy Sciences, Applied Mathematical Sciences, Los Alamos, NM, 1981.

Held August 5–7, 1981, at the National Security and Resources Study Center, Los Alamos National Laboratory.

Leonardo, Portrait of a Genius [motion picture], Armand Hammer Productions, Los Angeles, CA, Armand Hammer Productions, 1982.

Focuses on the life of Leonardo da Vinci and traces his career as it leads up to the compilation of the Codex Hammer (formerly called the Codex Leicester). Gives a description of the codex, a collection of scientific manuscripts dealing chiefly with the movement and control of water.

Lift, M.I.T. Center for Advanced Engineering Study, Cambridge, MA, © 1985. (Fluid dynamics, lecture 36) (M.I.T. video course)

Includes lift on an airfoil in a two-dimensional cascade, Blasius' theorem for the forces on a body in a streaming flow, virtual mass due to acceleration, Joukowsky law of lift, the Magnus effect, d'Alembert's paradox, shape of a subsonic airfoil, the Kutta condition, starting vortex, methods of airfoil design, the Joukowsky transformation, thin-airfoil theory, boundary-layer control for high lift, and induced drag for wings of finite span.

Liquid Helium II, The Superfluid [motion picture], Michigan State University, East Lansing, MI, 1963.

Records a transfer of liquid helium at 4.2 degrees Kelvin and describes the properties of helium I briefly. The liquid is then cooled by evaporation to the lambda point, demonstrating the transition to helium II. The viscosity paradox is proved by superleak and by the rotating cylinder method. The two-fluid model is discussed. The fountain effect is shown in two experiments, proving the zero entropy of the superfluid. Demonstrates the Rolin creeping film and second sound by the pulse technique.

Liquid Pressure and Buoyancy [motion picture], Iwanami Films, Tokyo, Santa Monica, CA, released in the U.S. by BFA Educational Media, 1972.

Demonstrates liquid pressure by using a transparent plastic container with thin rubber walls which is immersed in water and rotated. Shows that the greatest pressure is exerted on the lowest wall. Illustrates buoyancy by showing that an object which is lighter than water will not necessarily float in a tank of water unless there is water between the object and the tank.

Longitudinal Wave Formation [motion picture], Films Incorporated, Northbrook, IL, released by Hubbard Scientific Co., 1972. (Wave fundamentals)

Describes the propagation of longitudinal waves.

Low Reynolds Number Flows, produced by Educational Services Incorporated under the direction of the NCFMF, 1967.

Describes and discusses characteristics of the motion of fluids using a variety of models and demonstrates how, in low Reynolds number flows, a rough estimate is provided by the Reynolds number scale of the relative importance of inertia and viscosity.

Originally issued as a motion picture in 1966.

Lubrication, Ernest Rabinowicz, Massachusetts Institute of Technology, Cambridge, MA, 1978. (Friction, wear, and lubrication, 10-11)

Part 1 demonstrates fluid lubrication, boundary lubrication, and the types of lubrication in between. Part 2 shows the effect of reduction of surface energy on wear and discusses types of lubricants which are effective in this regard.

Also issued as a motion picture.

Magnetohydrodynamics, J. Arthur Shercliff, produced by Educational Services Incorporated, Chicago, IL, Encyclopaedia Britannica, 1989. (Fluid mechanics films)

Describes and illustrates interactions between electromagnetic and velocity fields which are central to magnetohydrodynamics; discusses changes in pressure fields due to irrotational JXB forces, changes of vorticity by rotational JXB forces, induced currents and time constants for their decay, and Hartmann, boundary layers, and Alfven waves.

Originally issued as a motion picture in 1966.

Mechanical Waves, Barr Films, director and writer, Roy Casden, Pasadena, CA, Barr Films, 1983.

Uses laboratory experiments and real life examples to explain the properties of mechanical waves which are defined as waves that require a material medium, such as air or water, to pass through. Also explores wave amplitude, frequency, wavelength, velocity, pulse, bore, periodic waves, standing waves, and the Doppler effect.

Mechanical Waves [motion picture], London, BBC-TV, 1969, released in the U.S. by Time–Life Films. (Waves no. 1)

Examines the nature of mechanical waves and how they travel.

The Momentum Theorem I, M.I.T. Center for Advanced Engineering Study, Cambridge, MA, © 1985. (Fluid dynamics, lecture 9) (M.I.T. video course)

Includes control-volume analysis using the momentum theorem; determination of drag by wake survey; derivation of Euler's equations by momentum theorem;

constant-area internal flows; and pressure differences due to changes in nonuniformity of velocity profile.

The Momentum Theorem II and the Theorem of Moment of Momentum I, M.I.T. Center for Advanced Engineering Study, Cambridge, MA, © 1985. (Fluid dynamics, lecture 10) (M.I.T. video course)

Includes further examples and applications of the momentum theorem and the theorem of moment of momentum.

The Navier–Stokes Equation, M.I.T. Center for Advanced Engineering Study, Cambridge, MA, © 1985. (Fluid dynamics, lecture 21) (M.I.T. video course)

Includes Cartesian tensor notation, the stress tensor, the deformation-rate tensor, modes of rheologic behavior, the Newtonian fluid, the Stokes fluid, dynamic equilibrium, and the Navier–Stokes equation and its boundary conditions.

Numerical Modeling of Reactive Flow Systems, Los Alamos Scientific Laboratory, Los Alamos, NM, 1980.

One-Dimensional, Steady, Compressible Flow I, M.I.T. Center for Advanced Engineering Study, Cambridge, MA, © 1985. (Fluid dynamics, lecture 15) (M.I.T. video course)

Includes differential equations for one-dimensional conservation of mass, momentum, and energy, adiabatic flows, isentropic (frictionless, adiabatic) flows, the critical state, choking at maximum flow per unit area, Mach number, detailed results for a liquid of constant compressibility, and detailed results (Mach number functions and tables) for a perfect gas.

One-Dimensional, Steady, Compressible Flow II, M.I.T. Center for Advanced Engineering Study, Cambridge, MA, © 1985 (Fluid dynamics, lecture 16) (M.I.T. video course)

Includes Mach number as an index of compressibility effects, nozzle shape required to accelerate fluid at subsonic and supersonic speed, differential equations for isentropic flow of a perfect gas, types of integral curves, choking, continuous passage from subsonic to supersonic flow, the normal shock wave in a perfect gas, and the supersonic pitot tube.

Operation at the Suction Side of a Pump, Tel-A-Train, Chattanooga, TN, 1994. (Industrial hydraulic technology)

Installation and maintenance of a hydraulic pump and symptoms of improper inlet vacuum, cavitation, and entrained air.

Petroleum Base Hydraulic Fluids and Fire Resistant Hydraulic Fluids, Tel-A-Train, Chattanooga, TN, 1994. (Industrial hydraulic technology)

Discusses lubrication, types of hydraulic fluids and the effects of temperature, pressures, and contamination.

The Physical World of a Machine, Tel-A-Train, Chattanooga, TN, 1994. (Industrial hydraulic technology)

Explanation and measurement of force, energy, inertia, resistance and power, and introduction of hydraulic systems.

Potential Flow I, M.I.T. Center for Advanced Engineering Study, Cambridge, MA, © 1985. (Fluid dynamics, lecture 34) (M.I.T. video course)

Includes irrotational flow, the velocity potential function, Bernoulli's integral for unsteady potential flow, three-dimensional, incompressible, irrotational flow, the stream function, geometry of potential lines and streamlines, Laplace's equation, solution by complex variables, and the principle of super-position, the sink-vortex.

Potential Flow II, M.I.T. Center for Advanced Engineering Study, Cambridge, MA, © 1985. (Fluid dynamics, lecture 35) (M.I.T. video course)

Includes various elementary flows, Hele–Shaw visualization, self-propulsion of a vortex pair, smoke rings, streaming flows, far-field disturbance, circular cylinder with circulation, and corner-type flows and their teachings.

Pressure Fields and Fluid Acceleration, Ascher H. Shapiro, produced by Educational Services incorporated, Chicago, IL, Encyclopaedia Britannica, 1988.

Presents steady-flow experiments in which the pressure gradient, as the main force affecting fluid acceleration, is demonstrated.

Pressure in Fluids at Rest [motion picture], Modern Film Rentals, New Hyde Park, NY, 1966.

Fluid pressure results from the freedom of molecular motion in fluids. Animation, models, and underwater photography show that pressure in a fluid is related to depth and density, and that this pressure acts in all directions, explaining buoyancy and Archimedes' principle. Pascal's law as applied to closed systems is illustrated in terms of both transmission of forces and multiplication of forces.

Principles of Lubrication, Antony Barrier Productions, Ltd., Chicago, IL, International Film Bureau, 1978.

Demonstrates two simple laws of friction which come into play when two solid surfaces come into contact without a fluid interposed. Discusses ways that friction and wear can be reduced, viscosity of lubricants, and modes of lubrication. Gives special attention to boundary lubrication.

First released in England in 1975. Also issued as a motion picture.

Principles of Lubrication, U.S. Office of Education, made by Graphic Films, Washington, D.C., distributed by National Audiovisual Center, 1979. (Engineering series)

Discusses the need for lubrication, properties of lubricants, action of lubricants, viscosity of lubricants, and conditions that determine proper viscosity.

Originally issued as a motion picture in 1945.

Quasi-Parallel Viscous Flows I, M.I.T. Center for Advanced Engineering Study, Cambridge, MA, © 1985. (Fluid dynamics, lecture 18) (M.I.T. video course)

Includes internal friction in fluids; Couette viscometer; Newtonian, shear-thinning, shear thickening, and viscoelastic fluids; viscosity; boundary conditions for real fluids; quasi-parallel flows; two-dimensional flow in a slit, driven by pressure or by wall motion; boundary layer development near a tube entrance; and Reynolds number.

Quasi-Parallel Viscous Flows II, M.I.T. Center for Advanced Engineering Study, Cambridge, MA, © 1985. (Fluid dynamics, lecture 19) (M.I.T. video course)

Includes fully-developed Poiseuille flow in a long, straight tube of circular cross section, unsteady motion of a plate in its own plane, order-of-magnitude analysis for impulsive start of the plate, exact solution for impulsive start, diffusion of vorticity, application to growth of steady boundary layer at leading edge of a plate, sinusoidal motion of the plate, and the Womersley solution for oscillatory flow in a circular tube.

Quasi-Parallel Viscous Flows III, M.I.T. Center for Advanced Engineering Study, Cambridge, MA, © 1985. (Fluid dynamics, lecture 20) (M.I.T. video course)

Includes inertia-free flow in a duct of small taper, quasi-Poiseuille flow, hydrodynamic lubrication, linear wedge bearing, and journal bearing.

Rarefied Gas Dynamics, produced by Education Development Center under the direction of the NCFMF, Chicago, IL, Encyclopaedia Britannica, 1988.

Shows the evolution of flow fields as density is varied from continuum levels to the rarefied levels of free-molecule flow. Demonstrates the interaction of free-molecular flows with solid surfaces and portrays the role of the molecular mean-free-path.

Reservoirs, Coolers and Filters, Tel-A-Train, Chattanooga, TN, 1994. (Industrial hydraulic technology)

Discusses maintenance of hydraulic systems, including differences between nominal and absolute ratings and effects of excessive heat, contamination, and entrained air.

Reverse Osmosis [motion picture], Kinescope, Berkeley, CA, available from University of California, Extension Media Center, 1970.

Examines reverse osmosis as an economical and efficient water purifying system. Discusses osmotic principles and techniques and present and future applications.

The Riemann Problem for Fluid Flow of Real Materials, Ralph Menikoff, Los Alamos, NM, Los Alamos National Laboratory, 1991. (Colloquium series)

Fellows Prize Recipient Colloquium presented March 19, 1991.

Rheological Behavior of Fluids, Hershal Markovitz, produced by Educational Services Incorporated, Chicago, IL, Encyclopaedia Britannica, 1989.

Illustrates the rheological behavior of fluids which do not obey the Navier–Stokes equations. Demonstrates non-Newtonian properties by nonlinear steady flows and by the time-dependent visco-elastic behavior of such materials. Contrasts the non-Newtonian behavior with Newtonian fluid behavior in similar experiments.

Rotary Drilling Fluids [motion picture], Petroleum Extension Service, University of Texas at Austin, Austin, TX, Visual Instruction Bureau of Film Library, Division of Extension, University of Texas, 1975.

Demonstrates the use, testing, and treatment of water base drilling fluids, and the role they play in the prevention of blowouts.

Rotating Flows, Dave Fultz, produced by Education Development Center, Chicago, IL, Encyclopaedia Britannica, 1989. (Fluid mechanics films)

Illustrates the phenomena associated with rotation of homogeneous fluids, including horizontal trajectories in surface gravity waves, low and high-Rossby-number flows around spheres, Taylor walls, normal modes of inertia oscillation, and Rossby waves in a cylindrical annulus.

Originally issued as a motion picture in 1970.

Schlieren, 3M Company, Minneapolis, MN, 1976.

Explains and provides examples of the Schlieren technique of photography.

Secondary Flow, produced by Education Development Center, Chicago, IL, Encyclopaedia Britannica, 1985. (Fluid mechanics films)

Introduces the concept of secondary flow, contrasting it with primary flow, and uses models to demonstrate various types.

Originally issued as a motion picture in 1968.

Secondary Flow in a Bend [motion picture], Encyclopaedia Britannica, Chicago, IL, 1976. (Fluid mechanics)

Demonstrates water flow in a duct, showing that transverse vorticity in the straight section upstream becomes streamwise vorticity as the flow rounds a bend in the channel, thus accounting for the secondary flow.

Similarities in Wave Behavior [motion picture], Pacific Telephone Film Library, San Francisco, CA, 1960.

Using specially built torsion wave machines, Dr. J. N. Shive of Bell Laboratories demonstrates and discusses various aspects of wave behavior.

Sounds in the Silent Deep [motion picture], Moody Institute of Science, Whittier, CA, 1972.

Explains the importance of sound for man in everyday life, and deals with factors that led to the erroneous conclusion that man cannot hear under water.

Source Moving at Speeds Below and Above Wave Speeds [motion picture], Encyclopaedia Britannica, Chicago, IL, 1976. (Fluid mechanics)

Small-amplitude waves on the free surface of a shallow water table illustrate the types of disturbance fields produced by a disturbance source that moves with respect to the fluid. Analogy to a source in subsonic, transonic, and supersonic compressible flows is demonstrated.

Standing Waves and the Principle of Superposition, Encyclopedia Britannica, Chicago, IL, 1971.

Production of standing waves by the superposition of two identical wave patterns traveling in opposite directions.

Stratified Flow, Robert R. Long, produced by Education Development Center, Chicago, IL, Encyclopaedia Britannica, 1988. (Fluid mechanics films)

Demonstrates the interface wave between fluids of varying densities, internal and lee waves of two-layer fluids, and the behavior and effects of continuously stratified fluids, especially as they flow over various obstacles.

Studying Fluid Behavior, Journal Films, Evanston, IL, 1979. (Scientific investigations series)

Investigates and explores the characteristics and behavior of fluids.

Superposition of Waves, Same Direction [motion picture], Films Incorporated, Northbrook, IL, released by Hubbard Scientific Co., 1972. (Wave fundamentals)

Demonstrates the process of superposition as waves converge on one medium from two others.

Surface Tension, M.I.T. Center for Advanced Engineering Study, Cambridge, MA, © 1985. (Fluid dynamics, lecture 3) (M.I.T. video course)

Includes the nature of surface tension and range of applications; pressure jump across an interface; bubbles and droplets; surfaces of minimum area; contact angle; wetting and nonwetting; motions produced by gradients of surface tension due to impurities, temperature gradients, and electric fields.

Surface Tension in Fluid Mechanics, produced by Educational Services Incorporated under the direction of the NCFMF, Chicago, IL, Encyclopaedia Britannica, 1988.

Presents a series of experiments to show that surfaces exert forces. Defines the fundamental boundary conditions governing the effects of these forces. Includes illustrations of nucleation, "wine tears," swimming bubbles, and high-speed pictures of the breakup of water sheets and soap films.

Originally issued as a motion picture in 1965.

Swimming Propulsion at Low Reynolds Number [motion picture], Encyclopaedia Britannica, Chicago, IL, 1976. (Fluid mechanics)

Bull spermatozoa are seen swimming in a manner resembling that of fish, but the mechanism is not the same, because the spermatozoa cannot impart rearward momentum to the fluid. One form of propulsion by which microscopic organisms swim is illustrated by experiments with cylindrical rods falling in syrup at different orientations, and with a mechanical swimming model.

Taylor Column in Rotating Flows at Low Rossby Number [motion picture], Encyclopaedia Britannica, Chicago, IL, 1976. (Fluid mechanics)

Experiments in a rotating tank of water illustrate the tendency of the motions in a slightly perturbed rotating flow to be two dimensional. Injected dye arranges itself into tall vertical sheets (Taylor walls). A solid sphere moving horizontally carries with it a vertical cylinder of fluid (Taylor column). A vertically moving sphere alters the vorticity in the Taylor columns above and below itself and encounters increased resistance.

Theorem of Movement of Momentum II, M.I.T. Center for Advanced Engineering Study, Cambridge, MA, © 1985. (Fluid dynamics, lecture 11) (M.I.T. video course)

Includes the sink-vortex, hurricanes, tornadoes, whirlpools, spiral diffusers, Euler's pump and turbine equation, performance of a simple axial-flow fan or pump, and the performance of a simple centrifugal fan or pump.

Transverse Wave Formation [motion picture], Films Incorporated, Northbrook, IL, Hubbard Scientific Co., 1972. (Wave fundamentals)

Describes the propagation of a transverse wave.

Troubleshooting Hydraulic Systems: Analyzing Component Faults, Tel-A-Train, Chattanooga, TN, 1986.

Demonstrates malfunctions in hydraulic components, including in pumps, actuators, relief valves, and pilot-operated directional control valves.

Turbulence, produced by Education Development Center, Encyclopaedia Britannica, Chicago, IL, 1988. (Fluid mechanics films).

Illustrates aspects of turbulence, including the effect of Reynolds number on inception and on turbulent flows increased pressure drop in pipe-flow, efficient mixing, turbulent transport of momentum and scalar properties, Reynolds stress, and effects of buoyancy.

Originally issued as a motion picture in 1969.

Turbulence, M.I.T. Center for Advanced Engineering Study, Cambridge, MA, © 1985. (Fluid dynamics, lecture 27) (M.I.T. video course)

Includes the origin, nature, and characteristics of turbulence—its effects on velocity profiles, wall shear stress, and frictional pressure drop; Reynolds stress tensor—why it delays boundary layer separation in adverse pressure gradients; and the turbulent energy cascade.

Turbulent Shear Flows, M.I.T. Center for Advanced Engineering Study, Cambridge, MA, © 1985. (Fluid dynamics, lecture 28) (M.I.T. video course)

Includes Reynolds stress; turbulence models; the closure problem; ''turbulent viscosity''; length and velocity scales of turbulence; mixing-length formulation; boundary layer structure; velocity distributions in the laminar sub-layer, wall region, wake region, and overlap region; Spalding's Law of the Wall; Coles' Law of the Wake; equilibrium profiles; and fully-developed turbulent pipe flow.

Underwater Sound—Basic Principles [motion picture], U.S. Navy Dept., Washington, D.C., released by National Audiovisual Center, 1969.

Illustrates the basic principles of the behavior of sound under water. Shows the effects of absorption, scattering, bottom loss, and sound refraction on sound transmissions in the sea.

Valves, Tel-A-Train, Chattanooga, TN, 1994.

The advantages, disadvantages, and applications for many different types of valves are discussed, including ball, butterfly, needle, globe and plug. Terms such as pressure drop and water hammer are explained.

Viscosity [motion picture], released by Hubbard Scientific Co., Northbrook, IL, 1971. (Properties of matter)

The differences in viscosity of various liquids and the causes of these differences are demonstrated.

Vorticity, Ascher H. Shapiro, produced by Educational Services Incorporated, Chicago, IL, Encyclopaedia Britannica, 1988.

Illustrates, using simple laboratory experiments, the concepts of vorticity and vortex motion.

Issued as a motion picture in 1961.

Vorticity, Circulation, and Vortices, M.I.T. Center for Advanced Engineering Study, Cambridge, MA, © 1985. (Fluid dynamics, lecture 30) (M.I.T. video course)

Includes vorticity as a measure of mean angular velocity; vortex lines, vortex tube, vortex filament; flux of vorticity; divergence-free nature of vorticity; circulation and its relationship to vorticity; vortex-type flows; and diverse examples of free vortices.

Wave Motion: Interference [motion picture], Gateway Educational Films, Ltd., Chicago, IL, distributed by International Film Bureau.

Includes a demonstration in teaching wave motion which shows true forms of interference patterns on the surface of the water. Special apparatus avoids the distortions usually seen in a simple ripple tank because of the lens-like action of the ripples themselves.

Wave Propagation I, M.I.T. Center for Advanced Engineering Study, Cambridge, MA, © 1985. (Fluid dynamics, lecture 12) (M.I.T. video course)

Includes the physics of wave propagation; analysis of one-dimensional propagation of compressibility waves, long gravity waves, and area waves in a compliant tube; the linearized wave equation and its solution; and physical interpretation of the mathematical solution.

Wave Propagation II, M.I.T. Center for Advanced Engineering Study, Cambridge, MA, © 1985. (Fluid dynamics, lecture 13) (M.I.T. video course)

Includes waves in compliant tubes, the speed of compressibility waves in liquids and gases, nonlinear changes in the shape of large-amplitude waves, calculation rules for small-amplitude waves, solution by step wavelets, the four elementary step wavelets, and types of wavelet events.

Wave Propagation III, M.I.T. Center for Advanced Engineering Study, Cambridge, MA, © 1985. (Fluid dynamics, lecture 14) (M.I.T. video course)

Includes examples and applications; start of flow from a reservoir, analyzed on three different time scales; nonlinear development of shock waves; and moving sources of waves.

Wave Reflection [motion picture], Films Incorporated, Northbrook, IL, released by Hubbard Scientific Co., 1972. (Wave fundamentals)

Presents a schematic illustration of reflection at the ends of a linear medium, distinguishing between reflection with a phase reversal of 180 degrees at a fixed end and reflection without phase reversal at an open end.

Waves in Fluids, produced by Educational Services Incorporated, Chicago, IL, Encyclopaedia Britannica, 1988. (Fluid mechanics films)

Introduces some of the basic phenomena of gravity waves, using models of water waves to demonstrate the behaviors of various types of one-dimensional waves. Illustrates how the concepts developed for gravity waves also apply to compressibility waves in gases.

Originally issued as a motion picture in 1964.

Waves on Water [motion picture], produced in collaboration with John S. Shelton, chief advisor, and Joseph S. Creager, Chicago, IL, Encyclopaedia Britannica, 1989.

With the help of large experimental tanks, explains how waves are created. Demonstrates that even though waves travel, the water does not move.

Note: The M.I.T. video course, *Fluid Dynamics*, by Ascher H. Shapiro is accompanied by a guide: v. 1, *Concepts, Principles, and Flow Phenomena Due to Inertia.* v. 2, *Viscous Behavior.* v. 3, *Deeper Insights.*

Suggested sources for videorecordings and films:

Aviation Video Library
Division of Bennett Video Group
730 Washington Street
Marina del Rey, CA 90292
(310) 821-3329
(800) 733-8862

Barr Films, Inc.
12801 Schabarum Avenue
P.O. Box 7878
Irwindale, CA 91706-7878
(818) 338-7878
(800) 234-7878

Burt Munk & Company
666 Dundee Rd.
Northbrook, IL 60062

Center for Instructional Technology
Wayne State University
5035 Woodward St.
Detroit, MI 48202

Center for Nonlinear Studies
Los Alamos National Laboratory
Los Alamos, NM 87544

Hubbard Scientific Co.
3101 Iris Ave., Suite 215
Boulder, CO 80301

International Film Bureau
332 S. Michigan Ave.
Chicago, IL 60604
(312) 427-4545

Journal Films
1560 Sherman Ave.
Evanston, IL 60201
(708) 328-6700
(800) 323-5448

Massachusetts Institute of Technology
Cambridge, MA 02139

Meridian Education Corp
236 E. Front Street
Bloomington, IL 61701
(309) 827-5455
(800) 727-5507

Michigan State University
IMC Marketing Division
Instructional Media Center
P.O. Box 710
East Lansing, MI 48826-0710
(517) 353-9229

Minnesota Mining and Manufacturing
   Co.
3-M Center
St. Paul, MN 55101

Moody Institute of Science
12000 E. Washington Blvd.
Whittier, CA 90606

NCFMF (National Committee for
   Fluid Mechanics Films)
distributed by Encyclopaedia Britannica
310 S. Michigan Ave.
Chicago, IL 60611

National Audiovisual Center
General Services Administration
Washington, DC 20409
(301) 763-1896

National Naval Medical Center
Bethesda, MD 20814

Sterling Educational Films, Inc.
241 East 34th St.
New York, NY 10016

Tel-A-Train
309 North Market Street
P.O. Box 4752
Chattanooga, TN 37405
(615) 251-0113
(800) 251-6018

Time–Life Films
Time Life Building
Rockefeller Center
New York, NY 10020

United States Film Corporation
Division of A. Hoffmann Group Co.
75-313 14th Green Drive
Indian Wells, CA 92210
(619) 773-9169

University of California
Extension Media Center
2223 Fulton St.
Berkeley, CA 94720

University of Iowa
Audiovisual Center/Marketing
C215 Seashore Hall
Iowa City, IA 52242
(319) 335-2539
(800) 369-4692

University of South Carolina
Columbia, SC 29208

Visual Instruction Bureau of Film
   Library
Division of Extension
University of Texas
Austin, TX 78712

18 Introduction to Computational Fluid Dynamics

KENNETH G. STEVENS, VICTOR L. PETERSON (Retired), and PAUL KUTLER

NASA Ames Research Center
Moffett Field, CA

CONTENTS

18.1 DIGITAL COMPUTERS FOR FLUID DYNAMICS CALCULATIONS
Kenneth G. Stevens

18.1.1 Computer Vocabulary

To understand the constraints and opportunities which digital computers offer to computational methods in fluid dynamics, one must first understand the vocabulary in which the constraints and opportunities are presented. The following glossary presents brief definitions and the succeeding sections will use the words defined in

Handbook of Fluid Dynamics and Fluid Machinery, Edited by Joseph A. Schetz and Allen E. Fuhs
ISBN 0-471-12598-9 Copyright © 1996 John Wiley & Sons, Inc.

this glossary to examine the constraints and opportunities which digital computers offer.

Arithmetic Logic Unit (ALU): The portion of a computer which performs the arithmetic and logic functions on user data.

Bank (of Memory): Independent memory module which can be addressed separately. A banked memory consists of multiple banks with addresses located in (address) mod (number of banks) bank.

Bit: A single binary digit.

Byte: 8 bits and used to store an alphanumeric character.

Cache: A small buffer memory which is significantly (i.e., 4 to 20 times) faster than main memory. Contiguous memory locations fetched are stored together in cache. Thus, accessing contiguous words can be done at near cache speeds while random words would be accessed at main memory speeds.

Floating Point: A binary representation of scientific notation which contains an exponent field and a mantissa field.

Millions of Instructions per Second (MIPS): A unit in which performance is measured.

Millions of Floating Point Operations per Second (MFLOPS): A unit in which performance is measured.

Multiple Instruction-Stream Multiple Data-Stream (MIMD): A computer architecture which utilizes several cooperating instruction streams to operate on multiple subsets of the problem data base.

Pipeline: An arithmetic logic unit which is organized in an assembly line so that multiple operands are being operated on at the same time and so that results can be produced at the end of the assembly line in a fraction of the time required to go through the full pipeline.

Pixel: A single point in a digitized image.

Random Access Memory (RAM): Memory which permits direct access to any particular word, byte, or bit with equal speed.

Raster Graphics: Similar to conventional television in that the image is a matrix of pixels.

Sequential Access: Can only be accessed sequentially like the bits on a magnetic tape reel.

Single Instruction-Stream Multiple Data-Stream (SIMD): A computer architecture which utilizes a single instruction stream to operate on multiple subsets of the data usually via pipelines and/or parallel processors.

Single Instruction-Stream Single Data-Stream (SIMD): A computer architecture which has a single instruction stream performing operations on a single stream of data. This is the first architecture developed and is the simplest architecture.

Virtual Memory: A mechanism which allows the program to refer to virtual (logical) addresses and permits the hardware to translate these to physical addresses. This permits the user to work in an address space which is larger than the physical memory.

Word: Usually 32 or 64 bits which are operated on as a single entity (floating point or integer number).

18.1.2 Performance of Computers

Performance of computers is governed by the speed of the computer and its memory capacity. In most cases, time (speed) and space (memory capacity) can be traded-off for one another. For example, one could choose to store a temporary variable and use it at a later time or one could save the space and recompute the temporary variable when it is needed again. To understand the implications of this time and space trade-off to the user of a computer and the designer of a computer, some discussion of computer architectures and technology is necessary.

There are four basic components which make up a computer. The first is the *control unit* which performs the functions of instruction fetch and instruction decode. The second is the *Arithmetic Logic Unit* (ALU) which performs the arithmetic and logical operations, such as add, shift, and logical, and on user data. The third component is the *memory* which is where the program and data are stored. The fourth component is the *Input/Output* (I/O) *Unit* which interfaces the computer with the rest of its peripherals and the outside world. Figure 18.1 shows these four components in diagram form. Each of these four components will be examined in some detail to indicate how the performance of computers has been improved architecturally and to indicate how the four components may affect the performance of a particular application.

(a) Control Unit. Figure 18.2 is a block diagram of the control unit. The control unit sequences through instructions unless a branch instruction is encountered. The instructions, after being fetched from memory (possibly through a cache or special instruction buffers), are decoded and a set of micro instructions are sent to the memory and ALU for actual execution. As the ALU complexity and/or memory complexity increases, the number of micro instructions will increase. When an instruction is fetched, it is fetched with a group of other words (a cache line in the case of microprocessors or the contents of an instruction buffer in the case of supercomputers). If there is a branch and the target address is not stored in the instruction cache/buffer then it must be fetched from memory which is a slower operation.

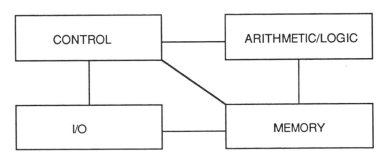

FIGURE 18.1 Basic computer design without concurrent functions.

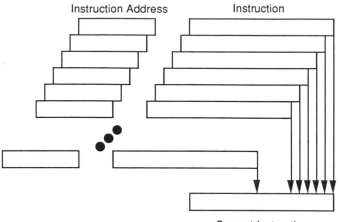

FIGURE 18.2 Block diagram of a control unit.

(b) Memory Unit. The memory unit consists of a sequentially addressed set of storage locations. The contents of each location may be a bit, byte, word, or larger set of bits. The control unit or the I/O unit places an address in a memory address register. If the instruction is a *store*, the contents of the register is stored in the address contained in the memory address register. If the instruction is a *fetch*, the contents of the memory location contained in the memory address register will be loaded into the memory interface register.

(c) Arithmetic Logic Unit. The ALU contains a set of registers, fast storage locations, and the hardware necessary to perform arithmetic operations (add, subtract, multiply, divide, compare, etc.) and logical operations (Boolean operations, shifts, rotates, etc.). This could be as simple as having a small number of general purpose registers and an independent floating point unit to as complicated as having a large number of special purpose registers (vector, mask, etc.) and many independent functional units. A block diagram is shown in Fig. 18.3.

(d) I/O Unit. The I/O unit contains buffer space and a connection network to peripherals (disks, tapes, networks, printers, etc.). For input, the control unit requests the I/O unit to access a peripheral device and read data from the device into the buffer which can be accessed by the control unit. For output, the control unit places the data to be written into the buffer and requests the I/O unit to write the contents of the buffer to a peripheral device.

Performance can also be improved by changing the architecture of the memory. Cray Research Inc. (CRI) uses multiple banked memories on its vector supercomputers. Each of the banks is an independent memory unit and contains addresses which are equal to a constant mod the number of banks. This memory architecture permits the fetch/store of a new word to begin every minor cycle if they are not in the same bank. If successive fetch/stores are required from the same bank, then the full memory cycle is required, and can range from a few clock cycles to tens of

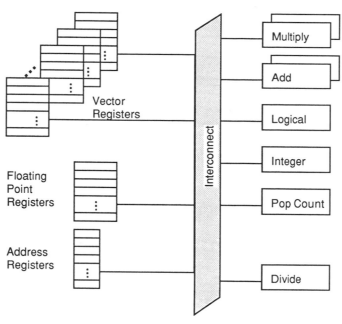

FIGURE 18.3 Complex ALU with multiple sets of registers and functional units.

clock cycles. The Cray 1 had only one port to memory. The Cray X/MP increased the number of ports to memory to three (two read and one write); it also added the complexity of having multiple CPUs accessing the same common memory. Variants of this architecture are used by CRI, Cray Computer Corp., and the Japanese vector supercomputer makers (Fujitsu, Hitachi, and NEC). When programming a computer with multiple banked memories, this extra delay can be avoided by not addressing memory locations which are separated by the number of banks, e.g.,

DIMENSION A(16,16)	DIMENSION A(17,16)
DO 1 I=1,16	DO 1 I=1,16
A(I,16) = A(I,1)	A(I,16) = A(I,1)
1 CONTINUE	1 CONTINUE
Bank Conflict	No Bank Conflict

If we assume 16 memory banks, the DO loop on the right will only access the same memory bank, while the one on the left will access different memory modules and not suffer the delay caused by bank conflict. Another technique of improving memory speed is to address multiple words simultaneously. When a machine with a cache memory makes a memory request for a word not already in cache, the main memory will be requested to supply multiple adjacent words (a cache line) to the cache where a single word is then extracted. Subsequent accesses to adjacent words which are now in cache will be significantly faster than if another main memory access is required.

Architectural changes may be made to the control unit to improve performance such as permitting the overlap of instruction execution. This is commonly done with a *score board* which schedules when various other resources of the computer (registers, ALU, memory, etc.) are busy and schedule subsequent instructions to begin as soon as all necessary resources become available. This is particularly important when one has instructions which require tens of clock cycles.

Thus far, we have discussed a class of machines which are referred to as Single Instruction-Stream Single Data-Stream (SISD), because they have a single control unit with instruction *decode* (single instruction-stream) and *perform* operations on one set of operands (single data-stream). In the early 1970's, another class of machines started being introduced which were called Single Instruction-Stream Multiple Data-Stream (SIMD) machines. These machines improved performance by performing the same operation on multiple sets of operands simultaneously. SIMD architectures take the form of parallel processors or pipelined processors.

A SIMD parallel processor which is currently available is the MasPar MP-2. The MP-2 has an array control unit and up to 16K processing elements (PEs) in the PE Array Unit. Each PE is a complete arithmetic logic unit, and all the PEs perform operations in lock step, e.g., when the array control unit requests the PE Array Unit to add, all the PEs perform an add simultaneously. Thus, operations of length 16K could be operated on component-wise in parallel rather than sequentially. However, to fully utilize the performance potential of a parallel processor like the MP-2, it is necessary to have identical operations to perform equal to the number of PEs throughout the program. This problem leads to *Amdahl's Law* which is exemplified by the following: Given a parallel processor with 100 processors and a code which has 10% sequential operations and 90% parallel operations, then the code will run only ten times faster on the parallel processor than on a single sequential processor with the same speed of a PE. Thus, when one is using a parallel processor, it is necessary not only to have a large fraction of the program executable in parallel but that very little (5% or less) must require sequential processing. Experience has shown that many of the finite difference and spectral methods used in computational fluid dynamics meet this requirement.

Pipelining is another architectural technique used in SIMD architectures to improve performance. A pipeline is like an automobile assemblyline in that the operation is subdivided into suboperations, and each suboperation is performed at a substation in the assemblyline. Thus, once the pipeline is full (each suboperation has a set of operands upon which to perform its suboperation), a completed result is created in the length of time required for the longest suboperation which is some fraction of the time required to perform the whole operation if a pipeline were not used. Pipelines, as they get longer (more substations), decrease the time required for each subsequent result once the pipeline is full. On the other hand, the longer the pipeline, the more time it takes to fill the pipeline. This time to fill the pipeline is referred to as startup time. The following equation expresses the length of time to perform a component-wise vector operation in a pipeline

$$\text{Time} = \text{Startup Time} + (\text{Vector Length} * \text{Cycle Time}) \qquad (18.1)$$

Thus, to maximize performance it is desirable to maximize vector length. Some vector supercomputers, e.g., CDC Cyber 205 and ETA 10, perform memory to

TABLE 18.1 Characteristics of Vector Super Computers

Model	Clock (MHz)	Processors (Max #)	Memory (GBytes)	Perf./Node (MFLOPS)	Max Perf. (GFLOPS)
Cray Research Y/MP	167	8	2	333	2.65
Cray Research C-90	240	16	8	960	15.36
Cray Computer C-3	500	16	8	1,000	16
Fujitsu VP2600	312	4 sc. & 2 vec.	2	1,600	5
Hitachi S-3800	500	4	2	8,000	32
NEC SX-3/44R	400	4	8	6,400	25.6

memory operations which permitted vector lengths up to 64K. Most of today's vector supercomputers perform vector register to vector register operations which limits vector length to the maximum vector register size. Some of the vector supercomputers utilize two (or four) pipelines to perform operations. In these cases, every other operand goes to a different pipe, and once both pipes are full they produce two results each clock cycle. However, using multiple pipelines effectively reduces the effective vector length. Table 18.1 lists some of the characteristics of contemporary vector supercomputers.

Notice that these vector supercomputers can come in multiprocessor configurations. Unlike the more massive parallel processors discussed subsequently, these processors share a common memory space. Only a small number of the applications run on these computers use multiple processors on a single job for improved turnaround. In most cases, unrelated jobs are run in the multiple processors to improve throughput of set of submitted jobs.

In the 1990's Multiple Instruction-Stream Multiple Data-Stream (MIMD) processors began to emerge using either microprocessors originally designed for workstations or chipsets customized for MIMD applications. Typical configurations are given in Table 18.2.

Note that the Cray Research uses the Digital Equipment Corp. Alpha processor, Convex uses the Hewlett Packard PA-RISC processor, Intel uses its own i860 processor, IBM uses its RISC processor, and Kendal Square and Thinking Machines use custom processors. Although not listed in the above table, networks of workstations can be used as a MIMD computer.

TABLE 18.2 Configurations of MIMD Processors

Model	Clock (MHz)	Processors (#)	Memory (GBytes)	Perf./Node (MFLOPS)	Max Perf. (GFLOPS)
Cray T3D	150	128	8	150	19.2
Convex SPP1	99	128	32	198	25
Intel Paragon XP/S	50	512	49	75	33.6
Kendal Square KSR-1	50	7200	8	40	10
Kendal Square KSR-2	100	64	230	80	576
IBM SP-1	62.5	512	16	125	8
IBM SP-2	66	512	262	264	135
Thinking Machines CM-5	32	1024	32	128	128

The MIMD architecture can be used as if it were an SIMD architecture in a data parallel fashion. They also can have unique programs on each processor and communicate via passing messages. There are currently no standard tools to program or debug MIMD configurations. One of the more popular programming tools is PVM which was developed by Oak Ridge National Laboratory and the University of Tennessee. Many implementations of PVM are available. Programming tools for vector computers often can permit utilizing 50% of the raw compute power provided by vector supercomputers for well-behaved algorithms found in CFD. On the other hand, it is not uncommon for MIMD computers utilizing message passing to only tap 5% of the available computer power given the current state of parallel systems software. It is hoped that this will improve significantly during the last 5 years of the century.

18.1.3 Graphics, Networks, and Other Non-Number Crunching Activities

The previous section discussed many of the aspects involved in the number crunching activities associated with supercomputers. This section will discuss the other aspects of the computing system which permit the researcher to interface with the number cruncher.

The users window on the number crunching activities is the terminal, personal computer, or workstation. A terminal running the X-Windows software is the minimal configuration recommended. This will permit you to display multiple windows on the CRT which could contain such things as the source code, Unix session, and graphics output. An X-terminal contains a microprocessor for maintaining the display but not for independent computing. The difference between personal computers and workstations is starting to blur. Traditionally, personal computers were based on Motorola 680X0 processors (Apple Macintosh) or Intel 80X86 (IBM PC and compatible) and workstations on more powerful microprocessors such as DEC Alpha, HP PA, IBM RISC 6000, MIPS R4400, etc. Some of this distinction is starting to blur with the introduction of the Pentium and PowerPC. Workstations often have custom accelerators for the graphics (e.g., translation, clipping, and rotation of images). Typical CRT resolution ranges from 640×480 pixels for personal computers to 1280×1024 pixels and larger for workstations. Associated with each pixel, there can be associated up to 24 bits for color and possibly additional bits for depth buffering.

To get data from the supercomputers, mass storage, or other workstations you will have to utilize networks which range from local area networks (LANs), wide area networks (WANs), to international configurations like the Internet. Table 18.3 lists some of the network standards and their communications rate.

The most common LANs are Ethernet. LANs have been moving up in speed to Fiber Distributed Data Interface (FDDI) and may move to a yet to be standardized *Fast Ethernet* which would have a speed comparable to FDDI. For WANs, T-1 or *bundles* of T-1s are common with some migration to T-3 from the phone companies. Beginning to emerge as WAN backbones is Asynchronous Transfer Mode (ATM) on OC-3 fiber. By the end of this century it is hoped that the *Information Super Highway* will be using OC-48 and providing gigabits per second transfer rates.

It is important when selecting a computing resource to have a balanced system which includes a computer with an architecture compatible with the algorithm, a

TABLE 18.3 Network Details

Ethernet	10 Mbps	
FDDI	800 Mbps	Fiber Distributed Data Interface
T-1	1.544 Mbps	Standard North American digital transmission techniques.
T-3	44.768 Mbps	Standard North American digital transmission techniques.
OC-1	51.84 Mbps	Optical carrier. A CCITT term used to define bit rate.
OC-3	155.52 Mbps	Optical carrier. A CCITT term used to define bit rate.
OC-12	622.08 Mbps	Optical carrier. A CCITT term used to define bit rate.
OC-48	2.49 Gbps	Optical carrier. A CCITT term used to define bit rate.
CCITT	International Consultative Committee for Telephony and Telegraphy and organization within the UN that develops international standards	

network which has sufficient bandwidth, sufficient storage for data, and a display device of sufficient capability to truly understand the results.

18.2 REQUIREMENTS FOR CALCULATION OF FLUID FLOWS
Victor L. Peterson

18.2.1 Introduction

The amount of computer power (speed and memory) required for computational fluid dynamics depends on three factors: 1) the complexity of the problem, which includes the fluid physics governing the problem to be solved and the complexity of the geometry about which the fluid moves, 2) the efficiency of the algorithm available for solving the governing equations numerically, and 3) the amount of computational time that can be invested to solve the problem at hand. The power of the largest currently available supercomputers (Class VI machines), such as the CYBER 205 and the CRAY X-MP, is far less than that required to solve the most complex problems involving the full Navier–Stokes equations and realistically shaped devices or vehicle geometries in a practical amount of computational time. This is true even though these machines can sustain computational rates greater than 100 MFLOPS (million floating point operations per second) and can have electronic memories in excess of 10 million words. Thus, the computational fluid dynamist is being limited by available computer power and must resort to approximations either to the physics, the geometry, or to both in order to obtain numerical solutions to his problem.

The relationships between problem complexity, algorithm efficiency, and computational time are reviewed in terms of the power of computers of the past and present. This will serve to highlight the factors influencing computer requirements and to provide some insight into the prospects for the future. Finally, some estimates of future requirements are presented and discussed.

18.2.2 Problem Complexity

The Navier–Stokes equations, although known to contain approximations when compared to the more basic nonlinear Boltzmann equation which governs molecular transport, are definitely suitable for solving fluid dynamic and aerodynamic problems in the continuum flow regime. The Navier–Stokes equations contain the physics governing all scales of turbulence of which between four and five decades of scales are of practical importance depending upon the Reynolds number. The equations are highly nonlinear and strongly coupled, and their solution is further complicated by the boundary conditions associated with complex shapes.

It is not possible to obtain closed-form solutions to these full equations for situations of interest in design. Even with the level of computational power available today, the complete equations cannot, within practical constraints, be solved numerically in their full form, except in highly simplified situations. Therefore, various degrees of approximation have been evoked over the years to obtain useful results. The four major levels of approximation that stand out in order of their evolution and complexity are discussed by Chapman (1979) and are summarized in Table 18.4.

Long before the advent of digital computers, mathematicians and fluid dynamicists devised methods for solving the linearized equations governing inviscid flows that are everywhere either subsonic or moderately supersonic. This was possible, because only three partial derivative terms are retained in the equations. The first solutions for two-dimensional airfoils were obtained about 1920 and for three-dimensional wings about 1940. With the advent during the 1960's of computers of the IBM-360 and CDC-6600 class, it became practical to compute inviscid flows about somewhat idealized, complete aircraft configurations. This level of approximation (level I) is still heavily used today, but because the equations neglect all viscous terms, as well as inviscid nonlinear terms, such calculations provide only limited help in the design process. For many years, it has been customary to increase the value of this level of approximation by making corrections for viscous effects based

TABLE 18.4 Major Levels of Approximation to the Navier–Stokes Equations and the Time Period for Initiation of Major Efforts to Treat Them Computationally

APPROXIMATION LEVEL	CAPABILITY	INITIATION TIME PERIOD	
		RESEARCH	APPLICATIONS
I LINEARIZED INVISCID	SUBSONIC/SUPERSONIC • PRESSURE DISTRIBUTIONS • VORTEX AND WAVE DRAG	1950s	1960s
II NONLINEAR INVISCID	ABOVE PLUS • TRANSONIC • HYPERSONIC	1960s	1970s
III RE-AVERAGED NAVIER-STOKES MODEL TURBULENCE	ABOVE PLUS • TOTAL DRAG • SEPARATED FLOW • STALL/BUFFET/FLUTTER	1970s	1980s
IV LARGE EDDY SIMULATION MODEL SUBGRID SCALE TURBULENCE	ABOVE PLUS • TURBULENCE STRUCTURE • AERODYNAMIC NOISE	1970s	1990s
EXACT FULL NAVIER-STOKES EQUATIONS	ABOVE PLUS • LAMINAR/TURBULENT TRANSITION • DISSIPATION	INCREASING INTENSITY OF RESEARCH 1970s ————————▶	

on the so-called boundary layer theory. This approach works reasonably well when flows do not encounter severe adverse pressure gradients and remain everywhere attached to surfaces over which the flows pass.

The advent of the computer brought the next level (level II) of approximation into the realm of practical usefulness. By including the nonlinear terms in the equations, it is possible to treat flows at all Mach numbers, including transonic and hypersonic, although viscous terms are still neglected. Hand-calculated solutions for the transonic flow over a nonlifting airfoil were obtained in the later 1940's and over a lifting airfoil with detached bow wave at transonic speeds in the mid-1950's. However, the first transonic solutions for a practical lifting airfoil with embedded shock wave required the use of a computer and were not obtained until about 1970. This latter achievement can be interpreted as the turning point in computational aerodynamics, for it marked the beginning of a series of advances that would not have been possible without computers. See Chap. 20 for a discussion of methods for inviscid flows.

Level III of approximation is now in the stages of vigorous development. This level of approximation does not neglect any terms in the Navier–Stokes equations. The basic equations are averaged over a time interval that is long relative to turbulent eddy fluctuations, yet small relative to macroscopic flow changes. Such a process introduces new terms representing the time-averaged transport of momentum and energy which must be modeled using empirical information. Computers are required to work with this level of approximation, but the potential advantages are enormous. Realistic simulations of separated flows, of unsteady flows (such as buffeting) and of total drag should be possible as the ability to model turbulence matures. Combined with computer optimization methods, these simulations should make it possible to develop aerodynamically optimum designs. Landmark advances in the decade of the 1970's include the investigation of shock-wave interaction with a laminar boundary layer, the treatment of high Reynolds-number transonic airfoil flows, and the first three-dimensional laminar flow over an inclined body of revolution. Relatively large amounts of computer time are required using the level III approximation. Although two-dimensional flows can be computed in a matter of minutes on a Class VI computer with current numerical methods, the routine computation of three-dimensional flows is not yet practical; three-dimensional flows require about 100 times more calculations than two-dimensional flows. Sections 21.1 through 21.4 discuss these approaches.

Level IV of approximation involves the direct numerical simulation of turbulent eddies over a range of scales sufficiently broad to capture the transport of nearly all momentum and energy. Only the small eddies that dissipate energy, transport little energy or momentum, tend toward isotropy and are nearly universal in character are modeled. Under such conditions, the computer results involve essentially no empiricism. This level of approximation is being used today on a limited basis to research the physics of turbulence at a level of detail not possible through experiment. Computer times for simulations of very basic flows, such as those in channels, range up to 80 hours on current supercomputers. Ultimately, this work should provide information that will lead to improved methods for modeling turbulence in the level III approximation, as well as to unlock the secrets of the generation of aerodynamic noise and the manipulation of friction drag. A description of this level of treatment is in Sec. 21.5.

The effect that increasing computer power has had on solving the various forms

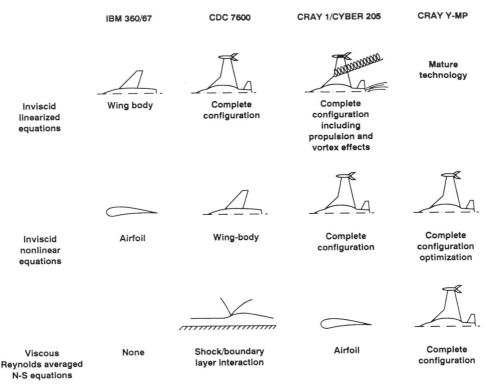

FIGURE 18.4   Examples illustrating impact of new computers on aerodynamic applications.

of the governing equations for flows about aerodynamic shapes is displayed pictorially in Fig. 18.4. Presently available machines are adequate for treating relatively complex configurations with the inviscid flow equations. Of course, the type of information derived from the computations is limited (e.g., no effects of flow separation). The viscous flow equations, being more complex and requiring fine computational meshes to resolve large flow gradients near surfaces and small scales of motion in regions of turbulence, demand substantially greater computational power to solve. Thus, the types of problems that can be solved with a given computer are necessarily less complex. In effect, a designer must make the choice between treating simple configurations with complex physics or treating complex configurations with simple physics. It is obvious, though, that in both the inviscid- and viscous-flow situations each new generation of computers has resulted in corresponding advances in the value of computational fluid dynamics as a design tool. The discipline will begin to mature when both complex configurations and complex physics can be treated together in a practical amount of time.

18.2.3   Algorithm Efficiency

Considerable effort is being expended to develop improved numerical methods for solving the governing equations, particularly for the II, III, and IV levels of approximation. Approaches based on finite difference algorithms (see Sec. 19.2) are the most popular, although work with the level I approximation is largely based on paneling methods (see Sec. 20.1) and that for the level IV approximation frequently involves consideration of spectral methods (see Sec. 19.9).

Finite difference methods require the use of a computational mesh engulfing the configuration and extending far enough in all directions to where outer boundary conditions can be expressed in terms of known quantities. Most meshes being used today are highly complicated nets that conform to geometric contours and involve stretching and clustering designed to tailor grid spacings commensurate with the physical detail that must be captured to provide an accurate solution (see Sec. 19.1). The concept is to solve the governing equations for all of the flow variables for the parcels of fluid within each little volume formed by the grid. Obviously, if computers had unlimited speeds and memories, the grid could be refined to the point of resolving all scales of motion without approximation.

The thrust of the work in algorithm development in the past has been aimed at increasing the degree of vectorization, economy of calculation at each time step, interval length between time steps, numerical stability, accuracy, and ease of computer memory management. Considerable attention is also being directed to developing mathematical representations of complex geometries and automated computational grid generators. Advancements in algorithm development are keeping pace with advances being made in computer speed. This is illustrated by the results shown in Fig. 18.5 where reductions in the cost of a computation owing to improvements in both computers and algorithms are compared. Improvements in computers have reduced the time and cost of a computation with a given algorithm by a factor of almost 100 in a period of 15 years. Improvements in algorithms have reduced the time and cost of a computation with a given computer by almost a thousandfold over the same period of time. These advances compound to result in an overall increase in cost effectiveness and reduction in time required to obtain solutions of nearly 10^5.

Improvements in algorithm efficiency cannot be expected to continue indefinitely, since there are absolute physical limits. At least one iteration will be required to resolve steady flows, and time accurate solutions of unsteady flows will require the selection of time steps commensurate with the frequencies involved. Considering these limits, only one- to two-orders of magnitude of further improvements can be expected. Other factors are beginning to emerge, however, to motivate optimism about further improvements in software efficiency beyond these limits.

Heretofore, improvements in the efficiency of algorithms for solving the various forms of the equations have resulted in almost corresponding reductions in the time required to solve problems of interest. This correspondence is beginning to become less direct as problems being tackled become more complex. Numerical representations of complex three-dimensional geometries, automated optimal grid generation and post-processing of the enormous result files for display and analysis are beginning to place significant burdens on the same computer that is used to solve the flow equations. These pre- and post-processing tasks are expected to account for a growing fraction of the computer power required to solve the important problems of the future. Because of these factors, we can expect that improvements in the more global problem solving methodology will continue to keep pace with improvements in computer power for some time to come, even though limits to improvements in algorithm efficiency are in sight.

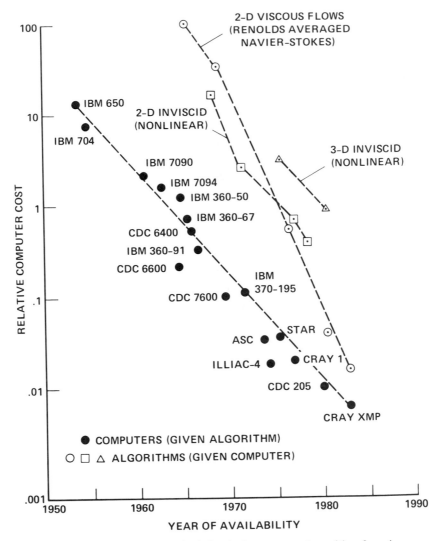

FIGURE 18.5 Comparison of numerical simulation cost trend resulting from improvements in algorithms with that owing to improvements in computers.

18.2.4 Practical Solution Times

A perspective on the computer time that is reasonable to invest in the solution of a problem can be obtained by reviewing the information in Table 18.5 on the historical use of computers for computational fluid dynamics. Experience has shown that computational methods are not routinely used in the design environment unless the time required to obtain a simulation is of the order of 10 to 15 minutes. Short computation times are required for practical sorting of many configurations early in the design cycle when fluid dynamic factors can have the largest effect on the shape of a new device. Simpler forms of the aerohydrodynamic equations, such as the two-dimensional inviscid nonlinear or three-dimensional inviscid linear, can be solved in 10 minutes or less on machines of the IBM-360 or CDC-6600 class. When machines

TABLE 18.5 Perspective on the Use of Computers for Computational Fluid Dynamics

Computer requirements		
Computer class	**Practical engineering computations (minutes of CPU time)**	**Research computation (code development) (hours of CPU time)**
IBM 360/67 CDC 6600	2-D Inviscid nonlinear 3-D Inviscid nonlinear	3-D Inviscid nonlinear
CDC 7600	3-D Inviscid nonlinear	2-D Reynolds-averaged Navier-Stokes
CRAY 1 CYBER 205	2-D Reynolds-averaged Navier-Stokes	3-D Reynolds-averaged Navier-Stokes
CRAY C-90	3-D Reynolds-averaged Navier-Stokes	Large eddy simulation

of this class were first made available to the research community, it was not uncommon to invest several hours of computer time to obtain a single solution; the IBM-360 and CDC-6600 class computers were used to pioneer solution methods for the more complicated three-dimensional inviscid nonlinear equations. Then, as more powerful machines of the CDC-7600 class became available, industry designers routinely used the two-dimensional inviscid nonlinear methods, while the researchers moved on to develop methods for solving the next higher level of approximation of the governing equations. Class VI computers such as the CRAY-1S and the CYBER-205 are adequate to solve routinely the two-dimensional Reynolds-averaged Navier–Stokes equations and to extend research to three dimensions; however the Class VI computers fall far short of making the three-dimensional viscous-flow simulations practical for design work.

18.2.5 Future Computer Requirements

More powerful computers are required to realize the full potential of computational aerodynamics. An estimate of the computational speed that is necessary to compute a flow with the three-dimensional Reynolds-averaged Navier–Stokes equations using 10^7 grid points can be obtained from the data in Fig. 18.6. These results, which are based on the use of improved algorithms projected to be available in 1995, show that a computer must perform at least 1 billion floating point operations per second on a sustained basis in order to complete the simulation in 10 minutes. It is interesting to note that the solution of the same problem would take about a month on an IBM-360, a day on a CDC-7600, and several hours on the Class-VI machines. Implicit in these results is the assumption that data transfer rates to and from memory are fast enough to keep pace with the processing units.

The other aspect of computational power that must be considered is memory. About 27 words of memory at each iteration are required for each grid point used in Reynolds-averaged Navier–Stokes simulations. Thus, a calculation involving 10 million grid points would require high-speed access to about 270 million words.

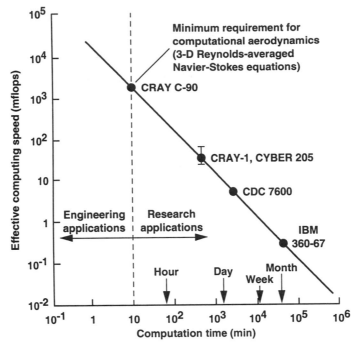

FIGURE 18.6   Relationship between the time required to solve the three-dimensional Reynolds-averaged Navier–Stokes equations and the speed of performing arithmetic operations: 1988 algorithms, 10^7 grid points.

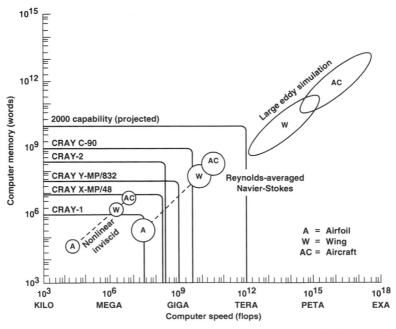

FIGURE 18.7   Computer speed and memory requirements compared with computer capabilities: 1988 algorithms, 15-minute runs.

A computer having a high-speed memory ranging in size from 250 million to 300 million words and a sustained computational rate of at least 1 billion floating point operations per second will bring a use of the Reynolds-averaged Navier–Stokes equations into the design environment. In addition, it would provide the tool required to research applications of direct numerical simulation of turbulent eddies, which is the highest level of approximation to the complete Navier–Stokes equations.

A composite summary of the computer speed and memory requirements for computational aerodynamics is shown in Fig. 18.7. These results are based on estimates presented in Chapman (1979) of the number of grid points required to resolve the flows to engineering accuracy for Reynolds numbers ranging from 10^6 to 10^8. Also shown for comparison are the speed and memory capacities of some existing and planned supercomputers. Computer requirements for simulation of the flow about a complete aircraft with the Reynolds-averaged Navier–Stokes equations should be met in the 1980's but probably not before the next century for corresponding calculations with the Large Eddy Simulation form of the governing equations.

18.3 MULTIDISCIPLINARY COMPUTATIONAL FLUID DYNAMICS
Paul Kutler

18.3.1 Introduction

The use of computational fluid dynamics (CFD) has steadily increased during the last several years. The primary reasons for this usage are the enhanced, validated applications software, available computer power to run that software and the past successful application of CFD to *real world* design problems. According to Holst *et al.* (1992), computer hardware execution speed has increased by a factor of about 15 over the past decade and by over 200 during the past two decades. It can safely be said that the increased use of computational simulations has improved the efficiency of aerospace vehicle performance while at the same time reducing their cost to design as one example of an area where CFD has been very heavily used.

A comparison of the advantages and disadvantages of three simulation techniques, analytical methods, computational fluid dynamics and experimentation is shown in Fig. 18.8. From that figure it can be seen that CFD offers significant advantages over the other two approaches, and this in part explains its rising popularity.

The design of aerospace vehicles requires improvements in such properties as thrust, weight, and stability that is pushing the state-of-the-art in these technologies. The same may be said for other fluid dynamic devices. As a result, development costs using conventional technology are becoming prohibitively expensive. To reduce these costs, more emphasis is being placed on computer simulations to provide data normally obtained from laboratory or large-scale testing. If numerical simulations are to begin supplementing such testing, they must include the complex physical processes that exist on actual aerospace vehicles. This requires modeling physical processes such as propulsion/airframe interaction, flutter, surge, engine unstart, and flameout.

The discipline of computational fluid dynamics has proven its value to many (for its simulation requirements, Boeing currently uses CFD 30% of the time versus 70%

TOOLS FOR AERODYNAMIC PREDICTION

ANALYTICAL
METHODS

COMPUTATIONAL
PROCEDURES

DATA

EXPERIMENTATION

ADVANTAGES

- CLOSED FORM SOLUTIONS (SIMILARITY RULES AND LAWS)
- MINIMAL AMOUNT OF COMPUTER TIME

- EASILY APPLIED
- FEWEST RESTRICTIVE ASSUMPTIONS
- OPTIMIZATION LINK POSSIBLE
- COMPLETE FLOW FIELD DEFINITION
- TREATMENT OF COMPLICATED CONFIGURATIONS
- NO MACH NO. OR REYNOLDS NO. LIMITATIONS
- COST EFFECTIVE

- REPRESENTATIVE OR ACTUAL CONFIGURATION
- REPRESENTATIVE AERODYNAMIC DATA
- OBSERVATION OF NEW FLOW PHENOMENA

DISADVANTAGES

- RESTRICTIVE SIMPLIFYING ASSUMPTIONS
- SIMPLISTIC CONFIGURATIONS
- LIMITED AERODYNAMIC CHARACTERISTICS

- INADEQUATE TURBULENCE MODELS
- LACK OF COMPUTER STORAGE AND SPEED
- ACCURACY OF FINITE – DIFFERENCE REPRESENTATIONS

- COSTLY MODELS AND TUNNEL TIME
- TUNNEL DEPENDENT FLOW CONDITIONS (WALLS, IMPURITIES, TURBULENCE, DISTORTION)
- LIMITED AMOUNT OF DATA
- ACCURACY OF DATA OBTAINED
- SCALING (VISCOUS EFFECTS, CHEMICAL NONEQUILIBRIUM, ETC.)

FIGURE 18.8   Tools for aerodynamic prediction.

for experimentation; it's projected that in the near future, the use of CFD will grow to 60%), and it is routinely being used to simulate flows about vehicles and components that in some instances don't yet fully represent *real* configurations. The overarching goal for computational fluid dynamics or more appropriately *multidisciplinary computational fluid dynamics* (MCFD) is to simulate the actual flow field in or about a computer optimized, realistic device or vehicle at true operating conditions in a reasonable amount of time on the computer. Phenomena such as shock and expansion waves, vortices, shear layers, separation, reattachment, and unsteadiness are part of the actual flow field that must be predicted. A *realistic configuration* consists of a structure that is geometrically accurate, flexes under loads, may have deflecting control surfaces that perturb the flow field, and might possess propulsion systems. A *reasonable amount of time* on the computer depends on the need for the data. Use for design studies requires a quick turn-around time, probably on the order of one hour. Use for design validation, that might supplement or complement an experiment or flight test, would not require such a short time.

Modeling the flow about realistic vehicles and devices will require computing capabilities that currently do not exist. Computer performance today is limited by circuit speeds and thus the speed of light. The advent of computers based on parallel architectures offers the computational scientist an opportunity to solve his problem in a reasonable amount of time and thus should have an impact on multidisciplinary problems. Pacing items for CFD problems thus include computer hardware, systems software, and applications software mapped efficiently onto the computer.

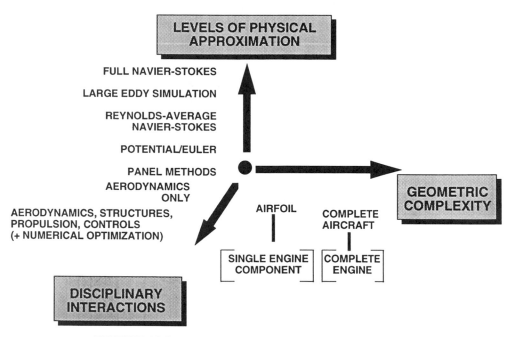

FIGURE 18.9 Dimensions of the computational challenge.

The simulation of an actual aircraft, as an example, is truly multidisciplinary in nature and offers numerous challenges to the CFD scientist. The dimensions of the computational challenge are depicted graphically in Fig. 18.9. Displayed on the vertical axis is the equation set complexity, displayed on the horizontal axis is the geometric complexity and displayed on the inclined axis are the disciplinary interactions. In the early stages of development of the CFD discipline, activity was focused at the origin of this graph, that is, simple gas-dynamic equation sets were being applied to simple geometries. As the discipline matured, more descriptive equation sets were being solved for more complex geometries. However, these were still only being applied to predict the aerodynamics. Similarly, other single computational disciplines were advancing in an analogous fashion such as structures, acoustics, propulsion, and controls. Computational fluid dynamicists today are capable of solving the Reynolds-averaged Navier–Stokes equations about complete aircraft to predict the viscous flow field. They are also capable of solving that same equation set coupled with a simple structural model for wings. This discipline is thus heading towards multidisciplinary numerical simulation.

To that end, NASA is currently involved in a national research program entitled High-Performance Computing and Communication (HPCCP). One part of the NASA element involves the solution of computational aeroscience grand challenge problems, that is, multidisciplinary problems, on massively parallel computers. The primary objective of the computational aeroscience applications portion of HPCCP is to develop robust computational methods and the associated pilot application computer programs that will enable integrated multidisciplinary analysis and design of advanced aerospace systems on massively parallel computers. This program is unique in that it will require teams of scientists working together with expertise in computer science, computational science, and numerous physical disciplinary sciences. One goal of the program is to make available computational power 1000 times greater than that existing today to solve the computational grand challenge problems. Research for this element of the program is focusing on the development of algorithms for multidisciplinary equation set coupling, algorithm/architecture mapping of the multidisciplinary models, multidisciplinary simulation and analysis, and application of these technologies to airframes, propulsion systems, and vehicle systems.

Challenges offered the CFD scientist include coupling the equations governing the necessary disciplines, developing the appropriate algorithms to solve that all-inclusive set of equations and efficiently programming those procedures on the computer. Solution of such a set of equations is estimated to require a teraFLOPS (10^{12}) computer and between 10 and 100 hr per solution.

18.3.2 Elements Pacing MCFD

Requirements for performing realistic simulations include multidisplinary application software, systems software, and computers powerful enough to run that software. The successful development of all three of these technologies is mandatory for a positive outcome of MCFD. There are a number of elements for MCFD that are pacing the progress not only of multidisciplinary problems but also single disciplinary problems. They include physical modeling (e.g., turbulence and transition modeling, chemistry modeling, other physics such as electromagnetics and structures), solution methodology (e.g., equation set coupling and algorithms), computer power, and multidisciplinary validation data. Secondary pacing technologies include pre- and post-data processing (e.g., surface, internal and structural definition, grid generation, and scientific visualization). These pacing elements are discussed below in more detail.

Solution Methodology. With the potential for enhanced computer power offered by machines with parallel architectures will come a search for algorithms that run effectively on those machines. Not only will existing algorithms have to be programmed and evaluated, but also *old* algorithms will have to revisited and tried to determine their effectiveness. In addition, the development of new algorithms specifically designed to take advantage of these architectures will have to be researched. These new algorithms should be those that can efficiently solve such equation sets as those governing gas-dynamics, structural dynamics, controls, and optimization as well as some combination of those equation sets.

Because there are different parallel architectures on the market, e.g., SIMD (single instruction, multiple data) and MIMD (multiple instruction, multiple data), there will be some algorithms, e.g., explicit and implicit, that operate more effectively on each of those parallel machines. Holst *et al.* (1992) briefly discuss some tradeoffs between algorithms and architectures. The ultimate goal, of course, is to produce the most cost-effective and accurate simulation possible, and this will involve tradeoffs among numerical algorithms, computer hardware, and physical disciplines.

Problems governed by multiple equation sets of different physical disciplines will require research as to the most optimum procedure for coupling those equation sets. Both indirect and direct procedures will have to be investigated. The indirect approach links the physical dependence of one discipline on another by sensitivity matrices. The direct approach mathematically links the governing equation sets. They can be linked in either a strong or weak form. For example, hypersonic problems involve the solution of both the gas dynamic and finite-rate chemistry equations. In the strongly coupled approach, the two equation sets are solved simultaneously by a large matrix inversion process for an implicit method. In the weakly coupled approach, each equation set is solved independently at each iteration, and the information from each equation set is then used to update the other equation set for the next iteration. The strongly coupled approach takes considerably more computer time than the weakly coupled approach. For hypersonic problems the weakly coupled approach has proved successful.

In addressing problems involving multi-equation sets, the *stiffness* of the governing equations might also dictate the solution procedure to be employed. For example, in computing aeroelastic problems using coupled procedures, the fluid equations are cast in an Eulerian system while the structural equations are in a Lagrangian system. The structural system is much stiffer than the fluid system. Therefore, it is numerically inefficient to solve both systems using a monolithic numerical scheme. In this case a domain decomposition approach can be employed in which the equation sets are coupled only at the boundaries between the fluids and structures.

One of the advantages that computer simulation possesses over either laboratory or field testing is that it can automatically enhance the design of the vehicle or component. The tedium of design-space parameter selection and the trial-and-error process of eliminating poor designs is performed much more efficiently by the computer. This is done using numerical optimization procedures. In this approach, an optimization algorithm is coupled with a flow solver. Design requirements and constraints are imposed on the problem, and the computer is employed to search the design space until the optimal configuration is found.

Physical Modeling. The numerical simulation of viscous flows about real vehicles and devices and their components is being attacked from both ends of the computational spectrum. At one end are Reynolds-averaged Navier–Stokes (RANS) solutions that employ suitable turbulence models (see Secs. 21.3 and 21.4) while at the other end are direct numerical solutions that don't require a turbulence model (see Sec. 21.5). As computer and algorithmic technology improves, these two approaches will merge [Kutler (1986)].

The development of suitable turbulence models for the RANS equations to date remains highly problem dependent. They are dependent on such elements of the problem as the free-stream conditions (for example, Mach and Reynolds number), configuration geometry (for example, deflected control surfaces and wing-body junctures) and flow field behavior (for example, shock boundary layer interaction). Accurate and numerically efficient turbulence models for numerical simulations, especially those types of models that can be used to obtain engineering answers for separated flow problems, must be developed.

One multidisciplinary area of application requiring advanced turbulence models is associated with combustion. In this area, turbulence and chemistry models must be developed to appropriately account for chemistry/turbulence interactions. Transition models, that predict the flow characteristics leading up to fully developed turbulent flows are also important, especially for laminar flow control (LFC) applications, e.g., those associated with the high-speed civil transport, and high-speed, high-altitude flows.

To improve the turbulence-model development process, it should be considered at the early stage of code development according to Marvin (1986). The first step in such a process is to identify those flows pacing the development of the aerodynamic computations. The second step is to develop models through a phased approach of building-block studies that combine theory, experiment, and computations. The final step is to provide verification and/or limits of the modeling through *benchmark* experiments over a practical range of Reynolds and Mach numbers.

Turbulence models that complete the above process must then be installed in the application software. This requires close collaboration between the model developer and software author. Issues associated with overall code stability must be addressed.

Computer Technology. Conventional serial computer architectures will probably not be capable of fulfilling the requirement of multidisciplinary computational fluid dynamics challenges in a reasonable amount of time. For that matter, some single discipline problems, such as direct simulation of turbulence, will also demand computer power beyond that offered by serial machines. Those machines are limited by the speed of light. To alleviate that problem, computers based on parallel architectures are being developed. Massively parallel computer architectures offer another avenue for improvement in computing speed. To obtain a numerical simulation on such a computer, the problem is divided among hundreds or even thousands of processors. It is important that the work be balanced across the processors and the communication between processors is minimized. By doing this, the time to compute a solution is dramatically reduced.

The goal for the multidisciplinary element within the HPCCP is a sustained execution speed of one teraFLOPS. This will require a peak speed of more than one teraFLOPS or approximately 10 teraFLOPS. The multidisciplinary performance goal was designed to push the current state of computing technology. The HPCCP, it is hoped, will lead to the commercial development of fully scaled systems complete with system software.

Parallel computers currently available at a major CFD center such as NASA Ames Research Center for performing calculations include: 1) the Connection Machine (CM-5) with its 128 nodes, 40 megahertz (MHz) Sparc processors, 4 vector units per processor, 32 megabytes of memory per node, 2) the Intel iPSC/860 Gamma with 128 nodes based on 40 MHz i860 XR superscalar RISC chips with 8 megabytes of memory per node, 3) the Intel Paragon with 227, 75 MHz, i860XP processors and 32 megabytes of memory per node, and 4) the IBM SP2 with 160, 66.6 MHz processors and 32 megabytes of memory per node with 6 nodes having 512 megabytes.

Turbulent flow numerical simulations require extremely large data bases and long run times. To solve problems such as these on the Intel machines, Wray (1992) first had to implement Vectoral (a highly optimized compiler), a very efficient Fast Fourier Transform and an efficient inter-processor communication scheme. Once this software was written, then the applications software could be easily ported to the machines. The first code ported simulates homogeneous turbulence in a three-dimensional box. For this code on the 128 processor Intel Gamma, it runs more than 11 times faster than on a single Cray Y-MP processor. However, that is about 25% its theoretical speed. If it were not for limitations imposed by the speed of memory accesses and of inter-processor communication, the code would run at 80% of the absolute upper bound of performance (80MFLOPS per processor in single precision). The performance on the Delta, using 512 processors, is about 30 times a Y-MP processor.

The results of these calculations indicate that highly parallel, distributed memory computers are well suited to large-scale turbulence simulations, memory speed degrades the i860 processor performance, memory sizes for the speed available are inadequate, and the interprocessor communication speeds are out of balance with the processor speeds.

Multidisciplinary Validation. Validation of multidisciplinary application software will be a key pacing item. It takes a long time for design engineers to gain confidence in applications software. That confidence is in part gained by comparison with experimental or field test data. It will be important to conceive and perform, in parallel with the application software development, companion experiments designed to validate the multidisciplinary computer programs being created. It is also important that these experiments be designed in concert with the MCFD scientists but at an early stage so that the data produced can impact the software development program in a timely fashion. In the past, such attempts resulted in a missed opportunity, because the experiments required such a long time to produce the validation data.

The multidisciplinary nature of the problem to be validated might preclude the validation experiment from being done in a laboratory because of its complicated nature. That is, the experiment could involve structures, controls and propulsion elements requiring a costly model and sophisticated instrumentation. The only alternative, therefore, would be field testing. It's expensive, but it produces realistic data. On the other hand, simple double-disciplinary experiments can be and have been performed in the laboratory. A typical example is aeroelasticity in which fluid and structure interactions are studied in a wind tunnel.

Secondary Pacing Technologies. A bottleneck technology for single disciplinary, numerical fluid dynamics simulation is grid generation (see Secs. 19.1 and 19.7). The goal for grid generation, of course, is to efficiently and effectively distribute grid points to generate the most accurate solution possible with the least number of points. Research in solution adaptive procedures for steady flows has resulted in substantial gains in the optimal placement of grid points. Research is still required to effectively apply these procedures to unsteady flows which will be important in multidisciplinary computational aerosciences problems.

For multidisciplinary problems, it is quite possible that additional grid points will be required to satisfy the accuracy requirements of the secondary disciplines. Thus, the grid generation process is a pacing technology for multidisciplinary problems. These problems involving multiple equation sets that describe each discipline obviously will require more time on the computer to solve. Because of this, scientists will tend to reduce the number of grid points from that required for an accurate flow solution in order to get their multidisciplinary problem solved in a reasonable amount of time on the computer. This impairs the overall solution accuracy and can bias the results. Thus, there is a need to develop solution adaptive grid procedures for unsteady problems in order to optimally utilize available grid points.

As an example, aeroelasticity problems require an unsteady flow solver to capture the physics generated by coupling the fluid and structural dynamic equations of motion. To do this for a simple wing-body configuration in a reasonable amount of time on existing computers limits the grid to only about half the size of that used for the fluids problem alone.

Numerically solved multidisciplinary problems will generate enormous amounts of data on massively parallel computers. To effectively display and digest this data, quite naturally, requires high-resolution, high-throughput computer graphics devices utilizing sophisticated software display packages.

REFERENCES

Chapman, D. R., "Computational Aerodynamics Development and Outlook," Dryden Lectureship in Research, AIAA Paper 79-0129, Jan. 1979.

Holst, T. L., Salas, M. D., and Claus, R. W., "The NASA Computational Aerosciences Program—Toward Teraflops Computing," AIAA-92-0558, 1992.

Kutler, P., "A Perspective of Computational Fluid Dynamics," NASA TM-88246, May 1986.

Marvin, J. G., "Future Requirements of Wind Tunnels for Computational Fluid Dynamics Code Verifications," AIAA Paper 86-0752-CP, 1986.

Wray, A. A. and Rogallo, R. S., "Simulation of Turbulence on the Intel Gamma and Delta," NASA Technical Memorandum, Apr. 1992.

19 Numerical Methods of Solving Fluid Dynamic Equations

S. NAKAMURA
Ohio State University
Columbus, OH

RAINALD LÖHNER
The George Mason University
Fairfax, VA

EVANGELOS HYTOPOULOS
Automated Analysis Corp.
Ann Arbor, MI

TAYFUN E. TEZDUYAR
MAREK BEHR
University of Minnesota
Minneapolis, MN

THOMAS J. R. HUGHES
Stanford University
Stanford, CA

JAMES L. THOMAS
W. KYLE ANDERSON
NASA Langley Research Center
Hampton, VA

H. C. YEE
NASA Ames Research Center
Moffett Field, CA

PETER R. EISEMAN
Program Development Corporation
White Plains, NY

PAUL N. SWARZTRAUBER
National Center for Atmospheric Research
Boulder, CO

Handbook of Fluid Dynamics and Fluid Machinery, Edited by Joseph A. Schetz and Allen E. Fuhs
ISBN 0-471-12598-9 Copyright © 1996 John Wiley & Sons, Inc.

ROLAND A. SWEET
University of Colorado at Denver
Denver, CO

JOHN KIM
University of California Los Angeles
Los Angeles, CA

CONTENTS

19.1 GRID GENERATION

The geometry of the flow domain in computational fluid analyses is usually complicated. To discretize the flow equations, two alternative approaches are commonly used: 1) to map the flow domain to a rectangular or parallel piped domain where a simple equispaced Cartesian grid is used, and 2) to cover the domain with a unstructured grid. The first approach is often preferred when the finite difference approximations of the flow equations are used. Although the coordinate transformation of a complicated geometry to a simple geometry increases the complexity of the equations, computational efficiency can be maintained to be high. The second approach also has advantages that, without cumbersome coordinate transformations, computational solution of flow equations can be applied to significantly more complicated geometries. First, we focus on basic aspects of theory and methods of coordinate transformation and grid generation for structured grids. This is followed by a section on unstructured grids. Adaptive grid generation is discussed in Sec. 19.7.

19.1.1 Coordinate Transformation and Structured Grid Generation
S. Nakamura

Transformation of Two-Dimensional Coordinates (Stationary Systems). The coordinates on the physical domain will be denoted by (x, y), and those of the computational domain by (ξ, η). The physical domain is bounded by the boundary D, while that of the computational domain is bounded by the boundary G (see Fig. 19.1). Any point, P, on the physical domain has a corresponding point on the com-

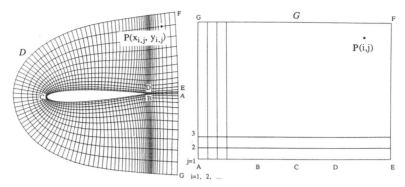

FIGURE 19.1 Mapping of grid between physical and computational domains.

putational domain, and *vice versa*. Thus, the relation between the physical domain and the computational domain can be written in the functional forms as [Thompson (1982i), (1988), Thompson *et al.* (1982), (1985), Ghia and Ghia (1983), and Knupp and Steinberg (1993)].

$$x = x(\xi, \eta)$$
$$y = y(\xi, \eta)$$
(19.1)

or equivalently as

$$\xi = \xi(x, y)$$
$$\eta = \eta(x, y).$$
(19.2)

For simple geometries, the transformation may be expressed analytically. The coordinate transformation that can be expressed by analytical functions is called *analytical transformation*. The conformal mapping [Ives (1982) and Anderson *et al.* (1982)] is one of the analytical transformations. However, in the usual computational fluid dynamics analysis, the geometry is so complicated that probably no analytical transformation can be found. In such cases, the only way is to express the coordinate transformation in a numerical form.

In the remainder of this section, however, we will first develop analytical transformation of derivatives. From Eq. (19.1), the derivatives $\partial/\partial x$ and $\partial/\partial y$ can be expressed by

$$\frac{\partial}{\partial x} = \xi_x \frac{\partial}{\partial \xi} + \eta_x \frac{\partial}{\partial \eta}$$
$$\frac{\partial}{\partial y} = \xi_y \frac{\partial}{\partial \xi} + \eta_y \frac{\partial}{\partial \eta}$$
(19.3)

From Eq. (19.2), the derivatives $\partial/\partial \xi$ and $\partial/\partial \eta$ can be written as

$$\frac{\partial}{\partial \xi} = x_\xi \frac{\partial}{\partial x} + y_\xi \frac{\partial}{\partial y}$$
$$\frac{\partial}{\partial \eta} = x_\eta \frac{\partial}{\partial x} + y_\eta \frac{\partial}{\partial y}$$
(19.4)

Solving Eq. (19.4) with respect to $\partial/\partial\xi$ and $\partial/\partial\eta$ gives

$$\frac{\partial}{\partial x} = J\left(y_\eta \frac{\partial}{\partial \xi} - y_\xi \frac{\partial}{\partial \eta}\right)$$

$$\frac{\partial}{\partial y} = J\left(-x_\eta \frac{\partial}{\partial \xi} + x_\xi \frac{\partial}{\partial \eta}\right)$$

(19.5)

where J is the Jacobian defined by $J = \xi_x y_y - \xi_y \eta_x = (x_\xi y_\eta - x_\eta y_\xi)^{-1}$.

Equation (19.5) thus obtained is equivalent to Eq. (19.3), and comparision of Eq. (19.3) and Eq. (19.5) gives us the following relations

$$\xi_x = y_\eta J$$

$$\xi_y = -x_\eta J$$

$$\eta_x = -y_\xi J$$

$$\eta_y = x_\xi J$$

(19.6)

The second partial derivatives on the physical domain may be transformed to the computational domain by repeating the tranformation of the first derivatives. The second derivative, with respect to x becomes

$$\frac{\partial^2}{\partial x^2} = \frac{\partial}{\partial x}\left(\xi_x \frac{\partial}{\partial \xi} + \eta_x \frac{\partial}{\partial \eta}\right)$$

$$= \left(\xi_x \frac{\partial}{\partial \xi} + \eta_x \frac{\partial}{\partial \eta}\right)\left(\xi_x \frac{\partial}{\partial \xi} + \eta_x \frac{\partial}{\partial \eta}\right)$$

$$= \xi_x \frac{\partial}{\partial \xi} \xi_x \frac{\partial}{\partial \xi} + \xi_x \frac{\partial}{\partial \xi} \eta_x \frac{\partial}{\partial \eta} + \eta_x \frac{\partial}{\partial \eta} \xi_x \frac{\partial}{\partial \xi} + \eta_x \frac{\partial}{\partial \eta} \eta_x \frac{\partial}{\partial \eta}$$

(19.7)

The second derivative with respect to y may be derived in a similar way. It is also possible to obtain a more convenient form after carrying out differentiation

$$\frac{\partial^2}{\partial x^2} = \xi_{xx} \frac{\partial}{\partial \xi} + \eta_{xx} \frac{\partial}{\partial \eta} + \xi_x^2 \frac{\partial^2}{\partial \xi^2} + 2\eta_x \xi_x \frac{\partial^2}{\partial \eta \partial \xi} + \eta_x^2 \frac{\partial^2}{\partial \eta^2}$$

$$\frac{\partial^2}{\partial y^2} = \xi_{yy} \frac{\partial}{\partial \xi} + \eta_{yy} \frac{\partial}{\partial \eta} + \xi_y^2 \frac{\partial^2}{\partial \xi^2} + 2\eta_y \xi_y \frac{\partial^2}{\partial \eta \partial \xi} + \eta_y^2 \frac{\partial^2}{\partial \eta^2}$$

(19.8)

In the actual numerical computations, the coordinate transformations are given in a discrete form. Figure 19.1 shows a simple example of the relation between the physical domain and the rectangular computational domain. The grid on the computational domain is equispaced in both directions. Each grid point on the computational grid has its corresponding point on the physical domain. The discrete coordinate transformation means that the coordinates (x, y) for the point $(\xi = i, \eta = j)$, is given as $(x_{i,j}, y_{i,j})$, but no analytical relation between the two coordinate systems is given, except when the transformation is based on an analytical relation.

The metrics on the grid points of the computational domain are evaluated by the difference approximations (see Sec. 19.2)

$$\xi_x = y_\eta J \approx \frac{y_{i,j+1} - y_{i,j-1}}{2} J$$

$$\xi_y = -x_\eta J \approx -\frac{x_{i,j+1} - x_{i,j-1}}{2} J$$

$$\eta_y = -y_\xi J \approx -\frac{y_{i+1,j} - y_{i-1,j}}{2} J \qquad (19.9)$$

$$\eta_y = x_\xi J \approx \frac{x_{i+1,j} - x_{i-1,j}}{2} J$$

$$J = (x_\xi y_\eta - x_\eta y_\xi)^{-1}$$

$$\approx \left[\frac{(x_{i+1,j} - x_{i-1,j})(y_{i,j+1} - y_{i,j-1}) - (x_{i,j+1} - x_{i,j-1})(y_{i+1,j} - y_{i-1,j})}{4} \right]^{-1}$$

where mesh interval sizes on the computational domain are assumed to be unity. Whenever the central difference approximations do not apply to grid points along the boundary, the three-point forward or backward difference approximation may be used. For example, ξ_x for $j = 1$ may be calculated by

$$\xi_x = y_\eta J \approx \frac{-y_{i,3} + 4y_{i,2} - 3y_{i,1}}{2} J \qquad (19.10)$$

where J is also evaluated by the three-point forward difference approximations.

Example 19.1: Transform the following equations onto the computational domain (ξ, η)

$$\frac{\partial u}{\partial x} + \frac{\partial v}{\partial y} = 0$$

$$u \frac{\partial u}{\partial x} + v \frac{\partial u}{\partial y} = -\frac{\partial p}{\partial x} \qquad (19.11)$$

$$u \frac{\partial v}{\partial x} + v \frac{\partial v}{\partial y} = -\frac{\partial p}{\partial y}$$

where $u = u(x, y)$ and $v = v(x, y)$ are velocity components and $p = p(x, y)$ is the pressure.

Introducing Eq. (19.5) into Eq. (19.11) yields

$$y_\eta \frac{\partial u}{\partial \xi} - y_\xi \frac{\partial u}{\partial \eta} - x_\eta \frac{\partial v}{\partial \xi} + x_\xi \frac{\partial v}{\partial \eta} = 0$$

$$u \left(y_\eta \frac{\partial u}{\partial \xi} - y_\xi \frac{\partial u}{\partial \eta} \right) + v \left(-x_\eta \frac{\partial u}{\partial \xi} + x_\xi \frac{\partial u}{\partial \eta} \right) = -y_\eta \frac{\partial p}{\partial \xi} + y_\xi \frac{\partial p}{\partial \eta} \qquad (19.12)$$

$$u \left(y_\eta \frac{\partial v}{\partial \xi} - y_\xi \frac{\partial v}{\partial \eta} \right) + v \left(-x_\eta \frac{\partial v}{\partial \xi} + x_\xi \frac{\partial v}{\partial \eta} \right) = x_\eta \frac{\partial p}{\partial \xi} - x_\xi \frac{\partial p}{\partial \eta}$$

where the equations have been multiplied by J^{-1}.

Example 19.2: The derivative normal to a curved boundary of the physical domain is given by

$$\frac{\partial}{\partial n} = l_x \frac{\partial}{\partial x} + l_y \frac{\partial}{\partial x} \tag{19.13}$$

where l_x and l_y are directional cosines of the unit vector normal to the boundary. Express the same derivative on the computational coordinates.

Introducing Eq. (19.5) into Eq. (19.13) yields

$$\frac{\partial}{\partial n} = c_1 \frac{\partial}{\partial \xi} + c_2 \frac{\partial}{\partial \eta} \tag{19.14}$$

where $c_1 = (l_x y_\eta - l_y x_\eta) J$ and $c_2 = (-l_x y_\xi + l_y x_\xi) J$.

Example 19.3: Transform the following compressible Euler equations onto the computational coordinates:

$$\frac{\partial Q}{\partial t} + \frac{\partial E}{\partial x} + \frac{\partial F}{\partial y} = 0 \tag{19.15}$$

$$Q = \begin{bmatrix} \rho \\ \rho u \\ \rho v \\ e \end{bmatrix}, \quad E = \begin{bmatrix} \rho u \\ \rho u^2 + p \\ \rho u v \\ (e + p)u \end{bmatrix}, \quad F = \begin{bmatrix} \rho v \\ \rho u v \\ \rho v^2 + p \\ (e + p)v \end{bmatrix} \tag{19.16}$$

$$p = (\gamma - 1) \left[e - \tfrac{1}{2}\rho(u^2 + v^2) \right] \tag{19.17}$$

Four conservation law equations are included in Eq. (19.15). The first equation is the mass conservation law explicitly written as

$$\frac{\partial \rho}{\partial t} = \frac{\partial \rho u}{\partial x} + \frac{\partial \rho v}{\partial y} \tag{19.18}$$

After the coordinate transformation, it becomes

$$\frac{\partial \rho}{\partial t} = \left(\xi_x \frac{\partial}{\partial \xi} + \eta_x \frac{\partial}{\partial \eta} \right) \rho u + \left(\xi_y \frac{\partial}{\partial \xi} + \eta_y \frac{\partial}{\partial \eta} \right) \rho v \tag{19.19}$$

Multiplying the foregoing equation by

$$J = \partial(\xi, \eta)/\partial(x, y) \tag{19.20}$$

yields

$$\frac{\partial J^{-1}\rho}{\partial t} = \left[\left(\xi_x \frac{\partial}{\partial \xi} + \eta_x \frac{\partial}{\partial \eta} \right) \rho u + \left(\xi_y \frac{\partial}{\partial \xi} + \eta_y \frac{\partial}{\partial \eta} \right) \rho v \right] J^{-1} \tag{19.21}$$

where the inverse Jacobian is placed within the time derivative, because it is time-independent. It is easy to show that this equation can be rewritten as

$$\frac{\partial J^{-1}\rho}{\partial t} = \frac{\partial}{\partial \xi} J^{-1}\rho(\xi_x u + \xi_y v) + \frac{\partial}{\partial \eta} J^{-1}\rho(\eta_x u + \eta_y v) \tag{19.22}$$

We define here new variables

$$U = \xi_x u + \xi_y v$$
$$V = \eta_x u + \eta_y v \tag{19.23}$$

which are named *contravariant velocities*. Then, Eq. (19.22) is written as

$$\frac{\partial J^{-1}\rho}{\partial t} = \frac{\partial}{\partial \xi} J^{-1}\rho U + \frac{\partial}{\partial \eta} J^{-1}\rho V \tag{19.24}$$

Three other conservation laws, namely two momentum equations plus the energy equation, can be rewritten in similar forms. The four equations thus obtained may be expressed as

$$\frac{\partial \hat{Q}}{\partial t} + \frac{\partial \hat{E}}{\partial \xi} + \frac{\partial \hat{F}}{\partial \eta} = 0 \tag{19.25}$$

where

$$\hat{Q} = \frac{Q}{J}, \quad \hat{E} = \frac{\xi_x E + \xi_y F}{J}, \quad \hat{F} = \frac{\eta_x E + \eta_y F}{J} \tag{19.26}$$

Equation (19.25) is called the strong conservation form of the Euler equation.

Transformation of Second Order Equations: The Laplace equation is a very important partial differential equation not only for fluid dynamics theory but also for co-ordinate transformation itself. Therefore, we focus our effort now on the Laplace equation

$$\frac{\partial^2 \phi}{\partial x^2} + \frac{\partial^2 \phi}{\partial y^2} = 0 \tag{19.27}$$

Introducing Eq. (19.8) into Eq. (19.27), we obtain

$$(\xi_x^2 + \xi_y^2) \frac{\partial^2 \phi}{\partial \xi^2} + 2(\eta_x \xi_x + \eta_y \xi_y) \frac{\partial^2 \phi}{\partial \eta \, \partial \xi} + (\eta_x^2 + \eta_y^2) \frac{\partial^2 \phi}{\partial \eta^2} + (\xi_{xx} + \xi_{yy}) \frac{\partial \phi}{\partial \xi}$$

$$+ (\eta_{xx} + \eta_{yy}) \frac{\partial \phi}{\partial \eta} = 0 \qquad (19.28)$$

The coefficients of Eq. (19.28) are in terms of derivatives of the ξ and η with respect to x and y, which are difficult to evaluate numerically. We wish to express the coefficients in terms of derivatives of x and y with respect to ξ and η. This can be done easily for the first three coefficients, using the relations of Eq. (9.6) in the preceding section, because they involve only first derivatives

$$A \equiv \xi_x^2 + \xi_y^2 = J^2(x_\eta^2 + y_\eta^2)$$

$$B \equiv \eta_x \xi_x + \eta_y \xi_y = -J^2(x_\xi x_\eta + y_\xi y_\eta) \qquad (19.29)$$

$$C \equiv \eta_x^2 + \eta_y^2 = J^2(x_\xi^2 + y_\xi^2)$$

Introducing the foregoing equations into Eq. (19.28) yields

$$A \frac{\partial^2 \phi}{\partial \xi^2} + 2B \frac{\partial^2 \phi}{\partial \eta \partial \xi} + C \frac{\partial^2 \xi}{\partial \eta^2} + (\xi_{xx} + \xi_{yy}) \frac{\partial \phi}{\partial \xi} + (\eta_{xx} + \eta_{yy}) \frac{\partial \phi}{\partial \eta} = 0 \quad (19.30)$$

Notice that Eq. (19.27) is satisfied if we set $\phi = x$ or $\phi = y$. Because Eq. (19.30) is equivalent to Eq. (19.27), it is also satisfied by $\phi = x$ and $\phi = y$. Thus, we obtain two equations as follows

$$A x_{\xi\xi} + 2B x_{\xi\eta} + C x_{\eta\eta} + (\xi_{xx} + \xi_{yy}) x_\xi + (\eta_{xx} + \eta_{yy}) x_\eta = 0$$
$$\qquad (19.31)$$
$$A y_{\xi\xi} + 2B y_{\xi\eta} + C y_{\eta\eta} + (\xi_{xx} + \xi_{yy}) y_\xi + (\eta_{xx} + \eta_{yy}) y_\eta = 0$$

Solving the two equations above for $(\xi_{xx} + \xi_{yy})$ and $(\eta_{xx} + \eta_{yy})$ yields

$$\xi_{xx} + \xi_{yy} = [x_\eta(A y_{\xi\xi} + 2B y_{\xi\eta} + C y_{\eta\eta}) - y_\eta(A x_{\xi\xi} + 2B x_{\xi\eta} + C x_{\eta\eta})]J$$
$$\qquad (19.32)$$
$$\eta_{xx} + \eta_{yy} = [y_\xi(A x_{\xi\xi} + 2B x_{\xi\eta} + C x_{\eta\eta}) - x_\xi(A y_{\xi\xi} + 2B y_{\xi\eta} + C y_{\eta\eta})]J$$

or more compactly

$$E \equiv \xi_{xx} + \xi_{yy} = (x_\eta H - y_\eta G)J$$
$$\qquad (19.33)$$
$$F \equiv \eta_{xx} + \eta_{yy} = (y_\xi G - x_\xi H)J$$

where

$$G = A x_{\xi\xi} + 2B x_{\xi\eta} + C x_{\eta\eta}$$
$$\qquad (19.34)$$
$$H = A y_{\xi\xi} + 2B y_{\xi\eta} + C y_{\eta\eta}$$

Thus, Eq. (19.30) is now written as

$$A \frac{\partial^2 \phi}{\partial \xi^2} + 2B \frac{\partial^2 \phi}{\partial \eta \partial \xi} + C \frac{\partial^2 \phi}{\partial \eta^2} + E \frac{\partial \phi}{\partial \xi} + F \frac{\partial \phi}{\partial \eta} = 0 \qquad (19.35)$$

If the coordinate transformation is based on the solution of

$$\begin{aligned} E &\equiv \xi_{xx} + \xi_{yy} = 0 \\ F &\equiv \eta_{xx} + \eta_{yy} = 0 \end{aligned} \qquad (19.36)$$

then Eq. (19.35) is simplified to

$$A \frac{\partial^2 \phi}{\partial \xi^2} + 2B \frac{\partial^2 \phi}{\partial \eta \partial \xi} + C \frac{\partial^2 \phi}{\partial \eta^2} = 0 \qquad (19.37)$$

If the coordinate transformation does not satisfy Eq. (19.36), the last two terms in Eq. (19.35) must be retained.

Example 19.4:   The incompressible Navier–Stokes equations in the vorticity and stream-function form are given by

$$\begin{aligned} \frac{\partial \psi}{\partial t} + u \frac{\partial \xi}{\partial x} + v \frac{\partial \psi}{\partial y} &= \frac{1}{Re} \left(\frac{\partial^2 \psi}{\partial x^2} + \frac{\partial^2 \psi}{\partial y^2} \right) \\ \frac{\partial^2 \phi}{\partial x^2} + \frac{\partial^2 \phi}{\partial y^2} &= \psi \end{aligned} \qquad (19.38)$$

where $\psi = \psi(x, y, t)$ and $\phi = \phi(x, y, t)$ are vorticity and stream function respectively. Transform the equations onto the computational domain.

The equations on the computational coordinates are

$$\frac{\partial \psi}{\partial t} + [(uy_\eta - vx_\eta)\psi_\xi + (vx_\xi - uy_\xi)\psi_\eta]J = \frac{1}{Re} \nabla^2_{\xi\eta} \psi$$

$$\nabla^2_{\xi\eta} \phi = \psi$$

$$\nabla^2_{\xi\eta} f = (Af_{\xi\xi} + 2Bf_{\xi\eta} + Cf_{\eta\eta})$$

$$+ [(y_\xi f_\eta - y_\eta f_\xi)G + (x_\eta f_\xi - x_\xi f_\eta)H]J$$

$$(19.39)$$

Transformation of Moving Coordinates in Two Dimensions.   Here, we consider the situation where the computational coordinate is not stationary but moving. In computations, this occurs when the grid points on the physical domain move with time. We consider the relations

$$x = x(\xi, \eta, t) \tag{19.40}$$
$$y = y(\xi, \eta, t)$$

where the (x, y) domain changes in time, while the (ξ, η) domain is fixed. Equation (19.40) may be equivalently expressed as

$$\xi = \xi(x, y, t) \tag{19.41}$$
$$\eta = \eta(x, y, t)$$

For any instant, the transformation of the spatial derivatives between the (x, y) domain and the (ξ, η) domain and *vice versa* are the same as for the stationary systems. However, the time derivative of the flow equation is affected when one of the coordinates moves. The time derivative of a physical quantity $A(x, y, t)$ on the moving (x, y) domain is related to that in the fixed (ξ, η) domain by

$$\left[\frac{\partial A}{\partial t}\right]_{(\xi, \eta)} = \left[\frac{\partial A}{\partial t}\right]_{(x, y)} + (\dot{x})_{(\xi, \eta)}\frac{\partial A}{\partial x} + (\dot{y})_{(\xi, \eta)}\frac{\partial A}{\partial y} \tag{19.42}$$

where the coefficients of the second and third terms on the right side are velocity components of the image of (ξ, η) moving on the (x, y) domain. Rewriting the equation above yields

$$\left[\frac{\partial A}{\partial t}\right]_{(x, y)} = \left[\frac{\partial A}{\partial t}\right]_{(\xi, \eta)} - (\dot{x})_{(\xi, \eta)}\frac{\partial A}{\partial x} - (\dot{y})_{(\xi, \eta)}\frac{\partial A}{\partial y} \tag{19.42a}$$

Furthermore, by introducing the transformation of the first derivatives with respect to x and y, we get

$$\left[\frac{\partial A}{\partial t}\right]_{(x, y)} = \left[\frac{\partial A}{\partial t}\right]_{(\xi, \eta)} - \left((\dot{x})_{(\xi, \eta)}\xi_x + (\dot{y})_{(\xi, \eta)}\xi_y\right)\frac{\partial A}{\partial \xi} - \left((\dot{x})_{(\xi, \eta)}\eta_x + (\dot{y})_{(\xi, \eta)}\eta_y\right)\frac{\partial A}{\partial \eta}$$

$$\tag{19.43}$$

The metrics, ξ_x, ξ_y, η_x and η_y, are time-dependent, so that in difference approximations these values are updated in each time step.

Example 19.5: Consider the two-dimensional, inviscid momentum equations:

$$u_t + uu_x + vu_y = -p_x \tag{19.44}$$
$$v_t + uv_x + vv_y = -p_y$$

where u and v are velocity components in the x and y directions, p is the pressure. Each of u, v, and p is a function of x, y, and t. Transform the equations above onto the (ξ, η) coordinates assuming the coordinate transformation is time-dependent.

By introducing the transformation of the derivatives, the equations become:

$$\frac{\partial u}{\partial t} + ((u - \dot{x})\xi_x + (v - \dot{y})\xi_y)\frac{\partial u}{\partial \xi} + ((u - \dot{x})\eta_x + (v - \dot{y})\eta_y)\frac{\partial u}{\partial \eta}$$

$$= -\xi_x\frac{\partial p}{\partial \xi} - \eta_x\frac{\partial p}{\partial \eta}$$

(19.45)

$$\frac{\partial v}{\partial t} + ((u - \dot{x})\xi_x + (v - \dot{y})\xi_y)\frac{\partial v}{\partial \xi} + ((u - \dot{x})\eta_x + (v - \dot{y})\eta_y)\frac{\partial v}{\partial \eta}$$

$$= -\xi_y\frac{\partial p}{\partial \xi} - \eta_y\frac{\partial p}{\partial \eta}$$

where u, v and p are functions of ξ, η, and t.

In the special case that the velocity of the (ξ, η) point is the same as the the fluid velocity, the second and third terms in the foregoing equations vanish. On the co-ordinate that moves with the fluid, the time derivative is named the *total derivative* or *the material derivative* and denoted by D/Dt. With this notation, Eq. (19.45) is written as

$$\frac{Du}{Dt} = -\xi_x\frac{\partial p}{\partial \xi} - \eta_x\frac{\partial p}{\partial \eta}$$

(19.46)

$$\frac{Dv}{Dt} = -\xi_y\frac{\partial p}{\partial \xi} - \eta_y\frac{\partial p}{\partial \eta}$$

Transformation of Three-Dimensional Coordinates. A three-dimensional coordi-nate transformation can be written as

$$x = x(\xi, \eta, \zeta)$$
$$y = y(\xi, \eta, \zeta)$$
$$z = z(\xi, \eta, \zeta)$$

(19.47)

or inversely

$$\xi = \xi(x, y, z)$$
$$\eta = \eta(x, y, z)$$
$$\zeta = \zeta(x, y, z)$$

(19.47a)

where (x, y, z) is the three-dimensional physical domain, while (ξ, η, ζ) is the three-dimensional computational domain. The first order partial derivative operators are found from Eq. (19.47) as

$$\begin{bmatrix} \dfrac{\partial}{\partial x} \\[2mm] \dfrac{\partial}{\partial y} \\[2mm] \dfrac{\partial}{\partial z} \end{bmatrix} = K \begin{bmatrix} \dfrac{\partial}{\partial \xi} \\[2mm] \dfrac{\partial}{\partial \eta} \\[2mm] \dfrac{\partial}{\partial \zeta} \end{bmatrix}$$

(19.48)

where

$$
K = \begin{bmatrix} \xi_x & \eta_x & \zeta_x \\ \xi_y & \eta_y & \zeta_y \\ \xi_z & \eta_z & \zeta_z \end{bmatrix}
\tag{19.49}
$$

An equivalent relation is also obtained from Eq. (14.47a) as

$$
\begin{bmatrix} \dfrac{\partial}{\partial \xi} \\[2ex] \dfrac{\partial}{\partial \eta} \\[2ex] \dfrac{\partial}{\partial \zeta} \end{bmatrix} = L \begin{bmatrix} \dfrac{\partial}{\partial x} \\[2ex] \dfrac{\partial}{\partial y} \\[2ex] \dfrac{\partial}{\partial z} \end{bmatrix}
\tag{19.50}
$$

where

$$
L = \begin{bmatrix} x_\xi & y_\xi & z_\xi \\ x_\eta & y_\eta & z_\eta \\ x_\zeta & y_\zeta & z_\zeta \end{bmatrix}
\tag{19.51}
$$

Because Eq. (19.48) and Eq. (19.50) are in the inverse relation, K must be the inverse of L. Therefore, the chain rules similar to Eq. (19.6) may be obtained by equating K to L^{-1} giving

$$
\begin{aligned}
\xi_x &= (y_\eta z_\zeta - y_\zeta z_\eta) J \\
\eta_x &= (y_\zeta z_\xi - y_\xi z_\zeta) J \\
\zeta_x &= (y_\xi z_\eta - y_\eta z_\xi) J \\
\xi_y &= (x_\zeta z_\eta - x_\eta z_\zeta) J \\
\eta_y &= (x_\xi z_\zeta - x_\zeta z_\xi) J \\
\zeta_y &= (x_\eta z_\xi - x_\xi z_\eta) J \\
\xi_z &= (x_\eta y_\zeta - x_\zeta y_\eta) J \\
\eta_z &= (x_\zeta y_\xi - x_\xi y_\zeta) J \\
\zeta_z &= (x_\xi y_\eta - x_\eta y_\xi) J
\end{aligned}
\tag{19.52}
$$

where J is the Jacobian and equals the inverse of the determinant of L, or equivalently the determinant of K.

Grid Generation in Two and Three Dimensions by Numerical Methods

Generating a Grid by Solving Two Heat Conduction Equations: A heat transfer model often helps to understand and develop an algorithm to generate grids. In this section, we discuss the concept of generating a two-dimensional grid by solving two heat conduction equations. Pros and cons of this approach, and the relation of this approach to elliptic grid generation will be discussed.

Suppose we have a flow domain as shown in Fig. 19.2(a) which is to be mapped onto the computational domain as shown in Fig. 19.2(b). The four corner points of the computational domain correspond to the four corners of the flow domain, although corner to corner correspondence is not necessarily required.

To generate one family of grid lines on the physical domain which maps to the vertical lines in the computational domain, we consider a heat conduction problem as follows. (The heat conduction problems we are going to discuss have no relation to the fluid flow analysis at all but are used just to develop a grid.) We consider the domain as a thin plate made of a homogeneous material with a uniform thickness that is insulated on both sides (front and back). We also assume that the temperature along the side AB is fixed at a constant temperature of 1 degree, that along the side CD is fixed to M degrees, and the two sides, BC and AD, are perfectly insulated. The temperature distribution, $T(x, y)$, for this problem satisfies the heat conduction equation given by

$$\frac{\partial^2 T(x, y)}{\partial x^2} + \frac{\partial^2 T(x, y)}{\partial y^2} = 0 \tag{19.53}$$

with

$$T = 1 \text{ along AB}$$

$$T = M \text{ along CD} \tag{19.54}$$

$$\frac{\partial T}{\partial n} = 0 \quad \text{along the boundaries BC and AD}$$

where $\partial/\partial n$ is the derivative normal to the boundary.

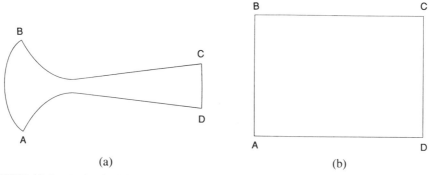

(a)   (b)

FIGURE 19.2 A physical flow domain and the computational domain. (a) Physical domain and (b) computational domain.

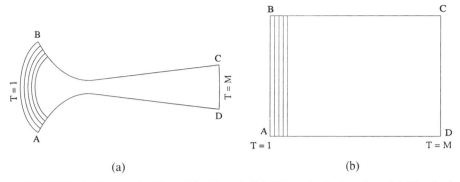

(a)   (b)

FIGURE 19.3 Isotherms for Eqs. (19.53) and (19.54) on both domains. (a) Physical domain and (b) computational domain.

If we plot the isotherms of the solution on Fig. 19.3, the isotherms for $T = 1$ and M degrees will be on the boundary curves AB and CD respectively, and the isotherms for $T = 2, 3, \ldots, M - 1$ will be located between the isotherms for $T = 1$ and $T = M$ in consecutive order. Isotherms never cross over each other, and they are orthogonal to the boundaries BC and AD (because of the boundary conditions).

We consider Eq. (19.53) with another set of boundary conditions

$$T = 1 \text{ along AD}$$

$$T = N \text{ along BC} \tag{19.55}$$

$$\frac{\partial T}{\partial n} = 0 \quad \text{along the boundaries AB and CD}$$

The isotherms of this second solution will become as in Fig. 19.4. The isotherm for $T = 1$ is on AD, the isotherm for $T = N$ is on BC. The isotherms for $T = 2, 3, \ldots, N - 1$ will be between BC and AD in consecutive order without any cross over. Thus, if we find the intersection of the ith isotherm from the first solution and jth isotherm from the second isotherm and denote the coordinates of the intersection by ($x_{i,j}$ and $y_{i,j}$), then the whole set of ($x_{i,j}$, $y_{i,j}$) will become a grid over the physical domain.

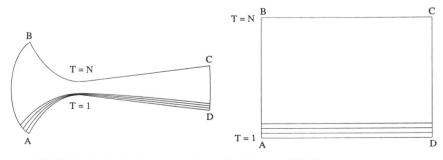

FIGURE 19.4 Isotherms for Eqs. (19.53) and (19.55) on both domains.

In generating a grid, often it becomes necessary to make the grid points denser in one region than another. For example, one may wish to have isotherms clustered toward the curve CD. This may be achieved by taking the isotherms of nonequispaced temperature values. For example, the increment of the temperature values toward the boundary CD is gradually decreased and smaller than unity, and just the opposite trend is given to the temperature increment toward the other boundary AB. However, an alternative way is to use the following equation in place of Eq. (19.53)

$$\frac{\partial}{\partial x} \Gamma(x, y) \frac{\partial T(x, y)}{\partial x} + \frac{\partial}{\partial y} \Gamma(x, y) \frac{\partial T(x, y)}{\partial y} = 0 \tag{19.56}$$

where $\Gamma(x, y)$ is the thermal conductivity which can be changed as desired.

It is well known that the temperature gradient in a good thermal conductor such as a metal is small, but, on the other hand, the temperature gradient in an insulating material such as wood is high. This means that if the thermal conductivity varies in space, the density of the isotherms becomes high where Γ is low, while the density of the isotherms becomes low where Γ is high. Therefore, if Γ is changed in such a way that Γ becomes smaller toward CD, then the isotherms for temperatures with a constant increment will be clustered toward CD.

A special application of the method described here is presented in [Nakamura *et al.* (1991)].

Elliptic Grid Generation Method: The method of generating grids by solving heat conduction equations has some disadvantages. First, it often needs a finite element solution because the boundaries are complicated. Second, after solving two heat conduction problems, intersections of the isotherms must be found. The *elliptic grid generation method* [Thompson *et al.* (1977)] eliminates these two disadvantages, although the analogy of the grid generation to the heat conduction problems still applies.

We now express the two heat conduction problems in the preceding section by writing two separate Laplace equations as

$$\frac{\partial^2 \xi(x, y)}{\partial x^2} + \frac{\partial^2 \xi(x, y)}{\partial y^2} = 0$$

$$\frac{\partial^2 \eta(x, y)}{\partial x^2} + \frac{\partial^2 \eta(x, y)}{\partial y^2} = 0 \tag{19.57}$$

The first equation is subject to the boundary conditions,

$$\xi = 1 \text{ along AB}$$

$$\xi = M \text{ along CD} \tag{19.58}$$

$$\frac{\partial \xi}{\partial n} = 0 \quad \text{along the boundaries BC and DA}$$

while the second equation is subject to boundary conditions,

$$\eta = 1 \text{ along BC}$$

$$\eta = N \text{ along DA} \qquad (19.59)$$

$$\frac{\partial \eta}{\partial n} = 0 \quad \text{along the boundaries AB and CD}$$

As discussed already in the prior section, the two sets of the isotherms form a coordinate system (ξ, η). Therefore, Eqs. (19.57) and (19.58) yields the relation in the form

$$\xi = \xi(x, y) \qquad (19.60)$$

$$\eta = \eta(x, y)$$

which may be also equivalently expressed by

$$x = x(\xi, \eta) \qquad (19.61)$$

$$y = y(\xi, \eta)$$

The expression of coordinate transformation in the form of Eq. (19.60) is not useful, because we need x and y values for sets of constant ξ and η. To obtain the transformation in the second form, we must exchange the role of (x, y) and (ξ, η).

Let us rewrite Eq. (19.28) to

$$A \frac{\partial^2 \phi}{\partial \xi^2} + 2B \frac{\partial^2 \phi}{\partial \eta \, \partial \xi} + C \frac{\partial^2 \phi}{\partial \eta^2} + (\xi_{xx} + \xi_{yy}) \frac{\partial \phi}{\partial \xi} + (\eta_{xx} + \eta_{yy}) \frac{\partial \phi}{\partial \eta} = 0 \quad (19.62)$$

where the coefficients are given by

$$A \equiv \xi_x^2 + \xi_y^2 = J^2(x_\eta^2 + y_\eta^2)$$

$$B \equiv \eta_x \xi_x + \eta_y \xi_y = -J^2(x_\xi x_\eta + y_\xi y_\eta) \qquad (19.63)$$

$$C \equiv \eta_x^2 + \eta_y^2 = J^2(x_\xi^2 + y_\xi^2)$$

Setting $\phi = x$ in the equation above, we obtain

$$A \frac{\partial^2 x}{\partial \xi^2} + 2B \frac{\partial^2 x}{\partial \eta \, \partial \xi} + C \frac{\partial^2 x}{\partial \eta^2} = 0 \qquad (19.64)$$

where the fourth and fifth terms vanish because of Eq. (19.57) [see also Eqs. (19.36) and (19.37)]. Similarly by setting $\phi = y$, we obtain

$$A \frac{\partial^2 y}{\partial \xi^2} + 2B \frac{\partial^2 y}{\partial \eta \, \partial \xi} + C \frac{\partial^2 y}{\partial \eta^2} = 0 \qquad (19.65)$$

The set of Eqs. (19.64) and (19.65) is named the *elliptic grid generation equations*. The adiabatic boundary conditions equivalent to Eqs. (19.58) and (19.58) may be

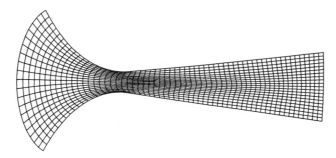

FIGURE 19.5   A grid to be generated by the elliptic grid generation method.

used. If so, the locus of (x, y) for ξ = constant or η = constant will correspond to an isotherm of the heat conduction problems. However, in the elliptic grid generation scheme, fixed value type boundary conditions are more often used than adiabatic conditions.

Suppose we wish to generate a grid as illustrated in Fig. 19.5. Along the boundary of the physical domain, the points are usually predetermined. So, the (x, y) values at the points on the computational grid are fixed.

Equations (19.64) and (19.65) are discretized using central difference formulas. (See Sec. 19.2). Denoting the value of x and y at point (i, j) by $x_{i,j}$ and $y_{i,j}$, the difference equations are written as

$$A(x_{i-1,j} - 2x_{i,j} + x_{i+1,j}) + \frac{B}{2}(x_{i-1,j-1} - x_{i+1,j+1} - x_{i-1,j+1} + x_{i+1,j+1})$$

$$+ C(x_{i,j-1} - 2x_{i,j} + x_{i,j+1}) = 0 \tag{19.66}$$

$$A(y_{i-1,j} - 2y_{i,j} + y_{i+1,j}) + \frac{B}{2}(y_{i-1,j-1} - y_{i+1,j-1} - y_{i-1,j+1} + y_{i+1,j+1})$$

$$+ C(y_{i,j-1} - 2y_{i,j} + y_{i,j+1}) = 0 \tag{19.67}$$

where the mesh spacing in both the i and j directions on the computational domain are assumed to be unity. The coefficients are also written in the finite difference form as

$$A = ((x_{i,j+1} - x_{i,j-1})^2 + (y_{i,j+1} - y_{i,j-1})^2)/4$$

$$B = -((x_{i+1,j} - x_{i-1,j})(x_{i,j+1} - x_{i,j-1})$$
$$- (y_{i+1,j} - y_{i-1,j})(y_{i,j+1} - y_{i,j-1}))/4 \tag{19.68}$$

$$C = ((x_{i+1,j} - x_{i-1,j})^2 + (y_{i+1,j} - y_{i-1,j})^2)/4$$

Here, all $x_{i,j}$ and $y_{i,j}$ are unknown except those that are prescribed along the boundaries.

A convenient way of solving Eqs. (19.66) and (19.67) is the *successive-over-relaxation* (SOR) scheme, which is written

$$x_{i,j}^t = \frac{\omega}{2(A + C)}\left[A(x_{i-1,j}^t + x_{i+1,j}^{t-1})\right.$$

$$+ \frac{B}{2}(x_{i-1,j-1}^t - x_{i+1,j-1}^t - x_{i-1,j+1}^{t-1} + x_{i+1,j+1}^{t-1})$$

$$\left. + C(x_{i,j-1}^t + x_{i,j+1}^{t-1})\right] + (1 - \omega)x_{i,j}^{t-1} \tag{19.69}$$

$$y_{i,j}^t = \frac{\omega}{2(A + C)}\left[A(y_{i-1,j}^t + y_{i+1,j}^{t-1})\right.$$

$$+ \frac{B}{2}(y_{i-1,j-1}^t - y_{i+1,j-1}^t - y_{i-1,j+1}^{t-1} + y_{i+1,j+1}^{t-1})$$

$$\left. + C(y_{i,j-1}^t + y_{i,j+1}^{t-1})\right] + (1 - \omega)y_{i,j}^{t-1}$$

where t is the iteration number and ω is the *successive-over-relaxation parameter*. The coefficients are computed for each point using the most updated values of $x_{i,j}$ and $y_{i,j}$ available. Because of the nonlinearity of the equations, the convergence history of SOR is significantly different from the linear elliptic problems. Indeed, during the early stages of SOR iteration process, the iteration parameter needs to be a low value satisfying $1 \le \omega < 2$, but it can be increased as the iteration number increases.

The elliptic grid generation scheme discussed so far is in its fundamental form, but it needs several modifications to provide grids useful for fluid flow analysis. Both *clustering* and *orthogonality control* of a grid may be achieved by adding first order derivative terms to the grid generation equations [Thomas (1981) and Steger and Sorenson (1979)].

$$A\left(\frac{\partial^2 x}{\partial \xi^2} + \phi\frac{\partial x}{\partial \xi}\right) + 2B\frac{\partial^2 x}{\partial \eta\,\partial \xi} + C\left(\frac{\partial^2 x}{\partial \eta^2} + \varphi\frac{\partial x}{\partial \eta}\right) = 0$$

$$A\left(\frac{\partial^2 y}{\partial \xi^2} + \phi\frac{\partial y}{\partial \xi}\right) + 2B\frac{\partial^2 y}{\partial \eta\,\partial \xi} + C\left(\frac{\partial^2 y}{\partial \eta^2} + \varphi\frac{\partial y}{\partial \eta}\right) = 0 \tag{19.70}$$

where ϕ and φ are parameters to control the clustering and orthogonality, and they can be set as functions of (ξ, η). In discretizing the equations above, the first order partial derivative terms are replaced by central difference approximations.

To consider the effect of ϕ on clustering first, we reduce Eq. (19.70) to a one-dimensional model by setting $B = 0$ and $C = 0$

$$\frac{\partial^2 x}{\partial \xi^2} + \phi\frac{\partial x}{\partial \xi} = 0 \tag{19.71}$$

The difference equation then becomes

$$(1 - \phi)x_{i-1} - 2x_i + (1 + \phi)x_{i+1} = 0 \tag{19.72}$$

Rewriting yields

$$x_i = \frac{(1 - \phi)x_{i-1} + (1 + \phi)x_{i+1}}{2} \tag{19.73}$$

If $\phi = 0$ in Eq. (19.73), x_i becomes the arithmetic average of the two neighbors. If ϕ approaches 1 on the other hand, x_i becomes closer to x_{i+1}, and if ϕ approaches -1, x_i becomes closer to x_{i-1}. The ϕ value should not violate, $-1 < \phi < 1$, because otherwise x_i becomes outside of $x_{i-1} < x < x_{i+1}$. The spatial distributions of ϕ and φ can be set in many different ways.

It can be shown that Eq. (19.70) may be derived by transforming the following equations

$$\frac{\partial^2 \xi(x, y)}{\partial x^2} + \frac{\partial^2 \xi(x, y)}{\partial y^2} = P(\xi, \eta)$$

$$\frac{\partial^2 \eta(x, y)}{\partial x^2} + \frac{\partial^2 \eta(x, y)}{\partial y^2} = Q(\xi, \eta) \tag{19.74}$$

where P and Q are prescribed functions related to ϕ and φ in Eq. (19.70) by $\phi = J^{-2}P/a$ and $\varphi = J^{-2}Q/c$ and J is the Jacobian.

Three-dimensional elliptic grid generation equations may be written as an extension of Eq. (19.70).

$$a_{11}(x_{\xi\xi} + \phi x_\xi) + a_{22}(x_{\eta\eta} + \varphi x_\eta) + a_{33}(x_{\varsigma\varsigma} + \lambda x_\varsigma)$$
$$+ 2a_{12}x_{\xi\eta} + 2a_{23}x_{\eta\varsigma} + 2a_{31}x_{\varsigma\xi} = 0$$

$$a_{11}(y_{\xi\xi} + \phi y_\xi) + a_{22}(y_{\eta\eta} + \varphi y_\eta) + a_{33}(y_{\varsigma\varsigma} + \lambda y_\varsigma) + 2a_{12}y_{\xi\eta}$$
$$+ 2a_{23}y_{\eta\varsigma} + 2a_{31}y_{\varsigma\xi} = 0$$

$$a_{11}(z_{\xi\xi} + \phi z_\xi) + a_{22}(z_{\eta\eta} + \varphi z_\eta) + a_{33}(z_{\varsigma\varsigma} + \lambda z_\varsigma)$$
$$+ 2a_{12}z_{\xi\eta} + 2a_{23}z_{\eta\varsigma} + 2a_{31}z_{\varsigma\xi} = 0 \tag{19.75}$$

Here, a_{ij} are defined by

$$a_{ij} = \sum_{m=1}^{3} A_{mi}A_{mj} \tag{19.76}$$

and A_{ij} is the *cofactor* of the (m, i) element in the matrix

$$A = \begin{bmatrix} x_\xi & x_\eta & x_\varsigma \\ y_\xi & y_\eta & y_\varsigma \\ z_\xi & z_\eta & z_\varsigma \end{bmatrix} \tag{19.77}$$

Numerical solution of Eq. (19.77) is essentially the same as for the two-dimensional elliptic grid generation. However, the boundary of the three-dimensional computational domain corresponds to a curved surface on the physical domain. Therefore, in order to predetermine the coordinates of the points on the boundary, a two-dimensional grid generation on the curved surface is necessary.

Hyperbolic Grid Generation Scheme: The two-dimensional *hyperbolic grid generation scheme* [Steger and Chausee (1980)] is based on one orthogonality relation and an additional equation to control the inverse Jacobian equated to a prescribed quantity V

$$x_\xi x_\eta + y_\xi y_\eta = 0 \tag{19.78}$$

$$\frac{\partial(x, y)}{\partial(\xi, \eta)} = V \tag{19.79}$$

where ξ and η are the coordinates on the computational domain, for which the grid indices of i and j, respectively, are used. We will assume that η is the direction nearly perpendicular to the surface to which the grid is desired to maintain orthogonality. The V in the equation above determines the size of the mesh cell and is a predetermined function in space. The set of Eqs. (19.78) and (19.79) may be more explicitly written as

$$x_\xi x_\eta + y_\xi y_\eta = 0 \tag{19.80}$$
$$x_\xi y_\eta - x_\eta y_\xi = V$$

Suppose we have the x and y that satisfy Eq. (19.80) along a given curve corresponding to $\eta = \eta_0$. Thus, knowing x and y along the given curve, we investigate the solution of Eq. (19.80) in the vicinity of that curve. Suppose $\Delta\eta$ is a small increment of η. Then, the solution of Eq. (19.80) along $\eta = \eta_0 + \Delta\eta$ may be expressed by

$$x(\xi, \eta_0 + \Delta\eta) = \tilde{x}(\xi) + \delta x(\xi) \tag{19.81}$$
$$y(\xi, \eta_0 + \Delta\eta) = \tilde{y}(\xi) + \delta y(\xi)$$

where $\tilde{x}(\xi) = x(\xi, \eta_0)$, and $\tilde{y}(\xi) = y(\xi, \eta_0)$ are the known solution along $\eta = \eta_0$.

If we introduce Eq. (19.81) into Eq. (19.80) and ignore the second-order terms with respect to δx and δy, we obtain the following equation in the matrix form

$$\begin{pmatrix} \tilde{x}_\eta & \tilde{y}_\eta \\ \tilde{y}_\eta & -\tilde{x}_\eta \end{pmatrix} \begin{pmatrix} \delta x_\xi \\ \delta y_\xi \end{pmatrix} + \begin{pmatrix} \tilde{x}_\xi & \tilde{y}_\xi \\ -\tilde{y}_\xi & \tilde{x}_\xi \end{pmatrix} \begin{pmatrix} \delta x_\eta \\ \delta y_\eta \end{pmatrix} = \begin{pmatrix} 0 \\ \delta V \end{pmatrix} \tag{19.82}$$

where the following relations are used

$$\tilde{x}_\xi \tilde{x}_\eta + \tilde{y}_\xi \tilde{y}_\eta = 0$$
$$\tilde{x}_\xi \tilde{y}_\eta - \tilde{x}_\eta \tilde{y}_\xi = \tilde{V} \tag{19.83}$$
$$\delta V = V - \tilde{V}$$

Equation (19.82) may be also rewritten as

$$
\begin{pmatrix} \tilde{x}_\eta & \tilde{y}_\eta \\ \tilde{y}_\eta & -\tilde{x}_\eta \end{pmatrix} \begin{pmatrix} x_\xi \\ y_\xi \end{pmatrix} + \begin{pmatrix} \tilde{x}_\xi & \tilde{y}_\xi \\ -\tilde{y}_\xi & \tilde{x}_\xi \end{pmatrix} \begin{pmatrix} x_\eta \\ y_\eta \end{pmatrix} = \begin{pmatrix} 0 \\ V + \tilde{V} \end{pmatrix}
\tag{19.84}
$$

Equation (19.84) may be proven easily by introducing Eq. (19.81).

The finite difference approximation of Eq. (19.81) is straightforward. Let i and j correspond to ξ and η, respectively, on the computational domain. We set $\Delta\eta = 1$ because the increment of η on the computational domain is unity. Assuming that the $x_{i,j}$ and $y_{i,j}$ are known along the jth grid line, the difference approximation for x and y may be written as

$$
x_\xi \approx \frac{x_{i+1,j} - x_{i+1,j}}{2}
$$
$$
y_\xi \approx \frac{y_{i+1,j} - y_{i-1,j}}{2}
\tag{19.85}
$$

where the central difference approximation is used with $\Delta\xi = 1$, and

$$
x_\eta \approx x_{i,j} - x_{i,j-1}
$$
$$
y_\eta \approx y_{i,j} - y_{i,j-1}
\tag{19.86}
$$

where the backward difference approximation is used with $\Delta\eta = 1$. Introducing Eqs. (19.85) and (19.86) into Eq. (19.84), a *block tridiagonal* equation is obtained as

$$
-A \begin{pmatrix} x_{i-1,j} \\ y_{i-1,j} \end{pmatrix} + B \begin{pmatrix} x_{i,j} \\ y_{i,j} \end{pmatrix} + A \begin{pmatrix} x_{i+1,j} \\ y_{i+1,j} \end{pmatrix} = \begin{pmatrix} 0 \\ V + \tilde{V} \end{pmatrix} + B \begin{pmatrix} x_{i,j-1} \\ y_{i,j-1} \end{pmatrix}
$$

$$
A = \tfrac{1}{2} \begin{pmatrix} \tilde{x}_\eta & \tilde{y}_\eta \\ \tilde{y}_\eta & -\tilde{x}_\eta \end{pmatrix}
$$

$$
B = \begin{pmatrix} \tilde{x}_\xi & \tilde{y}_\xi \\ -\tilde{y}_\xi & \tilde{x}_\xi \end{pmatrix}
\tag{19.87}
$$

Equation (19.87) may be solved by a block tridiagonal solution scheme with appropriate initial and boundary conditions.

We now consider the role and treatment of V in Eq. (19.87). The V for the grid point (i, j) is the desired volume of the grid cell, which we set as follows

$$
V = s\sqrt{x_\xi^2 + y_\xi^2}
\tag{19.88}
$$

where s is a user-defined parameter and the desired distance of the grid points between point (i, j) and point $(i, j - 1)$. Numerically, this can be set to

$$V = \frac{s}{2} \sqrt{(x_{i+1,j-1} - x_{i-1,j-1})^2 + (y_{i+1,j-1} - y_{i-1,j-1})^2} \qquad (19.89)$$

The boundary conditions for Eq. (19.87) may take different forms depending on the boundary type. One form is to fix the coordinates of the points on the side boundary. Another form is to leave the coordinates of the point undetermined except it is required to be on a specified boundary curve.

The algorithm developed so far is a straightforward implementation of the basic equation given by Eq. (19.83). However, it needs a few modifications in order to be useful in actual grid generation.

The hyperbolic grid generation scheme has no intrinsic mechanism to smooth out the distribution of points in the ξ direction. This causes two kinds of difficulties. First, the scheme may become unstable and yield cross-over of the grid lines. Second, if any irregularity such as a sudden change of the grid spacing or direction of the curve exists along the boundary, the irregularity may be amplified in an undesirable manner as the grid lines are generated in the η direction. To prevent these from occurring, a second-order or fourth-order diffusion term [Steger and Chausee (1980)] is added. We discuss how to use the second-order diffusion terms implicitly.

Equation (19.81) is written as

$$-B^{-1}A \begin{pmatrix} x_{i-1,j} \\ y_{i-1,j} \end{pmatrix} + \begin{pmatrix} x_{i,j} \\ y_{i,j} \end{pmatrix} + B^{-1}A \begin{pmatrix} x_{i+1,j} \\ y_{i+1,j} \end{pmatrix} = B^{-1} \begin{pmatrix} 0 \\ V + \tilde{V} \end{pmatrix} + \begin{pmatrix} x_{i,j-1} \\ y_{i,j-1} \end{pmatrix}$$

$$(19.90)$$

where the equation has been multiplied through by B^{-1}, and

$$B^{-1} = \frac{1}{|B|} \begin{pmatrix} \tilde{x}_\xi & -\tilde{y}_\xi \\ \tilde{y}_\xi & \tilde{x}_\xi \end{pmatrix}$$

$$B^{-1}A = \frac{1}{2|B|} \begin{pmatrix} \tilde{x}_\xi \tilde{x}_\eta - \tilde{y}_\xi \tilde{y}_\eta, & \tilde{x}_\xi \tilde{y}_\eta + \tilde{y}_\xi \tilde{x}_\eta \\ \tilde{y}_\xi \tilde{x}_\eta + \tilde{x}_\xi \tilde{y}_\eta, & \tilde{y}_\xi \tilde{y}_\eta - \tilde{x}_\xi \tilde{x}_\eta \end{pmatrix} \qquad (19.91)$$

or more compactly

$$B^{-1}A = \begin{pmatrix} a & b \\ b & -a \end{pmatrix} \qquad (19.91a)$$

with

$$a = \frac{\tilde{x}_\xi \tilde{x}_\eta - \tilde{y}_\xi \tilde{y}_\eta}{2|B|}$$

$$b = \frac{\tilde{x}_\xi \tilde{y}_\eta + \tilde{y}_\xi \tilde{x}_\eta}{2|B|} \qquad (19.92)$$

Now, we add a finite difference approximation for

$$\alpha \frac{d^2}{d\xi^2} \begin{pmatrix} x \\ y \end{pmatrix} \tag{19.93}$$

namely,

$$-\alpha \begin{pmatrix} x_{i-1,j} \\ y_{i-1,j} \end{pmatrix} + 2\alpha \begin{pmatrix} x_{i,j} \\ y_{i,j} \end{pmatrix} - \alpha \begin{pmatrix} x_{i+1,j} \\ y_{i+1,j} \end{pmatrix} \tag{19.94}$$

where α is a constant to control the intensity of the diffusion effect. Adding Eq. (19.94) to Eq. (19.90) yields

$$-\begin{pmatrix} a+\alpha & b \\ b & -a+\alpha \end{pmatrix} \begin{pmatrix} x_{i-1,j} \\ y_{i-1,j} \end{pmatrix} + \begin{pmatrix} 1+2\alpha & 0 \\ 0 & 1+2\alpha \end{pmatrix} \begin{pmatrix} x_{i,j} \\ y_{i,j} \end{pmatrix}$$
$$+ \begin{pmatrix} a-\alpha & b \\ b & -a-\alpha \end{pmatrix} \begin{pmatrix} x_{i+1,j} \\ y_{i+1,j} \end{pmatrix} = B^{-1} \begin{pmatrix} 0 \\ V+\tilde{V} \end{pmatrix} + \begin{pmatrix} x_{i,j-1} \\ y_{i,j-1} \end{pmatrix} \tag{19.95}$$

Notice that the matrix coefficient for the second term is always $1 + 2\alpha$. This has been made possible because Eq. (19.95) had been multiplied through by B^{-1}.

There is an additional advantage of adding second-order diffusion terms. That is, Eq. (19.95) may now be solved by an iterative solution rather than using the block tridiagonal scheme. Although the practical merit of an iterative solution is small in two-dimensional grid generation, the feasibility of using an iterative solution is a considerable benefit when the hyperbolic scheme is extended to three dimensions. It also benefits when the hyperbolic scheme is blended with the elliptic scheme as explained in more detail below.

Blending of the Hyperbolic Scheme with Elliptic Scheme: The significant advantage of the hyperbolic grid generation scheme is that the grid near the boundary becomes nearly orthogonal even when the shape of the boundary includes singularities. The disadvantage is that the outer boundary cannot be prescribed. By blending the hyperbolic scheme with the elliptic scheme [Nakamura *et al.* (1991) and Spradling *et al.* (1991)], however, the disadvantage is eliminated while adding the capability of orthogonalizing the grid in the elliptic scheme. Blending the hyperbolic grid generation equation with the elliptic grid generation equations is very similar to the blending of the former with the parabolic grid generation equations [Nakamura and Suzuki (1987)].

The elliptic scheme gives the diffusion effect to the hyperbolic scheme, so iterative solution of the blended scheme is possible as another advantage. To show how the two schemes can be blended, we first write Eqs. (19.90), (19.66), and (19.67) in the following form:

$$\begin{pmatrix} x_{i,j} \\ y_{i,j} \end{pmatrix} = B^{-1}A \begin{pmatrix} x_{i-1,j} - x_{i+1,j} \\ y_{i-1,j} - y_{i+1,j} \end{pmatrix} + B^{-1} \begin{pmatrix} 0 \\ V+\tilde{V} \end{pmatrix} + \begin{pmatrix} x_{i,j-1} \\ y_{i,j-1} \end{pmatrix} \tag{19.96}$$

$$
\binom{x_{i,j}}{y_{i,j}} =
\begin{bmatrix}
\dfrac{1}{2(\bar{a}+\bar{c})} \left[\bar{a}(x_{i-1,j}+x_{i+1,j}) + \dfrac{\bar{b}}{2}(x_{i-1,j-1}-x_{i+1,j-1}-x_{i-1,j+1}) \right. \\
\left. + x_{i+1,j+1}) + \bar{c}(x_{i,j-1}+x_{i,j+1}) \right] \\[2ex]
\dfrac{1}{2(\bar{a}+\bar{c})} \left[\bar{a}(y_{i-1,j}+y_{i+1,j}) + \dfrac{\bar{b}}{2}(y_{i-1,j-1}-y_{i+1,j-1}-y_{i-1,j+1}) \right. \\
\left. + y_{i+1,j+1}) + \bar{c}(y_{i,j-1}+y_{i,j+1}) \right]
\end{bmatrix}
$$

(19.97)

where $\bar{a}$, $\bar{b}$, and $\bar{c}$, respectively, denote A, B, and C in Eqs. (19.66) and (19.67).

To blend the equations above, we multiply Eq. (19.96) by c_h and Eq. (19.97) by c_e and add them together to yield

$$
\binom{x_{i,j}}{y_{i,j}} = c_h \left[B^{-1}A \binom{x_{i-1,j}-x_{i+1,j}}{y_{i-1,j}-y_{i+1,j}} + B^{-1} \binom{0}{V+\tilde{V}} + \binom{x_{i,j-1}}{y_{i,j-1}} \right]
$$

$$
+ c_e
\begin{bmatrix}
\dfrac{1}{2(\bar{a}+\bar{c})} \left[\bar{a}(x_{i-1,j}+x_{i+1,j}) + \dfrac{\bar{b}}{2}(x_{i-1,j-1}-x_{i+1,j-1}-x_{i-1,j+1}) \right. \\
\left. + x_{i+1,j+1}) + \bar{c}(x_{i,j-1}+x_{i,j+1}) \right] \\[2ex]
\dfrac{1}{2(\bar{a}+\bar{c})} \left[\bar{a}(y_{i-1,j}+y_{i+1,j}) + \dfrac{\bar{b}}{2}(y_{i-1,j-1}-y_{i+1,j-1}-y_{i-1,j+1}) \right. \\
\left. + y_{i+1,j+1}) + \bar{c}(y_{i,j-1}+y_{i,j+1}) \right]
\end{bmatrix}
$$

(19.98)

where c_h and c_e are the blending coefficients and satisfy

$$
c_e = 1 - c_h \tag{19.99}
$$

The hyperbolic scheme in this blended scheme may be viewed as a means to enhance the elliptic grid generation with respect to orthogonality near the boundary $j = 1$, so c_h should diminish as j increases. For this purpose we set c_h as

$$
c_h = \gamma \exp\left(-\beta(j-2)\right) \tag{19.100}
$$

where γ and β are *control parameters*. As β is increased, the contribution of the hyperbolic scheme diminishes faster as j increases. As a rule of thumb, the hyperbolic part in the blended scheme has a strong contribution if c_h is in the range $0.2 < c_h < 1$. A two-dimensional grid generated by the blended scheme is illustrated in Fig. 19.6.

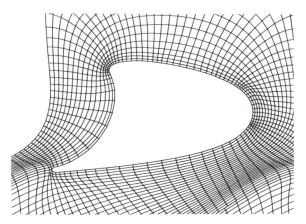

FIGURE 19.6 Illustration of a grid generated by blending elliptic and hyperbolic methods (hyperbolic method is made dominant near the blade surface).

The blending approach can be extended to three dimensions. The approach of blending the elliptic grid generation makes orthogonalization of the grid in the vicinity of sharp corners and singularities of the surface extremely easy compared with the traditional approach of orthogonalization by means of source terms of the elliptic grid generation scheme. Figure 19.7 shows the grid generated by the blended method for the upper part of an automobile [Spradling *et al.* (1991)].

Transfinite Interpolation: Transfinite interpolation [Eiseman (1982), Erickson (1985), Gordon and Hall (1973), Smith (1982), and Nakamura (1993)] is frequently used to generate grids in two and three dimensions. Consider a rectangular, two-dimensional space where the functional values along the external boundaries, as well as along the vertical and horizontal lines within the domain, are known. The traditional double interpolations apply when functional values are known only at the intersections of vertical and horizontal lines. In contrast to the double interpolations, the transfinite interpolation fits continuous functions specified along the horizontal and vertical lines.

Grid Generation on a Curved Surface. Generation of the grids for external flow calculations over a complex geometry such as an aircraft or automobile requires generation of a grid on the surface of the body, which is then used as a boundary condition for generation of a three-dimensional grid in the external fluid volume. For a detailed discussion of the surface grids see Nakamura *et al.* (1991).

19.1.2 Unstructured Grid Generation
Rainald Löhner

Introduction. An *unstructured grid* is one that lacks the logical indexing scheme of a *structured grid*, where given i, j, k one knows the identity of all the neighboring points in the grid. For an unstructured grid, a linked list must be supplied that identifies all the points and their neighbors.

Consider the task of generating an arbitrary unstructured mesh in a given com-

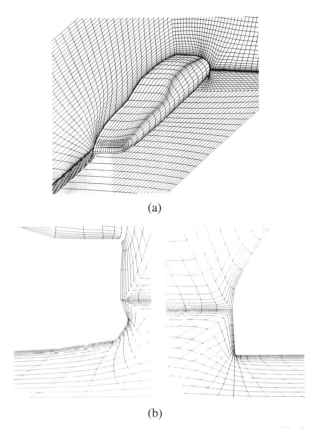

(a)

(b)

FIGURE 19.7 Illustration of a three-dimensional grid generated elliptic and hyperbolic methods (hyperbolic method is made dominant near the automotive surface). (a) A three-dimensional perspective view and (b) close views of the grid near edges.

putational domain. The information required to perform this task is: 1) a description of the bounding surfaces of the domain to be discretized, 2) a description of how the element size, shape and orientation should be in space, 3) the choice of element type, and 4) the choice of a suitable method to achieve the generation of the desired mesh. The most common ways to provide these four pieces of information are discussed in this section.

Description of the Domain to be Gridded. There are two possible ways of describing the surface of a computational domain: using analytic functions and via discrete data.

Analytic Functions: This is the preferred choice if a CAD–CAM database exists for the description of the domain. In this case, Splines, B-Splines, NURB Surfaces or other types of functions are used to describe the surface of the domain. An important characteristic of this approach is that the surface is continuous, i.e., there exist no *holes* in the information.

Discrete Data: Here, instead of functions, a cloud of points describes the surface of the computational domain. This choice may be attractive when no CAD–CAM database exists. Commercial digitizers can gather surface point information at speeds higher than 30,000 points/sec, allowing a very accurate description of a scaled model or the full configuration [Merriam (1991)]. The surface definition is completed with the triangulation of the cloud of points. The triangularization process is far from trivial and has been the subject of major research and development efforts [see, e.g., Choi (1988) and Hoppe (1992), (1993)]. This approach can lead to a discontinuous surface description. In order not to make any mistakes when discretizing the surface during mesh generation, only the points given in the cloud of points should be selected.

The current incompatibility of formats for the description of surfaces makes the surface definition by far the most man-hour intensive task of the CFD analysis process. Grid generation, flow solvers and visualization are processes that have been automated to a high degree. This is not the case with surface definition, and it may continue to be so for a long time. It may also have to do with the nature of analysis. For a CFD run, a vast portion of CAD–CAM data has to be filtered out (nuts, bolts, etc., are not required for the surface definition suitable for a standard CFD run), and the surface patches used by the designers seldom match, leaving gaps or overlap regions that have to be treated manually.

Variation of Element Size and Shape. Having described the surface of the domain to be gridded, the next task is to define how the element size and shape should vary in space. The parameters required to generate an arbitrary element are shown in Fig. 19.8. They consist of the side distance parameter, δ, commonly known as *element length*, two *stretching parameters*, S_1, S_2, and two associated *stretching directions* s_1, s_2. Furthermore, the assumption: $S_3 = 1$, $s_3 = s_1 \times s_2$ is made.

Internal Measures of Grid Quality: The idea here is to start from a given surface mesh. After the introduction of a new point or element, the quality of the current grid or front is assessed. Then, a new point or element is introduced in the most critical region. This process is repeated until either a mesh that satisfies a preset measure of quality is achieved [Holmes (1988)] or the number of faces in the front

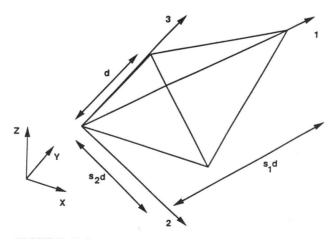

FIGURE 19.8   Parameters required for an arbitrary element.

has shrunk to zero [Huet (1990)]. This technique works well for equilateral elements, requiring minimal user input. On the other hand, it is not very general, as the surface mesh needs to be provided as part of the procedure.

Analytical Functions: In this case, the user codes the desired variation of element size, shape, and orientation in space in a small subroutine. Needless to say, this is the least general of all procedures, requiring new coding for every new problem. On the other hand, if the same problem needs to be discretized many times, an optimal discretization may be coded in this way. Although it may seem inappropriate to pursue such an approach within unstructured grids, the reader may be reminded that most current airfoil calculations are carried out using this approach.

Boxes: If all that is required are regions with uniform mesh sizes, one may define a series of boxes in which the element size is constant. For each location in space, the element size taken is the smallest of all the boxes containing the current location. When used in conjunction with surface definition via *quad/octrees*, one can automate the point distribution process completely in a very elegant way [Yerry (1984) and Shephard (1991)].

Point/Line/Surface Sources: A more flexible way that combines the smoothness of functions with the generality of boxes or other discrete elements is to define sources. The element size for an arbitrary location $\bar{x}$ in space is given as a function of the closest distance to the source $r(\bar{x})$. Point, line, and surface-sources have been used frequently in practice (see Fig. 19.9). The number of operations required to determine $r(\bar{x})$ is not considerable if one can pre-compute and store the geometrical parameters of the sources, and the loop over all sources can readily be vectorized. Having determined the distance from the source, the next step is to select a function that is general yet has a minimum of input to define the element size as a function of distance. Typically, a small element size is desired close to the sources, and a large element size away from them. Moreover, the element size should be constant (and small) in the vicinity $r < r_0$ of the source. An elegant way to satisfy these

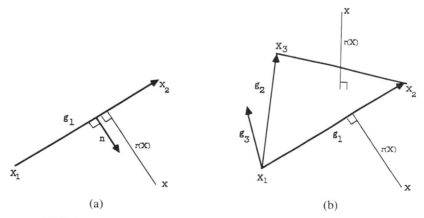

(a)   (b)

FIGURE 19.9 Sources. (a) Line source and (b) surface source.

requirements is to work with a function of the transformed variable

$$\rho = \max \left(0, \frac{r(\bar{x}) - r_0}{r_1} \right) \tag{19.101}$$

For obvious reasons, the parameter r_1 is called the *scaling length*. Commonly used functions of ρ to define the element size in space are:

  i) Power Laws given by expressions of the form [Löhner (1992)]

$$\delta(\bar{x}) = \delta_0[1 + \rho^\gamma] \tag{19.102}$$

  with the four input parameters δ_0, r_0, r_1, γ; typically, $1.0 \le \gamma \le 2.0$.
 ii) Exponential Functions which are of the form [Weatherill (1992)]

$$\delta(\bar{x}) = [\delta_0 e^{\gamma \rho}] \tag{19.103}$$

  with the four parameters δ_0, r_0, r_1, γ.
iii) Polynomial Expressions which avoid the high cost of exponents and logarithms by employing expressions of the form

$$\delta(\bar{x}) = \delta_0 \left[1 + \sum_{i=1}^{n} a_i \rho^i \right] \tag{19.104}$$

  with the $n + 3$ parameters δ_0, r_0, r_1, a_i; typically, quadratic polynomials are employed (i.e., $n = 2$, implying five free parameters).

Given a set of m sources, the minimum is taken whenever an element is to be generated

$$\delta(\bar{x}) = \min (\delta_1, \delta_1, \dots, \delta_m) \tag{19.105}$$

Some authors, notably Prizadeh (1992), have employed a smoothing procedure to combine background grids (see below) with sources. The effect of this smoothing is a more gradual increase in element size away from regions where small elements are required. Sources offer a very convenient and general way to define the desired element size in space. It is a simple matter to introduce sources interactively on a workstation with a mouse-driven menu once the surface data is available. Obviously, the number of sources should be kept small ($N_s < 50$) in order not to incur penalties in user set-up time and grid generation time.

Background Grids: Here, a coarse grid is provided by the user. At each of the nodes of this background grid, the element size, stretching and stretching direction are specified. Whenever a new element or point is introduced during grid generation, this background grid is interrogated to determine the desired size and shape of elements. While very general and flexible, the input of suitable background grids for complex 3-D configurations can become a tedious process. The main use of background grids is for adaptive remeshing (see Sec. 19.7). Given a first grid and a solution, the element size and shape for a mesh that is more suited for the problem at hand can be determined from an error indicator. With this information, a new grid can be generated by taking the current grid as the background grid. For this reason, background grids are still prevalent in most unstructed grid generators and are employed in conjunction with sources or other means of defining element size and shape.

Grid Size Attached to CAD Data: For problems that require gridding complex geometries, the specification of proper element sizes can become a tedious process. Conventional background grids would involve many tetrahedra, whose generation is a labor-intensive task. Point-, line- or surface-sources are not always appropriate either. Curved *ridges* between surface patches, as sketched in Fig. 19.10, may require many line-sources. Similarly, the specification of gridding parameters for surfaces with high curvature may require many surface sources. The net effect is that for complex geometries one is faced with excessive labor costs (background grids, many sources) and/or CPU requirements during mesh generation (many sources).

A better way to address these problems is to attach element size (or other gridding parameters) directly to CAD data. For many problems, the smallest elements are required close to the boundary. Therefore, if the element size for the points of the current front is stored, the next element size may be obtained by multiplying it with a user-specified increase factor, c_{incr}. The element size for each new point introduced is then taken as the minimum obtained from the background grid, δ_{bg}, the sources, δ_s, and the minimum of the point sizes corresponding to the face being deleted, multiplied by a user-specified increase factor c_i

$$\delta = \min (\delta_{bg}, \delta_s, c_i \min (\delta_A, \delta_B, \delta_C)) \tag{19.106}$$

Typical values for c_i are $c_i = 1.2–1.5$.

Element Type. Almost all current unstructured grid generators can only generate triangular or tetrahedral elements. If quad elements in 2-D are required, they can either be generated using an advanced front or *paving* [Blacker (1991)] technique, or by first generating a grid of triangles that is modified further. This last option can be summarized in the following five algorithmic steps:

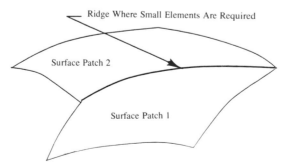

FIGURE 19.10 Specifying small element size for curved ridges.

i) Generate a triangular mesh with elements that are four times as big as the quad-elements required.

ii) Fuse as many pairs of triangles into quads as possible without generating quads that are too distorted. This process will leave some triangles in the domain.

iii) Smooth the mesh of triangles and quads.

iv) H-refine globally the mesh of triangles and quads. For the triangles, introduce an additional point in the element (see Fig. 19.11). In this way, the resulting mesh will only contain quads. Moreover, the quads will now be of the desired size.

v) Smooth the final mesh of quads.

The procedure outlined above will not work in 3-D, the problem being the face diagonals when matching tetrahedra with brick-elements. The face disappears in 2-D (only edges are present), making it possible to generate quad-elements from triangles.

Automatic Grid Generation Methods. There appear to be only the two following ways to fill space with an unstructured mesh.

Fill Empty, i.e., Not Yet Gridded Space: The idea here is to proceed into as yet ungridded space until the complete computational domain is filled with elements.

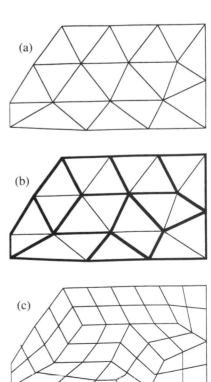

(a)

(b)

(c)

FIGURE 19.11 Generation of quad grids via triangles. (a) Mesh of triangles, (b) after triangle collapse, and (c) after H-refinement.

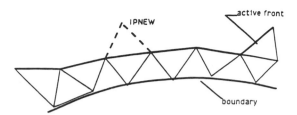

FIGURE 19.12 The advancing front method.

This is shown diagramatically in Fig. 19.12. The *front* denotes the boundary between the region in space that has been filled with elements and that which is still empty. The key step is the addition of a new volume or *element* to the ungridded space. Methods falling under this category are called *advancing front techniques* (AFTs).

Modify and Improve an Existing Grid: In this case, an existing grid is modified by the introduction of new points. After the introduction of each point, the grid is reconnected or reconstructed locally in order to improve the mesh quality. This procedure is sketched in Fig. 19.13. The key step is the addition of a new *point* to an existing grid. In most cases, the Delauney circumcircle or circumsphere criterion is used to reconnect the points. For this reason, the methods falling under this category have been called *Delaunay triangulation techniques* (DTT).

Given the number of ways for specifying element size and shape in space, as well as the method employed to generate the mesh, a number of combinations can be envisioned: AFT and internal measures of grid/front quality [Huet (1990)], DTT and internal measures of grid quality [Holmes 1988], AFT and boxes [Lo (1985)], DTT and boxes [Cavendish (1974), Baker (1989), Yerri (1984), and Shephard (1991)], AFT and background grids [Peraire (1987) and Löhner (1988)], DTT and background grids [Weatherhill (1992)], AFT and sources [Löhner (1993)], DTT and sources [Weatherhill (1992)], etc.

Before describing the two most commonly used unstructured grid generation procedures in CFD, the advancing front and the Delaunay triangulation techniques, a brief summary of other possible methods is given.

Other Grid Generation Methods. A number of other methods of generating meshes that are especially suited to a particular application have been developed. If the

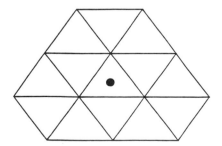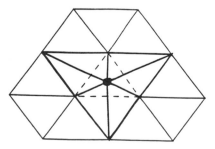

FIGURE 19.13 The Delaunay triangulation technique.

approximate answer to the problems being simulated is known (say we want to solve the same wing time and time again), the specialized development of an optimal mesh makes good sense. In many of these cases (e.g., an O-mesh for a subsonic/transonic inviscid steady-state airfoil calculation), grids generated by the more general methods listed above will tend to be larger than the specialized ones for the same final accuracy. There are three main methods falling under this specialized category.

Simple Mappings: In this case, it is assumed that the complete computational domain can be mapped into the unit square or cube. The distribution of points and elements in space is controlled either by an algebraic function, or by the solution of a partial differential equation in the transformed space [Thompson (1985)]. Needless to say, the number of points on opposing faces of the mapped quad or cube have to match line by line.

Macro-Element Approach: Here, the previous approach is applied on a local level by first manually or semi-manually discretizing the domain with large elements. These large elements are subsequently divided up into smaller elements using simple mappings [Zienkiewicz (1971)] (see Fig. 19.14).

In the aerospace community, this approach has been termed *multi-block*, and a whole service industry dedicated to the proper construction of grids has evolved [Thompson (1988), Steinbrenner (1990), and Allwright (1990)]. The macro-element approach is very general and has the advantage of being well suited for the generation of brick elements. However, it is extremely labor intensive, and efforts to automate the subdivision of space into macro-elements that precedes the actual grid generation have only shown limited success to date [Allwright (1990) and Dannenhoffer (1993)]. Generating these macro-blocks for complex geometries demands many man hours. As an example, it took a team of six engineers half a year to produce an acceptable grid for the flowfield around an F-16 fighter with two stores.

Uniform Background Grid: For problems that require uniform grids (e.g., Radar Cross-Section calculations, acoustics, homogeneous turbulence), most of the mesh covering the computational domain can readily be obtained from a uniform grid. This grid is also called a background grid, because it is laid over the computational

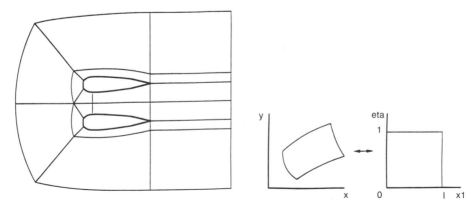

FIGURE 19.14   The macro-element gridding technique.

domain. At the boundaries, the regular mesh or the flow code have to be modified in order to take into account the boundaries. Options here are:

i) No Modifications. This is quite often done when the grids employed are very fine. The resulting grid exhibits *staircasing* at the boundaries (see Fig. 19.15), and for this reason codes that operate in such a manner are often denoted as *Legoland codes*.

ii) Modifying the Flow Solver. In this case, the change in volumes for the elements cut by boundaries is taken into account. Other modifications are required to avoid the timestep-limitations imposed by very small cells.

iii) Modifying the Mesh. Here, the points close to the surfaces are placed on the correct boundary. Some nontrivial logic is required to avoid problems at sharp corners, or when the surface curvature is such that more than one face of any given element is cut by the boundary. After modifying these surface elements, the mesh is smoothed in order to obtain a more uniform discretization close to the boundaries [Thacker (1980), Holmes (1986)]. The procedure is sketched in Fig. 19.15(a).

The Advancing Front Technique. Having described the general strategies currently available to generate unstructured grids, the advancing front technique will now be explained in more detail. The technique consists algorithmically of the following steps:

i) Define the boundaries of the domain to be gridded.

ii) Define the spatial variation of element size, stretchings, and stretching directions for the elements to be created. In most cases, this is accomplished with a combination of background grids and sources as outlined above.

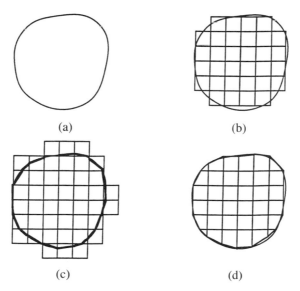

FIGURE 19.15 Uniform background grids. (a) Domain to be gridded, (b) Legoland representation, (c) improved Legoland, and (d) modified regular grid.

iii) Using the information given for the distribution of element size and shape in space and the line-definitions, generate sides along the lines that connect surface patches. These sides form an initial front for the triangulation of the surface patches.

iv) Using the information given for the distribution of element size and shape in space, the sides already generated, and the surface definition, triangulate the surfaces. This yields the initial front of faces.

v) Find the generation parameters (element size, element stretchings and stretching directions) for these faces.

vi) Select the next face to be deleted from the front; in order to avoid large elements crossing over regions of small elements, the face forming the smallest new element is selected as the next face to be deleted from the list of faces.

vii) For the face to be deleted:

    (a) Select a *best point* position for the introduction of a new point I PNEW.

    (b) Determine whether a point exists in the already generated grid that should be used in lieu of the new point. If there is such a point, set this point to I PNEW and continue searching [go to (b)].

    (c) Determine whether the element formed with the selected point I PNEW does not cross any given faces. If it does, select a new point as I PNEW and try again [go to (c)].

viii) Add the new element, point, and faces to their respective lists.

ix) Find the generation parameters for the new faces from the background grid and the sources.

x) Delete the known faces from the list of faces.

xi) If there are any faces left in the front, go to vi).

The complete grid generation of a simple 2-D domain using the advancing front technique is shown in Fig. 19.16.

The most important ingredient of the advancing front generator is a reliable and fast algorithm for checking whether two faces intersect each other. Experience from practical applications indicates that even slight changes in this portion of the generator greatly influence the final mesh. As with so many other problems in computational geometry, checking whether two faces intersect each other seems trivial for the eye, but is complicated to code. The problem is shown in Fig. 19.17. The checking algorithm is based on the following observation: two triangular faces do not intersect if no side of either face intersects the other face. The idea then is to build all possible side-face combinations between any two faces and check them in turn. If an intersection is found, then the faces cross. For each of the side-face combinations, a 3×3 matrix of covariant vectors has to be inverted, and additional vector multiplications have to be carried out to see if the side pierces through the face. Considering that on average about 40 close faces need to be checked, this way of checking the crossing of faces is very CPU intensive. If coded carelessly, it can easily consume more than 80% of the total CPU required for grid generation. In order to reduce the work load, a layered approach that consists of a simple min/max test, a local shape-function value check and the full side/face combination test was

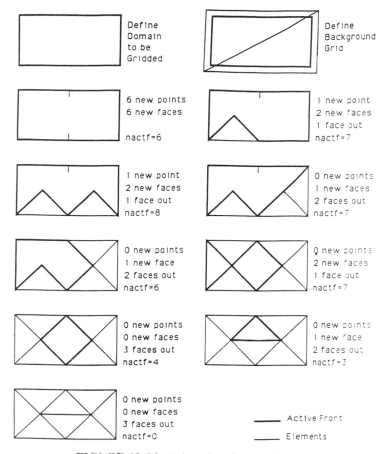

FIGURE 19.16 Advancing front technique.

proposed [Löhner (1988)]. Each of these three filters requires about an order of magnitude more CPU time than the preceding one. When implemented in this way, the face-crossing check requires only 25% of the total grid generation time.

The operations that could potentially reduce the efficiency of the algorithm to $O(N^{1.5})$ or even $O(N^2)$ are:

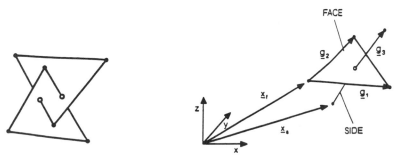

FIGURE 19.17 Intersection of faces.

1) finding the next face to be deleted;
2) finding the closest given points to a new point;
3) finding the faces adjacent to a given point; and
4) finding for any given location the values of generation parameters from the background grid and the sources. This is an interpolation problem on unstructured grids.

The verb *find* appears in all of these operations. The main task is to design the best data structures for performing these search operations as efficiently as possible. These data structures are typically *binary trees* or more complex trees. They were developed in the 1960's for Computer Science applications. Many variations are possible [see Williams (1964), Floyd (1964), Knuth (1973), Sedgewick (1983), and Yerry (1984)]. As with flow solvers, there does not seem to be clearly defined optimal data structure that all current grid generators use. For each of the data structures currently employed, one can find pathological cases where the performance of the tree-search degrades considerably. Data structures that have been used include:

1) heap-lists to find the next face to be deleted from the front,
2) quad-trees (2-D) and Octrees (3-D) to locate points that are close to any given location, and
3) n-trees, to determine which faces are adjacent to a point.

There are some additional techniques that can be used to improve the performance of the advancing front grid generator in terms of speed and reliability. The speed with which elements are generated can be improved by filtering techniques (i.e., working only with a minimum of nearest neighbors), by using *visibility cones* to remove close points that would form bad tetrahedra, by removing unused points from all pertinent data structures *on the fly*, by employing global *h*-refinement (i.e., generating coarser mesh that is subsequently refined) [Löhner (1992)], and by parallelization [Löhner (1992) and Shostko (1994)].

The quality of generated elements can be improved by not allowing bad elements to be created during the generation process, and by a so-called *sweep and retry* technique to enlarge and remesh again regions where elements could not be introduced [Löhner (1990), (1992)].

Typical 3-D advancing front generators construct grids at a rate of 12,000 tetrahedra per minute on the IBM-RISC/550 and 50,000 tetrahedra per minute on the CRAY–YMP. With one level of h-refinement, this last rate is boosted to 190,000 tetrahedra per minute. This rate is essentially independent of grid-size, but can decrease for very small grids.

Delaunay Triangulation. The Delaunay triangulation technique has a long history in mathematics, geophysics and engineering. In 1850, Dirichlet first proposed a method whereby a given domain could be systematically decomposed into a set of convex polyhedra. Given a set of points $\mathcal{P} := \bar{x}_1, \bar{x}_2, \ldots, \bar{x}_n$, one may define a set of regions or volumes $\mathcal{V} := v_1, v_2, \ldots, v_n$ assigned to each of the points, that satisfy the following property: any location within v_i is closer to $\bar{x}_i$ than to any other of the points

$$v_i := \mathcal{P} : \|\bar{\mathbf{x}} - \bar{\mathbf{x}}_i\| < \|\bar{\mathbf{x}} - \bar{\mathbf{x}}_j\| \ \forall \, j \neq i \qquad (19.107)$$

This set of volumes, $\mathcal{V}$, which covers the domain completely, is known as the *Dirichlet tesselation*. The volumes, v_i, are convex polyhedra and can assume fascinating shapes. They are referred to as *Voronoi regions*. Joining all the pairs of points $\bar{\mathbf{x}}_i$, $\bar{\mathbf{x}}_j$ across polyhedral boundaries results in a triangulation of the convex hull of $\mathcal{P}$. It is this triangulation that is commonly known as the *Delaunay triangulation*. The set of triangles (tetrahedra) that form the Delaunay triangulation satisfy the property that no other point is contained within the circumcircle (circumsphere) formed by the nodes of the triangle (tetrahedron). Although this and other properties of Delaunay triangulations have been studied for some time, practical triangulation procedures have only appeared in the last decade [Bowyer (1981), Watson (1981), Tanemura (1983), Kirkpatrick (1985), Cavendish (1985), Baker (1987, 1989), and Weatherhill (1992a,b)]. Most of these are based on Bowyer's algorithm.

The complete grid generation of a simple 2-D domain using the Delaunay triangulation algorithm is shown in Fig. 19.18. In this case, the point distribution was provided. The specification of a point distribution may be avoided by employing

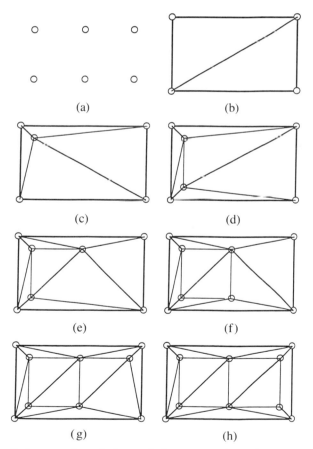

(a)   (b)

(c)   (d)

(e)   (f)

(g)   (h)

FIGURE 19.18 Delaunay triangulation technique. (a) Points, (b) external points and elements, (c) point 1, (d) point 2, (e) point 3, (f) point 4, (g) point 5, and (h) point 6.

sources or background grids to define in a more general way the desired distribution of element size and shape in space. The key idea, first proposed by Holmes (1988), is to check the discrepancy between the desired and the actual element shape and size of the current mesh. Points are then introduced in those regions where the discrepancy exceeds a user-defined tolerance.

The most important ingredient of the Delaunay generator is a reliable and fast algorithm for checking whether the circumsphere of a tetrahedron (triangle) contains the points to be inserted. A point $(\bar{\mathbf{x}}_p$ is within the radius R_i of the sphere centered at $\bar{\mathbf{x}}_c$ if

$$d_p^2 = (\bar{\mathbf{x}}_p - \bar{\mathbf{x}}_c) \cdot (\bar{\mathbf{x}}_p - \bar{\mathbf{x}}_c) < R_i^2 \qquad (19.108)$$

This check can be performed without any problems unless $|d_p - R_i|^2$ is of the order of the round-off of the computer. In such a case, an error may occur, leading to an incorrect rejection or acceptance of a point. Once an error of this kind has occurred, it is very difficult to correct, and the triangulation process breaks down. Baker (1987) has determined the following condition: Given the set of points $\mathscr{P} := \bar{\mathbf{x}}_1, \bar{\mathbf{x}}_2, \ldots, \bar{\mathbf{x}}_n$ with characteristic lengths $d_{max} = \max [\bar{\mathbf{x}}_i - \bar{\mathbf{x}}_j | \forall i \neq j$ and $d_{min} = \min |\bar{\mathbf{x}}_i - \bar{\mathbf{x}}_j| \forall i \neq j$, the floating point arithmetic precision required for the Delaunay test should be better than

$$\epsilon = \left(\frac{d_{min}}{d_{max}}\right)^2 \qquad (19.109)$$

Consider the generation of a mesh suitable for inviscid flow simulations for a typical transonic airliner (e.g., B-747). Taking the wing chord length as a reference length, the smallest elements will have a side length of the order of $10^{-3} L$, while far-field elements may be located as far as $10^2 L$ from each other. This implies that $\epsilon = 10^{-10}$, which is beyond the 10^{-8} accuracy of 32-bit arithmetic. For these reasons, unstructured grid generators generally operate with 64-bit arithmetic precision.

A related problem of degeneracy that may arise is linked to the creation of very flat elements or *slivers* [Cavendish (1985)]. The calculation of the circumsphere for a tetrahedron is given by the conditions:

$$(\bar{\mathbf{x}}_i - \bar{\mathbf{x}}_c) \cdot (\bar{\mathbf{x}}_i - \bar{\mathbf{x}}_c) = R^2, \qquad i = 1, 4 \qquad (19.110)$$

yielding four equations for the four unknowns $\bar{\mathbf{x}}_c, R$. If the four points of the tetrahedron lie on a plane, the solution is impossible ($R \to \infty$). In such a case, the point to be inserted is rejected and stored for later use (*skip and retry*).

The operations that could potentially reduce the efficiency of the algorithm to $O(N^{1.5})$ or even $O(N^2)$ are:

1) finding all tetrahedra whose circumspheres contain a point
2) finding all the external faces of the void that results due to the deletion of a set of tetrahedra
3) finding the closest new points to a point and
4) finding for any given location the values of generation parameters from the background grid and the sources.

As before, the verb *find* appears in all these operations, and the task is to design the best data structures for performing these search operations as efficiently as possible. Some of these data structures have already been discussed for the advancing front technique. The principal data structure required to minimize search overheads is the *element adjacent to element* or *element surrounding element* structure ESUEL(1:NFAEL,1:NELEM) that stores the neighbor elements of each element. This structure is used to march quickly through the grid when trying to find the tetrahedra whose circumstances contain a point. Once a set of elements has been marked for removal, the outer faces of this void can be obtained by interrogating ESUEL. As the new points to be introduced are linked to the elements of the current mesh, ESUEL can also be used to find the closest new points to a point. Furthermore, the equivalent ESUEL structure for the background grid can be used for fast interpolation of the desired element size and shape.

A major assumption that is used time and again to make the Delaunay triangulation process both unique and fast is the Delauney property itself, namely that no other point should reside in the circumsphere of any tetrahedron. This implies that in general during the grid generation process some of the tetrahedra will break through the surface. The result is a mesh that satisfies the Delauney property, but is not surface conforming (see Fig. 19.19). The most commonly used techniques to recover a surface conforming mesh are extra point insertion [Baker (1987), (1989)], and algebraic surface recovery [Weatherhill (1992a,b)].

There are some additional techniques that can be used to improve the performance of the Delaunay grid generator in terms of speed and reliability. The speed with which elements are generated can be improved by ordering the first set of boundary points such that contiguous points in the list are neighbors in space, by vectorization of the background grid and background source searches, and by employing global *h*-refinement. The reliability of the generator can be improved by avoiding bad elements, by being consistent in case *degenerate* point distributions are encountered (e.g., a perfectly cartesian point distribution) [Weatherhill (1992a,b)], and by introducing new points in an advancing front-like manner [Müller (1993), Merriam (1992), and Mavriplis (1992)].

Typical Delaunay grid generators construct grids at a rate of 60,000 tetrahedra per minute on the Sun Sparc-2 workstation. This is faster than the advancing front method, but the assumption is made that a surface grid was provided.

Grid Improvement. Practical implementations of either advancing front or Voronoi grid generators indicate that in certain regions of the mesh abrupt variations in element shape or size may be present. These variations appear even when trying to

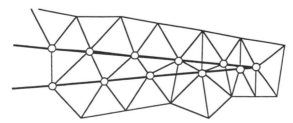

FIGURE 19.19 Delaunay triangulation with elements breaking through the surface.

generate perfectly uniform grids. In order to circumvent any possible problems these irregular grids may trigger for field solvers, the generated mesh is optimized further in order to improve the uniformity of the mesh. The three most popular ways of mesh optimization are: 1) removal of bad elements, 2) Laplacian smoothing, or 3) functional optimization [Cabello (1992)]. The most commonly used of these is the Laplacian smoothing.

Laplacian Smoothing: A number of smoothing techniques are lumped under this name. The edges of the triangulation are assumed to represent springs. These springs are relaxed in time using an explicit time stepping scheme, until an equilibrium of spring forces has been established. Because *globally* the variations of element size and shape are smooth, most of the nonequilibrium forces are local in nature. This implies that a significant improvement in mesh quality can be achieved rather quickly. The force exerted by each spring is proportional to its length and along its direction. Therefore, the sum of the forces exerted by all springs surrounding a point can be written as:

$$\bar{\mathbf{f}}_i = c \sum_{j=1}^{ns_i} (\bar{\mathbf{x}}_j - \bar{\mathbf{x}}_i) \tag{19.111}$$

where c denotes the spring constant, $\bar{\mathbf{x}}_i$ the coordinates of the point, and the sum extends over all the points surrounding the point. The time-advancement for the coordinates is accomplished as

$$\Delta x_i = \Delta t \frac{1}{ns_i} f_i \tag{19.112}$$

At the surface of the computational domain, no movement of points is allowed, i.e., $\Delta \bar{\mathbf{x}} = 0$. Usually, the timestep (or relaxation parameter) is chosen as $\Delta t = 0.8$, and 5–6 timesteps yield an acceptable mesh. The application of the Laplacian smoothing technique can result in inverted or negative elements. The presence of even one element with a negative Jacobian will render most field solvers inoperable. Therefore, these negative elements are eliminated. For the advancing front technique, it has been found advisable to remove not only the negative elements, but also all elements that share points with them. This element removal gives rise to voids or holes in the mesh, which are regridded using the advancing front technique. Another option, which can also be used for the Delaunay Technique, is the removal of negative elements.

Navier–Stokes Gridding Techniques. The need to grid complex geometries for the simulation of flows using the Reynolds-averaged Navier–Stokes (RANS) equations, i.e., including the effects of viscosity and the associated boundary or mixing layers, is encountered commonly in engineering practice. The difficulty of this task increases not only with the geometric complexity of the domain to be gridded, but also with the Reynolds number of the flow. For high Reynolds numbers, the proper discretization of the very thin, yet important boundary or mixing layers requires elements with aspect ratios well in excess of 1:1,000. This requirement presents formidable difficulties to general, *black-box* unstructured grid generators. The most

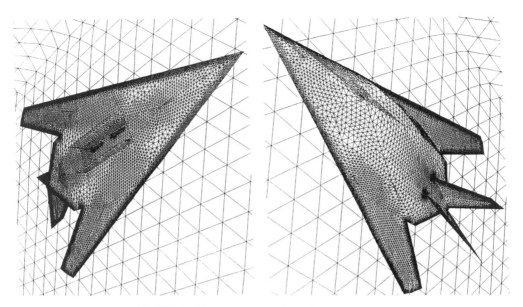

FIGURE 19.20 Unstructured mesh for an F-117.

common way to generate meshes suitable for RANS calculations for complex geometries is to employ a structured or semi-structured mesh close to wetted surfaces or wakes. This *Navier–Stokes region mesh* is then linked to an outer unstructured grid that covers the *inviscid* regions. In this way, the geometric complexity is solved using unstructured grids and the physical complexity of near-wall or wake regions is solved by semi-structured grids. This approach has proven very powerful in the past, as evidenced by many examples [Nakahashi (1987), (1988), Mavriplis (1990), Ramamurti (1990), Kallinderis (1992), Löhner (1993), Pirzadah (1993), (1994), Müller (1993), and Morgan *et al.* (1993)] and is still an area of active research.

Application Examples. We just include two examples to demonstrate the possibilities unstructured grids offer. The first example, shown in Fig. 19.20, is a mesh suitable for Euler simulations. The geometry is an F-117 with open bombbay, and four bombs inside. The mesh had approximately 1 Mtet elements. The second example, shown in Fig. 19.21, is a mesh suitable for Navier–Stokes simulations coupled with heat transfer in the solid. The geometry is an electronic chip. Four different material regions were gridded inside the chip in addition to the Navier–Stokes mesh for the fluid region. This mesh had approximately 1.5 Mtet elements.

19.2 INTRODUCTION TO FINITE DIFFERENCE METHODS
Evangelos Hytopoulos

19.2.1 Introduction

The numerical solution of the differential equations governing a complex fluid dynamics problem requires the introduction of a discretization method. Several methods have been developed and are currently in use. In the present section, the basic

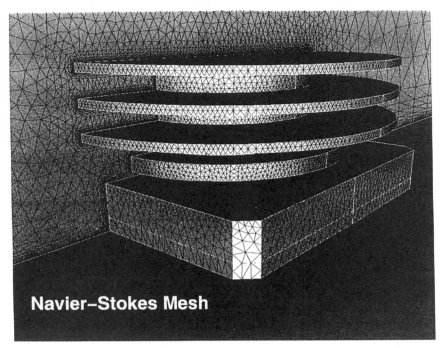

FIGURE 19.21 Unstructured mesh for an electronic chip.

ideas and techniques used in the development of the *finite difference method* will be presented. Other discretization methods are the *finite volume method* (see Sec. 19.3) and the *finite element method* (see Sec. 19.4). The finite difference method is the oldest method applied to the numerical solution of differential equations, and its development is based on the definition of the derivative and the properties of Taylor's series. In the finite difference method, the domain is discretized into a *mesh* or *grid* (Fig. 19.22), and the unknown variables exist only at discrete points called *nodes*. The derivatives are approximated by differences. Recall the definition of the

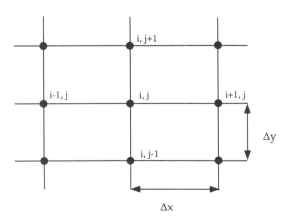

FIGURE 19.22 Typical finite difference mesh or grid.

derivative of a function $u(x)$ at the point x

$$\frac{\partial u}{\partial x} = \lim_{\Delta x \to 0} \frac{u(x + \Delta x) - u(x)}{\Delta x} \tag{19.113}$$

For a small but finite Δx, the expression on the right hand side is an approximation to the exact value of the derivative. Although the reduction of Δx leads to a better approximation for $\partial u/\partial x$, any finite value of the grid spacing introduces a truncation error that only tends to zero for Δx going to zero. Thus, the truncation error is defined as the difference between the derivative and its finite difference representation. The power of the leading term of the truncation error defines the accuracy of the approximation. For example, the finite difference approximation of the first derivative given by

$$\frac{\partial u}{\partial x} = \frac{u_{i+1} - u_{i-1}}{2\Delta x} \tag{19.114}$$

is said to be second-order accurate, since the truncation error tends to zero as $(\Delta x)^2$. The error can be formally derived for any finite difference approximation, using the Taylor's series expansion around the point x. By developing $u(x + \Delta x)$ in a Taylor's series about the point x, one obtains

$$u(x + \Delta x) = u(x) + \frac{\partial u}{\partial x}(x)\Delta x + \frac{\Delta x^2}{2}\frac{\partial^2 u}{\partial x^2}(x)$$

$$+ \frac{\Delta x^3}{6}\frac{\partial^3 u}{\partial x^3}(x) + \text{(higher order terms)} \tag{19.115}$$

We can then write

$$\frac{u(x + \Delta x) - u(x)}{\Delta x} = \frac{\partial u}{\partial x}(x) + \frac{\Delta x}{2}\frac{\partial^2 u}{\partial x^2}(x) + \cdots \tag{19.116}$$

Now assume that the x-axis has been discretized in such a way that the continuum is replaced by N mesh points x_i. For simplicity, we assume that the spacing between the mesh points is constant and equal to Δx. Without loss of generality, we can write $x_i = i\Delta x$, and we denote the value of the function $u(x)$ at the point x_i as u_i. Inserting these definitions in Eq. (19.116), one obtains the following expression for the first derivative

$$\frac{u_{i+1} - u_i}{\Delta x} = \frac{\partial u}{\partial x} + \frac{\Delta x}{2}\frac{\partial^2 u}{\partial x^2} + \cdots = \frac{\partial u}{\partial x} + O(\Delta x) \tag{19.117}$$

The symbol $O(\Delta x)$ indicates that the absolute value of the truncation error, for small Δx, is smaller than $a\Delta x$, where a is a positive number. Equation (19.117) gives a *forward* difference with respect to x_i. Taylor expansion of $u(x - \Delta x)$ about x provides a *backward difference* with truncation error of order Δx.

$$\frac{\partial u}{\partial x} = \frac{u_i - u_{i-1}}{\Delta x} + O(\Delta x) \tag{19.118}$$

The reader should notice that the second-order accurate, *central difference* formula in Eq. (19.114), can be obtained as an average of the Eqs. (19.117) and (19.118). Also, it is important to note that although Eq. (19.117) provides a first-order accurate expression for the first derivative at $x = x_i$, it provides a second-order accurate expression if considered as an approximation at $x = x_{i+1/2}$. The expressions presented so far represent only a few ways in which the first derivative of a function can be approximated. All of them involve only two nodal points, but this is not a restriction. Difference formulas for the first derivative can be constructed involving any number of adjacent points, and the accuracy of the approximation increases with the number of nodal points used. Later, a systematic way for obtaining approximations to any order derivatives based on the Taylor series will be presented.

An expression for the second derivative at $x = x_i$ can be obtained by using the Taylor expansions of u_{i+1} and u_{i-1}. By adding the two expressions and rearranging, one obtains

$$\frac{\partial^2 u}{\partial x^2} = \frac{u_{i+1} - 2u_i + u_{i-1}}{\Delta x^2} + O(\Delta x^2) \tag{19.119}$$

Forward and backward approximations for the second derivative are also possible. Before introducing the general methodology for obtaining finite difference approximation, a number of difference operators are presented. These operators make the expressions more compact, and they can be combined to represent more complicated formulas. The forward difference operator, δ^+, is defined as

$$\delta^+ u_i = u_{i+1} - u_i \tag{19.120}$$

The backward operator, δ^-, is defined as

$$\delta^- u_i = u_i - u_{i-1} \tag{19.121}$$

Two central difference operators can be defined

$$\delta = u_{i+1/2} - u_{i-1/2} \tag{19.122}$$

and

$$\bar{\delta} = u_{i+1} - u_{i-1} \tag{19.123}$$

The averaging operator, μ, is

$$\mu = \frac{u_{i+1/2} + u_{i-1/2}}{2} \tag{19.124}$$

Using these definitions, a number of relations between the operators can be developed. For example,

$$\bar{\delta} = \delta^+ + \delta^- \tag{19.125}$$

$$\delta^2 = \delta^+ - \delta^- = \delta^+ \delta^- = \delta^- \delta^+ \tag{19.126}$$

Let us now show a general way of obtaining finite difference formulas for any order derivative and with a given accuracy.

19.2.2 Finite Difference Formulas using the Taylor Expansion

The method is described for a uniform grid, but it can be used for nonuniform grids as well. Suppose that we want to find an expression for the first derivative that is second-order accurate. Since the term $(\partial u / \partial x)$, in the Taylor expansion [Eq. (19.115)] is multiplied Δx, and the truncation error must be $O(\Delta x^2)$, it is required that the quadratic term is eliminated from the final expression. A general way in which this can be accomplished is to write the first derivative as a linear combination of the nodal values

$$\left(\frac{\partial u}{\partial x} \right)_i = \frac{a_{i+k} u_{i+k} \cdots + a_{i+1} u_{i+1} + a_i u_i + a_{i-1} u_{i-1} + \cdots a_{i-l} u_{i-l}}{\Delta x}$$

$$\tag{19.127}$$

where k, l are integers and $i > l$. One can express the values, u_j, with $i - l \le j \le i + k$ as a Taylor expansion around u_i. Inserting these expressions in Eq. (19.127) leads to an expression, the terms of which are the terms of the Taylor expansion about the point x_i, but with coefficients that are functions of the unknown multipliers. Setting specific requirements that these coefficients must satisfy leads to a system of linear equations, the solution of which determines the unknown multipliers. It becomes obvious that this procedure does not indicate the number of points that must be used for a given accuracy or the nature of the difference (backward, central, forward). The latter is generally defined from the mathematical classification of the equation that one wants to solve. The number of points required can be found by finding the least number of conditions that must be satisfied. These conditions can be summarized as follows:

i) The coefficients of the derivatives of order lower than the one that is approximated must be zero.

ii) The coefficient of the derivative of interest is different than zero. It can conveniently be set equal to 1.

iii) If the truncation error is of order m and the derivative that is approximated is of order n, then all the coefficients of the derivatives of order k, with $n < k < m + n$ must be zero.

Conditions (i) and (iii) usually provide the required number of equations for calculating the unknown multipliers. Condition (ii) is used whenever two of the constraints required from (ii) and (iii) happen to be identical. This is the case in the calculation of the second derivative using a second-order accurate scheme on a uniform grid. Let us now apply the procedure to the example at hand. We can choose the number of points based on the number of requirements that must be satisfied.

From the first requirement, we have 0 points, from the second we have 1, and from the third we have 1, since the coefficient of the second derivative must be zero (order of derivative $= 1$, order of truncation error $= 2$, so $1 < k < 1 + 2$). So, if one chooses a central difference, the points that must be used are u_{i+1}, u_i, and u_{i-1}. Using the Taylor series for u_{i+1} and u_{i-1} and multiplying the u_{i+1} expression by a and the u_{i-1} expression by b, one obtains

$$au_{i+1} + bu_{i-1} = (a + b)u_i + (a - b)\frac{\partial u}{\partial x}\Delta x + (a + b)\frac{\partial^2 u}{\partial x^2}\frac{(\Delta x)^2}{2}$$

$$+ \frac{\partial^3 u}{\partial x^3}(a - b)\frac{(\Delta x)^3}{6} + \cdots \tag{19.128}$$

The aforementioned conditions require that

$$a - b = 1$$

$$a + b = 0 \tag{19.129}$$

The solution of the system is $a = 1/2$ and $b = -1/2$, which leads to the expression given by Eq. (19.114). A second-order accurate, backward difference for the same derivative can be derived in a similar way. Again, the reader can verify that three points are needed, namely u_i, u_{i-1}, u_{i-2}. The expansion of u_{i-2} about u_i gives

$$u_{i-2} = u_i - 2\Delta x\frac{\partial u}{\partial x} + \frac{(2\Delta x)^2}{2}\frac{\partial^2 u}{\partial x^2} - \frac{(2\Delta x)^3}{6}\frac{\partial^3 u}{\partial x^3} + \cdots \tag{19.130}$$

Multiplying the u_{i-1} by a and the u_{i-2} by b, one gets

$$au_{i-1} + bu_{i-2} = (a + b)u_i - (a + 2b)\Delta x\frac{\partial u}{\partial x} + (a/2 + 2b)(\Delta x)^2\frac{\partial^2 u}{\partial x^2}$$

$$- \frac{(a + 8b)}{6}(\Delta x)^3\frac{\partial^3 u}{\partial x^3} + \cdots \tag{19.131}$$

The aforementioned requirements lead to

$$a + 2b = 1$$

$$\frac{a}{2} + 2b = 0 \tag{19.132}$$

the solution of which is $a = 2$, $b = -1/2$. This leads to the following expression for the backward difference

$$\frac{\partial u}{\partial x} = \frac{(3u_i - 4u_{i-1} + u_{i-2})}{2\Delta x} \tag{19.133}$$

This is not the only method that can produce finite difference expressions in a systematic way. The interested reader can find derivations based on polynomial fitting in Anderson et al. (1984) and the operator technique in Hirsch (1990).

19.2.3 Implicit Finite Difference Formulas

It was mentioned that the accuracy of finite difference schemes can be improved by including more points. Increasing the number of points for increased accuracy leads to algebraic systems of equations with greater bandwidths, and they are thus more computationally expensive. So one can pose the question: Can we develop schemes that have a higher order of accuracy than the ones currently presented for a given number of points? The answer is yes. This goal can be accomplished by using *implicit* finite difference formulas. For example, the scheme

$$\frac{1}{6}\left(\left(\frac{\partial u}{\partial x}\right)_{i+1} + 4\left(\frac{\partial u}{\partial x}\right)_{i} + \left(\frac{\partial u}{\partial x}\right)_{i-1}\right)$$

$$= \frac{u_{i+1} - u_{i-1}}{2\Delta x} + O(\Delta x^4) \tag{19.134}$$

involves only three nodal points and it is fourth-order accurate. The corresponding explicit scheme involving the same three points is only second-order accurate. The accuracy of the scheme can be verified by expanding all the variables (both the nodal values and the derivatives) as Taylor expansions about u_i. This increased accuracy is not obtained for free. Expressions such as Eq. (19.134) do not permit the *explicit* evaluation of the numerical approximations to the derivative $(\partial u/\partial x)$ at $x = x_i$, because the derivatives of the adjacent nodes are present. Instead, an algebraic system of equations has to be solved for the unknowns $(\partial u/\partial x)_i$, $i = 1, 2, \ldots, N$. For example, the scheme given by Eq. (19.134) leads to a tridiagonal system of equations that can be solved using well known schemes like the *Thomas algorithm*. Thus, these methods are termed *implicit*. The increased accuracy of these schemes can be attributed to the coupling of the values at any point to the values at every other point of the domain. In a real problem, the values of the function, u_i, are also unknowns, so the system for the derivatives is augmented by the algebraic equations that come from the governing equations, and the global system has as unknowns both the values and the derivatives of the function at the nodal points. General implicit formulas can be derived by making use of the Taylor expansions. Again, an expression involving both nodal values and derivatives multiplied by constants, that are considered unknowns, is written down. The numerical values of the constants are calculated by means of an algebraic system that is the result of the requirements that the expression must satisfy. For example, if one chooses a three point representation for the second derivative of $u(x)$, a general expression that can be used for the derivation is

$$au_{i+1} + bu_i + cu_{i-1} + d\left(\frac{\partial u}{\partial x}\right)_{i+1} + e\left(\frac{\partial u}{\partial x}\right)_{i} + f\left(\frac{\partial u}{\partial x}\right)_{i-1}$$

$$+ g\left(\frac{\partial^2 u}{\partial x^2}\right)_{i+1} + h\left(\frac{\partial^2 u}{\partial x^2}\right)_{i} + l\left(\frac{\partial^2 u}{\partial x^2}\right)_{i-1} = 0 \tag{19.135}$$

Taylor expansion about x_i leads to a truncation error written in powers of Δx. In general, the conditions required for obtaining a specific order of truncation error will not uniquely determine all the unknowns. A parametric family of approximations is obtained, and extra conditions are required to uniquely determining all the constants. These extra conditions can be selected according to the number of derivatives and mesh points one wishes to maintain in the formula. It is also possible to determine all the coefficients by requiring that the formula has the highest degree of accuracy possible with the given number of nodal points. Extensive treatments of implicit formulas can be found in Peyret (1978) and Peyret and Taylor (1983).

19.2.4   Finite Difference Formulas on Nonuniform Grids

In many problems, it becomes important that a small grid spacing is used in some regions of the domain where variations are rapid (boundary layer regions, shock waves), while a coarser grid can be used elsewhere. This gives rise to grids with unequal spacing, and the finite difference approximations derived above are not valid on those grids. There exist several ways to account for that. One can use a transformation of coordinates which transforms the unequal spacing in the original system into an equal spacing in the new system. This procedure introduces *metric coefficients* that must be discretized in a consistent way. Another way is to derive finite difference formulas in a way similar to that used for the uniform grids. To illustrate this point, let us derive a second-order accurate expression for the first derivative using at most three points, i.e., x_{i-1}, x_i, and x_{i+1}. Define $\Delta x^+ = x_{i+1} - x_i$ and $\Delta x^- = x_i - x_{i-1}$. Then, the Taylor expansion about u_i can be written as follows

$$u_{i+1} = u_i + \left(\frac{\partial u}{\partial x}\right)_i (\Delta x^+) + \left(\frac{\partial^2 u}{\partial x^2}\right)_i \frac{(\Delta x^+)^2}{2} + \left(\frac{\partial^3 u}{\partial x^3}\right)_i \frac{(\Delta x^+)^3}{6} + \cdots \quad (19.136)$$

$$u_{i-1} = u_i - \left(\frac{\partial u}{\partial x}\right)_i (\Delta x^-) + \left(\frac{\partial^2 u}{\partial x^2}\right)_i \frac{(\Delta x^-)^2}{2} - \left(\frac{\partial^3 u}{\partial x^3}\right)_i \frac{(\Delta x^-)^3}{6} + \cdots \quad (19.137)$$

Multiplying the first equation by a and the second one by b and adding one gets

$$au_{i+1} + bu_{i-1} = (a + b)u_i + \left(a\frac{\Delta x^+}{\Delta x^-} - b\right)\Delta x^- \left(\frac{\partial u}{\partial x}\right)_i$$

$$+ \frac{(a(\Delta x^+)^2 + b(\Delta x^-)^2)}{2}\left(\frac{\partial^2 u}{\partial x^2}\right)_i$$

$$+ \frac{(a(\Delta x^+)^3 - b(\Delta x^-)^3)}{6}\left(\frac{\partial^3 u}{\partial x^3}\right)_i + \cdots \quad (19.138)$$

Applying the same requirements as for the case of the uniform grid, one obtains the following system of equations

$$a\frac{\Delta x^+}{\Delta x^-} - b = 1$$

$$a\Delta x^{+2} + b\Delta x^{-2} = 0 \quad (19.139)$$

The solution to this system is

$$a = \frac{\Delta x^{-2}}{\Delta x^+(\Delta x^+ + \Delta x^-)}, \quad b = -\frac{\Delta x^+}{\Delta x^- + \Delta x^+} \tag{19.140}$$

Thus, the final expression becomes

$$\left(\frac{\partial u}{\partial x}\right)_i = \frac{\Delta x^-}{\Delta x^+ + \Delta x^-}\frac{(u_{i+1} - u_i)}{\Delta x^+} + \frac{\Delta x^+}{\Delta x^+ + \Delta x^-}\frac{(u_i - u_{i-1})}{\Delta x^-}$$

$$- \frac{\Delta x^- \Delta x^+}{6}\left(\frac{\partial^3 u}{\partial x^3}\right)_i \tag{19.141}$$

In a similar way, one can derive backward and forward schemes for the first derivative. Now, turn attention to the second derivative. We have seen that on a uniform grid, a central difference representation involving the three points $i-1, i, i+1$ is second-order accurate. To check if this holds true for nonuniform grid, write the conditions that must be satisfied in this case. From Eq. (19.138) one obtains

$$a\frac{\Delta x^+}{\Delta x^-} - b = 0$$

$$a\left(\frac{\Delta x^+}{\Delta x^-}\right)^3 - b = 0 \tag{19.142}$$

For $\Delta x^+ \neq \Delta x^-$, the only solution of the system is $a = b = 0$, while for $\Delta x^+ = \Delta x^-$ the two conditions are identical. So, for unequal spacing the central difference scheme seems to be first order. Actually, second-order accuracy can be restored if $\Delta x^+/\Delta x^- = 1 + O(\Delta x)$. If the grid spacing varies abruptly though, then deterioration of the accuracy will result. This is a general property of the finite difference approximation, and more information can be found in Hoffman (1982).

19.2.5 Finite Difference Schemes in Two or More Dimensions

Almost all the fluid dynamics problems of engineering interest involve more than one dimension, so one must be able to write expressions for the partial derivatives in a way similar to the one described before. This is actually possible by making a use of the expressions derived in the previous sections and considering each variable separately. Let us illustrate this by applying the concept to the discretization of the Laplace operator on a two-dimensional domain. Suppose that the spacing in the x direction is Δx and the corresponding spacing in the y direction is Δy, where Δx and Δy are not equal. Then, $x_i = i\Delta x$, $y_j = j\Delta y$ and $u_{i,j} = u(x_i, y_j)$. To obtain a discrete representation of the Laplace operator in the two-dimensional space, we apply the expressions derived previously for the second derivative to either variable x, y, by changing the corresponding index i or j, separately. Choosing a central differencing scheme, the partial derivative with respect to x can be written as

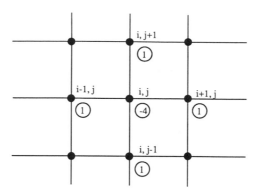

FIGURE 19.23   Computational molecule for the Laplace operator.

$$\left(\frac{\partial^2 u}{\partial x^2}\right)_{i,j} = \frac{u_{i+1,j} - 2u_{i,j} + u_{i-1,j}}{\Delta x^2} \tag{19.143}$$

and the corresponding partial derivative with respect to y as

$$\left(\frac{\partial^2 u}{\partial y^2}\right)_{i,j} = \frac{u_{i,j+1} - 2u_{i,j} + u_{i,j-1}}{\Delta y^2} \tag{19.144}$$

Thus, the discrete representation of the Laplace operator, with second-order accuracy in x and y can be written as (see Fig. 19.23)

$$\Delta u_{i,j} = \frac{\partial^2 u}{\partial x^2} + \frac{\partial^2 u}{\partial y^2} = \frac{u_{i+1,j} - 2u_{i,j} + u_{i-1,j}}{\Delta x^2} + \frac{u_{i,j+1} - 2u_{i,j} + u_{i,j-1}}{\Delta y^2} \tag{19.145}$$

This is not the only finite difference approximation for the Laplace operator. Other approximations resulting from combinations of the difference operators on the two spatial coordinates are possible [Hirsch (1990)].

19.2.6 Mixed Derivatives

Mixed derivatives often appear in differential equations, and their discretization can be performed following approaches similar to the ones used previously. The successive use of the one-dimensional operators yields expressions for these derivatives. Suppose that one wants to write a second order accurate representation of the mixed derivative $\partial^2 u/\partial x\,\partial y$. Successive application of the one-dimensional operators yields

$$\frac{\partial^2 u}{\partial x\,\partial y} = \frac{\partial}{\partial y}\left(\frac{\partial u}{\partial x}\right) = \frac{\bar{\delta}_y}{\Delta y}\frac{\bar{\delta}_x}{\Delta x}u_{i,j} = \bar{\delta}_y\frac{1}{2\Delta x}(u_{i+1,j} - u_{i-1,j})$$

$$= \frac{1}{4\Delta x\Delta y}(u_{i+1,j+1} - u_{i+1,j-1} - u_{i-1,j+1} + u_{i-1,j-1}) + O(\Delta x^2, \Delta y^2) \tag{19.146}$$

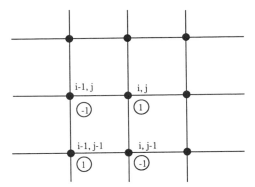

FIGURE 19.24   Computational molecule for the mixed derivative (backward differencing).

Other expressions can be obtained by mixing forward, backward and central differencing operators. Applying the backward difference operators in both directions leads to the following first order scheme (see Fig. 19.24)

$$\left(\frac{\partial^2 u}{\partial x\, \partial y}\right)_{i,j} = \frac{1}{\Delta x \Delta y}\,(u_{i-1,j-1} - u_{i-1,j} - u_{i,j-1} + u_{i,j}) + O(\Delta x, \Delta y) \quad (19.147)$$

In a similar way, one can use forward difference operators to derive another first order accurate formula with truncation error equal to that of Eq. (19.147), but with opposite sign. Summation of the two formulas leads to a second-order accurate scheme that involves the value at the point i, j. This may be helpful in some cases for increasing the diagonal dominance.

19.2.7   Application of Finite Difference Methods to Model Equations

In this section, we describe the general concepts that are involved in the application of finite difference methods to differential equations. For simplicity, and without loss of generality, we will discuss only simple model equations. Application to the full equations governing fluid dynamics are found in later sections and chapters of this handbook.

We start by defining the truncation error of the finite difference representation of a partial differential equation. Suppose that a numerical solution of the differential wave equation

$$\frac{\partial u}{\partial t} + a\,\frac{\partial u}{\partial x} = 0 \quad\quad (19.148)$$

is sought using the following finite difference scheme (called *upwind*)

$$\frac{u_j^{n+1} - u_j^n}{dt} + a\,\frac{u_j^n - u_{j-1}^n}{dx} = 0 \quad\quad (19.149)$$

By approximating the derivatives with finite differences, one introduces an error. Inserting the Taylor expansion for the different terms, one obtains the following expression

$$\frac{\partial u}{\partial t} + a\frac{\partial u}{\partial x} = -\frac{dt}{2}\frac{\partial^2 u}{\partial t^2} + a\frac{dx}{2}\frac{\partial^2 u}{\partial x^2} + \cdots \tag{19.150}$$

The left-hand side of the equation is the partial differential equation (PDE) that we are interested in solving. If the finite difference representation were exact, the right hand side should have been zero. This expression is the leading term of the error. Thus, the *truncation error* is the difference between the PDE and the finite difference approximation to it. The lowest powers of dt and dx that appear in the truncation error give the accuracy of the scheme. In the present example the scheme is first-order accurate both in space and time.

The truncation error also reveals the answer to another important question that must be posed, i.e., "is the numerical scheme that was invented an acceptable approximation to the original differential equation?" The answer to this question introduces the concept of consistency. Let us try to make this point clear by using the heat equation as our model, i.e.,

$$\frac{\partial u}{\partial t} = a\frac{\partial^2 u}{\partial x^2} \tag{19.151}$$

The equation will be approximated by the use of the DuFort and Frankel (1953) scheme

$$\frac{u_i^{n+1} - u_i^{n-1}}{2\,dt} = a\frac{u_{i+1}^n - u_i^{n+1} - u_i^{n-1} + u_{i-1}^n}{\Delta x^2} \tag{19.152}$$

where n is the current time step and i represents the position along the x axis. Expanding all the terms into a Taylor's series about u_i^n and substituting into the numerical scheme, one obtains

$$\frac{\partial u}{\partial t} - a\frac{\partial^2 u}{\partial x^2} = -\left(\frac{\partial^3 u}{\partial t^3}\right)_i^n \frac{dt^2}{6} + \frac{a}{12}\left(\frac{\partial^4 u}{\partial x^4}\right)_i^n$$

$$\cdot\, \Delta x^2 - a\left(\frac{\partial^2 u}{\partial t^2}\right)_i^n \left(\frac{dt}{dx}\right)^2 + \cdots \tag{19.153}$$

Thus, the leading order of the truncation error is $O(dx^2, dt^2, (dt/dx)^2)$. The numerical scheme tends to the exact differential equation only if dt tends to zero faster than dx. A numerical scheme is *consistent*, if the discretized equations tends to the PDE that it approximates when dt and dx tend to zero. According to the definition, the DuFort–Frankel method, when applied to the heat equation, is consistent only under the condition that dt goes to zero faster than dx.

We need to introduce one more concept related to the behavior of the error during the numerical solution. That is the concept of *stability*. A scheme is said to be stable when it does not permit errors to grow as the numerical solution proceeds. The last two concepts are related to the concept of convergence through *Lax's Equivalence*

Theorem [Lax and Richtmyer (1956) and Richtmyer and Morton (1967)]. The theorem states that for a properly posed initial value problem and a discrete representation that satisfies consistency, stability is the necessary and sufficient condition for convergence. The theorem strictly applies only to linear PDE's, but it seems to hold true for the majority of the cases that involve nonlinear ones, although no proof exists for these cases. By *convergence*, one means that the numerical solution tends to the exact solution for any given point in space and time as the mesh is refined.

There are several methods developed for the analysis of the stability of numerical schemes. Here, the one developed by von Neumann is presented in detail. The interested reader can find more information concerning those methods in the references at the end of the chapter. Let us start by noting that the exact solution, u_i^n, of the difference equation can be written as the sum of the actual numerical solution, $u_i'^n$, (obtained with a finite number of digits) and an error ϵ_i^n (round-off error, error in the initial data, etc.)

$$u_i^n = u_i'^n + \epsilon_i^n \tag{19.154}$$

Now, suppose we use the following numerical scheme for the heat equation

$$\frac{u_i^{n+1} - u_i^n}{\Delta t} = a \frac{u_{i+1}^n - 2u_i^n + u_{i-1}^n}{\Delta x^2} \tag{19.155}$$

Inserting Eq. (19.154) into Eq. (19.153) and noting that the exact solution should satisfy the difference equation, we obtain

$$\frac{\epsilon_i^{n+1} - \epsilon_i^n}{\Delta t} = a \frac{\epsilon_{i+1}^n - 2\epsilon_i^n + \epsilon_{i-1}^n}{\Delta x^2} \tag{19.156}$$

which indicates that the error satisfies the same difference equation that the exact solution satisfies. So, studying the growth of the error during the numerical solution is equivalent to studying the growth of the numerical solution. Suppose that the error in one-dimensional space can be written as a Fourier series

$$\epsilon(x, t) = \sum_n a_n(t) e^{ik_n x} \tag{19.157}$$

where the fundamental frequency ($n = 1$) corresponds to the maximum wavelength of $2L$. For a domain of length L, the wave number k_n can be expressed as follows

$$k_n = n \frac{\pi}{L}, \quad n = 0, 1, \ldots, N \tag{19.158}$$

where N is the number of uniform subdivisions, Δx, used to discretize the domain. For linear equations, since superposition can be used, examining one of the terms of the series is adequate for determining the behavior of the error. Each term can be expressed in the following form

$$\epsilon_j^{n+1} = b_j^{n+1} e^{i\beta j} \tag{19.159}$$

where b_j^{n+1} is the amplitude of the jth term at time $t + \Delta t$, and $\beta = k_j \Delta x$. A scheme is *stable* if the amplitude of every harmonic, b_j^{n+1}, does not grow in time, i.e.,

$$|G| = |b_j^{n+1}/b_j^n| \leq 1 \; \forall \beta \tag{19.160}$$

The factor G is called *amplification factor*, and it is a function of the time and space mesh sizes as well as the frequency. This analysis assumes that periodic boundary conditions have been imposed. Clearly, the method is not suitable for situations where the influence of the boundary conditions is important.

In the case of nonlinear governing equations or equations with non-constant coefficients, only local stability limits can be obtained by linearizing the formulation. This can be accomplished by *freezing* the nonlinear non-constant coefficients. We can now demonstrate the method by applying it to some model equations.

Model Hyperbolic Equation: Linear Convection Equation.

Seek a numerical representation of the equation

$$\frac{\partial u}{\partial t} + c \frac{\partial u}{\partial x} = 0 \tag{19.161}$$

Let us propose the following scheme

$$\frac{u_j^{n+1} - u_j^n}{dt} + c \frac{u_{j+1}^n - u_{j-1}^n}{2\,dx} = 0 \tag{19.162}$$

or by setting $\nu = c\,dt/dx$, one obtains

$$u_j^{n+1} = u_j^n - \frac{\nu}{2}(u_{j+1}^n - u_{j-1}^n) \tag{19.163}$$

To check the stability of the scheme, introduce Eq. (19.159) into the numerical scheme (the error satisfies the difference equation) and, after some algebra, obtain

$$\frac{b_j^{n+1}}{b_j^n} = G = 1 - \nu i \sin \beta \tag{19.164}$$

The amplification factor is equal to

$$|G|^2 = 1 + \nu^2 \sin^2 \beta \geq 1 \tag{19.165}$$

Thus, the errors will grow, and the scheme is *unconditionally unstable*.

Suppose now that we evaluate the spatial derivative at time $t = t^{n+1}$, i.e.,

$$\frac{u_j^{n+1} - u_j^n}{dt} + c \frac{u_{j+1}^{n+1} - u_{j-1}^{n+1}}{2\,dx} = 0 \tag{19.166}$$

Applying the von Neumann stability analysis, we obtain

$$G = \frac{1}{1 + i\nu \sin \beta} \tag{19.167}$$

so the amplification factor is equal to

$$|G|^2 = \frac{1}{1 + \nu^2 \sin^2 \beta} \leq 1 \tag{19.168}$$

The amplification factor is always smaller than unity, and the scheme is *unconditionally stable*. This means that there is no restriction concerning the selection of *dt* for satisfying stability. The stability of the scheme should not be confused with the accuracy. Although, the stability is not affected by an increase in *dt*, the accuracy of the scheme would decrease.

Finally, analyze the following scheme, where the spatial derivative is approximated by a backward difference

$$\frac{u_j^{n+1} - u_j^n}{dt} + c \frac{u_j^n - u_{j-1}^n}{dx} = 0 \tag{19.169}$$

Applying the von Neumann analysis, one obtains

$$G = 1 - \nu(1 - \cos \beta) - i\nu \sin \beta \tag{19.170}$$

The amplification factor is equal to

$$|G|^2 = 1 - 4\nu(1 - \nu) \sin^2 \frac{\beta}{2} \tag{19.171}$$

The requirement for stability ($|G| \leq 1$) is satisfied if $0 \leq \nu \leq 1$. This condition is called the Courant–Friedrichs–Levy (CFL) condition, and it sets a limit on the speed with which the numerical scheme propagates the information from one level to the other. The scheme is called *conditionally stable*, since disturbances will die out only for combinations of *dt* and *dx* that satisfy the CFL condition.

A Model Parabolic Equation: Linear Convection-Diffusion Equation. Consider the following equation

$$\frac{\partial u}{\partial t} + c \frac{\partial u}{\partial x} = \epsilon \frac{\partial^2 u}{\partial x^2} \tag{19.172}$$

and propose the following numerical representation

$$u_j^{n+1} = u_j^n - \frac{\nu}{2} (u_{j+1}^n - u_{j-1}^n) + \kappa(u_{j+1}^n - 2u_j^n + u_{j-1}^n) \tag{19.173}$$

Applying the von Neumann analysis, we obtain

$$G = 1 - i\nu \sin \beta - 2\kappa(1 - \cos \beta) \tag{19.174}$$

which leads to the following amplification factor

$$|G|^2 = (1 - 2\kappa(1 - \cos \beta))^2 + \nu^2 \sin^2 \beta \tag{19.175}$$

For $\beta = \pi$, the following condition must be satisfied for stability

$$(1 - 4\kappa)^2 \leq 1 \tag{19.176}$$

or $0 \leq \kappa \leq 1/2$. For $\beta = \pi/2$, the condition for stability requires that $(\nu^2/\kappa) \leq 2$. The reader can verify that the ratio of the two numbers, ν/κ, defines a *mesh Reynolds number*, which expresses the ratio of convection to diffusion. Depending on the relative magnitude of the two numbers that appears in the stability requirements, the behavior of the equation can be primarily of the hyperbolic type, the parabolic type or a mixture of the two. That can cause severe problems of accuracy for the numerical method.

This example concludes the presentation of the finite difference methods. A number of references are included at the end of the chapter. The attempt was to give the most important elements of the derivation and application of the finite difference method to the numerical solution of the PDEs. Extensive description and analysis of methods applied to the solution of the governing fluid flow equations can be found later in this Handbook.

19.3   FINITE ELEMENT METHODS
Tayfun E. Tezduyar, Marek Behr, and Thomas J. R. Hughes

19.3.1   Introduction

Finite element methods (FEM), with their capability of handling intricate geometries of practical interest, have become a significant area of computational fluid dynamics (CFD). The basic finite element procedure can be outlined as follows [see also Fletcher (1984)]:

i) *Conversion to a weak formulation.* The classical form of the boundary-value problem is converted to an equivalent weak, or variational, form. In this form, the order of differentiation is typically reduced and some of the boundary conditions appear in the integral form.

ii) *Selection of the function spaces.* The trial solution and weighting function spaces appearing in the weak formulation are infinite-dimensional. In a finite element method, they are approximated by finite-dimensional space. To construct these function spaces, the problem domain is discretized into subdomains (elements), and a simple functional form is assumed to approximate the solution and weighting functions on each element. The collection of element-level functions give rise to globally defined functions.

iii) *Solution of the discretized equations.* The finite-dimensional weighting function space is then used to produce a system of either algebraic, or ordinary differential, equations. Due to the local support of the function space bases, the coupling between equations is limited to neighboring elements. These equations are the solved either directly or iteratively.

In a variational formulation, selecting the weighting functions from the same class that the interpolation functions are selected from, leads to a *Galerkin formulation.* When applied to differential equation systems with symmetric operators, (e.g., diffusion equations, most structural and solid mechanics problems) Galerkin formulations produce solutions with a *best approximation* property. That is, the error is minimized with respect to a certain norm. For systems with nonsymmetric operators (e.g., most fluid dynamics problems), however the Galerkin formulation does not possess a best approximation property. This, in some cases, may result in solutions with spurious node-to-node oscillations. In fact, this problem is not limited to Galerkin formulations. It also arises for finite difference schemes when non-symmetric operators are approximated centrally.

Instead of using weighting functions which lead to a Galerkin formulation, one can employ a *Petrov–Galerkin formulation* by modifying, in a way, those weighting functions according to an optimal rule, leading to a series of additional element-level integrals in the variational formulation. The basic idea is to minimize the spurious oscillations without introducing excessive diffusion to the solution.

An optimal *Streamline–Upwind/Petrov–Galerkin* (SUPG) formulation for advection-dominated flows was developed in late 1970's by Hughes and Brooks (1979) and was successfully applied to the solution of advection-diffusion and viscous, incompressible flow problems. Similar ideas were applied to the first-order hyperbolic systems, particularly compressible Euler equations by Tezduyar and Hughes (1983).

This successful application of a stabilization technique which preserves the weighted residual character of the original Galerkin formulation was followed by similar treatment of many other shortcomings of the early applications of the Galerkin method. The advent of *stabilized* formulations has been one of the most important recent developments in the finite element CFD.

19.3.2 Stabilized Methods

Today, with the benefit of hindsight, it is easy to understand that the early disappointments with the Galerkin method as applied to fluid mechanics problems stemmed from two main sources of potential numerical instabilities.

One is due to the presence of advection terms in the governing equations, and its effect can appear as spurious node-to-node oscillations, primarily in the velocity field. This oscillatory behavior can rapidly lead to the divergence of the iterative algorithm used for solving the nonlinear equation system. The problem becomes particularly apparent for high Reynolds number, advection-dominated flows and flows which involve shocks or sharp gradients in the solution. The other source of the numerical instability is most often encountered in the incompressible flow modeling and is due to the incompatibility (discussed later in this section) of the interpolations used for the velocity and pressure fields. The oscillations can appear primarily in the pressure field. Several of the most convenient choices of the interpolations for the two fields, including the equal-order interpolation, would be vulnerable to this instability.

The approaches listed in the remainder of this section represent ways of neutralizing these and other difficulties encountered with the standard Galerkin method.

Streamline–Upwind/Petrov–Galerkin Method. The Streamline–Upwind/Petrov–Galerkin (SUPG) technique is designed to stabilize the Galerkin formulation in the presence of dominant advection terms. It was introduced and developed by Hughes and Brooks (1979), Brooks and Hughes (1982), and Tezduyar and Hughes (1983). The SUPG method relies on the addition to the Galerkin formulation of a series of element-level integrals, all of which have the residual of the momentum equation as a weighted term. Thus, the weighted residual character of the original formulation is preserved. There has been a growing appreciation of the SUPG qualities, i.e., stability and accuracy. The initial numerical appraisal of the method has been now firmly placed on mathematical foundations, developed by Johnson [see e.g., Johnson and Saranen (1986)], and others.

Pressure–Stabilizing/Petrov–Galerkin Method. The *Pressure–Stabilizing/Petrov–Galerkin* (PSPG) approach is a way of circumventing the compatibility condition, also known as the *Babuška–Brezzi condition*, which is imposed on the interpolation functions for velocity and pressure in incompressible flow models. Some of the most appealing elements that do not satisfy the compatibility condition are the $Q1P0$ (linear velocity, constant discontinuous pressure) and the $Q1P1$ (linearly velocity and pressure) elements. It was shown by Brezzi and Pitkäranta (1984) and Hughes *et al.* (1986a) that these convenient elements can be used to solve the Stokes problem, provided a proper stabilization is used. This stabilization again includes a series of element-level integrals which do not disturb the weighted residual properties of the Galerkin formulation. The PSPG stabilization, introduced by Tezduyar *et al.* (1992d), is the generalization of that concept to Navier–Stokes equations.

Galerkin/Least-Squares Method. The *Galerkin/Least-Squares* (GLS) stabilization combines, in a conceptually simple form, the SUPG and PSPG techniques. This approach has been introduced for Stokes problems by Hughes and Franca (1987), for compressible flows by Hughes *et al.* (1989) and for incompressible, Navier–Stokes problems by Hansbo and Szepessy (1990). In GLS, the added stabilizing terms include the sum of the variations of the squared residual of the governing equations integrated over the element domains.

Discontinuity Capturing. Even the stabilizing methods listed above, while providing robustness, do not entirely eliminate oscillatory behavior which is strictly local to the discontinuities and sharp layers in the solution. Control over gradients in those areas can be provided by nonlinear discontinuity capturing (DC) operators. These operators were developed for the advection–diffusion equations by Hughes *et al.* (1986d), for advection-diffusion-reaction equations by Tezduyar and Park (1986), and by Hughes and Mallet (1986c) for the formulation of compressible flows based on the entropy variables (see the next subsection). Subsequently, an equivalent stabilization was given by LeBeau and Tezduyar (1991) for the formulation of compressible flows based on the conservation variables.

Entropy Variables. To obtain the properties needed to establish stability proofs and accuracy estimates, the conservation variables in the standard formulations of compressible Navier–Stokes equations have been converted to entropy variables (see Hughes *et al.* (1986b)). This transformation symmetrizes the Navier–Stokes equations and makes their Galerkin formulation dimensionally consistent. It is this framework that allowed the design of the nonlinear DC operators, which were only later adapted for use with the conservation variables.

19.3.3 Methods for Deforming Domains

Another important development has been the growing interest in the modeling of problems which involve moving boundaries and interfaces, i.e., problems in which the domain taken up by the fluid is deforming. A factor in the increasing application of finite element methods to such problems has been the popularity of an *Arbitrary Lagrangian–Eulerian* (ALE) approach [see Hughes *et al.* (1981)]. More recently, space–time methods further extended the capability to deal with problems involving deforming domains.

Space–Time Methods. The space–time methods for problems involving fixed domains have been developed and extensively analyzed, mainly by Johnson *et al.* (1984), Hansbo and Szepessy (1990), and Hughes and Hulbert (1988). The potential of the space–time treatment to handle problems with moving boundaries has been exploited in the *Deforming-Spatial-Domain/Stabilized-Space–Time* (DSD/SST) method developed by Tezduyar *et al.* (1992b, 1992c) and Hansbo (1992). In the space–time approach, both the spatial and temporal domains of the problem are discretized using finite elements. The deformation of the spatial domain is reflected simply in the deformation of the elements of the space-time mesh.

Mesh Moving Strategies. The space–time methods give substantial freedom as far as movement of the spatial mesh is concerned. This movement may be constrained at the mesh boundaries, e.g., at a free surface, the mesh boundary should move in such a way that the flux through that boundary vanishes. On the other hand, the movement of the mesh in the interior of the domain is largely arbitrary. In the *Characteristic Streamline Diffusion* (CSD) method [see Hansbo (1992)] the movement of the nodes in the mesh is governed by the local fluid velocity field, i.e., the nodes move along the flow characteristic lines. At each time step, this is followed by remeshing which is needed to recover the acceptable shape of the finite elements.

In order to avoid remeshing and associated numerical diffusion, other mesh moving schemes can be developed. If the deformation of the domain is somewhat predictable, the original mesh can be generated in such a way that the displacement of the interior modes is expressed explicitly as an algebraic function of the node displacement at the boundary. For more general deformations, an automatic mesh moving scheme such as the one described by Johnson and Tezduyar (1994) can be employed. In an automatic mesh moving scheme, the motion of the interior nodes is governed by the equations of linear elasticity, with the boundary conditions determined by the motion of the boundary nodes. The fictitious elasticity coefficients in the elasticity equations is typically adjusted to decrease the deformation of small elements at the expense of the large ones.

19.3.4 Parallel Computations

In recent years, the computation speed delivered by a single processing unit has begun to stagnate, and so did the performance of the traditional scalar and vector supercomputers. In contrast, massively parallel computers compensate for the limited speed of a single processing node by linking hundreds or thousands of nodes together with fast communication channels. With this approach, computational speeds on the order of trillion floating point operations per second (TeraFLOPS) are within grasp, with sustained rates of 10–50 GigaFLOPS achievable now.

To take advantage of these new capabilities, however, algorithm design and programming techniques have to be extensively modified. In the future, this problem may be alleviated by advanced parallelizing compilers. But for the moment, special programming approaches have to be employed, and some of them are outlined in the following subsections. Note that these techniques are relevant to finite element implementations which rely on solution of coupled systems of equations on general unstructured meshes.

Generalized Minimum Residual Iterative Solver. The difficulty of efficiently implementing a direct solver on a distributed memory machine, as well as the rapid increase in the computational cost of a direct method with the problem size, fuels the popularity of iterative solvers as a mean of solution of the large, coupled systems of equations arising from implicit finite element implementations. Among these, the *Generalized Minimum Residual* (GMRES) solver introduced by Saad and Schultz (1986), is the method of choice for nonsymmetric systems stemming from the Navier–Stokes or Euler equations. In the GMRES procedure, the full system is projected onto a much smaller Krylov subspace, and the residual is minimized on that subspace. The size of the Krylov space required for convergence is usually smaller than 15 for compressible flow problems [see Shakib *et al.* (1989)]. This requirement increases for incompressible flow problems and finer meshes.

Element-by-Element Storage. In iterative algorithms such as the GMRES method, the interaction with the system matrix takes place exclusively through matrix–vector product. Thus, the computation of the matrix-vector product is possible directly from the element-level matrices, without the global sparse matrix ever being assembled. This element-by-element storage mode is also sufficient to construct the element-by-element (EBE) preconditioner [see Hughes *et al.* (1983)]. Employing the element-by-element storage, the matrix-vector product $\mathbf{A}\bar{\mathbf{x}}$ is actually computed as $\mathbf{A}\,(\mathbf{a}^e\bar{\mathbf{x}}^e)$, using the fact that $\mathbf{A} = \mathbf{A}\,\mathbf{a}^e$, where $\mathbf{a}^e$ are the element-level matrices, $\bar{\mathbf{x}}^e$ is the vector $\bar{\mathbf{x}}$ distributed (gathered) to the element level, and $\mathbf{A}$ represents the finite element assembly operator.

Matrix-Free Techniques. The matrix-vector product can be also approximated by two residual evaluations, eliminating the need to store even the element-level matrices and resulting in substantial memory savings. This approach, known as *matrix-free algorithm*, is described, e.g., in Johan *et al.* (1991). Here, the matrix-vector product $\mathbf{A}(\bar{\mathbf{u}})\bar{\mathbf{x}}$ can be written as $(\mathbf{R}(\bar{\mathbf{u}} + \epsilon\bar{\mathbf{x}}) - \mathbf{R}(\bar{\mathbf{u}}))/\epsilon$, where ϵ is a suitably small number. In the matrix-free computations, the complex analytical task of finding the Jacobian matrix $\mathbf{A}(\bar{\mathbf{u}})$ of the nonlinear system is also avoided.

Domain Decomposition. A good quality decomposition of the finite element domain among the processing nodes of a multi-processor computer is a necessary requirement for all message-passing parallel implementations, and it is also very advantageous in data parallel codes. In the latter, the domain decomposition serves to minimize interprocessor communication between various sets of data, as it is done in the 2-step gather/scatter algorithm developed for Connection Machine computers by Johan *et al.* (1994). Effective techniques have been developed in order to obtain a good decomposition of a general, unstructured finite element mesh. A *recursive spectral bisection* (RSB) technique developed by Pothen *et al.* (1990), has compared favorably with other techniques and has been implemented on a data parallel computer by Johan (1992).

Preconditioning. At the moment, the choice of preconditioning for the iterative solver on parallel machines is limited. The diagonal and nodal block-diagonal preconditioning [see, e.g., Shakib (1988)] work well for compressible flow problems. In incompressible flow simulations, diagonal preconditioning performs adequately, but parallel implementations of more complex preconditioners, such as the *clustered-element-by-element* (CEBE) preconditioner and its derivatives [see Tezduyar *et al.* (1992a)], are also clearly needed.

19.3.5 Governing Equations

In this section, we state the problem in the form of first the incompressible, and then compressible, Navier–Stokes equations. In the following, $\Omega_t \subset R^{nsd}$ will denote a bounded region at time $t \in (0, T)$, with boundary Γ_t, where n_{sd} is the number of space dimensions. The time index indicates that the domain may be deforming. The symbols $\rho(\bar{\mathbf{x}}, t)$, $\bar{\mathbf{u}}(\bar{\mathbf{x}}, t)$, $p(\bar{\mathbf{x}}, t)$ and $e(\bar{\mathbf{x}}, t)$ will represent the density, velocity, pressure and the total (internal plus kinetic) energy fields, respectively. The external forces, such as the gravity, will be represented by $\mathbf{f}(\bar{\mathbf{x}}, t)$.

Incompressible Flows. The Navier–Stokes equations for incompressible flows are

$$\rho\left(\frac{\partial \bar{\mathbf{u}}}{\partial t} + \bar{\mathbf{u}} \cdot \nabla \bar{\mathbf{u}} - \bar{\mathbf{f}}\right) - \nabla \cdot \bar{\bar{\sigma}} = \mathbf{0} \quad \text{on } \Omega_t \quad \forall t \in (0, T) \quad (19.177)$$

$$\nabla \cdot \bar{\mathbf{u}} = 0 \quad \text{on } \Omega_t \quad \forall t \in (0, T) \quad (19.178)$$

where ρ is assumed to be constant. For the Newtonian flows under consideration here, the stress tensor for a fluid with dynamic viscosity μ is defined as follows

$$\bar{\bar{\sigma}} = -p\bar{\bar{\mathbf{I}}} + \bar{\bar{\tau}}, \quad \bar{\bar{\tau}} = 2\mu\bar{\bar{\varepsilon}}(\bar{\mathbf{u}}) \quad (19.179)$$

where $\bar{\bar{\tau}}$ is the Newtonian stress-tensor and $\bar{\bar{\varepsilon}}$ is the strain rate tensor. This equation set is completed by suitable boundary conditions and an initial condition consisting of a divergence-free velocity field specified over the entire domain

$$\bar{\mathbf{u}}(\bar{\mathbf{x}}, 0) = \bar{\mathbf{u}}_0, \quad \nabla \cdot \bar{\mathbf{u}}_0 = 0 \quad \text{on } \Omega_0 \quad (19.180)$$

Compressible Flows. The Navier–Stokes equations for compressible flows can be written in the vector (or array) form

$$\frac{\partial \mathbf{U}}{\partial t} + \frac{\partial \mathbf{F}_i}{\partial x_i} - \frac{\partial \mathbf{E}_i}{\partial x_i} = \mathbf{0} \quad \text{on } \Omega_t, \quad \forall t \in (0, T) \tag{19.181}$$

where $\mathbf{U} = (\rho, \rho u_1, \rho u_2, \rho u_3, \rho e)$, is the vector of conservation variables, $\mathbf{F}_i$ and $\mathbf{E}_i$ are, respectively, the Euler and viscous flux vectors defined as

$$\mathbf{F}_i = \begin{pmatrix} u_i \rho \\ u_i \rho u_1 + \delta_{i1} p \\ u_i \rho u_2 + \delta_{i2} p \\ u_i \rho u_3 + \delta_{i3} p \\ u_i(\rho e + p) \end{pmatrix} \tag{19.182}$$

$$\mathbf{E}_i = \begin{pmatrix} 0 \\ \tau_{i1} \\ \tau_{i2} \\ \tau_{i3} \\ -q_i + \tau_{ik} u_k \end{pmatrix} \tag{19.183}$$

and q_i are the components of the heat flux. Here, the equation of state is modeled with the ideal gas equation.

Alternatively, Eq. (19.181) can be written in the following form

$$\frac{\partial \mathbf{U}}{\partial t} + \mathbf{A}_i \frac{\partial \mathbf{U}}{\partial x_i} - \frac{\partial}{\partial x_i} \left(\mathbf{K}_{ij} \frac{\partial \mathbf{U}}{\partial x_j} \right) = \mathbf{0} \quad \text{on } \Omega_t, \quad \forall t \in (0, T) \tag{19.184}$$

where

$$\mathbf{A}_i = \frac{\partial \mathbf{F}_i}{\partial \mathbf{U}} \tag{19.185}$$

$$\mathbf{K}_{ij} \frac{\partial \mathbf{U}}{\partial x_j} = \mathbf{E}_i \tag{19.186}$$

It is assumed that appropriate sets of boundary and initial conditions are specified with Eq. (19.184).

19.3.6   Finite Element Formulations

In a space-time formulation, the space-time domain is first divided into a sequence of space-time slabs Q_n, and each slab is decomposed into space-time elements Q_n^e. A slab Q_n is located between the time levels t_n and t_{n+1}. The integration of a functional over a slab will include integration over both the spatial domain Ω_t and the temporal one (t_n, t_{n-1}). Since many of the functions introduced in the following subsection will be discontinuous across slab interfaces, we will employ the notation $(\cdot)_n^-$ and $(\cdot)_n^+$ to indicate the values at t_n as it is approached from below and above, respectively. The number of elements in slab n is written as $(n_{\text{el}})_n$.

Incompressible Flows. The finite element formulation begins with choosing appropriate trial solution $[(\mathcal{S}_{\mathbf{u}}^h)_n$ and $(\mathcal{S}_p^h)_n]$ and weighting function $[(\mathcal{V}_{\mathbf{u}}^h)_n$ and $(\mathcal{V}_p^h)_n = (\mathcal{S}_p^h)_n]$ spaces for velocity and pressure. In our computations, we employ piecewise linear functions for all fields.

The stabilized space-time formulation for deforming domains can be written as follows: given $(\bar{\mathbf{u}}^h)_n^-$, find $\bar{\mathbf{u}}^h \in (\mathcal{S}_u^h)_n$ and $p^h \in (\mathcal{S}_p^h)_n$ such that $\forall \bar{\mathbf{w}}^h \in (\mathcal{V}_u^h)_n$ and $\forall q^h \in (\mathcal{V}_p^h)_n$

$$
\int_{Q_n} \bar{\mathbf{w}}^h \cdot \rho \left(\frac{\partial \bar{\mathbf{u}}^h}{\partial t} + \bar{\mathbf{u}}^h \cdot \nabla \bar{\mathbf{u}}^h - \bar{\mathbf{f}} \right) dQ + \int_{Q_n} \bar{\bar{\varepsilon}}(\bar{\mathbf{w}}^h) : \bar{\bar{\sigma}}(p^h, \bar{\mathbf{u}}^h) \, dQ
$$

$$
+ \int_{Q_n} q^h \nabla \cdot \bar{\mathbf{u}}^h \, dQ + \int_{\Omega_n} (\bar{\mathbf{w}}^h)_n^+ + \rho((\bar{\mathbf{u}}^h)_n^+ - (\bar{\mathbf{u}}^h)_n^-) \, d\Omega
$$

$$
+ \sum_{e=1}^{(n_{\text{el}})_n} \int_{Q_n^e} \tau_{\text{MOM}} \frac{1}{\rho} \left[\rho \left(\frac{\partial \bar{\mathbf{w}}^h}{\partial t} + \bar{\mathbf{u}}^h \cdot \nabla \bar{\mathbf{w}}^h \right) - \nabla \cdot \bar{\bar{\sigma}}(q^h, \bar{\mathbf{w}}^h) \right]
$$

$$
\cdot \left[\rho \left(\frac{\partial \bar{\mathbf{u}}^h}{\partial t} + \bar{\mathbf{u}}^h \cdot \nabla \bar{\mathbf{u}}^h - \bar{\mathbf{f}} \right) - \nabla \cdot \bar{\bar{\sigma}}(p^h, \bar{\mathbf{u}}^h) \right] dQ
$$

$$
+ \sum_{e=1}^{(n_{\text{el}})_n} \int_{Q_n^e} \tau_{\text{CONT}} \nabla \cdot \bar{\mathbf{w}}^h \rho \nabla \cdot \bar{\mathbf{u}}^h \, dQ = \int_{(P_n)_\mathbf{h}} \bar{\mathbf{w}}^h \cdot \bar{h}^h \, dP \qquad (19.187)
$$

Here $\bar{h}^h$ represents the Neumann boundary condition imposed, $(P_n)_h$ is the part of the slab boundary with such conditions, and τ_{MOM} and τ_{CONT} are the stabilization parameters. The solution to Eq. (19.187) is obtained sequentially for all space-time slabs $Q_1, Q_2, \ldots, Q_{N-1}$, and the computations start with

$$
(\bar{\mathbf{u}}^h)_0^- = \bar{\mathbf{u}}_0^h \qquad (19.188)
$$

The deformation of the mesh is reflected in the deformation of space-time elements and is automatically accounted for when computing the transport terms. In the formulation given by Eq. (19.187), the first four integrals, together with the right-hand side, represent the time-discontinuous Galerkin formulation of Eqs. (19.177)–(19.178). The fourth integral enforces, weakly, the continuity of the velocity field in time. The two series of element-level integrals in the formulation are the least-squares stabilization terms. The reader can refer to Tezduyar *et al.* (1992b, 1992c) and Behr and Tezduyar (1994) for further details regarding the space–time formulation for incompressible flows, including definitions of the stabilization parameters. For problems not involving moving boundaries and interfaces, a semidiscrete formulation derived from the Eq. (19.187) can be written by dropping the fourth integral and the term $\partial \bar{\mathbf{w}}^h / \partial t$, and by converting all space-time integrations to spatial integrations.

Compressible Flows. In the finite element formulation of compressible flows, we define the function spaces $\mathcal{S}_n^h$ and $\mathcal{V}_n^h$ corresponding to the trial solutions and weighting functions, respectively. Again, we use first-order polynomials as interpolation functions. Globally, these functions are continuous in space but discontinuous in time.

The DSD/SST formulation of Eq. (19.184) can be written as follows: given $(\mathbf{U}^h)_n^-$, find $\mathbf{U}^h \in \mathcal{S}_n^h$ such that $\forall \mathbf{W}^h \in \mathcal{V}_n^h$

$$
\int_{Q_n} \mathbf{W}^h \cdot \left(\frac{\partial \mathbf{U}^h}{\partial t} + \mathbf{A}_i^h \frac{\mathbf{U}^h}{\partial x_i} \right) dQ + \int_{Q_n} \left(\frac{\partial \mathbf{W}^h}{\partial x_i} \right) \cdot \left(\mathbf{K}_{ij}^h \frac{\partial \mathbf{U}^h}{\partial x_j} \right) dQ
$$

$$
+ \int_{\Omega_n} (\mathbf{W}^h)_n^+ \cdot ((\mathbf{U}^h)_n^+ - (\mathbf{U}^h)_n^-) \, d\Omega
$$

$$
+ \sum_{e=1}^{(n_{el})_n} \int_{Q_n^e} \tau (\mathbf{A}_k^h)^T \left(\frac{\partial \mathbf{W}^h}{\partial x_k} \right) \cdot \left[\frac{\partial \mathbf{U}^h}{\partial t} + \mathbf{A}_i^h \frac{\mathbf{U}^h}{\partial x_i} - \frac{\partial}{x_i} \left(\mathbf{K}_{ij}^h \frac{\partial \mathbf{U}^h}{\partial x_j} \right) \right] dQ
$$

$$
+ \sum_{e=1}^{(n_{el})_n} \int_{Q_n^e} \delta \left(\frac{\partial \mathbf{W}^h}{\partial x_i} \right) \cdot \left(\frac{\partial \mathbf{U}^h}{\partial x_i} \right) dQ = \int_{(P_n)_h} \mathbf{W}^h \cdot \mathbf{H}^h \, dP \qquad (19.189)
$$

The solution to (13) is obtained sequentially for $Q_1, Q_2, \ldots, Q_{N-1}$, commencing with

$$
(\mathbf{U}^h)_0^- = \mathbf{U}_0^h \qquad (19.190)
$$

where $\mathbf{U}_0$ is the initial value of the vector $\mathbf{U}$.

In the formulation shown in Eq. (19.189), the first three integrals, together with the right-hand side, represent the time-discontinuous Galerkin formulation of Eq. (19.184). The third integral enforces, weakly, the continuity of the conservation variables in time. The first series of element-level integrals are the SUPG stabilization terms, and the second series are the shock-capturing terms added to the formulation. The definition of τ and δ are given in Aliabadi and Tezduyar (1993). For problems not involving moving boundaries and interfaces, a semidiscrete formulation derived from Eq. (19.184) can be written by dropping the third integral, and by converting all space-time integrations to spatial integrations.

19.3.7 Examples

In this section, we describe two numerical examples that illustrate the power of FEM. In both problems, we employ a GMRES iterative solver. All computations were carried out on the Thinking Machines Corporation Connection Machines CM-5 and CM-200.

First we simulate flow around a Los Angeles-class submarine. We use the stabilized semidiscrete velocity-pressure formulation. This example is especially demonstrative of the ability of finite element methods to handle complex geometries. The finite element mesh was generated with the aid of a finite octree mesh generator developed by Shephard and Georges (1991). The Reynolds number based on the free-stream velocity and the hull length is 1×10^6. A Smagorinsky turbulence model

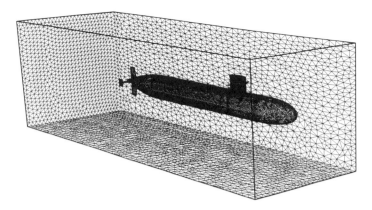

FIGURE 19.25 Flow around a submarine. Surface of the submarine mesh.

[see Kato and Ikegawa (1991)] was used in this unsteady computation, which was restarted from a steady-state solution at Reynolds number 1×10^5. This problem is described in more detail in Kennedy *et al.* (1994). We employed a spatial mesh consisting of 86,111 nodes and 428,157 tetrahedral elements. This has resulted in 293,642 equations solved using GMRES iterations. Figure 19.25 shows selected surfaces of the finite element mesh, while Fig. 19.26 shows the pressure field on the submarine hull. This problem was computed on the CM-5.

Next consider simulation of the air intake of a jet engine at Mach 2 and Reynolds number 0.8 million. This axisymmetric computation demonstrates the potential of the DSD/SST formulation to model intricate compressible flows involving interactions between boundary layers, shocks and moving surfaces. This type of flow is encountered in the air intake of a jet engine with adjustable spool. The efficiency of these engines at supersonic speeds can be improved by moving the spool back and forth and thus adapting the shock. The free-stream Mach number is 2, and the Reynolds number based on the free-stream values and the gap size is 0.8 million. We use 48,450 quadrilateral elements and 49,091 nodes. At each time step, 386,974 nonlinear equations are solved simultaneously. Computations start with the free-stream values as an initial condition, and after a prescribed period of time the spool starts to move forward.

The mesh moving strategy used for this problem is such that the connectivity of the mesh remains unchanged throughout the simulation. This eliminates the projec-

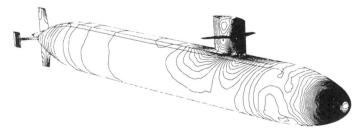

FIGURE 19.26 Flow around a submarine. Pressure distribution on the submarine surface.

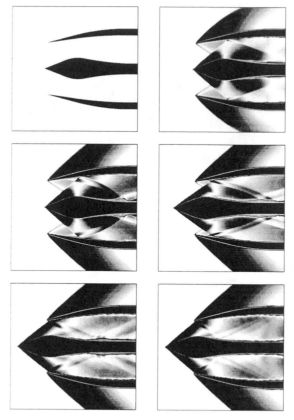

FIGURE 19.27 Axisymmetric simulation of a jet engine with adjustable spool. The images show the Mach number at six different instants during the motion of the spool.

tion errors associated with remeshing and also eliminates the parallelization overhead associated with remeshing. The images in Fig. 19.27 show the Mach number at six different instants during the motion of the spool. A full problem description and discussion can be found in Aliabadi and Tezduyar (1994). This simulation was carried out on the CM-200.

19.4 FINITE VOLUME METHODS
James L. Thomas and W. Kyle Anderson

The *finite volume method* (FVM) of solving the fluid dynamic equations is derived from the integral form of the conservation law equations,

$$\frac{d}{dt} \iiint_V \mathbf{w} \, dV + \iint_S \bar{\mathbf{f}} \cdot \bar{\mathbf{n}} \, dS \tag{19.191}$$

where $\bar{\mathbf{n}}$ is the unit normal outward from a surface S bounding a volume V, and $\bar{\mathbf{f}}$ is the net flux across a cell. The vector (or array) $\mathbf{w}$ represents density, Cartesian mass flux, and total energy per unit volume; the equations represent conservation of mass, momentum, and energy. The net flux is composed generally of convective, shear stress, and heat transfer terms. Using this integral approximation, numerical formulations can be determined by arbitrary divisions of the computational domain into discrete cells. The computational cell in a generalized coordinate transformation, or structured-grid, approach corresponds to a hexahedral (quadrilateral) cell in three (two) dimensions defined by coordinates of constant generalized coordinates. The finite volume method is equivalent to a finite difference (see Sec. 19.2) solution to the strong conservation law form of the governing equations, subject to proper interpretation of the metric derivatives. The generalized coordinate approach is illustrated in Fig. 19.28(a), in which a series of body-fitted block-structured grids are shown

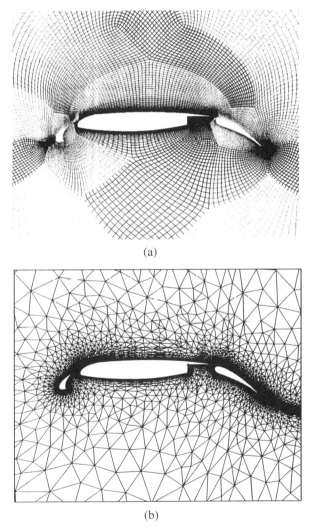

(a)

(b)

FIGURE 19.28 Generalized coordinates for a multi-element airfoil. (a) Structured grids, (b) unstructured grids, (c) Mach number contours, and (d) generalized coordinate decomposition for a hexahedral cell.

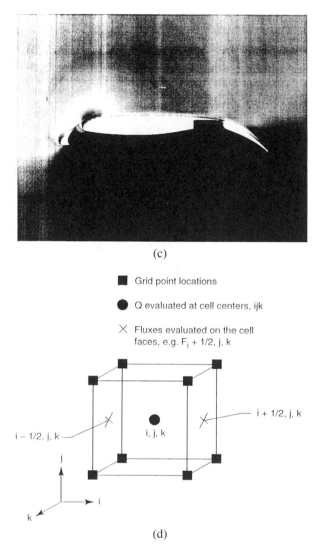

(c)

■ Grid point locations

● Q evaluated at cell centers, ijk

✕ Fluxes evaluated on the cell
faces, e.g. $F_{i + 1/2, j, k}$

(d)

FIGURE 19.28 *(Continued)*

for a multi-element two-dimensional airfoil. This approach has the distinct advantage that the body-fitted grids can be stretched in the direction normal to the body, allowing for a excellent numerical representation of the thin viscous layers occurring at high Reynolds numbers. Also, a number of different implicit schemes have been developed and are in common use which rely on the underlying structure of the generalized coordinate transformation. The disadvantage of this approach is that the grids for complex geometry can be very labor intensive to construct.

Numerical schemes which rely on arbitrary arrangements of cells are termed *unstructured grid* (see Sec. 19.1) methods and have their roots in the finite element area (see Sec. 19.3). The most common arrangements are tetrahedrons (triangles) in three (two) dimensions. A typical two-dimensional unstructured grid is shown in

Fig. 19.28(b), for the same airfoil shown previously. A flow solution showing Mach number contours at low speed for this multi-element airfoil is shown in Fig. 19.28(c). The unstructured grid methods are increasing in popularity, because the grid generation task is easier and locally adaptive methods are much easier to implement. In an adapted grid method, the grids are locally clustered to resolve important flow features in order to distribute the local truncation error uniformly over the mesh and thus minimize the computational work (see Sec. 19.7). The drawback to these methods is that additional bookkeeping must be done to keep track of the connections between the arbitrary arrangements of cells.

Using the integral formulation in Eq. (19.191), the time evolution of the average value of the flow variable at the center of the element, defined by the surrounding grid points, is determined by a flux balance across the bounding surfaces. This process is illustrated in Fig. 19.28(d) for a structured-grid (generalized coordinate) approach. The defining points for the grid are constructed by a transformation from Cartesian coordinates to generalized coordinates, defined here as i, j, k indices. The fluxes at the cell faces in the i-direction are illustrated; the average flow variable is at the center of the hexahedral cell defined by the eight surrounding grid points. The basic aspects of one class of time-stepping schemes are described below for finite volume schemes. The accuracy at steady state of such schemes is determined by the type of spatial differencing used; a number of different discretizations are described with special attention paid to the numerical errors that arise in practice which must be small for consistent engineering calculations. The last section shows some of the ideas and approaches for determining appropriate physical and numerically stable boundary conditions. The subsequent discussion is directed towards finite volume methods, but it also applies equally well to most finite difference methods.

19.4.1 Semi-Discrete Time Advancement Schemes

The time discretization methods used to solve the Euler equations can be typed into two classes: 1) coupled space-time methods, such as the method of MacCormack (1969), and 2) semidiscrete algorithms. In the latter approach, the spatial discretization is decoupled from the temporal discretization by first differencing the spatial derivative terms; the partial differential equations are thus transformed to a system of first-order ordinary differential equations in time. For steady flows, the time-rate-of-change of the spatial residual equation can be driven to zero, and the resulting physical solution is independent of the particular path taken to convergence or the time step used to advance the equations. As a consequence, the solution is only dependent on the spatial differencing approximation.

Explicit and implicit methods are discussed below for this class of semidiscrete algorithms. For a diffusion dominated flow, the allowable explicit time step scales on the square of the distance between the grid points; for the highly clustered grids required for the resolution of viscous flows at high Reynolds numbers, the maximum allowable time step of purely explicit schemes is prohibitively small. Implicit schemes have a less restrictive time step limitation and are generally more versatile and efficient, especially for time dependent computations. However, the implicit schemes entail more arithmetic operations, since the solution of a coupled system of equations at each time step is generally required.

Explicit Runge–Kutta Schemes. Time stepping procedures can be based upon the schemes available in the literature for the solution of ordinary differential equations. Single step methods can be used based upon Euler and Taylor series methods. The Runge–Kutta method is commonly used for time advancement of both the inviscid and viscous compressible equations, although for the latter especially, convergence acceleration methods must be used to overcome the explicit stability limitations of the scheme. The discretization of spatial operators in the equations leads to a system of coupled ordinary differential equations of the form

$$\frac{d\mathbf{w}}{dt} + \mathbf{R}(\mathbf{w}) = 0 \tag{19.192}$$

where $\mathbf{R}(\mathbf{w})$ is the vector of the residuals corresponding to the discrete evaluation of the spatial derivative terms. Let $\mathbf{w}^n$ be the numerical result after n time steps; the formulation to advance the solution to time level $n + 1$ is

$$\mathbf{w}^{(0)} = \mathbf{w}^n$$
$$\mathbf{w}^{(1)} = \mathbf{w}^{(0)} - \alpha_1 \Delta t \, \mathbf{R}^{(0)}$$
$$\cdots$$
$$\mathbf{w}^{(m-1)} = \mathbf{w}^{(0)} - \alpha_{m-1} \Delta t \, \mathbf{R}^{(m-2)}$$
$$\mathbf{w}^{(m)} = \mathbf{w}^{(0)} - \Delta t \, \mathbf{R}^{(m-1)}$$
$$\mathbf{w}^{n+1} = \mathbf{w}^{(m)} \tag{19.193}$$

where Δt is the time step. The residual in the $q + 1$ stage is evaluated as

$$\mathbf{R}^{(q)} = \sum_{r=0}^{q} \beta_{qr} \mathbf{R}(\mathbf{w}^{(r)}); \quad \sum_{r=0}^{q} \beta_{qr} = 1 \tag{19.194}$$

Typically, 3, 4, or 5 stage schemes have been used. The number of stages and the Runge–Kutta parameters can be varied to reduce the required storage, improve the temporal accuracy, or improve the high-frequency damping of the scheme in multigrid applications. The scheme is conditionally stable and, without alteration, is stable for a Courant (CFL) number of 2.8 for purely inviscid simulations for the most commonly used 4-stage scheme. It is straightforward to construct the stability region for which the scheme is stable. Figure 19.29 shows the stability region for commonly used Runge-Kutta methods. The time step is selected so that the eigenvalues of the Jacobian matrices of the linearized equations lie within the stability region. The eigenvalues for purely inviscid simulations lie along the imaginary axis, and those for purely viscous simulations lie along the negative real axis. The stage parameters can be adjusted to optimize the stability and/or damping given the eigenvalue range of the spatial operators. A number of possibilities have been exploited by Jameson (1985a).

Implicit Schemes. The prototype implicit algorithm considered is the backward time differencing equation, given as

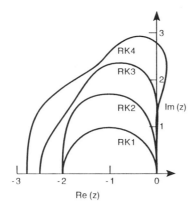

FIGURE 19.29 Stability regions for explicit Runge–Kutta methods.

$$\frac{\Delta\mathbf{w}^n}{\Delta t} + \mathbf{R}^{n+1} = 0 \qquad (19.195)$$

where $\mathbf{R}^{n+1}$ is the discrete residual evaluated at time level $(n + 1)\,\Delta t$, and $\Delta\mathbf{w}^n \equiv \mathbf{w}^{n+1} - \mathbf{w}^n$ is the change in the dependent variables over a time step. The scheme is first-order accurate in time. The *Trapezoidal scheme* can be written as

$$\frac{\Delta\mathbf{w}^n}{\Delta t} + \frac{1}{2}\,[\mathbf{R}^n + \mathbf{R}^{n+1}] = 0 \qquad (19.196)$$

and the three-point backward time scheme can be written as

$$\frac{1}{\Delta t}\left[\frac{3}{2}\,\Delta\mathbf{w}^n - \frac{1}{2}\,\Delta\mathbf{w}^{n-1}\right] + \mathbf{R}^{n+1} = 0 \qquad (19.197)$$

both of which are second-order accurate in time. The three schemes above can be considered as examples from the well-known class of linear multistep methods developed for solving ordinary differential equations. The stability of such methods can be determined from an analysis of the eigenvalues of the coefficient matrix arising from linearization of the nonlinear terms. For discrete solutions to the Euler equations, these eigenvalues generally lie in the left half of the complex plane. A method without any time step stability limit in such a case is referred to as an *A-stable method.* It is known from a theorem of Dahlquist that:

i) The order of an A-stable method cannot exceed two.
ii) The second-order A-stable scheme with lowest truncation error is the Trapezoidal scheme.

Generally, the three-point backward time scheme is preferred for second-order accuracy, since the Trapezoidal method, also known as Crank–Nicholson, is susceptible to an odd-even decoupling in time of the highest frequencies in the solution.

Either Eq. (19.195) or (19.197) represents a nonlinear system of equations to be solved at each time step and can be written generically as

$$\hat{\mathbf{R}} = 0 \qquad (19.198)$$

where the hatted notation denotes that the residual contains both temporal and spatial discretization terms. Applying Newton's method for the root of a nonlinear system of equations gives a linear system to be solved iteratively,

$$\hat{\mathbf{R}}^{i+1} \approx \hat{\mathbf{R}}^i + \left[\frac{\partial \hat{\mathbf{R}}}{\partial \mathbf{w}}\right]^i (\mathbf{w}^{i+1} - \mathbf{w}^i) \equiv 0, \quad i = 1, 2, 3, \ldots \quad (19.199)$$

The linearization is about an estimate $\mathbf{w}^i$, which can be taken initially as $\mathbf{w}^n$, and the solution converges to the solution at the new time level $\mathbf{w}^{n+1}$. The requirement to solve a linear equation arises from the linearization of the nonlinear spatial residual terms at the new time level. In that respect, the treatment of second-order accuracy in time is similar to that of the first-order scheme, since the additional terms involved are all evaluated at time levels which have already been computed. Hence, restricting the discussion below to the first-order backward time scheme, the linear system is written as

$$\left[\mathbf{I} + h\frac{\partial \mathbf{R}}{\partial \mathbf{w}}\right]^i (\mathbf{w}^{i+1} - \mathbf{w}^i) = -[\mathbf{w}^i - \mathbf{w}^n + h\mathbf{R}(\mathbf{w}^i)],$$

$$i = 1, 2, 3, \ldots \quad (19.200)$$

where $h = \Delta t$. For each iteration, Eq. (19.200) requires the solution of a block-banded linear system of equations with the property that quadratic convergence can be attained at each iteration if the approximation is sufficiently near the root of the equation. For large time steps, Newton's method is recovered for the solution of the steady-state residual equation. A general block-matrix equation solved with *Gauss elimination* requires

$$O(N^3 m^3) \text{ operations} \quad (19.201)$$

where N is the total number of equations and m is the block size. A general block-banded matrix equation can be solved in

$$O(Nb^2 m^3) \text{ operations}; \quad b = \max{(p, q)} \quad (19.202)$$

where p is the number of nonzero off-diagonals in the matrix at or above the main diagonal and q is the number at or below the main diagonal of the matrix to be solved.

The computational work for typical structured grid solvers can be estimated assuming an implicit computational stencil which spans three points in each coordinate direction. For a three-dimensional ordering of the unknowns by generalized coordinate directions,

$$N = J \cdot K \cdot L; \quad p = q = J \cdot K \quad (19.203)$$

where J, K, L is the number of points in each of the three coordinate directions, respectively. For a two-dimensional case with ordering by rows.

$$N = J \cdot K, \quad p = q = J \quad (19.204)$$

Hence, the computational work scales as

$$O(N^{7/3}m^3) \text{ operations: } 3 - D$$
$$O(N^2 m^3) \text{ operations: } 2 - D \tag{19.205}$$

assuming an equal number of points in each coordinate direction. In two dimensions, the relative bandwidth is smaller which results in a smaller operation count. Since both N and m are usually smaller in this case, direct solutions of the linear system are possible with available computers [Bailey and Beam (1991)] at least for steady-state solutions.

Specialized versions of Gaussian elimination are used for sparse linear systems to minimize storage costs and reduce operation counts [Duff *et al.* (1986)]. Banded-matrix direct solvers are perhaps the most common approach to reducing the storage and operation count of a full Gaussian elimination. Banded solvers store diagonal entries of the matrix as vectors and, hence, store all coefficient elements of the matrix out to the last diagonal entry that has a nonzero coefficient. In large part, most work on banded direct solvers has been done in the structural finite element field. Consequently, most banded solvers are specialized for symmetric positive-definite matrices.

Because of the number of operations and storage involved in a direct solution of the linear system at each iteration, the complexity which can arise in linearizing the equations exactly, and the realization that quadratic convergence is only obtained when the approximation is near to the exact solution, the method as outlined above is rarely used in application. A number of approximate techniques have been devised, including approximate factorizations, relaxation schemes, and hybrid iterative schemes. Approximate linearizations can be used: 1) to reduce the bandwidth of the linear system or 2) to reduce the complexity associated with an exact linearization. The equations can be cast in *delta form*, so that the nonlinear equation on the right-hand side will be satisfied as long as the sequence of iterates converge.

Within the framework of *approximate factorization* (AF) methods, implicit schemes which factor spatially the unsplit matrix equation into a sequence of simpler matrix equations are known as *alternating direction implicit* (ADI) schemes and have been widely used. For the compressible Euler and Navier–Stokes equations, Beam and Warming (1978) and Briley and MacDonald (1980) laid the foundations of the current ADI algorithms which are generalizations of the alternating-direction implicit algorithms developed in the 1950's for solving parabolic equations. In addition to the classical spatially factored scheme, a number of alternative schemes are possible by factoring the implicit operator according to the eigenvalues of the split Jacobian matrices and using type-dependent differencing [Steger and Warming (1981)]. These alternative factorizations can be used to split the full operator into a lower, L, and an upper, U, factor independent of the number of spatial dimensions of the problem, thereby increasing the allowable time step based on stability considerations and/or decreasing in the number of operations. The largest deficiency of the spatially factored approach is that the factored operator incurs a splitting error in three dimensions which is proportional to the cube of the time step, and the resulting algorithm is only conditionally stable at best. The LU schemes, on the other hand, incur only $(\Delta t)^2$ splitting errors in two and three dimensions and can attain longer stability bounds.

With the development and use of upwind discretizations for the Euler equations, Chakravarthy (1984), and Van Leer and Mulder (1984), observed that the linearized implicit equations can be solved efficiently with classical relaxation methods. Also, for supersonic flows, relaxation schemese can be constructed to recover efficient space-marching schemes. Hybrid relaxation-factorization schemes have also been developed to improve the stability characterizations of the three-factor ADI scheme [Thomas and Walters (1985)]. Relaxation is applied along one coordinate direction only; the approximate factorization is applied in the other two directions.

A class of algorithms that can be very effective and is applicable to both structured and unstructured grid methods are minimum-residual methods. For nonsymmetric matrices, methods such as the *Generalized Minimum Residual Method* (GMRES) [Saad and Schultz (1986)] are often used. The GMRES procedure is usually applied to a *preconditioned* system of equations. The role of the preconditioners is to obtain a more favorable distribution of eigenvalues than the original system in order to obtain faster convergence. In practice, the success of using GMRES depends very strongly on the effectiveness of the preconditioners.

19.4.2   Space Discretization—Dissipation

All numerical schemes used for obtaining solutions to the fluid dynamics equations must contain a certain level of dissipation to prevent odd-even point decoupling, to maintain stability at discontinuities, and to eliminate nonphysical solutions such as expansion shocks. The dissipation may be explicitly added on top of a naturally nondissipative scheme such as pure central differencing, or it may arise naturally from the spatial discretization such as occurs with upwind differencing algorithms. The exact form of dissipation has a large impact on the accuracy of the scheme, as well as on the stability and robustness of the overall algorithm. While some discretizations provide dissipation due to the coupling of the space-time discretizations, these schemes generally have a steady state that depends on the time step and are not discussed. Below, methods of explicitly adding dissipation to central differencing schemes are discussed, as well as methods of achieving naturally dissipative schemes through upwind differencing. Also discussed are methods of capturing sharp discontinuities without introducing nonphysical oscillations into the computations. While some of the discussion is particularly focused on the compressible flow equations, the need for artificial viscosity for solving incompressible flow is not eliminated, and the methodology for obtaining good dissipation without compromising accuracy is similar.

Artificial Dissipation Models for Central-Difference Schemes. Because the basic numerical scheme uses central differences to represent spatial derivatives, the artificial dissipation required to avoid spurious oscillations in the vicinity of shocks and to stabilize the scheme is implemented in a convenient manner by modifying the convective fluxes:

$$\hat{\mathbf{F}}_{i+(1/2)} = \tfrac{1}{2}(\hat{\mathbf{F}}_i + \hat{\mathbf{F}}_{i+1}) - \mathbf{d}_{i+(1/2)} \tag{19.206}$$

The $\mathbf{d}_{i+(1/2)}$ term represents the dissipative term in the i direction. Although many variations in dissipation models are presented in the literature, only two specific forms are discussed here.

Scalar Dissipation Model: The basic dissipation model is a nonisotropic model, where the dissipative terms are functions of the spectral radii of the Jacobian matrices associated with the appropriate coordinate directions. The details of the model vary with specific researchers, and no attempt is made here to describe the many variations. However, the essential ingredients are described below, and more details can be found in many references [e.g., Radespiel, Rossow, and Swanson (1989), Vatsa and Wedan (1990), Martinelli (1987), Pulliam (1985), Jameson, Schmidt, and Turkel (1981), Jameson (1992), Swanson and Turkel (1987)].

For clarity, a description of the dissipative terms for the *i* direction is given as

$$\mathbf{d}_{i+(1/2),j,k} = \lambda_{i+(1/2),j,k}[\epsilon^{(2)}_{i+(1/2),j,k}(\mathbf{W}_{i+1,j,k} - \mathbf{W}_{i,j,k})$$

$$+ \epsilon^{(4)}_{i+(1/2),j,k}(\mathbf{W}_{i+2,j,k} - 3\mathbf{W}_{i+1,j,k}$$

$$+ 3\mathbf{W}_{i,j,k} - \mathbf{W}_{i-1,j,k})] \tag{19.207}$$

In the above expression, the coefficients $\epsilon^{(2)}$ and $\epsilon^{(4)}$ are related to the pressure gradient parameter ν_i

$$\nu_i = \frac{|p_{i+1,j,k} - 2p_{i,j,k} + p_{i-1,j,k}|}{p_{i+1,j,k} + 2p_{i,j,k} + p_{i-1,j,k}} \tag{19.208}$$

$$\epsilon^{(2)}_{i+(1/2),j,k} = \kappa^{(2)} \max(\nu_{i+1}, \nu_i) \tag{19.209}$$

$$\epsilon^{(4)}_{i+(1/2),j,k} = \max\{0, (\kappa^{(4)} - \epsilon^{(2)}_{i+(1/2),j,k})\} \tag{19.210}$$

where $\kappa^{(2)}$ and $\kappa^{(4)}$ are constants with typical values of 1/2 and 1/64, respectively and the variable $\mathbf{W}$ is related to the solution vector $\mathbf{w}$ by the equation

$$\mathbf{W} = \mathbf{w} + [0, 0, 0, 0, p]^T \tag{19.211}$$

where

$$\mathbf{w} = \begin{Bmatrix} \rho \\ \rho u \\ \rho v \\ \rho w \\ E \end{Bmatrix} \tag{19.212}$$

is the density, the momentum in the *x*-, *y*-, and *z*-directions, and the total energy per unit volume. The term $\lambda_{i+(1/2),j,k}$ is the scaling factor associated with the ξ-coordinate.

The pressure switches given in Eqs. (19.208), (19.209), and (19.210) serve to increase the second difference dissipation while switching off the fourth differences near discontinuities. This formulation is given by Jameson, Schmidt, and Turkel (1981) and increases the sharpness of the shock wave without the introduction of unwanted oscillations due to differencing across the shock.

Matrix-Valued Dissipation: The dissipation model described above is not optimal in the sense that the same dissipation scaling is used for all the governing equations in a given coordinate direction. Reduced artificial dissipation can be obtained by individually scaling the dissipation contribution to each equation, as is done implicitly in upwind schemes [Turkel (1988) and Swanson and Turkel (1990)]. The scalar coefficients used in the artificial dissipation model are replaced by the modulus (absolute values) of flux Jacobian matrices. The matrices **A**, **B**, and **C** have very few nonzero elements and can be found in the cited reference in their entirety. The absolute value of these matrices, illustrated here for the matrix **A**, is defined as

$$|\mathbf{A}| = \mathbf{T}_\xi |\mathbf{A}_\xi| \mathbf{T}_\xi^{-1} \tag{19.213}$$

where

$$|\mathbf{\Lambda}_\xi| = \begin{bmatrix} |\lambda_1| & 0 & 0 & 0 & 0 \\ 0 & |\lambda_2| & 0 & 0 & 0 \\ 0 & 0 & |\lambda_3| & 0 & 0 \\ 0 & 0 & 0 & |\lambda_3| & 0 \\ 0 & 0 & 0 & 0 & |\lambda_3| \end{bmatrix} \tag{19.214}$$

is a diagonal matrix with the absolute values of the eigenvalues along the diagonals.

In practice, λ_1, λ_2, λ_3 cannot be simply the eigenvalues of the Jacobian matrix. Near stagnation points, λ_3 approaches zero; near sonic lines, λ_1 or λ_2 approaches zero. Since zero artificial viscosity can create numerical difficulties, these values are usually limited away from zero.

A comparison of results obtained with both scalar and matrix dissipation is shown in Fig. 19.30. These results have been obtained with the code described by Swanson and Turkel (1990). The case shown is the inviscid flow over an NACA 0012 airfoil at a free-stream Mach number of 0.8 and an angle of attack of 1.25°. This flow field

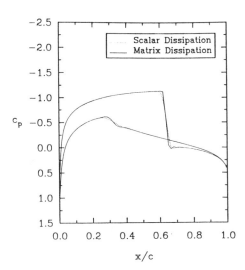

FIGURE 19.30   Central-difference scheme results with scalar and matrix dissipation for NACA 0012 with $M_\infty = 0.8$ and $\alpha = 1.25°$.

is characterized by a moderate strength shock on the upper surface and a much weaker shock on the lower surface. As seen in the figure, the results obtained with both the scalar and matrix dissipation are very similar. However, the solution obtained with matrix dissipation exhibits slightly sharper shocks on both the upper and lower surfaces. It should be pointed out that while the given example is for an inviscid flow, significant improvements in using matrix dissipation are also observed for viscous flows [Van Leer, Thomas, Roe, and Newsome (1987)].

Upwind Differencing

Flux–Vector Splitting: For flux–vector splitting, the fluxes in generalized coordinates, $\hat{\mathbf{F}}$, $\hat{\mathbf{G}}$, and $\hat{\mathbf{H}}$ are split into forward and backward contributions according to the signs of the eigenvalues of the Jacobian matrices and are differenced accordingly. For example, the flux in the ξ direction can be differenced as

$$\delta_\xi \hat{\mathbf{F}} = \delta_\xi^- \hat{\mathbf{F}}^+ + \delta_\xi^+ \hat{\mathbf{F}}^- \tag{19.215}$$

because $\hat{\mathbf{F}}^+$ has all nonnegative eigenvalues, and $\hat{\mathbf{F}}^-$ has all nonpositive eigenvalues. Two methods of flux–vector splitting are discussed below.

The first method is the technique outlined by Steger and Warming (1981). Since the flux vectors are homogeneous functions of degree one in $\mathbf{w}$, they can be expressed in terms of their Jacobian matrices. For example, considering the flux vector in the ξ direction, $\hat{\mathbf{F}}$ can then be written as

$$\hat{\mathbf{F}} = \hat{\mathbf{A}}\hat{\mathbf{w}} = \frac{\partial \hat{\mathbf{F}}}{\partial \hat{\mathbf{w}}} \hat{\mathbf{w}} \tag{19.216}$$

A similarity transformation allows Eq. (19.216) to be written as

$$\hat{\mathbf{F}} = \hat{\mathbf{A}}\hat{\mathbf{w}} = \mathbf{T}_\xi \mathbf{\Lambda}_\xi \mathbf{T}_\xi^{-1} \hat{\mathbf{w}} \tag{19.217}$$

where the matrix $\mathbf{\Lambda}_\xi$ is a diagonal matrix composed of the eigenvalues of $\hat{\mathbf{A}}$. The eigenvalues can then be decomposed into nonnegative and nonpositive components

$$\lambda_i = \lambda_i^+ + \lambda_i^- \tag{19.218}$$

where

$$\lambda_i^\pm = \frac{\lambda_i \pm |\lambda_i|}{2} \tag{19.219}$$

Similarly, the eigenvalue matrix $\mathbf{\Lambda}_\xi$ can be decomposed into

$$\mathbf{\Lambda}_\xi = \mathbf{\Lambda}_\xi^+ + \mathbf{\Lambda}_\xi^- \tag{19.220}$$

where $\mathbf{\Lambda}_\xi^+$ is made up of the nonnegative contributions of λ_i^+, and $\mathbf{\Lambda}_\xi^-$ is constructed of the nonpositive contributions of λ_i^-. This splitting of the eigenvalue matrix, combined with Eq. (19.217), allows the flux vector $\hat{\mathbf{F}}$ to be rewritten as

$$\hat{\mathbf{F}} = \mathbf{T}_\xi (\mathbf{\Lambda}_\xi^+ + \mathbf{\Lambda}_\xi^-) \mathbf{T}_\xi^{-1} \hat{\mathbf{w}}$$
$$= (\mathbf{A}^+ + \mathbf{A}^-) \hat{\mathbf{w}} = \hat{\mathbf{F}}^+ + \hat{\mathbf{F}}^- \tag{19.221}$$

The flux vector $\hat{\mathbf{F}}$ has three distinct eigenvalues and can therefore be written as a sum of three subvectors, each of which has a distinct eigenvalue as a coefficient [Janus (1984)]

$$\hat{\mathbf{F}} = \hat{\mathbf{F}}_1 + \hat{\mathbf{F}}_2 + \hat{\mathbf{F}}_3 \tag{19.222}$$

For supersonic and sonic flow in the ξ direction (i.e., $|M_\xi| = |\bar{u}/a| \geq 1$), where $\bar{u} = q_\xi/|\nabla\xi|$ represents the velocity normal to a ξ = constant face, the fluxes in this direction become

$$\hat{\mathbf{F}}^+ = \hat{\mathbf{F}} \quad \hat{\mathbf{F}}^- = 0 \quad (M_\xi \geq 1)$$
$$\hat{\mathbf{F}}^+ = 0 \quad \hat{\mathbf{F}}^- = \hat{\mathbf{F}} \quad (M_\xi \leq -1) \tag{19.223}$$

The split fluxes in the other two directions are easily obtained by interchanging η or ζ in place of ξ.

The fluxes split in the above manner are not continuously differentiable at zeros of the eigenvalues (i.e., sonic and stagnation points) [Buning and Steger (1982)]. The lack of differentiability of the split fluxes has been shown in some cases to cause small oscillations at sonic points that are rarely noticeable for most aerodynamic applications but can be remedied by biasing the eigenvalues near zero to a small value [Buning and Steger (1982)]. Another method of splitting the flux vector has been proposed by Van Leer (1982). Here, the fluxes are split so that the forward and backward flux contributions blend smoothly at eigenvalue sign changes (i.e., near sonic and stagnation points). Just as for the Steger–Warming splitting, the Jacobian matrices $\partial\hat{\mathbf{F}}^+/\partial\hat{\mathbf{w}}$ must have nonnegative eigenvalues, and $\partial\hat{\mathbf{F}}^-/\partial\hat{\mathbf{w}}$ must have nonpositive eigenvalues so that upwind differencing can be used for the spatial derivatives. In addition, both Jacobians have one zero eigenvalue for subsonic Mach numbers, which leads to steady transonic shock structures with at most only two transition zones [Van Leer (1982)]. In practice, when second-order spatial differencing is used, shocks with only one interior zone are usually obtained [Anderson, Thomas, and Van Leer (1986)]. This feature is not observed with the Steger–Warming flux splitting.

An interesting numerical consequence of splitting the fluxes (since the continuity and energy equations are differenced in different manners) is that although mass, momentum, and energy are conserved by the numerical algorithm, the total enthalpy is not conserved. Hänel *et al.* (1987) modified the energy formulation for the Van Leer flux–vector splitting so that the total enthalpy is preserved. In addition, because the total enthalpy is a constant at steady state, the energy flux remains degenerate so that the shock-capturing capabilities are not compromised.

Godunov's Method: A very successful scheme for solving the Euler equations that has led to significant improvements in the accuracy of modern numerical algorithms is due to Godunov [see Hirsch (1990)]. For this scheme, a piecewise constant approximation of the data in each cell is obtained, which represents the average of the data in the cell

$$\mathbf{w}_i = \frac{1}{\Delta x} \int_{i-(1/2)}^{i+(1/2)} \mathbf{w}(x, t)\, dx \qquad (19.224)$$

Each cell interface, located at $i + 1/2$, is then considered to separate two regions of constant properties in the same manner as a diaphragm separates regions of high and low pressure gas in a shock tube. Because an exact solution exists for that problem, the evolution of the flow field can be easily determined by solving for the interaction of the resulting wave system, provided that waves from neighboring cells do not interact. Afterwards, the solution in each cell is averaged, and the process is repeated. The process described above is summarized in Fig. 19.31.

To advance the solution in time, the one-dimensional, time-dependent Euler equations are integrated over both space and time to yield an equation that describes the time evolution of the cell average in each cell. For example, in cell i,

$$\mathbf{w}_i^{n+1} = \mathbf{w}_i^n - \frac{\Delta t}{\Delta x} [\mathbf{F}(\mathbf{w}_{i+(1/2)}) - \mathbf{F}(\mathbf{w}_{i-(1/2)})] \qquad (19.225)$$

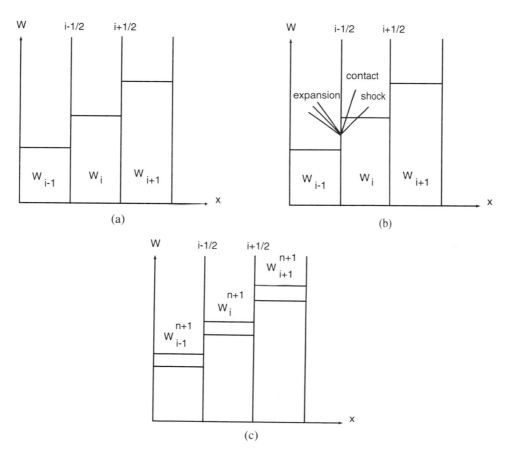

FIGURE 19.31 Illustration of Godunov's method. (a) Average state at time $t = n$, (b) solution of local Riemann problem, and (c) average state at time $t = n + 1$.

Here, $\mathbf{F}(\mathbf{w}_{i + (1/2)})$ represents the time average of the flux between times n and $n +$ 1. Recall that in advancing the solution in time, Δt is chosen so that there is no interaction of the waves from neighboring cell faces. Therefore, the solution at the interface is constant over the time interval of interest. The solution can then be advanced in time by forming the fluxes on the faces from the data obtained by solving the Riemann problem and advancing the solution using Eq. (19.225).

This process can be broken down into a *projection* and an *evolution* stage as described by Van Leer (1977). In the *projection stage*, the behavior of the data in each cell is reconstructed, whereas the *evolution stage* refers to the solution of the Riemann problem. In Godunov's method, the data in the cell are reconstructed by assuming it to be piecewise contant, which leads to first-order spatial accuracy. By replacing this approximation by a piecewise linear representation of the data, the accuracy of the scheme can be raised to second order [Van Leer (1979)].

Osher's Scheme: In Godunov's technique, the solution of the Riemann problem requires an interative procedure at each interface whenever a shock wave is present. To circumvent the iterative process, an approximate solution to the Riemann problem can be obtained by replacing a shock wave with an overturned rarefaction [Osher and Solomon (1982) and Chakravarthy (1985)]. Therefore, because all nonlinear waves are expressed in terms of rarefactions, explicit relations are obtained for the intermediate state variables connected by each wave.

Flux-Difference Splitting: For *flux-difference splitting* [Roe (1981)], the solution of the Riemann problem is again considered. However, the solution of the Riemann problem is foregone in favor of an exact solution to an approximate problem that does not require any iteration. More specifically, for one space dimension, data are advanced in time through a linearized version of the Euler equations given by

$$\frac{\partial \mathbf{w}}{\partial t} + \tilde{\mathbf{A}} \frac{\partial \mathbf{w}}{\partial x} = 0 \qquad (19.226)$$

where $\tilde{\mathbf{A}}$ is a specially constructed constant matrix that satisfies the property that for any $\mathbf{w}_L$ and $\mathbf{w}_R$ (which represent the left and right state variables on either side of a cell face),

$$\tilde{\mathbf{A}}(\mathbf{w}_L, \mathbf{w}_R) \, \Delta \mathbf{w} = \Delta \mathbf{F} \qquad (19.227)$$

where $\Delta(\cdot) = (\cdot)_L - (\cdot)_R$ (i.e., the jump across an interface). Note that the tilde ($\sim$) denotes that the matrix is constructed with a specific averaging procedure that is described in Roe (1981).

Equation (19.227) can also be written as

$$\tilde{\mathbf{T}}\tilde{\mathbf{A}}\tilde{\mathbf{T}}^{-1}\Delta \mathbf{w} = \Delta \mathbf{F} \qquad (19.228)$$

Since the eigenvalues represent wave speeds of individual waves, $\Delta \mathbf{q} = \tilde{\mathbf{T}}^{-1} \, \Delta \mathbf{w}$ represent jumps in the variables due to the influence of each wave. Hence, the change in flux between the left and right states is expressed in terms of the jumps in these states projected onto the eigenvectors

$$\Delta \mathbf{F} = \sum (\tilde{\lambda}_k \Delta \mathbf{q}) \tilde{\mathbf{T}}_k \tag{19.229}$$

By considering the backward-moving ($\tilde{\lambda}_k < 0$) and forward-moving waves ($\tilde{\lambda}_k > 0$) separately, the flux on the face at $i + \frac{1}{2}$ in Fig. 19.31(b) can be determined through any of the following equations (all of which are equivalent)

$$\mathbf{F}_{i+(1/2)} = \mathbf{F}_L + \sum_{\tilde{\lambda} < 0} (\tilde{\lambda}_k \Delta \mathbf{q}) \tilde{\mathbf{T}}_k \tag{19.230}$$

$$\mathbf{F}_{i+(1/2)} = \mathbf{F}_R - \sum_{\tilde{\lambda} > 0} (\tilde{\lambda}_k \Delta \mathbf{q}) \tilde{\mathbf{T}}_k \tag{19.231}$$

$$\mathbf{F}_{i+(1/2)} = \tfrac{1}{2}(\mathbf{F}_L + \mathbf{F}_R) - \tfrac{1}{2} \sum_k (|\tilde{\lambda}_k| \Delta \mathbf{q}) \tilde{\mathbf{T}}_k \tag{19.232}$$

The last form can be considered to represent a central-difference term plus a dissipative term.

Many researchers currently use the flux-difference splitting technique described above. Further, Van Leer *et al.* (1987) demonstrate that for viscous flows, this flux function is more accurate than central-difference formulations with scalar dissipation, as well as upwind formulations based on flux-vector splitting. The explanation lies in the consideration of Eq. (19.232) as a central-difference flux with an added dissipative term. By considering the influence of the individual waves, it is apparent that as an eigenvalue associated with the wave vanishes, the corresponding dissipation also vanishes. This mechanism is the means through which the exact solution to a single discontinuity is obtained. For viscous flows, the boundary layer is considered to consist of a series of shear waves normal to the body. Because the velocity in this direction is small, the corresponding dissipation is also small; the result is that boundary layers are captured with very high accuracy.

Observe that vanishing eigenvalues (hence artificial viscosity for the wave) occur at shock waves, where $\tilde{\lambda}_k = \bar{u} - a$ passes through zero, as well as contact discontinuities, where $\tilde{\lambda} = \bar{u} = 0$. Unfortunately, vanishing eigenvalues also occur at sonic points at which the flow transitions from subsonic to supersonic flow and $\tilde{\lambda}_k = \bar{u} - a$ is again zero. This is a consequence of replacing the full nonlinear problem with a linearized version in Eq. (19.226) which considers an expansion to be described by a single wave instead of a series of waves. This can lead to expansion shocks in which the flow transitions from subsonic to supersonic in a discontinuous manner. To remedy the situation, the eignvalue $\tilde{\lambda}_k = \bar{u} - a$ is often modified slightly so that it does not vanish at these points. As a result, a small amount of dissipation is added to the scheme, so the wave is spread slightly over several mesh points. Although many forms of this modification exist, the most common implementation is to modify the eigenvalue according to [Harten (1984)]

$$|\tilde{\lambda}| \equiv \begin{cases} |\tilde{\lambda}|_{i+(1/2)} & \text{if } |\tilde{\lambda}|_{i+(1/2)} \geq \epsilon \\ \dfrac{1}{2}\left(\dfrac{\tilde{\lambda}^2_{i+(1/2)}}{\epsilon} + \epsilon\right) & \text{if } |\tilde{\lambda}|_{i+(1/2)} \leq \epsilon \end{cases} \tag{19.233}$$

Multidimensional Upwind Methods: The upwind techniques discussed above are applicable strictly to one-dimensional problems. To apply these techniques to two- and three-dimensional problems, the usual procedure is to assume that waves propagate normal to grid lines, which allow Riemann solvers to be applied in a one-dimensional manner separately in each coordinate direction. This approach leads to quite satisfactory solutions when features such as shock waves and shear waves are essentially aligned with the mesh. However, severe degradation of accuracy can occur when the features are oblique to the grid lines, because they are interpreted by the Riemann solver to be composed of pairs of grid-aligned waves instead of a single wave. The result is that shocks and shear waves may be severely smeared. In recent years, several research efforts have been aimed at overcoming the possible loss of accuracy attributable to the dimension-by-dimension approach to upwind differencing. A summary of some of the more promising techniques is given in Van Leer (1992) and Struijs, Deconinck, DePalma, Roe, and Powell (1991).

Higher Order Schemes: Up to this point in the discussion of upwind schemes, the determination of the left and right state variables on either side of a cell face has been, for the most part, left unspecified. Recall that for Godunov's method, if a piecewise constant approximation of the data in each cell is assumed, then the resulting scheme is first-order accuracy in space. The accuracy of the approximation can be raised to higher order by replacing the piecewise constant approximation of the data with a piecewise polynomial approximation [Van Leer (1979) and Colella and Woodward (1984)]. For instance, the state variables on the left and right side of the cell interface located at $i + 1/2$ in Fig. 19.32 can be determined as

$$\mathbf{w}_L = \mathbf{w}_i + \tfrac{1}{4}[(1 - \kappa)\mathbf{\Delta}_- + (1 + \kappa)\mathbf{\Delta}_+]_i$$

$$\mathbf{w}_R = \mathbf{w}_{i+1} - \tfrac{1}{4}[(1 - \kappa)\mathbf{\Delta}_+ + (1 + \kappa)\mathbf{\Delta}_-]_{i-1} \qquad (19.234)$$

where

$$(\mathbf{\Delta}_+)_i \equiv \mathbf{w}_{i+1} - \mathbf{w}_i$$

$$(\mathbf{\Delta}_-)_i \equiv \mathbf{w}_i - \mathbf{w}_{i-1} \qquad (19.235)$$

Equation (19.234) represents a one-parameter family of schemes. A fully one-sided approximation of the data is obtained by inserting $\kappa = -1$, while $\kappa = 1/3$ leads to a third-order upwind-biased approximation, and $\kappa = +1$ yields a second-order central-difference scheme. All the upwind-biased approximations use the same number of cells for the residual computation as the fully one-sided scheme and may be implemented with only a slight increase in computational effort. The third-order scheme is strictly third-order accurate only in one-dimensional calculations. To obtain a third-order scheme in two or three dimensions, computation of the flux across a cell face on the basis of an averaged state is not sufficient, because the difference

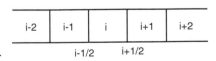

FIGURE 19.32 Higher order interpolation stencil.

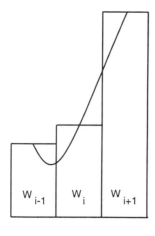

FIGURE 19.33 Introduction of new extrema using upwind-biased interpolation.

between that average flux and the flux computed from the averaged states is a term of second order and vanishes only for a linear system of conservation laws. Nevertheless, by switching from a fully upwind approximation ($\kappa = -1$) to the third-order ($\kappa = 1/3$) scheme, the accuracy of smooth solutions can be increased [Thomas and Walters (1985)].

A deficiency in using Eq. (19.234) for reconstructing the data at the cell faces is that new extrema can be introduced even when the original data is monotone. For example, in Fig. 19.33, a non-monotone interpolation is obtained between cells $i - 1$ and i. If this profile is convected and the cell averages are then reconstructed, nonphysical oscillations can result in the solution.

For determining acceptable limits on the slopes, the data in each cell are first represented by a Taylor's series expansion about the center of the cell. For example, the data on the boundaries of cell i in Fig. 19.33 can be determined as

$$\mathbf{w}_{i+(1/2)} = \mathbf{w}_i + \frac{1}{2}\left(\frac{\partial \mathbf{w}}{\partial x}\right)\Delta \mathbf{x} = \mathbf{w}_i + \frac{1}{2}\delta \mathbf{w} \qquad (19.236)$$

$$\mathbf{w}_{i-(1/2)} = \mathbf{w}_i - \frac{1}{2}\left(\frac{\partial \mathbf{w}}{\partial x}\right)\Delta x = \mathbf{w}_i - \frac{1}{2}\delta \mathbf{w} \qquad (19.237)$$

where Δx is the width of the cell and

$$\delta \mathbf{w} = \tfrac{1}{2}[(1 - \kappa)\mathbf{\Delta}_- + (1 + \kappa)\mathbf{\Delta}_+]_i \qquad (19.238)$$

For monotone increasing data, a sufficient bound on the size of $\mathbf{\Delta}_-$ and $\mathbf{\Delta}_+$ is obtained by requiring that the interpolated data on the cell faces does not exceed the values in the surrounding cells. This limit is achieved provided that

$$\mathbf{w}_i + \tfrac{1}{2}\delta \mathbf{w} \leq \mathbf{w}_{i+1} \Rightarrow \delta \mathbf{w} \leq 2\mathbf{\Delta}_+$$

$$\mathbf{w}_i - \tfrac{1}{2}\delta \mathbf{w} \geq \mathbf{w}_{i-1} \Rightarrow \delta \mathbf{w} \leq 2\mathbf{\Delta}_- \qquad (19.239)$$

In order to ensure a monotone interpolation, the magnitude of δw may have to be limited to be no larger than either $2\Delta_+$ or $2\Delta_-$

$$(\delta w) \leq \min (2\Delta_+, 2\Delta_-) \tag{19.240}$$

Equation (19.24) provides a guideline for reducing the magnitude of any gradient that would result in a nonmonotone interpolation. Following Van Leer (1987), the value of δw that will maintain monotone interpolation will be referred to as $(\delta w)_{limited}$ and can be written as

$$(\delta w)_{limited} = R(\phi)\delta w \tag{19.241}$$

where $\phi = (\Delta_+/\Delta_-)$ and $R(\phi)$ serves to limit the size of the original gradient. From Eqs. (19.238), (19.240), and (19.241), $R(\phi)$ is written as

$$R(\phi) = \min \left[\frac{4}{(1 - \kappa) + (1 + \kappa)\phi}, \frac{4\phi}{(1 - \kappa) + (1 + \kappa)\phi} \right] \tag{19.242}$$

In Fig. 19.34, a plot of $R(\phi)$ from Eq. (19.242) is shown, where κ has been assumed to be zero ($\kappa = 0$) in Eq. (19.238). In the figure, the area that lies inside the curve is the region for which monotone interpolation is obtained.

For $\phi = 1$, Δ_+ is equal to Δ_-, and the data in the cell appears linear. Because a second-order scheme should reconstruct linear data exactly, a basic requirement on the limiter function is that it pass through unity for $\phi = 1$, which can easily be achieved by modifying Eq. (19.242) to read

$$R(\phi) = \min \left[\frac{4}{(1 - \kappa) + (1 + \kappa)\phi}, 1, \frac{4\phi}{(1 - \kappa) + (1 + \kappa)\phi} \right] \tag{19.243}$$

In this manner, the slopes as calculated from Eq. (19.238) are left unchanged, provided the interpolations remain monotone.

Many variations for $R(\phi)$ have appeared in the literature that preserve monoton-

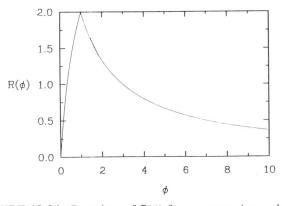

FIGURE 19.34   Boundary of $R(\phi)$ for monotone interpolation.

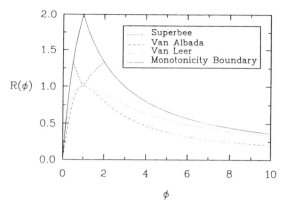

FIGURE 19.35 Superbee, Van Albada, and Van Leer limiters.

icity and are second-order accurate. Some of the most widely used are those of Van Albada *et al.* (1982), Van Leer (1974), and Roe's *superbee* limiter [see for example Sweby (1984)]. A plot of the Van Albada, Van Leer, and Roe's superbee limiters is shown in Fig. 19.35. Note that all the limiters shown pass through unity when $\phi = 1$ which maintains second-order accuracy.

In the previous discussion, it has been assumed that the data are monotonically increasing; similar arguments hold for monotonically decreasing data. For non-monotone data in which the sign of Δ_+ and Δ_- differ, $(\delta\mathbf{w})_{\text{limited}}$ can be simply set to zero to make sure that any extrema are not magnified [Van Leer (1987)]. In the work of Spekreijse (1987), (1988), general conditions are derived for the limiter function that will maintain second-order accuracy as well as monotone steady-state solutions that do not set the gradient to zero at the extrema.

Another method that is useful to the design of nonoscillatory schemes is based on the definition of total variation. For a discrete one-dimensional scalar solution on an infinite domain, the *total variation* is defined at time level n as

$$TV(W^n) = \sum_{-\infty}^{-\infty} |W_{i+1} - W_i| \tag{19.244}$$

In the top part of Fig. 19.36, for a monotone grid function, the total variation is determined strictly by the endpoints (i.e., $|W_\infty^n - W_{-\infty}^n|$). However, if a new extrema is introduced as in the lower half of the figure, the total variation will increase. Hence, for a scheme to remain nonoscillatory, the total variation should remain the same or decrease as the solution is updated

$$TV(W^{n+1}) \leq TV(W^n) \tag{19.245}$$

Schemes derived from these guidelines are appropriately called *total variation diminishing* (TVD).

Sufficient conditions for constructing TVD schemes were first developed by Harten (1983), and many investigators have examined criteria for constructing TVD schemes [e.g., Osher and Chakravarthy (1984) and Jameson and Lax (1986)]. Many

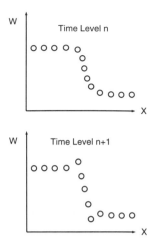

FIGURE 19.36 Example of increasing total variation.

schemes have been developed and applied to difficult aerodynamic problems that rely heavily on the TVD concepts discussed above. A few examples can be found in Chakravarthy (1987), Harten (1984), Chakravarthy and Osher (1985), Yee (1987), and Jameson (1985b).

The methodologies described above only achieve higher order accuracy for smoothly varying meshes. To obtain higher order accuracy on general meshes, as well as to extend the accuracy beyond second order, an extensive amount of research has been conducted for both structured and unstructured grids. An important class of such schemes is the *essentially nonoscillatory* (ENO) schemes. Although no details are given here, the essential ingredients include a polynomial reconstruction of the data from cell-averaged data that is exact up to a specified order of accuracy. Further details can be found in [Harten *et al.* (1986), Chakravarthy *et al.* (1986a), (1986b), Casper and Atkins (1993), and Godfrey *et al.* (1992)].

Another technique to obtain higher order accuracy that has been particularly useful for unstructured grids is the so-called *k-exact* method. Here, the conservation of the mean is enforced, and the reconstruction is such that a polynomial of degree k or less is reconstructed exactly [Barth (1992)]. In the implementations of Godfrey *et al.* (1992), Barth and Frederickson (1990), and Mitchell and Walters (1993), the stencil generally remains fixed, which results in a somewhat lower computational expense. However, in order to avoid oscillations, a limiting procedure must be applied where steep gradients are present in order to avoid nonphysical oscillations. This tends to reduce the order of accuracy in these regions. Algorithms have been proposed that incorporate stencil-varying techniques into high-order ENO methods for unstructured grids [Harten and Chakravarthy (1991) and Abgrall (1991)].

High-order ENO schemes can also be implemented in a finite-difference manner in which the reconstruction operator acts upon pointwise fluxes [Shu and Osher (1988)].

19.4.3 Boundary Conditions

Numerical computations must necessarily be done on a finite mesh, thus information which is incoming and outgoing with respect to the computational domain crosses the boundary of the mesh. Straightforward application of the interior point algorithm at the boundary requires information from outside the domain, which is generally not completely available. Since the boundary conditions are what distinguishes one solution of the equations from another, this additional information is critical to the proper numerical simulation of a physical problem.

The type of boundary condition to be specified is problem dependent and as Moretti (1981) noted: "A physically consistent model of the outside world must be provided." For instance, the upstream boundary conditions to be specified for a subcritical, two-dimensional, converging-diverging nozzle might consist of the recognition that the fluid comes from a uniform reservoir condition corresponding to specified total pressure and entropy conditions as well as a specification of the direction of the velocity.

For near-field boundaries, the physical boundary conditions are usually augmented by conditions derived from field equations evaluated on the boundary. For example, the viscous no-slip physical condition at the surface is that the velocity components be zero and either the temperature or its gradient be prescribed at the surface. This leaves one degree of freedom at the boundary, and usually a numerical boundary specification of the pressure is formulated by satisfying the normal momentum equation evaluted at the surface.

For many cases, the boundary is located where viscous effects can be neglected; in this case, the time-dependent Euler equations are an effective model. These equations are a hyperbolic system of equations and the number of boundary conditions to be specified at the boundary comes from a straightforward characteristic analysis. This characteristic decomposition indicates that the equations representing outgoing waves can be differenced using information from the computational domain. The equations representing incoming waves cannot be stably differenced using only information available in the interior, hence those equations need to be replaced by boundary conditions.

Characteristic Equations

Linearized Equations: The linearized form of the time-dependent Euler equations can be written as

$$a \frac{\partial \mathbf{q}}{\partial x} + b \frac{\partial \mathbf{q}}{\partial y} + c \frac{\partial \mathbf{q}}{\partial z} = 0 \qquad (19.246)$$

where $\mathbf{a}, \mathbf{b}, \mathbf{c}$ are Jacobian matrices. The choice of variables is not unique and is generally selected to make the Jacobian matrices as simple as possible [Roe (1983) and Thompson (1987)]. A common choice is the set

$$\mathbf{q} = \begin{bmatrix} \rho \\ u \\ v \\ w \\ p \end{bmatrix} \quad \mathbf{a} = \begin{bmatrix} u & \rho & 0 & 0 & 0 \\ 0 & u & 0 & 0 & 1/\rho \\ 0 & 0 & u & 0 & 0 \\ 0 & 0 & 0 & u & 0 \\ 0 & \rho a^2 & 0 & 0 & u \end{bmatrix} \qquad (19.247)$$

from which the eigenvalues of **a** can be easily computed as $\{u, u, u, u + a, u - a\}$, where u is velocity and a the speed of sound. All of the Jacobian matrices have real eigenvalues and a set of linearly independent eigenvectors and each, individually, can be diagonalized, although not simultaneously since the Jacobian matrices do not have the same eigenvectors. The characteristic equations result from diagonalizing the Jacobian matrices as

$$\frac{\partial \mathbf{q}}{\partial t} + \mathbf{T}_x \mathbf{\Lambda}_x \mathbf{T}_x^{-1} \frac{\partial \mathbf{q}}{\partial x} + \mathbf{T}_y \mathbf{\Lambda}_y \mathbf{T}_y^{-1} \frac{\partial \mathbf{q}}{\partial y} + \mathbf{T}_z \mathbf{\Lambda}_z \mathbf{T}_z^{-1} \frac{\partial \mathbf{q}}{\partial z} = 0 \qquad (19.248)$$

where $\mathbf{\Lambda}_x$, $\mathbf{\Lambda}_y$, $\mathbf{\Lambda}_z$ are diagonal matrices.

Considering a plane boundary as coincident with a surface of constant x, the derivatives in the two directions tangent to the boundary (y, z) can be determined from information on the boundary. In general, the computation of the derivative normal to the boundary requires information about the state vector at locations outside the computational domain. Defining the terms corresponding to derivatives in the plane of the boundary as a source term $\mathbf{S}$, the equations may be written as

$$\mathbf{T}^{-1} \frac{\partial \mathbf{q}}{\partial t} + \mathbf{\Lambda} \mathbf{T}^{-1} \frac{\partial \mathbf{q}}{\partial x} + \mathbf{T}^{-1} \mathbf{S} = 0 \qquad (19.249)$$

where the x-subscript notation has been dropped. This can also be written in terms of each component of the equation as

$$\mathbf{l}_i \frac{\partial \mathbf{q}}{\partial t} + \lambda_i \mathbf{l}_i \frac{\mathbf{q}}{\partial x} + \mathbf{l}_i \mathbf{S} = 0 \qquad (19.250)$$

where $\mathbf{l}_i$ is the left eigenvector of the Jacobian matrix **a,** corresponding to the ith eigenvalue (and also forms the ith row of $\mathbf{T}^{-1}$).

Characteristic Variables: If a characteristic variable, V_i, can be defined to satisfy the so-called compatibility equations below [Roe (1983) and Hirsch (1990)]

$$dV_i = \mathbf{l}_i \, d\mathbf{q} + \mathbf{l}_i \mathbf{S} \, dt \qquad (19.251)$$

then Eq. (19.250) reduces to a set of wave equations

$$\frac{\partial V_i}{\partial t} + \lambda_i \frac{\partial V_i}{\partial x} = 0 \qquad (19.252)$$

for which the characteristic variable is constant along characteristic curves defined in the $x - t$ plane as $dx/dt = \lambda_i$. The characteristic directions are sketched in Fig. 19.37 for subsonic and supersonic flow. Positive u is indicated; outflow conditions correspond to the exterior boundary defined by $x > 0$. The characteristics are traced back from the new time level t^{n+1} for three characteristic directions, C^0, C^+, C^-, corresponding to the repeated eigenvalue λ_1 and λ_4, λ_5.

The construction of V_i is generally not possible for the Euler equations without

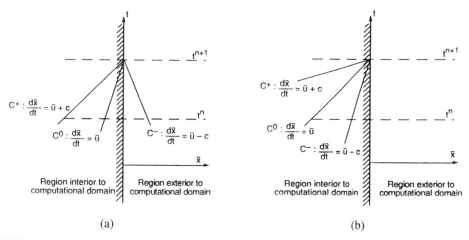

FIGURE 19.37 Sketch of characteristic directions at the boundary. Outflow (inflow) corresponds to the exterior domain described by $\bar{x} > 0$ ($\bar{x} < 0$), for a positive normal velocity $\bar{u}$. (a) Subsonic flow—outflow ($\bar{u} > 0$) and (b) supersonic flow—outflow ($\bar{u} > 0$).

assuming the diagonalizing matrices are constant [Roe (1983)]. Assuming the exterior domain is described by $x > 0$, the characteristic form, either Eq. (19.250) or (19.252), indicates that outgoing waves are described by equations with $\lambda_i \geq 0$ and depend on information at and within the boundary. Incoming waves, representing information reaching the boundary from the exterior, are described by equations with $\lambda_i < 0$. These latter wave equations cannot be differenced stably using just interior and boundary information, since the numerical domain of dependence would not include the physical domain of dependence. Hence, these equations need to be replaced with boundary conditions.

Diagonal Equations: Assuming no variations in the plane of the boundary and the diagonalizing matrices to be constant, the equations can be reduced to a set of diagonal equations

$$\frac{\partial \overline{\mathbf{w}}}{\partial t} + \mathbf{\Lambda} \frac{\partial \overline{\mathbf{w}}}{\partial x} = 0 \tag{19.253}$$

where the linearized characteristic variable is defined as

$$\overline{\mathbf{w}} \equiv \mathbf{T}_0^{-1}\mathbf{q} = \begin{bmatrix} \rho - p/a_0^2 \\ v \\ w \\ u + p/\rho_0 a_0 \\ -u + p/\rho_0 a_0 \end{bmatrix} \tag{19.254}$$

and the subscript notation denotes evaluation at a nearby reference value.

Homentropic Equations: Further assuming locally homentropic flow (i.e., that the entropy is uniform everywhere), the equations can be reduced to

$$\frac{d}{dt}(v) = 0 \quad \text{along} \quad \frac{dx}{dt} = u$$

$$\frac{d}{dt}(w) = 0 \quad \text{along} \quad \frac{dx}{dt} = u$$

$$\frac{d}{dt}(R^{\pm}) = 0 \quad \text{along} \quad \frac{dx}{dt} = u \pm a \tag{19.255}$$

where the Riemann variables are defined as

$$R^{\pm} = u \pm [2a/(\gamma - 1)] \tag{19.256}$$

and x is the local normal pointing out of the domain and γ the ratio of specific heats. The equations are in a form very similar to one-dimensional, unsteady flow, except that the tangential velocities, in addition to the entropy, are convected along the particle path.

Numerical Procedures. The equations which replace the characteristic equations for the incoming waves are generally referred to as the *physical boundary conditions.* The procedures to determine the remaining variables at the boundary, which should be as compatible as possible with the outgoing characteristic equations, are sometimes referred to as *numerical boundary conditions*, but should be more properly termed *numerical treatments at the boundary* [Moretti (1981)].

The numerical procedures at the boundary are different than those used at the interior scheme. Thus, two factors of the coupled system need to be taken into account—accuracy and stability. Gustafsson (1975) has pointed out that the accuracy of the numerical procedure for a linear equation can be one order lower than the order of the interior scheme without adversely influencing the global accuracy of the solution. The stability of the boundary procedure can be analyzed in many cases using the analysis of Gustafsson, Kreiss, and Sundstrom (1972). Generally, the closer the numerical scheme is coupled to the characteristic equations, the more well-behaved the numerical procedure.

Homentropic Subsonic Flow: Assuming a locally orthogonal coordinate system where $\bar{x}$ is the local normal pointing out of the domain (sketched in Fig. 19.38), then the homentropic equations, Eqs. (19.255)–(19.256), can be used to update the equations along the boundary at the new time level. For subsonic flow, for instance, R^- can be evaluated from the far field, corresponding to conditions outside the boundary, and R^+ can be evaluated locally from the interior of the domain. Then, the normal velocity and speed of sound can be evaluated as

$$\bar{u} = (R^+ + R^-)/2$$

$$a = (R^+ - R^+)(\gamma - 1)/4 \tag{19.257}$$

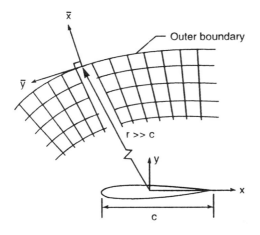

FIGURE 19.38 Sketch of local coordinate system at the boundary used for characteristic method in homentropic flow.

Depending on the sign of the normal velocity, the entropy and tangential velocities are extrapolated from the exterior or interior of the domain. Thus, the three velocity components, entropy, and speed of sound can be constructed at the new time level.

The influence of the computed airfoil (body) is felt at large distances upstream and downstream, thus the assumption of uniform flow at the boundary necessitates the construction of a grid which extends quite far from the airfoil. By including the first-order effect of the circulation imposed by the airfoil to the state variable vector at locations exterior to the boundary, the boundary need not extend as far, thus the computations can be restricted to a smaller domain with fewer grid points and/or less stretching. An example [Thomas and Salas (1986)] of including this effect on the lift coefficient of an airfoil is shown in Fig. 19.39. Subcritical and supercritical cases are shown. Using free-stream conditions to evaluate far-field boundary con-

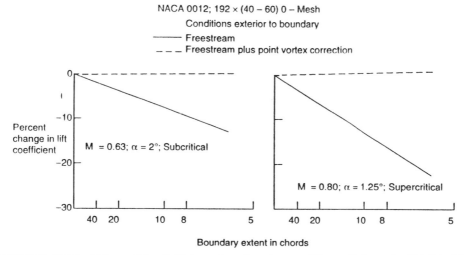

FIGURE 19.39 Effect of far-field boundary location on lift coefficient for NACA 0012 airfoil at subcritical and supercritical conditions.

tributions, the lift coefficient shows an inverse radial dependence on the boundary extent, which is the same functional dependence as the leading-order term in the far-field expansion. Updating the boundary conditions with the far-field contribution corrected as above, the sensitivity of the solution is dramatically reduced. The supercritical case shows a stronger dependence on the outer boundary extent, as expected, due to the increased lateral extent of the disturbances at the higher Mach number. The correction term, which scales on the lift, is effective in both cases.

Other Methods: Other methods include the characteristic methods of Chakravarthy (1983) and the so-called *nonreflecting*, or *radiation*, *boundary conditions* of Hedstrom (1979) and Thompson (1987). Further developments in nonreflecting boundary conditions are given by Giles (1989) and Atkins and Casper (1993). An alternative analytic procedure has been developed by Verhoff *et al.* (1992). It is a consistent method for coupling linearized analytic solutions with nonlinear numerical solutions in the far field through the computational boundary condition.

Simulation of Propulsion Systems and Boundary Layers: In many applications, it is not necessary to simulate the full details of the propulsion system. Rather, the propulsion system can be considered as a *black box* across or through which the solution can change in a manner consistent with that of a complete modeling. The additional energy and/or swirl added by a jet engine can be specified at a location which might represent a faired-over representation of the actual geometry as a function of engine thrust, for example, and thus avoid the considerable cost and complexity associated with a full simulation [Hartwich and Frink (1992)].

Boundary conditions can also be used to simulate the displacement effects of the viscous boundary layer near the surface in inviscid simulations [Lighthill (1958)].

19.5   TIME ACCURATE FINITE DIFFERENCE APPROACH
H. C. Yee

19.5.1   Introduction

When a finite difference method is used to approximate a partial differential equation (PDE), the domain of the continuous problem is *discretized* so that the dependent variables are considered to exist only at discrete points. Partial derivatives are approximated by finite difference quotients resulting in an algebraic representation of the PDE. The nature of the resulting algebraic system depends on the character of the original PDE and the type of difference method used to approximate the partial derivatives. Finite difference approximations can be classified into two types according to the nature of the algebraic system. When the numerical value at the next time step can be obtained directly in terms of the current known quantities without solving a set of simultaneous equations, the finite difference method is called *explicit*, otherwise, the method is called *implicit*. The composite of several considerations determines whether the solution so obtained will be a good approximation to the exact solution of the original PDE. Typical considerations are *truncation error (order of accuracy)*, *consistency*, *stability*, and *convergence* [Richtunger and Morton (1967), Hildebrand (1965), and Mitchell (1976)].

The purpose of this section is to review the fundamentals of finite difference methods for time dependent PDE's, i.e., PDE's of the hyperbolic and parabolic type. In particular, the analysis of hyperbolic equations is stressed here, since initial boundary value problems of this type are more difficult to analyze. The intention is to give an overview of the material and provide references for those who wish to pursue these ideas.

19.5.2 One-Dimensional Model Equations

In one spatial dimension, the nonconservative, compressible Navier–Stokes equations can be written

$$\frac{\partial \mathbf{U}}{\partial t} + \mathbf{A}(\mathbf{U}) \frac{\partial \mathbf{U}}{\partial x} = \frac{\partial}{\partial x} \mathbf{B}(\mathbf{U}) \frac{\partial \mathbf{U}}{\partial x} \tag{19.258}$$

where the components of the vector $\mathbf{U} = (\mathbf{U}_1, \mathbf{U}_2, \mathbf{U}_3)$ are mass, momentum, and energy per unit volume, and $\mathbf{A}$ and $\mathbf{B}$ are 3×3 matrices. If the viscous term on the right-hand side of Eq. (19.258), are neglected then the quasi-linear equation

$$\frac{\partial \mathbf{U}}{\partial t} + \mathbf{A}(\mathbf{U}) \frac{\partial \mathbf{U}}{\partial x} = 0 \tag{19.259}$$

is defined to be *hyperbolic* if there exists a similarity transformation Q such that

$$Q^{-1}\mathbf{A}Q = \begin{bmatrix} c_1 & 0 & 0 \\ 0 & c_2 & 0 \\ 0 & 0 & c_3 \end{bmatrix} \tag{19.260}$$

where the eigenvalues c_l are real. For fluid dynamics $c_1 = u$, $c_2 = u + a$, and $c_3 = u - a$, where u is the velocity and a is the sound speed. If the matrix A is assumed to be locally constant, Eq. (29.259) can be transformed into an uncoupled system of scalar hyperbolic equations

$$\frac{\partial u_l}{\partial t} + c_l \frac{\partial u_l}{\partial x} = 0, \quad l = 1, 2, 3 \tag{19.261}$$

by defining a new vector $(u_1, u_2, u_3)^T = Q^{-1}U$. When analyzing the stability properties of a linearized version of an algorithm for a PDE associated with Eq. (19.259), we normally only apply the scheme to the scalar equation Eq. (19.261) provided the boundary conditions are applied to the characteristic variable, u_l. Hereafter, for simplicity, drop the subscript l and refer to Eq. (19.259) as the *hyperbolic model equation*.

Likewise, if the convective term of Eq. (19.258) is dropped then the resulting equation

$$\frac{\partial \mathbf{U}}{\partial t} = \frac{\partial}{\partial x} \mathbf{B}(\mathbf{U}) \frac{\partial \mathbf{U}}{\partial x} \tag{19.262}$$

is defined to be *parabolic* if there exists a similarity transformation P that diagonalizes $\mathbf{B}$, i.e.,

$$P^{-1}\mathbf{B}P = \begin{bmatrix} \sigma_1 & 0 & 0 \\ 0 & \sigma_2 & 0 \\ 0 & 0 & \sigma_3 \end{bmatrix} \tag{19.263}$$

where the eigenvalues $\sigma_l \geq 0$ are the transport coefficients. Again for locally constant B, the system decouples to

$$\frac{\partial u_l}{\partial t} = \sigma_l \frac{\partial^2 u_l}{\partial x^2}, \quad l = 1, 2, 3. \tag{19.264}$$

Dropping the subscript l, refer to Eq. (19.264) as the *parabolic model equation*.

19.5.3   Finite Difference Methods for Time-Dependent Partial Differential Equations—Initial Value Problem

Let $u(x)$ be a function defined in the interval $0 \leq x \leq l$. The interval is discretized by the equally spaced grids, $x_0 = 0, x_1, \cdots, x_J = l$. Let the values u_j be some discrete approximation of $u(x)$. The mth derivative of $u(x)$ at point x_j can be approximated in the form

$$\frac{d^m u(x_j)}{dx^m} \approx \sum_{k=-J_1}^{J_2} \alpha_k u_{j+k} \tag{19.265}$$

where the α_k's are determined by means of Taylor expansions, and J_1, J_2 are integers depending on the order m of the derivative under consideration and also on the degree of accuracy of the approximation. The degree of accuracy p, or the error of the approximation, is normally denoted by $O(\Delta x^p)$. For $m = 1, J_1 = J_2 = 1$, Eq. (19.265) can be rewritten as

$$\frac{du(x_j)}{dx} \approx \frac{(1 - \alpha)u_{j+1} + 2\alpha u_j - (1 + \alpha)u_{j-1}}{2\Delta x} \tag{19.266}$$

where α is a constant. By specifying the value of α, obtain the standard difference:

Centered: $\alpha = 0$

$$\frac{du(x_j)}{dx} \approx \frac{u_{j+1} - u_{j-1}}{2\Delta x} \equiv D_x u_j, \quad \text{error} = O(\Delta x^2) \tag{19.266a}$$

Backward: $\alpha = 1$

$$\frac{du(x_j)}{dx} \approx \frac{u_j - u_{j-1}}{\Delta x} \equiv D_x^- u_j, \quad \text{error} = O(\Delta x) \tag{19.266b}$$

Forward: $\alpha = -1$

$$\frac{du(x_j)}{dx} \approx \frac{u_{j+1} - u_j}{\Delta x} \equiv D_x^+ u_j, \quad \text{error} = O(\Delta x) \qquad (19.266c)$$

Now, if we choose $J_1 = 2$ and $J_2 = 0$ and employ a Taylor's series to find the α_j's, we obtain the second-order accurate backward approximation

$$\frac{du(x_j)}{dx} \approx \frac{3u_j - 4u_{j-1} + u_{j-2}}{2\Delta x} \qquad (19.266d)$$

Similarly, if $J_1 = 0$, $J_2 = 2$, we obtain the second-order accurate forward approximation.

$$\frac{du(x_j)}{dx} \approx \frac{-3u_j + 4u_{j+1} - u_{j+2}}{2\Delta x} \qquad (19.266e)$$

A fourth-order accurate approximation can be obtained if $J_1 = J_2 = 2$, i.e.,

$$\frac{du(x_j)}{dx} \approx \frac{-u_{j+2} + 8u_{j+1} - 8u_{j-1} + u_{j-1}}{12\Delta x} \qquad (19.266f)$$

We can approximate higher derivatives in the same manner. For example, the classical difference approximation of the second derivative is

$$\frac{d^2u(x_j)}{dx^2} \approx \frac{u_{j+1} - 2u_j + u_{j-1}}{\Delta x^2} \equiv D_{xx}u_j \qquad (19.267)$$

In the following sections, the function u will be a function of more than one variables (time and space).

The Method of Lines. In the *method of lines*, spatial derivatives are approximated by finite difference formulae to generate a system of ordinary differential equations (ODE's) in time The resulting system of ODE's is then integrated by standard numerical techniques for ODE's. The appropriate time integration scheme depends on the eigenvalue spectrum of the resulting ODE system [Lambert (1979) and Gear (1971)]. There are several reasons why one prefers to use this approach: 1) the properties of the resulting systems of ODE's will give insight into the desirable properties of time discretization techniques, 2) numerical methods for ODE's are more well established and many user-oriented general ODE software packages are available, and 3) the concepts related to discretization and covergence estimates are more *transparent* than for the fully discretized systems.

Let u_j be the numerical solution of $u(j\Delta x, t)$, where Δx is the spatial increment and $x = j\Delta x$. Some of the properties of the ODE system resulting from the method of lines can be best illustrated by the following examples:

Consider the semidiscrete approximation

$$\frac{du_j}{dt} = -\frac{(u_{j+1} - u_{j-1})}{2\Delta x} \tag{19.268}$$

for the PDE $\partial u/\partial t + \partial u/\partial x = 0$ with periodic boundary condition, with u_x approximated by a centered difference formula. In matrix notation, this semi-discrete approximation can be rewritten as

$$\frac{d\bar{u}}{dt} = G\bar{u} \tag{19.269}$$

The eigenfunction associated with the matrix G are simply complex exponentials $e^{i\omega x}$, and the eigenvalues are $i(\sin \omega \Delta x/\Delta x)$. Here $\bar{u}$ is a vector and ω is the frequency. Thus, if we want to solve the resulting system of ODE's, we should use a method which is appropriate for a system of ODE's which has purely imaginary eigenvalues [Lambert (1979)].

Now consider the parabolic equation $\partial u/\partial t = \partial^2 u/\partial x^2$, with spatial differencing

$$\frac{du_j}{dt} = \frac{(u_{j+1} - 2u_j + u_{j-1})}{(\Delta x)^2} \tag{19.270}$$

The discrete operator has real eigenvalues $-4(\sin^2 \omega \Delta x/(\Delta x)^2)$ and eigenvectors $e^{i\omega x}$. The eigenvalues here are real and non-positive. In this case, we need an ODE method that is good for systems with widely varying eigenvalues. Such systems are called *stiff* and the appropriate ODE solvers to use are *stiff solvers* [Gear (1971)].

Next, consider the hyperbolic–parabolic model equation

$$\frac{\partial u}{\partial t} + c\frac{\partial u}{\partial x} = \sigma\frac{\partial^2 u}{\partial x^2}. \tag{19.271}$$

If we apply the method of lines to this equation, we will obtain an ODE system with eigenvalues

$$\lambda = -4\sigma\frac{\sin^2 \omega \Delta x}{(\Delta x)^2} - ic\frac{\sin \omega \Delta x}{\Delta x} \tag{19.272}$$

The eigenvalues lie in the left half of the complex λ plane. In this case, again we need a stiff ODE solver.

Full Discretization. Not all difference methods can be separated into two different composite difference operators (time and space) as in the method of lines. As a matter of fact, many of the classical approaches use *full discretization* [Richtunger and Morton (1967)]. The method of lines is only a special subclass of the *full discretization* difference scheme.

In general, there is no best method for obtaining approximating difference formulae. The only requirement is that the difference formula must pass certain tests for accuracy, consistency, stability, and convergence. Difference formulas can be classified into two types: *explicit* and *implicit*. An explicit formula involves only one grid point at the advanced time level $t = (n + 1)\Delta t$, with Δt the time step. An implicit formula involves more than one grid point at the advanced time level $t = (n + 1)\Delta t$.

The *Lax–Wendroff method* is an example of a fully discretized difference scheme. Let u_t^n be the numerical solution of $u(j\Delta x, n\Delta t)$ and consider a hyperbolic equation

$$\frac{\partial u}{\partial t} + c\,\frac{\partial u}{\partial x} = 0 \tag{19.273}$$

The Lax–Wendroff method starts from a Taylor series in t

$$u_j^{n+1} = u_j^n + \Delta t\left(\frac{\partial u}{\partial t}\right)_j^n + \frac{\Delta t^2}{2}\left(\frac{\partial^2 u}{\partial t^2}\right)_j^n + \cdots \tag{19.274}$$

The t-derivatives are then replaced by x-derivatives by means of Eq. (19.273), namely

$$\frac{\partial^2 u}{\partial t^2} = -\frac{\partial}{\partial t}\,c\,\frac{\partial u}{\partial x} = -c\,\frac{\partial}{\partial x}\,\frac{\partial u}{\partial t} = c^2\,\frac{\partial^2 u}{\partial x^2} \tag{19.275}$$

The x-derivatives are approximated by a 3-point central difference approximation resulting in the following form

$$u_j^{n+1} = u_j^n - c\Delta t D_x(u_j^n) + \frac{(c\Delta t)^2}{2}\,D_{xx}(u_j^n) \tag{19.276}$$

The *MacCormack (1969) method* is a two step variant of the Lax–Wendroff scheme and can be written as

$$u_j^* = u_j^n - v(u_{j+1}^n - u_j^n) \tag{19.277}$$

$$u_j^{n+1} = \frac{1}{2}(u_j^n + u_j^*) - \frac{v}{2}(u_j^* - u_{j-1}^*)$$

Next, we briefly define some basic terminology for difference schemes.

Consistency: A finite difference representation of a PDE is said to be consistent if the difference between the PDE and the finite difference equation vanishes as the mesh is refined.

Truncation Error: Truncation error is defined as the difference between the PDE and the difference approximation to it.

Convergence: Convergence means that the solution to the finite difference equation approaches the true solution to the PDE having the same initial and boundary conditions as the mesh is refined.

Stability: A numerical scheme is said to be stable if the numerical solution is bounded as the mesh is refined (i.e., $\Delta x \to 0$) for a fixed value of $n\Delta t$ and a fixed relation between Δt and Δx. There are usually two kinds of stable schemes: conditional stability and unconditional stability. The conditional stability usually amounts to a restriction on the permissible size of Δt in terms of the sizes of the other increments, whereas unconditional stability amounts to no restriction on the size of Δt and usually is associated with implicit methods. Unconditionally stable methods are most suitable for problems with widely varying characteristic speeds and/or length scales (stiff systems, e.g., unsteady transonic flow).

Lax's Equivalence Theorem: Given a well-posed initial value problem and a finite difference approximation to it that satisfies the consistency condition, stability is the necessary and sufficient condition for convergence [Richtunger and Morton (1967)].

For most of the difference methods (for scalar equations), the *von Neumann test* is necessary and sufficient for stability of the initial value problem. In the von Neumann analysis, one considers a pure initial value problem and assumes that the solution of the difference equation is spatially periodic

$$u_j^n = z^n(\omega)e^{i\omega j \Delta x} \tag{19.278}$$

where z^n is the Fourier coefficient, ω is the Fourier variable and $i = \sqrt{-1}$. The amplification factor $\mathcal{G}$ defined by

$$z^{n+1} = \mathcal{G}(\omega)z^n \tag{19.279}$$

is determined by inserting (19.278) for u_j^n and u_j^{n+1} into the difference equation.

By using the von Neumann analysis, one can obtain, for example, the stability bound of the Lax–Wendroff method for the hyperbolic model equation as

$$|c| \frac{\Delta t}{\Delta x} \leq 1, \tag{19.280}$$

whereas, the *Crank–Nicholson scheme*

$$\left[1 + \frac{c\Delta t}{2} D_x\right] u_j^{n+1} = \left[1 - \frac{c\Delta t}{2} D_x\right] u_j^n \tag{19.281}$$

is unconditionally stable. It is customary to refer to $|c|(\Delta t/\Delta x)$ as the *CFL number* named after Courant, Friedrichs, and Lewy. Let $\nu = c(\Delta t/\Delta x)$, $\gamma = \sigma[\Delta t/(\Delta x)^2]$, and denote the accuracy of the scheme by (q_1, q_2) where q_1 is the order of accuracy in space and q_2 is the order of accuracy in time. The following is a list of stability bounds and order of accuracy for a few classical consistent schemes for hyperbolic and parabolic model equations.

For the hyperbolic model equation:

i) Leap-frog method

$$u_j^{n+1} = u_j^{n-1} - 2c\Delta t D_x(u_j^n) \tag{19.282}$$
$$\text{Stable if } |\nu| \leq 1$$
$$\text{Accuracy: } (2, 2)$$

ii) Crank–Nicholson method

$$\left[1 + \frac{c\Delta t}{2} D_x\right] u_j^{n+1} = \left[1 - \frac{c\Delta t}{2} D_x\right] u_j^n \qquad (19.283)$$

Unconditionally stable
Accuracy: (2, 2)

iii) Lax–Wendroff method

$$u_j^{n+1} = u_j^n - c\Delta t D_x(u_j^n) + \frac{(c\Delta t)^2}{2} D_{xx}(u_j^n) \qquad (19.284)$$

Stable if $|\nu| \leq 1$
Accuracy: (2, 2)

iv) MacCormack method

$$u_j^* = u_j^n - \nu(u_{j+1}^n - u_j^n)$$

$$u_j^{n+1} = \frac{1}{2}(u_j^n + u_j^*) - \frac{\nu}{2}(u_j^* - u_{j-1}^*) \qquad (19.285)$$

Stable if $|\nu| \leq 1$
Accuracy: (2, 2)

v) Forward Euler (time) method

$$u_j^{n+1} = u_j^n - c\Delta t D_x(u_j^n) \qquad (19.286)$$

Unconditionally unstable
Accuracy: (2, 1)

vi) Backward Euler (time) method

$$(1 + c\Delta t D_x)u_j^{n+1} = u_j^n \qquad (19.287)$$

Unconditionally stable
Accuracy: (2, 1)

For the parabolic model equation:

i) Leap-frog method

$$u_j^{n+1} = u_j^{n-1} + 2\sigma\Delta t D_{xx}(u_j^n) \qquad (19.288)$$

Unconditionally unstable
Accuracy: (2, 2)

ii) Crank–Nicholson method

$$\left(1 - \frac{\sigma\Delta t}{2} D_{xx}\right) u_j^{n+1} = \left(1 + \frac{\sigma\Delta t}{2} D_{xx}\right) u_j^n \qquad (19.289)$$

Unconditionally stable
Accuracy: (2, 2)

iii) Forward Euler (time) method

$$u_j^{n+1} = u_j^n + \sigma \Delta t D_{xx}(u_j^n) \tag{19.290}$$
$$\text{Stable if } |\gamma| \leq 1/2$$
$$\text{Accuracy: } (2, 1)$$

19.5.4 Effect of Numerical Boundary Conditions on the Stability and Accuracy of Finite Difference Schemes

When a finite difference approximation is used to obtained a solution of the PDE, it is essential to start with a well-posed problem. The PDE is defined only when a proper set of initial and/or boundary conditions is given. We cannot expect our difference approximations to yield a reasonable solution if the PDE does not have a reasonable solution. In many instances, a good understanding of the theory of well-posed problems can guide us to exclude many boundary conditions which might look physically reasonable. A basic requirement for a *well-posed initial boundary value problem* (IBVP) is to not overspecify or underspecify the boundary conditions with given smooth initial data. In order to mathematically define a well-posed IBVP, we have to establish the existence and uniqueness of the solution and its continuous dependence on the initial and boundary data or to establish the existence of certain *a priori* estimates or energy inequalities. From here on we assume all problems under consideration are well-posed. For general references, see Oliger and Sundstrom (1978) and Kreiss and Oliger (1973), and for applications to gas dynamic problems, see Yee (1981).

The main purpose of this section is to discuss the stability of finite difference approximations to the IBVP for the hyperbolic model equation [Osher (1969), Gustafsson *et al.* (1992), Gustafsson and Oliger (1982), Yee *et al.* (1982), Beam *et al.* (1982), and Warming *et al.* (1983)]. In general, stability analysis for IBVP of the parabolic type are less complicated [Kreiss and Oliger (1973)].

Consider the hyperbolic model equation

$$\frac{\partial u}{\partial t} + c \frac{\partial u}{\partial x} = 0, \quad 0 \leq x \leq l, \quad t \geq 0 \tag{19.291a}$$

where c is a real constant. Initial data are given at $t = 0$

$$u(x, 0) = f(x), \quad 0 \leq x \leq l \tag{19.291b}$$

and the problem is well-posed if boundary values are prescribed at $x = 0$

$$u(0, t) = g(t), \quad \text{for} \quad c > 0 \tag{19.291c}$$

or at $x = l$

$$u(l, t) = g(t), \quad \text{for} \quad c < 0 \tag{19.291d}$$

The boundary conditions required for a well-posed problem are generally referred to as *analytical boundary conditions*.

If one considers a difference approximation to a hyperbolic IBVP, the numerical scheme may require more boundary conditions than the analytical boundary conditions needed for the PDE (a typical example is the use of spatially centered schemes). These additional boundary conditions for finite-difference equations are often called *numerical boundary conditions*. The numerical boundary conditions cannot be imposed arbitrarily but are determined, in general, using interior information, e.g., by extrapolation or uncentered approximations. The numerical procedure used to provide a numerical boundary condition will be called a *boundary scheme*, and the numerical scheme used for the interior points will be called an *interior scheme*.

In principle, one would like to avoid difference methods that require boundary schemes. An example of a scheme which does is the first-order upwind scheme

$$u_j^{n+1} = u_j^n - c \frac{\Delta t}{\Delta x} (u_j^n - u_{j-1}^n), \quad c > 0. \tag{19.292}$$

However, for nonlinear coupled systems in conservation law form with eigenvalues of both positive and negative signs, upwind schemes generally require boundary schemes. In practice, almost all numerical algorithms for hyperbolic systems require some special treatment at the boundaries.

During the 1960's and early 1970's Godunov and Ryabenkii [see Richtmyer and Morton (1967)], Kreiss and Oliger (1973), Osher (1969), and Gustafsson *et al.* (1972) developed a stability theory for difference approximations to IBVP's. The algebraic condition to check for stability evolved from this theory is sometimes called the *normal mode analysis*. Since the theory is fairly complicated to explain, interested readers should consult references given above. Application of the theory to model equations can be found in Gustafsson and Oliger (1982), Yee *et al.* (1982), Beam *et al.* (1982), and Warming *et al.* (1983). Here we cite two results obtained by applying the normal mode analysis to analyze the stability of difference approximations for the model hyperbolic IBVP, Eq. (19.291a).

First, we consider the leap-frog scheme, Eq. (19.282), as the interior scheme with spatial grid indexing 0, 1, 2, . . . , J. Since this is a three point scheme, we need a numerical boundary condition on the left boundary. Let us consider space extrapolation

$$u_0^n = u_1^n \tag{19.293}$$

as a boundary scheme. Normal mode analysis shows that the combined scheme (interior scheme + boundary scheme) is unstable. However, if instead of Eq. (19.293) we consider the space–time extrapolation

$$u_0^n = u_1^{n-1} \tag{19.294}$$

as the boundary scheme, then the resulting combined scheme will be stable for the IBVP for $-c(\Delta t/\Delta x) \leq 1$.

Next, we consider the Crank–Nicholson scheme (implicit scheme), Eq. (19.281), with the space-time extrapolation, Eq. (19.294). Normal mode analysis shows that the combined scheme has a stability bound of $|c|(\Delta t/\Delta x) \leq 2$, although the corresponding initial value problem ($-\infty \leq x \leq \infty$) is unconditionally stable. If instead of Eq. (19.294) we use the space extrapolation, Eq. (19.293), then the combined scheme is unconditionally stable.

We emphasize that if one combines an unconditionally stable interior scheme with a boundary scheme, the combined scheme may be unstable, conditionally stable, or unconditionally stable for the IBVP.

As for accuracy of the combined scheme, Gustafsson (1975) has proved that the boundary scheme, in general, can be one order lower than the interior scheme without reducing the global order of accuracy. Thus, if the interior scheme is second-order accurate in space and the boundary scheme is first-order accurate, then the combined scheme can retain second-order global accuracy in space.

Therefore, from the above examples, we can see that improper treatment of the numerical boundary conditions can lead to instability and/or inaccuracy, even though one starts with a stable and consistent interior scheme. Consequently, it is essential that stability and accuracy analyses include both the interior scheme and the boundary scheme. All of the above theories are based on linear stability analysis. In general, the theory does not apply to nonlinear problems with discontinuous solutions.

19.5.5   Fractional Step and Alternating Direction Implicit (ADI) Methods for Problems in More Than One Space Dimension

For the method of lines approach or a two level time difference scheme, it is quite straightforward to extend these methods from one space dimension to two or three dimensions. However, in practice it is advisable not to do so for implicit methods. The reason is that the resulting systems of equations become unreasonably expensive to solve, and application to practical problems is almost impossible. Therefore, other methods have been devised. The earliest such methods were *fractional step (splitting)* methods [Yanenko (1971) and LeVeque (1982)] and *alternating direction implicit* (ADI) methods [Peaceman and Rachford (1955), Douglas (1955), Steger and Warming (1981), and Warming and Beam (1977)]. The method of fractional steps is applicable to both explicit and implicit methods. One can also use the fractional step approach to split the convection and diffusion terms or to split an equation with both stiff and nonstiff parts [LeVeque (1982)]. We will discuss the fractional step approach first.

Fractional Step Method.   Consider an equation

$$\frac{\partial u}{\partial t} = (A + B)u \tag{19.295}$$

where A and B may be matrices if the method of lines is used. (Also, A and B can be either differential operators $\partial/\partial x$ and $\partial/\partial y$ or $\partial/\partial x$ and $\partial^2/\partial x^2$ depending on whether dimensional splitting or the splitting of the convection and diffusion terms is desired.) Let $h = \Delta t$ and assume A and B commute, then we can write the solution as

$$u(t + h) = e^{Ah}e^{Bh}u(t). \tag{19.296}$$

If $\mathcal{L}_A^h$ and $\mathcal{L}_B^h$ are operators of approximate methods for the reduced problems $\partial u/\partial t = Au$ and $\partial u/\partial t = Bu$, respectively, we can approximate Eq. (19.295) by

$$u^{n+1} = \mathcal{L}_A^h \mathcal{L}_B^h u^n \qquad (19.297)$$

based on Eq. (19.296). Here, we assume the time differencing is two level, and we have used the notation $u(t + h)$ to be the exact solution to Eq. (19.295) at $t + h$ and u^{n+1} to be the approximate solution to $u(t + h)$. The stability of Eq. (19.297) follows immediately from the stability of the component methods. However, if A and B do not commute, Eq. (19.296) results in an error of $\mathbf{O}(h^2)$. Strang [see LeVeque (1982)] has observed that $e^{Ah/2}e^{Bh}e^{Ah/2}u(t)$ is one order more accurate. Thus, to retain the original time accuracy of the scheme, we have to use

$$u^{n+1} = \mathcal{L}_A^{h/2} \mathcal{L}_B^h \mathcal{L}_A^{h/2} u^n. \qquad (19.298)$$

Noting that $e^{Ah/2}e^{Ah/2} = e^{Ah}$, we have

$$u(t + 2h) = e^{Ah/2}e^{Bh}e^{Ah}e^{Bh}e^{Ah/2}u(t) \qquad (19.299)$$

So, if one handles the intermediate boundary conditions correctly, one only needs to do the *half steps* at the begining and immediately before printout, i.e.,

$$u^{n+2} = \mathcal{L}_A^{h/2} \mathcal{L}_B^h \mathcal{L}_A^h \cdots \mathcal{L}_B^h \mathcal{L}_A^{h/2} u^0 \qquad (19.300)$$

where u^0 is the initial condition.

The method of fractional steps is more apparent if we look at the following example. Let Δy be the spatial increments in the y direction, and let $u_{j,k}$ approximate the solution of u at $t = n\Delta t$, $x = j\Delta x$, and $y = k\Delta y$. Consider the hyperbolic equation

$$\frac{\partial u}{\partial t} + \frac{\partial u}{\partial x} + \frac{\partial u}{\partial y} = 0 \qquad (19.301)$$

Using the Crank–Nicholson approximation for both $\mathcal{L}_A$ and $\mathcal{L}_B$, we have

$$\left[1 + \frac{\Delta t}{4} D_x \right] u_{j,k}^{1/2} = \left[1 - \frac{\Delta t}{4} D_x \right] u_{j,k}^0$$

$$\left[1 + \frac{\Delta t}{2} D_y \right] u_{j,k}^{3/2} = \left[1 - \frac{\Delta t}{2} D_y \right] u_{j,k}^{1/2}$$

$$\left[1 + \frac{\Delta t}{2} D_x \right] u_{j,k}^{5/2} = \left[1 - \frac{\Delta t}{2} D_x \right] u_{j,k}^{3/2}$$

$$\vdots$$

$$\left[1 + \frac{\Delta t}{2} D_y \right] u_{j,k}^{n+3/2} = \left[1 - \frac{\Delta t}{2} D_y \right] u_{j,k}^{n+1/2}$$

$$\left[1 + \frac{\Delta t}{2} D_x \right] u_{j,k}^{n+2} = \left[1 - \frac{\Delta t}{4} D_x \right] u_{j,k}^{n+3/2} \qquad (19.302)$$

where $D_x u_{j,k} = (1/2\Delta x) (u_{j+1,k} - u_{j-1,k})$ and $D_y u_{j,k} = (1/2\Delta y) (u_{j,k+1} - u_{j,k-1})$.

ADI Method. The ADI methods also reduce the problem to a sequence of one-dimensional operators so that a more practical recursive solution process can be employed for the implicit methods. The basic idea of the ADI method is due to Peaceman and Rachford (1955) and Douglas (1955) who developed alternating-direction methods for parabolic equations. The method is best illustrated by an example. Consider the parabolic equation

$$\frac{\partial u}{\partial t} = a_1 \frac{\partial^2 u}{\partial x^2} + a_2 \frac{\partial^2 u}{\partial y^2}, \quad a_1, a_2 > 0 \tag{19.303}$$

As a notational convenience, we keep spatial derivatives continuous; in practice some appropriate difference formula will be used with $u_{j,k}^n$ denoting the discrete approximation of u. For the time differencing, we use the *trapezoidal method*. Then, an ADI method of Eq. (19.303) can be written as

$$\left[1 - \frac{a_1 \Delta t}{2} \frac{\partial^2}{\partial x^2} \right] u^{n+1/2} = \left[1 + \frac{a_2 \Delta t}{2} \frac{\partial^2}{\partial y^2} \right] u^n \tag{19.304a}$$

$$\left[1 - \frac{a_2 \Delta t}{2} \frac{\partial^2}{\partial y^2} \right] u^{n+1} = \left[1 + \frac{a_1 \Delta t}{2} \frac{\partial^2}{\partial x^2} \right] u^{n+1/2} \tag{19.304b}$$

Equation (19.304) can be written in a single step by eliminating $u^{n+1/2}$ from Eq. (19.304a) to obtain

$$\left[1 - \frac{a_1 \Delta t}{2} \frac{\partial^2}{\partial x^2} \right]\left[1 - \frac{a_2 \Delta t}{2} \frac{\partial^2}{\partial y^2} \right] u^{n+1} = \left[1 + \frac{a_1 \Delta t}{2} \frac{\partial^2}{\partial x^2} \right]\left[1 + \frac{a_2 \Delta t}{2} \frac{\partial^2}{\partial y^2} \right] u^n$$

$$\tag{19.305}$$

Equation (19.305) can be rewritten as

$$\left[1 - \frac{a_1 \Delta t}{2} \frac{\partial^2}{\partial x^2} \right]\left[1 - \frac{a_2 \Delta t}{2} \frac{\partial^2}{\partial y^2} \right] (u^{n+1} - u^n) = \Delta t \left[a_1 \frac{\partial^2 u}{\partial x^2} + a_2 \frac{\partial^2 u}{\partial y^2} \right]^n \tag{19.306}$$

The solution procedure is to solve first

$$\left[1 - \frac{a_1 \Delta t}{2} \frac{\partial^2}{\partial x^2} \right] u^* = \Delta t \left[a_1 \frac{\partial^2 u}{\partial x^2} + a_2 \frac{\partial^2 u}{\partial y^2} \right]^n \tag{19.307a}$$

for u^*, and then to solve

$$\left[1 - \frac{a_2 \Delta t}{2} \frac{\partial^2}{\partial y^2} \right] (u^{n+1} - u^n) = u^* \tag{19.307b}$$

for u^{n+1}.

Extension of the ADI methods to systems of parabolic and hyperbolic PDE's following the same procedure. See for example Steger and Warming (1981) and

Warming and Beam (1977). Application of difference methods to gas dynamics can be found in Roach (1972), Peyret and Taylor (1983), Beam and Warming (1982), and Turkel (1982).

19.5.6 Difference Methods for Nonlinear Hyperbolic Conservation Laws

So far, we have only discussed the stability of difference schemes for constant coefficient PDE's [i.e., the model equations, Eq. (19.259) and Eq. (19.264)]. In practical problems the PDE's are often nonlinear. It is standard practice that the stability theory for linear difference equation is of use in checking the *local stability* of linearized equations obtained from truly nonlinear equations. Practical experience shows that instabilities usually start as local phenomena, at least for mildly nonlinear problems. However, in many instances when strong discontinuities are present, local stability is neither necessary nor sufficient for the nonlinear problems. One traditional remedy is to introduce *linear numerical dissipation* or *artificial viscosity* or a *smoothing term* into the difference schemes. One can do so by designing the scheme so that the eigenvalues, λ, of the amplication matrix satisfy an inequality of the form

$$|\lambda| \leq 1 - \delta|\omega\Delta x|^{2r}, \quad \text{when } |\omega\Delta x| \leq \pi \qquad (19.308)$$

for some real constant $\delta > 0$ and positive integer r. A difference approximation that satisifies (19.308) is called *dissipative of order 2r*. For example, the Lax–Wendroff scheme is dissipative of order 2.

The majority of fluid dynamic practical applications before the mid-1980s was based on this traditional approach of adding an additional dissipation term [Roe (1983) and Beam and Warming (1982)] to their scheme to improve nonlinear instability. However, this approach alone will not guarantee convergence to a physically correct solution in the nonlinear case.

Here, we give a few words on the conservative form of the Navier–Stokes equations before discussing schemes that are more appropriate for the calculation of shocks. At the present stage of development, numerical computation of the Navier–Stokes equations of gas dynamics is concentrated on improving the method for the inviscid part of the equations; i.e., the Euler equations. The reason is that mathematical modelling of the viscous terms for most physical problems is not in a well developed stage. Also, it seems that the slow convergence rate in solving both the Euler and the Navier–Stokes equations is associated with the inviscid or Euler terms when strong viscous effects are not present. Therefore, we will concentrate on the analysis of the inviscid part. The Euler equations, if written in conservative form, are a highly nonlinear system of hyperbolic conservation laws.

There are three major difficulties in using schemes which are developed under the guidelines of linear theory to compute weak solutions of a nonlinear hyperbolic system, such as the compressible Euler equations:

i) Schemes that are second- (or higher) order accurate may produce oscillations wherever the solution is not smooth.

ii) Nonlinear instabilities may develop in spite of the stability in the constant coefficient case.

iii) The scheme may select a nonphysical solution.

Lax (1972) showed that the limit solution of any finite difference scheme in a conservation form which is consistent with the conservation laws satisfies the jump conditions across a discontinuity *automatically*. This was a conceptual breakthrough which enabled the direct discretization of the conservation laws by introducing the notion of numerical dissipation. However, weak solutions (solutions with shocks and contact discontinuities) of hyperbolic conservation laws are not uniquely determined by their initial values; an entropy condition is needed to select the physically relevant solutions. The question arises whether finite difference approximations converge to this particular solution. It is shown in Harten *et al.* (1976) and Crandall and Majda (1980) that in the case of a single conservation law, *monotone schemes* (to be defined later), when convergent, always converge to the physically relevant solution. If the scheme is not monotone, then the scheme must be consistent with an entropy inequality for the insurance of convergence to a physically relevant solution.

Thus, monotone schemes are guaranteed to not have difficulties ii) and iii) above. However, monotone schemes are only first-order accurate. Consequently, they are not accurate enough to compute complicated flow fields unless extremely small grid spacing is used.

To overcome the above difficulties, a new class of schemes that is more appropriate for the computation of weak solutions of nonlinear hyperbolic conservation laws is considered. These schemes are required to be: 1) total variation diminishing in the nonlinear scalar case and the constant coefficient system case, 2) consistent with the conservation law and an entropy inequality [see Harten *et al.* (1976)], and 3) at least second-order accurate away from shocks. The first property guarantees that the scheme does not generate spurious oscillations. We refer to schemes with this property as *Total Variation Diminishing* (TVD) schemes. The second property guarantees that the weak solutions are physical ones. Schemes in this class are guaranteed to avoid difficulties i) to iii) mentioned above (for one-dimensional, nonlinear scalar conservation laws and constant coefficient hyperbolic systems).

The class of TVD schemes contains monotone schemes, but is significantly larger as it can include second-order (or higher) accurate schemes. Existence of second-order accurate TVD schemes was demonstrated in van Leer (1979), Collela and Woodward (1982), Harten (1983), Roe and Baines (1982), and Osher (1984). Unlike monotone schemes, TVD schemes are not automatically consistent with the entropy inequality. Consequently, some mechanism may have to be explicitly added to a TVD scheme in order to enforce the selection of the physical solution. Harten (1983) and Harten and Hyman (1980) have demonstrated a way of modifying a TVD scheme to be consistent with an entropy inequality. The following is an introduction to monotone, explicit and implicit TVD schemes.

Monotone and Explicit TVD Schemes. Consider the scalar hyperbolic conservation law

$$\frac{\partial u}{\partial t} + \frac{\partial f(u)}{\partial x} = 0 \tag{19.309}$$

where $a(u) = \partial f/\partial u$ is the characteristic speed. A general three-point two-level explicit difference scheme in conservation form can be written as

$$u_j^{n+1} = u_j^n - \kappa(\bar{f}_{j+1/2}^n - \bar{f}_{j-1/2}^n) \tag{19.310}$$

where $\bar{f}_{j+1/2}^n = \bar{f}(u_j^n, u_{j+1}^n)$, $\kappa = \Delta t/\Delta x$. Here, $\bar{f}$ is commonly called a *numerical flux function*. The numerical flux function, $\bar{f}$, is required to be consistent with the conservation law in the following sense

$$\bar{f}(u_j, u_j) = f(u_j) \tag{19.311}$$

Rewrite Eq. (19.310) as

$$u_j^{n+1} = H(u_{j-1}^n, u_j^n, u_{j+1}^n) \tag{19.312}$$

The numerical scheme, Eq. (19.312), is said to be *monotone* if H is a monotonic increasing function of each of its arguments. Examples of monotone schemes are: *Lax-Friedrick scheme, Godunov scheme*, [see Richtuyer and Morton (1967)], and *Engquist and Osher (1980) scheme*.

Consider a numerical scheme with numerical flux functions of the following form

$$\bar{f}_{j+1/2} = \tfrac{1}{2}[f_j + f_{j+1} - Q(a_j + \tfrac{1}{2})\Delta_{j+1/2}u] \tag{19.313}$$

where $f_j = f(u_j)$, $\Delta_{j+1/2}u = u_{j+1} - u_j$ and

$$a_{j+1/2} = \begin{cases} (f_{j+1} - f_j)/\Delta_{j+1/2}u & \Delta_{j+1/2}u \neq 0 \\ a(u_j) & \Delta_{j+1/2}u = 0 \end{cases} \tag{19.314}$$

Here, Q is a function of $a_{j+1/2}$ and κ. The function Q is sometimes referred to as the *coefficient of numerical viscosity*. Three familiar schemes with the numerical fluxes of the form Eq. (19.313) are:

i) A form of the Lax–Wendroff (L–W) scheme with

$$\bar{f}_{j+1/2} = \tfrac{1}{2}[f_j + f_{j+1} - \kappa(a_{j+1/2})^2\Delta_{j+1/2}u] \tag{19.315}$$

where $Q(a_{j+1/2}) = \kappa(a_{j+1/2})^2$.

ii) Lax–Friedrichs (L–F) scheme with

$$\bar{f}_{j+1/2} = \frac{1}{2}\left[f_j + f_{j+1} - \frac{1}{\kappa}\Delta_{j+1/2}u\right] \tag{19.316}$$

where $Q(a_{j+1/2}) = 1/\kappa$

iii) A generalization of the *Courant–Isaacson–Rees* (GCIR) scheme with

$$\bar{f}_{j+1/2} = \tfrac{1}{2}[f_j + f_{j+1} - |a_{j+1/2}|\Delta_{j+1/2}u] \tag{19.317}$$

where $Q(a_{j+1/2}) = |a_{j+1/2}|$

The total variation of a mesh function, u, is defined to be

$$TV(u) = \sum_{j=-\infty}^{\infty} |u_{j+1} - u_j| = \sum_{j=-\infty}^{\infty} |\Delta_{j+1/2}u| \tag{19.318}$$

The numerical scheme, Eq. (19.310) is said to be TVD if

$$TV(u^{n+1}) \leq TV(u^n) \tag{19.319}$$

It can be shown [Harten (1984)] that a sufficient condition for Eq. (19.310), together with (19.313), to be a TVD scheme is

$$\frac{\kappa}{2} [-a_{j+1/2} + Q(a_{j+1/2})] \geq 0 \tag{19.320a}$$

$$\frac{\kappa}{2} [a_{j+1/2} + Q(a_{j+1/2})] \geq 0 \tag{19.320b}$$

$$\kappa Q(a_{j+1/2}) \leq 1 \tag{19.320c}$$

Applying the above conditions to the three examples, it can be easily shown that the L–W scheme is not a TVD scheme, and the latter two schemes are TVD schemes. Note that there is a further distinction between the L–F scheme and GCIR scheme, the L–F scheme is consistent with an entropy inequality, whereas the GCIR is not. In fact, the L–F scheme is a monotone scheme. Note that entropy satisfying first-order TVD schemes are not necessarily monotone schemes.

It should be emphasized that condition, Eq. (19.320), is only a sufficient condition; i.e., schemes that fail this test might still be TVD. The L–W scheme, besides failing the condition in Eq. (19.320), does not satisfy Eq. (19.319).

Implicit TVD Schemes. Now we consider a one parameter family of three-point conservative schemes of the form

$$u_j^{n+1} + \kappa\eta(\bar{f}_{j+1/2}^{n+1} - \bar{f}_{j-1/2}^{n+1}) = u_j^n - \kappa(1 - \eta)(\bar{f}_{j+1/2}^n - \bar{f}_{j-1/2}^n) \tag{19.321}$$

where η is a parameter, $\bar{f}_{j+1/2}^n = \bar{f}(u_j^n, u_{j+1}^n)$, $\bar{f}_{j+1/2}^{n+1} = \bar{f}(u_j^{n+1}, u_{j+1}^{n+1})$, and $\bar{f}(u_j, u_{j+1})$ is the numerical flux, Eq. (19.313). This one-parameter family of schemes contains implicit as well as explicit schemes. When $\eta = 0$, Eq. (19.321) reduces to Eq. (19.310), the explicit method. When $\eta \neq 0$, Eq. (19.321) is an implicit scheme. For example: if $\eta = 1/2$, the time differencing is the trapezoidal formula and if $\eta = 1$, the time differencing is the backward Euler method. To simplify the notation, we will rewrite Eq. (19.321) as

$$L \cdot u^{n+1} = R \cdot u^n, \tag{19.322}$$

where L and R are the following finite-difference operators

$$(L \cdot u)_j = u_j + \kappa\eta(\bar{f}_{j+1/2} - \bar{f}_{j-1/2}) \tag{19.323a}$$

$$(R \cdot u)_j = u_j - \kappa(1 - \eta)(\bar{f}_{j+1/2} - \bar{f}_{j-1/2}) \tag{19.323b}$$

A sufficient condition for Eq. (19.321) to be a TVD scheme is that

$$TV(R \cdot v) \le TV(v) \tag{19.324a}$$

and

$$TV(L \cdot v) \ge TV(v) \tag{19.324b}$$

A sufficient condition for Eq. (19.324) is the CFL-like restriction

$$|\kappa a_{j+1/2}| \le \kappa Q(a_{j+1/2}) \le \frac{1}{1-\eta} \tag{19.325}$$

where $a_{j+1/2}$ is defined in Eq. (19.314). For a detailed proof of Eq. (19.323) and Eq. (19.324), see Harten (1984). Observe that the backward Euler implicit scheme, $\eta = 1$ in Eq. (19.321) is unconditionally TVD, while the trapezoidal formula, $\eta = 1/2$ is TVD under the CFL-like restriction of 2. The forward Euler explicit scheme, $\eta = 0$ or Eq. (19.310), is TVD under the CFL restriction of 1. We remark that three-point conservative TVD schemes of the form, Eq. (19.321), are generally first-order accurate in space. When $\eta = 1/2$, the scheme is second-order accurate in time.

Generalization of first order TVD schemes to second-order accurate ones is somewhat involved. Interested readers should refer to Harten (1984), Yee and Harten (1987), Yee (1985), (1986), (1987), (1989), Yee and Shinn (1989), Montagne *et al.* (1987), (1988), and Yee *et al.* (1990). Entropy satisfying second-order accurate TVD schemes have the properties that they not generate spurious oscillations across discontinuities and that the weak solutions are physical ones (for nonlinear scalar hyperbolic conservation laws and the constant coefficient hyperbolic systems). The goal of constructing these highly nonlinear schemes is to simulate complex flow fields more accurately. TVD schemes are usually rather expensive to use compared with the conventional shock-capturing methods such as variants of the Lax–Wendroff scheme.

Generalization and application of the TVD schemes to the Euler and Navier–Stokes equations of gas dynamics can be found in Harten (1983), Yee *et al.* (1982), (1985), and the references just cited above. For an overview of the state-of-the-art development of TVD schemes for computational fluid dynamics, see the author's lecture notes [Yee (1989)].

Since the development of higher than first-order TVD schemes, essentially oscillatory schemes (ENO) were developed. ENO schemes are less restricted schemes in the sense that they allow spurious oscillations across discontinuities up to the order of the truncation error. See Yee (1989) for a discussion and references cited therein.

19.6 CONVERGENCE ACCELERATION
W. Kyle Anderson and James L. Thomas

The convergence rate of numerical computations is an important limiting factor when conducting grid convergence studies to verify the numerical accuracy of computational solutions. An integral part of this process is the uniform refinement of the grid in each direction until little or no variation in the solution is observed with increased grid size. Unfortunately, without convergence acceleration, the convergence rate of iterative methods severely degrades as the grid spacing is decreased through the grid refinement process. To mitigate the penalties associated with the use of fine grids, several methods of convergence acceleration have been introduced and are discussed below. These methods are especially important in three dimensions, where an eight-fold increase in the number of grid points occurs when the points are doubled in all three coordinate directions.

19.6.1   Local Time Stepping

Many iterative schemes for obtaining steady-state solutions can be cast in terms of a time-marching algorithm. One of the simplest and most commonly used methods of convergence acceleration in such cases is *local time stepping* [Li (1973)]. Each cell is advanced at its local stability limit so that a spatially varying time step size is used. The time step is generally based on a combination of the flow variables in each cell as well as the cell size.

Perhaps the most commonly used method of local time stepping for inviscid flows is to base the time step in each cell on a local Courant–Fredrichs–Lewy (CFL) number in which the time step in each cell is determined by

$$\Delta t_{\text{cell}} = \text{CFL} \cdot \Delta t_{(\text{CFL}=1)} \tag{19.326}$$

where $\Delta t_{(\text{CFL}=1)}$ is the time step required for a CFL of unity and may be determined with a variety of definitions for multidimensional problems. One form that is commonly used is given by

$$\Delta t_{(\text{CFL}=1)} = \frac{2V}{\displaystyle\int (|\bar{\mathbf{v}} \cdot \bar{\mathbf{n}}| + a)\, dA} \tag{19.327}$$

where the integral is evaluated over the surface of each control volume, $\bar{\mathbf{v}} \cdot \bar{\mathbf{n}}$ is the velocity normal to a cell face, and a is the speed of sound.

An example of the effectiveness of local time stepping is shown in Fig. 19.40, where an explicit method is used to compute the flow around an NACA 0012 airfoil at a Mach number of 0.8 and an angle of attack of 1.25°. The effect of local time stepping is dramatic. By using local time stepping, the residual is reduced 10 orders of magnitude in approximately 3000 iterations; without local time stepping, little progress towards convergence is achieved. The effect is even more dramatic on the lift coefficient, where, with local time stepping, the final lift value is obtained in about 500 iterations. The solution obtained without local time stepping has failed to reach a steady state after 3000 iterations. This technique is very simple and easily implemented and offers a clear advantage toward acceleration of the solution to a steady state.

19.6.2   Residual Smoothing

To accelerate the convergence of explicit algorithms, one methodology that has been extremely effective is *residual smoothing* [Lerat (1979) and Jameson (1983b)]. For

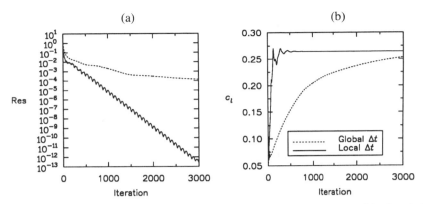

FIGURE 19.40   Effect of local time stepping on convergence rate. (a) Residual and (b) lift.

this method, the steady-state residual calculated at each step is modified in such a way that the support of the scheme is enhanced, which increases the implicitness of the algorithm. In practice, this technique has been particularly effective for central-differencing schemes when used in conjunction with multistage time stepping, although recent improvements for upwind discretizations have been reported [Blazek *et al.* (1991)]. For this reason, the general procedure is outlined below for a four-stage Runge–Kutta type of algorithm, applied to a one-dimensional model problem with central differencing. The effect of residual smoothing on the stability is examined through the application of a Fourier analysis.

Consider the model problem given by

$$u_t + u_x + \mu \Delta x^3 u_{xxxx} = 0 \qquad (19.328)$$

A four-stage Runge–Kutta type method is given by

$$u^{(0)} = u^{(n)}$$
$$u^{(1)} = u^{(0)} - \alpha_1 \Delta t R^{(0)}$$
$$u^{(2)} = u^{(0)} - \alpha_2 \Delta t R^{(1)}$$
$$u^{(3)} = u^{(0)} - \alpha_3 \Delta t R^{(2)}$$
$$u^{(4)} = u^{(0)} - \Delta t R^{(3)}$$
$$u^{(n+1)} = u^{(4)} \qquad (19.329)$$

where $R^{(i)} = \delta_x u^{(i)} + \mu \Delta x^3 \delta_{xxxx} u^{(i)}$ denotes the discretized steady-state residual formed from data at stage level i. Note than in the present form this scheme is one in which the dissipative term is evaluated at each stage. It is possible (and more economical) to use schemes in which the dissipation is only evaluated periodically, for example on the first and third stages [Jameson (1983b)]. However, for illustrative purposes, the dissipation will be evaluated on each stage.

To determine the stability of the current scheme, a Fourier mode is substituted for u

$$u = \hat{U} e^{i\beta x} \qquad (19.330)$$

The Fourier symbol for $\Delta t R^{(i)}$ is now written as

$$\Delta t \hat{R}^{(i)} = -\hat{U}^{(i)} e^{i\beta x} Z \tag{19.331}$$

where

$$Z = -\lambda(i \sin \xi + 4\mu(1 - \cos \xi)^2) \tag{19.332}$$

and $\lambda = \Delta t / \Delta x$ is the Courant number and $\xi = \beta \Delta x$. Substitution of Eqs. (19.330), (19.331), and (19.332) into Eq. (19.329) yields an equation for the amplification factor

$$g = 1 + Z + \alpha_3 Z^2 + \alpha_2 \alpha_3 Z^3 + \alpha_1 \alpha_2 \alpha_3 Z^4 \tag{19.333}$$

which indicates the extent to which errors decrease (or grow) from one iteration to the next. Stability of the scheme requires that $|g| \le 1$ for all Z. By cycling through values of $0 \le \xi \le \pi$, with Eq. (19.332) used in Eq. (19.333), the amplification factor can be obtained for a fixed value of λ and a dissipation coefficient μ. With a *standard* set of coefficients given by $\alpha_1 = 1/4$, $\alpha_2 = 1/3$, $\alpha_3 = 1/2$, and $\mu = 1/32$, the scheme is stable for values of λ up to about 2.8.

To obtain a stable algorithm for higher values of λ, the support of the scheme may be increased by replacing the residuals $(\cdots, R_{i-1}, R_i, R_{i+1}, \cdots)$ at each point with an average of the residuals on either side:

$$\overline{R}_i = \epsilon R_{i-1} + (1 - 2\epsilon)R_i + \epsilon R_{i+1} = (1 + \epsilon\delta_{xx})R_i \tag{19.334}$$

With this modification to the residual, the Fourier symbol Z is now given by

$$Z = -\lambda[i \sin \xi + 4\mu(1 - \cos \xi)^2][1 - 2\epsilon(1 - \cos \xi)] \tag{19.335}$$

With the addition of the second factor, the value of λ may be increased. However, a disadvantage of this procedure is that for certain values of ϵ, $\overline{R}_i$ computed with Eq. (19.334) will be zero if the residuals exhibit an odd-even type of behavior so that no update of the dependent variables at each grid point would occur regardless of the value of R_i at that point. To overcome this difficulty and to allow arbitrary values of ϵ, an average residual may be calculated from an implicit relation given by

$$-\epsilon\overline{R}_{i-1} + (1 + 2\epsilon)\overline{R}_i + \epsilon\overline{R}_{i+1} = (1 - \epsilon\delta_{xx})\overline{R}_i = R_i \tag{19.336}$$

In this manner, the support of the scheme can be made to extend over the entire grid thus relaxing the time step limitation. The Fourier symbol of the resulting scheme is given by

$$Z = \frac{-\lambda[i \sin \xi + 4\mu(1 - \cos \xi)^2]}{[1 + 2\epsilon(1 - \cos \xi)]} \tag{19.337}$$

The denominator is greater than 1 for all values of $\xi > 0$ and reduces the magnitude of Z so that larger time steps can be taken. Venkatakrishnan (1986) shows that in

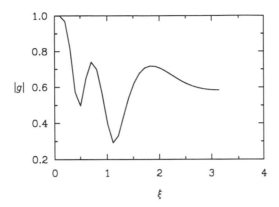

FIGURE 19.41 Amplification factor for standard four-stage scheme with implicit residual smoothing; $\lambda = 7$, $\epsilon = 1\frac{3}{8}$, $\alpha_1 = 1/4$, $\alpha_2 = 1/3$, $\alpha_3 = 1/2$, and $\mu = 1/32$.

the absence of dissipation, stability is maintained for any value of λ if ϵ is chosen so that

$$\epsilon \geq \frac{1}{4}\left[\left(\frac{\lambda}{\lambda^*}\right)^2 - 1\right] \qquad (19.338)$$

where λ^* is the limit for the original scheme without residual averaging.

The success of this technique is demonstrated in Fig. 19.41, where the value of λ^* is assumed to be 2.8, which is the stability limit of the standard four-stage scheme without added dissipation ($\mu = 0$). For example, to achieve stability for $\lambda = 7$, Eq. (19.338) indicates that a value of ϵ of $1\frac{3}{8}$ is appropriate. As shown in the figure, the amplification factor remains below unity for all values of $0 \leq \xi \leq \pi$.

19.6.3 Vector Sequence Extrapolation

Vector-sequence extrapolation is a well-known technique for accelerating the convergence rate of sequences. An example of this is the well-known Aitken-δ^2 method, in which a new sequence is derived from the original sequence, which hopefully converges much faster than the original one [Greenberg (1978)]. Although many variants of this technique and many related algorithms exist, concentration below focuses on one particular method, commonly referred to as *Wynn's ϵ algorithm*. First, a brief discussion of the essential ingredients of vector-sequence extrapolation methods is presented.

For the Aitken-δ^2 method, the sequence is derived by assuming that the original members of the sequence s_n can be adequately described as

$$s = s_n + \kappa\rho^n \qquad (19.339)$$

where s is the limiting value of the sequence and κ and ρ are constants. By evaluating Eq. (19.339) at n, $n + 1$, and $n + 2$, the limiting value of the sequence may be obtained from the solution of the set of simultaneous equations for s, κ, and ρ. If the original sequence is accurately described by Eq. (19.339), then the exact answer will be obtained. If, on the other hand, Eq. (19.339) does not provide a precise description of s_n, then application of the procedure will be only approximate, but may still provide a better estimate for s than is currently available. This estimate for the value of s is given by

$$s_n^{(1)} = \frac{s_{n+2}^{(0)} s_n^{(0)} - (s_{n+1}^{(0)})^2}{s_{n+2}^{(0)} + s_n^{(0)} - 2s_{n+1}^{(0)}} \tag{19.340}$$

where the superscript (0) is used to denote terms in the original sequence. This new sequence may now be used to define another sequence given by

$$s_n^{(2)} = \frac{s_{n+2}^{(1)} s_n^{(1)} - (s_{n+1}^{(1)})^2}{s_{n+2}^{(1)} + s_n^{(1)} - 2s_{n+1}^{(1)}} \tag{19.341}$$

which may supply a further improvement to the limiting value of the sequence. This procedure can be applied repeatedly, increasing the accuracy each time over the previous estimate.

An example of this procedure, borrowed from Greenberg (1978), is given in Table 19.1. The original sequence is taken to be the first nine terms in the series given by

$$\sum_{k=1}^{\infty} \frac{(-1)^{k+1}}{k} \tag{19.342}$$

which is the Taylor's series expansion for $\ln(1 + x)$ evaluated at $x = 1$. In the current example, the solution is obtained to seven-digit accuracy ($\ln(2) = 0.6931472$) at the end of the extrapolation procedure with only the first nine partial sums. Note that to obtain similar accuracy from simply summing the series directly would require approximately 10 million terms [Greenberg (1978)]. The acceleration procedure is clearly very useful in this case.

A shortcoming of the above technique lies in the underlying assumption that the sequence behaves similarly to that described by Eq. (19.339). For this reason, this extrapolation procedure is most effective on geometric series and becomes less effective as the series deviates from this behavior. To overcome this shortcoming, Shanks (1955) derived other extrapolations based on the assumption that the se-

TABLE 19.1   Illustration of Aitken-δ^2 Method

n	s_n	$s_n^{(1)}$	$s_n^{(2)}$	$s_n^{(3)}$	$s_n^{(4)}$
1	1.0	0.7	0.6932773	0.6931489	0.6931472
2	0.5	0.6904762	0.6931058	0.6931467	
3	0.8333333	0.6944444	0.6931633	0.6931474	
4	0.5833333	0.6924242	0.6931399		
5	0.7833333	0.6935897	0.6931508		
6	0.6166667	0.6928571			
7	0.7595238	0.6933473			
8	0.6345238				
9	0.7456349				

quence may be described in the more general form

$$s = s_n + \kappa_1 \rho_1^n + \cdots + \kappa_N \rho_N^n \qquad (19.343)$$

which is referred to as the *Nth order Shank's transformation*. Equation (19.343) is evaluated for five values of n resulting in a set of equations that can be used to solve for s in much the same manner as to obtain Eq. (19.340). Although this higher order transformation may provide a more accurate representation of a general sequence, the implementation in this manner can be inefficient for higher order transforms.

Wynn (1966) describes an efficient algorithm for computing the higher order extrapolations. The extension and application of Wynn's algorithm to systems of equations such as those that arise in Euler solvers is given by Hafez *et al.* (1987). In Fig. 19.42, examples are presented from Hafez *et al.* (1987) in which this procedure is applied to the computation of a NACA 0012 airfoil at transonic conditions. For the computation, an explicit, multigrid, multistage time-stepping scheme is used, and only five terms are included in the initial sequence. Shown in Fig. 19.42(a) is the convergence history obtained by using the ϵ algorithm after 250 time steps. As seen, the residual drops dramatically at this point, which indicates the effectiveness of the acceleration procedure. Note, however, that at the point the acceleration is applied, the residual has been reduced by about 6 orders of magnitude, which should be more than sufficient for obtaining global quantities such as lift and drag. An attempt to apply the procedure earlier in the iteration history is shown in Fig. 19.42(b). As seen, the effectiveness of the current algorithm, when applied after 100

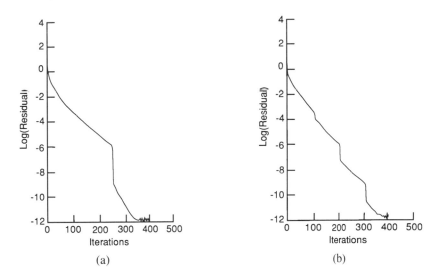

(a)   (b)

FIGURE 19.42 Effect of vector-sequence acceleration for NACA 0012 airfoil; $M_\infty = 0.8$ and $\alpha = 1.25°$. (a) Applied after 250 time steps. (Reprinted with permission from AIAA Paper No. 87-1143, "Applications of Wynn's e-Algorithm to Transonic Flow-Calculations," Hafez, M., Falaniswamy, S., Kuruvils, G., and Salas, M. D., 1987.) (b) Applied after 100, 200, and 300 time steps. (Reprinted with permission from AIAA Paper No. 85-1494, "GMRES Acceleration of Computational Fluid Dynamics Codes," Wigton, L. B., Yu, N. J., and Young, D. P., 1985.)

iterations, is minimal. After 200 iterations, however, a sudden drop in the residual is observed, and a further drop is seen at 300 iterations, where the algorithm is applied once again.

Although the ϵ algorithm can be very effective at achieving dramatic reductions in the residuals, systematic knowledge of when to apply the algorithm and how often is not clear. In the results shown above, the acceleration is impressive, but the implementation requires storage of twice the number of terms included in the initial sequence; for the example given above in which five terms in the initial sequence are used, the solution would need to be stored 10 times. For very large problems, this requirement could become prohibitive.

19.6.4   Multigrid Acceleration

Introduction to Multigrid.   One of the most successful and widely used methods of convergence acceleration is multigrid. The greatest benefit of this method is that the convergence rate (i.e., the spectral radius that measures the ratio of errors at successive time steps) remains constant, independent of the mesh spacing. In this way, solutions can be obtained in $O(N)$ operations (i.e., the computational cost varies linearly with the total number of grid points N). Without convergence acceleration, the computational cost is considerably higher, because of the penalty associated with a deterioration of the convergence rate as the mesh spacing decreases. Although most of the existing theory on multigrid methods pertains specifically to elliptic equations, a number of references [for example, Jameson (1983b), Anderson *et al.* (1988), Ni (1982), Mulder (1985), and Jameson and Baker (1984)] have shown that the multigrid method can greatly accelerate the convergence rate of numerical schemes used to solve the Euler and Navier–Stokes equations.

Description of Multigrid.   The multigrid method most widely used for accelerating the convergence of iterative methods for nonlinear equations, such as the Euler and Navier–Stokes equations, is the *full-approximation scheme* (FAS) that appears in many references [Hackbush (1985) and Brandt (1977), (1982), (1980)] and is summarized below. First, consider the solution of a general nonlinear system of equations

$$L(Q) = S \qquad (19.344)$$

where L is a geneal nonlinear operator, Q is the solution vector of unknowns, and S represents a forcing function. Equation (19.344) is solved numerically by dividing the domain into discrete cells that yield a system of equations to be solved simultaneously at each point as

$$L_N(Q_N) = S_N \qquad (19.345)$$

where Q_N is the exact solution to the discretized system and L_N is the discrete analog of the operator L. If initial conditions are close enough to the final solution, Eq. (19.345) could be solved iteratively with Newton iteration. This approach, however, may be prohibitively expensive if the number of unknowns is large which typically occurs in multidimensional problems. Many other iterative schemes have, therefore, been devised that require significantly fewer operations. After a few iterations, however, these methods generally exhibit a slow convergence rate, which reduces the residuals by a very small amount each time [Brandt (1977)]. The reason for the slow asymptotic convergence rate is the inadequate damping of the low-frequency errors [Stuben and Trottenberg (1982)].

The *multigrid method* efficiently damps the low-frequency errors by using a sequence of grids G_0, G_1, $\ldots$, G_N, where G_N denotes the finest grid, from which successively coarser grids G_{N-1}, G_{N-2}, $\ldots$, G_0 are created. In a structured-grid, full-coarsening algorithm, the coarser grids are constructed by deleting every other grid line in each coordinate direction. In practice, and particularly for unstructured grids, the coarser grids need not be constructed as a subset of the finest grid, (i.e., they can be created independently of the finest grid). In this context, the high-frequency error components on a given grid are those that cannot be resolved on the next coarser mesh because of the increased grid spacing. If an iterative method is chosen that quickly damps the high-frequency errors on a given grid, then after a few iterations, the remaining errors will be those associated with the smoother, low-frequency error components. Because these components appear as higher frequencies on coarser meshes, a sequence of coarser grids can be effectively used to accelerate the convergence rate on the finest grid. Therefore, the low-frequency errors on the fine grid that are usually responsible for slow convergence can be efficiently damped on the coarser grids. These computations are relatively inexpensive so the total overhead of the method is not excessively high. For example, the work required to solve the equations on all the grids in relation to that required to solve on just the finest grid can be estimated by

$$1 + \tfrac{1}{4} + \tfrac{1}{16} + \tfrac{1}{64} + \cdots \approx 1\tfrac{1}{3} \qquad (19.346)$$

for two dimensions and

$$1 + \tfrac{1}{8} + \tfrac{1}{64} + \tfrac{1}{512} + \cdots \approx 1\tfrac{1}{7} \qquad (19.347)$$

for three-dimensional calculations. Note that the above estimates are for structured grids from which coarser grids are formed by removing every other mesh line in all directions. Also, the estimates only account for the number of unknowns on the various grids and do not consider the extra residual computations necessary to compute the relative truncation error.

To use the coarser grids, an equation must be obtained on the fine mesh that can be accurately represented by the coarser mesh. Neither the solution nor the high-frequency error components on the fine grid can generally be resolved on a coarser grid. The high-frequency errors, however, can be sufficiently damped on a fine grid by using iterative schemes specifically designed to damp high-frequency errors in the solution, so that the remaining errors will be composed of only low-frequency components that can be adequately represented on coarser meshes. Because only the low-frequency errors may be represented well on coarser meshes, it is necessary to obtain an equation on the fine mesh in terms of the errors.

To solve iteratively, Eq. (19.345) is solved approximately at each step as

$$L_N(q_N^c) = S_N + R_N \qquad (19.348)$$

where q_N^c is the most current approximation to Q_N and R_N is the residual that will be zero only when $q_N^c = Q_N$. Hence, the exact discrete solution is obtained. Equation 19.348 is subtracted from Eq. (19.345) to yield an equation on the finest grid in terms of the residual

$$L_N(Q_N) - L_N(q_N^c) = -R_N \tag{19.349}$$

With the assumption that the high-frequency errors have been previously smoothed, the fine-grid residual, Eq. (19.349), can be adequately approximated on a coarser mesh as

$$L_{N-1}(Q_{N-1}) = \hat{I}_N^{N-1}(-R_N) + L_{N-1}(I_N^{N-1}q_N^c) \tag{19.350}$$

where I_N^{N-1} and $\hat{I}_N^{N-1}$ are restriction operators for transferring both the dependent variables and the residual from the fine grid to the coarse grid, respectively. Here, $I_N^{N-1}q_N^c$ serves as an initial approximation to the solution on the coarse mesh; Q_{N-1} is the exact solution of the coarse-grid problem and is the sum of the initial approximation and a correction [Brandt (1980)]. Because the full solution is computed and stored on each grid level (as opposed to only the corrections, which is all that is required for a linear equation), this process is referred to as the FAS.

On a sufficiently coarse grid, Eq. (19.350) can be solved exactly with a variety of numerical techniques to obtain Q_{N-1}, from which the coarse-grid correction can be formed as

$$V_{N-1} = Q_{N-1} - I_N^{N-1}q_N^c \tag{19.351}$$

This can then be transferred to the fine grid and used as a correction to q_N^c, which is replaced by its previous value plus the prolongated correction

$$q_N^c \leftarrow q_N^c + \hat{I}_{N-1}^N V_{N-1} \tag{19.352}$$

This process yields a simple FAS two-level algorithm where the operations on the coarse grid, Eqs. (19.350)–(19.352) used to update the fine-grid solution are termed the *coarse-grid correction*. Normally, the exact solution of Eq. (19.350) can be expensive to obtain. Also, because the correction on the coarse grid serves only as an approximation to the fine-grid correction, the exact solution of Eq. (19.350) is not required. Therefore, instead of solving Eq. (19.350) to completion, several iterations can be carried out to get a reasonable approximation of Q_{N-1}. For an approximate solution q_{N-1}^c of Eq. (19.350), a corresponding coarse-grid residual R_{N-1} can be defined from

$$L_{N-1}(q_{N-1}^c) = \hat{I}_N^{N-1}(-R_N) + L_{N-1}(I_N^{N-1}q_N^c) + R_{N-1} \tag{19.353}$$

whose solution differs from the solution of Eq. (19.350) only by the residual term R_{N-1}, which will be zero when $q_{N-1}^c = Q_{N-1}$. If the errors are smooth, then subtraction of Eq. (19.353) from Eq. (19.350) yields an equation that can be well represented on a still coarser mesh G_{N-2}. If this equation is written on G_{N-2}, then

$$L_{N-2}(Q_{N-2}) = \hat{I}_{N-1}^{N-2}(-R_{N-1}) + L_{N-2}(I_{N-1}^{N-2}q_{N-1}^c) \tag{19.354}$$

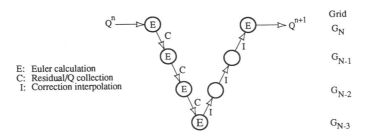

FIGURE 19.43 Multigrid V-cycle.

where Eq. (19.353) is used to determine R_{N-1}. The solution may be obtained in one of three ways: 1) by solving Eq. (19.354) exactly, 2) by approximating by several iterations, or 3) by introducing more coarse-grid levels. On all coarse grids, one or more FAS cycles (smoothing followed by coarse-grid correction) are completed. In this manner, each of the coarse meshes is used to obtain a correction for the solution on the next finest mesh. Because only the equations for smooth error components may be represented well on coarser grids, only the corrections (and not the full solution) must be passed from a coarse grid to the next finest grid.

When only one FAS cycle is carried out for each of the coarser grids, the resulting global strategy is termed a *V-cycle*, which is depicted in Fig. 19.43 for an application of multigrid to the Euler equations. Another cycling strategy of interest, which is shown in Fig. 19.44, is termed a *W-cycle* and results when two FAS cycles are used on each of the coarse meshes. In both of the figures, the corrections from the coarser grids are prolongated to the next finest mesh with trilinear interpolation, and no additional iteration steps are taken between meshes.

Note that Eq. (19.350) can be recast by using Eq. (19.348) as

$$I_{N-1}(Q_{N-1}) = S_{N-1} + \tau_{N-1} = P_{N-1} \tag{19.355}$$

where

$$S_{N-1} = \hat{I}_N^{N-1} S_N \tag{19.356}$$

$$\tau_{N-1} = L_{N-1}(I_N^{N-1} q_N^c) - \hat{I}_N^{N-1}(L_N(q_N^c)) \tag{19.357}$$

Here, τ_{N-1} is the *relative truncation error* (or *defect correction*) between the grids, so that the solution on the coarse grid is driven by the fine grid, and the defect

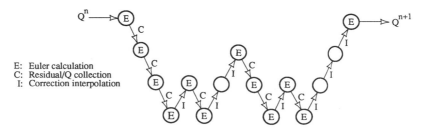

FIGURE 19.44 Multigrid W-cycle.

correction accounts for the difference in the truncation error between the coarse and fine grids [Brandt (1977)]. The analogous equation for grid G_{N-2} is given by

$$L_{N-2}(Q_{N-2}) = S_{N-2} + \tau_{N-2} \tag{19.358}$$

$$S_{N-2} = \hat{I}_{N-1}^{N-2} S_{N-1} \tag{19.359}$$

and

$$\tau_{N-2} = L_{N-2}(I_{N-1}^{N-2} q_{N-1}^c) - \hat{I}_{N-1}^{N-2}[L_{N-1}(q_{N-1}^c)] + \hat{I}_{N-1}^{N-2}\tau_{N-1} \tag{19.360}$$

Note that the relative truncation error on the $N-2$ grid is the sum of the relative truncation error between grids N and $N-1$, as well as $N-1$ and $N-2$. Thus, the equations solved on the coarser meshes, Eqs. (19.355) and (19.358), for example, appear exactly as the original equation, except that a forcing function appears on the coarser meshes. The result is that the coarse meshes can be updated with the same scheme that is used on the fine mesh, with only a slight modification to the right-hand side.

For solving problems in fluid dynamics, both explicit and implicit algorithms have been for used with success. Examples of explicit algorithms that have been used include the pioneering work done by Jameson (1983a), (1983b), (1985) and Jameson and Baker (1984) in which a multistage Runge–Kutta scheme was used along with implicit residual smoothing. In addition, the coefficients of the Runge–Kutta algorithm have been chosen so that the damping of the high-frequency errors is enhanced. Various researchers who have used implicit algorithms include Anderson et al. (1988), Mulder (1985), Jameson and Yoon (1986), and Spekreijse (1988).

An example of the application of multigrid for the Euler equations is given here for a three-dimensional transonic flow computation over the ONERA M6 wing. The wing consists of symmetrical airfoil sections with a planform swept at $30°$ along the leading edge, an aspect ratio of 3.8, and a taper ratio of 0.56 [Schmitt and Charpin (1979)]. A solution is obtained on a C-H mesh, which has a C type of mesh topology around the airfoil profile and an H type mesh in the spanwise direction. The effectiveness of multigrid acceleration is demonstrated for a computation at transonic conditions with a Mach number of 0.84 and an angle of attack of $3.06°$. Figure 19.45(a) and (b) show the effect of using multigrid on the residual and lift-coefficient histories for a mesh with over 210,000 points. The mesh is a $193 \times 33 \times 33$ C-H mesh that has 193 points along the airfoil and wake (110 of which are on the airfoil), 33 points approximately normal to the airfoil, and 33 points in the spanwise direction (17 of which are on the wing planform). For this calculation, an implicit, upwind differencing method is used for smoothing the errors, with a V-cycle and four grid levels (one fine and three coarser grids). The multigrid method is very effective in accelerating convergence of both the residual and lift coefficients. The residual is reduced to machine zero in 400 cycles, whereas the single-grid method has reduced the residual only between 1 and 2 orders of magnitude. The benefit of multigrid is especially pronounced in the lift-coefficient history where the lift-coefficient value is obtained to within 0.1 percent of its final value in 41 cycles. This is a dramatic improvement over the single-grid result, which required more than 400 cycles to converge to the same level of accuracy for the lift coefficient.

The multigrid procedure has also been implemented into multiblock versions of

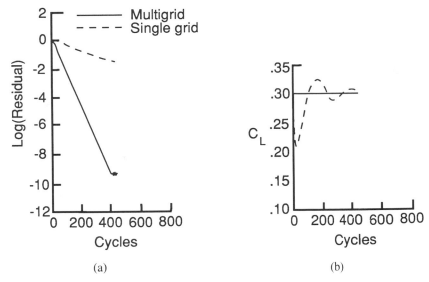

FIGURE 19.45 Effect of multigrid ONERA M6 wing with $M_\infty = 0.84$ and $\alpha = 3.06°$. (a) Residual history and (b) lift history.

several codes to handle complex geometries and viscous flows [Vatsa *et al.* (1993), Atkins (1991), Rossow (1992), Thomas *et al.* (1989), and Cannizzaro *et al.* (1990)]. In addition, several examples of the application of multigrid to reduce the computational times required for time-accurate calculations can be found in [Jameson (1991), Jespersen (1985), Anderson *et al.* (1989), and Melson *et al.* (1993)].

19.6.5 Generalized Minimum Residual Algorithm

The *generalized minimal residual* (GMRES) [Saad and Schultz (1986)] algorithm for solving a nonsymmetric linear system of equations has been extended to nonlinear problems and applied to the Euler calculations by Wigton *et al.* (1985). In this implementation, the equation considered for solution is written as

$$\mathbf{R}(\mathbf{w}) = 0 \tag{19.361}$$

where, for the Euler equations, $\mathbf{R}(\mathbf{w}) = 0$ represents the steady-state residual. The differential of $\mathbf{R}(\mathbf{w}) = 0$ in a general direction $\mathbf{p}$ is denoted by $\tilde{\mathbf{R}}(\mathbf{w}; \mathbf{p})$ and is given by

$$\tilde{\mathbf{R}}(\mathbf{w}; \mathbf{p}) = \lim_{\epsilon \to 0} \frac{\mathbf{R}(\mathbf{w} + \epsilon\mathbf{p}) - \mathbf{R}(\mathbf{w})}{\epsilon} \tag{19.362}$$

Analogous to the procedure for linear systems, the GMRES algorithm first obtains k orthonormal search directions $\mathbf{p}_1, \mathbf{p}_2, \ldots, \mathbf{p}_k$ and then updates the solution as

$$\mathbf{w}^{n+1} = \mathbf{w}^n + \sum_{j=1}^{k} a_j p_j \tag{19.363}$$

where the a_j are chosen to minimize

$$\|\mathbf{R}(\mathbf{w}^{n+1})\|^2 = \left\| \mathbf{R}\left(\mathbf{w}^n + \sum_{j=1}^{k} a_j \mathbf{p}_j \right) \right\|^2$$

$$\approx \left\| \mathbf{R}(\mathbf{w}^n) + \sum_{j=1}^{k} a_j \tilde{\mathbf{R}}(\mathbf{w}^n; \mathbf{p}_j) \right\|^2$$

(19.364)

The orthogonal search directions are determined by a Gram–Schmidt process

$$\mathbf{p}_1 = \frac{\mathbf{R}(\mathbf{w}^n)}{\|\mathbf{R}(\mathbf{w}^n)\|}$$

(19.365)

For $j = 1, 2, \ldots, k - 1$,

$$\mathbf{p}_{j+1} = \tilde{\mathbf{R}}(\mathbf{w}^n; \mathbf{p}_j) - \sum_{i=1}^{j} b_{ij} \mathbf{p}_i$$

(19.366)

$$\mathbf{p}_{j+1} = \frac{\mathbf{p}_{j+1}}{\|\mathbf{p}_{j+1}\|}$$

(19.367)

where b_{ij} is the projection of $\tilde{\mathbf{R}}(\mathbf{w}^n; \mathbf{p}_j)$ in the direction of $\mathbf{p}_i$

$$b_{ij} = [\tilde{\mathbf{R}}(\mathbf{w}^n; \mathbf{p}_j), \mathbf{p}_i]$$

(19.368)

In practice, the above process is actually applied to a *preconditioned* equation that has the same solution as the original problem, but has a more favorable distribution of eigenvalues. For the problem $\mathbf{R}(\mathbf{w}) = 0$, most computer codes generate an improved approximation to the current estimate of the solution as

$$\mathbf{w}^{n+1} = \mathbf{M}(\mathbf{w}^n)$$

(19.369)

where $\mathbf{M}$ represents some methodology such as *line relaxation*, *approximation factorization*, or *Runge–Kutta time stepping*. Convergence is achieved when $\mathbf{w}^{n+1} = \mathbf{w}^n$. Therefore, the solution of Eq. (19.361) can be replaced by the equation

$$\mathbf{R}'(\mathbf{w}) = \mathbf{w} - \mathbf{M}(\mathbf{w})$$

(19.370)

for which GMRES is much more effective. Note, however, that every evaluation of $\mathbf{R}'(\mathbf{w})$ involves an evaluation of $\mathbf{M}$.

An example of results with GMRES to accelerate the convergence of an existing flow solver for a transonic calculation is shown in Fig. 19.46. Here, GMRES is applied to a two-dimensional, central-differenced implicit Euler code denoted as ARC2D [Pulliam (1984)]. As seen in the figure, the use of GMRES can result in a significant increase in the convergence rate. After 400 function calls [where one function call is one evaluation of Eq. (19.370)], the residual is reduced about 4 orders of magnitude over that without GMRES.

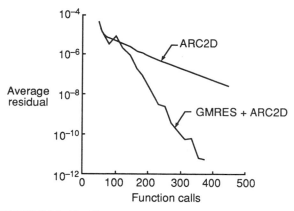

FIGURE 19.46 Convergence acceleration with GMRES.

19.6.6 Preconditioning

Recent work has been undertaken to accelerate the convergence rate of iterative schemes by essentially multiplying the time derivative by a matrix that allows faster convergence, but does not alter the steady state. The motivation for this is easily seen by examining the one-dimensional Euler equations

$$\frac{\partial \mathbf{w}}{\partial t} + \frac{\partial \mathbf{F}}{\partial x} = 0 \tag{19.371}$$

After this equation is linearized and a similarity transformation is used, it can be recast into the form

$$\frac{\partial \mathbf{q}}{\partial t} + \mathbf{\Lambda} \frac{\partial \mathbf{q}}{\partial x} = 0 \tag{19.372}$$

where $\mathbf{\Lambda}$ is a diagonal matrix whose entries are the eigenvalues of the flux Jacobian

$$\mathbf{\Lambda} = \begin{bmatrix} u & 0 & 0 \\ 0 & u+a & 0 \\ 0 & 0 & u-a \end{bmatrix} \tag{19.373}$$

The equations given by Eq. (19.372) are now uncoupled, so that each equation can be approximated separately. For example, simple explicit time differencing can be used in conjunction with first-order spatial accuracy, where each equation is differenced according to the sign of the eigenvalue. The allowable time step for stability depends on the size of the maximum eigenvalue as well as on the grid spacing through the CFL number (defined as the product of the convective speed and the time step divided by the grid spacing). If a CFL is maintained less than unity for the simple explicit scheme considered, then the numerical characteristics completely enclose the physical ones. If all equations are advanced in time with the same Δt, then the CFL number for the equation whose convective speed (eigenvalue) is smallest may not be advanced nearly as fast as the stability criteria allows. For example, for $u = 0.5$ and $a = 1$, the limiting condition corresponds to the largest eigenvalue, which is $u + a = 1.5$. This corresponds to determining the time step by the speed of an acoustic wave that is moving to the right with a speed $u + a$. Note, however, that choosing the time step based on this eigenvalue means that the first equation (associated with the eigenvalue $\lambda_1 = u$) is advanced at a time step somewhat lower than the stability criteria requires. For the conditions chosen above, this restriction is not too prohibitive. However, for a low-speed flow, the wide disparity in the size of the eigenvalues can lead to slow convergence unless a time step is used separately for each equation, based on the individual eigenvalues.

The basic premise of *preconditioning* is to advance each equation with an optimum time step for each. For the one-dimensional case, this can be easily achieved by multiplying the right-hand side of Eq. (19.371) by a matrix, so that when the equation is diagonalized, all eigenvalues are equal

$$\frac{\partial \mathbf{w}}{\partial t} + \mathbf{P}\,\frac{\partial \mathbf{F}}{\partial x} = 0 \tag{19.374}$$

Note that the preconditioning matrix does not change the steady state. Also note that $\mathbf{P}$ should be a positive definite matrix. Otherwise, the nature of the flow could be changed, as it would if $\mathbf{P} = -\mathbf{I}$, which would correspond to marching backward in time. For the one-dimensional case, this matrix is given by

$$\mathbf{P} = |\mathbf{A}^{-1}| = \left| \mathbf{T}\left(\frac{1}{\mathbf{\Lambda}}\right)\mathbf{T}^{-1} \right| \tag{19.375}$$

where $\mathbf{T}$ and $\mathbf{T}^{-1}$ are the right and left eigenvectors of the matrix $\mathbf{A} = \partial \mathbf{F}/\partial \mathbf{w}$ and $1/\mathbf{\Lambda}$ is a diagonal matrix whose entries are the inverse of the eigenvalues of $\mathbf{A}$.

In the above example, the criteria used in determining the optimum time step for each equation is based solely on taking the largest allowable time step for each equation. Other criteria can be used, such as selection of a time step to provide maximum damping of certain frequencies for use in a multigrid algorithm. Also, note that if all the eigenvalues are of comparable size, such as for hypersonic flow, then no significant benefit would be expected.

The simplicity of the matrix in Eq. (19.375) is attributable to the fact that the one-dimensional Euler equations are easily diagonalized. The complexity of devising a preconditioner increases in multidimensions, because the equations cannot be simultaneously diagonalized (with the exception of supersonic flow). However, recent work at preconditioning the equations in multidimensions has been undertaken with good success [Choi and Merkle (1985), Turkel (1986), (1992), van Leer *et al.* (1991), and Godfrey (1992)].

19.7 GRID ADAPTATION
Peter R. Eiseman

19.7.1 Overview

Almost all computational fluid dynamic algorithms are constructed relative to some disection of physical space. That disection is called a *grid*. The beginning of grid adaptivity is clearly the need to capture the region geometry. This is achieved when either the grid conforms exactly to the geometric boundaries or represents those boundaries by a fine enough resolution. Because boundary conformity means that the boundary is represented by actual grid points rather than nearby grid points, the fluid dynamic simulation is generally more accurate and simpler to execute. As a consequence, the majority of applications use boundary conforming grids. This is evident from surveys, texts, and conference proceedings on grid generation [Eiseman (1985), Eiseman *et al.* (1989), (1990), Thompson *et al.* (1985), George (1991), Knupp and Steinberg (1993), Castillo (1991), Häuser and Taylor (1986), Sengupta *et al.* (1988), Arcilla *et al.* (1991), and Weatherill *et al.* (1994)].

Once the boundaries of the flow region are represented by the grid, the next adaptive requirement is the clustering of grid points to capture the requisite physics. This can appear either before or during the simulation. Before a simulation, the use of prior knowledge or estimates are applied. For example, with Navier–Stokes simulations, a clustering for attached boundary layers is given based upon the Reynolds Number. When resolution must be given at locations that are not predictable in advance, the only recourse is to give the resolution during the simulation. As the fluid dynamic solution evolves, the features in need of resolution appear in their respective locations. If the aim is for a steady state solution, then there is the opportunity to adjust the resolution to meet the needs at a small number of time-like stations in the evolution toward convergence. In this instance, the resolution may be supplied in either a manual or an automatic fashion. If a time accurate simulation is sought, the resolution must be supplied in a dynamic enough fashion to both resolve and track the motion of the requisite physical features. The resolution for the time accurate case must be supplied automatically.

19.7.2 Detectability

The issues of solution adaptivity are the detectability of the solution features, the creation of a new grid to resolve those features, and the adjustment of the solver to use the new grid. The detectability is focused upon determining a means to identify features and their strengths. This can vary from a visual assessment to analytically formulated measures. When automatic action is required, the measures are created from error estimates, physical features, or the governing equations. The error estimates are derived from the numerical solution method as applied to the governing equations. The advantage of these is that there is no need to identify certain physical attributes as being important enough to drive adaptive concerns. Because of the mathematical complexity involved in getting tight enough estimates, usually weaker estimates are used. To extract features directly from the governing equations, one can consider the *ad hoc* balance between various terms, characteristic directions, or invariants. The invariants will lead to the most methodical approach but are also the most complex. In formal terms, one finds the group of continuous transformations of independent and dependent variables that leave the governing equations in an unchanged form. This group is a *Lie Group*. The associated *Lie Algebra* is the collection of tangent fields to the Lie Group, and from them one knows the direction for which action is occurring. To couple this formalism to the discrete solution algorithm, one must use a modified equation approach to find the continuous system that is being solved by the algorithm. Details on this type of procedure can be found in Shokin (1983). A direct use of terms in modified equations has also been considered by Klopfer and McRae (1981).

19.7.3 Monitor Surfaces

From the above discussion, it is clear that the price for the more universal adaptive measures is an increase in complexity. Also, since it is the physical features that are important in the simulation, they are usually readily identified. As a consequence, most solution adaptive procedures use those features directly. In fact, it is common practice to take feature variables such as pressure or density for shock wave identification, velocity magnitudes for boundary layers, and so forth. In the direct application of such parameters, the first and sometimes second derivatives are inserted directly into weight functions which are used to increase the point density with increasing weight.

This insertion practice is, however, not a very coordinated approach. What is needed is the means to represent this data in a more consistent manner. With this motivation, Eiseman (1985) introduced the idea of a *monitor surface*. In a subsequent study, Liseikin (1992), (1993) reintroduced it independently. To form it, one takes the indentified field parameters with proper scaling and forms a surface directly from them. A surface point is given by appending to the spatial coordinates further coordinates which represent the identified features. As such, this is a surface which is the same dimension of physical space, sits on top of the physical space, and is viewed as being embedded in a higher dimensional space. While such a surface expression may seem uncomfortably abstract, one should bear in mind that the fluid dynamic solution which is understood in concrete terms can also be viewed as a similar surface in a higher dimensional space. In fact, the solution surface itself could be taken as a monitor surface. This, however, may not be efficient or desirable.

The motivation behind the formulation of the monitor surface is to consolidate all of the important basic field parameters into one geometric object. It is those parameters which will most directly reflect the features of rapid variation that need to be resolved. The components are generally taken from solution components or derived combinations of them. Since the aim is to capture the variations in the physical solution, the final step in forming the monitor surface is to filter it. Filtering is needed to remove solution generated wiggles or other irrelevant properties. In some cases, clipping is also required prior to the smoothing filter. An example of clipping occurred in the study of shock-vortex interaction by Bockelie and Eiseman (1992). The clipping and filtering actions provide a smooth enough surface to insure well defined discrete derivatives. It is important to remember that this surface exists solely for the purpose of driving the generation of an adapted grid. That is, while solution data may be used for its formulation, that solution itself is left alone.

19.7.4 Grid Quality and Structure

With the requisite formulation of adaptive data, the next step is the use of it to generate a new grid. As with all grid generation issues, grid quality is an important factor in the process. A loss of quality can often lead to a loss of accuracy just when the adaptivity is supposed to enhance it. While quality can also be interpreted as a reasonable resolution of the severe solution behavior as represented by the monitor surface, there are a number of basic metric measures as well as other factors. The metric measures include cell aspect ratios, lengths, angles, and sizes. It also includes the variations of these items as we go to their neighbors. This is just the smoothness in the distribution of these properties. In the case of structured grids, plain geometric quality is seen as smooth variations in cell properties with angles that are nearly perpendicular. In the case of unstructured grids, it is seen as smoothness together with nearly equilateral triangles in 2-D or similarly contrived tetrahedra in 3-D as in Baker (1989). For unstructured paving (quadralaterals) and plastering (hexahedrons) as the Blacker and Stephenson (1991) and Blacker and Meyers (1993), it is smoothness and orthogonality.

19.7.5 Methods of Generation

With an assumed level of quality, adaptive grids can be generated by refinement, movement, or both. In addition, the solver accuracy can also be adapted. In the case of finite element methods, this is called the *p method* while the actions of grid refinement and movement are respectively referred to as the *h and r methods*. Altogether, there can be any mixture of grid and solver adaptivity.

In the case of refinement (and de-refinement), the overall grid is fixed, while only localized alterations are performed. This means that certain adaptively flagged cells are split (or merged). The splitting may be accomplished either isotropically or homogeneously (i.e., with or without a directional bias). An advantage with this approach is that there is a local transfer of solution data from the old to the new grid. This can help preserve the accuracy in the nonadapted portions of grid. A further advantage is that the often challenging process of boundary conforming grid generation is done only once, at the beginning. All adaptivity is then relative to the initially generated grid. A disadvantage of this approach is the need for a more complex data structure and management thereof as well as the possible need to deal with hanging nodes or to circumvent them by adding further transitional structures.

When there is a desire to maintain a given data structure, a grid of the same type is generated. With unstructured grids, a new boundary conforming grid is created along with the adaptive data. This has a different number of points and is created from a global view of the action. It also possesses the advantage of being able to readily deal with a moving boundary. Unlike the refinement and de-refinement case, there is a need to globally transfer data from old to new grids. This transfer can be diffusive if it is applied in a time accurate setting. In a similar spirit, the use of adaptive grid motion requires the global generation of a new grid and can readily track moving boundaries. Unlike the unstructured actions, pure movement maintains the number of grid points and directly uses the old grid to evolve towards the new one. This means that the boundary conforming properties of the previous grid are not discarded but rather are used as a basis to go to the new grid.

19.7.6 Adaptive Grid Movement

While adaptive grid movement is applicable to grids of any structure, the main application is for structured grids. A comprehensive review of this was given in Eiseman (1987). The main additions to adaptive grid movement since then have been the use of harmonic mappings between manifolds as initiated by Dvinsky (1991), the use of deformations of the identity map as represented by Liao and Su (1992), and interactive adaptivity with control points as represented by Lu and Eiseman (1992) and Choo and Henderson (1992). The 1987 survey is more than just an enumeration of various techniques: it is synthesis of methodology which is presented in a *how to do it* instructive style.

Adaptive grid movement for structured grids is particularly attractive when it is desired to maintain the simplicity that comes with a regular coordinate line pattern of grid points. These patterns vary from the use of a single coordinate transformation up to the application of many. The assembly of coordinate transformations can vary from overlapped collections to those which are fit together with full continuity. The overlapped case trades relative simplicity in grid generation for data management and transfer problems. The full continuity case relieves the data treatment problems and produces grids which are referred to as *multiblock grids*. Here, each block is a separate coordinate grid which is attached to its neighbors with some level of continuity. This provides the best arrangement of grid points for general application. Flow solvers for them are generally more efficient and more available.

19.7.7 Traditional Multiblock Grid Generation

Until recently, however, multiblock grids have been generally harder to create. The main problem has been a bottom-up approach using interactive graphics to assemble a rather large amount of input data. That process is a manual one that requires expert judgement, albeit the appearance of a high technology approach through sophisticated graphics.

In building from the bottom up, the user forms coordinate blocks by working from lower dimensional parts to higher dimensional ones. This process typically starts with the creation of a surface grid which is to be generated over a region's geometric boundaries. The first part of surface grid generation is the preparation of boundaries to find the curves which represent the intersection between surfaces. The next part is to segment the boundary into further curves which can be used to define the block faces which lie on the boundary. Each of these curves is put into place based upon the user's judgement as to the best location for them as well as the best shape. The next step for judgement and action is to put grid points on the curves and to give them an appropriate distribution. For every set of four curves that give the perimeter of a face, the process continues by filling it in with a patch of grid. This is typically done with transfinite interpolation and is automatic when given perimeter grid points. Since this patch of grid is generally close to, but not exactly on the surface, a projection must also be performed to get it there. While such a surface grid can be used to directly form the progression into a volume grid, it may not be smooth enough. To improve the smoothness, elliptic grid generation is applied and then projected as necessary.

In the course of surface grid generation, there are a number of other technical considerations. One such item is the need to use parameterized surface geometries that are not correctly set up for grid generation purposes. This situation arises when the coordinate grid extends over the boundaries of geometry defining patches. In this instance, there is a need to transition between respective patch parameter spaces.

Once the grid is generated on the bounding surfaces, the volume grid must be constructed. In anticipation for this, the pattern of surface grid lines and their concentration has been chosen with appropriate judgement. The next level of judgement and construction is to define the block edges away from the boundaries. To start, one must determine the corner locations where block edges will come together. Once such points are put somewhere in 3-D space, the next step is to create the connecting curves which emanate from the corresponding surface corners as well as all other curves which connect the corners in space. With the geometry of these connectors (block edges) in place, grid points must be placed upon them and appropriately distributed. As in the case of surface grids, the rest is fairly automatic. Every pair of four edges defines a face by transfinite interpolation and every set of six faces defines a block by transfinite interpolation. While the generation process often ends with the algebraically generated grid, there is always the option to add smoothness by applying elliptic grid generation methods.

The theme of traditional multiblock grid generation occurs in many forms with various distinguishing features. These can vary with the organization of the data and the sequencing of events as well as the recording of such events. The event record is sometimes available and is called a *journal file*. It provides the opportunity to edit and replay the grid generation process. When packaged in a suitable fashion, it is called a *template*. For geometries sufficiently close to the original one, this allows grid generation by journal file replay.

19.7.8   Software

Here, we discuss one software package that is currently available—GridPro. There are two main lines of the GridPro software. One is for multiblock grid generation with automatic zoning, and the other is for dynamic grid modeling. The blueprint for both lines was established in early 1989. At several stages, there were studies in grid quality. In addition, grid modeling was combined with the traditional multiblock grid generation code EAGLE from Eglin AFB. EAGLE was the first general purpose multiblock code which now has assumed the form of EAGLEView with the addition of graphics (GUI). The result was a grid modeling within EAGLE as well as the elliptic motion of control points to form the EAGLE/CPF code. The control point form of algebraic grid generation (CPF) was started by Eiseman (1988), (1992). The grid modeler, GridPro™/sb3020 and the EAGLE/CPF code are discussed by Eiseman, Lu, Jiang, and Thompson (1994).

Grid Modeling and Adaptivity. The combination of grid modeling and grid quality measures showed the elements of interactive grid adaptivity. In that combination, not only were the metrical measures for grid quality used but also monitor surface data was deployed. This involved monitor surface metrics and directional elements. On the interactive screen, the user would see an overlay of items. First, the background would show the smooth continuous plot of colors for a chosen scalar variable. Then on top of that, there was a grid plot and a plot of the control points over it all. As a further option, a vector field (e.g., velocity field) can also be shown at each grid point. The action occurs when the user moves a control point. When this happens, the grid concentrations and alignments dynamically appear in response. If the plotted data is imported from a solution, it is static relative to this process and does not change with the change in the control points. If the plotted data is a grid quality measure, then the user sees the plots change dynamically with the grid and continuously sees the changes in grid quality. Thus, he can move the control point in such a way as to improve quality as measured by the displayed measure. Moreover, if there is a need to distinguish some fine grain details that cannot be witnessed clearly enough with the current color map, there is the opportunity to interactively adapt the color map as well. At the present time, this interactive adaptive code is only in 2-D. It, however, has shown the fundamental value of this action and has set the stage for entry into the 3-D grid modeling software at some time in the future.

The 3-D grid modeling product is called GridPro™/sb3020 and operates upon a single block of grid at a time. It does not matter which multiblock or single block grid generator is used. A block of grid is merely imported from anywhere and is then modeled in real time. With the import, the user is asked to define the portion of grid to be modeled as well as the amount of control to be applied in each direction. Then, an initial control net is automatically constructed to start the action. The action progresses with the motion of one or more control points. During this process, the boundaries are respected. Initially, they are specified on a point-by-point basis. They can be selectively opened for free form manipulation as well. Moreover, there are separate control nets for each face so that the face grid can be manipulated while preserving the face geometry. In addition, there are several levels of elliptic motion of control points. These are for the motion of interior control points by fixing anywhere from 0 to 3 layers of them from the boundary.

Multiblock Grid Generation With Automatic Zoning. The 3-D multiblock grid generation product is called GridPro™/az3000 and represents a radical departure from the traditional methods. Unlike the others, there is no manual assembly of the bits and pieces of grid. Instead, the action is automatic once the user supplies a relatively small amount of input. That data is also more flexibly given. The input consists of the geometric boundaries, a coarse wire frame to define the pattern of grid lines, and a small number of parameters to define or control such items as the number of grid points and the clustering intensities.

Of the inputs, the geometric boundaries are common to any grid generation technique. Its use in GridPro™/az3000 is more flexible, since all surfaces are referenced directly for their trace of locations in terms of (x, y, z). The traditional techniques more often than not will treat surfaces in terms surface parameterizations. In so doing, there are technical problems such as the management of parameter spaces for neighboring patches and the correct treatment of metric properties. In contrast, the view of boundaries for only their trace in (x, y, z) space removes the detailed problems associated with how the surfaces were constructed. At a very rudimentary level, the only concern is the evaluation of a point on a surface. This can be done with wide variety of surface definitions.

In this spirit, a natural definition is one which comes directly from specifying a function $f(x, y, z)$. A level surface of f is one for which $f = c$ for some constant c. For simplicity, c is taken to be 0. This is quite general, since f could always be replaced by $f - c$ in order to make this happen. As it turns out, this is a very general means to give surface definitions. In fact, it is the primary way surfaces are defined in algebraic geometry. Such definitions are more general than the parametric ones, since they include topological configurations that cannot be represented by parametrically defined surfaces. For example, the surfaces can easily have a number of holes in them. In terms of modern surface modeling tools, the NURBS surfaces are parametrically defined surfaces and have been shown to be a special case of level surfaces, although this path to NURBS requires the extraction of the appropriate parameters.

In addition to the level surface definitions, there is the use of locally defined surfaces such as represented by multiple parametric patches or unstructured data sets. The evaluation of points on these surfaces requires a search to find the appropriate patch and then uses the local patch definition to find the desired point. Continuing, there are still more surface types. One is the periodic boundary, common in turbomachinery. Another is the use of copious data which is an IGES entity type. In particular, GridPro™/az3000 will accept surfaces defined by unstructured triangular and quadrilateral elements. In each of these, the user has a choice between a NASTRAN format and a natural one which is given in the manual. In fact, the structure of GridPro™/az3000 will admit a growth path that will include all of the entity types of modern CAD (e.g., NURBS). The reason is the direct use of positions in physical space.

The pattern of points is given by a coarse wire frame. Each wire frame cell corresponds to a block in the grid to be generated. In a sense, these coarse wire frame cells look like macro elements in an unstructured hexahedral grid. The represented pattern is called the *grid topology*. The user defines it in a regular systematic fashion. That process employs a Topology Input Language (TIL). This contains subprograms which communicate by passing variables between each other. The basic variables are the coarse wire frame corners and the defined surfaces. The surface definitions appear here by reference. For example, a surface given by a data file will be put into TIL with one line that will give the surface type and the name of the data file. The programs and subprograms are called *COMPONENTS*. This term is appropiate because the various objects really appear like components. The *COMPONENT* is reusable and can be called any number of times. The result is an efficient assembly process which can take advantage of recurrent structure. Moreover, a collection of components can be put into a separate file that can be included in a component with an *include* command. The *include* is particularly convenient for making libraries of components.

Once the TIL code is finished and put into a file, the next task is to run it. This is done by creating a schedule file for the run. The schedule can vary from a straight run to one which is dynamically changed. The option to change the course of a run allows the user to optimize the run. For example, the run could be started with a fairly modest number of grid points which are then increased in stages as the grid is relaxed towards its equilibrium that progressively satisfies the underlying variational scheme. As one could imagine, there are a number of parameters in the schedule that provide the user helpful options. One of these provides the bounds for curvature clustering. This is in effect a geometry adaptive process. Upon activation, the volumetric grid points distribute themselves about the boundaries in proportion to the boundary curvature.

Some Examples. A simple, 2-D application is shown first in Fig. 19.47 for a grid inside of a turbine blade. A 3-D application is shown next in Fig. 19.48. The case is for a fuel tank of an automobile, generated with only one page of TIL code. The curvature clustering is quite evident by examining the locations where the surface changes its direction relative to being planar or nearly so. The fuel tank geometry is given by unstructured triangular data in TRIA NASATRAN format. The surface for the fuel level is given by the built-in analytic planar type. In addition, parameters for the general size and shape of a fuel tank are first computed with a FORTRAN

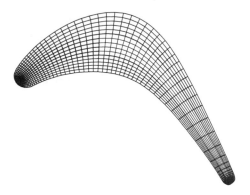

FIGURE 19.47 A 2-D grid inside of a turbine blade.

program which creates a file of them. The TIL code then reads that file and automatically generates the grid. Thus, grids are rapidly created for a large number of fuel tank cases. In comparison with the traditional approach for this problem, grid generation was reduced from about one month per grid to about 15 minutes per grid on a modern workstation.

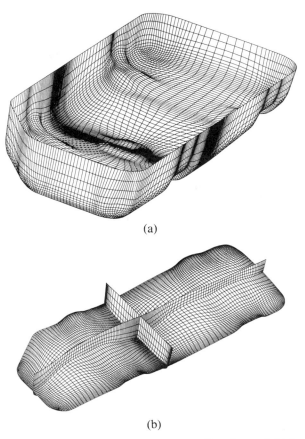

(a)

(b)

FIGURE 19.48 A grid for a fuel tank. (a) Exterior and (b) interior.

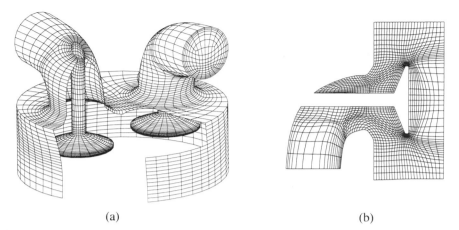

(a)   (b)

FIGURE 19.49 A grid for a dual port cylinder. (a) Overall view and (b) cut through valve.

More examples are shown in the paper by Eiseman, Cheng, and Häuser (1994). Among those cases is the configuration for the flow through a bent pipe and about a rod which intersects the pipe at the elbow bend. On a subsequent occasion, the TIL code for this case (a little over one page) was used as input for the unrelated application of a port cylinder configuration. Figure 19.49 shows one of the grids. To examine the effect of adaptive grid movement, Figure 19.50 shows a grid for shock vortex interaction. This case involved dynamic tracking. Figure 19.51 shows a 3-D adaptive clustering for a spherical disturbance within a region with curved boundaries.

19.7.9 Adaptivity and Multiblock Grids

The consequence of GridPro™/az3000 is that multiblock grid generation is automatic. Given the relatively small demands for user input data, this automation is much more automatic than any other multiblock grid generator and rivals that which was promised by the unstructured approach. With this advance in the technology of multiblock grid generation, the user gets both the automation and the quality that certainly enhances his CFD procedure. The quality includes not only the proper ordering and hexahedral structure that is common with the multiblock approach, but also a grid which is smooth and nearly orthogonal even in 3-D. It may also be set up for a straight-forward application of multigrid techniques as in Vatsa *et al.* (1994).

If it is desired to strictly maintain the high quality structure for multiblock grids, the adaptivity must come through grid point movement entirely. There are, however, a rather rich supply of options. One can also split cells in various ways as in Schön-feld and Rudgyard (1994). One may also do this in the FEM approach with the *h* or *hp* approach as represented by Oden *et al.* (1990). One may also consider embedded patches of grid as in Berger and Colella (1989) or a combination of movement as in Arney and Flaherty (1986). With pure adaptive movement, some examples are given by Benson and McRae (1994), Bockelie and Eiseman (1992), and Eiseman and Tokahashi (1994).

The ease with which very high quality multiblock grids can be generated with automatic zoning technology points to a renewed emphasis on grid adaptivity within

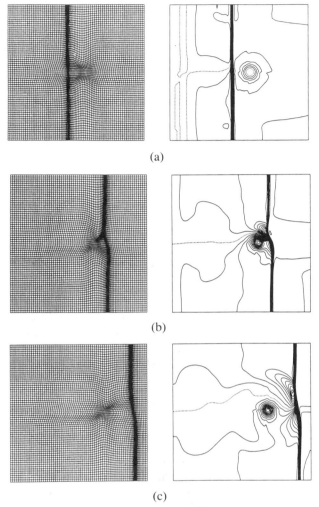

FIGURE 19.50   Adaptive grid for shock-vortex interaction. (a) $t = 0.30$, (b) $t = 0.70$, and (c) $t = 1.00$.

this environment. It is an environment which emphasizes quality under the constraint of structure. This leaves room for grid refinement and movement. While the refinement builds an unstructured form within the global multiblock framework, movement preserves it exactly when it is not applied concurrently with refinement. Each form of adaptivity will inherit the quality of the underlying multiblock grid.

19.8   FOURIER AND CYCLIC REDUCTION METHODS FOR POISSON'S EQUATION
Paul N. Swarztrauber and Roland A. Sweet

19.8.1   Introduction

Mathematical models of fluids frequently require the solution of Poisson's *equation* often to solve for the pressure or vorticitiy. Here, we will review two direct methods

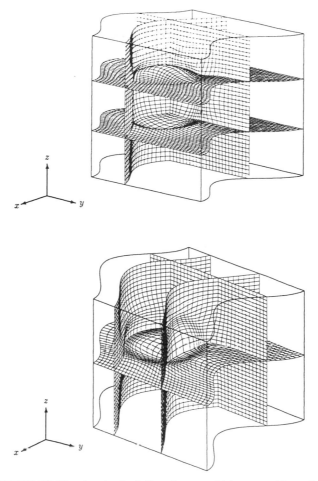

FIGURE 19.51 A spherical disturbance within curved boundaries.

for solving Poisson's equation-namely, the Fourier and cyclic reduction methods. Direct methods are distinguished by the property that a solution is obtained in a finite number of operations, and the accuracy is determined by the accumulation of roundoff errors. The accuracy of an iterative method is determined by the convergence test which is usually inferior to the accuracy of the direct method because the convergence test must be maximized to minimize computing time. This tradeoff between accuracy and computing time presents a nontrivial problem associated with iterative methods in the context of time-dependent fluid models.

In this section, we will review these methods and note that further details can be found in Buzbee *et al.* (1970) and Swarztrauber (1984). In what follows, we will determine an approximate solution to Poisson's equation in Cartesian coordinates

$$\frac{\partial^2 u}{\partial x^2} + \frac{\partial^2 u}{\partial y^2} = f(x, y) \qquad (19.376)$$

To solve Poisson's equation, one must specify the region as well as the appropriate conditions on the solution on the boundaries of the region. Although fast, direct methods can be used to facilitate the solution on irregular regions [Buzbee *et al.* (1971) and Proskurowski and Widlund (1976)] here, for clarity of exposition, we assume a rectangle $a \leq x \leq b$ and $c \leq y \leq d$. The region is assumed rectangular in the coordinate system but not necessarily in physical space. For example, in cylindrical coordinates, a fast solution can be obtained on the region $a \leq r \leq b$ and $c \leq z \leq d$. In fluid models, one of the following boundary conditions can be specified on each of the four boundaries of the rectangle:

  i) The solution is specified on the boundary.
  ii) The derivative of the solution, normal to the boundary, is specified on the boundary.
  iii) The solution is periodic.

Specifying combinations of the latter two boundary conditions leads to a problem having two important characteristics.

First, a solution will exist only if the boundary condition and f satisfy an auxiliary condition. For example, suppose we wish to solve with i) on the rectangle given above with the condition that the solution is periodic in both x and y. If a solution exists, it satisifes i) which, if integrated over the rectangle, yields

$$\int_a^b \int_c^d f(x, y) \, dy \, dx = 0 \tag{19.377}$$

where we have made use of the periodic boundary conditions. Therefore, if a solution exists, the mean of f must be zero, or, conversely, if the mean of f is not zero, a solution does not exist. Second, if a solution to this problem exists, then it is not unique. That is, if u is a solution, then $u + c$ is also a solution for any constant c. This is usually not a problem for fluid models, because the gradient of u is used in the time-dependent equations that describe the motion of the fluid. This situation is not uncommon in fluid models and is the subject of a later subsection.

Any combination of derivative and periodic boundary conditions may define a problem without a solution. However, it is usually the case that a solution can be found to a perturbed problem. That is, if the problem arises from a particular physical application it is likely that observational or computational errors will produce a small but nevertheless nonzero right side of Eq. (19.377), which implies a solution does not exist. Nevertheless, if f is perturbed so that its mean is zero, then a solution exists which is a *least squares* solution to the unperturbed problem.

In the remaining sections we develop the large linear system of equations that arise from the second-order, central finite difference approximation to Eq. (19.376). We also present two algorithms that provide reliable and fast solutions, discuss extensions to more general problems, and describe software that can be used to solve Poisson's equation subject to the boundary conditions listed above. Finally, we discus the implementation on multiprocessor computers.

19.8.2 Finite Difference Approximation

To illustrate fast, direct methods, consider the development of a finite difference approximation to a typical problem; namely, that of finding a solution u to Poisson's equation, Eq. (19.376), on the rectangle $a \le x \le b$, $c \le y \le d$, assuming the solution is periodic in both x and y. We begin with the selection of positive integers M and N that define grid spacings $\Delta x = (b - a)/M$ and $\Delta y = (d - c)/N$ and grid points $x_i = a + i\Delta x$ for $i = 0, \ldots, M$ and $y_j = c + j\Delta y$ for $j = 0, \ldots, N$. We wish to determine an approximate solution $v_{i,j}$ to Eq. (19.376) at the grid points x_i and y_j. To this end, we require $v_{i,j}$ to satisfy the following second-order centered finite difference approximation to that equation

$$\frac{1}{\Delta x^2} (v_{i-1,j} - 2v_{i,j} + v_{i+1,j}) + \frac{1}{\Delta y^2} (v_{i,j-1} - 2v_{i,j} + v_{i,j+1}) = f_{i,j} \quad (19.378)$$

where $f_{i,j} = f(x_i, y_j)$. The periodic boundary conditions imply

$$v_{i,j} = v_{i+M,j+N} \text{ for all } i \text{ and } j \quad (19.379)$$

Equation (19.378) plus boundary conditions in Eq. (19.379) can be written

$$\mathbf{A}v_0 + \mathbf{v}_1 + \mathbf{v}_{N-1} = \mathbf{g}_0 \quad (19.380a)$$

$$\mathbf{v}_{j-1} + \mathbf{A}v_j + \mathbf{v}_{j+1} = \mathbf{g}_j \quad \text{for} \quad j = 1, \ldots, N - 2 \quad (19.380b)$$

$$\mathbf{v}_0 + \mathbf{v}_{N-2} + \mathbf{A}v_{N-1} = \mathbf{g}_{N-1} \quad (19.380c)$$

where

$$\mathbf{v}_j^T = (v_{0,j}, v_{2,j}, \ldots, v_{M-1,j})$$

$$\mathbf{g}_j^T = \Delta y^2 (f_{0,j}, f_{2,j}, \ldots, f_{M-1,j})$$

and

$$\mathbf{A} = \rho^2 \begin{bmatrix} -2\alpha & 1 & & & 1 \\ 1 & -2\alpha & 1 & & \\ & 1 & -2\alpha & & 1 \\ & & 1 & \ddots & 1 \\ 1 & & & 1 & -2\alpha \end{bmatrix} \quad (19.381)$$

is a $M \times M$ matrix, with $\rho = \Delta y/\Delta x$ and $\alpha = (1 + \rho^2)/\rho^2$. The problem becomes one of solving the linear system (5a–c) for the approximate solution, $v_{i,j}$, of Poissons equation. Before we describe the two methods of solution, we note that if the $M \times N$, Eq. (19.378) or Eqs. (19.380) are added, the left side of the sum vanishes and we obtain

$$0 = \sum_{i=0}^{M-1} \sum_{j=0}^{N-1} f_{i,j} \quad (19.382)$$

This is the discrete version of Eq. (19.377), in which the integral is approximated by a trapezoidal quadrature. Therefore, the solvability requirement for Poisson's equation has a finite difference analog that must be satisfied if the system Eqs. (19.380) has a solution. The accuracy of Eq. (19.382) relative to Eq. (19.377) meets or exceeds the accuracy of Eq. (19.378) relative to Eq. (19.376).

19.8.3 The Fourier Method

Here, we will use the Fourier method to solve the finite difference equations and note only that the Fourier method can also be used to solve Poisson's equation directly. The latter approach is called the *spectral method* [Haldvogel and Zang (1979)] which is more accurate, however the treatment of nonhomogeneous boundary conditions is somewhat more complicated.

The Fourier method for solving the finite difference equations comes under the more general heading of *matrix decomposition* for reasons that will become evident. The matrix decomposition method can be applied to problems with nonzero boundary conditions. The resulting solution has the accuracy of the finite difference method, which is usually consistent with the accuracy of the other approximations in the fluid model, but it is not as accurate as the spectral method. In this section, we will describe the matrix decomposition method for solving the large, sparse system of Eqs. (19.380).

Matrix decomposition consists of the following major steps:

i) Derive a new system of equations that is easy to solve and whose variables are the Fourier coefficients of the solution. This step includes the transformation of the equations from physical space to Fourier space, which is called the *Fourier analysis phase*.

ii) Solve the new system of equations for the Fourier coefficients.

iii) Compute the solution of Poisson's equation from its Fourier coefficients. This step includes the transformation of the solution from Fourier space back to physical space which is called the *Fourier synthesis phase*.

The Fourier method is quite straightforward, and its implementation is significantly assisted by software. In particular, the transforms required to implement the Fourier method are in FFTPACK 5.0 which is available from the National Center for Atmospheric Research (NCAR). Let $\mathbf{Q}$ be an orthogonal matrix; i.e., $\mathbf{QQ}^T = \mathbf{I}$, yet to be defined, and compute

$$\hat{\mathbf{g}}_j = \mathbf{Q}^T \mathbf{g}_j \quad j = 0, \ldots, N - 1 \tag{19.383a}$$

Further, define

$$\hat{\mathbf{v}}_j = \mathbf{Q}^T \mathbf{v}_j \quad j = 0, \ldots, N - 1 \tag{19.383b}$$

Multiplying Eq. (19.380b) by $\mathbf{Q}^T$ we obtain

$$\hat{\mathbf{v}}_{j-1} + \mathbf{Q}^T \mathbf{A} \mathbf{Q} \hat{\mathbf{v}}_j + \hat{\mathbf{v}}_{j+1} = \hat{\mathbf{g}}_j \tag{19.384}$$

If $\mathbf{Q}$ is chosen such that $\mathbf{Q}^T \mathbf{A} \mathbf{Q}$ is a diagonal matrix, i.e.,

$$\mathbf{Q}^T \mathbf{A} \mathbf{Q} = \text{diag}(\lambda_0, \lambda_1, \ldots, \lambda_{M-1}) \tag{19.385}$$

the resulting system of equations decouples into M independent periodic tridiagonal systems

$$\lambda_k \hat{v}_{k,0} + \hat{v}_{k,1} + \hat{v}_{k,N-1} = \hat{g}_{k,0} \tag{19.386a}$$

$$\hat{v}_{k,j-1} + \lambda_k \hat{v}_{k,j} + \hat{v}_{k,j+1} = \hat{g}_{k,j}, \quad \text{for} \quad j = 1, \ldots, N-2 \tag{19.386b}$$

$$\hat{v}_{k,0} + \hat{v}_{k,N-2} + \lambda_k \hat{v}_{k,N-1} = \hat{g}_{k,N-1} \tag{19.386c}$$

for $k = 0, \ldots, M - 1$. The Fourier coefficients $\hat{v}_{k,j}$ can easily be determined by solving these systems, and the solution can then be computed from the inverse of Eq. (19.383b).

Using the results we can now summarize the matrix decomposition method:

i) Given the tabulation of the right-hand side $\mathbf{g}_j$, then, using the *fast Fourier transform* (FFT), [Cooley and Tukey (1965) and Swarztrauber (1982)], we first compute

$$\hat{\mathbf{g}}_j = \mathbf{Q}^T \mathbf{g}_j \tag{19.387}$$

ii) Next, we solve the M independent periodic tridiagonal systems, Eqs. (19.386), for the Fourier coefficients $\hat{v}_{k,j}$.

iii) Finally, we compute the solution $\mathbf{v}_j$ from the inverse of Eq. (19.383b) or

$$\mathbf{v}_j = \mathbf{Q}\hat{\mathbf{v}}_j \tag{19.388}$$

The FFT is the key to the efficiency of this method. The matrix $\mathbf{Q}$ depends on the form of the matrix $\mathbf{A}$, which in turn depends on the boundary conditions. For periodic boundary conditions, the $M \times M$ matrix $\mathbf{A}$ has the form given in Eq. (19.381) For M even, the Fourier transform, Eq. (19.387), is given by

$$\hat{g}_{0,j} = \frac{2}{M} \sum_{i=0}^{M-1} g_{i,j}$$

$$\hat{g}_{M-1,j} = \frac{2}{M} \sum_{i=0}^{M-1} (-1)^i g_{i,j} \tag{19.389a}$$

and for $k = 1, \ldots, M/2 - 1$

$$\hat{g}_{2k-1,j} = \frac{2}{M} \sum_{i=0}^{M-1} g_{i,j} \cos ik2\pi/M$$

$$\hat{g}_{2k,j} = \frac{2}{M} \sum_{i=0}^{M-1} g_{i,j} \sin ik2\pi/M \tag{19.389b}$$

A similar formula can be given for the case when M is an odd integer. The inverse Fourier transform, Eq. (19.388), is given by

$$v_{i,j} = \tfrac{1}{2}\hat{v}_{0,j} + \tfrac{1}{2}(-1)^i \hat{v}_{M-1,j}$$

$$+ \sum_{k=1}^{M/2-1} (\hat{v}_{2k-1,j} \cos ik2\pi/M + \hat{v}_{2k,j} \sin ik2\pi/M) \tag{19.390}$$

The eigenvalues λ_k, for use in Eqs. (19.380), are

$$\lambda_0 = -2$$

$$\lambda_{2k-1} = \lambda_{2k} = -2(1 + 2\rho^2 \sin^2 \pi k/M), \quad k = 1, \ldots, M/2 - 1 \quad (19.391)$$

and

$$\lambda_{M-1} = -2(1 + 2\rho^2)$$

Consider now the asymptotic number of operations required to solve Eqs. (19.380a–c) where an operation is defined as one multiplication plus an addition. Steps i) and iii) of the Fourier method require $2N$ matrix–vector products. Ordinarily, such a matrix–vector would require M^2 operations. However, the matrices Q and Q^T are such that the product can be realized in $5M \log_2 M$ operations, using the fast Fourier transform. Therefore, the total number of operations for steps i) and iii) is $10MN/\log_2 M$. Step ii) requires the solution of M periodic tridiagonal systems, each of which takes $5N$ operations, with the result that the total operation count for step ii) is $5MN$. Since this last term is of lower order, it is not included in the asymptotic count. Hence, for large M and N, the matrix decomposition method requires about $10MN \log_2 M$ operations.

19.8.4 Cyclic Reduction

The second fast direct method we present is the cyclic reduction algorithm [Buzbee *et al.* (1970), Swarztrauber (1974)]. This algorithm is a recursive scheme that eliminates half of the unknowns at each step until there remains a single equation that can be solved. The remaining unknowns are computed easily by a back-substitution method.

To describe this algorithm, we present the reduction and back-substitution for the case $N = 8 = 2^3$ from which the general case for $N = 2^p$ will become evident. Although not a requirement, for the purpose of exposition we will assume that N is a power of two. The case of general N is treated in Sweet (1974). For the case $N = 8$ the system in Eqs. (19.380) has the form:

$$
\begin{aligned}
\mathbf{A v_0} + \mathbf{v_1} \qquad\qquad\qquad\qquad\qquad + \mathbf{v_7} &= \mathbf{g_0} \\
\mathbf{v_0} + \mathbf{A v_1} + \mathbf{v_2} \qquad\qquad\qquad\qquad &= \mathbf{g_1} \\
\mathbf{v_1} + \mathbf{A v_2} + \mathbf{v_3} \qquad\qquad\qquad &= \mathbf{g_2} \\
\mathbf{v_2} + \mathbf{A v_3} + \mathbf{v_4} \qquad\qquad &= \mathbf{g_3} \\
\mathbf{v_3} + \mathbf{A v_4} + \mathbf{v_5} \qquad &= \mathbf{g_4} \\
\mathbf{v_4} + \mathbf{A v_5} + \mathbf{v_6} \qquad &= \mathbf{g_5} \\
\mathbf{v_5} + \mathbf{A v_6} + \mathbf{v_7} &= \mathbf{g_6} \\
\mathbf{v_0} + \qquad\qquad\qquad\qquad\qquad \mathbf{v_6} + \mathbf{A v_7} &= \mathbf{g_7} \quad (19.392)
\end{aligned}
$$

Multiply the second equation by $-\mathbf{A}$ and add to it the first and third equations. The result is the equation

$$(2\mathbf{I} - \mathbf{A}^2)\mathbf{v}_1 + \mathbf{v}_3 + \mathbf{v}_7 = -\mathbf{A}\mathbf{g}_1 + \mathbf{g}_0 + \mathbf{g}_2 \qquad (19.393)$$

Repeating this procedure on the fourth, sixth, and eighth equation we obtain the reduced system

$$
\begin{aligned}
\mathbf{A}^{(1)}\mathbf{v}_1 + \quad \mathbf{v}_3 \quad\quad\quad + \quad \mathbf{v}_7 &= \mathbf{g}_1^{(1)} \\
\mathbf{v}_1 + \mathbf{A}^{(1)}\mathbf{v}_3 + \quad \mathbf{v}_5 \quad\quad &= \mathbf{g}_3^{(1)} \\
\mathbf{v}_3 + \mathbf{A}^{(1)}\mathbf{v}_5 + \quad \mathbf{v}_7 &= \mathbf{g}_5^{(1)} \\
\mathbf{v}_1 + \quad\quad\quad \mathbf{v}_5 + \mathbf{A}^{(1)}\mathbf{v}_7 &= \mathbf{g}_7^{(1)}
\end{aligned}
\qquad (19.394)
$$

where

$$\mathbf{A}^{(1)} = 2\mathbf{I} - \mathbf{A}^2 \text{ and } \mathbf{g}_{2j-1}^{(1)} = \mathbf{A}\mathbf{g}_{2j-1} + \mathbf{g}_{2j-2} + \mathbf{g}_{2j} \qquad (19.395)$$

The system of Eqs. (19.394) has half the number of original unknowns. Furthermore, it has exactly the same form as Eq. (19.392), so the process can be repeated and the unknowns can be halved again. Multiply the second equation by $-\mathbf{A}^{(1)}$ and add to it the first and third equations. Multiply the fourth equation by $-\mathbf{A}^{(1)}$ and add to it the third and first equations to get the further reduced system

$$
\begin{aligned}
\mathbf{A}^{(2)}\mathbf{v}_3 + 2\mathbf{v}_7 &= \mathbf{g}_3^{(2)} \\
2\mathbf{v}_3 + \mathbf{A}^{(2)}\mathbf{v}_7 &= \mathbf{g}_7^{(2)}
\end{aligned}
\qquad (19.396)
$$

where

$$\mathbf{A}^{(2)} = 2\mathbf{I} - (\mathbf{A}^{(1)})^2 \text{ and } \mathbf{g}_{4j-1}^{(2)} = -\mathbf{A}^{(1)}\mathbf{g}_{4j-1}^{(1)} + \mathbf{g}_{4j-3}^{(1)} + \mathbf{g}_{4j+1}^{(1)} \qquad (19.397)$$

Next we eliminate $\mathbf{v}_3$ from system of Eqs. (19.396) by multiplying the second equation by $-\mathbf{A}^{(2)}$ and adding twice the second equation to obtain

$$\mathbf{A}^{(3)}\mathbf{v}_7 = \mathbf{g}_7^{(3)} \qquad (19.398)$$

where

$$\mathbf{A}^{(3)} = 4\mathbf{I} - (\mathbf{A}^{(2)})^2 \text{ and } \mathbf{g}_7^{(3)} = -\mathbf{A}^{(2)}\mathbf{g}_7^{(2)} + \mathbf{g}_3^{(2)} \qquad (19.399)$$

Once we have solved Eq. (19.398) for $\mathbf{v}_7$, we can solve for $\mathbf{v}_3$ from the first of Eq. (19.396) and solve for $\mathbf{v}_1$ and $\mathbf{v}_5$ from the first and third of Eq. (19.394); we obtain $\mathbf{v}_0$, $\mathbf{v}_2$, $\mathbf{v}_4$, and $\mathbf{v}_6$ from Eq. (19.392). There remains now the task of solving Eq. (19.398) for $\mathbf{v}_7$. Note that the auxiliary matrices $\mathbf{A}^{(k)}$ that were created in the reduction process are polynomials of degree 2^k in the original matrix $\mathbf{A}$. These polynomials have known real roots $\lambda_i^{(k)}$ so we can rewrite Eq. (19.398), using the factored form of the polynomial for $\mathbf{A}^{(3)}$, as

$$\prod_{i=0}^{7} (\mathbf{A} - \lambda_i^{(7)}\mathbf{I})\mathbf{v}_7 = \mathbf{g}_7^{(3)} \tag{19.400}$$

which can be solved by the algorithm

1. set $\mathbf{z}_0 = \mathbf{g}_7^{(3)}$

2. for $i = 1, \ldots, 8$, solve

$$(\mathbf{A} - \lambda_i^{(7)}\mathbf{I})\mathbf{z}_i = \mathbf{z}_{i-1} \tag{19.401}$$

The last vector obtained, $\mathbf{z}_8$, is the solution $\mathbf{v}_7$, of Eq. (19.384).

The algorithm presented above is known as *cyclic odd–even reduction* and is numerically unstable. The elements of the matrices $\mathbf{A}^{(k)}$ grow exponentially as a function of k, with the result that all significance can be lost in the calculation of the right sides $\mathbf{g}_j^{(k)}$. The Buneman (1969) variant stabilizes the calculation by assuming that each of the $\mathbf{g}_j^{(k)}$ can be written as

$$\mathbf{g}_j^{(k)} = \mathbf{A}^{(k)}\,\mathbf{p}_j^{(k)} + \mathbf{q}_j^{(k)} \tag{19.402}$$

and developing recurrence relations for $\mathbf{p}_j^{(k)}$ and $\mathbf{q}_j^{(k)}$ which involve solving a linear system of equations with coefficient matrix $\mathbf{A}^{(k-1)}$.

We have illustrated the algorithm for the case $N = 8 = 2^3$. In general, for $N = 2^p$, p reduction steps are required to reduce the original system to a single equation for the unknown $\mathbf{v}_N$. At each step of the reduction phase, there are $N/2$ tridiagonal systems to be solved. Solving for $\mathbf{v}_N$ requires the solution of N periodic tridiagonal systems. Finally, there are p back-substitution steps to find the remaining unknowns, at each step of which $N/2$ periodic tridiagonal systems are solved. The total number of tridiagonal systems to be solved is, therefore, $N(p + 1) = N \log_2 N + N$. Each periodic tridiagonal system requires $5M$ multiplications, so the total operation count is about $5MN \log_2 N$. This count is less than the count for the Fourier method, however, in practice, the stable Buneman variant is about 25% *slower* than the Fourier method on a vector computer. The relative speed of these two methods will vary depending on the implementation and the computer.

19.8.5   Least Squares Solutions

Perhaps the most unique attribute of direct methods is the ability to determine least squares solutions to Poisson's equation. As mentioned earlier, it is relatively common in fluid models to pose problems that do not have a computational solution. This difficulty results from computational and/or observational errors and would not be a problem for the continuous fluid model with exact data. That is, roundoff, truncation, and observational errors likely produce an inconsistent system of Eqs. (19.380a–c) in which the constraint in Eq. (19.382) is not satisfied even though constraint in Eq. (19.377) is satisfied with an exact right side $f(x, y)$. In this section, we show how to perturb the right side of Poisson's equation so that either the Fourier or cyclic reduction method can be used to compute a solution to the perturbed system. Solutions to the perturbed system are least squares solutions to the unperturbed system.

If we replace $f_{i,j}$ with a perturbed right side

$$g_{i,j} = f_{i,j} - \frac{1}{MN} \sum_{i=0}^{M-1} \sum_{j=0}^{N-1} f_{i,j} \qquad (19.403)$$

then $g_{i,j}$ satisfies the constraint in Eq. (19.382). In Swarztrauber and Sweet (1975) it is shown that system of Eqs. (19.380a–c) has a solution if $f_{i,j}$ is replaced with $g_{i,j}$. Further, it is shown that the solution is a least squares solution to the unperturbed problem. That is, the solution minimizes the l_2 norm of the residual in Eq. (19.380a–c).

A slight modification of the direct methods is necessary for the doubly periodic case under consideration. Using the Fourier method, the periodic tridiagonal system of Eqs. (19.386a–c) is singular for $k = 0$. Using a fast variant of Gauss elimination, a zero pivot is computed for $j = N - 1$. However, because $g_{i,j}$ satisfies the constraint in Eq. (19.382), it can be shown that the right side of the zero pivot equation is also zero yielding an equation of the form $0 \cdot \hat{v}_{0,N-1} = 0$. Therefore, $\hat{v}_{0,N-1}$ can be set to any value which demonstrates that the solution is not unique. In practice, the zeros in the pivot equation are not identically zero, but on the order of roundoff error. Hence, proceeding with Gauss elimination will compute a $\hat{v}_{0,N-1}$ that is on the order of one. Therefore, the only modification of the direct method is an identically zero test in which case $\hat{v}_{0,N-1}$ can be set to any value, say, $\hat{v}_{0,N-1} = 1$.

The constraint in Eq. (19.382) is valid only for the doubly periodic case considered here. It will vary depending on the boundary conditions and the coordinate system under consideration. The general form of the constraint is

$$0 = \sum_{i-i_s}^{i_f} \sum_{j=f_{js}}^{j_f} w_i w_j f_{i,j} \qquad (19.404)$$

where i_s, i_f, j_s, j_f and the weights w_i, and w_j vary depending on the problem. These quantities are tabulated in Swarztrauber and Sweet (1975) for the boundary conditions listed in Sec. 19.8.1 and for Poisson's equation in Cartesian, cylindrical, and spherical coordinate systems.

Finally, we include an important comment about the least squares solutions. The method will formally work for any right side $f_{i,j}$. However, if the perturbation $g_{i,j}$ is significantly different from $f_{i,j}$, then the least squares solution provides the exact solution to a problem that is significantly different from the original problem. In this case, an effort should be made to understand why the original problem did not have a solution.

19.8.6 Software

The advent of fast direction methods has significantly reduced computation time for fluid models. In addition, they have also reduced the development time since the methods have been implemented in a public domain FORTRAN package called FISHPACK [Swarztrauber and Sweet (1975), (1979)]. This package automatically provides complete second-order finite difference approximatons to two-dimensional Poisson equations in several frequently used coordinate systems (Cartesian, cylindrical, and spherical), and the three-dimensional Poisson equation in Cartesian coordinates. The software incorporates the given boundary data and calculates the correct approximation at coordinate singularities, e.g., the origin $r = 0$ in spherical coordinates. When the problem specified is singular, the software checks the correct condition to determine whether a solution exists and, if it does not, subtracts the appropriate constant from the data to guarantee that a weighted least-squares solution to the original equation exists, and finds it.

In addition, the software includes subroutines for more general separable elliptic equations that have the added capability of providing fourth-order accurate solutions. The most general equation solved by the package is the separable elliptic equation

$$a(x) \frac{\partial^2 u}{\partial x^2} + b(x) \frac{\partial u}{\partial x} + c(x)u + d(y) \frac{\partial^2 u}{\partial y^2} + e(y) \frac{\partial u}{\partial y} + f(y)u = g(x, y) \quad (19.405)$$

that is solved with a software implementation of the generalized cyclic reduction algorithm given in Swarztrauber (1974). Except for the three-dimensional Cartesian solver, the codes are written for two space dimensions. Nevertheless, three-dimensional versions are available in private domain software CRAYFISHPACK [Sweet (1992)]. Software is also available for irregular regions [Buzbee *et al.* (1970), Proskurowski (1983)].

FISHPACK was developed at the National Center for Atmospheric Research (NCAR) and was tested by a group from five federal laboratories [Steuerwalt (1979)]. The testing included compilation of routines, verification of example programs, verification that the input error detection code works correctly, and the construction and running of various test problems. Finally, the Fourier method described in Sec. 19.8.3 can be implemented directly with the assistance of FFTPACK 5.0 which includes sine, cosine, and the quarter wave transforms as well as the traditional real and complex periodic transforms. Both FISHPACK and FFTPACK are distributed by NCAR. Distribution information can be obtained from either author.

19.8.7 Parallel Computation

With the advent of parallel and distributed computing, it is of interest to develop methods for the implementation of the direct solvers on these machines. The solution of Poisson's equation on a parallel computer is discussed in some detail in Swarztrauber and Sweet (1989). The focus of that paper is on the distribution of the various computational segments of the direct solver. That is, parallel algorithms were developed for the FFT and the solution of the tridiagonal systems. Here, we describe a different approach in which the traditional scalar algorithms are used, and parallelism is obtained from the multiplicity of sequences that must be transformed or the multiplicity of tridiagonal systems that must be solved.

For example, consider the development of weather/climate models on massively parallel multiprocessors. These models contain a number of computations that can be performed in parallel. On each latitude, a Fourier transform is performed in the longitudinal direction. These transforms can be performed in parallel if the data on each latitude are in the same processor; that is, if the data are distributed in latitude but not longitude. The Legendre transform in the latitudinal direction can similarly be performed in parallel if the data are distributed in longitude but not in latitude.

There are two approaches to performing these computations on a multiprocessor. First, a fixed distribution of the data could be selected; that is, the data could be distributed in longitude but not latitude. Then, the Legendre transform could be performed in-processor without interprocessor communication. However, the Fourier transform in longitude would then require interprocessor communication and the development of a distributed or parallel FFT algorithm. As an alternative, the data could be dynamically redistributed (transposed) between these computational modules, and both the Legendre and Fourier transforms could be done in-processor. This approach is called the *transpose method*.

There are several reasons why the transpose method is preferred. First and foremost, it requires less communication. Using current parallel transpose algorithms, the communication complexity is proportional to the data complexity; the transposition of a $P \times N$ array can be performed in $O(N)$ time on P processors. However, if the data are not reconfigured, then the communication complexity of the FFT's in the longitudinal direction may be proportional to its computational complexity. If N is the number of longitudinal points, the computational (and communication) complexity of P distributed FFT's of length N is $O(N \log N)$. In general, where applicable, it is preferable to move the data into a processor with a communication complexity proportional to the data complexity, rather than implement a distributed algorithm in which the communication complexity may be proportional to the computational complexity.

A second reason to prefer the transpose method is the resulting simplicity of the approach. The method uses existing efficient algorithms and software without the need to develop what are often less efficient parallel or distributed algorithms. Finally, the porting of a model is simplified, because, apart from the addition of the transposition, much of the original code can remain relatively unchanged. Hence, the problem of weather/climate modeling on a multiprocessor is reduced to finding the most efficient parallel algorithm for transposing an array on a multiprocessor.

19.9 SPECTRAL AND PSEUDO-SPECTRAL METHODS
John Kim

Spectral methods, which had been used widely in mathematical analyses before the advent of computers [e.g., Lanczos (1956)], have recently regained popularity because of the development of fast transform methods that allow the efficient implementation of spectral methods. These methods have become a powerful tool for numerical simulations of transitional and turbulent flows where high accuracy is needed for a realistic realization of these flows. In this section, we present by using simple examples the basic concept involved with spectral methods. Detailed description of these methods can be found in Gottlieb and Orszag (1977) and Canuto *et al.* (1987). We also assume the reader to be familiar with the basic concepts of Fourier series and Sturm-Liouville problems; otherwise, they should refer to mathematical references such as Courant and Hilbert (1953) and Arfken (1970).

In spectral methods, the solution to a problem is represented by a truncated series of known functions. The proper choice of the expansion functions depends upon the particular problem, especially the problem geometry and the boundary conditions. A wide variety of the eigenfunctions of Sturm–Liouville problems, such as trigonometric functions and Chebyshev, Legendre, and Jacobi polynomials can be used. Fourier series and Chebyshev polynomials are most commonly used. For problems with periodic boundary conditions, Fourier series are the natural choice. However, for nonperiodic problems, Fourier series converge too slowly because of the well-known Gibb's phenomenon at boundaries [e.g., Arfken (1970)]. To avoid this slow convergence due to the Gibb's phenomenon, one has to resort to the eigenfunctions of singular Strum–Liouville problems. Chebyshev polynomials [Fox and Parker (1968)] are eigenfunctions of this kind and are generally used for nonperiodic boundary conditions. For both Fourier series and Chebyshev expansions, Fast Fourier Transforms (FFT) [Cooley and Tuckey (1965)] can be used for the efficient implementation of the spectral methods. Other choices of eigenfunctions such as Legendre and Jacobi polynomials are also possible, but the lack of a fast transform method decreases the efficiency of the methods, although this is not considered a serious problem, especially for multidimensional problems [Orszag (1980)].

Spectral methods are very accurate numerical differentiators. This is the major advantage of spectral methods. To illustrate this point and to show how a spectral method is used in practice to obtain numerical differentiations, we consider the following example.

Example 19.6:   Compare how a finite difference and a spectral method differentiate a simple wave, $f(x) = e^{ikx}$, $0 \le x \le 2\pi$.

Consider the second-order central differencing scheme

$$\frac{\delta f}{\delta x} = \frac{f_{j+1} - f_{j-1}}{2\Delta}$$

$$= \frac{e^{ik(x+\Delta)} - e^{ik(x-\Delta)}}{2\Delta}$$

$$= \frac{e^{ik\Delta} - e^{-ik\Delta}}{2\Delta} e^{ikx}$$

$$= i\frac{\sin k\Delta}{\Delta} f = ik'f \tag{19.406}$$

where $\Delta = 2\pi/(N-1)$ and N is the number of computational points. Since the exact answer to the problem is ikf, the accuracy of the differencing scheme can be measured by how well $k' = \sin k\Delta/\Delta$ approximates k. Expanding Eq. (19.406) in terms of Taylor series,

$$k' = \frac{\sin k\Delta}{\Delta} = k - \frac{1}{6}k^3\Delta^2 + \cdots = k + O(\Delta^2) \tag{19.407}$$

Note that the leading error term is $O(\Delta^2)$ as expected. It is also a function of the wave number k, and the error becomes greater for high wave numbers (a high wave number is a rapidly varying function in space). To obtain the spectral approximation of $\partial f/\partial x$, we first represent $f(x)$ in terms of truncated Fourier series (we use Fourier expansions since $f(x)$ is periodic)

$$f(x) = \sum_{n=0}^{N-1} a_n e^{i\omega_n x}.$$ (19.408)

where $\omega_n = n - N/2$, $n = 0, \ldots, N - 1$. Then, by differentiating Eq. (19.408) term by term, we obtain

$$\frac{\partial f}{\partial x} = \sum_{n=0}^{N-1} i\omega_n a_n e^{i\omega_n x}$$ (19.409)

Thus from Eq. (19.409), $i\omega_n a_n$ is the Fourier transform of $\partial f/\partial x$. In actual computation to get the spectral approximation of the first derivative, we: 1) Fourier transform $f(x)$ to get a_n, 2) compute $i\omega_n a_n$, and 3) inverse transform $i\omega_n a_n$ to get $\partial f/\partial x$. In this procedure, FFT can be used for efficient implementation of the transformations. The transformation can be computed with $O(N \ln N)$ arithmetic operations in contrast to $O(N^2)$ required for the case of a straightforward sum. For the present example, the above procedure will give the exact answer, that is $\delta f/\delta x = ikf$, if $-N/2 \le k \le N/2 - 1$. This is the main distinction between spectral and finite difference methods; spectral methods give high accuracy at high wave numbers, while finite difference methods fail badly. In addition to the superior accuracy, spectral methods introduce no extra difficulty near boundaries in contrast to higher order finite difference methods which require special treatment near boundaries.

The above example was a rather fortuitous case, though, since the trial function could be represented exactly by the Fourier expansion (in fact, one of the components of the expansion). In a more general case, for *infinitely differentiable functions*, the truncation error in spectral methods goes to zero faster than any power of $1/N$ as $N \to \infty$, while in finite difference approximation, the error goes to zero with $O(1/N^p)$ for a finite p. This fast convergence rate is guaranteed, however, only when a proper expansion is used with regard to boundary conditions. Otherwise, the convergence rate depends on the smoothness of the function *and* the boundary conditions. This is why one has to use the eigenfunctions of singular Sturm–Liouville problems for general boundary conditions as mentioned earlier. In this case, the convergence rate depends only on the smoothness of the function.

There are basically three different techniques that can be used to implement the spectral methods to solve a differential equation—*Galerkin*, *tau*, and *collocation methods*. Galerkin and tau methods are generally called *spectral methods* while the collocation method is called *pseudo-spectral* method [Orszag (1971)]. The difference among these methods can be summarized by showing how the solution to a problem is sought in terms of truncated series. Consider $u(x, t)$, which satisfies a differential equation and certain boundary conditions. With the Galerkin method, the solution is assumed in the form of

$$u(x, t) = \sum_{n=1}^{N} a_n(t)\phi_n(x)$$ (19.410)

where ϕ_n's are assumed to be linearly independent and satisfy given boundary conditions. Therefore $u(x, t)$ automatically satisfies all the necessary boundary conditions. Substituting Eq. (19.410) into the governing equation for $u(x, t)$ and using the orthogonal properties of $\phi_n(x)$, we obtain differential equations for the coefficients $a_n(t)$ to be solved. Once we have $a_n(t)$, we can get $u(x, t)$ from Eq. (19.410). In the tau method, which was developed by Lanczos [see Lanczos (1956)], the solution is sought in the form of

$$u(x, t) = \sum_{n=1}^{N+k} a_n(t)\phi_n(x) \tag{19.411}$$

where ϕ_n need not satisfy the boundary conditions, and k is the number of independent boundary constraints. The same procedure as the Galerkin method follows except that the k boundary constraints are imposed together with the given equations, thus giving $(N + k)$ equations for $(N + k)$ *unknown* coefficients. With the collocation method, the same representation as the Galerkin approximation is used, but this is enforced only at selected N points (collocation points)

$$u(x_j, t) = \sum_{n=1}^{N} a_n(t)\phi_n(x_j), \quad j = 1, \ldots, N \tag{19.412}$$

Furthermore, this spectral approximation is only used to compute derivatives. With the collocation method, the truncation error in Eq. (19.412) is zero at the collocation points.*

The difference between the Galerkin and tau methods is minor, but the difference between *spectral* and *pseudo-spectral* becomes significant especially in computing nonlinear terms. The main advantage of the pseudo-spectral method lies in the fact that it allows computation in either spectral or physical space, whichever is more convenient. This is particularly useful when there are nonconstant coefficients or nonlinear terms. This will be illustrated in the next example.

Example 19.7:   Consider *Burger's equation* with a periodic boundary condition.

$$\frac{\partial u}{\partial t} + u\frac{\partial u}{\partial x} = \nu\frac{\partial^2 u}{\partial x^2}, \quad 0 \leq x \leq L \tag{19.413}$$

$$u(0, t) = u(L, t) \tag{19.414}$$

First, we use spectral methods. With the periodic boundary condition, Fourier-spectral is appropriate. We look for the solution in terms of

$$u(x, t) = \sum_{n=0}^{N-1} a_n(t)e^{ik_n x} \tag{19.415}$$

---

*In the Galerkin method, the error in approximating Eq. (19.410) is forced to be orthogonal to ϕ_n, $n = 1, \ldots, N$.

where $k_n = (2\pi/L)(n - N/2)$, $n = 0, \ldots, N - 1$. By substituting Eq. (19.419) into Eq. (19.413) and using the orthogonal property of $e^{ik_n x}$, we obtain

$$\frac{\partial a_n}{\partial t} + \sum_{\substack{k_p + k_q = k_n \\ 0 \le p, q \le N-1}} ik_n a_p a_q = -\nu k_n^2 a_n, \quad n = 0, \ldots, N - 1 \quad (19.416)$$

The second term on the left side is due to the nonlinear term. This convolution sum requires $O(N^2)$ operations for each time step.

Now, use the Pseudo-Spectral (Collocation) method. We enforce Eq. (19.413) at N selected points with the spectral approximation,

$$u(x_j, t) = \sum_{n=1}^{N-1} a_n(t) e^{ik_n x_j} \quad (19.417)$$

where $x_j = (j - 1)L/N$, $j = 1, \ldots, N$. Thus, we have to solve

$$\frac{\partial u(x_j, t)}{\partial t} + u(x_j, t) \frac{\partial u(x_j, t)}{\partial x} = -\nu \frac{\partial^2 u(x_j, t)}{\partial x^2} \quad (19.418)$$

for each j. In solving Eq. (19.418), we use the spectral approximation, Eq. (19.417) to evaluate $u_x(x_j, t)$ and $u_{xx}(x_j, t)$, that is

$$\mathcal{F}(u) \to a_n; \quad \mathcal{F}^{-1}(ik_n a_n) \to u_x$$
$$; \quad \mathcal{F}^{-1}(-k_n^2 a_n) \to u_{xx} \quad (19.419)$$

where $\mathcal{F}$ and $\mathcal{F}^{-1}$ represent the Fourier and its inverse transforms. Again, the transformation between spectral and physical space can be carried out by using FFT with $O(N \ln N)$ operations. In general, a pseudo-spectral method is much cheaper and easier to implement than a spectral method. The only disadvantage of the former is that it contains *aliasing* error, and this could lead to a serious numerical instability unless a special care is taken [Patterson and Orszag (1971)]. Most spectral codes solving fluid mechanics problems use a pseudo-spectral method with some kind of aliasing control.

So far, we have discussed only Fourier-spectral methods for periodic problems. Other boundary conditions require different spectral expansions to ensure rapid convergence of the series as mentioned earlier. The next example shows how Chebyshev polynomials are used for this purpose.

Example 19.8: We consider a one-dimensional wave equation with a nonperiodic boundary condition:

$$u_t(x, t) + uu_x(x, t) = 0 \quad -1 \le x \le 1, \quad t > 0,$$

$$u(-1, t) = 0, \quad u(x, 0) = f(x) \quad (19.420)$$

We use the Chebyshev-pseudo-spectral method to obtain a numerical solution to the problem. Use $\phi_n(x) = T_n(x) - (-1)^n T_0$ as the expansion function. Here, T_n represents the nth order Chebyshev polynomial and note that we have chosen ϕ_n such that it satisfies the boundary condition. Choose the collocation points as extrema of the Chebyshev polynomial $T_N(x)$, they are $x_j = \cos(\theta_j)$ where $\theta_j = \pi j / N$, $j = 0, \ldots$, $N - 1$. Thus, we have

$$u(x_j, t) = \sum_{n=1}^{N} a_n(t) \phi_n(x_j) \qquad (19.421)$$

By defining $a_0 = -\sum_{m=1}^{N} (-1)^m a_m$, we can rewrite Eq. (19.421) in terms of Chebyshev polynomials,

$$u(x_j, t) = \sum_{n=0}^{N} a_n(t) T_n(x_j) \qquad (19.422)$$

Therefore, we solve

$$u_t(x_j, t) + u(x_j, t) u_x(x_j, t) = 0 \quad \text{for} \quad x_j = \cos\left(\frac{\pi j}{N}\right) \qquad (19.423)$$

We use the following relationship to compute $u_x(x_j, t)$

$$\frac{\partial u}{\partial x} = \sum_{n=0}^{N} a_n(t) T_n'(x_j)$$

$$= \sum_{n=0}^{N} b_n(t) T_n(x_j) \qquad (19.424)$$

From the properties of Chebyshev polynomials [Fox and Parker (1968)], we obtain

$$c_{n-1} b_{n-1} - b_{n+1} = 2n a_n, \quad n \geq 1$$

where c_n is defined such that $c_0 = 2$, $c_n = 1$ for $n \geq 2$. Also not that $b_N = 0$ from Eq. (19.424) and $b_{N+1} = 0$ is implied in Eq. (19.425). With the above choice of the collocation points, we can use FFT to compute a_n, since $T_n(x) = T_n(\cos \theta) = \cos n\theta$ with $\theta = \cos^{-1} x$. Therefore, given $u(x_j, t)$, we can compute $a_n(t)$ using FFT. We then compute $b_n(t)$ using Eq. (19.425), and the inverse transform of Eq. (19.424) with FFT will give back $u_x(x_j, t)$. Thus, it requires two evaluations of FFT for each time step to advance Eq. (19.423) in time.

In this section, we have shown how spectral/pseudo-spectral methods can be used to find a numerical solution to differential equations. For simple geometries, well-designed spectral codes can produce very accurate solutions with little extra cost. Application of spectral methods to complex geometries is, however, a current research topic [Orszag (1980)]. For large scale computations, in addition to the extra computations involved with the spectral methods, one needs more elaborate management of databases, since the spectral approximations require all the data points (global) in contrast to finite difference methods which require only the adjacent data points (local). Whether one needs to use spectral methods with the extra effort of computations and coding should be decided depending upon the accuracy requirement and the smoothness of a solution to the problem. For sufficiently smooth solutions, spectral methods provide higher accuracy with a much lower number of computational nodes. But, the advantage of spectral methods as an accurate differentiator diminishes for nonsmooth solutions. One study shows roughly twice as many computational nodes are needed for a second-order finite difference method to achieve the same accuracy as a spectral method in a turbulent simulation [Herring *et al.* (1974)]. For some problems, one can also develop mixed spectral and finite difference methods to obtain the required accuracy and efficiency for special purposes [Moin and Kim (1982)]. In this case, one or more directions are selected for spectral expansions, and the rest are represented by finite difference approximations. One has to be careful about the balance between the resolutions of different directions, since the overall accuracy of the solution is influenced by the lowest order approximation.

Further information can be found in Collaz (1960), Elliot (1961), Fox and Orszag (1973), Leonard and Wray (1982), Orszag (1969), (1972), Orszag and Israeli (1974), and Canuto *et al.* (1987).

REFERENCES

Abgrall, R., "Design of an Essentially Non-Oscillatory Reconstruction Procedure on Finite-Element Type Meshes," ICASE Report 91-84, 1991.

Aliabadi, S. and Tezduyar, T. E., "Parallel Fluid Dynamics Computations in Aerospace Applications," *Int. J. Num. Meth. Fluids*, 1995.

Aliabadi, S. and Tezduyar, T. E., "Space-Time Finite Element Computation of Compressible Flows Involving Moving Boundaries and Interfaces," *Computer Methods Appl. Mech. and Eng.*, Vol. 107, pp. 209–224, 1993.

Allwright, S., "Multiblock Topology Specification and Grid Generation for Complete Aircraft Configurations," AGARD-CP-464, 1990.

Anderson, D. A., Tannehill, J. C., and Pletcher, R. H., *Computational Fluid Mechanics and Heat Transfer*, Hemisphere, New York, 1984.

Anderson, O. L., Davis, R. T., Hankins, G. B., and Edwards, D. E., "Solution of Viscous Internal Flows on Curvilinear Grids Generated by the Schwarts–Christoffel Transformation," *Numerical Grid Generation*, North-Holland, 1982.

Anderson, W. K., Thomas, J. L., and Rumsey, C. L., "Extension and Application of Flux-Vector Splitting to Calculations on Dynamic Meshes," *AIAA J.*, Vol. 27, No. 6, 1989.

Anderson, W. K., Thomas, J. L., and van Leer, B., "A Comparison of Finite Volume Flux Vector Splittings for the Euler Equations," *AIAA J.*, Vol. 24, No. 9, pp. 1453–1460, 1986.

Anderson, W. K., Thomas, J. L., and Whitfield, D. L., "Three-Dimensional Multigrid Algorithms for the Flux-Split Euler Equations," NASA, Tech. Rep. 2829, 1988.

Arcilla, A. S., Häuser, J., Eiseman, P. R., and Thompson, J. F. (Eds.), *Numerical Grid Generation in Computational Fluid Dynamics and Related Fields—The Third International Conference*, North-Holland, 1991.

Arfken, G., *Mathematical Methods for Physicists*, 2nd ed., Academic Press, New York, 1970.

Arney, D. C. and Flaherty, J. E., "A Moving-Mesh Finite Element Method With Local Refinement for Parabolic Partial Differential Equations," *Comp. Meths. Appl. Mech. and Engr.*, Vol. 55, pp. 3–26, 1986.

Atkins H. L. and Casper, J., "Nonreflective Boundary Conditions for High-Order Methods," AIAA 93-0152, 1993.

Atkins, H. L., "A Multiblock Multigrid Method for the Solution of the Euler and Navier–Stokes Equations for Three-Dimensional Flows," AIAA 91-0101, 1991.

Bailey, H. E. and Beam, R. M., "Newtons Method Applied to Finite-Difference Approximations for the Steady-State Compressible Navier–Stokes Equations," *J. Comp. Phys.*, Vol. 93, pp. 108–127, 1991.

Baker, T. J., "Developments and Trends in Three-Dimensional Mesh Generation," *Appl. Num. Math.*, Vol. 5, pp. 275–304, 1989.

Baker, T. J., "Element Quality in Tetrahedral Meshes," *Finite Elements in Fluids*, Chung, T. J. (Ed.), University of Alabama in Huntsville Press, 1989.

Baker, T. J., "Three-Dimensional Mesh Generation by Triangulation of Arbitrary Point Sets," AIAA-CP-87-1124, 1987.

Barth, T. J. and Frederickson, P. O., "Higher Order Solution of the Euler Equations on Unstructured Grids Using Quadratic Reconstruction," AIAA 90-0013, 1990.

Barth, T. J., "Aspects of Unstructured Grids and Finite-Volume Solvers for the Euler and Navier–Stokes Equations," *Unstructured Grid Methods for Advection Dominated Flows*, AGARD Report 787, 1992.

Beam, R. M. and Warming, R. F., "An Implicit Finite-Difference Algorithm for Hyperbolic Systems in Conservation-Law Form," *J. Comp. Phys.*, Vol. 22, pp. 87–110, 1976.

Beam, R. M. and Warming, R. F., "Implicit Numerical Methods for the Compressible Navier–Stokes and Euler Equations," *von Karman Institute for Fluid Dynamics Lecture Series: Computational Fluid Dynamics*, Belgium, 1982.

Beam, R. M., Warming, R. F., and Yee, H. C., "Stability Analysis of Numerical Boundary Conditions and Implicit Difference Approximations for Hyperbolic Equations," *J. Comp. Phys.*, Vol. 48, No, 2, pp. 200–222, 1982.

Behr, M. and Tezduyar, T. E., "Finite Element Solution Strategies for Large-scale Flow Simulations," *Computer Methods in Appl. Mech. and Eng.*, Vol. 112, pp. 3–24, 1994.

Benson, R. A. and McRae, S. D., "Time Accurate Simulations of Unsteady Flows With a Dynamic Solution-Adaptive Grid Algorithm," *Numerical Grid Generation in Computational Fluid Dynamics and Related Fields*, Weatherill, N. P., Eiseman, P. R., Häuser, J., and Thompson, J. F. (Eds.), Pineridge Press, 1994.

Berger, M. J. and Colella, P., "Local Adaptive Mesh Refinement for Shock Hydrodynamics," *J. Comp. Phys.*, Vol. 82, pp. 64–84, 1989.

Blacker, T. D. and Stephenson, M. B., "Paving: A New Approach to Automated Quadrilateral Mesh Generation," *Int. J. Num. Method. Eng.*, Vol. 32, pp. 811–847, 1992.

Blacker, T. D. and Meyers, R. J., "Seams and Wedges in Plastering: A 3-D Hexahedral Mesh Generation Algorithm," *Engr. with Computers*, Vol. 9, pp. 83–93, 1993.

Blacker, T. D. and Stephenson, M. B., "Paving: A New Approach to Automated Quadrilateral Mesh Generation," *Int. J. Num. Meth. Engr.*, Vol. 32, pp. 811–847, 1991.

Blazek, J., Kroll, N., Radespiel, R., and Rossow, C. C., "Upwind Implicit Residual Smoothing Method for Multi-Stage Schemes," AIAA 91-1533-CP, 1991.

Bockelie, M. J. and Eiseman, P. R., "Adaptive Grid Method for Unsteady Flow Problems," *Intl. J. Numer. Meth. Heat and Fluid Flow*, Vol. 2, No. 2, pp. 171–190, 1992.

Bonet, J. and Peraire, J., "An Alternate Digital Tree Algorithm for Geometric Searching and Intersection Problems," *Int. J. Num. Meth. Eng.*, 1992.

Bowyer, A., "Computing Dirichlet Tesselations," *The Computer Journal*, Vol. 24, No. 2, pp. 162–167, 1981.

Brandt, A., "Guide to Multigrid Development," *Multigrid Methods: Volume 960 of Lecture Notes in Mathematics*, Springer-Verlag, 1982.

Brandt, A., "Multi-Level Adaptive Solutions to Boundary-Value Problems," *Math. Comput.*, Vol. 31, No. 138, pp. 333–390, 1977.

Brandt, A., "Multilevel Adaptive Computations in Fluid Dynamics," *AIAA J.*, Vol. 18, No. 10, pp. 1165–1172, 1980.

Brezzi, F. and Pitkäranta, J., "On the Stabilization of Finite Element Approximations of the Stokes Problem," *Efficient Solutions of Elliptic Systems, Notes on Numerical Fluid Mechanics*, Vol. 10, Hackbusch, W. (Ed.), Vieweg, Wiesbaden, 1984.

Briley, W. R. and McDonald, H., "On the Structure and Use of Linearized Block Implicit Schemes," *J. Comp. Phys.*, Vol. 34, pp. 54–73, 1980.

Brooks, A. N. and Hughes, T. J. R., "Streamline Upwind/Petrov–Galerkin Formulations for Convection Dominated Flows With Particular Emphasis in the Incompressible Navier–Stokes Equations," *Computer Methods in Appl. Mech. and Eng.*, Vol. 32, pp. 199–259, 1982.

Buneman, O., "A Compact Noniterative Poisson Solver," Rept. 294, Stanford Univ. Inst. for Plasma Research, 1969.

Buning, P. G. and Steger, J. L., "Solution of the Two-Dimensional Euler Equations with Generalized Coordinate Transformation Using Flux Vector Splitting," AIAA 82-0971, 1982.

Buzbee, B., Dorr, F., George, J., and Golub, G., "The Direct Solution of the Discrete Poisson Equation on Irregular Regions," *SIAM J. Numer. Anal.*, Vol. 8, pp. 722–736, 1971.

Buzbee, B., Golub, G., and Nielson, C., "On Direct Methods for Solving Poisson's Equation," *SIAM J. Numer. Anal.*, Vol. 7, pp. 627–656, 1970.

Cabello, J., Löhner, R., and Jacquotte, O-P., "A Variational Method for the Optimization of Two- and Three-Dimensional Unstructured Meshes," AIAA-92-0450, 1992.

Cannizzaro, F. E., Elimiligui, A., Melson, N. D., and von Lavante, E., "A Multiblock Multigrid Method for the Solution of the Three-Dimensional Euler Equations," AIAA 90-0105, 1990.

Canuto, C., Hussaini, M. Y., Quarteroni, A., and Zang, T. A., *Spectral Methods in Fluid Dynamics*, Springer-Verlag, 1987.

Casper, J. and Atkins, H., "A Finite-Volume High-Order ENO Scheme for Two-Dimensional Hyperbolic Systems," *J. Comp. Phys.*, Vol. 106, pp. 62–76, 1993.

Castillo, J. E., *Mathematical Aspects of Numerical Grid Generation*, SIAM, 1991.

Cavendish, J. C., Field, D. A., and Frey, W. H., "An Approach to Automatic Three-Dimensional Finite Element Generation," *Int. J. Num. Meth. Eng.*, Vol. 21, pp. 329–347, 1985.

Cavendish, J. C., "Automatic Triangulation of Arbitrary Planar Domains for the Finite Element Method," *Int. J. Num. Method. Eng.*, Vol. 8, pp. 679–696, 1974.

Chakravarthy, S. R., "Euler Equations—Implicit Schemes and Boundary Conditions," *AIAA J.*, Vol. 21, pp. 699–706, 1983.

Chakravarthy, S. R., "Relaxation Methods for Unfactored Implicit Upwind Schemes," AIAA 84-0165, 1984.

Chakravarthy, S. R., "Development of Upwind Schemes for the Euler Equation," NASA CR 4043, 1987.

Chakravarthy, S. R. and Osher, S., "A New Class of High Accuracy TVD Schemes for Hyperbolic Conservation Laws," AIAA 85-0363, 1985.

Chakravarthy, S. R., Harten, A., and Osher, S., "A New Class of High Accuracy TVD Schemes for Hyperbolic Conservation Laws," AIAA 86-0363, 1986.

Chakravarthy, S. R., Harten, A., and Osher, S., "Essentially Non-Oscillatory Shock-Capturing Schemes of Arbitrarily High Accuracy," AIAA 86-0339, 1986.

Choi, B. K., Chin, H. Y., Loon, Y. I., and Lee, J. W., "Triangulation of Scattered Data in 3D Space," *Comp. Aided Geom. Des.*, Vol. 20, pp. 239–248, 1988.

Choi, Y. H. and Merkle, C. L., "Application of Time-Iterative Schemes to Incompressible Flows," *AIAA J.*, Vol. 23, No. 10, 1985.

Choo, Y. K. and Henderson, T. L., "Interactive Solution-Adaptive Grid Generation," *Software Systems and Surface Modeling and Grid Generation*, Smith, R. E. (Ed.), NASA CP 3143, pp. 347–361, 1992.

Colella, P. and Woodward, P. R., "The Piecewise Parabolic Method (PMM) for Gas-Dynamical Simulations," *J. Comp. Phys.*, Vol. 54, pp. 174–201, 1984.

Colella, P. and Woodward, P. R., "The Piecewise-Parabolic Method (PPM) for Gas-Dynamical Simulations," LBL Report No. 14661, 1982.

Collaz, L., *The Numerical Treatment of Differential Equations*, Springer-Verlag, Berlin, 1960.

Cooley, J. and Tukey, J., "An Algorithm for the Machine Calculation of Complex Fourier Series," *Math. Comp.*, Vol. 19, pp. 297–301, 1965.

Courant, R. and Hilbert, D., *Methods of Mathematical Physics*, Vol. 1, Wiley–Interscience, New York, 1953.

Crandall, M. G. and Majda, A., "Monotone Difference Approximations for Scalar Conservation Laws," *Math. Comp.*, Vol. 34, No. 149, pp. 1–21, 1980.

Dekoninck, W. and Barth, T. (Eds.), AGARD Rep. 787, *Proc. Special Course on Unstructured Grid Methods for Advection Dominated Flows*, VKI, Belgium, 1992.

Douglas, J., "On the Numerical Integration of $u_{xx} + u_{yy} = u_t$ by Implicit Methods," *J. Soc. Indust. Appl. Math.*, Vol. 3, pp. 42–65, 1955.

Duff, I. S., Erisman, A. M., and Reid, J. K., *Direct Methods for Sparse Matrices*, Monographs on Numerical Analysis, Clarendon Press, Oxford, 1986.

DuFort, E. C. and Frankel, S., "Stability Conditions in the Numerical Treatment of Parabolic Differential Equations," *Math Tables and Other Aids to Computation*, Vol. 7, pp. 135–153, 1953.

Dvinsky, A. S., "Adaptive Grid Generation from Harmonic Maps on Riemannian Manifolds," *J. Comp. Phys.*, Vol. 95, pp. 450–476, 1991.

Eiseman, P. R., "Automatic Algebraic Coordinate Generation," *Numerical Grid Generation*, North-Holland, 1982.

Eiseman, P. R., "Grid Generation for Fluid Mechanics Computations," *Ann. Review of Fluid Mech.*, Vol. 17, pp. 487–522, 1985.

Eiseman, P. R., "Adaptive Grid Generation," *Computer Meth. Appl. Mech. and Eng.*, Vol. 64, Nos. 1–3, pp. 321–376, 1987.

Eiseman, P. R., "A Control Point Form of Algebraic Grid Generation," *Int. J. Num. Meth. in Fluids*, Vol. 8, pp. 1165–1181, 1988.

Eiseman, P. R., "Control Point Grid Generation," *Computers Math. Appl.*, Vol. 24, No. 5/6, pp. 57–67, 1992.

Eiseman, P. R. and Erlebacher, G., "Grid Generation for the Solution of Partial Differential Equations," *State-of-the-Art Surveys on Computational Mechanics*, by Noor, A. K. and Oden, T. (Eds.), ASME AMD, Vol., 1989.

Eiseman, P. R., Cheng, Z., and Häuser, J., "Applications of Multiblock Grid Generation With Automatic Zoning," *Numerical Grid Generation in Computational Fluid Dynamics and Related Fields*, Weatherill, N. P., Eiseman, P. R., Häuser, J., and Thompson, J. F. (Eds.), Pineridge Press, 1994.

Eiseman, P. R., Lu, N., Jiang, M.-Y., and Thompson, J. F., "Algebraic-Elliptic Grid Generation," *Numerical Grid Generation in Computational Fluid Dynamics and Related Fields*, Weatherill, N. P., Eiseman, P. R., Häuser, J., and Thompson, J. F. (Eds.), Pineridge Press, 1994.

Eiseman, P. R., Peraire, J., Thompson, J. F., and Weatherill, N. P., *Von Karman Institute Lecture Series on Numerical Grid Generation*, 1990.

Eiseman, P. R. and Takahashi, S., "Grid Generation and Adaptivity for Complex Boundaries," *The Third World Congress on Computational Mechanics*, Chiba, Japan, p. 917, 1994.

Elliot, D., "A Method for the Numerical Integration of One-Dimensional Heat Equation Using Chebyshev Series," *Proc. Cambridge Philos. Soc.*, Vol. 57, p. 823, 1961.

Engquist, B. and Osher, S., "Stable and Entropy Satisfying Approximations for Transonic Flow Calculations," *Math. Comp.*, Vol. 34, pp. 45–75, 1980.

Erickson, L., "Practical Three-Dimensional Mesh Generation Using Transfinite Interpolation," *SIAM J. Sci. Comput.*, Vol. 6, 1985.

Fletcher, C. A. J., *Computational Galerkin Methods*, Springer-Verlag, New York, 1984.

Fox, D. G. and Orszag, S. A., "Pseudospectral Approximation to Two-Dimensional Turbulence," *J. Comp. Phys.*, Vol. 11, p. 612, 1973.

Fox, L. and Parker, I. B., *Chebyshev Polynomials in Numerical Analysis*, Oxford University Press, London, 1968.

Gear, C. W., *Numerical Initial Value Problems in Ordinary Differential Equations*, Prentice-Hall, New Jersey, 1971.

George, P. L., *Automatic Mesh Generation*, John Wiley & Sons, New York, 1991.

Ghia, K. N. and Ghia, U. (Eds.), *Advances in Grid Generation*, ASME FED, Vol. 5, 1983.

Giles, M. B., "Nonreflecting Boundary Conditions for Euler Equation Calculations," AIAA 89-1942, 1989.

Godfrey, A. G., "Topics on Spatially High-Order Accurate Methods and Preconditioning for the Navier–Stokes Equations with Finite-Rate Chemistry," Ph.D. thesis, Virginia Polytechnic Institute and State University, Blacksburg, VA, 1992.

Gordon, W. J. and Hall, C. A., "Construction of Curvilinear Coordinate Systems and Application to Mech Generation," *Int. J. Num. Meth. Eng.*, Vol. 7, pp. 461–477, 1973.

Gottlieb, D. and Orszag, S. A., *Numerical Analysis of Spectral Methods: Theory and Application*, CBMS-NSF Monograph No. 26, Soc. Ind. and Appl. Math., Philadelphia, 1977.

Greenberg, M. D., *Foundations of Applied Mathematics*, Prentice-Hall, New Jersey, 1978.

Gustaffson, B., Kreiss, H. O., and Sundstrom, A., "Stability Theory of Difference Approximations for Mixed Initial Boundary Value Problems, II," *Math. of Computations*, Vol. 26, 1972.

Gustaffson, B., "The Convergence Rate for Difference Approximations to Mixed Initial Value Problems," *Math. of Computations*, Vol. 29, 1975.

Gustafsson, B. and Oliger, J., "Stable Boundary Approximations for Implicit Time Discretization for Gas Dynamics," *SIAM J. Sci. and Stat. Computing*, Vol. 3, No. 4, pp. 408–421, 1982.

Gustafsson, B., Kreiss, H.-O., and Sundstrom, A., "Stability Theory of Difference Approximations for Mixed Initial Boundary Value Problems. II," *Math. Comp.*, Vol. 26, pp. 649–686, 1972.

Gustafsson, B., "The Convergence Rate for Difference Approximations to Mixed Initial Boundary Value Problems," *Math. Comp.*, Vol. 29, pp. 396–406, 1975.

Hackbush, W., *Multi-Grid Methods and Applications*, Springer-Verlag, 1985.

Hafez, M., Palaniswamy, S., Kuruvila, G., and Salas, M. D., "Applications of Wynn's Epsilon Algorithm to Transonic Flow Calculations," AIAA 87-1143, 1987.

Haidvogel, D. and Zang, T., "The Accurate Solution of Poisson's Equation By Expansion in Chebyshev Polynomials," *J. Comput. Phys.*, Vol. 30, pp. 167–180, 1979.

Hänel, D., Schwane, R., and Seider, G., "On the Accuracy of Upwind Schemes for the Solution of the Navier–Stokes Equations," AIAA 87-1105, 1987.

Hansbo, P. and Szepessy, A., "A Velocity-Pressure Streamline Diffusion Finite Element Method for the Incompressible Navier–Stokes Equations," *Computer Methods in Appl. Mech. and Eng.*, Vol. 84, pp. 175–192, 1990.

Hansbo, P., "The Characteristic Streamline Diffusion Method for the Time-Dependent Incompressible Navier–Stokes Equations," *Computer Methods in Appl. Mech. and Eng.*, Vol. 99, pp. 171–186, 1992.

Harten, A., "A High Resolution Scheme for the Computation of Weak Solutions of Hyperbolic Conservation Laws," *J. Comp. Phys.*, Vol. 49, pp. 357–393, 1983.

Harten, A., "High Resolution Schemes for Hyperbolic Conservation Laws," *J. of Comp. Phys.*, Vol. 49, pp. 357–393, 1983.

Harten, A., "On a Class of High Resolution Total-Variation-Stable Finite-Difference Schemes," *SIAM J. Num. Anal.*, Vol. 21, pp. 1–23, 1984.

Harten, A. and Chakravarthy, S., "Multidimensional ENO Schemes for General Geometries," NASA CR 187637, also ICASE Report 91-76, 1991.

Harten, A. and Hyman, J. M., "A Self-Adjusting Grid for the Computation of Weak Solutions of Hyperbolic Conservation Laws," Los Alamos Nat. Lab. Report LA9105, 1981.

Harten, A., Engquist, B., Osher, S., and Chakravarthy, R., "Uniformly High Order Accurate Essentially Non-Oscillatory Schemes III," ICASE Report 86-22, 1986.

Harten, A., Hyman, J. M., and Lax, P. D., "On Finite-Difference Approximations and Entropy Conditions for Shocks," *Comm. Pure Appl. Math.*, Vol. 29, pp. 297–322, 1976.

Hartwich, P. M. and Frink, N. T., "Estimation of Propulsion-Induced Effects on Transonic Flows Over a Hypersonic Configurations," AIAA 92-0523, 1992.

Hauser, J. and Taylor, C. (Eds.), *Numerical Grid Generation in Computational Fluid Dynamics—The First International Conference*, Pineridge Press, 1986.

Hedstrom, G. W., "Nonreflecting Boundary Conditions for Nonlinear Hyperbolic Systems," *J. Comp. Phys.*, Vol. 30, pp. 222–227, 1979.

Herring, J. R., Orszag, S. A., Kraichnan, R. H., and Fox, D. G., "Decay of Two-Dimensional Homogeneous Turbulence," *J. Fluid Mech.*, Vol. 66, p. 417, 1974.

Hestenes, M. and Stiefel, E., "Methods of Conjugate Gradients for Solving Linear Systems," *J. Res. Nat. Bur. Stand.*, Vol. 49, pp. 409–436, 1952.

Hildebrand, F. B., *Introduction to Numerical Analysis*, McGraw-Hill, New York, 1965.

Hirsch, C., *Numerical Computation of Internal and External Flows, Vol. 1—Fundamentals of Numerical Discretization*, John Wiley & Sons, New York, 1988.

Hirsch, C., *Numerical Computation of Internal and External Flows, Vol. 2—Computational Methods for Inviscid and Viscous Flows*, John Wiley & Sons, New York, 1990.

Hoffman, J. D., "Relationship Between the Truncation Errors of Centered Finite Difference Approximation on Uniform and Nonuniform Meshes," *J. Comp. Phys.*, Vol. 46, pp. 469–474, 1982.

Holmes, D. G. and Lamson, S. C., "Compressible Flow Solutions on Adaptive Triangular Meshes," Open Forum AIAA Reno'86 Meeting, 1986.

Hoppe, H., DeRose, T., Duchamp, T., McDonald, J., and Stuetzle, W., "Surface Reconstruction from Unorganized Points," *Comp. Graph.*, Vol. 26, No. 2, pp. 71–78, 1992.

Hoppe, H., DeRose, T., Duchamp, T., McDonald, J., and Stuetzle, W., "Mesh Optimization," *Proc. Comp. Graph. Ann. Conf.*, pp. 19–26, 1993.

Huet, F., "Generation de Maillage Automatique dans les Configurations Tridimensionelles Complexes Utilization d'une Methode de 'Front'," AGARD-CP-464, 1990.

Hughes, T. J. R., Franca, L. P., and Balestra, M., "A New Finite Element Formulation for Computational Fluid Dynamics: V. Circumventing the Babuška–Brezzi Condition: A Stable Petrov–Galerkin Formulation of the Stokes Problem Accommodating Equal-Order Interpolations," *Computer Methods in Appl. Mech. and Eng.*, Vol. 59, pp. 85–99, 1986a.

Hughes, T. J. R. and Brooks, A. N., "A Multi-Dimensional Upwind Scheme With No Crosswind Diffusion," *Finite Element Methods for Convection Dominated Flows*, Hughes, T. J. R. (Ed.), AMD, Vol. 34, pp. 19–35, ASME, New York, 1979.

Hughes, T. J. R. and Franca, L. P., "A New Finite Element Formulation for Computational Fluid Dynamics: VII. The Stokes Problem With Various Well-Posed Boundary Conditions: Symmetric Formulations That Converge For All Velocity/Pressure Spaces," *Computer Methods in Appl. Mech. and Eng.*, Vol. 65, pp. 85–96, 1987.

Hughes, T. J. R. and Hulbert, G. M., "Space-time Finite Element Methods for Elastodynamics: Formulations and Error Estimates," *Computer Methods in Appl. Mech. and Eng.*, Vol. 66, pp. 339–363, 1988.

Hughes, T. J. R. and Mallet, M., "A New Finite Element Formulation for Computational Fluid Dynamics: IV. A Discontinuity-Capturing Operator for Multidimensional Advective-Diffusive Systems," *Computer Methods in Appl. Mech. and Eng.*, Vol. 58, pp. 329–339, 1986c.

Hughes, T. J. R., Franca, L. P., and Hulbert, G. M., "A New Finite Element Formulation for Computational Fluid Dynamics: VIII. The Galerkin/Least-Squares Method For Advective-Diffusive Equations," *Computer Methods in Appl. Mech. and Eng.*, Vol. 73, pp. 173–189, 1989.

Hughes, T. J. R., Franca, L. P., and Mallet, M., "A New Finite Element Formulation for Computational Fluid Dynamics: I. Symmetric Forms of the Compressible Euler and Navier–Stokes Equations and the Second Law of Thermodynamics," *Computer Methods in Appl. Mech. and Eng.*, Vol. 54, pp. 223–234, 1986b.

Hughes, T. J. R., Levit, I., and Winget, J., "An Element-By-Element Solution Algorithm for Problems of Structural and Solid Mechanics," *Computer Methods in Appl. Mech. and Eng.*, Vol. 36, pp. 241–254, 1983.

Hughes, T. J. R., Liu, W. K., and Zimmermann, T. K., "Lagrangian–Eulerian Finite Element Formulation for Incompressible Viscous Flows," *Computer Methods in Appl. Mech. and Eng.*, Vol. 29, pp. 329–349, 1981.

Hughes, T. J. R., Mallet, M., and Mizukami, A., "A New Finite Element Formulation for Computational Fluid Dynamics: II. Beyond SUPG," *Computer Methods in Appl. Mech. and Eng.*, Vol. 54, pp. 341–355, 1986d.

Ives, D. C., "Conformal Grid Generation," *Numerical Grid Generation*, North-Holland, 1982.

Jameson, A., "Solution of the Euler Equations by a Multigrid Method," *Appl. Math. and Computations*, Vol. 13, pp. 345–356, 1983.

Jameson, A., "Solution of the Euler Equations for Two Dimensional Transonic Flow by a Multigrid Method," MAE Report, Princeton Univ., 1983.

Jameson, A., "A Non-Oscillatory Shock Capturing Scheme Using Flux Limited Dissipation," *Lectures in Applied Mathematics*, Engquist, B. E., Osher, S., and Sommerville, R. C. J. (Eds.), also MAE Report 1653, Princeton University, 1985.

Jameson, A., "Multigrid Algorithms for Compressible Flow Calculations," *Lecture Notes in Mathematics. Proceedings of the 2nd European Conference on Multigrid Methods*, 1985.

Jameson, A., "Using Euler Schemes," *AIAA Professional Study Series*, Snowmass, CO, 1985.

Jameson, A., "Time Dependent Calculations Using Multigrid with Applications to Unsteady Flows Past Airfoils and Wings," AIAA 91-1596, 1991.

Jameson, A., "Computational Algorithms for Aerodynamic Analysis and Design," MAE Report 1966, Princeton University, 1992.

Jameson, A. and Lax, P., "Conditions for the Construction of Multi-Point Total Variation Diminishing Difference Schemes," Tech. Rep. 178076, ICASE, 1986.

Jameson, A., Schmidt, W., and Turkel, E., "Numerical Solutions of the Euler Equations by Finite Volume Methods Using Runge–Kutta Time-Stepping Schemes," AIAA 81-1259, 1981.

Jameson, A. and Baker, T., "Multigrid Solution of the Euler Equations for Aircraft Configurations," AIAA 84-0093, 1984.

Jameson, A. and Yoon, S., "LU Implicit Schemes with Multiple Grids for the Euler Equations," AIAA 86-0105, 1986.

Jameson, A., Baker, T. J., and Weatherhill, N. P., "Calculation of Inviscid Transonic Flow over a Complete Aircraft," AIAA-86-0103, 1986.

Janus, J. M., "The Development of a Three-Dimensional Split Flux Vector Euler Solver With Dynamic Grid Applications," Master's thesis, Mississippi State University, 1984.

Jespersen, D., "A Time-Accurate Multiple-Grid Algorithm," AIAA 85-1493, 1985.

Johan, Z., Hughes, T. J. R., and Shakib, F., "A Globally Convergent Matrix-Free Algorithm for Implicit Time-Marching Schemes Arising in Finite Element Analysis in Fluids," *Computer Methods in Appl. Mech. and Eng.*, Vol. 87, pp. 281–304, 1991.

Johan, Z., Mathur, K. K., Johnsson, S. L., and Hughes, T. J. R., "An Efficient Communications Strategy for Finite Element Methods on the Connection Machine CM-5 System," *Computer Methods in Appl. Mech. and Eng.*, Vol. 113, pp. 363–387, 1994.

Johan, Z., *Data Parallel Finite Element Techniques for Large-Scale Computational Fluid Dynamics*, Ph.D. Thesis, Dept. of Mech. Eng., Stanford University, 1992.

Johnson, A. A. and Tezduyar, T. E., "Mesh Update Strategies in Parallel Finite Element Computations of Flow Problems With Moving Boundaries and Interfaces," *Computer Methods in Appl. Mech. and Eng.*, Vol. 119, pp. 73–94, 1994.

Johnson, C. and Saranen, J., "Streamline Diffusion Methods for the Incompressible Euler and Navier–Stokes Equations," *Math. of Comp.*, Vol. 47, pp. 1–18, 1986.

Johnson, C., Navert, U., and Pitkäranta, J., "Finite Element Methods for Linear Hyperbolic Problems," *Computer Methods in Appl. Mech. and Eng.*, Vol. 45, pp. 285–312, 1984.

Kallinderis, Y. and Ward, S., "Prismatic Grid Generation with an Efficient Algebraic Method for Aircraft Configurations," AIAA-92-2721, 1992.

Kato, C. and Ikegawa, M., "Large Eddy Simulation of Unsteady Turbulent Wake of a Circular Cylinder Using the Finite Element Method," *Advances in Numerical Simulation of Turbulent Flows*, Celik, I., Kobayashi, T., Ghia, K. N., and Kurokawa, J. (Eds.), FED-Vol. 117, pp. 49–56, ASME, New York, 1991.

Kennedy, J. G., Behr, M., Kalro, V., and Tezduyar, T. E., "Implementation of Implicit Finite Element Methods for Incompressible Flows on the CM-5," *Computer Methods in Appl. Mech. and Eng.*, Vol. 119, pp. 95–111, 1994.

Kirkpatrick, R. C., "Nearest Neighbor Algorithm," *Springer Lecture Notes in Physics 238*, Fritts, M. J., Crowley, W. P., and Trease, H. (Eds.), Springer-Verlag, 1985.

Klopfer, G. H. and McRae, D. S., "The Nonlinear Modified Equation Approach to Analyzing Finite Difference Schemes," AIAA 5th-Computational Fluid Dynam. Conf., 1981.

Knupp, P. and Steinberg, S., *Fundamentals of Grid Generation*, CRC Press, 1993.

Kreiss, H.-O. and Oliger, J., "Method for the Approximate Solution of Time-Dependent Problems," GARP Publ. Series No. 10, Global Atmospheric Research Program, 1973.

Lambert, J. D., *Computational Methods In Ordinary Differential Equations*, John Wiley & Sons, Chichester, 1979.

Lanczos, C., *Applied Analysis*, Prentice-Hall, Englewood Cliffs, NJ, 1956.

Lax, P. D. and Richtmyer, R. D., "Survey of the Stability of Linear Finite Difference Equations," *Comm. Pure Appl. Math.*, Vol. 9, pp. 267–293, 1956.

Lax, P. D., "Hyperbolic Systems of Conservation Laws and Mathematical Theory of Shock Waves," *SIAM*, Philadelphia, 1972.

LeBeau, G. J. and Tezduyar, T. E., "Finite Element Computation of Compressible Flows With the SUPG Formulation," *Advances in Finite Element Analysis in Fluid Dynamics*, Dhaubhadel, M. N., Engelman, M. S., and Reddy, J. N. (Eds.), FED-Vol. 123, pp. 21–27, ASME, New York, 1991.

Lee, D. T. and Schachter, B. J., "Two Algorithms for Constructing a Delaunay Triangulation," *Int. J. Comp. Inf. Sci.*, Vol. 9, No. 3, pp. 219–242, 1980.

Leonard, A. and Wray, A. A., "A New Numerical Method for the Simulation of Three-Dimensional Flow in a Pipe," *Proc. 8th Int. Conf. on Numerical Methods in Fluid Dynamics, Lecture Notes in Physics*, Vol. 170, Springer-Verlag, 1982.

Lerat, A., "Une Class de Scheme aux Difference Implicites pour les Systems Hyperboliques de Lois de Conservation," *Compte Rendus Acad. Sciences Paris*, Vol. 288A, 1979.

LeVeque, R. J., "Time-Split Methods for Partial Differential Equations," Ph.D. thesis, Stanford University, Stanford, CA, 1982.

Li, C. P., "Numerical Solution of Viscous Reacting Blunt Body Flows of a Multicomponent Mixture," AIAA 73-202, 1973.

Liao, G. and Su, J., "Grid Generation via Deformation," *Appl. Math. Letters*, Vol. 5, No. 3, pp. 27–29, 1992.

Lighthill, M. J., "On Displacement Thickness," *J. Fluid Mech.*, Vol. 4, pp. 383–392, 1958.

Liseikin, V. D., "On a Variational Method of Generating Adaptive Grids on n-Dimensional Surfaces," *Soviet Math. Dokl.*, Vol. 44, No. 1, pp. 149–152, 1992.

Liseikin, V. D., "On Some Interpretations of a Smoothness Functional Used in Constructing Regular and Adaptive Grids," *Russian J. Numer. Math. Modeling*, No. 6, pp. 507–518, 1993.

Löhner, R. and Parikh, P., "Three-Dimensional Grid Generation by the Advancing Front Method," *Int. J. Num. Meth. Fluids*, Vol. 8, pp. 1135–1149, 1988.

Löhner, R., Camberos, J., and Merriam, M., "Parallel Unstructured Grid Generation," *Comp. Meth. Appl. Mech. Eng.*, Vol. 95, pp. 343–357, 1992.

Löhner, R., "Matching Semi-Structured and Unstructured Grids for Navier–Stokes Calculations," AIAA-93-3348-CP, 1993.

Löhner, R., "Some Useful Data Structures for the Generation of Unstructured Grids," *Comm. Appl. Num. Meth.*, Vol. 4, pp. 123–135, 1988.

Löhner, R., "Three-Dimensional Fluid-Structure Interaction Using a Finite Element Solver and Adaptive Remeshing," *Comp. Syst. Eng.*, Vol. 1, Nos. 2–4, pp. 257–272, 1990.

Lo, S. H., "A New Mesh Generation Scheme for Arbitrary Planar Domains," *Int. J. Num. Meth. Eng.*, Vol. 21, pp. 1403–1426, 1985.

Lu, N. and Eiseman, P. R., "A Grid Quality Manipulation System," *Numerical Grid Generation in Computational-Fluid Dynamics and Related Fields—The Third International Conference*, North-Holland, 1991.

MacCormack, R. W., "The Effect of Viscosity in Hypervelocity Impact Cratering," AIAA 69-354, 1969.

Marchant, M. J. and Weatherhill, N. P., "The Construction of Nearly Orthogonal Multiblock Grids for Compressible Flow Simulation," *Comm. Appl. Num. Meth.*, Vol. 9, pp. 567–578, 1993.

Martinelli, L., "Calculation of Viscous Flows with Multigrid Methods," Ph.D. thesis, Princeton University, 1987.

Mavriplis, D., "Euler and Navier–Stokes Computations for Two-Dimensional Geometries Using Unstructured Meshes," ICASE Rep. 90-3, 1990.

Melson, N. D., Sanetrik, M. D., and Atkins, H. L., "Time-Accurate Navier–Stokes Calculations with Multigrid Acceleration," AIAA 93-3392, 1993.

Merriam, M. and Barth, T., "3D CFD in a Day: The Laser Digitizer Project," AIAA-91-1654, 1991.

Mitchell, A. R., *Computation Methods in Partial Differential Equations*, John Wiley & Sons, London, 1976.

Mitchell, C. R. and Walter, R. W., "K-Exact Reconstruction for the Navier–Stokes Equations on Arbitrary Grids," AIAA 93-0536, 1993.

Moin, P. and Kim, J., "On the Numerical Solution of Time-Dependent Viscous Incompressible Fluid Flows Involving Solid Boundaries," *J. Comp. Physics*, Vol. 35, p. 381, 1980.

Moin, P. and Kim, J., "Numerical Investigation of Turbulent Channel Flow," *J. Fluid Mech.*, Vol. 118, p. 341, 1982.

Montagné, J.-L., Yee, H. C., Klopfer, G. H., and Vinokur, M., "Hypersonic Blunt Body Computations Including Real Gas Effects," *Proc. 2nd Int. Conf. Hyperbolic Problems*, 1988 (also NASA TM-100074, 1988).

Montagné, J.-L., Yee, H. C., and Vinokur, M., "Comparative Study of High-Resolution Shock-Capturing Schemes for a Real Gas," *AIAA J.*, Vol. 27, No. 10, pp. 1332–1345, 1989.

Moretti, G., "A Physical Approach to the Numerical Treatment of Boundaries in Gas Dynamics," *Numerical Boundary Condition Procedures*, NASA CP-2201, 1981.

Morgan, K., Probert, J., and Peraire, J., "Line Relaxation Methods for the Solution of Two-Dimensional and Three-Dimensional Compressible Flows," AIAA-93-3366, 1993.

Mulder, W., "Multigrid Relaxation for the Euler Equations," *J. of Comp. Phys.*, Vol. 60, No. 2, pp. 235–252, 1985.

Müller, J.-D., "Proven Angular Bounds and Stretched Triangulations with the Frontal Delaunay Method," AIAA-93-3347-CP, 1993.

Nakahashi, K. and Obayashi, S., "Viscous Flow Computations Using a Composite Grid," AIAA-CP-87-1128, 1987.

Nakahashi, K., "FDM-FEM Zonal Approach for Viscous Flow Computations over Multiple Bodies," AIAA-87-0604, 1987.

Nakahashi, K., "Optimum Spacing Control of the Marching Grid Generation," AIAA-88-0515, 1988.

Nakamura, S. and Suzuki, M., "Noniterative Three-Dimensional Grid Generation Using a Parabolic Hyperbolic Hybrid Scheme," AIAA 87-0277, 1987.

Nakamura, S., Fradl, D. P., Spradling, M. L., and Kuwahara, K., "Mapping Curved Surfaces onto a Side Boundary of the Three-Dimensional Computational Grid Using Two-Elliptic Partial Differential Equations," *Grid Generation in Computational Fluid Dynamics and Related Fields*, North-Holland, 1991.

Nakamura, S., Spradling, M., Fradl, D., and Kuwahara, K., "A Grid Generating System for Automobile Aerodynamic Analysis," Paper No. SP-855, SAE International Congress, 1991.

Nakamura, S., *Applied Numerical Methods in C*, pp. 578–581, Prentice Hall, New Jersey, 1993.

Ni, R.-H., "A Multiple-Grid Scheme for Solving the Euler Equations," *AIAA J.*, Vol. 20, No. 11, 1982.

Oden, J. T., Demkowicz, L., Liska, T., and Rachowicz, W., "h-p Adaptive Finite Element Methods for Compressible and Incompressible Flows," *Computing Systems in Eng.*, Vol. 1, Nos. 2–4, pp. 523–534, 1990.

Oliger, J. and Sundstrom, A., "Theoretical and Practical Aspects of Some Initial Boundary Value Problems in Fluid Dynamics," *SIAM J. Appl. Math.*, Vol. 35, No. 3, 1978.

Orszag, S. A. and Israeli, M., "Numerical Simulation of Viscous Incompressible Flows," *Ann. Rev. Fluid Mech.*, Vol. 5, 1974.

Orszag, S. A., "Comparison of Pseudospectral and Spectral Approximation," *Studies in Appl. Math.*, Vol. 51, p. 253, 1972.

Orszag, S. A., "Numerical Methods for the Simulation of Turbulence," *Phys. of Fluids*, Supplement II, p. II-250, 1969.

Orszag, S. A., "Numerical Simulation of Incompressible Flows Within Simple Boundaries. I. Galerkin (Spectral) Representations," *Studies in Appl. Math.*, Vol. L, No. 4, p. 293, 1971.

Orszag, S. A., "Spectral Methods for Problems in Complex Geometries," *J. Comp. Physics*, Vol. 37, p. 70, 1980.

Osher, S., "Stability of Difference Approximations of Dissipative Type for Mixed Initial-Boundary Value Problems. I," *Math. Comp.*, Vol. 23, pp. 335–340, 1969.

Osher, S., "Shock Modeling in Transonic and Supersonic Flow," *Advances in Computational Transonics*, Habashi, W. G. (Ed.), Pineridge Press, 1984.

Osher, S. and Chakravarthy, S., "Very High Order Accurate TVD Schemes," Tech. Rep. 84-44, ICASE, 1984.

Osher, S. and Solomon, F., "Upwind Difference Schemes for Hyperbolic Systems of Conservation Laws," *Math. of Computations*, Vol. 38, No. 158, pp. 339–374, 1982.

Patterson, G. S. and Orszag, S. A., "Spectral Calculations of Isotropic Turbulence: Efficient Removal of Aliasing Interactions," *Phys. Fluids*, Vol. 14, p. 2538, 1971.

Peaceman, D. W. and Rachford, H. H., "The Numerical Solution of Parabolic and Elliptic Differential Equations," *J. Soc. Indust. Appl. Math.*, No. 3, pp. 28–41, 1955.

Peraire, J., Morgan, K., and Peiro, J., "Unstructured Finite Element Mesh Generation and Adaptive Procedures for CFD," AGARD-CP-464, 1990.

Peraire, J., Vahdati, M., Morgan, K., and Zienkiewicz, O. C., "Adaptive Remeshing for Compressible Flow Computations," *J. Comp. Phys.*, Vol. 72, pp. 449–466, 1987.

Peyret, R., "A Hermitian Finite Difference Method for the Solution of the Navier–Stokes Equations," *Proc. 1st Int. Cong. on Num. Meth. in Laminar and Turbulent Flows*, John Wiley & Sons, New York, pp. 43–54, 1978.

Peyret, R. and Taylor, T. D., *Computational Methods for Fluid Flow*, Springer-Verlag, New York, 1983.

Pirzadeh, S., "Unstructured Viscous Grid Generation by Advancing-Layers Method," AIAA-93-3453, 1993.

Pirzadeh, S., "Viscous Unstructured Three-Dimensional Grids by the Advancing-Layers Method," AIAA-94-0417, 1994.

Pothen, A., Simon, H. D., and Liou, K. P., "Partitioning Sparse Matrices With Eigenvectors of Graphs," *SIAM J. Matrix Analysis and Applications*, Vol. 11, pp. 430–452, 1990.

Proskurowski, W. and Widlund, O., "On the Numerical Solution of Helmholtz's Equation by the Capacitance Matrix Method," *Math. Comput.*, Vol. 30, pp. 433–468, 1976.

Proskurowski, W., "Algorithm 593—A Package for the Helmholtz Equation in Nonrectangular Planar Regions," *ACM Trans. Math. Software*, Vol. 9, pp. 117–124, 1983.

Pulliam, T. H., "Euler and Thin-Layer Navier–Stokes Codes: ARC2D, ARC3D," Computational Fluid Dynamics Workshop, Univ. Tennessee Space Institute, 1984.

Pulliam, T. J., "Artificial Dissipation Models for the Euler Equations," AIAA 85-0438, 1985.

Radespiel, R., Rossow, C., and Swanson, R. C., "An Efficient Cell-Vertex Multigrid Scheme for the Three-Dimensional Navier–Stokes Equations," AIAA 89-1953, 1989.

Ramamurti, R. and Löhner, R., "Simulation of Subsonic Viscous Flows Using Unstructured Grids and a Finite Element Solver," AIAA-90-0702, 1990.

Richtmyer, R. D. and Morton, K. W., *Difference Methods for Initial-Value Problems*, Wiley–Interscience, New York, 1967.

Roach, P. J., *Computational Fluid Dynamics*, Hermosa Publishers, Albuquerque, NM, 1972.

Roe, P. L., "Approximate Riemann Solvers, Parameter Vectors, and Difference Schemes," *J. Comp. Phys.*, Vol. 43, pp. 357–372, 1981.

Roe, P. L., "An Introduction to Numerical Methods Suitable for the Euler Equations," *von Karman Institute for Fluid Dynamics Lecture Series: Introduction to Computational Fluid Dynamics*, Belgium, 1983.

Roe, P. L., "Upwind Schemes Using Various Formulations of the Euler Equations," *Numerical Methods for the Euler Equations of Fluid Dynamics*, SIAM, 1983.

Roe, P. L. and Baines, M. J., "Algorithms for Advection and Shock Problems," *Proc. 4th GAMM Conference on Numerical Methods in Fluid Mechanics*, Viviand, H. (Ed.), Vieweg, 1982.

Rossow, C. C., "Efficient Computation of Inviscid Flow Fields Around Complex Configurations Using a Multiblock Multigrid Method," *Comm. in Appl. Num. Meth.*, Vol. 8, pp. 735–747, 1992.

Saad, Y. and Schultz, M. H., "GMRES: A Generalized Minimum Residual Algorithm for Solving Nonsymmetric Linear Systems," *SIAM J. Sci. Stat. Comp.*, Vol. 7, No. 3, pp. 856–869, 1986.

Schmitt, V. and Charpin, F., "Pressure Distributions on the ONERA-M6 Wing at Transonic Mach Numbers," *Experimental Data Base for Computer Program Assessment*, AGARD-AR-138, 1979.

Schönfeld, T. and Rudgyard, M., "A Cell-Vertex Approach to Local Mesh Refinement for the 3-D Euler Equations," AIAA-94-0318, 1994.

Sengupta, S., Häuser, J., Eiseman, P. R., and Thompson, J. F. (Eds.), *Numerical Grid Generation in Computational Fluid Dynamics—The Second International Conference*, Pineridge Press, 1988.

Shakib, F., *Finite Element Analysis of the Compressible Euler and Navier–Stokes Equations*, Ph.D. Thesis, Dept. of Mech. Eng., Stanford University, 1988.

Shakib, S., Hughes, T. J. R., and Johan, Z., "A Multi-Element Group Preconditioned GMRES Algorithm for Nonsymmetric Systems Arising in Finite Element Analysis," *Computer Methods in Appl. Mech. and Eng.*, Vol. 75, pp. 415–456, 1989.

Shanks, D., "Non-linear Transformation of Divergent and Slowly Convergent Sequences," *J. of Math. and Physics*, XXXIV, pp. 1–42, 1955.

Shenton, D. N. and Cendes, Z. J., "Three-Dimensional Finite Element Mesh Generation Using Delaunay Tesselation," *IEEE Trans. Mag. MAG-21*, pp. 2535–2538, 1985.

Shepard, M. S. and Georges, M. K., "Automatic Three-Dimensional Mesh Generation by the Finite Octree Technique," *Int. J. Num. Meth. Eng.*, Vol. 32, pp. 709–749, 1991.

Shokin, Yu I., *The Method of Differential Approximation*, Springer-Verlag, 1983.

Shostko, A. and Löhner, R., "Three-Dimensional Parallel Unstructured Grid Generation," AIAA-94-0418, 1994.

Shu, C. and Osher, S., "Efficient Implementation of Essentially Non-Oscillatory Shock-Capturing Schemes," *J. Comp. Phys.*, Vol. 77, No. 2, pp. 439–471, 1988.

Sloan, S. W. and Houlsby, G. T., "An Implementation of Watson's Algorithm for Computing 2-Dimensional Delaunay Triangulations," *Adv. Eng. Software*, Vol. 6, No. 4, pp. 192–197, 1984.

Smith, R. E., "Algebraic Grid Generation," *Numerical Grid Generation*, North-Holland, 1982.

Sorenson, R. L. and Steger, J. L., "Numerical Grid Generation of Two-Dimensional Grids by the Use of Poisson Equations with Grid Control," NASA CP 2166, 1980.

Spekreijse, S. P., "Multigrid Solution of Monotone Second-Order Discretizations of Hyperbolic Conservation Laws," *Math. Comp.*, Vol. 49, pp. 135–155, 1987.

Spekreijse, S. P., "Multigrid Solution of the Steady Euler Equations," *CWI-tract 46*, Center for Mathematics and Computer Science, Amsterdam, 1988.

Spradling, M. L., Nakamura, S., and Kuwahara, K., "Application of Elliptic Grid Generation Equations Blended with Hyperbolic Method to Three-Dimensional Grids for Vehicle Aerodynamic Analysis," *Grid Generation in Computational Fluid Dynamics and Related Fields*, North-Holland, 1991.

Steger, J. L. and Chausee, D. S., "Generation of Body-Fitted Coordinates Using Hyperbolic Partial Differential Equations," in *SIAM J. Sci. Stat. Comput.*, Vol. 1, pp. 431–437, 1980.

Steger, J. F. and Sorenson, R. L., "Automatic Mesh-Point Clustering near a Boundary in Grid Generation with Eilliptic Partial Differential Equations," *J. Comp. Phys.*, Vol. 33, pp. 405–410, 1979.

Steger, J. L. and Warming, R. F., "Flux Vector Splitting of the Inviscid Gasdynamic Equations with Application to Finite-Difference Methods," *J. Comp. Phys.*, Vol. 40, No. 2, pp. 263–293, 1981.

Steinbrenner, J. P., Chawner, J. R., and Fouts, C. L., "A Structured Approach to Interactive Multiple Block Grid Generation," AGARD-CP-464, 1990.

Steuerwalt, M., "Certification of Algorithm 541—Efficient Fortran Subroutines for the Solution of Separable Elliptic Partial Differential Equations," *ACM Trans. Math. Software*, Vol. 5, pp. 365–371, 1979.

Struijs, R., Deconinck, H., DePalma, P., Roe, P., and Powell, K., "Progress on Multidimensional Upwind Euler Solvers for Unstructured Grids," AIAA 91-1550-CP, 1991.

Stuben, K. and Trottenberg, U., "Multigrid Methods: Fundamental Algorithm, Model Problems Analysis and Applications," *Multgrid Methods: Volume 960 of Lecture Notes in Mathematics*, Springer-Verlag, 1982.

Swanson, R. C. and Turkel, E., "Artificial Dissipation and Central Difference Schemes for the Euler and Navier–Stokes Equations," AIAA 87-1107, 1987.

Swanson, R. C. and Turkel, E., "On Central-Difference and Upwind Schemes," *J. Comp. Phys.*, Vol. 101, pp. 292–306, 1992 (also ICASE Report 90-44, 1990).

Swartrauber, P. N. and Sweet, R. A., "Efficient FORTRAN Subprograms for the Solution of Elliptic Partial Differential Equations," Tech. Note TN/IA-109, Nat. Ctr. for Atmospheric Res., Boulder, CO, 1975 (available from the Nat. Tech. Inform. Service as document PB263 498/AS).

Swarztrauber, P. N. and Sweet, R. A., "Algorithm 541—Efficient Fortran Sub-Programs for the Solution of Separable Elliptic Partial Differential Equations," *ACM Trans. Math. Software*, Vol. 5, pp. 352–364, 1979.

Swarztrauber, P. N. and Sweet, R. A., "Vector and Parallel Methods for the Direct Solution of Poisson's Equation," *J. Comp. Applied Math.*, Vol. 27, pp. 241–263, 1989.

Swarztrauber, P. N., "A Direct Method for the Discrete Solution of Separable Elliptic Equations," *SIAM J. Numer. Anal.*, Vol. 11, pp. 1136–1150, 1974.

Swarztrauber, P. N., "Fast Poisson Solvers," *Studies in Numerical Analysis*, MAA Studies in Mathematics, Vol. 24, Math. Assoc. of America, 1984.

Swarztrauber, P. N., "Vectorizing the FFT's," *Parallel Computations*, Rodrigue, G. (Ed.), Academic Press, New York, 1982.

Sweby, P. K., "High Resolution Schemes Using Flux Limiters For Hyperbolic Conservation Laws," *SIAM J. Numer. Anal.*, Vol. 21, No. 5, pp. 995–1011, 1984.

Sweet, R. A., "A Cyclic Reduction Algorithm for Block Tridiagonal Systems of Arbitrary Dimension," *SIAM J. Numer. Anal.*, Vol. 11, pp. 506–520, 1974.

Sweet, R. A., "CRAYFISHPAK: A Vectorized Fortran Package to Solve Helmholtz Equations," *Recent Developments in Numerical Methods and Software for ODEs/ DAEs/PDEs*, Byrne, G. D. and Schiesser, W. E. (Eds.), World Scientific, Singapore, 1992.

Tanemura, M., Ogawa, T., and Ogita, N., "A New Algorithm for Three-Dimensional Voronoi Tesselation," *J. Comp. Phys.*, Vol. 51, pp. 191–207, 1983.

Temperton, C., "Self-Sorting Mixed-Radix Fast Fourier Transforms," *J. Comp. Phys.*, Vol. 52, pp. 1–23, 1983.

Tezduyar, T. E. and Hughes, T. J. R., "Finite Element Formulations for Convection Dominated Flows With Particular Emphasis on the Compressible Euler Equations," AIAA 83-0125, 1983.

Tezduyar, T. E. and Park, Y. J., "Discontinuity Capturing Finite Element Formulations for Nonlinear Convection-Diffusion-Reaction Problems," *Computer Methods in Appl. Mech. and Eng.*, Vol. 59, pp. 307–325, 1986.

Tezduyar, T. E., Behr, M., Aliabadi, S. K., Mittal, S., and Ray, S. E., "A New Mixed Preconditioning Method for Finite Element Computations," *Computer Methods in Appl. Mech. and Eng.*, Vol. 99, pp. 27–42, 1992a.

Tezduyar, T. E., Behr, M., and Liou, J., "A New Strategy for Finite Element Computations Involving Moving Boundaries and Interfaces—The Deforming-Spatial-Domain/Space-Time Procedure: I. The Concept and the Preliminary Tests," *Computer Methods in Appl. Mech. and Eng.*, Vol. 94, pp. 339–351, 1992b.

Tezduyar, T. E., Behr, M., Mittal, S., and Liou, J., "A New Strategy for Finite Element Computations Involving Moving Boundaries and Interfaces—The Deforming-Spatial-Domain/Space-Time Procedure: II. Computation of Free-Surface Flows, Two-Liquid Flows, and Flows with Drifting Cylinders," *Computer Methods in Appl. Mech. and Eng.*, Vol. 94, pp. 353–371, 1992c.

Tezduyar, T. E., Mittal, S., Ray, S. E., and Shih, R., "Incompressible Flow Computations With Stabilized Bilinear and Linear Equal-Order-Interpolation Velocity-Pressure Elements," *Computer Methods in Appl. Mech. and Eng.*, Vol. 95, pp. 221–242, 1992d.

Tezduyar, T., Aliabadi, S., Behr, M., Johnson, A., and Mittal, S., "Parallel Finite-Element Computation of 3D Flows," *IEEE Computer*, Vol. 26, No. 10, pp. 27–36, 1993.

Thacker, W. C., Gonzalez, A., and Putland, G. E., "A Method for Automating the Construction of Irregular Computational Grids for Storm Surge Forecast Models," *J. Comp. Phys.*, Vol. 37, pp. 371–387, 1980.

Thomas, J. L. and Salas, M. D., "Far-Field Boundary Conditions for Transonic Lifting Solutions to the Euler Equations," *AIAA J.*, Vol. 24, No. 7, pp. 1074–1080, 1986.

Thomas, J. L. and Walter, R. W., "Upwind Relaxation Algorithms for the Navier–Stokes Equations," AIAA 85-1501, 1985.

Thomas, J. L., Walters, R. W., Reu, T., Ghaffari, F., Weston, R. P., and Luckring, J. M., "A Patched-Grid Algorithm for Complex Configurations Directed Toward the F/A-18 Aircraft," AIAA 89-0121, 1989.

Thomas, P. D., "Contruction of Composite Three Dimensional Grids from Subregion Grids Generated by Elliptic Systems," Proc. AIAA Computational Fluid Dynamics Conference, Palo Alto, CA, 1981.

Thompson J. F. (Ed.), *Numerical Grid Generation*, North-Holland, 1982.

Thompson, J. F., Thames, F. C., and Mastin, C. W., "Boundary-Fitted Curvilinear Coordinate Systems for Solutions of Partial Differential Equations on Fields Containing Any Arbitrary Two-Dimensional Bodies," NASA CR 2729, 1977.

Thompson, J. F., Wasri, Z. U. A., and Mastin, C. W., *Numerical Grid Generation—Foundations and Applications*, North-Holland, 1985.

Thompson, J. F., Wasri, Z. U. A., and Mastin, C. W., "Boundary-Fitted Coordinate Systems for Numerical Solution of Partial Differential Equations—A Review," *J. Comp. Phys.*, Vol. 47, p. 1, 1982.

Thompson, J. F., "A Composite Grid Generation Code for General 3D Regions—The EAGLE Code," *AIAA J.*, Vol. 26, No. 3, 1988.

Thompson, J. F., "Grid Generation," *Handbook of Numerical Heat Transfer*, Wiley–Interscience, New York, 1988.

Thompson, K. W., "Time Dependent Boundary Conditions for Hyperbolic Systems," *J. Comp. Phys.*, Vol. 68, pp. 1–24, 1987.

Turkel, E., "Improving the Accuracy of Central Difference Schemes," *11th International Conference on Numerical Methods in Fluid Dynamics, Springer-Verlag Lecture Notes in Physics*, Vol. 323, pp. 586–591, 1988.

Turkel, E., "Preconditioned Methods for Solving The Incompressible and Low Speed Compressible Equations," ICASE Report No. 86-14, 1986.

Turkel, E., "Progress in Computational Physics," ICASE Report No. 82-23, 1982.

Turkel, E., "Review of Preconditioning Methods for Fluid Dynamics," ICASE Report No. 92-47, 1992.

van Albada, G. D., van Leer, B., and Roberts, W. W., "High Resolution Schemes Using Flux Limiters For Hyperbolic Conservation Laws," *J. Astron. Astroph.*, p. 108, 1982.

van Leer, B., "Towards the Ultimate Conservative Difference Scheme. II: Monotonicity and Conservation Combined in a Second Order Scheme," *J. Comp. Phys.*, Vol. 14, pp. 361–370, 1974.

van Leer, B., "Towards the Ultimate Conservative Difference Scheme. III: Upstream-Centered Finite Difference Schemes for Ideal Compressible Flow," *J. Comp. Phys.*, Vol. 23, pp. 263–275, 1977.

van Leer, B., "Towards the Ultimate Conservative Difference Scheme. V. A Second-Order Sequel to Godunov's Method," *J. Comp. Phys.*, Vol. 32, pp. 101–136, 1979.

van Leer, B., "Flux Vector Splitting for the Euler Equations," *Lecture Notes in Physics*, Vol. 170, pp. 501–512, 1982.

van Leer, B., "Numerical Fluid Dynamics II," ICASE Int. Rep. 36, 1987.

van Leer, B., "Progress in Multi-Dimensional Upwind Differencing," ICASE Rep. 92-43, 1992.

van Leer, B. and Mulder W., "Relaxation Methods for Unfactored Implicit Schemes," Rep. 84-20, Delft University of Technology, 1984.

van Leer, B., Lee, W., and Roe, P., "Characteristic Time-Stepping or Local Preconditioning of the Euler Equations," AIAA 91-1552-CP, 1991.

van Leer, B., Thomas, J. L., Roe, P. L., and Newsome, R. W., "A Comparison of Numerical Flux Formulas for the Euler and Navier–Stokes Equations," AIAA 87-1104, 1987.

Vatsa, V. N. and Wedan, B. W., "Development of an Efficient Multigrid Code for 3-D Navier–Stokes Equations," AIAA 89-1791, 1990.

Vatsa, V. N., Sanetrik, M. D., and Parlette, E. B., "Development of a Flexible and Efficient Multigrid-Based Multiblock Flow Solver," AIAA 93-0677, 1993.

Vatsa, V. N., Sanetrik, M. D., Parlette, E. B., Eiseman, P. R., and Cheng, Z., "Multiblock Structured Grid Approach for Solving Flows Over Complex Aerodynamic Configurations," AIAA-94-0655, 1994.

Venkatakrishnan, V., "Computation of Unsteady Transonic Flows Over Moving Airfoils," Ph.D. thesis, Princeton University, Princeton, NJ, 1986.

Verhoff, A., Stookesberry, D., and Agrawal, S., "Far-Field Computational Boundary Conditions for Two-Dimensional External Flow Problems," *AIAA J.*, Vol. 30, No. 11, pp. 2585–2594, Nov. 1992.

Warming, R. F. and Beam, R. M., "On the Construction and Application of Implicit Factored Schemes for Conservation Laws," *SIAM-AMS Proceedings*, Vol. 11, Symposium on Comp. Fluid Dynamics, New York, 1977.

Warming, R. F., Beam, R. M., and Yee, H. C., "Stability of Difference Approximations for Initial-Boundary-Value Problems," *Proc. 3rd Int. Symp. Numerical Methods for Eng.*, Paris, France, 1983.

Watson, D. F., "Computing the N-Dimensional Delaunay Tesselation with Application to Voronoi Polytopes," *The Computer Journal*, Vol. 24, No. 2, pp. 167–172, 1981.

Weatherill, N. P. and Hassan, O., "Efficient Three-Dimensional Grid Generation Using the Delaunay Triangulation," *Proc. First Europ. CFD Conf.*, Hirsch, Ch. (Ed.), Brussels, Belgium, 1992.

Weatherill, N. P., Eiseman, P. R., Häuser, J., and Thompson, J. F., *Numerical Grid Generation in Computational Fluid Dynamics and Related Fields*, Pineridge Press, 1994.

Weatherill, N. P., "Delaunay Triangulation in Computational Fluid Dynamics," *Comp. Math. Appl.*, Vol. 24, No. 5/6, pp. 129–150, 1992.

Wigton, L. B., Yu, N. J., and Young, D. P., "GMRES Acceleration of Computational Fluid Dynamics Codes," AIAA 85-1494 1985.

Wynn, P., "On the Convergence and Stability of the Epsilon Algorithm," *J. Siam Num. Anal.*, Vol. 3, No. 3, pp. 91–122, 1966.

Yanenko, N. N., *The Method of Fractional Steps*, Springer-Verlag, 1972.

Yee, H. C. and Harten, A., "Implicit TVD Schemes for Hyperbolic Conservation Laws in Curvilinear Coordinates," *AIAA J.*, Vol. 25, No. 2, pp. 266–274, 1987.

Yee, H. C. and Shinn, J., "Semi-Implicit and Fully Implicit Shock-Capturing Methods for Hyperbolic Conservation Laws with Stiff Source Terms," *AIAA J.*, Vol. 27, No. 1, pp. 299–307, 1989.

Yee, H. C., Beam, R. M., and Warming, R. F., "Boundary Approximations for Implicit Schemes for One-Dimensional Inviscid Equation of Gasdynamics," *AIAA J.*, Vol. 20, No. 9, pp. 1203–1211, 1982.

Yee, H. C., Klopfer, G., and Montagné, J.-L., "High-Resolution Shock-Capturing Schemes for Inviscid and Viscous Hypersonic Flows," *J. of Comput. Phys.*, Vol. 88, No. 1, pp. 31–61, 1990.

Yee, H. C., Warming, R. F., and Harten, A., "A High-Resolution Numerical Technique for Inviscid Gas-Dynamic Problems with Weak Solutions," *Proc. 8th Int. Conf. Numerical Methods in Fluid Dynamics, Lecture Notes in Physics 170*, Krause, E. (Ed.), Springer-Verlag, 1982.

Yee, H. C., Warming, R. F., and Harten, A., "Implicit Total Variation Diminishing (TVD) Schemes for Steady-State Calculations," *J. Comput. Phys.*, Vol. 57, No. 3, 1985.

Yee, H. C., "A Class of High-Resolution Explicit and Implicit Shock-Capturing Methods," *VKI Lecture Notes in Computational Fluid Dynamics*, von Karman Institute for Fluid Dynamics, Belgium (also NASA TM-101088, 1989).

Yee, H. C., "Construction of Explicit and Implicit Symmetric TVD Schemes and Their Applications," *J. Comp. Phys.*, Vol. 68, pp. 151–179, 1987.

Yee, H. C., "Construction of Explicit and Implicit Symmetric TVD Schemes and Their Application," *J. Comput. Phys.*, Vol. 68, No. 1, pp. 180–190, 1987.

Yee, H. C., "Generalized Formulation of a Class of Explicit and Implicit TVD Schemes," *J. Comput. Phys.*, Vol. 68, No. 1, pp. 151–179, 1987.

Yee, H. C., "Linearized Forms of Implicit TVD Schemes for the Multidimensional Euler and Navier–Stokes Equations," *Int. J. Computers and Math. with Applications*, Vol. 12A, No. 4/5, pp. 413–432, 1986.

Yee, H. C., "Numerical Approximation of Boundary Conditions with Applications to Inviscid Equations of Gas Dynamics," NASA TM-81265, 1981.

Yerry, M. A. and Shepard, M. S., "Automatic Three-Dimensional Mesh Generation by the Modified-Octree Technique," *Int. J. Num. Meth. Eng.*, Vol. 20, pp. 1965–1990, 1984.

Zienkiewicz, O. C. and Phillips, D. V., "An Automatic Mesh Generation Scheme for Plane and Curved Surfaces by Isoparametric Co-Ordinates," *Int. J. Num. Meth. Eng.*, Vol. 3, pp. 519–528, 1971.

20 Computational Methods for Inviscid Flow

J. L. HESS (Retired)
McDonnell Douglas Corporation
Long Beach, CA

K. KUWAHARA
Institute of Space and Astronautical Science
Sagamihara, Japan

M. D. SALAS
NASA Langley Research Center
Hampton, VA

TERRY L. HOLST
THOMAS H. PULLIAM
NASA Ames Research Center
Moffett Field, CA

CONTENTS

Handbook of Fluid Dynamics and Fluid Machinery, Edited by Joseph A. Schetz and Allen E. Fuhs
ISBN 0-471-12598-9 Copyright © 1996 John Wiley & Sons, Inc.

20.1 PANEL METHODS
John L. Hess

20.1.1 General Remarks

If any technique can be called the *workhorse* of fluid dynamic analysis it is the *panel method*. While there are situations to which it does not apply, its geometric generality, ease of use, and low computational cost have led to its virtual daily employment in a variety of facilities concerned with the design of aircraft and ships and its occasional employment in a much broader context, such as pollution control and snow accretion. The basis of the panel-method approach is the superposition of fundamental flow solutions due to source, doublet, and vorticity singularities. The superposition principle is discussed in Sec. 1.3, and the point singularity formulas are presented in Sec. 1.3.3. Section 1.3.6 discusses the underlying ideas of panel methods and, for a particular type of panel method, gives a two-dimensional example in detail and verbal description of a three-dimensional example. Thus, Sec. 1.3.6 provides an introduction to the present section which will duplicate material presented there as little as possible.

As may be inferred from the above, a panel method is applicable to inviscid, incompressible flow (although under certain circumstances, compressibility effects may be accounted for approximately). Despite this apparent restriction, panel-method predictions have been found to agree well with experiment over a surprisingly large range of flow conditions [Hess and Smith (1966)]. Even when the calculated results fail to give the proper experimental values, they are frequently useful in predicting the incremental effect of a proposed design change or in ordering various designs in terms of effectiveness. Clearly, for a panel method to be applicable, the flow must be subsonic and free of catastrophic separation. Most good designs of aircraft and ships strive to minimize separation and, thus, are precisely the shapes for which panel methods are valid. If wings and fuselages were circular cylinders and spheres, panel methods would never have been developed. However, they are quite compatible with the streamlined shapes actually used.

Since the theoretical formulation of the panel method is very general geometrically, it is customary for investigators, including the present author, to claim that their newly developed panel method can handle completely arbitrary configurations. This is true in principle, but not in practice. When bodies having some new fluid-dynamic or geometric feature are encountered, a method will frequently fail, because some implicit assumption in the numerical formulation is no longer valid. After a particular method has been applied to a large variety of problems, many such difficulties will have arisen and one-by-one overcome by suitably improving the procedure. A panel method that has been heavily used in design for several years can fairly claim to have encountered almost every such difficulty and, indeed, to be applicable to *arbitrary* bodies. A new method, or one that has been applied only occasionally, would have little or no chance of successfully calculating a case such as that of Fig. 1.34.

The discussion of this section is restricted to three-dimensional problems, because space is limited and because most current interest lies in such problems. Hopefully, the two-dimensional case is covered adequately in Sec. 1.3.6. Even with this relatively tight focus, the present article is necessarily terse and relies on the references for details.

20.1.2 Mathematical Formulation

If a general three-dimensional body S is immersed in an incompressible flow with an onset flow velocity $\overline{U}$ (usually a uniform stream), the velocity $\overline{V}$ at any point is the sum of $\overline{U}$ and a perturbation velocity $\overline{v}$, which is irrotational, i.e.,

$$\overline{V} = \overline{U} + \overline{v} = \overline{U} + \nabla\phi \tag{20.1}$$

where ϕ is the perturbation potential which satisfies the Laplace equation (see Fig. 20.1). The boundary condition at infinity is that $\overline{v}$ vanishes, while on the body the total normal velocity must be zero, i.e.,

$$\overline{v} \cdot \overline{n} = \frac{\partial\phi}{\partial n} = -\overline{U} \cdot \overline{n} \quad \text{on } S \tag{20.2}$$

where $\overline{n}$ is the unit normal vector pointing into the flow. It is known [Lamb (1932)] that the perturbation potential can be expressed in terms of surface distributions of sources and doublets

$$\phi(x, y, z) = \iint [\phi_{\text{source}}\sigma(x_0, y_0, z_0) + \phi_{\text{doublet}}\mu(x_0, y_0, z_0)]\, dS \tag{20.3}$$

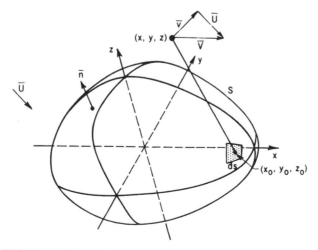

FIGURE 20.1 A three-dimensional body in an onset flow.

where the point source and doublet potentials in the integrand are those of Sec. 1.3.3. The potential of Eq. (20.3) satisfies the Laplace equation and the condition at infinity for any surface variation of source strength, σ, and doublet strength, μ. Thus, these functions are available to satisfy Eq. (20.2). Since there are two functions and only one condition there is a *degree of freedom* in the problem, and one additional condition on σ and/or μ may be prescribed arbitrarily. This problem of *singularity mix* is discussed below.

The domain of integration in Eq. (20.3) has been deliberately left vague. If singularities are distributed only on or inside S, the resulting flow is nonlifting. Lift can be obtained only with the presence of a trailing vortex wake issuing from the *trailing edge* of the lifting portion of the body (Fig. 1.33). The extent and location of the trailing edge is to some extent a matter of legislation by the user [Hess (1974)]. It is certain, however, that an additional condition, the *Kutta condition*, is required along the trailing edge and that it is this condition that fixes the lift. The location of the wake is unknown, which theoretically introduces a nonlinearity into the problem. In principle, the location is determined from the condition of zero normal velocity on and zero pressure jump through the wake. Since at any stage the panel method requires the locations of all singularity surfaces to be known, the wake location must be determined iteratively. In the majority of cases the wake location is simply assumed.

Of the many possible *singularity mixes*, two have been investigated extensively almost to the exclusion of the others: the *source method* and the *Green's Identity Method*. In the source method the form of the doublet distribution on or inside S is assumed in terms of a relatively small number of adjustable parameters used to satisfy the Kutta condition. The variation of source strength, σ, over S is determined via an integral equation resulting from application of the normal velocity boundary condition, Eq. (20.2) to the potential, Eq. (20.3). In the Green's Identity Method, the source strength on S is

$$\sigma = -\overline{\mathbf{n}} \cdot \overline{\mathbf{U}} \qquad (20.4)$$

which is known. The doublet strength, μ, is determined from an integral equation that expresses both the normal velocity boundary condition and the Kutta condition. Application of the normal velocity boundary condition is done indirectly by means of the equivalent condition that the perturbation potential, ϕ, be constant inside S.

No matter how the singularities are chosen, the exterior flow field is theoretically unique independently of the individual values of source and doublet strength. The *singularity mix*, i.e., the particular relation of σ and μ chosen to eliminate the above non-uniqueness, is given principal emphasis in many theoretical presentations. The implication is that this choice is the most important determinant of the efficiency and accuracy of the resulting method. In fact, it is less important than other theoretical considerations, such as the Kutta condition, and even certain details of the numerical implementation.

20.1.3   Numerical Implementation

In the numerical implementation, the body surface and wake are represented by four-sided *panels* that give these methods their name. (Compare Figs. 20.2 and 1.33.)

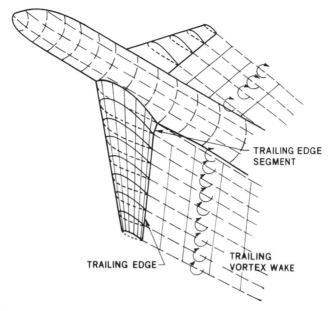

FIGURE 20.2 Discretization of a three-dimensional lifting configuration.

The singularity strengths σ and μ are either taken constant over each panel or are represented by first- or second-degree, two-variable polynomials whose coefficients are related numerically to the values on adjacent panels. The result is a number of unknown singularities σ or μ equal to the panel number plus the number of trailing-edge segments (Fig. 20.2). These strengths are adjusted by means of a set of linear equations (which comprise the numerical approximation of the integral equation mentioned above) to satisfy the normal velocity condition at a single control point of each panel and the Kutta condition at each trailing-edge segment.

The quantities basic to this approach are the influences of the panels on each other's control points, the so-called *aerodynamic influence coefficients*. Specifically, the source method require velocities induced by polynomial source and doublet distributions over a panel at a general point in space while a Green's Identity Method requires the corresponding potentials. These quantities are obtained analytically by integration over flat panels. (Curved panels are treated by expansion about a flat panel.) The complexity of the resulting formulas depends on the complexity of the underlying formulation, with the simplest formulas corresponding to constant singularity on a flat panel. Even these last formulas are rather complicated, because they include the effects of every detail of the panel shape, e.g., vertex angle, side lengths, etc. When the control point in question is far from the panel, such precision is inappropriate, and approximate expressions are used to reduce computing cost. Different approximations are used for different ranges of distance, and, above some distance, simple point singularity formulas are used. This is crucial, because the number of panel influences that must be computed is N^2, where N is the number of panels used to define the body and may run into the thousands (Fig. 1.34). In large cases, well over 95% of the panel influences are calculated by point singularity formulas, but, even so, this portion of the calculation is one of the two major contributors to computing cost. If approximations were not used, the computing cost would be greatly increased.

The second major portion of the calculation is the solution of the linear equations that express the normal velocity boundary condition together with the Kutta condition. Existing methods use either direct Gaussian elimination or iterative techniques. As is well known, the computing effort involved in direct elimination is proportional to N^3 and is independent of the matrix or the number of solutions obtained as long as this is small compared to N. The computing effort for an iterative solution is proportional to $I \times F \times N^2$, where I is the number of iterations required for convergence, and F is the number of solutions (flows) being obtained. If I is independent of N, then, for sufficiently large panel numbers, the iterative solution requires far less computation.

It turns out that the development of efficient iterative techniques for complicated geometries such as that of Fig. 1.34 is far from trivial. Block iteration is necessary where the strong panel influences are contained within blocks, which are inverted directly. Moreover, for unfavorable bodies, some form of convergence acceleration must be applied. A robust procedure has been developed by Clark (1985), which obtains convergence in 10–20 iterations.

In lifting flows, the circulation distributions on the lifting portions drive the entire solution. Accordingly, accurate determination of these circulations is crucial. Values of circulation are determined mainly from the Kutta condition along the trailing edge, thus specification of the Kutta condition is more important than any other detail of the numerical implementation.

The theoretical form of the Kutta condition states that the velocity shall remain finite all along the sharp trailing edge of a lifting portion, e.g. a wing (Fig. 1.33). Such a condition cannot be imposed in a numerical method, and an equivalent alternative condition must be found. Physically, it is evident that the limiting values of the pressure as the trailing edge is approached along the upper surface and along the lower surface must be equal. This equal pressure serves very well as a numerical form of the Kutta condition and is used in the source method [Hess (1974) and Hess *et al.* (1985)]. Since pressure contains the square of velocity, which is linear in the source and doublet strengths, the equal-pressure Kutta condition is quadratic in these strengths. This nonlinearity has led to its rejection by most investigators, who have selected a linear alternative to the Kutta condition so that the entire set of equations to be solved will be linear. A variety of linear conditions involving wake direction or doublet continuity have been used in the various Green's Identity Methods [Johnson (1980) and Maskew (1982)]. Some are clearly approximate, but all can lead to a nonphysical pressure mismatch at the trailing edge. When the nonlinear equal pressure conditions are used, at each stage of the iterative solution they are linearized by the well-known Newton–Raphson procedure. Thus, the Kutta-condition equations may be thought of as linear equations whose coefficients vary as the iteration proceeds.

Once the singularity strengths have been determined, velocities and thus pressures are calculated at the control points. The source method obtains velocities using the vector matrix of aerodynamic influence coefficients. The Green's Identity Method obtains the velocity potentials directly as the solution of the linear equations. Moreover, the only matrix involved is the scalar potential influence matrix, which requires only one-third as much storage as the matrix used by the source method. This is the

main advantage claimed for this approach. An obvious objection to this formulation
is that, since the solution consists only of values of the potential at the control points,
the velocity itself must be obtained by the equivalent of numerical differentiation.
However, no difficulties due to this cause have been reported.

Of the methods currently in use, the source method has been developed by the
author and his colleagues [Hess and Smith (1966), Hess (1974), Hess *et al.* (1985)].
Present Green's Identity Methods, which were originally formulated by Morino
(1974), are due to Johnson (1980) and Maskew (1982). An extensive bibliography
and an outline of the historical development is contained in Hess (1986).

20.1.4 Current Use of Panel Methods

As mentioned above, the panel method has attained the status of an essential design
tool whose use is so routine it is no longer remarked upon. Moreover, there seems
to be no possibility that it will be replaced by any other method. In major aircraft
companies, cases such as that of Fig. 1.34 are run a dozen times a day. Furthermore,
the complexity of these cases is increasing. The configuration of Fig. 1.34 uses 8000
panels to represent the body − 4000 on each side of the symmetry plane. Mathematical efficiencies and especially increases in computer speed now permit much
larger panel numbers. Users may successively refine the panelling to demonstrate
convergence. Use of 20,000 panels on each side of the symmetry plane is not un-
usual.

Concurrent developments in computer technology have put panel-method calculations in reach of the individual investigator, instead of being restricted to large
companies. One particular personal computer, whose approximate price is
$5,000.00, has the following characteristics: speed, 16 MFLOPS (million floating
point operations per second); core, 32 megabytes RAM (8 million words random
access memory); and disc, one gigabyte (250 million words of storage). Such a
machine can calculate a 10,000 panel case in 18 hours of dedicated running and a
case like that of Fig. 1.34 in 45 minutes.

20.2 VORTEX METHODS
K. Kuwahara

20.2.1 Introduction

Many incompressible flows at high Reynolds numbers are characterized by regions
of concentrated vorticity imbedded in an irrotational flow. By the theorems of Helmholtz and Kelvin, the inviscid motion of the vorticity in these regions is given by
the local fluid velocity which in turn is determined kinematically from the vorticity
field (see Chap. 1). Thus, it is often very convenient to consider inviscid fluid dynamics in terms of parcels of vorticity which introduce motion on each other as an
alternative to pressure-velocity considerations. By this mutual induction process a
vortex ring propels itself along its axis, and a pair of aircraft-trailing vortices induce
downward motion, each upon the other. Vortex methods simulate flows of this type
by discretizing the vorticity-containing regions and tracking this discretization in a
Lagrangian reference frame. The required local velocities are computed as a solution

to a Poisson equation for the velocity field often in terms of a Green's function or Biot-Savart integration. Typically, the Lagrangian coordinates of this discretization satisfy a nonlinear system of ordinary differential equations giving the time evolution of the coordinates. See Chap. 1 for definition and discussion of Lagrangian coordinates.

The use of vortex methods for unsteady, two-dimensional flow in free space [Rosenhead (1931), Chistiansen (1973), Aston (1976), Ashurst (1979), (1983), Nakamura *et al.* (1982), and Inoue (1984)] or past a simple body [Kuwahara (1973), (1978), Chorin (1973), Sarpkaya (1975), Kiya and Arie (1977), Sarpkaya and Schoaff (1979), Horiuti *et al.* (1980), Ono *et al.* (1980), Spalart and Leonard (1981), Oshima *et al.* (1983), and Izumi and Kuwahara (1983)] has significantly increased. There are a number of review papers available [Cheer (1983), Clemerts and Maull (1975), and Leonard (1980)]. Many attempts were made to prove their convergence to the Euler equations. Under certain restrictions, the convergence has been proved [Oshima and Oshima (1982), Hald and Mauceri Del Prete (1978), Hald (1979), and Beale and Majda (1982a)].

Excellent background material may be found in Saffman (1992) and Ogawa (1992) on vortex dynamics in general.

20.2.2 Discretized Vortex Equations in Two Dimensions

The rotational region in an inviscid two-dimensional incompressible flow is divided into a number of small subdomains. Each of the subdomains is replaced by a point vortex located at its center, and the strength of each point vortex is set equal to the circulation around the boundary of the domain. The flow field is in the x–y plane and perpendicular to the direction of the vorticity. Let (x_i, y_i) be the position of the ith point vortex at time t, and γ_i be its strength. Each vortex is subjected to the velocity field induced by all other vortices. Thus, the equations of the motion of the ith vortex are

$$\frac{dx_i}{dt} = -\sum_{i \neq j} \gamma_j \frac{y_i - y_j}{r_{ij}^2}, \quad \frac{dy_i}{dt} = \sum_{i \neq j} \gamma_j \frac{x_i - x_j}{r_{ij}^2} \tag{20.5}$$

where

$$r_{ij}^2 = (x_i - x_j)^2 + (y_i - y_j)^2$$

These are the basic equations of the vortex method in its original form.

Equations (20.5) can be derived directly from the two-dimensional Euler equations of a constant density. The vorticity ω satisfies the transport equation

$$\frac{\partial \omega}{\partial t} + u \frac{\partial \omega}{\partial x} + v \frac{\partial \omega}{\partial y} = 0 \tag{20.6}$$

where (u, v) is the velocity computed as the solution of the Poisson equations

$$\Delta u = -\frac{\partial \omega}{\partial y}$$
$$\Delta v = \frac{\partial \omega}{\partial x} \tag{20.7}$$

If the two-dimensional flow has no interior boundaries and the fluid is at rest at infinity, the solution to Eq. (20.7) may be written as the Biot–Savart integral,

$$
u = -\frac{1}{2\pi} \int \frac{(y - y') \cdot \omega(x', y', t)\, dx'\, dy'}{r^2}
$$

$$
v = \frac{1}{2\pi} \int \frac{(x - x') \cdot \omega(x', y', t)\, dx'\, dy'}{r^2}
\tag{20.8}
$$

$$
r^2 = (x - x')^2 + (y - y')^2
$$

For the numerical purposes, one must discretize the vorticity field. It might be assumed that ω has the representation of point vortices, i.e.,

$$
\omega(x, y, t) = 2\pi \sum_j \gamma_j\, \delta(x - x_j(t),\, y - y_j(t))
\tag{20.9}
$$

where δ is the two-dimensional Dirac delta function, (x_j, y_j) are the locations of the vortices, and $2\pi\gamma_j$ are their respective circulations, constant in time. Putting Eq. (20.9) into Eq. (20.8), Eq. (20.5) is obtained.

Equation (20.5) has singularities at point vortices. This makes the motion of point vortices quite random. To remove this effect, Kuwahara and Takami (1973) proposed a viscous core, using a strict Navier–Stokes solution of a single point vortex; γ_j in Eq. (20.5) is replaced by

$$
\gamma_j = \kappa_j \left\{ 1 - \exp\left(-\frac{r_{ij}^2}{4\nu t} \right) \right\}
\tag{20.10}
$$

Chorin (1973) introduced a core of constant radius as follows

$$
\gamma_j =
\begin{cases}
\kappa_j & r > \sigma \\
\kappa_j r_{ij} & r < \sigma
\end{cases}
\tag{20.11}
$$

The core radius σ can be modified to be consistent with Eq. (20.6), if it is time-dependent as $\sigma = 2.24\sqrt{\nu t}$. Spalart and Leonard (1981) use a simpler core

$$
\gamma_j = \frac{\kappa_j r_{ij}^2}{r_{ij}^2 + \sigma^2}
\tag{20.12}
$$

These types of cores have almost the same effect on stabilization of the motion of vortices. The vortex method using nonsingular discrete vortices has been called the *vortex blob method*. However, this is not always needed. Some computations using simple point vortex without any core (*point vortex method*) give very good agreements with experiments [Horiuti *et al.* (1980), Ono *et al.* (1980), and Izumi and Kuwahara (1983)].

The above vortex blob methods introduce some viscous effects, but they are not directly based on molecular diffusion. Chorin (1973) applied a random walk concept to the motion of a point vortex, which simulates viscous diffusion of vorticity. Using an Euler approximation of Eq. (20.5), one obtains

$$x_i^{n+1} = x_i^n + u_i \Delta t + \eta_1$$
$$y_i^{n+1} = y_i^n + v_i \Delta t + \eta_2$$

(20.13)

where η_1 and η_2 are independent random variables with a Gaussian distribution of zero mean and variance $2\Delta t/R$, and (u_i, v_i) is the induced velocity. This method is sometimes called the *random vortex method*.

20.2.3 Numerical Error and the Diffusion of Vorticity

Integration of the equations of motion of point vortices is always accompanied by some numerical errors. These errors are not small when the time integration scheme is the first-order-accurate Euler type. These errors are nearly random and make the calculated flow pattern by the vortex method somewhat similar to viscous flow. This *viscosity effect* of numerical errors is almost always larger than that of the actual molecular viscosity of the fluid under consideration. The *effective viscosity* is large especially where the concentration of vortices is strong. As a matter of fact, it is rather close to that of turbulent viscosity [Kuwahara and Takami (1983)]. This fact is the main reason why such a simple method of simulation is so effective for the computation of air flow past, say, large buildings. The striking point is that a first-order scheme sometimes gives much better results than a higher-order scheme like Runge–Kutta method.

It is essential to evaluate the diffusion of vorticity due to numerical errors. The time derivative of the second moment of the vorticity is proportional to the viscosity

$$\frac{d}{dt} \iint \omega r^2 \, dx \, dy = 4\nu \iint \omega \, dx \, dy$$

(20.14)

where $r^2 = x^2 + y^2$, and ω is the vorticity. If the fluid is inviscid, the left-hand side should be exactly equal to zero. In numerical computation, however, this term does not vanish because of the time integration error. Following Nakamura *et al.* (1982), this may be estimated by calculating the effective viscosity

$$\nu_{\text{eff}} = \frac{d}{4 \, dt} \left(\sum_{n=1}^{N} \gamma_n r_n^2 \right) \bigg/ \sum_{n=1}^{N} \gamma_n$$

(20.15)

The effective viscosity, γ_{eff}, is proportional to Δt for the first-order Euler method, $(\Delta t)^2$ for the second-order predictor corrector method and $(\Delta t)^5$ for the Runge–Kutta method. Similar dependence on the strength of the vorticity is easily derived from that dependence on time step. The above definition of the effective viscosity is based on the equation of molecular diffusion, that is, the diffusion coefficient is constant everywhere. However, in computations, it depends on space and time. Therefore, care must be taken when using this definition for the diffusion.

To illustrate the diffusion process mentioned above, the dependence of the diffusion of vorticity on the parameters and integration scheme is investigated in a circular eddy replaced by a number of point vortices. Total circulation of the eddy is fixed to be 2π and the diameter, 0.5.

Figure 20.3 shows the dependence of the diffusion on the number of point vortices. The times integration scheme is Euler's. This clearly shows that the diffusion does not depend on the number of point vortices N that represent the vortex region. This fact is very important, because it means that the number of point vortices to simulate the vortical flow need not be large so long as the diffusion is concerned.

Figure 20.4 shows the diffusion of vorticity by the effect of molecular viscosity. This computation was done by a fourth-order Runge–Kutta method with a random walk representation of molecular viscosity. The difference between the molecular diffusion and the numerical diffusion is clearly seen.

Figure 20.5 shows the diffusion of vorticity by the Euler scheme in the case of Gaussian distribution of initial vorticity. In the region where the vorticity is stronger, its diffusion is also stronger. Therefore, the distribution of vorticity quickly becomes uniform in the eddy.

Although this type of diffusion simulates the turbulent diffusion very well, it is not perfectly controllable. It may be possible to model turbulence by modifying the random variables in Eq. (20.13) and by using a higher order integration scheme such

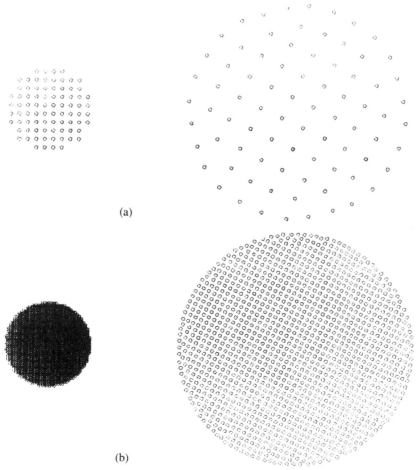

(a)

(b)

FIGURE 20.3 Dependence of diffusion on the number of point vortices ($\Delta t = 0.1$). (a) $N = 80$ and (b) $N = 1264$.

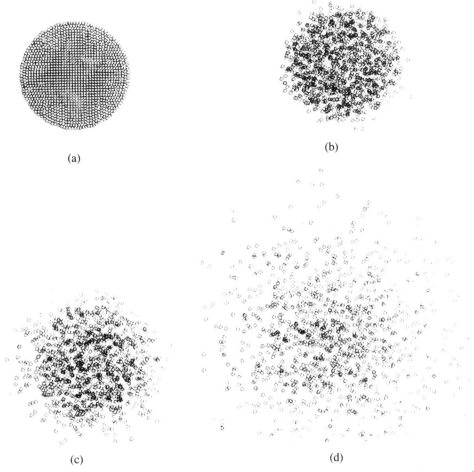

FIGURE 20.4   Effect of viscous diffusion of vorticity ($\Delta t = 0.1$). (a) $R = \infty$, (b) $R = 10^3$, (c) $R = 10^2$, and (d) $R = 10$.

as Runge–Kutta. If these random variables are Gaussian distributed, molecular diffusion will be better modeled.

20.2.4   Flow Past a Body

There are two different ways to treat solid boundaries in vortex methods. The first is to use mirror images of vortices; the body under consideration is mapped onto a circle and every calculation is accomplished on the mapped plane. Mirror images of vortices are set to match the boundary condition. Although this method can take the exact boundary condition into account, it is not easy to map an arbitrary two-dimensional shape onto a circle.

The second approach that has been developed [Spalart and Leonard (1981), Izumi and Kuwahara (1992), and Kuwahara and Takami (1983)] is based essentially on the panel method (see Sec. 20.1) using singularities on the boundary. The simplest version of the vortex method without conformal mapping as well as the method based on the mirror images of vortices are described below.

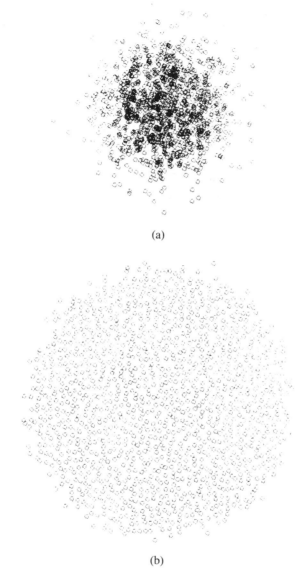

(a)

(b)

FIGURE 20.5 Effect of vorticity diffusion by the Euler scheme ($\Delta t = 0.05$). (a) $t = 0$ and (b) $t = 1$.

Another important problem when computing a flow past a body is the generation of vortices. Two methods are described here. One is the no-slip condition for a smooth boundary [Kuwahara (1978)], and the other is the Kutta condition for a sharp edge [Kuwahara (1973)].

Circular Cylinder. For a circular cylinder, a mirror image system of vortices may be employed for the normal boundary condition, and the no-slip condition can be used for the generation of vorticies. At $t = 0$, the flow field is assumed to be inviscid and irrotational, that is without vorticity in the flow field. However, there must be an infinitely thin boundary layer along the surface of the cylinder which is considered a vortex layer. This layer is divided into a finite number of corresponding segments and replaced by the same number of nascent vortices the circulations of which are assumed to be equal to those of the segments. This condition means that net flow along the surface does not occur.

It is convenient to use complex variables for the derivation of the vortex method. The equation of motion is

$$\frac{d\bar{z}_j}{dt} = \left(\frac{df_j}{dz}\right)_{z=z_j}$$

(20.16)

in the complex form, where z_j is the position of the jth point vortex, f_j is the complex velocity potential of the flow in which the term due to the jth vortex is excluded. The over-bar indicates the complex conjugate. The complex velocity potential f is given by

$$f(z) = z + \frac{1}{z} + i \sum_{j=1}^{n} \kappa_j \log (z - z_j) - i \sum_{j=1}^{n} \kappa_j \log (z - z_j^*)$$

(20.17)

$$z_j^* = \frac{1}{\bar{z}_j}$$

where κ_j is the strength of the jth point vortex, and n is the number of point vortices in the flow field. The first two terms represent the initial potential flow; the third, the flow due to the generated vortices; and the fourth, the flow due to the image vortices to match the boundary condition. The complex velocity potential f can be written as

$$f(z) = \Phi(x, y) + i\Psi(x, y)$$

$$z = x + iy, \quad \bar{z} = x - iy$$

(20.18)

where $\Phi(x, y)$ and $\Psi(x, y)$ represent the velocity potential and the stream function respectively.

The strengths of the m nascent vortices are determined by the following no-slip condition

$$\Phi_1 = \Phi_2 = \cdots = \Phi_m, \quad \kappa_n + \kappa_{n-1} + \cdots + \kappa_{n-m+1} = 0$$

$$\Phi_j = \left(\cos 2\pi \frac{j}{m}, \sin 2\pi \frac{j}{m}\right) = \text{Re } f(e^{2\pi i (j/m)}), \quad j = 1, \ldots, m$$

(20.19)

The positions of the nascent vortices at this moment are given by

$$z_{n-m+j} = (1 + \epsilon)e^{2\pi i [j - (1/2)]/m}$$

(20.20)

Each point vortex is convected by the velocity field due to the other vortices, their images and the initial potential flow. After one time step, the vortices which have compensated the circulation caused by the initial boundary layer can no longer compensate the new circulation. Therefore, new vortices must be introduced. A large

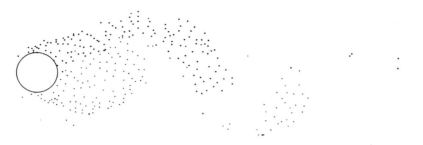

FIGURE 20.6 Flow pattern after a cylinder at $t = 25$ (nondimensional time).

number of point vortices seems to be desirable, but considering the fact that the computation time is nearly proportional to the square of the number of the point vortices, the number can not be too large. The smallest number to obtain reasonable results is two. Setting up at the initial instant, a larger number is better, say, 36. Figure 20.6 shows the flow pattern at nondimensional time $t = 25$.

The forces on the body are calculated from the *generalized Blasius formula*

$$X - iY = i\rho \oint \frac{\partial f}{\partial t} \, dz + \frac{1}{2} i\rho \oint \left(\frac{df}{dz}\right)^2 dz \qquad (20.21)$$

where X, Y are the drag and the lift respectively, and ρ is the density. Using the complex velocity potential, the forces can be calculated algebraically.

Flat Plate. In this case, the computation should be done using comformal mapping to transform the flat plate to a circle. The generation of vortices is determined by the Kutta condition. The mapping which transforms the two edges of the plate, $z = \pm e^{-i\alpha}$ into the points $\zeta = \pm 1$ on the circle is given by

$$z = \frac{1}{2} \exp\left(-i\alpha\right) \cdot \left(\zeta + \frac{1}{\zeta}\right) \qquad (20.22)$$

The complex velocity potential in the ζ-plane is given by

$$f = \frac{1}{2} \exp\left(-i\alpha\right) \cdot \left(\zeta + \frac{\exp\left(2i\alpha\right)}{\zeta}\right) - i \sum_{j=1}^{2n} \kappa_j \log\left(\zeta - \zeta_j\right)$$

$$+ i \sum_{j=1}^{2n} \kappa_j \log\left(\zeta - \zeta_j'\right) \qquad (20.23)$$

where $\zeta_j' = 1/\bar{\zeta_j}$ and κ_j, ζ_j are the strength and position of the jth point vortex respectively, and n is the number of point vortices shed from one edge of the plate.

The equation of motion in the ζ-plane is given by

$$\frac{d\bar{\zeta_i}}{dt} = \frac{df_i}{d\zeta} \cdot \left\|\frac{d\zeta}{dz}\right\|^2 \qquad (20.24)$$

The strength of the vortex κ_j is determined using the Kutta condition

$$\left(\frac{df}{d\varsigma}\right)_{\varsigma = \pm 1} = 0 \tag{20.25}$$

This leads to

$$\sin \alpha + \sum_{j=1}^{2n} \kappa_j \left(\frac{1}{1 - \varsigma_j} - \frac{1}{1 - \varsigma_j'}\right) = 0$$
$$\sin \alpha + \sum_{j=1}^{2n} \kappa_j \left(\frac{1}{-1 - \varsigma_j} - \frac{1}{-1 - \varsigma_j'}\right) = 0 \tag{20.26}$$

The computation can be done almost in the same way as for the case of circular cylinder. Figure 20.7 shows the flow pattern at $t = 17$.

Treatment of Boundary Condition without Conformal Mapping. The above method based on conformal mapping can treat the boundary condition exactly, but it is not always easy to map a complex geometry to a circle. A vortex method without conformal mapping is more desirable, although the boundary condition is satisfied only approximately. Here, the calculation of the flow past an arbitrary geometry is shown with the simplest version of the method. Point vortices are placed on the boundary, and the condition that the flow does not penetrate the boundary is imposed to determine the vortex strengths. This method can treat a two-dimensional flow around any shape. The most important point is that it can be easily extended to three-dimensional flow, while the extension of the circular cylinder and flat plate cases to 3D are very difficult. The boundary conditions are satisfied better as the number of vortices on the boundary is increased.

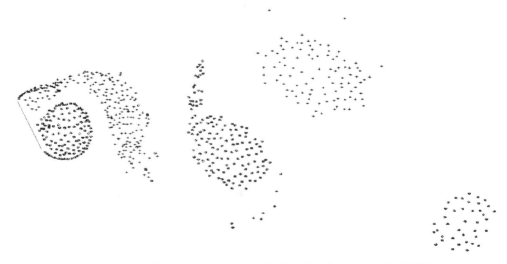

FIGURE 20.7   Flow pattern behind a flat plate at $t = 21.1495$.

The complex velocity potential f is given by

$$f(z) = z + i \sum_{j=1}^{m} \kappa_j \log (z - z_j) + i \sum_{j=m+1}^{n} \kappa_j \log (z - z_j) \qquad (20.27)$$

where κ_j is the strength of the jth point vortex, m is the number of point vortices on the boundary, and n is the total number of vortices. The first term represents the uniform flow. The point vortices are divided into two groups. The first group involves the vortices on the boundary including the *nascent vortices*. The second group represents the vorticity in the wake. The strength of each vortex in the first group is determined at each time step, while that of the vortex in the second group does not change. The strength of a vortex in the first group is determined by the following *no penetration condition*, that is, net flow through the boundary does not occur between any two point vortices on the boundary

$$\Psi(z_1) = \Psi(z_2) = \Psi(z_3) = \cdots = \Psi(z_m) \qquad (20.28)$$

The centers of the vortices are singular points, so this condition is not strictly applicable. Therefore, the control points are shifted slightly from the centers of the vortices. In the following calculation, the centers are shifted by 0.1 percent of the size of the body all in the same direction. This is acceptable, because the singularity of the logarithmic function is very weak. The number of conditions thus obtained is $m - 1$. The last condition which determines the strengths of the vortices is the conservation of the total circulation of the flow

$$\kappa_1 + \kappa_2 + \kappa_3 + \cdots + \kappa_n = 0 \qquad (20.29)$$

The vortices on the edges of the boundary of the body are used as the nascent vortices, because the vortices on the boundary can be thought to approximate the boundary layer. The edge vortices separate the flow from the edges.

Using this method, the flow past a circular arc cylinder was simulated and compared with the corresponding experiment. For the time integration, the Euler and Runge–Kutta schemes were both used and compared in Fig. 20.8. It was found that high Reynolds number flows are more accurately simulated by the Euler scheme than by the Runge–Kutta scheme. The larger numerical error of the former method provided enough turbulent diffusion effect so that the intrinsic error of the vortex method caused by the inviscid assumption fortunately was compensated. The amount of this turbulent diffusion effect can be controlled by the time step size. Ironically, the Runge–Kutta method, which is numerically more accurate than the Euler method, is less capable of yielding the diffusion effect.

20.2.5 Three-Dimensional Vortex Methods

Extensions of vortex methods to three-dimensional flows have been tried by many authors, e.g., Leonard (1980), Hama (1962), Chorin (1981), and Nakamura *et al.* (1983). In three dimensions, the rotational region is replaced by a number of *vortex filaments*. Each vortex filament is divided into a number of segments. The velocity of each segment is induced by the segment itself and the other segments, and it is

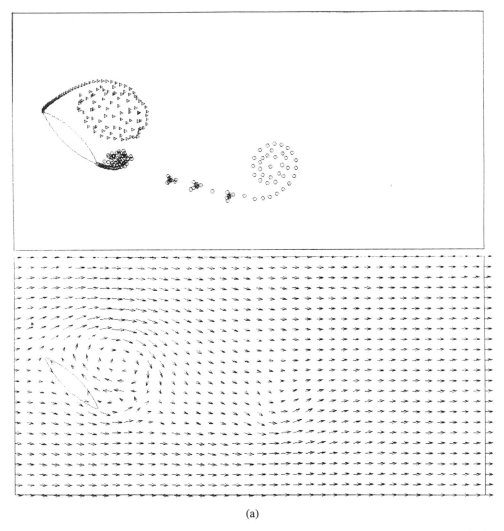

(a)

FIGURE 20.8   Flow pattern behind an airfoil (a) Euler scheme and (b) Runge–Kutta method.

calculated by *Biot–Savart law*,

$$\delta v = \frac{1}{2} \frac{\kappa \sin \theta \, ds}{r^2} \cdot \mathbf{e} \tag{20.30}$$

$$\bar{e} = \frac{\overline{\omega}}{\omega} \times \frac{\bar{r}}{r} \tag{20.31}$$

where $2\pi\kappa$ is the circulation of the vortex filament, $\overline{\omega}$ is the vorticity vector, and θ is the angle between $\bar{r}$ and $\overline{\omega}$. In two dimensions, the self induced velocity is zero, that is the velocity of the vortex filament induced by itself is zero. In three dimensions, however, the self induced velocity is not only not zero, but it plays an im-

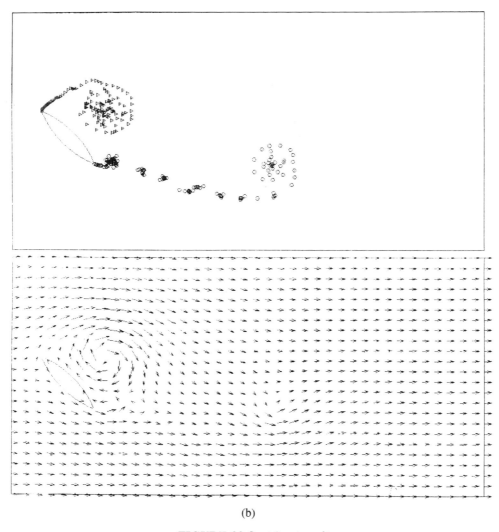

(b)

FIGURE 20.8 (*Continued*)

portant role [Lewis (1981)]. It becomes even infinite if the core radius of the filament is zero, but its curvature is not zero. Therefore, a core must be always assumed. Another difficult problem is vortex stretching. In two dimensions, vortex filaments can not stretch, but stretching should be duly treated in any three dimensional vortex method.

Although most vortex methods in three dimensions are based on the Biot-Savart law, many different methods can be designed, and many versions are being tested. A very promising method which is simple and does not use vortex filaments was developed by Beale and Majda (1982a), (1982b) with a proof of convergence to the Euler equations, but computational results are not available yet.

At this stage, only limited success has been obtained in three-dimensional vortex methods, but this area is being most actively investigated [Oshima and Kuwahara (1984), Shirayama *et al.* (1985), and Shirayama and Kuwahara (1986)].

20.2.6   Conclusions

The vortex method with an Euler time-integration scheme effectively simulates real vortical flows, although it has a strong diffusive effect. This effect suggests that numerical diffusion in the vortex method is similar to turbulent diffusion.

The vortex method is not only an approximation to the Euler equation as widely believed, but it is a good approximation to the Navier–Stokes equations at high Reynolds number because of the inevitable diffusive effect due to numerical errors. The vortex method has built-in turbulence modeling and can treat effectively a turbulent vortical flow such as a turbulent shear flow or a wake of a bluff body. Usually a first-order time integration scheme is appropriate, and a small number of vortices is sufficient to simulate turbulent flows.

For relatively low Reynolds number flows, molecular diffusion can be simulated effectively by a random walk, but a large number of point vortices is required.

Three-dimensional vortex methods are an attractive field of research.

20.3   NUMERICAL TREATMENT OF SHOCK WAVES
M. D. Salas

In a real fluid, a shock wave consists of a very thin region, of the order of a few molecular mean free paths, where the convective steepening tendencies of the wave are balanced by viscosity and heat dissipation effects. The experimentally-observed structure of shock waves has been found to be in good agreement with the structure predicted by the Navier–Stokes equations [Gilbarg and Paolucci (1953)]. For an inviscid fluid, the shock wave structure is replaced by a surface of zero thickness across which the flow properties pressure, temperature, density, and velocity change discontinuously. Rankine (1870) and some years later Hugoniot (1889) obtained the conservation relations that determine the jump of the flow properties across a shock. As might be expected, the discontinuous behavior of flow properties across a shock presents some serious problems to the numerical calculation of flows containing these surfaces. Two numerical approaches have evolved to handle these problems. The first restricts application of the differential equations governing the flow to smooth regions. This approach requires knowledge of the location in time and space of all surface discontinuities in order to connect, using the Rankine–Hugoniot relations, the piecewise smooth regions of the flow. The second approach generalizes the concept of a solution to admit discontinuities. It then numerically *captures* these discontinuities as very thin regions of high gradients, somewhat mimicking the shock wave structure of a real fluid. Both approaches have their advantages and disadvantages. The first method results in very accurate and efficient solution algorithms, because the numerical integration of the differential equations is limited to smooth regions. It is ideally suited to problems where some fundamental questions about the behavior of the flow is being investigated, and it is necessary to model the governing equations as closely as possible. Its major shortcoming is the coding complexity needed to track all the discontinuities and their interactions, particularly in three-dimensional applications. The advantage of the second method is its simplicity of coding, since little or no special treatment is given to discontinuities. Its main disadvantages are that it smears the discontinuities over several mesh points, reduces the accuracy of the numerical scheme in the vicinity of the shock, and requires many mesh points in order to keep the shock thickness small and calculate the high gradients within it.

Begin by examining the second approach. The equations governing the motion of a perfect inviscid flow may be written in the following divergence form

$$\frac{\partial \rho}{\partial t} + \nabla \cdot (\rho \overline{V}) = 0$$

$$\frac{\partial \rho \overline{V}}{\partial t} + \nabla \cdot (\rho \overline{V}\, \overline{V} + p\overline{\overline{I}}) = 0 \tag{20.32}$$

$$\frac{\partial \rho E}{\partial t} + \nabla \cdot (\rho \overline{V} H) = 0$$

where ρ is the density, $\overline{V}$ is the velocity vector, p is the pressure, E is the specific stagnation energy, and H is the specific stagnation enthalpy. To simplify the notation, the momentum equation, the second equation above, is written using tensor notation where $\nabla \cdot (\rho \overline{V}\, \overline{V})$ is the divergence of a dyadic product defined by

$$\nabla \cdot (\rho \overline{V}\, \overline{V}) = \overline{V} \nabla \cdot (\rho \overline{V}) + \rho \overline{V} \nabla \cdot \overline{V} \tag{20.33}$$

and $\overline{\overline{I}}$ is the metric tensor, such that

$$\nabla \cdot (p\overline{\overline{I}}) = \nabla p \tag{20.34}$$

The governing equations can be conveniently written in the form

$$\frac{\partial U}{\partial t} + \nabla \cdot \overline{\overline{F}}(U) = 0 \tag{20.35}$$

where

$$U = [\rho, \rho \overline{V}, \rho E]^T \tag{20.36}$$

and $\overline{\overline{F}}$ is the flux tensor. One says that U is a *weak solution* of Eq. (20.35) if it satisfies the integral relation [Lax (1954)]

$$\iint \left(U \frac{\partial \phi}{\partial t} + \overline{\overline{F}} \cdot \nabla \phi \right) d\overline{x}\, dt - \int \phi_0 U_0\, d\overline{x} = 0 \tag{20.37}$$

for every continuously-differentiable test function ϕ which vanishes outside a subdomain of $\overline{x}$, and where ϕ_0 and U_0 are initial conditions at time zero. Furthermore, if Σ is a smooth surface and U is a continuously differentiable solution of Eq. (20.35) on either side of Σ and discontinuous across Σ, then U satisfies on Σ the compatibility relation

$$\frac{\partial \Sigma}{\partial t} [U] + \overline{n} [\overline{\overline{F}}] = 0 \tag{20.38}$$

Here, the brackets denote the jumps of the quantities within brackets across the surface Σ, $\bar{n}$ is the unit normal to Σ and $\partial\Sigma/\partial t$ is the speed with which Σ moves in the $\bar{n}$ direction. It is easy to show that the solution of Eq. (20.35) includes, in addition to the Rankine–Hugoniot jump conditions for shock waves and slip surfaces, a nonphysical expansion shock. In order to exclude the latter, the mathematical model represented by Eqs. (20.35) and (20.38) is augmented by requiring that fluid particles crossing a discontinuity should undergo an increase in entropy. This is called the *entropy condition* and is equivalent to requiring that the discontinuity be crossed by the forward drawn (i.e., in the direction of increasing time) characteristics originating from either side of the discontinuity. In numerical calculations, the entropy condition is usually implemented by adding an artificial dissipation operator to the right-hand side of Eq. (20.35). This term usually has a form similar to the viscous terms of the Navier–Stokes equations. For more details on the *entropy condition* and artificial dissipation consult Lax (1954). Depending on how the artificial dissipation term is written and on the specific details of the discretization of Eq. (20.35), many shock-capturing methods can be constructed. The success of a particular method depends on a large number of factors, such as the treatment of boundary conditions, the numerical scheme used, the suitability of the mesh, and the complexity of the problem being solved. Examples abound in the literature which show both the success and failure of particular shock-capturing applications. Very impressive results have been presented by Colella (1982) for a very difficult double Mach reflection problem using a windward differencing shock capturing scheme.

Two variations of the first approach have been developed. In the first variation, the shock surface is mapped through some stretching of coordinates onto a computational boundary of the problem investigated. This method is generally known as *shock fitting*. This method can practically handle only a few discontinuities, because the topology of the computational mesh is tied to the shape of the discontinuities. It is best applied to problems where the investigator has *a priori* knowledge of the approximate shape and location of the discontinuity. The best known application is the classical problem of a blunt body in a supersonic stream [Moretti (1968)], but more complex flow fields have also been calculated with this method as, for example, the supersonic flow past a wing-body configuration [Marconi and Salas (1973)]. In the second variation, the computational mesh is defined independently of the discontinuities that may occur in the flow. This is the so-called *floating shock-fitting method*. It has the advantage of treating more complex shock patterns, but it requires somewhat more complicated programing logic to keep track of the discontinuities and is not quite as accurate as the previously mentioned variation. An example of its application to very complex scramjet inlet flows and to the cylindrical implosion of a thin tube is given in Salas (1976) and de Neef and Hechtman (1978), respectively. With some differences in implementation details, both shock-fitting methods use the Rankine–Hugoniot jump conditions and the compatibility relation along the characteristic reaching the shock on the high pressure side to obtain an equation describing the acceleration of the shock surface Σ. The acceleration is then integrated to obtain the shock speed, $\partial\Sigma/\partial t$, and the shock speed is integrated to obtain the new shock position, which defines the location of Σ at a new time level. It is not possible in this short exposition to give more details, but they can be found in Moretti (1972), where various shock-fitting methods are compared, and in de Neef (1978) where a simple strategy to obtain the shock acceleration is presented.

In general, shock-fitting codes require some method of detecting the formation of shock waves from originally smooth data. In one approach [Moretti (1972)], the formation of a shock is detected by monitoring the pressure distribution throughout the flow field by fitting it locally with a cubic polynomial. It is then assumed that a shock forms when the pressure gradient at one point in the field becomes infinite. In a second approach [Salas (1976)], the formation of a shock is detected by monitoring the coalescence of characteristics of the same family. When shocks are fitted as boundaries, the detection of a new shock requires a repartitioning of the mesh to accommodate the new shock, while with *floating shock fitting*, only a reordering of the tracking logic is required. Slip surfaces and other discontinuities can be treated in a similar way.

The use of *shock fitting* removes the need for using the divergence form of the governing equations, and it reduces the need for artificial dissipation. Shock fitting however requires the proper handling of the interactions that occur between discontinuities and rigid surfaces. *Shock capturing*, on the other hand, handles this problem automatically, but in most cases the results are rather poor, and considerable information is lost at the interaction point. In general, when discontinuities are fitted, the treatment of an interaction is reduced to a locally two-dimensional problem by treating the interaction in the plane normal to both interacting surfaces. In most practical problems, the interaction process is well understood, and the theory is well developed [Courant and Friedrichs (1967), and Ferri (1949)]. There are, however, some problems of current interest where the interaction mechanism is not well understood. One such problem is the Mach reflection of a weak shock wave [Sichel (1971)] for which under certain conditions it appears as if the Rankine–Hugoniot jumps are not satisfied. Another problem is the determination of the exact conditions at which a regular reflection makes a transition to a Mach reflection [Marconi (1983)] and yet another is related to the stability of weak and strong shock waves. Although these problems present some difficulties to their treatment with shock-fitting techniques, it is clear that shock-fitting techniques are better suited to provide an understanding of the basic underlying fluid dynamic phenomena than shock-capturing techniques.

20.4 TRANSONIC POTENTIAL METHODS
Terry L. Holst

20.4.1 Full Potential and Transonic Small Disturbance Potential Equations

Two transonic potential formulations have been used in many practical calculations in both two and three dimensions. These formulations are based on the full or exact potential equation and the transonic small disturbance (TSD) potential equation (See Chap. 9). The full potential formulation is derived from the Euler equations by assuming the flow to be irrotational and isentropic. Under these assumptions, the mass and momentum conservation laws associated with the Euler equations become identical, that is the equations can be expressed mathematically as one equation (for example, conservation of mass). The energy equation reduces to a single algebraic condition. Thus, the full potential formulation consists of a scalar partial differential equation, whereas the Euler equations are a set of partial differential equations.

The TSD equation is derived from the full potential equation by further assuming the geometry of interest (airfoils or wings) to be thin with respect to some characteristic length, for example, the airfoil chord. In addition, the *thin* body must have a small angle of inclination with respect to the freestream. Next, properly formulated series expansions are substituted into the full potential equation and truncated in a consistent manner. This process utilizes the thin-body assumption and requires freestream Mach numbers to be near unity. Several different forms of the TSD equation can be derived by this process. The various forms along with details of the derivation process are presented in Holst *et al.* (1982) and van der Vooren *et al.* (1976).

The TSD equation expressed in two-dimensional Cartesian coordinates (x, y) is given by

$$[1 - M_\infty^2 - M_\infty^2(\gamma + 1)\varphi_x]\, \varphi_{xx} + \varphi_{yy} = 0 \tag{20.39}$$

where M_∞ is the freestream Mach number, γ is the ratio of specific heats (equal to 1.4 for air), and φ is the small-disturbance or perturbation velocity potential defined by

$$\nabla \varphi = \overline{\mathbf{q}} - \overline{\mathbf{q}}_\infty \tag{20.40}$$

In Eq. (20.40), $\overline{\mathbf{q}}$ and $\overline{\mathbf{q}}_\infty$ are local and freestream velocity vectors defined by

$$\overline{\mathbf{q}} = u\overline{\mathbf{i}} + v\overline{\mathbf{j}}, \quad q_\infty = u_\infty\overline{\mathbf{i}} \tag{20.41}$$

where $\overline{\mathbf{i}}$ and $\overline{\mathbf{j}}$ are the unit vectors along the x and y directions, respectively. Thus, the perturbation velocity components are given by

$$\varphi_x = u - u_\infty, \quad \varphi_y = v \tag{20.42}$$

Boundary conditions for a typical *thin* airfoil used in conjunction with Eq. (20.39) are given by

$$\varphi_y(x, 0^\pm) = u_\infty \frac{dg^\pm}{dx} \tag{20.43}$$

where $g^+(x)$ and $g^-(x)$ define the upper and lower airfoil surfaces, respectively. Application of this boundary condition, which is made at the airfoil *slit* (or *chord line*), simulates the required flow tangency boundary condition at the airfoil surface to an accuracy consistent with small disturbance theory. An auxillary relation, usually used in this formulation to define the airfoil surface pressure coefficient, is given by

$$C_p^\pm = -2\varphi_x(x, 0^\pm) \tag{20.44}$$

where C_p^+ and C_p^- correspond to the upper and lower surfaces, respectively.

Equation (20.39) is hyperbolic when

$$\varphi_x > \frac{1 - M_\infty^2}{(\gamma + 1)M_\infty^2} \tag{20.45}$$

and elliptic when

$$\varphi_x < \frac{1 - M_\infty^2}{(\gamma + 1)M_\infty^2} \tag{20.46}$$

In other words, the sign of the coefficient for the first term determines the equation type. If this coefficient is positive, the local flow is subsonic, and if negative, the local flow is supersonic. The nonlinearity of the first term is essential for describing the mixed character of transonic flow and is the mechanism by which shock waves are formed.

The full potential equation written for steady flow in conservative form for two-dimensional Cartesian coordinates (x, y) is given by

$$\frac{\partial}{\partial x} (\rho \phi_x) + \frac{\partial}{\partial y} (\rho \phi_y) = 0 \tag{20.47a}$$

where the density ρ is defined by

$$\rho = \left[1 - \frac{\gamma - 1}{\gamma + 1} (\phi_x^2 + \phi_y^2) \right]^{1/\gamma - 1} \tag{20.47b}$$

In Eq. (20.47b), ϕ is the full or exact velocity potential given by

$$\nabla \phi = \phi_x \mathbf{i} + \phi_y \mathbf{j} = u\mathbf{i} + v\mathbf{j} \tag{20.48}$$

where ϕ_x and ϕ_y are the velocity components in the x and y directions, respectively. The density and velocity components appearing in Eq. (20.47) are nondimensionalized by the stagnation density ρ_s and the critical speed of sound, a_*. Thus, at a stagnation point

$$\phi_x = 0, \qquad \phi_y = 0, \qquad \rho = 1 \tag{20.49}$$

and at a sonic point (for $\gamma = 1.4$)

$$\phi_x^2 + \phi_y^2 = 1, \qquad \rho = \left[1 - \frac{\gamma - 1}{\gamma + 1} \right]^{1/\gamma - 1} = 0.633938145 \ldots \tag{20.50}$$

The latter condition is quite useful in providing a simple test for supersonic flow. If $\rho < 0.6339 \ldots$, the flow is supersonic; if $\rho > 0.6339 \ldots$, the flow is subsonic.

The density expression shown in Eq. (20.47b) has been derived from three auxillary relations that are given as follows:

Bernoulli equation (energy equation)

$$\frac{q^2}{2} + \frac{a^2}{\gamma - 1} = \frac{1}{2} \frac{\gamma + 1}{\gamma - 1} \tag{20.51}$$

Isentropic equation of state (perfect gas)

$$\frac{p}{\rho^\gamma} = \frac{\gamma + 1}{2\gamma} \tag{20.52}$$

Speed of sound definition

$$a^2 = \frac{dp}{d\rho} = \frac{\gamma p}{\rho} \tag{20.53}$$

where the second equality in Eq. (20.53) is obtained using the isentropic equation of state. Equations (20.51) and (20.52) have been written using the same nondimensionalization as used for the full potential equation.

The full potential equation given by Eq. (20.47a) is valid for isentropic, irrotational flows about arbitrary shapes, that is the full potential equation is not restricted to thin shapes as is the TSD equation. However, to obtain physically realistic results, the full potential equation is restricted to flows ranging from incompressible ($M_\infty = 0$) to transonic ($M_\infty \sim 1$). These potential formulations are valid in transonic flow (even though the formulations are, strictly speaking, isentropic) because weak (or transonic) shock waves can be reasonably approximated by an isentropic formulation. A comparison of the isentropic shock jump relation with the exact Euler shock jump relations (Rankine–Hugoniot relations) is given in Fig. 20.9 [Steger and Bald-

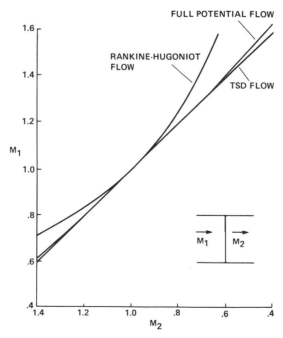

FIGURE 20.9 A comparison of the isentropic shock-jump relation and the Euler shock-jump relations. (Reprinted with permission of Steger, J. L. and Baldwin, B. S., "Shock Waves and Drag in the Numerical Calculation of Isentropic Transonic Flow," NASA TND-6997, 1972.)

win (1972)]. For a local Mach number (M_1) at or below 1.3 a reasonable approximation is obtained by the isentropic assumption.

Although the potential equation formulation generally requires the isentropic assumption (as mentioned above), several nonisentropic potential formulations have been proposed [Klopfer and Nixon (1983), Hafez and Lovell (1983)]. The basic idea behind this approach is to explicitly add entropy at shock waves so as to model the Euler shock jump relations. This effectively weakens the potential shock jump and extends the range of application for potential formulations. Good shock approximations can be obtained for local Mach numbers far in excess of 1.3 for the non-isentropic formulation.

20.4.2   Operator Notation

In general, the notation $u_{i,j}^n$ will be used to represent the nth iterate of the discrete dependent variable u at a position in the finite-difference mesh given by $x = i\Delta x$ and $y = j\Delta y$. Definitions for the various difference operators used in this section are given by

First central difference (second-order accurate)

$$\delta_x(\ )_{i,j} = \frac{1}{2\Delta x}[(\ )_{i+1,j} - (\ )_{i-1,j}] \qquad (20.54)$$

First backward difference (first-order accurate)

$$\delta_x^-(\ )_{i,j} = \frac{1}{\Delta x}[(\ )_{i,j} - (\ )_{i-1,j}] \qquad (20.55)$$

First forward difference (first-order accurate)

$$\delta_x^+(\ )_{i,j} = \frac{1}{\Delta x}[(\ )_{i+1,j} - (\ )_{i,j}] \qquad (20.56)$$

Second central difference (second-order accurate)

$$\delta_{xx}(\ )_{i,j} = \frac{1}{\Delta x^2}[(\ )_{i+1,j} - 2(\ )_{i,j} + (\ )_{i-1,j}] \qquad (20.57)$$

Forward shift operator

$$E_x^{+1}(\ )_{i,j} = (\ )_{i+1,j} \qquad (20.58)$$

Backward shift operator

$$E_x^{-1}(\ )_{i,j} = (\ )_{i-1,j} \qquad (20.59)$$

Central average operator

$$\mu_x(\ )_{i,j} = \tfrac{1}{2}[(\ )_{i+1,j} + (\ )_{i-1,j}] \qquad (20.60)$$

Forward average operator

$$\mu_x^+(\ )_{i,j} = \tfrac{1}{2}[(\ )_{i+1,j} + (\ )_{i,j}] \tag{20.61}$$

Backward average operator

$$\mu_x^-(\ )_{i,j} = \tfrac{1}{2}[(\ )_{i,j} + (\ )_{i-1,j}] \tag{20.62}$$

These definitions have been written using the x direction. Of course, operators can be defined for any space or time coordinate similar to those for the x direction. Certain identities associated with these operators exist and can be useful. For example, all difference and averaging operators can be expressed using shift operator notation

$$\delta_x^- = \frac{1}{\Delta x}(1 - E_x^{-1})$$

$$\mu_x^- = \frac{1}{2}(1 + E_x^{-1})$$

$$\delta_x^-\delta_x^- = \frac{1}{\Delta x}(1 - E_x^{-1})\frac{1}{\Delta x}(1 - E_x^{-1}) = E_x^{-1}\delta_{xx} \tag{20.63}$$

$$\delta_x^-\delta_x^- = \frac{1}{\Delta x^2}(1 - 2E_x^{-1} + E_x^{-2})$$

Notice that the symmetrical combination of first-order-accurate, one-sided difference operators creates second-order-accurate centered difference operators.

$$\delta_{xx} = \frac{1}{\Delta x^2}(E_x^{+1} - 2 + E_x^{-1}) = \frac{E_x^{+1} - 1}{\Delta x}\frac{1 - E_x^{-1}}{\Delta x}$$
$$= \delta_x^-\delta_x^+ = \delta_x^+\delta_x^- \tag{20.64}$$

This fact will be useful in the application of fully implicit approximate factorization schemes for solving the full potential and Euler equations.

20.4.3   von Neumann Stability Analysis

Any general iteration scheme can be investigated for numerical stability by using the Fourier or von Neumann test for numerical stability. This scheme was developed by von Neumann in the early 1940's and was first discussed in detail by O'Brien *et al.* (1951). Additional details can be found in Smith (1978), Ames (1977) or Mitchell (1969).

The propagation of numerical error is studied by substituting a suitable solution

$$u_{i,j}^n = e^{\lambda t}e^{i\alpha x}e^{i\beta y} \tag{20.65}$$

into the finite-difference scheme. In Eq. (20.65), α and β are integer wave numbers associated with the Fourier series which represents the solution at $t = 0$, λ is (in general) a complex constant, and the superscript i is $\sqrt{-1}$. The resulting expression is solved for

$$G = \frac{e^{\lambda(t + \Delta t)}}{e^{\lambda t}} = e^{\lambda \Delta t} \qquad (20.66)$$

This quantity, G, is called the amplification factor and provides an indication of error growth or decay. If $G > 1$ for any set of values of α and β, the scheme is numerically unstable. If $G \leq 1$ for all values of α and β, the scheme is stable.

Several important limitations regarding the von Neumann stability test exist and are now discussed. First, the test applies only to linear difference schemes with constant coefficients. If the difference scheme in question has variable coefficients, the method can still be applied locally with a good chance for predicting accurate numerical stability characteristics. Second, the effect of boundary conditions is neglected by the method (that is, the boundary conditions are assumed to be periodic). Another type of stability analysis, referred to as the matrix method, is preferable when the effect of boundary conditions needs to be analyzed [Smith (1978), Ames (1977), and Mitchell (1969)].

20.4.4 Transonic Flow Potential Formulations

The details of transonic flow field physics and the associated governing equations have been previously presented in Chap. 9, Transonic Flow; see also Sec. 20.4.2. In this section, details associated with the spatial differencing and iteration schemes are presented. Schemes for both the TSD and full potential formulations are discussed.

One of the most successful spatial differencing schemes for solving Eq. (20.39) is due to Murman (1974). This scheme uses difference operators which are dependent on the local flow type and can be expressed by considering

$$f_x + g_y = 0 \qquad (20.67)$$

where

$$f = (1 - M_\infty^2)\varphi_x - \tfrac{1}{2}M_\infty^2(\gamma + 1)\varphi_x^2, \; g = \varphi_y \qquad (20.68)$$

The f and g quantities represent mass fluxes (or more precisely freestream mass flux perturbations) in the x and y directions, respectively. Because all variables are inside the differentiation, Eq. (20.67) is said to be written in *strong conservation-law form*. The sign of f determines the flow type. A differencing scheme valid for both subsonic and supersonic regions of flow is given by

$$\frac{1}{\Delta x}(\overline{f}_{i+1/2,j} - \overline{f}_{i-1/2,j}) + \frac{1}{\Delta y}(g_{i,j+1/2} - g_{i,j-1/2}) = 0 \qquad (20.69)$$

where the modified flux $\overline{f}$ is defined by

$$\overline{f}_{i+1/2,j} = \mu_i f_{i+1/2,j} + (1 - \mu_i) f_{i-1/2,j} \tag{20.70}$$

and the *switching function*, μ_i, is given by

$$\mu_i = \begin{cases} 1, & M_{i,j} \leq 1 \\ 0, & M_{i,j} > 1 \end{cases} \tag{20.71}$$

In the above expression, $M_{i,j}$ is the local Mach number computed at point (i, j). Contained within this differencing scheme are four differencing operators: 1) subsonic operator, $\mu_i = 1$, $\mu_{i-1} = 1$, 2) supersonic operator, $\mu_i = 0$, $\mu_{i-1} = 0$, 3) sonic-point operator, $\mu_i = 0$, $\mu_{i-1} = 1$, and 4) shock-point operator, $\mu_i = 1$, $\mu_{i-1} = 0$.

Variations of this scheme have been extended to three dimensions and used to compute the transonic flow about wings, wing/fuselage combinations, and other three-dimensional configurations [Van der Vooren et al. (1976), and Bailey and Ballhaus (1975)]. Example calculations will be presented in the subsequent section that discusses transonic relaxation schemes.

Spatial differencing schemes for the full potential equation are now presented. The finite-volume, spatial difference scheme of Caughey and Jameson (1979) was first used to solve the conservative full potential equation in 1979. Since then, many applications of this scheme have been made in both two and three dimensions [Holst et al. (1982)].

The artificial density, spatial differencing scheme for the full potential equation has been independently presented in several different forms [Eberle, (1978) Holst and Ballhaus, (1978) and Hafez et al. (1979)]. These forms, while not identical, have certain similarities to the earlier work of Jameson (1975). Jameson's work is characterized by a scheme with an explicitly added artificial viscosity term. This term biases the spatial difference scheme in the upwind direction for supersonic regions of flow but does not affect the centrally-differenced scheme in subsonic regions. The three schemes of Eberle (1978), Holst and Ballhaus (1979), and Hafez et al. (1979) use this approach with one basic simplification. Namely, the upwind bias is accomplished by an upwind evaluation of the density coefficient. All three procedures compute this upwind or artificial density quantity in different ways.

In the procedure of Holst and Ballhaus (1979) the finite-difference approximation for the full potential equation written in Cartesian coordinates [see Eq. (20.47a)] is given by

$$L\phi_{i,j} = (\delta_x^- \tilde{\rho}_i \delta_x^+ + \delta_y^- \rho_{i,j+1/2} \delta_y^+)\phi_{i,j} \tag{20.72}$$

where the density coefficient $\tilde{\rho}_i$ is defined by

$$\tilde{\rho}_i = [(1 - \nu)\rho]_{i+1/2,j} + \nu_{i+1/2,j}\rho_{i-1/2,j} \tag{20.73}$$

The switching or transition function, ν, depends on the local Mach number, $M_{i,j}$, and the flow direction and is defined by

$$\nu_{i+1/2,j} = \max\left[(M_{i,j}^2 - 1)C, 0\right] \tag{20.74}$$

The quantity C is a user-specified constant usually set to a value between one and two.

The density calculation is performed in a straightforward manner by using a discretized version of Eq. (20.47b). Values of the density are computed and stored at cell centers (i.e., at $i + 1/2, j + 1/2$). Values of ϕ_x and ϕ_y required for computing the density are given by

$$\phi_x\big|_{i+1/2,j+1/2} \simeq \mu_y^+ \delta_x^+ \phi_{i,j}, \quad \phi_y\big|_{i+1/2,j+1/2} \simeq \mu_x^+ \delta_y^+ \phi_{i,j} \tag{20.75}$$

Values of the density required at $i + 1/2, j$ or $i, j + 1/2$ are computed using simple averages.

With the spatial differencing scheme just outlined, an upwind influence in supersonic regions is achieved without the explicit addition of an artificial viscosity term. Instead, the stabilizing influence is produced by the upwind evaluation of the density in an otherwise centrally-differenced scheme. This represents a simple technique for stabilizing the supersonic region associated with transonic flow fields and contributes to the stability and reliability of the present algorithm for many difficult test cases.

The next subject of discussion involves the iterative process by which the initial solution is evolved into the final solution. The iteration scheme is primarily responsible for the amount of computational work associated with each algorithm through the number of iterations required for convergence. In this section, three iteration schemes will be discussed. Each scheme will be presented in a standard *two-level correction form* (sometimes called *delta form*) given by

$$NC_{i,j}^n + \omega L\phi_{i,j}^n = 0 \tag{20.76}$$

where $C_{i,j}^n$ is the nth iteration correction defined by

$$C_{i,j}^n = \phi_{i,j}^{n+1} - \phi_{i,j}^n \tag{20.77}$$

The quantity $L\phi_{i,j}^n$ is the nth iteration residual, which is a measure of how well the finite-difference equation is satisfied by the nth level solution, $\phi_{i,j}^n$, and ω is a relaxation parameter. The general iteration scheme given by Eq. (20.76) can be considered as an iteration in time, i.e., pseudotime; the scheme does not, in general, apply to a physical time-dependent differential equation. This consideration allows the n superscript to be regarded as a time index, i.e., $C \sim \Delta t \phi_t$. The operator N determines the type of iteration scheme, therefore it is the only quantity from Eq. (20.76) to change with the iteration scheme.

A classical *relaxation* scheme, the *successive-line overrelaxation* (SLOR) iteration scheme, is still used today for some applications in both two and three dimensions for solving the TSD and full potential equations. The SLOR algorithm can be stated in simple terms for solving the full potential equation as follows

$$\left(-\frac{\tilde{\rho}_i^n}{\Delta x^2} - \frac{\tilde{\rho}_{i-1}^n}{\Delta x^2}\delta_x^- + \delta_y^- \rho_{i,j+1/2}\delta_y^+\right) C_{i,j}^n + \omega L\phi_{i,j}^n = 0 \tag{20.78}$$

where the L operator is defined by Eq. (20.72), and ω is a relaxation parameter. For $\omega < 1$ the scheme is said to be underrelaxed, and for $\omega > 1$ the scheme is overrelaxed. A standard von Neumann stability analysis reveals that $0 \leq \omega \leq 2$ is required for numerical stability. Values of ω approaching two in the subsonic region generally produce best convergence especially as the mesh if refined. Values of $\omega > 1$ in the supersonic region of flow generally produce poorer convergence. Therefore, setting $\omega = 1$ in supersonic regions is a general characteristic of most transonic relaxation schemes [Jameson (1974)].

The SLOR algorithm as defined in Eq. (20.78) contains the entire y operator. Thus, a tridiagonal matrix equation along the y direction must be inverted at each x location in the finite-difference mesh. As a result, this type of SLOR algorithm is sometimes called *vertical successive-line overrelaxation* (VSLOR). Because the N operator contains an upwind contribution due to the δ_x^- term, the order of inversion of these tridiagonal matrix equations must proceed from the inflow boundary downstream to the outflow boundary. This N-operator construction produces a time-like damping term (ϕ_{xt}), which provides a stabilizing effect to the relaxation process in supersonic regions of flow [Jameson (1974)].

Another relaxation scheme, which is considerably faster in convergence than SLOR, is the *Alternating Direction Implicit* or ADI scheme. An ADI scheme suitable for solving the conservative full potential equation for transonic flow can be stated by writing the standard N operator as follows

$$NC_{i,j}^n = -\frac{1}{\alpha}(\alpha - \delta_x^- \tilde{\rho}_i^n \delta_x^+)(\alpha - \delta_y^- \rho_{i,j+1/2}^n \delta_y^+)C_{i,j}^n \qquad (20.79)$$

where α is a convergence acceleration parameter cycled over a sequence of values. The ADI scheme of Eq. (20.79) is implemented in a two step format given by

Step 1:

$$(\alpha - \delta_x^- \tilde{\rho}_i^n \delta_x^+)f_{i,j}^n = \alpha \omega L \phi_{i,j}^n \qquad (20.80a)$$

Step 2:

$$(\alpha - \delta_y^- \rho_{i,j+1/2}^n \delta_y^+)C_{i,j}^n = f_{i,j}^n \qquad (20.80b)$$

where $f_{i,j}^n$ is an intermediate result stored over the entire finite-difference mesh. The residual, $L\phi_{i,j}^n$, is defined by Eq. (20.72). Step 1 consists of inverting a set of tridiagonal matrix equations along the x-direction, and step 2 consists of inverting a set of tridiagonal matrix equations along the y-direction. The construction of this ADI scheme does not automatically provide the necessary ϕ_{xt} term to stabilize supersonic regions. However, this type of term can be included by adding

$$\beta \delta_x^- \qquad (20.81)$$

inside the parentheses of the first step. The β coefficient is a constant specified by the user according to need.

Another very effective scheme in the approximate factorization family is the AF2 (*approximate factorization scheme* 2) algorithm. The AF2 scheme was first presented in Ballhaus and Steger (1975) for solving the low-frequency (unsteady) TSD equation. This algorithm was subsequently applied to the steady TSD equation [Ballhaus *et al.* (1978)] and the conservative full potential equation [Holst and Ballhaus (1978) and Holst (1980)]. The AF2 fully implicit scheme can be expressed by choosing the standard N operator as follows

$$NC_{i,j}^n = -\frac{1}{\alpha}(\alpha - \delta_x^+ \tilde{\rho}_{i-1}^n)(\alpha\delta_x^- - \delta_y^- \rho_{i,j+1/2}^n \delta_y^+)C_{i,j}^n \qquad (20.82)$$

The AF2 scheme is implemented in a two-step format given by

Step 1:

$$(\alpha - \delta_x^+ \tilde{\rho}_{-1}^n)f_{i,j}^n = \alpha\omega L\phi_{i,j}^n \qquad (20.83a)$$

Step 2:

$$(\alpha\delta_x^- - \delta_y^- \rho_{i,j+1/2}^n \delta_y^+)C_{i,j}^n = f_{i,j}^n \qquad (20.83b)$$

where α is a convergence acceleration parameter cycled over a sequence of values [Holst and Ballhaus (1978) and Ballhaus *et al.* (1978)], and $f_{i,j}^n$ is an intermediate result stored over the entire finite-difference mesh. Step 1 consists of inverting a set of bidiagonal matrix equations along the x-direction, and step 2 consists of inverting a set of tridiagonal matrix equations along the y-direction. With the AF2 factorization, the x-difference approximation is split between the two steps. This generates a ϕ_{xt} term, which is useful to the iteration scheme as time-like dissipation. The split x term also places a sweep direction restriction on both steps, namely in the negative x-direction for the first step and in the positive x-direction for the second step. No sweep direction restrictions are placed on either of the two sweeps due to flow direction.

Numerical results comparing the convergence characteristics of the three iteration schemes just outlined (SLOR, ADI, and AF2) are now presented. All three iteration schemes have been applied to the same artificial-density, spatial-differencing scheme for the conservative form of the full potential equation. A two-dimensional, 10% thick circular-arc airfoil with small-disturbance boundary conditions is used as a test case. The finite-difference grid is Cartesian with variable spacing in both the x- and y-directions. Both subcritical and supercritical cases are considered ($M_\infty = 0.7$ and 0.84, respectively). Pressure coefficient distributions for these two cases are displayed in Fig. 20.10. Note the perfect symmetry associated with the subcritical case and the existence of a moderate strength shock at about 80% of chord for the supercritical case. For more details about these calculations see Holst and Ballhaus (1978).

Convergence characteristics for the subcritical case are displayed in Fig. 20.11. All of the convergence parameters for each scheme have been selected by a trial-and-error optimization process. Based on a six order of magnitude reduction in the maximum residual, the ADI scheme is about twice as fast as the AF2 scheme, and about 16 times faster than SLOR. These speed ratios are in terms of iteration count.

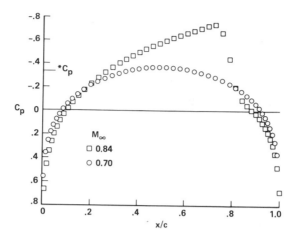

FIGURE 20.10 Subcritical and supercritical solutions used for the convergence history comparisons of Figs. 20.11–13, 10% circular-arc airfoil, nonlifting. (Reprinted with permission of Holst, T. L. and Ballhaus, W. F., Jr., "Fast Conservative Schemes for the Full Potential Equation Applied to Transonic Flow," NASA TM-78469, 1978.)

The ADI and AF2 schemes require about 50% and 30% more CPU time per iteration, respectively than SLOR. This should be considered when speed ratios based on the total amount of computational work are desired.

Convergence characteristics for the supercritical case are displayed in Fig. 20.12. Again, the convergence parameters have been optimized by a trial-and-error process. Based on a six order of magnitude reduction in the maximum residual and in terms of iteration count, AF2 is slightly more than twice as fast as ADI, and about 11

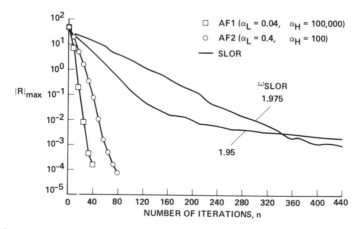

FIGURE 20.11 Maximum residual convergence history comparison, subcritical case, M_∞ = 0.7. (Reprinted with permission of Holst, T. L. and Ballhaus, W. F., Jr., "Fast Conservative Schemes for the Full Potential Equation Applied to Transonic Flow," NASA TM-78469, 1978.)

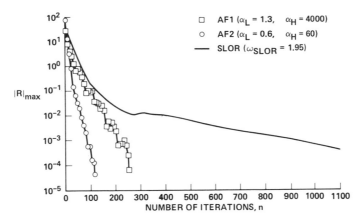

FIGURE 20.12   Maximum residual convergence history comparison, supercritical case, M_∞ = 0.84. (Reprinted with permission of Holst, T. L. and Ballhaus, W. F., Jr., "Fast Conservative Schemes for the Full Potential Equation Applied to Transonic Flow," NASA TM-78469, 1978.)

times faster than SLOR. The number of supersonic points (NSP) plotted versus iteration number for the supercritical case is shown in Fig. 20.13. The AF2, ADI, and SLOR schemes reach the final value of NSP in 29, 103, and 320 iterations, respectively.

The AF2 iteration scheme was relatively consistent in terms of convergence speed for both cases. The ADI iteration scheme, on the other hand, displayed remarkable speed for the subcritical case but was a disappointment for the supersonic case. This is because the ϕ_{xt} type error term produced by the AF2 factorization is more suitable for supersonic regions than the ϕ_t-type error term resulting from the ADI factorization. In fact, the ϕ_t-type error term has been shown to be destabilizing in the supersonic region [Jameson (1975)].

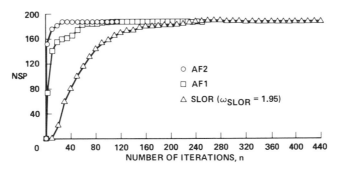

FIGURE 20.13   Development of the supersonic region, supercritical case, M_∞ = 0.84. (Reprinted with permission of Holst, T. L. and Ballhaus, W. F., Jr., "Fast Conservative Schemes for the Full Potential Equation Applied to Transonic Flow," NASA TM-78469, 1978.)

20.5 IMPLICIT METHOD FOR THE EULER EQUATIONS
Thomas H. Pulliam

20.5.1 The Euler Equations

The starting point is the strong conservation law form of the two-dimensional Euler equations in generalized coordinates. Details of the transformations from the Cartesian form can be found in Steger (1978) and Pulliam (1983). The strong conservation law form is chosen in order to admit shock capturing (see Sec. 20.3). The equations in nondimensional form are

$$\partial_\tau \hat{Q} + \partial_\xi \hat{E} + \partial_\eta \hat{F} = 0 \tag{20.84}$$

$$
\hat{Q} = J^{-1} \begin{bmatrix} \rho \\ \rho u \\ \rho v \\ e \end{bmatrix}, \quad
\hat{E} = J^{-1} \begin{bmatrix} \rho U \\ \rho u U + \xi_x p \\ \rho v U + \xi_y p \\ U(e + p) - \xi_t p \end{bmatrix},
$$

$$
\hat{F} = J^{-1} \begin{bmatrix} \rho V \\ \rho u V + \eta_x p \\ \rho v V + \eta_y p \\ V(e + p) - \eta_t p \end{bmatrix} \tag{20.85}
$$

where $U = \xi_t + \xi_x u + \xi_y v$ and $V = \eta_t + \eta_x u + \eta_y v$ are the contravariant velocities. Pressure is related to the conservative flow variables, $\hat{Q}$, by the equation of state

$$p = (\gamma - 1)(e - \tfrac{1}{2}\rho(u^2 + v^2)) \tag{20.86}$$

where γ is the ratio of specific heats, generally taken as 1.4 for air.

The choice of nondimensional parameters is arbitrary. Here, the variables ρ (density), u and v (the Cartesian velocities), and e (the total energy) are scaled as

$$\tilde{\rho} = \frac{\rho}{\rho_\infty}, \quad \tilde{u} = \frac{u}{a_\infty}, \quad \tilde{v} = \frac{v}{a_\infty}, \quad \tilde{e} = \frac{e}{\rho_\infty a_\infty^2} \tag{20.87}$$

where ∞ refers to free stream quantities. The speed of sound is a which for ideal gases, $a^2 = \gamma p/\rho$.

Assuming a reference length, l (usually taken as some characteristic physical length such as chord of an airfoil), time t scales as $\tilde{t} = ta/l$. For the remainder of this development the $\sim$ and $\hat{}$ signs will be dropped for simplicity.

The transformation variables (typically called *metrics*) are

$$\xi_x = Jy_\eta, \qquad \xi_y = -Jx_\eta, \qquad \xi_t = -x_\tau\xi_x - y_\tau\xi_y$$

$$\eta_x = -Jy_\xi, \qquad \eta_y = Jx_\xi, \qquad \eta_t = -x_\tau\eta_x - y_\tau\eta_y \tag{20.88}$$

where J is defined to be the metric Jacobian and $J^{-1} = (x_\xi y_\eta - x_\eta y_\xi)$.

20.5.2 Implicit Finite Difference Algorithm

The algorithm to be presented is an implicit finite difference scheme (see Sec. 19.2) with approximate factorization which can be either first- or second-order-accurate in time. Local time linearizations are applied to the nonlinear terms, and an approximate factorization of the two-dimensional implicit operator is used to produce locally one-dimensional operators. This results in block tridiagonal matrices which are easy to solve. The spatial derivative terms are approximated with second-order central differences. A spatially variable time step is used to accelerate convergence for steady-state calculations. A diagonal form of the algorithm is also discussed, which produces a computationally efficient modification of the standard algorithm where the diagonalization produces scalar tridiagonal operators in place of the block operators.

Implicit Time Differencing. Consider Eq. (20.84) and apply an implicit three point time differencing scheme of the form [see Warming and Beam (1978)]

$$\Delta Q^n = \frac{\vartheta\Delta t}{1 + \varphi} \frac{\partial}{\partial t} (\Delta Q^n) + \frac{\Delta t}{1 + \varphi} \frac{\partial}{\partial t} Q^n + \frac{\varphi}{1 + \varphi} \Delta Q^{n-1}$$

$$+ O\left[\left(\vartheta - \frac{1}{2} - \varphi\right)\Delta t^2 + \Delta t^3\right] \tag{20.89}$$

where $\Delta Q^n = Q^{n+1} - Q^n$ and $Q^n = Q(n\Delta t)$. The parameters ϑ and φ can be chosen to produce different schemes of either first- or second-order accuracy in time.

For $\vartheta = 1$ and $\varphi = 0$, Eq. (20.89) yields the first order Euler implicit scheme, and for $\vartheta = 1$ and $\varphi = 1/2$, the three point implicit scheme.

Restrict the example to the first-order scheme [although all of the subsequent development can easily be extended to any second-order scheme formed from Eq. (20.89)]. Applying Eq. (20.89) to Eq. (20.84), results in

$$Q^{n+1} - Q^n + h(E_\xi^{n+1} + F_\eta^{n+1}) = 0 \tag{20.90}$$

with $h = \Delta t$.

Local Time Linearizations. We wish to solve Eq. (20.90) for Q^{n+1} given Q^n. The flux vectors E and F are nonlinear functions of Q, therefore Eq. (20.90) is nonlinear in Q^{n+1}. The nonlinear terms are linearized in time about Q^n by a Taylor series such that

$$E^{n+1} = E^n + A^n\Delta Q^n + O(h^2)$$

$$F^{n+1} = F^n + B^n\Delta Q^n + O(h^2) \tag{20.91}$$

where $A = \partial E/\partial Q$ and $B = \partial F/\partial Q$ are the flux Jacobians (see the Appendix to this section), and ΔQ^n is $O(h)$.

Note that the linearizations are second-order-accurate, and, if a second-order time scheme had been chosen, the linearizations would not degrade the time accuracy.

Applying Eq. (20.91) to Eq. (20.90) and combining the ΔQ^n terms produces the *delta form* of the algorithm

$$[I + h\partial_\xi A^n + h\partial_\eta B^n]\Delta Q^n = -h(\partial_\xi E^n + \partial_\eta F_n) \qquad (20.92)$$

This is the unfactored form of the block algorithm. Call the right-hand side of Eq. (20.92) the *explicit* part and the left-hand side the *implicit* part of the algorithm.

Space Differencing. The next step is to take the continuous differential operators ∂_ξ and ∂_η and approximate them with finite difference operators on a discrete mesh. Introducing a grid of mesh points (j, k), variables are defined at mesh points as

$$u_{j,k} = u(j\Delta\xi, k\Delta\eta) \qquad (20.93)$$

The grid spacing in the computational domain is chosen to be unity so that $\Delta\xi = 1$ and $\Delta\eta = 1$. Second-order central difference operators can be used, for example,

$$\delta_\xi u_{j,k} = (u_{j+1,k} - u_{j-1,k})/2 \text{ and } \delta_\eta u_{j,k} = (u_{j,k+1} - u_{j,k-1})/2 \qquad (20.94)$$

The choice of the type and order of the spatial differencing is important both in terms of accuracy and stability. In most applications, second-order accuracy has proven to be sufficient provided the grid resolution is reasonable.

Approximate Factorization. The integration of the full, two-dimensional operator is too expensive. One way to simplify the solution process is to introduce an approximate factorization of the two-dimensional operator into two one-dimensional operators. The implicit side of Eq. (20.92) can be written as

$$[I + h\delta_\xi A^n + h\delta_\eta B^n]\Delta Q^n = [I + h\delta_\xi A^n] [I + h\delta_\eta B^n]\Delta Q^n - h^2\delta_\xi A^n\delta_\eta B^n\Delta Q^n$$

$$(20.95)$$

The cross term is second-order-accurate since ΔQ^n is $O(h)$. It can therefore be neglected without degrading the time accuracy of any second-order scheme which may be chosen.

The resulting factored form of the algorithm is

$$[I + h\delta_\xi A^n] [I + h\delta_\eta B^n]\Delta Q^n = -h[\delta_\xi E^n + \delta_\eta F^n] \qquad (20.96)$$

Two implicit operators have been identified, each of which is block tridiagonal. The solution algorithm now consists of two, one-dimensional sweeps—one in the ξ and one in the η direction. Each step requires the solution of a linear system involving a block tridiagonal which is solved by block LUD (lower-upper decomposition). The resulting solution process is much more economical than the unfactored algorithm in terms of computer storage and CPU time.

Implicit and Explicit Nonlinear Artificial Dissipation. Implicit and explicit artificial dissipation terms are added to the implicit algorithm to enhance the nonlinear stability, especially for flows with shock waves. Details for the form and choices of coefficients can be found in numerous articles, e.g., Pulliam (1983).

Explicit fourth-order smoothing

$$-\epsilon_e \Delta t J^{-1} \left[(\nabla_\xi \Delta_\xi)^2 + (\nabla_\eta \Delta_\eta)^2 \right] JQ^n \tag{20.97a}$$

is added to the right-hand side of Eq. (20.96), and implicit second-order smoothing

$$-\epsilon_i \Delta t J^{-1} \nabla_\xi \Delta_\xi J, \quad -\epsilon_i \Delta t J^{-1} \nabla_\eta \Delta_\eta J \tag{20.97b}$$

is inserted into the respective implicit block operators. The difference operators are defined as

$$\nabla_\xi q_{j,k} = q_{j,k} - q_{j-1,k}, \quad \Delta_\xi q_{j,k} = q_{j+1,k} - q_{j,k}$$

$$\nabla_\eta q_{j,k} = q_{j,k} - q_{j,k-1}, \quad \Delta_\eta q_{j,k} = q_{j,k+1} - q_{j,k} \tag{20.98}$$

The parameter ϵ_e is chosen to be $O(1)$, and $\epsilon_i = 2\epsilon_e$. The smoothing terms are scaled with Δt which makes the steady state independent of the time step.

It is important to realize that these added terms modify the original partial differential equation, and the coefficients used should be kept as small as possible while still maintaining stability. Typically, it is a matter of experience which decides the values for the coefficients.

Variable Time Step. Note that for the delta form of the algorithm (either factored or unfactored) the steady-state solution is independent of the time step, h. (There are numerical schemes where this is not the case, such as Lax–Wendroff.) Therefore, the time step path to the steady-state does not affect the final solution, and one can envision using time step sequences or spatially variable time steps to accelerate convergence. One particular form of a spatially variable time step which has been successfully applied to the implicit factored algorithm is to replace h in Eq. (20.92) with

$$\Delta t_0 \left(\frac{1}{1 + \sqrt{J}} \right) \tag{20.99}$$

with Δt_0 chosen to be $O(1)$. The use of the local time step can improve convergence time by a factor of two to three.

Diagonal Form of Implicit Algorithm. Another way to improve the efficiency of a numerical scheme is to modify or simplify the algorithm so that the computation work is decreased. Most of the computational work for the implicit algorithm is tied to the block tridiagonal solution process. One way to reduce that work would be to reduce the block size for the tridiagonals. This can be accomplished by reducing the equation set from four variables (density, x-momentum, y-momentum, and energy) to three variables (density and the two momenta) by assuming constant total enthalpy, $H = (e + p)/\rho = H_0$ or similar thermodynamic approximations.

The computational work can also be decreased by introducing a diagonalization of the blocks in the implicit operators as developed by Pulliam and Chaussee (1981). Here, the factored algorithm is taken in delta form, Eq. (20.96) and A and B are replaced with their eigensystem decompositions. See the Appendix of this section for the definitions of the eigenvector matrices and other terms.

$$[T_\xi T_\xi^{-1} + h\delta_\xi(T_\xi\Lambda_\xi T_\xi^{-1})] [T_\eta T_\eta^{-1} + h\delta_\eta(T_\eta\Lambda_\eta T_\eta^{-1})]\Delta Q^n$$

$$= \text{the explicit right-hand side of Eq. (20.96)} = R^n. \qquad (20.100)$$

At this point, Eq. (20.96) and (20.100) are exactly equivalent. A modified form of Eq. (20.100) can be obtained by factoring the T_ξ and T_η eigenvector matrices outside the spatial derivative terms, δ_ξ and δ_η. The eigenvector matrices are functions of ξ and η, therefore this modification reduces the time accuracy to at most first-order in time. The resulting equations are

$$T_\xi[I + h\delta_\xi\Lambda_\xi]N[I + h\delta_\eta\Lambda_\eta] T_\eta^{-1} \Delta Q^n = R^n \qquad (20.101)$$

where $N = T_\xi^{-1}T_\eta$; see the Appendix.

The explicit side of the diagonal algorithm (the steady-state finite difference equations) is exactly the same as in the original algorithm, Eq. (20.96). The modifications are restricted to the implicit side, so, if the diagonal algorithm converges, the steady-state solution will be identical to one obtained with the unmodified algorithm. In fact, linear stability analysis would show that the diagonal algorithm has exactly the same unconditional stability as the original algorithm. (This is because the linear stability analysis assumes constant coefficients and diagonalizes the blocks to scalars; the diagonal algorithm then reduces to the unmodified algorithm.) The modification (pulling the eigenvector matrices outside the spatial derivatives) of the implicit operator does affect the time accuracy of the algorithm. It reduces the scheme to at most first-order in time and also gives time accurate shock calculations a non-conservative feature, i.e., errors in shock speeds and shock jumps. But, remember that the steady-state is fully conservative, since the steady-state equations are unmodified. Also, computational experiments by Pulliam and Chaussee (1981) have shown that the convergence and stability limits of the diagonal algorithm are identical to that of the unmodified algorithm.

The diagonal algorithm reduces the block tridiagonal inversion to 4×4 matrix multiplies and scalar tridiagonal inversions. The operation count associated with the implicit side of the full block algorithm is 410 multiplies, 356 adds, and 10 divides, a total of 776 operations, while the diagonal algorithm requires 233 multiplies, 125 adds, and 26 divides or 384 operations. Adding in the explicit side and other overhead such as I/O (input/output) and initialization, the overall savings in computational work can be as high as 40%.

Appendix to Section 20.5

The flux Jacobians of the transformed fluxes E, F of Eq. (20.88) can be written in the general form

$$
A \text{ or } B = \begin{bmatrix}
\kappa_t & \kappa_x & \kappa_y & 0 \\
-u\theta + \kappa_x\phi^2 & \kappa_t + \theta - (\gamma - 2)\kappa_x u & \kappa_y u - (\gamma - 1)\kappa_x v & (\gamma - 1)\kappa_x \\
-v\theta + \kappa_y\phi^2 & \kappa_x v - (\gamma - 1)\kappa_y u & \kappa_t + \theta - (\gamma - 2)\kappa_y v & (\gamma - 1)\kappa_y \\
\theta[2\phi^2 - \gamma(e/\rho)] & \kappa_x[\gamma(e/\rho) - \phi^2] - (\gamma - 1)u\theta & \kappa_y[\gamma(e/\rho) - \phi^2] - (\gamma - 1)v\theta & \gamma\theta + \kappa_t
\end{bmatrix}
$$

(A.1)

with $\theta = \kappa_x u + \kappa_y v$, $\phi^2 = \frac{1}{2}(\gamma - 1)(u^2 - v^2)$, and $\kappa = \xi$ or η for A or B, respectively.

The flux Jacobian matrix has eigenvalues and a complete set of eigenvectors of the general form

$$
A = T_\xi \Lambda_\xi T_\xi^{-1} \text{ and } B = T_\eta \Lambda_\eta T_\eta^{-1} \tag{A.2}
$$

where

$$
\Lambda_\xi = \begin{bmatrix}
U & & & \\
& U & & \\
& & U + a\sqrt{\xi_x^2 + \xi_y^2} & \\
& & & U - a\sqrt{\xi_x^2 + \xi_y^2}
\end{bmatrix}, \tag{A.3a}
$$

$$
\Lambda_\eta = \begin{bmatrix}
V & & & \\
& V & & \\
& & V + a\sqrt{\eta_x^2 + \eta_y^2} & \\
& & & V - a\sqrt{\eta_x^2 + \eta_y^2}
\end{bmatrix}, \tag{A.3b}
$$

with

$$
T_k = \begin{bmatrix}
1 & 0 & \alpha & \alpha \\
u & \tilde{\kappa}_y\rho & \alpha(u + \tilde{\kappa}_x a) & \alpha(u - \tilde{\kappa}_x a) \\
v & -\tilde{\kappa}_x\rho & \alpha(v + \tilde{\kappa}_y a) & \alpha(v - \tilde{\kappa}_y a) \\
\dfrac{\phi^2}{(\gamma - 1)} & \rho(\tilde{\kappa}_y u - \tilde{\kappa}_x v) & \alpha\left(\dfrac{\phi^2 + a^2}{(\gamma - 1)} + a\tilde{\theta}\right) & \alpha\left(\dfrac{\phi^2 + a^2}{(\gamma - 1)} - a\tilde{\theta}\right)
\end{bmatrix}
$$

(A.4)

$$
T_\kappa^{-1} = \begin{bmatrix}
1 - \dfrac{\phi^2}{a^2} & (\gamma - 1)a^{-2}u & (\gamma - 1)a^{-2}v & -(\gamma - 1)a^{-2} \\
-\rho^{-1}(\tilde{\kappa}_y u - \tilde{\kappa}_x v) & \tilde{\kappa}_y\rho^{-1} & -\tilde{\kappa}_x\rho^{-1} & 0 \\
\beta(\phi^2 - a\tilde{\theta}) & \beta[\tilde{\kappa}_x a - (\gamma - 1)u] & \beta[\tilde{\kappa}_y a - (\gamma - 1)v] & \beta(\gamma - 1) \\
\beta(\phi^2 + a\tilde{\theta}) & -\beta[\tilde{\kappa}_x a + (\gamma - 1)u] & -\beta[\tilde{\kappa}_y a + (\gamma - 1)v] & \beta(\gamma - 1)
\end{bmatrix}
$$

(A.5)

and $\alpha = \rho/(\sqrt{2}a)$, $\beta = 1/(\sqrt{2}\rho a)$, $\tilde{\theta} = \tilde{\kappa}_x u + \tilde{\kappa}_y v$, and, for example, $\tilde{\kappa}_x = \kappa_x/\sqrt{\kappa_x^2 + \kappa_y^2}$.

Relations exist between T_ξ and T_η of the form

$$N = T_\xi^{-1} T_\eta, \qquad N^{-1} = T_\eta^{-1} T_\xi \tag{A.6}$$

where

$$N = \begin{bmatrix} 1 & 0 & 0 & 0 \\ 0 & m_1 & -\mu m_2 & \mu m_2 \\ 0 & \mu m_2 & \mu^2(1 + m_1) & \mu^2(1 - m_1) \\ 0 & -\mu m_2 & \mu^2(1 - m_1) & \mu^2(1 + m_1) \end{bmatrix} \tag{A.7a}$$

and

$$N^{-1} = \begin{bmatrix} 1 & 0 & 0 & 0 \\ 0 & m_1 & \mu m_2 & -\mu m_2 \\ 0 & -\mu m_2 & \mu^2(1 + m_1) & \mu^2(1 - m_1) \\ 0 & \mu m_2 & \mu^2(1 - m_1) & \mu^2(1 + m_1) \end{bmatrix} \tag{A.7b}$$

with $m_1 = (\tilde{\xi}_x \tilde{\eta}_x + \tilde{\xi}_y \tilde{\eta}_y)$, $m_2 = (\tilde{\xi}_x \tilde{\eta}_y - \tilde{\xi}_y \tilde{\eta}_x)$ and $\mu = 1/\sqrt{2}$.

REFERENCES

Acton, E., "The Modeling of Large Eddies in a Two-Dimensional Shear Layer," *J. Fluid Mech.*, Vol. 76, pp. 561–592, 1976.

Albone, C. M., Hall, M. G., and Joyce, G., "Numerical Solutions for Transonic Flows Past Wing-Body Configurations," *Symposium Transsonicum II*, Springer–Verlag, New York, 1976.

Ames, W. F., *Numerical Methods for Partial Differential Equations*, Academic Press, New York, 1977.

Ashurst, W. T., "Large Eddy Simulation via Vortex Dynamics," AIAA paper 83-1879-CP, 1983.

Ashurst, W. T., "Numerical Simulation of Turbulent Mixing Layers via Vortex Dynamics," *Turbulent Shear Flows I*, Springer, Berlin, pp. 402–413, 1979.

Bailey, F. R. and Ballhaus, W. F., Jr., "Comparisons of Computed and Experimental Pressures for Transonic Flows about Isolated Wings and Wing-Fuselage Configurations," NASA SP-347, Aerodynamic Analysis Requiring Advanced Computers, March 1975.

Ballhaus, W. F., Jr. and Steger, J. L., "Implicit Approximate Factorization Schemes for the Low-Frequency Transonic Equations," NASA TM X-73082, 1975.

Ballhaus, W. F., Jr., Jameson, A., and Albert, J., "Implicit Approximate Factorization Schemes for the Efficient Solution of Steady Transonic Flow Problems," *AIAA J.*, Vol. 16, pp. 573–579, 1978.

Beale, J. T. and Majda, A., "Vortex Method I: Convergence in Three Dimensions," *Math. Comput.*, Vol. 39, pp. 1–27, 1982.

Beale, J. T. and Majda, A., "Vortex Methods II: Higher Order Accuracy in Two and Three Dimensions," *Math Comput.*, Vol. 39, pp. 29–52, 1982.

Boppe, C. W. and Stern, M. A., "Simulated Transonic Flows for Aircraft with Nacelles, Pylons, and Winglets," AIAA Paper No. 80-130, 1980.

Caughey, D. A. and Jameson, A., "Numerical Calculation of Transonic Potential Flow About Wing-Body Combinations," *AIAA J.*, Vol. 17, pp. 175–181, 1979.

Cheer, A. Y., Rept. C.P.A.M., University of California, Berkeley, PAM135, 1983.

Chorin, A. J., "Estimates of Intermittency, Spectra and Blow-Up in Developed Turbulence," *Commun. Pure Appl. Math.*, Vol. 34, pp. 853–866, 1981.

Chorin, A. J., "Numerical Study of Slightly Viscous Flow," *J. Fluid Mech.*, Vol. 57, pp. 785–796, 1973.

Christiansen, J. P., "Numerical Simulation of Hydrodynamics by the Method of Point Vortices," *J. Comp. Phys.*, Vol. 13, pp. 363–379, 1973.

Clark, R. W., "A New Iterative Matrix Solution Procedure for Three-Dimensional Panel Method," AIAA Paper No. 85-0176, 1985.

Clements, R. R. and Maull, D. J., "The Representation of Sheets of Vorticity by Discrete Vortices," *Prog. Aerospace Sci.*, Vol. 16, pp. 129–214, 1975.

Colella, Ph. and Glaz, H. M., "Numerical Modelling of Inviscid Shocked Flows of Real Gases," *8th Int. Conf. on Num. Methods in Fluid Dyn.*, p. 175, 1982.

Courant, R. and Friedrichs, K. O., *Supersonic Flow and Shock Waves*, Springer–Verlag, New York, 1967.

de Neef, T. and Hechtman, C. "Numerical Study of the Flow Due to a Cylindrical Implosion," *Computers and Fluids*, Vol. 6, p. 185, 1978.

de Neef, T., "Treatment of Boundaries in Unsteady Inviscid Flow Computations," Delft University of Technology, Rep. LR-262, 1978.

Eberle, A., "A Finite Volume Method for Calculating Transonic Potential Flow Around Wings from the Pressure Minimum Integral," NASA TM-75324, 1978. [Translated from "Eine Methode der Finiten Volumen zur Berechnung der Transsonischen Potentialstromung um Flugel aus dem Druckmininumintegral," MBB-UFE1407(0), Feb. 1978.]

Ferri, A., *Elements of Aerodynamics of Supersonic Flows*, Macmillan, New York, 1949.

Gilbarg, D. and Paolucci, D., "The Structure of Shock Waves in Continuum Theory of Fluid," *J. Rat. Mech. Anal.*, Vol. 2, p. 617, 1953.

Hafez, M. M. and Lovell, D., "Entropy and Vorticity Corrections for Transonic Flows," AIAA Paper No. 83-1926, 1983.

Hafez, M. M., Murman, E. M., and South, J. C., Jr., "Artificial Compressibility Methods for Numerical Solution of Transonic Full Potential Equation," AIAA Paper No. 78-1148, 1978. (See also *AIAA J.*, Vol. 17, pp. 838–844, 1979.)

Hald, O. H. and Mauceri Del Prete, V., *Math. Comp.*, Vol. 32, p. 791, 1978.

Hald, O. H., "Convergence of Vortex Methods for Euler's Equations II," *SIAM J. Numer. Anal.*, Vol. 16, pp. 726–755, 1979.

Hama, F. R., "Progressive Deformation of a Curved Vortex Filament by its Own Induction," *Phys. Fluids*, Vol. 5, pp. 1156–1162, 1962.

Hess, J. L. and Smith, A. M. O., "Calculation of Potential Flow about Arbitrary Bodies," *Progress in Aeronautical Sciences*, Vol. 8, Pergamon Press, New York, 1966.

Hess, J. L., "Calculation of Potential Flow About Arbitrary Three-Dimensional Lifting Bodies," *Computer Methods in Applied Mechanics and Engineering*, Vol. 4, No. 3, pp. 283–319, 1974.

Hess, J. L., "Review of the Source Panel Technique for Flow Computation," *Innovative Numerical Methods in Engineering*, Shaw, R. P., Periaux, J., Chaudonet, A., Wu, J., Marine, C., and Brebbia, C. (Eds.), pp. 197–210, Springer–Verlag, Berlin/Heidelberg, 1986.

Hess, J. L., Friedman, D. M., and Clark, R. W., "Calculation of Compressible Flow About Three-Dimensional Inlets with Auxilliary Inlets, Slats, and Vanes by Means of a Panel Method," NASA CR-174975, 1985.

Holst, T. L. and Ballhaus, W. F., Jr., "Fast Conservative Schemes for the Full Potential Equation Applied to Transonic Flow," NASA TM-78469, 1978. (See also *AIAA J.*, Vol. 17, pp. 145–152, Feb. 1979.)

Holst, T. L., "A Fast, Conservative Algorithm for Solving the Transonic Full-Potential Equation," *AIAA J.*, Vol. 18, pp. 1431–1439, 1980.

Holst, T. L., Slooff, J. W., Yoshihara, H., and Ballhaus, W. F., Jr., "Applied Computational Transonic Aerodynamics," AGARDograph No. 266, Aug. 1982.

Horiuti, K., Kuwahara, K., and Oshima, Y., "Study of Two-Dimensional Flow Past an Elliptic Cylinder by Discrete-Vortex Approximation," *Proc. 7th ICNMFD*, Springer, Berlin, 1980.

Hugoniot, A., "A Memoir on the Propagation of Motion in Bodies and Particularly in Perfect Gases," *J. Ec. Polyt., Paris*, Vol. 58, p. 1, 1889.

Inoue, O., "A Numerical Simulation of Turbulent Mixing Layer by Discrete Vortex Method," AIAA paper 84-0434, 1984.

Izumi, K. and Kuwahara, K., "Unsteady Flow Field, Lift and Drag Measurements of Impulsively Started Elliptic Cylinder," AIAA paper 83-1711, 1983.

Jameson, A., "Iterative Solution of Transonic Flows Over Airfoils and Wings, Including Flows at Mach 1," *Commun. Pure Appl. Math.*, Vol. 27, pp. 283–309, 1974.

Jameson, A., "Transonic Potential Flow Calculations Using Conservative Form," *Proc. AIAA 2nd Computational Fluid Dynamics Conf. AIAA*, pp. 148–155, June 1975.

Johnson, F. T., "A General Panel Method for the Analysis and Design of Arbitrary Configurations in Incompressible Flows," NASA CR-3079, 1980.

Kiya, M. and Arie, M., "A Contribution to an Inviscid Vortex-Shedding Model for an Inclined Flat Plate in Uniform Flow," *J. Fluid Mech.*, Vol. 82, pp. 223–243, 1977.

Klopfer, G. H. and Nixon, D., "Non-Isentropic Potential Formulation for Transonic Flows," AIAA Paper No. 83-0375, 1983.

Kuwahara, K. and Takami, H., "Numerical Studies of Two-Dimensional Vortex Motion by a System of Point Vortices," *J. Phys. Soc. (Japan)*, Vol. 34, p. 247, 1973.

Kuwahara, K. and Takami, H., "Study of Turbulent Wake Behind a Bluff Body by Vortex Method," *Turbulence and Chaotic Phenomena in Fluids*, North-Holland, 1984.

Kuwahara, K., "Numerical Study of Flow Past an Inclined Flat Plate by an Inviscid Model," *J. Phys. Soc. (Japan)*, Vol. 35, pp. 1545–1551, 1973.

Kuwahara, K., "Study of Flow Past a Circular Cylinder by an Inviscid Model," *J. Phys. Soc. (Japan)*, Vol. 45, p. 292, 1978.

Lamb, H., *Hydrodynamics*, Cambridge University Press, London, 1932.

Lax, P. D., "Weak Solutions of Nonlinear Hyperbolic Equations and their Numerical Computations," *Comm. Pure Appl. Math.*, Vol. 7, p. 159, 1954.

Leonard, A., "Vortex Methods for Flow Simulation," *J. Comp. Phys.*, Vol. 37, pp. 289–335, 1980.

Lewis, R. I., "Surface Vorticity Modeling of Separated Flows," *J. Mech. Eng. Sci.*, Vol. 23, pp. 1–12, 1981.

MacCormack, R. W. and Paullay, A. J., "The Influence of the Computational Mesh on Accuracy for Initial Value Problems with Discontinuous or Non-Unique Solutions," *Computers and Fluids*, Vol. 2, p. 339, 1974.

Majda, A. and Osher, S., "Numerical Viscosity and the Entropy Condition," *Comm. Pure Appl. Math.*, Vol. 32, p. 797, 1979.

Marconi, F. and Salas, M. D., "Computation of Three-Dimensional Flows About Aircraft Configurations," *Computers and Fluids*, Vol. 1, p. 185, 1973.

Marconi, F., "Shock Reflection Transition in Three-Dimensional Steady Flow About Interfering Bodies," *AIAA J.*, Vol. 21, pp. 707, 1983.

Maskew, B., "Prediction of Subsonic Aerodynamic Characteristics: A Case For Low-Order Panel Methods," *J. Aircraft*, Vol. 19, pp. 157–163, 1982.

Mitchell, A. R., *Computational Methods in Partial Differential Equations*, John Wiley & Sons, New York, 1969.

Moretti, G., "Inviscid Blunt Body Shock Layers," Polyt. Inst. Bkln., PIBAL Rep. 68-15, 1968.

Moretti, G., "Thoughts and Afterthoughts About Shock Computations," Polyt. Inst. Bkln., PIBAL Rep. 72-37, 1972.

Morino, L. and Luo, C. C., "Subsonic Potential Aerodynamics for Complex Configurations. A General Theory," *AIAA J.*, Vol. 8, No. 2, Feb. 1974.

Murman, E. M., "Analysis of Embedded Shock Waves Calculated by Relaxation Methods," *AIAA J.*, Vol. 12, pp. 626–633, 1974.

Nakamura, Y., Leonard, A., and Spalart, P. R., "Numerical Simulation of Vortex Breakdown by the Vortex-Filament Method," *Proc. AGARD Symp., Aerodynamics of Vortical Type Flows in Three Dimensions*, Loughton, 1983.

Nakamura, Y., Leonard, A., and Spalart, P. R., "Vortex Simulation of an Inviscid Shear Layer," AIAA paper 82-0948, 1982.

O'Brien, C. G., Hyman, M. A., and Kaplan, S., "A Study of the Numerical Solution of Partial Differential Equations," *J. Math. Phys.*, Vol. 29, pp. 223–251, 1951.

Ogawa, A., *Vortex Flow*, CRC Press, Boca Raton, FL, 1992.

Ono, K., Kuwahara, K., and Oshima, K., "Numerical Analysis of Dynamic Stall Phenomena of an Oscillating Airfoil by the Discrete-Vortex Approximation," *Proc. 7th ICNMFD*, Springer, Berlin, 1980.

Oshima, Y. and Kuwahara, K., "Experimental and Numerical Study of Vortex Interaction," AIAA paper 84-1546, 1984.

Oshima, K. and Oshima, Y., *Proc. 8th ICNMFD*, Springer, Berlin, 1982.

Oshima, Y., Izutsu, N., Oshima, K., and Kuwahara, K., "Autorotation of an Elliptic Airfoil," AIAA paper 83-0130, 1983.

Pulliam, T. and Chaussee, D., "A Diagonal Form of an Implicit Approximate Factorization Algorithm," *J. Comp. Phys.*, Vol. 39, No. 2, pp. 347–363, 1981.

Pulliam, T., "Implicit Finite-Difference Methods for the Euler Equations," *Advances in Computational Transonics*, Habashi, W. G. (Ed.), Pineridge Press Ltd., 1983.

Rankine, W. J. M., "On the Thermodynamic Theory of Waves of Finite Longitudinal Disturbance," *Phil. Trans. Roy. Soc. (London)*, Vol. 160, p. 277, 1870.

Rosenhead, C., "The Formation of Vortices from a Surface of Discontinuity," *Proc. Roy. Soc. (London)*, Vol. A134, p. 170, 1931.

Saffman, P. G., *Vortex Dynamics*, Cambridge University Press, Cambridge, MA, 1992.

Salas, M. D. and Morgan, B., "Stability of Shock Waves Attached to Wedges and Cones," *AIAA J.*, Vol. 21, p. 1611, 1982.

Salas, M. D., "Shock Fitting Method for Complicated Two-Dimensional Supersonic Flows," *AIAA J.*, Vol. 14, p. 583, 1976.

Sarpkaya, T. and Schoaff, R. L., "Inviscid Model of Two-Dimensional Vortex Shedding by a Circular Cylinder," *AIAA J.*, Vol. 17, pp. 1193–1200, 1979.

Sarpkaya, T., "An Inviscid Model of Two-Dimensional Vortex Shedding for Transient and Asymptotically Steady Separated Flow Over an Inclined Plate," *J. Fluid Mech.*, Vol. 68, pp. 109–128, 1975.

Shirayama, S., Kuwahara, K., and Mendez, R. H., "A New Three-Dimensional Vortex Method," AIAA paper 85-1488, 1985.

Shirayama, S. and Kuwahara, K., "Computation of Flow Past a Parachute by a Three-Dimensional Vortex Method," AIAA paper 86-0350, 1986.

Sichel, M., "Two-Dimensional Shock Structure in Transonic and Hypersonic Flow," *Adv. Appl. Mech.*, Vol. 11, p. 131, 1971.

Smith, G. D., *Numerical Solution of Partial Differential Equations: Finite Difference Methods*, 2nd ed., Clarendon Press, Oxford, 1978.

Spalart, P. R. and Leonard, A., "Computation of Separated Flows by a Vortex Tracing Algorithm," AIAA paper 81-1246, 1981.

Steger, J. L. and Baldwin, B. S., "Shock Waves and Drag in the Numerical Calculation of Isentropic Transonic Flow," NASA TN D-6997, 1972.

Steger, J., "Implicit Finite-Difference Simulation of Flow about Arbitrary Two-Dimensional Geometries," *J. Comp. Phys.*, Vol. 16, pp. 679–686, 1978.

Van der Vooren, J., Slooff, J. W., Huizing, G. H., and van Essen, A., "Remarks on the Suitability of Various Transonic Small Perturbation Equations to Describe Three-Dimensional Transonic Flow-Examples of Computations Using a Fully Conservative Rotated Difference Scheme," *Symposium Transsonicum II*, Springer–Verlag, New York, 1976.

Warming, R. and Beam, R., "On the Construction and Application of Implicit Factored Schemes for Conservation Laws," *SIAM-AMS Proc.*, Vol. 11, pp. 85–129, 1978.

21 Computational Methods for Viscous Flow

UNMEEL B. MEHTA
NASA Ames Research Center
Moffett Field, CA

U. GHIA
K. N. GHIA
University of Cincinnati
Cincinnati, OH

SAAD RAGAB
Virginia Polytechnic Institute and State University
Blacksburg, VA

CONTENTS

Handbook of Fluid Dynamics and Fluid Machinery, Edited by Joseph A. Schetz and Allen E. Fuhs
ISBN 0-471-12598-9 Copyright © 1996 John Wiley & Sons, Inc.

21.1   PHYSICAL ASPECTS OF COMPUTING VISCOUS FLUID FLOW

Unmeel B. Mehta

21.1.1   Introduction

One of the main themes in fluid dynamics at present is computing of viscous fluid flow with the primary focus on the determination of drag, flow separation, vortex flows, and unsteady flows. A computation of the flow of a viscous fluid requires an understanding and consideration of the physical aspects of the flow. This is done by identifying the flow regimes, the scales of fluid motion, and the sources of vorticity. Discussions of flow regimes deal with conditions of incompressibility, transitional and turbulent flows, Navier–Stokes and non-Navier–Stokes regimes, shock waves, and strain fields. Discussions of the scales of fluid motion consider transitional and turbulent flows, thin- and slender-shear layers, triple- and four-deck regions, viscous–inviscid interactions, shock waves, strain rates, and temporal scales. Discussions of the sources of vorticity present the significance and generation of vorticity. These physical aspects mainly guide computations of the flow of a viscous fluid. The aforementioned discussions are based on a technical memorandum by Mehta (1984).

Brief Historical Background. In the eighteenth century, the Euler equations [Euler (1755)] governing incompressible inviscid fluid flows were developed. As a consequence of the assumption of inviscid flow, these equations led to the *paradox of d'Alembert* [d'Alembert (1752)], that is, contrary to the situation encountered in the flow of a real fluid, the drag of a solid body in a uniform flow was predicted as zero. The idea that real fluids exert some sort of friction was associated with the concept of viscosity for which Newton (1726) had presented a crude model. In the nineteenth century, the main theoretical motivation was to provide scientific reasons why drag existed in a fluid at all. Great efforts were made to explain how a vanishingly small frictional force in the fluid can have a significant effect on the flow properties. Navier (1826), Poisson (1831), St. Venant (1843), and Stokes (1847) provided rigorous theoretical explanations for viscosity. The paradox was that progress in aerodynamics depended on this explanation of viscosity, but its presence complicated the mathematics so as to make the prediction of drag impossible at speeds of practical interest. The mathematical difficulties of integrating the Navier–Stokes equations could be eliminated by neglecting the nonlinear terms. This approximation, valid only for slow motions, at least gave nonzero drag, but it yielded drag too small in magnitude for faster motions.

Helmhotz (1868), Kirchhoff (1869), and Rayleigh (1876) dealt with the problem of flow separation from a sharp edge, again to reconcile real flows and inviscid theory. The mathematical model permitted free-streamlines to exist, across which the fluid characteristics change discontinuously, thereby avoiding the paradox of d'Alembert. This free-streamline model led to the concept of a *vortex sheet*. This model was, however, of limited use because of the difficulties of predicting the shape of the vortex sheet and of determining the pressure on surfaces in the separated region.

Prandtl (1905) provided a conceptual breakthrough. His idea was that for large Reynolds numbers, the effect of viscosity is confined to a thin layer of fluid adjacent to a solid surface. In this layer, a rapid change from virtually inviscid flow to viscous flow takes place. In this transitional or boundary layer region, an approximate form of the Navier–Stokes equations may be used; outside this region, the Euler equations apply. The small thickness of the boundary layer simplifies the Navier–Stokes equations to the Prandtl boundary layer equations. Using these boundary layer equations on a flat plate, Blasius (1907) was able to deduce a law for the laminar skin friction that agreed with the experimental data. See Secs. 1.4 and Chap. 4.

The boundary layer concept, originally developed for laminar, incompressible flow along a solid boundary, has been extended to the corresponding cases of turbulent, incompressible flow, compressible flow, and also to boundary-free shear flow occurring in wakes and jets. This concept, having changed the type of the governing equation from elliptic to parabolic and reduced its order in the streamwise direction, sometimes gave rise to difficulties when attempts were made for improvement because a systematic approach was lacking. Stüper (1933) first accounted for the effect of displacement thickness to improve the boundary layer solution of flow past an airfoil. But a systematic approach, called the *method of matched asymptotic expansions*, was only developed much later [Van Dyke (1965)]. Improvements in boundary layer solutions were found only for flows without separation. The evolution of three-dimensional boundary layers under a prescribed external velocity field is still imperfectly understood [Stewartson (1981)].

A consequence of the boundary layer concept is that when the skin–friction coefficient vanishes, the boundary layer no longer has only a small effect on the inviscid flow outside. Prandtl introduced the phrase *flow separation* to describe the phenomenon of the inviscid flow leaving the neighborhood of the surface in an adverse pressure gradient. Goldstein (1948) showed that the solution of the boundary layer equations could not be continued downstream of separation. Whenever the external velocity (or pressure) is prescribed, a singularity arises in the boundary layer calculations. A two-layer model of the boundary layer was proposed by Lighthill (1953) and independently by Müller (1953) to overcome the impasse caused by this separation singularity. Further, it was shown by Lees and Reeves (1964) and by Catherall and Mangler (1966) that it is possible to construct external velocity distributions that do not lead to the singularity. These studies culminated into the formulation of the localized theory of the triple-deck developed independently by Neiland (1969), Stewartson (1969), and Messiter (1970). During the 1970's considerable progress was made in further development and application of this theory for two-dimensional flows, particularly for laminar flows. The development of a rational theory for three-dimensional, and possibly unsteady two-dimensional flows, is considered to be in its infancy. The limiting form of the structure of turbulent separation and of a large separated region has not been developed (see also Chap. 4).

In the past, fluid dynamicists were primarily concerned with aircraft, ships and airship-like shapes and were interested in streamline shapes that prevented or limited separation [Jones (1929)]. In the cruise condition, this is still the aim for aircraft of moderate-to-high aspect ratio. But at present, there are significant other applications in which the effects of controlled separation and vortex flows are incorporated in design of aircraft and ocean vehicles.

The importance of unsteady aerodynamics in helping with an understanding of aircraft stability was recognized during the First World War [Jones (1963)]. *Flutter*, an important problem caused by the interaction of aerodynamics and structural design, was identified. There was interest in how the *circulation* grew or was modified as an airfoil was started from rest, or changed its incidence quickly. Farren (1935) studied the dynamic stall of a pitching airfoil, a phenomenon of importance in the design of helicopter rotors.

Future Emphasis. Three-dimensional regions of separation generally exist on aerodynamic and underwater hydrodynamic configurations. Some examples from aerodynamics are the following. On an aircraft of moderate-to-high aspect ratio, at high lift and low speeds (for example, in the landing mode), there could be substantial regions of separation. Controlled separation from sharp edges is the hallmark of low-aspect-ratio wings. Controlled vortex flow can be used to clean up a detrimental separated flow, as for example, the vortex flow from a *strake* on a low-aspect-ratio aircraft such as F-16 or F/A-18. Further, the requirement for high rates of turning by military aircraft and missiles precludes maintenance of attached flows on smooth surfaces. Inviscid theory cannot determine the location at which a three-dimensional boundary layer detaches from the surface and forms a well-organized vortex. This requires a viscous theory.

There are many situations in which time-dependent aspects can be neglected. But, in a sense, it is steady flow that is peculiar. Fundamental understanding of the nature of transition from laminar to turbulent flow and of the detailed characteristics of turbulent flows requires study of unsteady flows. Practical areas in which unsteady phenomena predominate are increasing and will continue to increase significantly. Generally, separated flows and vortical flows are inherently time-dependent.

Unsteady aerodynamics is no longer a topic of limited applications. For example, the structural dynamics of a helicopter rotor interact with unsteady aerodynamic aspects that are manifested during the forward motion of the helicopter. The flow over each blade is greatly influenced by a very complex wake shed from the other blades. There are three-dimensional unsteady transonic effects on the advancing blade and, generally, dynamic stall phenomena on the retreating blade. This helicopter rotor problem is somewhat analogous to the aerodynamics of flight in a gusty atmosphere. Shock waves that terminate in the vicinity of boundary layers are seldom steady, particularly at transonic speeds. Flutter prediction and prevention and the stability of aircraft and missiles depend on the unsteady interactions between largely viscous-oriented aerodynamics and structural dynamics. The elimination of *buffet*, *stall*, and to some extent control-surface *buzz* depends on unsteady viscous effects. The evaluation of heat transfer to the stagnation region of a blunt body executing oscillations in pitch is an example of an unsteady boundary-layer problem. In turbomachines, unsteady effects are important in determining performance, blade flutter, stall surge, and system response to inlet distortions. The reduction in aerodynamic noise, in part, depends on the dissipation of acoustic energy by viscosity; this requires the study of wave motions in fluids (see also Chaps. 6, 12, and 27).

There are other areas of fluid dynamics which contain unsteady viscous effects. For example, industrial aerodynamics is concerned with the oscillatory lift and drag forces that accompany unsteady separation and vortex shedding from bluff bodies such as buildings, bridges, and automobiles. Problems in hydrodynamics are somewhat analogous to those in aerodynamics. In *hemodynamics*, the flow of blood downstream of constrictions and at branching junctions of the aorta and the large arteries also exhibits the phenomena of unsteady separation and vortex shedding.

The future aerodynamic design needs are no different from those in the past: airframe efficiency, control, and maneuverability. But, future emphasis is likely to be on determining effects of viscosity as it affects drag, flow separation, vortex flows, and unsteady flows. The effects of viscosity on nonboundary layer types of fluid flow were studied in the past largely indirectly. For example, a separation bubble is often implied from the measurements of surface-pressure distribution. Presently, viscosity effects are studied both directly, through computational fluid dynamics and advances in experimental techniques, and indirectly through experimental techniques. Computers will allow these effects to be studied more directly, primarily, via computational fluid dynamics, however experiments will always be needed to validate both computational and theoretical fluid dynamics and to provide answers when these disciplines cannot. Most of the progress in theoretical fluid dynamics has been already made. Further conceptual progress in theoretical fluid dynamics will be made, as in understanding the nature of turbulence. Like experiments, theoretical fluid dynamics will support and check computational fluid dynamics. However, one of the dominant themes in fluid dynamics for the future will be computational fluid dynamics.

Scope. The basic equations governing viscous fluid motion are essentially nonlinear (more precisely, quasi-linear) and coupled, and therefore, exact (analytical) solutions are rare. The only recourse available is to solve the equations for a physical problem by a numerical procedure, a process which is approximate and time-consuming. In this process, it is, therefore, necessary: 1) to determine the appropriate governing equations, including a model for turbulence, 2) to determine the computational form of these equations, the scaling laws, and the coordinate system, and 3) to choose the computational methods, space-time grid systems, and accuracy requirements. To achieve all this, it is imperative that the flow regimes and scales of fluid motion be identified. Primarily, the flow regimes are characterized by the relative magnitude of various terms in the equations of motion and flow parameters. The scales of motion are evaluated by experimental observations or by assumptions (or conjectures) when there is no experimental information. These are spatial scales and temporal scales that generally differ in different parts of the flow field. Along with the study of a kinematic property of flow of fluids, namely, vorticity, the flow regimes, and scales of motion also provide an understanding of flow structure.

The discussion of physical aspects starts with a brief identification of some of the flow regimes. This is followed by a discussion of scales of motion, both spatial and temporal. This section is concluded with a brief indication of the importance of vorticity and a description of how the vorticity is generated. It is intended that this section be a broad reference work on physical aspects of computing the flow of a viscous fluid. Because the subject is vast, an attempt is made to identify and briefly describe these aspects.

21.1.2 Flow Regimes

The classification scheme considered here for viscous-flow regimes is primarily a series of binary subclassifications displayed in Fig. 21.1. These are the following: 1) incompressible and compressible, 2) steady and unsteady, 3) laminar and turbulent, 4) attached and separated flows, 5) Navier–Stokes and non-Navier–Stokes, 6) weak and strong shock/boundary layer interactions, and 7) simple and complex strain fields. The first four classifications may be considered as based on physical aspects, and the rest may be considered as based on analytical aspects. There are also intermediate regimes such as transitional, subsonic, transonic, supersonic, and hypersonic flows. Most of these are standard and well known. Comments will be given on computationally significant physical aspects of most of these flow regimes.

Conditions of Incompressibility. The classical Navier–Stokes equations are applicable when the following assumptions are valid. First, when the velocity distribution is approximately *solenoidal*. In other words, when the changes in density of a material element resulting from pressure variations are negligible. (This is, strictly speaking, the condition of *incompressibility*.) The density variations are negligible in the absence of body forces when: 1) the Mach number, M, is less than about 0.3 everywhere in the flow field, and 2) when the flow is slowly varying, that is, when the product of the dominant frequency of unsteadiness and characteristic length scale is much smaller than the speed of sound. Second, when the viscosity is constant.

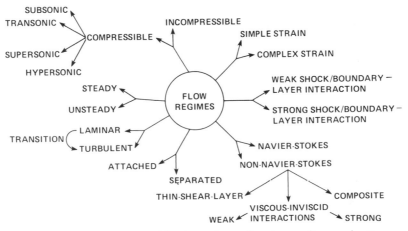

FIGURE 21.1 Classification scheme for viscous-flow regimes.

Third, when the thermodynamic processes of the fluid are irrelevant. (That is, when the heat added by dissipation and conduction is negligible, and when the enthalpy changes are negligible owing to pressure variations.) Therefore, when the temperature differences in the flow are not large, the heat-transfer problem can be treated separately or in parallel with the flow problem. In compressible flows, these two problems are intimately interconnected.

Transitional Flows. The phenomenon of transition from laminar to turbulent flow in vortical two- or three-dimensional, external-, internal-, or free-shear layers has so far eluded the completely rational understanding that is needed in predicting practical applications [Morkovin (1978), (1993) and Reshotko (1994)]. Reynolds (1883) and later Rayleigh (1892) hypothesized that transition to turbulent flow is a consequence of the instability of the laminar flow. When vorticity, entropy, and sound disturbances penetrate a laminar shear layer, these facets of the flow initiate in it forced and free response (Fig. 21.2). If these initiating disturbances are large, the disturbances grow by the forcing mechanism to (three-dimensional) nonlinear levels and lead to a turbulent shear layer, as explained below. This is the high-intensity *bypass process*. If the disturbances are small, these initiating perturbations excite and linearly amplify free disturbances [such as, *Tollmien–Schlichting (TS) waves* or *Görtler waves*] in the laminar shear layers. This *receptivity* of shear layers to small disturbances is influenced by *operation modifiers* such as pressure gradient, curvature, surface waviness, roughness and temperature, angle of attack, angle of yaw, leading-edge sweep, Mach number, suction, and blowing. The TS waves amplify and cease to follow the predictions of the linearized theory, and three-dimensionality of flow motion and vorticity stretching emerge. The initialization of three-dimensional, high frequencies associated with secondary instabilities and characterized by a sudden spike in the streamwise velocity signal is called the K-breakdown [Herbert and Morkovin (1979)] of laminar flow. Subsequently, turbulent spots are formed in external and internal shear layers, high straining is produced in rotating flows and pairing motion of rolled-up vortices is observed in free-shear layers. Growth of these

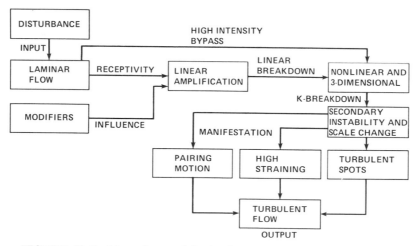

FIGURE 21.2 Many facets of the laminar-turbulent transition process.

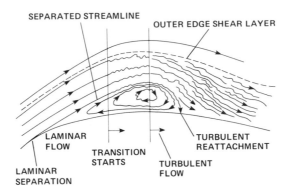

FIGURE 21.3   Transition in a free (separated) shear layer (transitional separation bubble).

flow features leads to turbulent shear layers. It is viscosity which limits the growth of disturbances. This process of transition from laminar to turbulent flow is not a well-defined problem because of the sensitivity to poorly defined initiating disturbances (see also Sec. 4.3). An example of a transitional flow is the leading-edge separation bubble, sometimes, observed on airfoils and illustrated in Fig. 21.3.

The only tool that can provide information, unobtainable from experiments and theory, to the nonlinear–laminar-to-turbulent transition process is the numerical solution of the compressible Navier–Stokes equations. The main requirements are that: 1) the discretization errors do not contaminate physical phenomena such as instabilities, that is, the finest scales in the transition process are adequately resolved in space and time, 2) the introduction of artificial boundaries owing to the limited size of the computational domain does not interfere with the physical upstream influence which is due to the ellipticity of the Navier–Stokes equations, and 3) the initiating disturbances are properly represented.

Turbulent Flows.   As in the case of the transition simulations, direct numerical simulation (DNS) of turbulence represents an attempt to resolve all scales of motion accurately both in time and space. This means that the only errors made are numerical ones, and there is no need for *turbulence modeling*. The principal problems are to encompass a broad enough range of scale and to compute a large enough sample so that the simulated event is meaningful. Computer resources limit such simulations to much-lower-than-normal Reynolds numbers and to unbounded, homogeneous, and inhomogeneous turbulent flows. See Sec. 21.5 for further discussion of DNS.

At normal (high) Reynolds numbers, viscosity plays a small part in the behavior of turbulence, except in the viscous sublayer and superlayer of a turbulent boundary layer (see Fig. 21.4). Generally, large-scale motions dominate transport processes, and small-scale structures dissipate energy by the action of viscosity. The large-scale structures are not universal. Except possibly in flows with strain or shear [Reynolds (1981)] (the flows of aerodynamic interest), the small-scale structures are nearly universal or isotropic at very high Reynolds numbers, and the spectrum of small-scale fluctuations can be parameterized by the local rate of energy dissipation and the kinematic viscosity. This spectral behavior leads to the concept of so-called

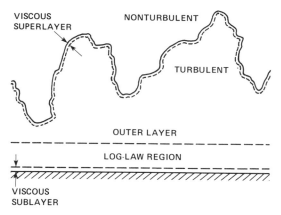

FIGURE 21.4   Sketch of instantaneous turbulent boundary layer and defined statistical regions.

large-eddy simulation (LES)* where the larger eddies are computed directly and the very small eddies are modeled. This idea originated in the works of Smagorinsky (1963), Lilly (1967), and Deardorff (1970a), (1970b), (1973); the Stanford–Ames LES program further developed it, providing extensive demonstrations [see for example, Ferziger, Mehta, and Reynolds (1977), and Moin, Reynolds, and Ferziger (1978)].

In the LES approach, one derives equations for the grid-resolvable field from the Navier–Stokes equations by averaging over spatial grid-volumes. These equations involve terms associated with the grid-unresolvable structures which have to be modeled. By definition, turbulence is necessarily unsteady and three-dimensional, therefore, in this approach, time-dependent and three-dimensional computations are necessary. Modeling of scales which are not resolved by the grid system, is done by phenomenological models of the simple eddy-viscosity type when the small-scale structures are homogeneous and isotropic. These models seem to be inadequate for inhomogeneous flows [Deardorff (1973)]. In addition, in regions in which viscous effects are important, these models are less likely to be satisfactory. In such cases, complex modeling, for example equations for the Reynolds stresses, may be required. The degree of complexity of modeling turbulence is essentially the same whether the averaging is done over spatial dimensions or over the time dimension, provided the computations are done in three dimensions and in an unsteady mode.

The LES approach is in its infancy for computations of viscous flows. The approach has been primarily applied to low Reynolds number, fully developed turbulent flows of incompressible fluids with periodic boundary conditions in the stream-

---

*The adjective *large* is misleading when the important scales to be resolved are also small as, for example, near solid surfaces. Further, the word *eddy* is also a misnomer when the simulation or the identification is not in a spectral space, as this word was used by Townsend (1956) to describe the stochastic picture of turbulent fluid through the Eulerian spatial-correlation tensor (a statistical concept), not a physical structural descriptor. Early use of the phrase *large eddy* related to measurements of the turbulence energy balance in the wake of a cylinder [Townsend (1993)].

wise and spanwise directions, that is, to extremely simple flow fields. The only flow that has been extensively studied is the channel flow [see for example, Deardorff (1970a), Schumann (1973), (1975) Moin, Reynolds, and Ferziger (1978), and Moin and Kim (1982)]. The computations of the turbulence field in practical applications almost always are done with the time-averaged equations of motion. Section 21.5 describes the LES method in more detail.

Navier–Stokes Regime. When a shear layer is not thin or slender (Fig. 21.5), the boundary layer equations do not apply and the Navier–Stokes equations are necessary. (A variation of velocity in the direction normal to the direction of the velocity itself is called a *rate-of-shear strain*. The region over which this variation takes place is called a *shear layer*.) The upstream influence of both the pressure field and streamwise gradients of viscous and turbulent stresses are then important. The upstream influence of the pressure field is much more pronounced than in a thin-shear layer owing to significant normal pressure gradients which occur, for example, in large separated regions, in the trailing-edge region of an airfoil and in interactions of a shock wave and a boundary layer. Further, the streamwise gradients of stresses cannot be neglected: 1) in low-Reynolds-number flows, 2) when the velocity field encounters rapid variations in streamwise direction, such as past the trailing edge of an airfoil, across a shock wave within a boundary layer, and around a point of flow separation, 3) when transition or turbulence is computed without modeling, and 4) when viscous LES simulations are done.

Non-Navier–Stokes Regime. The subclassification of viscous flow regimes in terms of non-Navier–Stokes and Navier–Stokes is based on the form of governing equations that are used to compute the flow fields. In the former category, the following flows occur: 1) the thin-shear-layer theory, 2) the strong viscous–inviscid interaction theory, and 3) the composite theory. The first is traditionally known as the boundary layer theory. It includes both the classical and higher-order formulations. The second includes theories such as the interacting boundary layer theory [Melnik (1980)] and the triple-deck theory or the asymptotic theory of free interactions [Adamson and Messiter (1980)]. The last includes the composite thin- and slender-shear-layer theory, and it also contains the parabolized Navier–Stokes theory.

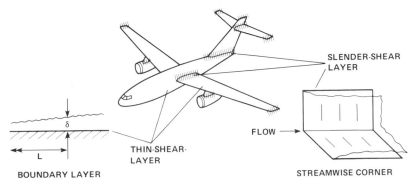

FIGURE 21.5 Thin-shear layers and slender-shear layers (shaded) on a commercial transport plane in a cruise condition.

Thin-Shear-Layer Theory: In the absence of flow separation (that is, departure of the shear layer from a solid surface), a steady, asymptotic solution of the Navier–Stokes equation at large Reynolds number is determined from the classical boundary layer theory. This solution is the first approximation to the solution of the Navier–Stokes equations. The second-order corrections can be classified into two categories, according to the appearance of curvature terms in the differential equations or through the interaction with the flow external to the boundary layer. It is convenient to subdivide each of these categories further. Curvature effects are due to longitudinal or transverse curvature or both. In incompressible fluids, the interaction effects are due to displacement by the wall-bounded viscous layer of the external flow, the external vorticity, and the external gradient of the stagnation temperature. In compressible fluids, the interaction effects result from displacement of the external flow, external gradient of entropy, and the external gradient of enthalpy. A further classification arises in this case, namely, effects caused by noncontinuum surface phenomena [Van Dyke (1965)]. The boundary layer theory is the most developed for laminar flows, for steady motions, and for plane or axisymmetric flows. Although the classical boundary layer equations are nonlinear, all equations of higher-order corrections are linear, consequently the above second-order corrections are additive.

The boundary layer theory is based on the assumption that the shear layer is thin and on the condition that this layer grows only slowly in the general direction of flow. This theory is applicable not only to layers on a boundary but also to layers in jets and wakes. Therefore, it is also called the *thin-shear-layer theory*. As a consequence of the above assumption and condition, the upstream influence within the thin layer is not accounted for by the first-order theory and is only partially accounted for by the second-order theory. This is the fundamental limitation of the thin-shear-layer theory.

The upstream influence is provided by the pressure field in subsonic regions, by transport of momentum in the upstream direction in separated flows, and by streamwise gradients of viscous and turbulence stresses, as indicated by the direction of the arrows in Fig. 21.6. The first or the last of these influences makes the flow field and the governing equations elliptic in the spatial dimensions. The correction owing

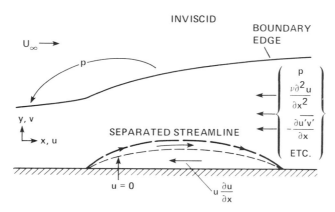

FIGURE 21.6 Upstream influence of pressure field (p), transport of momentum in separated flows, and streamwise gradients of viscous and turbulent stresses (boundary layer growth is exaggerated).

to the displacement effect can indirectly transmit the upstream influence of the pressure field via the external flow but not through the shear layer. A correction owing to displacement thickness in the *inviscid* flow at a station leads to a correction in the pressure distribution upstream of this station. This procedure is a part of the viscous–inviscid interaction theory. On the other hand, the correction owing to the curvature effect accounts for the variation of the pressure field across the shear layer but not for the upstream influence of this field through this layer. In addition, this correction is limited to shear layers that are thin, since the curvature of streamline varies substantially across thick-shear-layers. The upstream influence of the velocity field can be numerically accounted for by methods, such as the downstream–upstream iterative (DUIT) scheme [Williams (1975)], which march both upstream and downstream alternately. However, these methods are limited to flows in which there is significant upstream influence of the velocity field relative to that of the pressure field within the shear layer so that the latter can be neglected. This is usually the case in high-Re, laminar separated flows. The thin-shear-layer theory does not account for the upstream influence of streamwise gradients of viscous and turbulence stresses. These stresses are generally negligible at normal (high) Reynolds numbers. In a shock wave, the viscous diffusion is not negligible (provided the Navier–Stokes equations are valid, see the discussion of thickness of shock waves in Sec. 21.1.3), and, in a shock layer formed by the interaction of the shock wave with a shear layer, both viscous and turbulent streamwise diffusion may be significant.

The thin-shear-layer equations are singular when the normal momentum (or the error in conserving it) is not negligible. In a steady flow for an incompressible fluid, this singularity occurs at the location where the flow separates. At a singular point or a line, the direct (pressure-specified) computational procedures fail, but the inverse (displacement-thickness or wall-shear specified) procedures usually work when the upstream influence of the pressure field is negligible compared to that of the velocity field. The inverse procedures simply ignore the significant upstream influence of the pressure through the shear layer at and in the vicinity of the separation location [see also Bradshaw (1978) and Bradshaw (1982a)]. A further discussion of the thin-shear-layer theory is given by Telionis (1981).

Viscous–Inviscid Interaction Theory: If different regions of a flow field can be studied almost independently with good approximation, these regions are said to have a *weak interaction*. When the different regions cannot be treated separately because of strong coupling, the flow is classified as one with a *strong interaction*. This is explained again by considering the displacement effect of a viscous layer on the inviscid flow. First, the inviscid pressure distribution is determined independent of the viscous layer. This first inviscid approximation is used to obtain the first viscous approximate solution. Then, the second inviscid approximation is computed taking into account the displacement effect caused by the viscous region resulting in the second inviscid pressure distribution. This, in turn, is used to determine the second approximate solution of viscous region. If the second approximate displacement surface is nearly the same as the first one, then this iterative procedure is stopped, and the viscous–inviscid interaction is commonly known as a weak interaction [see, for instance, Lock and Firmin (1981) and Le Balleur (1981)]. If however, the iterative procedure must be repeated a large number of times, then the interaction is considered to be strong.

The thin-shear-layer theory fails when there are nonuniformities or abrupt changes in the surface boundary conditions caused by flow separation, a corner, strong injection, a trailing edge, or by a shock wave intersecting the boundary layer. In the neighborhood of such nonuniformities, the normal pressure gradients will be significant. For example, once laminar flow separates, it may not be possible to use the irrotational, inviscid solution as a basis for determining a uniformly valid first approximation to the flow as Re tends to infinity. Flow separation that occurs in an adverse pressure gradient is accompanied by a sharp increase in displacement thickness. This dramatic change in the boundary layer structure has a significant effect on the external flow. During this interaction of the boundary layer and the inviscid flow, the viscous effects are confined to the lower part of the boundary layer; the rest of the boundary layer behaves virtually as if it were inviscid. These two parts of the boundary layer are usually referred to as the *lower* and the *main decks*, respectively, [Lighthill (1953) and Müller (1953)]. The region of the external flow field just above the two decks is most affected by the rapid changes in these decks; it is called the *upper deck*. This structure is given in Fig. 21.7. This multilayered structure of the boundary layer and the associated analytical procedures for laminar flows have been called the *triple-deck theory* [Stewartson (1974)]. This theory is applicable if the streamwise length of nonuniformity is relatively short. Another example of dramatic change of the boundary-layer structure affecting the external flow is that of a weak shock wave sharply increasing the displacement thickness even in the absence of flow separation.

The multilayered structure of a turbulent boundary layer caused by nonuniformities is different from that of a laminar boundary layer. This difference is caused by the basic difference in structure of the turbulent and laminar boundary layers. In the main deck, the viscous stress gradients are usually neglected in laminar flows, whereas the Reynolds stresses are considered *frozen* along the streamlines in turbulent flows. Further, the turbulent boundary layer develops a two-layer structure described by the law of the wall and the law of the wake, as formulated by Coles (1956) for incompressible fluid and as modified by Maise and McDonald (1968) for compressible fluids. In this case, the asymptotic theory leads to a mismatch in both the Reynolds stress and streamwise velocity between the main and inner decks. This mismatch is resolved by introducing a *blending layer* [Melnik (1980)] or *Reynolds-stress sublayer* [Adamson and Feo (1975)], resulting in the *four-deck* theory. However, the nonasymptotic theory [Inger and Mason (1976)] does not use this blending layer. There are also qualitative differences in interactive response between laminar

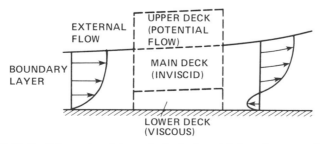

FIGURE 21.7 Schematic of three decks in the vicinity of a separation point.

and turbulent flow. In the latter case, there is much smaller upstream influence of pressure, and the boundary layer is less prone to separate. Note that a model has not been developed for turbulent separated flows [Stewartson (1982)].

Composite Theory: The basic ingredient of the interaction theories, for high-Reynolds-number problems, is the coupling of the external inviscid and thin- or slender-shear-layer flows. Instead of achieving this coupling through matching boundary conditions, it can be achieved by considering a set of equations, less complex than the Navier–Stokes equations, that are valid for both these flows. The basic idea is similar to that of Oseen (1910) who corrected the nonuniformity in Stokes's approximation for plane flow at low Reynolds numbers by partially including the neglected convective terms. If a set of equations includes the inviscid equations and the viscous and turbulence stress terms (for example, from the thin-shear-layer equations), which are absent in the inviscid equations, then the resulting equations automatically achieve the coupling between inviscid and viscous flows. These equations are called *composite equations* in *singular-perturbation theory* [Van Dyke (1975)].

The composite equations for thin-shear-layers contain leading terms up to second-order in the viscous–inviscid interaction theory, provided the dependent variables are defined in the same coordinate system as the independent variables. These equations are called the composite thin-shear-layer equations.* In a body-fitted or optimal [Kaplan (1954) and Davis (1974)] coordinate system, a composite set of equations exhibits upstream influence of the pressure field through the shear layer, but not upstream influence of viscous and turbulent stresses. In the absence of these stresses, the parabolic form for the velocities in streamwise direction has led to the expression *parabolized Navier–Stokes equations.* These equations may be used to solve problems that particularly need to be treated by the strong viscous–inviscid interaction theory. In addition, these equations are necessary for studying upstream influence of the pressure field through shear layers and flow fields such as the strong-interaction region downstream of the sharp leading edge of a flat plate in hypersonic rarefied flow, a mixing region with a strong transverse pressure gradient, and a blunt body in a supersonic flow at high altitudes.

Shock Waves. In the absence of viscosity, the interaction of a shock wave with a solid surface would result in a surface-pressure discontinuity and possibly an irregular reflection. In reality, viscosity makes the surface pressure variation continuous. The interaction of a shock wave with a boundary layer is a complex phenomenon, see, for instance, Van Dyke (1982). It can occur in the presence of an incident shock, oblique or normal, or be caused by an irregularity in surface shape. The external pressure discontinuity across the shock wave is transmitted through the boundary layer toward the subsonic part adjacent to the surface. This subsonic flow cannot support such a discontinuity; as a result, pressure disturbances are emitted both upstream and downstream of the shock impingement, developing a gradual streamwise pressure rise at the surface, along with pressure gradients normal to the

---

*These equations are commonly known as the *thin-layer Navier–Stokes equations*. It is, however, misleading to describe these equations as the thin-layer approximation of the Navier–Stokes equations, because the thin-(shear)-layer approximation of the Navier–Stokes equations gives the thin-shear-layer (boundary layer) equations.

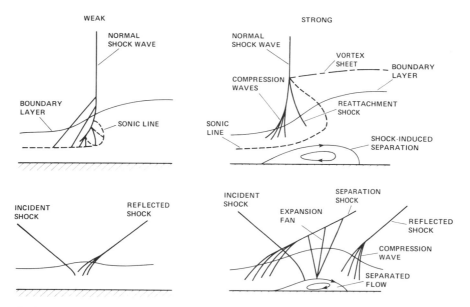

FIGURE 21.8 Schematics of weak and strong shock/boundary layer interactions.

surface. The pressure rise upstream of the shock causes thickening of the boundary layer, which in turn generates compression waves emanating from the sonic line toward the external flow as shown in Fig. 21.8. These compression waves converge and coalesce into the incident shock wave or a different external shock wave. Thus, there are both local effects, such as the smearing of discontinuous pressure rise and a different wave pattern in the external flow, and global effects, such as substantial thickening of the boundary layer downstream of the interaction region. If the adverse pressure gradient caused by the compression in front of the shock wave causes flow separation, then the interaction of a shock wave with a boundary layer is labeled as a strong-interaction problem. If, on the other hand, there is no shock-induced separation, the interaction is considered to be weak.

There is a pronounced difference between laminar and turbulent shock/boundary layer interactions. The pressure rises more rapidly for turbulent than for laminar flows. The displacement thickness through the shock increases considerably less for turbulent than for laminar flows. Reynolds number has almost no effect in turbulent flows, but it has a strong effect in laminar flows. In turbulent flows, a large pressure rise is required for the flow to separate, and the separation point moves upstream more slowly as the shock wave becomes stronger. The opposite is the case in laminar flows. Also, in laminar attached flows, the spread of the surface-pressure distribution at the foot of the shock wave, depends strongly on the Reynolds number, increasing with decreasing Reynolds number.

Strain Fields. Modeling of turbulence is largely governed by strain fields, which are classified as being simple or complex [Bradshaw (1973), (1982b)]. Simple shear layers have monotonic velocity profiles and nearly straight streamlines. In these layers, the only significant rate-of-strain component is the variation of streamwise velocity in the normal direction, $\partial U/\partial n$, which is of the same sign everywhere or

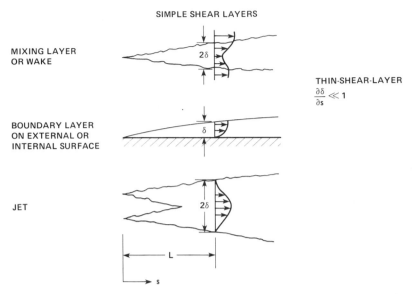

FIGURE 21.9 Condition for a shear layer to be thin in two-dimensional or axisymmetric flows.

changes sign only at an axis of symmetry. Simple strain fields are noticed in simple shear layers such as symmetric wakes and jets, and plane boundary layers (see Fig. 21.9). All other flows that are significantly affected by additional rates of strain are considered complex. In two-dimensional or axisymmetric flows, these extra rates of strain appear in shear layers that experience streamwise curvature, lateral divergence, bulk dilatation, buoyancy, or system rotation; see, for example, Fig. 21.10. This distinction among strain fields is made because extra rates of strain can have a large effect on turbulence, which in turn influences modeling of turbulence. Eddy-viscosity or mixing-length models can give good results when strain fields are simple but not when the fields are complex.

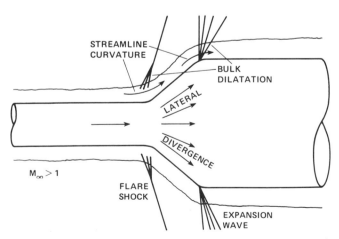

FIGURE 21.10 Supersonic flow over a cylinder-flare with extra rates of strain produced by streamwise curvature, lateral divergence of flow, and bulk dilatation.

21.1.3 Spatial and Temporal Scales

In this section, some of the spatial and temporal scales are presented that are observed in viscous flows. Also, flow structures in transitional flows and in turbulent flows in the vicinity of a wall are described. The spatial scales of the following physical phenomena are discussed: 1) mean turbulent boundary layer profile, 2) thin- and slender-shear layers, 3) flow separation, flow past the trailing edge of a flat plate and weak shock-wave boundary layer interactions, 4) weak and strong viscous–inviscid interactions, and 5) shock waves. Further, based on magnitudes of strain rates, simple and complex shear layers are classified. The temporal scales are discussed in terms of the *Strouhal number*, *diffusive time-scales*, and *turbulence frequencies*. These spatial and temporal scales help determine the computational space-time resolution requirements.

Flow Structures in Transitional Flows. * In a transitional flow from a laminar to a fully developed turbulent flow, the scales of motion depend on the Reynolds number of the flow and on the details of the shape of the body. Viscosity determines the relevant small scales. These scales are those in the fully-developed turbulent flow and are given in the next section.

The primary structure observed in a wall-bounded, external transitional flow is the turbulent *spot*, the existence of which was first observed by Emmons (1951). Each turbulent spot is surrounded by nonturbulent fluid. In plan view, it is like a rounded-off elongated delta, with blunt vertex pointing downstream as depicted in Fig. 21.11. The upstream and downstream edges travel at constant, but not equal, velocities. Near the wall, the propagation speeds (*celerities*) of the upstream and downstream interface, are typically 0.5 and 0.9, respectively, of the free-stream velocity [Schubauer and Klebanoff (1955)]. These spot propagation velocities are larger than the velocity of the undisturbed laminar flow. Therefore, the fluid down-

---

*Reproduced with permission from "Some Physical and Numerical Aspects of Computing and Effects of Viscosity of Fluid Flow," *Computational Methods in Viscous Flows*, Vol. 3, Pineridge Press Ltd., Swansea, U.K., 1984.

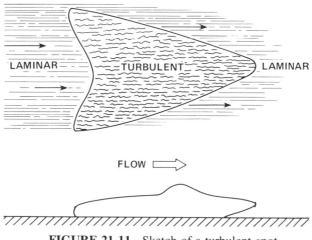

FIGURE 21.11 Sketch of a turbulent spot.

stream of the spot is overtaken by the spot, becomes turbulent, and then again laminar after the passage of the spot. The velocity (rate) of spot growth is about 0.2 of the free-stream velocity. The motion within the spot is similar to that in a turbulent boundary layer with streamwise streaks next to the wall. The ensemble-averaged vorticities within the spot are essentially a single horseshoe vortex superimposed on small-scale motions [Coles and Barker (1975) and Wygnanski, Sokolov, and Freidman (1976)].

The main structure in an internal transitional flow is somewhat different from that in an external flow. For example, in a pipe, a turbulent *slug* is observed. It is different from a turbulent spot. The *turbulent slugs* occur in smooth or slightly disturbed inlets for Re > 3200 based on the diameter of the pipe [Wygnanski and Champagne (1973)]. These structures occupy the entire cross section of the pipe and are elongated in the streamwise direction. The leading edge and the trailing edge of the slugs travel at the same speed, relative to the maximum turbulent velocity in the pipe, as the celerities of edges in the case of turbulent spots. Another type of intermittent turbulent flow, referred to as a *puff* is also found in pipes. A *turbulent puff* occurs when a large disturbance is introduced at the entrance of the pipe for 2000 ≤ Re ≤ 2700. Puffs correspond to an incomplete relaminarization process.

Scales in Turbulent Flows. Inside the viscous sublayer, the large scales are the friction velocity, v^*, and the length scale determined by v/v^*. Here v^* is given by $(\tau_w/\rho)^{1/2}$, τ_w is the shear at the wall, and v is the kinematic viscosity. Outside the viscous sublayer, the large scales are the fluctuating velocity relative to the mean flow, u', and the length scale is given by u'^3/ϵ, where ϵ is the mean dissipation rate [Corrsin and Kistler (1954)]. Inside and outside the viscous sublayer, the small length scales are identified with the *Kolmogorov microscale of length*, η, determined by $(v^3/\epsilon)^{1/4}$, and the corresponding *Kolmogorov scales of time* and *Kolmogorov scales of velocity*. Actually, there is significant flow structure that is smaller than this microscale of length, both outside and inside the sublayer.

If $\mathcal{L}$ is a macroscopic length scale, then outside and inside the viscous sublayer, the ratio of length scales are respectively, $\mathcal{L}/\eta = R_\lambda^{3/2}/15^{3/4}$ and $\mathcal{L}/\eta = O(R_\lambda^{-1/2})$, where R_λ is a Reynolds number based on the Taylor microscale ($\lambda^2 = 15vu'^2/\epsilon$) and fluctuating velocity relative to the mean flow. In terms of Re, the former ratio is $O(\text{Re}^{3/4})$. The Kolmogorov microscale of time is $\tau = (v/\epsilon)^{1/2}$, and the frequencies corresponding to macroscales are of $O(U/\mathcal{L})$, where U is the mean velocity. This gives $\tau/\mathfrak{I} \sim \text{Re}^{-(1/2)}$ where $\mathfrak{I} \sim \mathcal{L}/U$. Therefore, the total number of degrees of freedom in a volume of fluid of $O(L^3)$ is $O(\text{Re}^{9/4})$.

Organized Structures in Wall-Bounded Turbulent Flows.* Townsend (1956) was the first to identify organized** structures in turbulent flows. He described a *double structure* of fully developed turbulence consisting of more or less organized large

---

*Reproduced with permission from ''Some Physical and Numerical Aspects of Computing and Effects of Viscosity of Fluid Flow,'' *Computational Methods in Viscous Flows*, Vol. 3, Pineridge Press Ltd., Swansea, U.K., 1984.

**Organized structures are also known as *coherent structures*. Coherent means having the quality of cohering, of being logically consistent, or having a definite phase relationship (Webster's Dictionary, 1976). Since organized flow structures are not necessarily coherent, these structures are not called coherent structures here.

eddies. The concept of an *eddy* is a mathematical abstraction useful in constructing simplified models of turbulence. Different techniques are available for the identification of eddy structures [Bonnet and Glauser (1993)]. Organized structures are usually identified by flow-visualization, conditional-sampling, and the Eulerian space-time correlation methods. The identification of organized structures is also made by a technique based on conventional statistical approaches using the proper orthogonal decomposition theorem of probability theory. It appears that there is some form of organized structure present in all types of turbulent shear flows. However, in general, these structures contain a small part of the total energy and play a secondary role in controlling transport [Lumley (1981)]. Reviews of *coherent* structures in turbulent flows are given by, for example, Lumley (1981) and Hussain (1986).

Although there is a lack of consensus at present concerning the organized structures in a turbulent boundary layer, the definitions or identifications and the physical dimensions and origins and certain characteristics of these structures are beginning to be established. These structures and their scales are relevant in any unsteady, three-dimensional computation in which most of the scales of turbulence are resolved with an appropriate space-time grid, whether it uses finite space-averaging or short time-averaging.

It appears that there are five main parts of the turbulent boundary layer, and they are illustrated in Fig. 21.12: 1) streamwise vortices and streaks next to the wall, 2) pockets of energetic fluid near the wall, 3) large-scale motions in the main (or outer) part of boundary layer, 4) mushroom-shaped intermediate-scale motions of energetic fluid near both the outer and inner edges of the outer-region of the boundary layer, and 5) the viscous *superlayer*. The outer-region exists for $y^+ > 100$, where $y^+ = u_\tau y/\nu$ which is a Reynolds number related to the energy-containing turbulence. (Any variable, which is given the plus sign superscript, has been made dimensionless in this way, and is said to be in *wall units*.) In the following paragraphs, these structures are briefly identified for a simple boundary layer, that is, in the absence of streamwise curvature, pressure gradient and roughness. Further review of these structures can be found in Willmarth (1975), Cantwell (1981), and Falco (1983).

Next to the wall, the flow structure consists of a fluctuating array of counter-rotating vortices [Bakewell and Lumley (1967)] and of an alternating array of high-

FIGURE 21.12 Schematic of organized structures in a turbulent boundary layer.

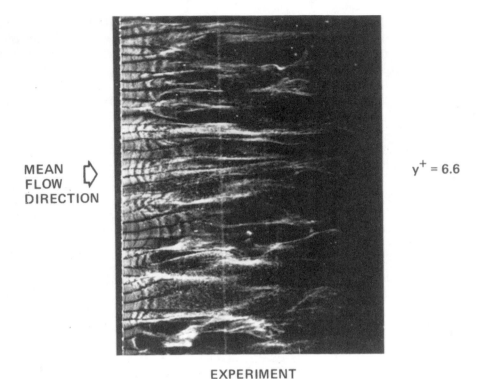

MEAN FLOW DIRECTION

$y^+ = 6.6$

EXPERIMENT

FIGURE 21.13   Structure of turbulent boundary layer visualized by passive markers introduced along a horizontal wire. (Courtesy of S. J. Kline.)

and low-speed regions called *streaks*, both aligned in the streamwise direction (see the flow visualization in Fig. 21.13). Streaks are found in the regions between vortices. High-speed fluid moving from the outer region into the region next to the wall, called a *sweep*, and low-speed fluid moving out of this region, called an *ejection*, are primary sources of the continual production of turbulence in the wall layer. Observations of the length of these structures vary from $x^+ = 100$ to $x^+ = 2000$ [Blackwelder and Eckelmann (1979)]. The vertical extent varies from $y^+ = 20$ to $y^+ = 50$, and the mean spanwise extent between vortex centers is 50 wall units, with a possible dependence on Reynolds number [Gupta, Laufer, and Kaplan (1971)]. The streamwise streaks interact with the outer flow when they break up, following lift-up and oscillation, during a *bursting process* [Kline, Reynolds, Schraub, and Runstadler (1967)]. Formerly, the mean dimensionless time between bursts was considered to be determined by the outer scaling variables, $U_\infty T/\delta \approx 6$ [Rao, Narasimha, and Narayanan (1971)]. Subsequently, it has been shown that the inner variables are considered to be the appropriate scaling variables [Falco (1983) and Blackwelder and Haritonidis (1983)].

The bursting process leaves a *pocket* of energetic, intermediate-scale motions in a region between $y^+ \approx 5$ and 40. Pockets lead to the formation of streamwise vortices and hairpin vortices. The period of formation of pockets is $(v^*)^2 T/\nu \approx 25$, and their dimensions are not easy to identify. Cantwell (1981) estimates the streamwise length of an energetic pocket to range from 20 to 40 wall units, and its normal extent to vary from 15 to 20 wall units. However, Falco (1983) reports streamwise scales between 50 and 90 wall units. There is no direct information about the spanwise extent. These energetic pockets persist over a distance between 0.5δ and 1.5δ and travel at about $(0.65 \pm 0.05)U_\infty$. The instantaneous maximum Reynolds shear stress is very large compared with the local mean value during bursting; for example, at $y^+ \approx 30$, it can exceed 60 times the local mean value [Willmarth and Lu (1972)].

In the outer layer, flow is fully turbulent, but intermittent for $y > 0.4\delta$. Two different structures have been observed in this layer. One occurs in its interior and the other at its inner and outer edges. In the interior, relatively large-scale energetic motions are observed. At a height of about 0.8δ, the streamwise and spanwise extent of the large-scale motions varies between δ and 2δ, and between 0.5δ and δ, respectively. Further, the centers of these motions are spaced about 2 to 3δ in the spanwise direction. The distance over which the centers persist is about 1.8δ, and the centers travel with a speed of about $0.85U_\infty$. The instantaneous maximum shear stress owing to fluctuating velocities may exceed 10 times the local mean value.

Along the edges of the outer flow, mushroom-shaped, intermediate-scale, energetic regions are observed. These are called *typical eddies* by Falco (1977). They have the streamwise and normal length, respectively, of about 200 and of $O(100)$ wall units between $730 \le \text{Re}_\theta \le 3.9 \times 10^4$. The scales of typical eddies are Reynolds-number dependent. At very low Reynolds numbers, the scales are $O(\delta)$ and appear to be the large-scale motions. These regions exist approximately over a distance 5 times their own streamwise extent, and the regions also travel at about $0.85U_\infty$. The energetic motions that are just above the wall layer interact with it leading to the formation of energetic pockets. Lifted fluid filaments in the wall layer may evolve into mushroom-shaped energetic regions moving into the outer region. These regions scale with the Taylor microscale and account for a major part of the Reynolds stresses.

The outer layer is bounded by the laminar-like viscous superlayer, outside of which vorticity fluctuations are zero but the velocity fluctuations are not. The superlayer is of the order of Kolmogorov length scale, η. The propagation velocity of the superlayer normal to itself is proportional to the Kolmogorov velocity scale, $v_e \equiv (\nu\epsilon)^{(1/4)}$ [Corrsin and Kistler (1954) and Willmarth (1975)].

From the above description of three-dimensional, unsteady turbulent boundary layer, it is concluded that the wall layer and the outer energetic regions are important. These regions need appropriate treatment in any direct or large-eddy simulation. As a turbulent boundary layer is an elliptic region, the wall layer and the outer layer interact, and the cause and effect cannot be separated.

Structure of the Mean Turbulent Boundary Layer.*
For a laminar velocity profile, the concept of Reynolds similarity can be used, but such a concept does not hold for a turbulent mean-velocity profile. These two velocity profiles greatly differ in both external and internal flows. The time-averaged turbulent velocity profile is made of two layers, an inner layer and an outer or core layer shown in Fig. 21.14 (see also Sec. 4.5). The inner layer is adjacent to the wall and is divided into two layers,

*Reproduced with permission from ''Some Physical and Numerical Aspects of Computing and Effects of Viscosity of Fluid Flow,'' *Computational Methods in Viscous Flows*, Vol. 3, Pineridge Press Ltd., Swansea, U.K., 1984.

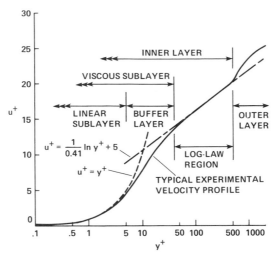

FIGURE 21.14   Mean turbulent velocity profile in wall-layer nomenclature.

the viscous *sublayer* and the *log-law region* or the inertial sublayer. Further, the viscous sublayer is composed of the linear sublayer and the *buffer* layer. The outer layer is also subdivided into two regions. Proceeding away from the wall, these are the *wake-law layer* and the viscous superlayer. The velocity and length scales, based on measurements and dimensional analysis, are identified below for a constant-density, two-dimensional mean flow over an impermeable smooth wall.

The inner layer of a wall shear layer extends about 15% of the boundary-layer thickness. The characteristic velocity scale and length scale are, respectively, $v*$ and $(v/v*)$, where $v*$ is the friction velocity. These are known as the inner scales. The thicknesses of the linear sublayer, the buffer layer, and the log-law region are, respectively, $0 < y^+ \leq 5$, $5 \leq y^+ \approx 40$, and from $y^+ \approx 40$ to 500–1000 wall units. These bounds are, of course, approximate. Actually, the logarithmic law of the wall can be used from $y^+ = 30$, but the eddy viscosity assumes a linear variation with y^+ only above $y^+ \approx 50$, when the direct effect of viscosity is negligible [Hinze (1975)]. The thickness of the inner layer is about 0.1δ ($y^+ \approx 500$) at the lowest Reynolds number at which turbulent flow can be maintained, that is, at $\text{Re}_\theta \approx 320$, where Re_θ is a *momentum-thickness Reynolds number* [Preston (1958)]. The linear sublayer is of the order of 0.001 to 0.01δ, depending on the Reynolds number.

In the linear sublayer, the Reynolds stresses are negligible, the pressure gradient has little effect, and the averaged velocity varies linearly. In the buffer layer, both viscous and Reynolds shear stresses are of the same order of magnitude, and there is no simple variation of velocity with the normal distance. In the log-law region, the viscous stresses are negligible compared with the Reynolds stresses. Both the buffer and inertial layer are affected by the streamwise pressure gradient. At the end of this region, the mean velocity is about 70% of the velocity at the edge of outer layer in small-pressure-gradient flows.

The outer layer of an external shear layer contains about 85% of the boundary layer thickness. In the wake-law region, the characteristic velocity scales and length scale are, respectively, $v*$ and U_∞ and δ for zero pressure gradient flows. This region is bounded by the viscous superlayer. The laminar-like superlayer is likely to be at least as thick as the turbulent viscous sublayer. In fully developed internal flows, there is no superlayer. With decreasing Reynolds number, both the viscous sublayer and the superlayer become thick, and the wake region ultimately disappears at $Re_\theta \approx 500$ [Coles (1962)]. Further, the mixing length and the dissipation-length parameter increase as Reynolds number decreases; see for example, Murlis, Tsai, and Bradshaw (1962). The ratio of the viscous superlayer length-scale to δ becomes independent of Reynolds number above $Re_\theta \approx 5000$ [Murlis, Tsai, and Bradshaw (1962)].

The above scales are valid, for both internal and external flows, provided the total shear stress varies slowly or is nearly constant in the normal direction. These scales are affected by surface nonuniformities and by the surface pressure gradient, if the latter is not small relative to the normal shear stress gradient at the wall, $\partial\tau_w/\partial y$.

In a computation, it is possible to use a *law of the wall* as a boundary condition instead of resolving the viscous sublayer. However, one of the conclusions of the 1980–81 AFOSR-HTTM-Stanford Conference on Complex Turbulent Flows was that it is better to integrate the governing equations representing a turbulent boundary layer all the way to the wall rather than to assume a law of the wall.

Thin- and Slender-Shear Layers. With s as the streamwise coordinate, a shear layer, whose thickness is δ at a distance L from its origin, is considered to be thin if $\partial\delta/\partial s$ is of order δ/L and is much smaller than unity, that is $\partial\delta/\partial s \ll 1$ (see Fig. 21.9). Note that it is the local rate of change that matters. In this case, the Navier–Stokes equations reduce to the thin-shear-layer equations. These latter equations are useful only if they are in a coordinate system such that s is in the direction of the thin-shear-layer. When this thin-shear-layer approximation is applicable, the ratio of neglected to retained stress gradients is $O[(\partial\delta/\partial s)^2]$ for laminar flows and $O(\partial\delta/\partial s)$ for turbulent flows [see, for example, Bradshaw (1973)]. Since $(\partial\delta/\partial s)$ is generally larger in turbulent flows than in laminar flows, this approximation is less accurate in turbulent flows. Typically, the neglected streamwise viscous diffusion and turbulent diffusion are, respectively, $O(5.3\ Re^{-1})$ and $O(0.37\ Re^{-1/5})$ for $Re \le 10^7$.

A shear layer is considered to be *slender*, if it is thin in two directions normal to the streamwise direction. In slender-shear-layers, the Reynolds shear-stress gradients dominate the streamwise momentum equation, and the Reynolds normal-stress gradients affect any secondary flow.* Only the streamwise gradients of viscous and Reynolds stresses can be neglected in the governing equations. This approximation gives the slender-shear-layer equations.

Triple-Deck Region. In laminar flows, the triple-deck theory is applicable in the vicinity of the separation point on a smooth surface, at the trailing edge of a flat plate, and in the neighborhood of a weak shock wave interacting with a boundary layer. This asymptotic theory is valid over a short distance. It gives the extent of this distance to be $O(Re^{-3/8}\ L)$ in the streamwise direction as shown in Fig. 21.15. Here, the Reynolds number, Re, is based on L and local inviscid conditions, and L is a representative length of the boundary layer. The lateral scales of the lower, main and upper decks are, respectively, $O(Re^{-5/8}\ L)$, $O([Re^{-1/2} - Re^{-5/8}]L)$, and

---

*The motion in the plane perpendicular to the streamwise direction is called the *secondary flow*. In general, the streamwise component of vorticity is nonzero in secondary flows.

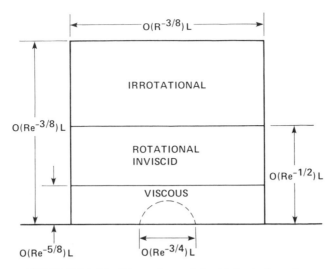

FIGURE 21.15   Dimensions of the triple-deck region.

$O([\text{Re}^{-3/8} - \text{Re}^{-1/2}]L)$. Note that $O(\text{Re}^{-1/2} L)$ is the order of the boundary layer thickness. The upstream influence of the streamwise viscous stress gradient is felt only over a region of $O(\text{Re}^{-3/4} L)$ within the lower deck and centered, for example, on the trailing edge of the flat plate. On the other hand, the upstream influence of the pressure field is noticed over the length of this deck. It would be prudent to resolve the above asymptotic length scales in numerical computations of problems that are amenable to the triple-deck theory. Burggraf *et al.* (1979) demonstrated this requirement on grid-spacing.

In the above examples, the contribution of the streamwise viscous gradient in the lower deck is relatively negligible. Therefore, the thin-shear-layer approximation is valid, and the composite thin-shear-layer equations are applicable to small separated regions. In catastrophic or large separation regions, in which the flow is not laminar everywhere at moderate or high Reynolds numbers, the relative upstream influence of streamwise stress gradients needs to be quantified.

Shock Wave/Turbulent Boundary Layer Interaction and Trailing Edge Regions. When the turbulent boundary layer remains unseparated, the asymptotic length scales for shock wave/boundary layer interactions and trailing-edge flows have been reviewed by Melnik (1980). The overall structure of the former case with a weak shock wave is similar to that of the latter case. In case of a weak normal shock wave/boundary layer interaction, the streamwise length of interaction is $O(\kappa^{3/2})$ where $\kappa \equiv (\ln \text{Re})^{-1} \approx 0.05$ (typically) normalized by L (see Fig. 21.16). The lateral dimension of the lower deck, blending deck, main deck, and outer deck are, respectively, $O(\kappa\hat{\kappa})$, $O(\kappa^{5/2})$, $O(\kappa)$, and $O(\kappa^{-1/2})$, where $\hat{\kappa} \equiv (\kappa^2 \text{Re})^{-1}$. In case of a turbulent flow past the trailing edge of a flat plate or a cusped trailing edge of an airfoil, the streamwise length of interaction is $O(\kappa)$. This is shown in Fig. 21.17. The lateral dimensions of the lower deck, blending deck, main deck, and outer deck are, respectively, $O(\kappa\hat{\kappa})$, $O(\kappa^2)$, $O(\kappa)$, and $O(\kappa^{-1/2})$. In addition, there is an inner

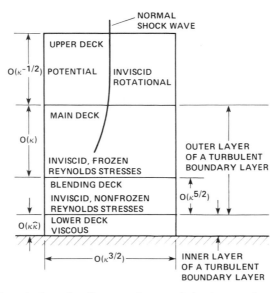

FIGURE 21.16 Four-deck region for a weak normal shock wave interacting with a turbulent boundary layer. (From Melnik, 1980.)

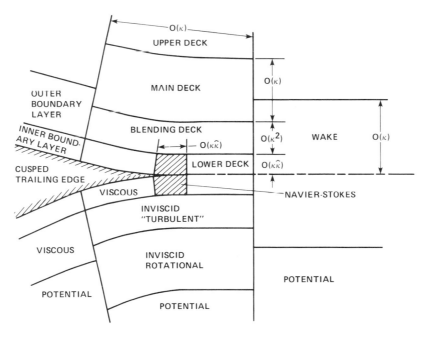

FIGURE 21.17 Asymptotic flow-field structure around a cusped trailing edge of an airfoil in a turbulent flow. (From Melnik, 1980.)

wall layer within the wall layer, whose streamwise length is $O(\kappa\hat{\kappa})$, and the lateral width is the same as that of the wall layer. In this region, the Navier–Stokes equations are required. The nonasymptotic length scales for shock wave/boundary layer interactions are discussed by Inger (1982).

Classification of Viscous–Inviscid Interactions. The nature of a *viscous–inviscid interaction* is identified by the thickness of the shear layer or the displacement thickness. When a shear layer suddenly thickens, the interaction is classified as a strong viscous–inviscid interaction. For example, on a smooth surface and in negligible pressure gradient, if the boundary layer thickness is almost either

$$\delta = 5.3 \ \text{Re}^{-1/2} L \tag{21.1}$$

for laminar flow or

$$\delta = 0.37 \ \text{Re}^{-1/5} L, \qquad 5 \times 10^5 \leq \text{Re} \leq 10^7$$

$$\delta = \left(\frac{0.0598}{\log_{10} \text{Re} - 3.170}\right) L, \qquad \text{Re} > 10^7 \tag{21.2}$$

for turbulent flow [Granville (1959)], before, during, and after the viscous-inviscid interaction, then it is considered to be *weak*. When a shear layer sharply thickens and subsequently becomes larger than this magnitude, the interaction is considered to be a *strong interaction*.

Thickness of Shock Waves. The effects of viscosity and heat conductivity develop a continuous change of flow variables through a shock wave, which is an internal transitional or *boundary* layer. The shock wave is quite thin outside any viscous region. The thickness of the shock wave increases when the wave begins to interact with the viscous boundary layer giving rise to a shock layer. This layer may even lose its identity as it penetrates into the viscous region. A turbulent boundary layer thickens the shock wave more than a laminar boundary layer.

At a Mach number of 1.05 and a Reynolds number of 10^7 with $L = 1$, the shock thickness in air is almost the same as the thickness of the linear sublayer of a turbulent boundary layer on a smooth flat surface. Another estimate is given by Taylor and Maccoll (1963) for a weak shock wave, in which the total speed change is small compared with the speed of sound. They give the thickness of the region within which 80% of the change from U_1 to U_2 occurs as $T = 1/(U_1 - U_2)$ cm for air. When the change in velocity through a shock wave is one tenth the speed of sound, that is, 30 m/sec, the shock thickness is 3×10^{-4} cm which is comparable to the thickness of the linear sublayer at the above Reynolds number. [Lagerstrom (1964) also states that the thickness of the shock wave is of the order of 2×10^{-6} m for air.] From measurements of shock wave structure, Sherman (1955) found that the Navier–Stokes equations are valid within the shock wave up to a Mach number of 2. This conclusion is considered to be somewhat cautious by Lagerstrom (1964).

Scales of Strain Rates. Frequently, turbulence in thin-shear-layers is modeled by eddy-viscosity formulas such that turbulent shear stress is proportional to averaged rate of shear strain, $\partial u_s / \partial n$, where u_s is the velocity in the streamwise direction s and the normal direction is represented by n (see Sec. 4.5). This model satisfactorily predicts only simple shear layers in which any extra strain rate, e, does not affect the turbulence structure. When $e/(\partial u_s / \partial n) < 0.001$ (approximately), Bradshaw (1973) considers any additional strain rate unimportant, but in complex shear layers, the extra strain rates affect the turbulence structure. Although the thin shear layer approximation is valid when $e/(\partial u_s / \partial n) < 0.01$, the eddy-viscosity model must be modified. If $0.01 < e/(\partial u_s / \partial n) < 0.1$, the thin shear layer approximation is also not valid, and some of the neglected terms of the Navier–Stokes equations are approximated in the governing equations. When $e/(\partial u_s / \partial n) > 0.1$, the flow is classified as a *strongly distorted flow*.

Temporal Scales. * Temporal scales are determined by the frequency of any unsteadiness, which may be either externally induced or self-induced. Externally induced unsteadiness may be due to fluctuations of boundaries or of the free stream or to initial conditions. Self-induced unsteadiness is developed by hydrodynamic instability, for instance, transitional flows, turbulent flows, and vortex shedding from bluff bodies. Typically, laminar flow problems may involve externally imposed unsteadiness, whereas turbulent flow problems may contain externally or self-induced unsteadiness, or both. There may be a coupling between externally induced and self-induced unsteadiness. For example, unsteady forces and moments owing to vortex shedding from a bluff body may force the body to deform, which in turn may change the shedding frequency. Unsteadiness can be completely stochastic, highly organized, or a combination of random and periodic components which under certain conditions may interact with each other. Further, there may be a single narrow-band frequency or a broad band of frequencies. Transitional and turbulent flows are characterized by a wide range of frequencies as discussed above. But, the externally induced unsteadiness is usually driven by a discrete frequency, which determines the dominant temporal scale. From a computational point of view, the highest significant frequency must be resolved. This fixes the computational time-step.

When a flow field is oscillating with a dominant frequency, f, the *Strouhal number*, $\mathrm{St} = fL/U_\infty$, is used to identify the time-scale of importance. This number characterizes the relative importance of the local velocity (acceleration) with respect to the convective velocity (acceleration), that is, the time-scale of physical motion to the basic fluid dynamic time-scale. For slender body oscillations, the reduced frequency, $k = \omega L/U_\infty$, is the relevant nondimensional parameter where $f = \omega/2\pi$. If St is referred to the speed of sound instead of U_∞, a necessary condition for the fluid to behave as if it were incompressible is that $\mathrm{St}^2 \ll 1$.

When the imposed unsteadiness is gradual so that the viscous diffusion is able to keep pace with the change, there is a *quasi-steady* flow. But if changes are rapid relative to the diffusion process, the unsteadiness is confined to a layer, called the *Stokes layer*, thinner than the boundary layer. Obviously, the flow cannot then be treated as quasi-steady. The *diffusive time-scale* is $O(\delta^2/\nu)$ for diffusion of vorticity and momentum outward from the body surface. For a body moving from rest, the

*Reproduced with permission from ''Some Physical and Numerical Aspects of Computing and Effects of Viscosity of Fluid Flow,'' *Computational Methods in Viscous Flows*, Vol. 3, Pineridge Press Ltd., Swansea, U.K., 1984.

boundary layer thickness is $O[(\nu t)^{1/2}]$ at small times. For an oscillatory body, the boundary layer thickness is $O[(\nu/\omega)^{1/2}]$.

When the flow variables are decomposed into a mean field and a fluctuating field, with fluctuating velocity components that satisfy the Reynolds conditions [Monin and Yaglom (1971)] as constraints, the time-averaged equations of motion are appropriate for computation of unsteady flows, provided the averaging time-interval is large compared with the periods characteristic of time-scales that cannot be resolved computationally, but small compared with the period of mean unsteady motion. This double decomposition of the velocity field is satisfactory for many physical problems involving relatively small-amplitude and low frequency oscillations. The above decomposition has limitations, however, as any harmonics higher than the first may become indistinguishable from the random fluctuations caused by turbulence. Further, the mean effects of small scales on large scales and *vice versa* are not discernible. Broad ranges of scales, therefore, require at least a triple decomposition of the velocity field with three different velocity, length, and time scales.

When the turbulence structure is unaffected by externally induced unsteadiness, steady flow turbulence models can be used to predict unsteady turbulent flows. Above a critical frequency, the turbulence structure is affected by externally imposed unsteadiness, and steady flow models cannot be used. In zero pressure gradient flows, the value of this critical frequency ranges from 20% to 100% of $f_b = U_\infty/(5\delta)$, which is roughly the *burst frequency* based on outer scales [see Acharya and Reynolds (1975) and Ramaprian and Tu (1980)]. In adverse pressure gradient flows, the critical frequency is well below this burst frequency. Values of 6% to 28% of f_b have been reported [see Cousteix and Houdeville (1983) and Simpson, Chew, and Shivaprasad, 1980)].

Refer to Chap. 12 for further discussion of unsteady flows.

21.1.4   Significance and Generation of Vorticity

The concept of a vortex has been used since prehistoric times to describe natural phenomena. In historic times, the analysis of vortices was applied to explain such phenomena as the evolution of the universe, the motions of celestial bodies, and atomic structure of matter. These efforts were unsuccessful and ended. But, with the development and application of Helmholtz's vortex theory [Serrin (1959a)] within the realm of classical mechanics, this concept has had an enormous effect on fluid dynamics (see Sec. 13.1 for more details).

Significance of Vorticity. A mathematical term used by Euler and d'Alembert, which later was identified as *vorticity* [Truesdell (1954)] has had a profound effect on the study of fluid motions. In an inviscid fluid, a vortex is defined as a finite volume or area of rotational fluid bounded by irrotational fluid or solid walls. Viscosity smooths out discontinuities or singularities. Therefore, in a viscous fluid, a vortex is identified by closed equivorticity lines with a relative extremum of vorticity inside the innermost closed loop [Mehta and Lavan (1975) and Mehta (1977)], as seen in Fig. 21.18. This identification depends on rotation of the reference frame and on stratification. Therefore, Lugt (1983) has defined a vortex as the rotating motion of a multitude of material particles around a common center.

Vortices and vortex motions are *the sinews and muscles of fluid motions* [Küchmann (1965)]. Vortices are omnipresent in separated flows and in many types of

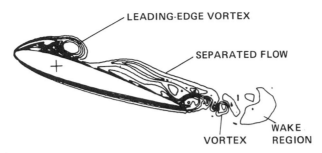

FIGURE 21.18   Equivorticity lines around an oscillating airfoil. (From Mehta, 1977.)

turbulent flows [see Lugt (1983) and Peake and Tobak (1980)]. In many cases, vortices are formed by an ejection, at a salient edge or from a smooth surface, of the vortical boundary layer fluid into the main body of fluid where viscosity plays a negligible role. In some cases, vortices are formed next to a boundary by shear in the external flow. Turbulent flow structures, organized or not organized, are essentially vortices. For example, a mixing layer consists largely of a row of quasi-two-dimensional organized structures [Roshko (1981)]. Also, such structures may be found in jets, wakes, and boundary layers. Organized vortices are ordered structures of fluid motion.

Some vortex patterns and interactions may be represented by a combination of inviscid and viscous models, by suitably dividing the flow domain, and some may be represented by only viscous-flow models. When circulation cannot be convected out of a closed separated region (a separation bubble), viscosity plays an important role along with convection and possibly vortex stretching.

The significance of the vorticity vector, which is defined as the curl of the velocity vector, is really mathematical rather than physical. This significance is specifically for the understanding and mathematical description of the study of fluid flow of a uniform, incompressible fluid or a *homentropic fluid* when the direct effect of viscosity may be neglected. In general, the nature and structure of the laminar or turbulent flow are controlled by the vorticity field. Vorticity considerations illuminate the detailed development of any rotational flow just as clearly as do momentum considerations and allow a compact description of the kinematics and dynamics of fluid motions. At high Reynolds numbers, the vorticity field is confined to thin layers or fine tubes, whereas the velocity vector is present in the whole flow field. In addition, study of this field is the most enlightening in understanding how the nonlinearity of the equation of motion generates highly complicated solutions that represent turbulent flows.

Generation of Vorticity. Although one cannot see the vorticity field and although its measurement is difficult, its presence in aerodynamic applications is easily detected by the determination of circulation, Γ, which is defined as the line integral of the velocity field around any closed curve. Kelvin's theorem of circulation [Kelvin (1869)]

$$\frac{D\Gamma}{Dt} = \oint \left(-\frac{\nabla p}{\rho} \right) \cdot d\bar{\mathbf{l}} + \oint \frac{1}{\rho} (-\nabla \times [\mu\bar{\mathbf{\Omega}}]$$

$$+ \nabla[(4/3)\mu\nabla \cdot \bar{\mathbf{U}}]) \cdot d\bar{\mathbf{l}} + \oint \bar{\mathcal{F}} \cdot d\bar{\mathbf{l}} \qquad (21.3)$$

shows that the time rate of change of circulation associated with a closed curve $\bar{l}$, always made up of the same fluid particles, is governed by the torques produced as a result of pressure forces, viscous forces, and body forces. In the above equation, $\rho, p, \overline{\Omega}, \mu, \overline{U}$, and $\overline{\mathfrak{F}}$ are, density, pressure, vorticity vector, dynamic viscosity with the second coefficient of viscosity as $-(2/3)\mu$, velocity vector, and body force. Irrotational body forces (i.e., conservative forces with single-valued force potential) do not produce circulation. Since fluid dynamic problems, in general, contain such body forces, the effects of the body force term has been omitted in the following discussion.

If isobars and isochors in a fluid particle are parallel, the situation is described as barotropic. In this case, the density field is function of the pressure field alone, and the pressure forces do not produce circulation. In an inviscid incompressible fluid of uniform density and in an inviscid compressible fluid, provided the flow field is homentropic, internal sources of vorticity do not exist. On the other hand, in a compressible, nonhomentropic fluid, pressure forces provide internal sources of vorticity. When all the torque-producing agents are absent, the dynamics of the fluid is governed by Helmholtz's vortex laws (see Sec. 13.1).

If the sources of vorticity owing to pressure are absent, vorticity flux or circulation cannot be created in the interior of the fluid [see Batchelor (1967)]. In such cases, vorticity is generated by viscous forces at a solid boundary or at a free surface.

Hadamard (1903) was first to point out that vortices are generated by shock waves and that the flow is no longer barotropic after shock waves. This is demonstrated by the *Crocco–Vazsonyi equation* [Vazsonyi (1945)] for steady flow of an inviscid fluid

$$\nabla H = T\nabla S + \overline{U} \times \overline{\Omega} \tag{21.4}$$

where H, T, and S are, respectively, total enthalpy, temperature, and entropy. The vorticity vector, therefore, depends on the rates of change of entropy and enthalpy normal to streamlines. When all streamlines have the same H but different entropy, there is production of vorticity. This is the situation downstream of a curved shock wave because the entropy increase across a shock wave is determined by the local angle of the shock. Even if the flow upstream of a shock wave were to be irrotational, the flow downstream of a curved shock wave is rotational.

The vorticity field in a Newtonian fluid is governed by the vorticity equation,

$$\frac{D\overline{\Omega}}{Dt} = (\overline{\Omega} \cdot \nabla)\overline{U} + \nabla \times \left(-\frac{\nabla p}{\rho}\right) + \nabla \times \left[\frac{1}{\rho}(-\nabla \times (\mu\overline{\Omega}) + \nabla[(4/3)\mu\nabla \cdot \overline{U}])\right] \tag{21.5}$$

which is obtained by taking the curl of both sides of the equation of motion. This equation shows that the time rate of change of vorticity as one follows the fluid is produced by a nonlinear, inviscid extension or contraction and tilting of existing vorticity by the strain rate, a source of vorticity resulting from the pressure force and diffusion of vorticity by viscosity. Tilting of a vortex line or tube changes the relative magnitude of the vorticity components and its stretching concentrates the vorticity in a smaller volume. When a flow is two-dimensional, the vorticity stretching and tilting term is absent, and, when the fluid is incompressible, the source term is absent.

Production of circulation generally means generation of vorticity, as circulation is a measure of integrated vorticity field, but a change in vorticity does not necessarily change circulation. This is readily noticed by comparing Eq. (21.3) (without the body force term) and Eq. (21.5). Both have pressure and viscous terms, but the term $(\overline{\Omega} \cdot \nabla)\overline{U}$ in Eq. (21.5) has no counterpart in Eq. (21.3). The vorticity of a material element increases when it is extended in the direction of the local vortex line. If the above torque-producing elements are absent, the circulation is the same for every contour enclosing the vortex line, the ratio of vorticity to the product of density and the length of the line remains constant and vortex lines remain located to the same elements of the fluid forever. But for the action of viscosity, the intensification of vorticity by extension of vortex lines continues. The total amount of vorticity in a body of fluid increases, until thorough change of the vorticity field, the smoothing effect of viscosity is able to balance or exceed this gain. This is one of the most remarkable characteristics of turbulent flows.

When a solid body passes through a viscous fluid that is void of vorticity, vorticity is imparted to the fluid through the mechanism of the viscosity, and it is diffused into the interior of the fluid, again by the action of viscosity. The no-slip boundary condition applied to the Navier–Stokes equations gives the instantaneous local rate of generation of positive and negative vorticity per unit area, per unit time, in terms of the strength of the vorticity source given by $(\nabla p)/\rho$. The pressure gradient along the surface of the body creates vorticity tangential to the surface. The maximum value of absolute vorticity need not be at the surface, but it could be in the interior of the fluid near the surface. This mechanism of vorticity generation is altogether absent in free shear layers. In case of a free surface, a nonzero jump in velocity derivative is responsible for the generation of vorticity.

The mean level of vorticity in a turbulent flow is also changed because of the production of vorticity by the Reynolds stresses arising from fluctuating velocity and vorticity fields. This is shown here for an incompressible fluid. Instantaneous values $\overline{U}$ and $\overline{\Omega}$ are decomposed into mean and fluctuating values. This is expressed by the relation

$$f = \overline{f} + f' \tag{21.6}$$

where the overbar designates a mean or averaged value, and the prime denotes the remaining fluctuating value. After substituting such relations into Eq. (21.5), taking a short-time average of all terms in the equation constrained with the Reynolds conditions [Monin and Yaglom (1971)], multiplying by $\overline{\Omega}_i$, and rearranging terms, we obtain

$$\frac{D}{Dt}\left(\frac{\overline{\Omega}_i \overline{\Omega}_i}{2}\right) = \overline{\Omega}_i \overline{\Omega}_j S_{ij} - \overline{\Omega}_i \epsilon_{ijk} \frac{\partial^2 \,(\overline{u'_i u'_k})}{\partial x_j\, \partial x_l} + \nu\, \frac{\partial^2 \overline{\Omega}_i}{\partial x_j\, \partial x_j} \tag{21.7}$$

where S_{ij}, ϵ_{ijk}, u'_i, and ν are, respectively, mean strain rate, alternating tensor, fluctuating velocity, and kinematic viscosity; the Cartesian tensor summation convention is used. The first term on the right-hand side is the stretching or contracting and tilting term, the second contains amplification or attenuation within the averaging time of the mean square vorticity caused by the stretching of fluctuating vorticity through fluctuating strain rates, and the last term represents viscous transport and dissipation of $\overline{\Omega}_i \overline{\Omega}_i$.

It is a fundamental property of turbulent flows that nearby points tend to move apart because of the diffusive action of turbulence. As a result, on the average, vortex lines and vortex sheets tend to be stretched rather than contracted. Therefore, the above stretching terms represent a growth of vorticity, limited only by viscous dissipation. The continued process of vortex stretching implies a cascade of energy from large-scale vortex motions to small-scale motions resulting in larger and larger velocity gradients.

21.2 ANALYTICAL ASPECTS OF COMPUTING VISCOUS FLUID FLOW

Unmeel B. Mehta

21.2.1 Introduction

The analytical aspects of computing the effects of viscosity on fluid flow include the equations of motion, turbulence models, initial conditions, and boundary conditions.* This section begins with a compact expression for the Navier–Stokes equations in two or three dimensions. It also contains some simplified approximate forms of the compressible Navier–Stokes equations for cases in which parts of the flow are inviscid, other parts are viscous, and the two are interacting strongly with one another, namely, the composite thin- and slender-shear layer equations. It further includes some discussion of the modeling of turbulence and some simple turbulent models. It concludes with a discussion of boundary conditions. The objective of this section is to highlight some of the current and future analytical aspects of computing the flow of a viscous fluid. Please refer to Secs. 4.2 and 4.6 for discussions of analytical aspects of the thin-shear layer (boundary layer) theory.

21.2.2 Navier–Stokes Equations

Continuum, incompressible fluid mechanics is described by combining the classical Navier–Stokes equations with an equation governing conservation of mass. Compressible fluid mechanics is described by these classical equations, properly modified to take into account variations in density, temperature and properties, together with equations governing conservation of mass and energy and an equation of state (often taken as that for a perfect gas) from equilibrium thermodynamics. Herein, both of these systems are referred to generically as the Navier–Stokes equations. These equations are considered to be the proper fundamental equations for viscous flows as long as the scales of interest are many times larger than the mean free paths of fluid molecules. It is generally assumed that solutions of these equations, subject to appropriate initial and boundary conditions, do exist and are unique. However, only under special conditions are existence and uniqueness theorems available [Temam (1979)]. [See also the local existence theorems in two- and three-dimensional problems for compressible fluids given by Solonnikov and Kazhikhov (1981).] Further,

---

*Please note that a discussion of analytical aspects of viscous flow presented by Lomax and Mehta (1984) was based on an early manuscript of this section, and that the discussion on boundary conditions is adapted from Mehta and Lomax (1982).

nonuniqueness is not a common occurrence. An example of nonuniqueness for an incompressible fluid is given by Stewartson, Smith, and Kaups (1982).

Problems for incompressible fluids can be formulated in terms of the vorticity vector, $\overline{\boldsymbol{\Omega}}$, the velocity vector, $\overline{\mathbf{U}}$, and the stream function vector, $\overline{\boldsymbol{\Psi}}$. This has the advantage of eliminating the pressure. The momentum equations of the Navier–Stokes equations can be represented by the vorticity equation. This equation is associated with an equation for the solenoidal, stream-function vector, $\overline{\boldsymbol{\Psi}}$, such that

$$\overline{\mathbf{U}} \equiv \nabla \times \overline{\boldsymbol{\Psi}} \tag{21.8}$$

This definition of velocity automatically satisfies the equation of conservation of mass. The equation satisfied by $\overline{\boldsymbol{\Psi}}$ is formed by applying the curl operator to this definition of velocity vector and using the definition of the vorticity vector ($\overline{\boldsymbol{\Omega}} \equiv \nabla \times \overline{\mathbf{U}}$). The Navier–Stokes equations for an incompressible fluid in term of vorticity and stream-function are

$$\frac{D\overline{\boldsymbol{\Omega}}}{Dt} = (\overline{\boldsymbol{\Omega}} \cdot \nabla)\overline{\mathbf{U}} + \nu\nabla^2\overline{\boldsymbol{\Omega}}$$

$$\nabla^2\overline{\boldsymbol{\Psi}} = -\overline{\boldsymbol{\Omega}} \tag{21.9}$$

Problems for turbulent flows, compressible or incompressible, are usually formulated in terms of quantities representing mean and fluctuating parts of a variable. This is expressed by Eq. (21.6). Monin and Yaglom (1971) present a general space-time averaging procedure for functions $f(\overline{\mathbf{x}}, t)$ given by the equation

$$\overline{f}(\overline{\mathbf{x}}, t) = \int_{-\infty}^{\infty} \int_{-\infty}^{\infty} f(\overline{\mathbf{x}} - \overline{\boldsymbol{\zeta}}, t - \tau)g(\overline{\boldsymbol{\zeta}}, \tau) \, d\overline{\boldsymbol{\zeta}} \, d\tau \tag{21.10}$$

where $\overline{\mathbf{x}}$ and $\overline{\boldsymbol{\zeta}}$ are vectors. The non-negative weighting function g satisfies the normalizing condition

$$\int_{-\infty}^{\infty} \int_{-\infty}^{\infty} g(\overline{\boldsymbol{\zeta}}, \tau) \, d\overline{\boldsymbol{\zeta}} \, d\tau = 1 \tag{21.11}$$

The choice of this weighting function determines the significance of the averaged quantities. The term $\overline{f}(\overline{\mathbf{x}}, t)$ is referred to as a time-averaged quantity if $g(\overline{\boldsymbol{\zeta}}, \tau)$ is set equal to $g(\tau)\delta(\overline{\boldsymbol{\zeta}})$ and if $g(\tau)$ is constant over some segment of time and zero elsewhere. (In this section, δ represents the Dirac delta function of its argument.) Space-averaged equations follow from setting $g(\overline{\boldsymbol{\zeta}}, \tau) = g(\overline{\boldsymbol{\zeta}})\delta(\tau)$ and letting $g(\overline{\boldsymbol{\zeta}})$ be a piecewise, continuously differentiable function which tends to zero at least as fast as $1/\zeta^4$. The system of equations obtained by applying the above time-averaging technique constrained with the Reynolds conditions [Monin and Yaglom (1971)] gives rise to *Reynolds-Averaged Navier–Stokes equations* (RANS). On the other hand, the system of equations determined by applying the above space-averaging technique is used in large-eddy simulation procedures (see Sec. 21.1.2).

For an incompressible fluid without external forces, both space-averaged and time-averaged Navier–Stokes equations can be written in dimensional form as

$$\frac{\partial \bar{u}_i}{\partial x_i} = 0$$

$$\frac{\partial \bar{u}_i}{\partial t} - \epsilon_{ijk}\overline{\bar{u}_j \omega_k} + \frac{1}{2}\frac{\partial \overline{\bar{u}_j \bar{u}_j}}{\partial x_i} = -\frac{1}{\rho}\frac{\partial \bar{p}}{\partial x_i} + \frac{\partial \sigma_{ij}}{\partial x_j}$$

(21.12)

Taken together, the second and third terms on the left-hand side are equivalent to $\partial \overline{\bar{u}_i \bar{u}_j}/\partial x_j$. The quantity σ_{ij} is referred to as the stress tensor and is usually broken down into two terms

$$\sigma_{ij} = 2\nu S_{ij} - R_{ij}$$

(21.13)

This includes contributions from both molecular and turbulent transport. It is composed of the mean flow rate-of-strain tensor, S_{ij}, and tensor $-\bar{\rho}R_{ij}$, which for convenience is called the Reynolds-stress tensor. These are related to the mean and fluctuating terms by

$$S_{ij} = \frac{1}{2}\left(\frac{\partial \bar{u}_i}{\partial x_j} + \frac{\partial \bar{u}_j}{\partial x_i}\right)$$

(21.14)

$$R_{ij} = \overline{u_i' u_j'} + \overline{\bar{u}_i u_j'} + \overline{u_i' \bar{u}_j}$$

(21.15)

No approximations have been made if we express the Navier–Stokes equations in terms of the mean and fluctuating quantities using the stress and strain tensors defined above. The Reynolds-averaged Navier–Stokes equations are an approximating subset of this system found when it is assumed that

$$\overline{\bar{u}_i \bar{u}_j} \to \bar{u}_i \bar{u}_j$$

$$\overline{\bar{u}_i u_j'} = \overline{u_i' \bar{u}_j} \to 0$$

(21.16)

For compressible fluids, the time-averaged Navier–Stokes equations are presented herein. These contain moments—second-order moments, such as $\overline{\rho' u'}$, and third-order moments, such as $\overline{\rho' u' u'}$—owing to fluctuations in the fluid density. In order to remove such moments, mass-weighted, time-averaged equations have been formulated. In this formulation, a mass-weighted velocity, $\bar{u}_i$, is defined such that

$$\bar{u}_i = \overline{\rho u_i}/\bar{\rho}$$

(21.17)

There are similar definitions for the other quantities. This averaging procedure eliminates the moment terms such as $\overline{\rho' u'}$ from the equations, but it does not, of course, remove the effect of density fluctuations on the turbulence. Mass-weighting was first used in the study of atmospheric turbulence by Hesselberg (1926). A comprehensive discussion of the procedure is presented by Favre (1965). Henceforth, the equations derived from this type of averaging are combined with the assumptions given in Eq. (21.16), and the result is referred to as the compressible, Reynolds-averaged Navier–Stokes equations. These equations can be written as

$$\frac{\partial \overline{\rho}}{\partial t} + \frac{\partial (\overline{\rho})\overline{u}_i}{\partial x_i} = 0$$

$$\frac{\partial (\overline{\rho})\overline{u}_i}{\partial t} + \frac{\partial (\overline{\rho})\overline{u}_j \overline{u}_i}{\partial x_j} = -\frac{\partial \overline{p}}{\partial x_i} + \frac{\partial \overline{\rho}\sigma_{ij}}{\partial x_j}$$

$$\frac{\partial \overline{\rho}\overline{h}}{\partial t} + \frac{\partial (\overline{\rho})\overline{u}_j \overline{h}}{\partial x_j} = \frac{\partial \overline{p}}{\partial t} + \overline{u}_j \frac{\partial \overline{p}}{\partial x_j} + \overline{\rho}\sigma_{ij} \frac{\partial \overline{u}_i}{\partial x_j} - \frac{\partial \overline{\rho}q_j}{\partial x_j} \qquad (21.18)$$

where $\overline{h}$ and q_j are the static enthalpy and heat-flux vector. These equations exhibit a term-by-term correspondence with the equations for incompressible fluids. They are also identical to the equations used to determine laminar flows, except for the Reynolds-stress tensor and turbulent heat-flux vector. The specific stress tensor and heat flux vector are given by

$$\sigma_{ij} = 2\nu \left(S_{ij} - \frac{1}{3}\delta_{ij}\frac{\partial \overline{u}_k}{\partial x_k} \right) - \Re_{ij}$$

$$q_j = -\frac{\nu}{\mathrm{Pr}_l}\frac{\partial \overline{h}}{\partial x_j} + \frac{\overline{\rho u_j' h'}}{\overline{\rho}} \qquad (21.19)$$

where

$$\Re_{ij} = \frac{\overline{\rho u_i' u_j'}}{\overline{\rho}} \qquad (21.20)$$

Here δ_{ij} and $\mathrm{Pr}_l \equiv c_p \mu / k$ are the *Kronecker delta* and the *molecular Prandtl number*, respectively. The mass-weighted average equivalent of R_{ij} is now denoted as $\Re_{ij}$. It is generally assumed that the molecular dynamic viscosity and thermal conductivity have the same functional relationship with temperature for gases [Lagerstrom (1964)].

The incompressible Reynolds-averaged Navier–Stokes equations are presented above in Cartesian coordinates $(\overline{\mathbf{x}}, t)$. The equations can also be expressed in arbitrary curvilinear coordinates $(\overline{\xi}, \tau)$ and in a conservation law form. As an example of representing the equations in $(\overline{\xi}, \tau)$, a compact form of the compressible Reynolds-averaged Navier–Stokes equations in conservative Cartesian variables is presented below. These equations are made nondimensional by constructing characteristic values of all the variables entering these equations from reference values of length, velocity, density, the coefficients of viscosity and thermal conductivity, and of temperature and entropy such that the product of these two is the square of the reference velocity. Consequently, the dimensional and nondimensional equations are identical, except for the appearance of the Reynolds number (Re $\equiv U_\infty L / \nu$) and Prandtl number in the nondimensional equations. These equations are written as

$$\frac{\partial \mathcal{D}Q_m}{\partial \tau} + \sum_{i=1}^{d} \frac{\partial C_{mi}}{\partial \xi_i} = \frac{1}{\mathrm{Re}} \sum_{i=1}^{d} \frac{\partial V_{mi}}{\partial \xi_i} \qquad (21.21)$$

where d is the number of spatial dimensions, m is the equation number, varying from 1 to $d + 2$, and Q_m, C_{mi}, and V_{mi} are as follows

$$Q_m \equiv [\bar{\rho}, (\bar{\rho})\bar{u}_1, \ldots, (\bar{\rho})\bar{u}_d, \bar{e}]^T, \qquad C_{mi} = \mathfrak{D} Q_m \mathfrak{U}_i + \bar{p}\Phi_{mi}$$

$$V_{mi} \equiv \mathfrak{D} \sum_{j=1}^{d} \frac{\partial \xi_i}{\partial x_j} \left(0, \sigma_{j1}, \ldots, \sigma_{jd}, \sum_{k=1}^{d} \bar{u}_k \sigma_{jk} - q_j\right)^T$$

$$\mathfrak{U}_i = \frac{\partial \xi_i}{\partial t} + \sum_{j=1}^{d} \bar{u}_j \frac{\partial \xi_i}{\partial x_j}$$

$$\Phi_{mi} \equiv \mathfrak{D} \sum_{j=1}^{d} \frac{\partial \xi_i}{\partial x_j} (0, \delta_{j1}, \ldots, \delta_{jd}, \bar{u}_j)^T, \qquad \mathfrak{D} \equiv J\left(\frac{x_1, \ldots, x_d}{\xi_1, \ldots, \xi_d}\right)$$

$$e_I = \frac{\bar{e}}{\bar{\rho}} - \sum_{k=1}^{d} \frac{\bar{u}_k \bar{u}_k}{2}, \qquad \bar{p} = (\gamma - 1)\bar{\rho} e_I$$

$$\frac{\partial \xi_i}{\partial t} = -\sum_{j=1}^{d} \frac{\partial \xi_i}{\partial x_j} \frac{\partial x_j}{\partial \tau}, \qquad \frac{\partial \xi_i}{\partial x_j} \equiv \frac{1}{\mathfrak{D}} J\left(\frac{x_{j+1}, x_{j+2}}{\xi_{i+1}, \xi_{i+2}}\right) \qquad (21.21a)$$

In the previous expression, subscripts $(i, i + 1, i + 2)$ and $(j, j + 1, j + 2)$ vary in a cyclic order, $(1, 2, 3)$, $(2, 3, 1)$, etc. The symbol J represents the Jacobian, and the superscript T denotes a transpose of a vector. The Stokes hypothesis, $(3\lambda + 2\mu = 0)$, of local thermodynamic equilibrium has been used. The total energy per unit volume and the internal energy per unit mass are represented by e and e_I, respectively. Further, the stress tensor and heat flux vector are given by

$$\sigma_{ij} = \mu \sum_{k=1}^{d} \left(\frac{\partial \xi_k}{\partial x_j} \frac{\partial \bar{u}_i}{\partial \xi_k} + \frac{\partial \xi_k}{\partial x_i} \frac{\partial \bar{u}_j}{\partial \xi_k} - \frac{2}{3} \sum_{n=1}^{d} \frac{\partial \xi_n}{\partial x_k} \frac{\partial \bar{u}_k}{\partial x_n} \delta_{ij}\right) - \mathfrak{R}_{ij}$$

$$q_i = -\frac{\gamma\mu}{\mathrm{Pr}_I} \sum_{j=1}^{d} \frac{\partial \xi_j}{\partial x_i} \frac{\partial e_I}{\partial x_j} + \frac{\overline{\rho u_i' h'}}{\bar{\rho}} \qquad (21.22)$$

Unfortunately, the turbulence quantities R_{ij}, Eq. (21.15), and $\mathfrak{R}_{ij}$ and $\overline{\rho u_i' h'}/\bar{\rho}$, Eq. (21.22), cannot be expressed in terms of the other variables. This leads to the classical *closure* problem and the necessity to construct turbulence models.

Detailed representations of the Navier–Stokes equations and their derivations in various coordinate systems are given by Lagerstrom (1964), Bird, Stewart, and Lightfoot (1960), Serrin (1959a), and Anderson, Tannehill, and Pletcher (1984).

21.2.3 Composite Equations for Thin- and Slender-Shear-Layers

Many viscous flow problems that cannot be solved by the thin- or slender-shear-layer approximations via thin- or slender-shear-layer equations, can be solved with an intermediate set of equations that fall between these simple equations and the complex Navier–Stokes equations. This intermediate set of equations is applicable to both inviscid and viscous regions, just as the Navier–Stokes equations are, therefore these equations can be used to compute strong interactions between the two. Since the equations couple the viscous and inviscid approximations to a fluid flow, they have been referred to as *composite equations* [Van Dyke (1975)]. The main feature of these equations is the presence of a nonzero normal pressure gradient, a necessary term for coupling and simultaneously solving the viscous and inviscid regions. Relative to the Navier–Stokes equations, these composite equations require less computational effort, because they contain fewer terms. Quite often a steady flow can be computed by a boundary layer type of marching procedure in the streamwise direction that requires very little computing work.

There are several forms of the composite equations. The various forms are referred to by various names, which differ primarily by the adjective used to describe them—*thin layer, slender-layer, parabolized, parabolic, partially parabolized, thin-layer parabolized,* and *conical*; an example is the parabolized Navier–Stokes equations. The terms *thin-layer, slender-layer,* and *conical* are based on physical considerations; the terms *parabolic* and *parabolized* are derived from analytical characteristics. A set called the *viscous shock-layer equations* also belongs to this class. These equations are parabolic in the flow direction only when there is no influence of the pressure field in the direction opposite to this direction. If the extent of the upstream influence is small, then sometimes the governing equations are modified locally to maintain the parabolic or marching characteristic. When the upstream effects are large, then the time or pseudotime dimension is introduced. The governing equations then are partially parabolic in time, and a marching numerical procedure in time is used.

The composite thin-shear-layer equations are generally referred to in the literature as *Thin-Layer Navier–Stokes equations* (TLNS). These equations are obtained by neglecting all streamwise and spanwise derivatives of the viscous and turbulence stress and heat-flux terms and any term involving mixed derivatives. This is justified either by physical order of magnitude analysis or by a computational accuracy argument. The latter argument amounts, in simple terms, to the observation that since the neglected terms cannot be computed correctly with the available grid resolution anyway, why keep them? On the other hand, the *composite slender-layer Navier–Stokes equations* are derived by neglecting only the streamwise derivatives of the viscous and turbulence stress and heat-flux terms, and mixed-derivative terms involving the streamwise derivative. (Note that if the slender-shear layer is axisymmetric, then the use of cylindrical coordinates leaves only one large gradient of these terms in the radial direction.) For a compressible fluid, these sets of equations may be written as

$$\frac{\partial \mathcal{D} Q_m}{\partial \tau} + \sum_{i=1}^{d} \frac{\partial C_{mi}}{\partial \xi_i} = \frac{\mathcal{D}}{\text{Re}} \sum_{i=2}^{l} \sum_{j=1}^{d} \frac{\partial \xi_i}{\partial x_j} \frac{\partial \mathcal{V}_{mj}}{\partial \xi_i} \tag{21.23}$$

in which the composite thin-shear layer equations and slender-shear layer equations are, respectively, given by the value of l equal to 2 for $d > 1$ and to 3 for $d > 2$, if the streamwise direction is represented by ξ_1 and if the directions normal to the shear layer are represented by ξ_2 and ξ_3. By the same arguments, the upper and lower limits for the k and j summations in Eq. (21.22) are replaced by l and 2, respectively. The truncated viscous term is evaluated using

$$\mathcal{V}_{mj} = \left(0,\ v_{j1},\ \ldots,\ v_{jd},\ \sum_{k=1}^{d} \bar{u}_k v_{jk} - \theta_{mj}\right)^T$$

$$v_{ij} = \mu \sum_{k=2}^{l} \left(\frac{\partial \xi_k}{\partial x_j} \frac{\partial \bar{u}_i}{\partial \xi_k} + \frac{\partial \xi_k}{\partial x_i} \frac{\partial \bar{u}_j}{\partial \xi_k} - \frac{2}{3} \sum_{n=1}^{d} \frac{\partial \xi_n}{\partial x_k} \frac{\partial \bar{u}_k}{\partial x_n} \delta_{ij}\right) - \mathcal{R}_{ij}$$

$$\theta_i = -\frac{\gamma \mu}{\mathrm{Pr}_l} \sum_{j=2}^{l} \frac{\partial \xi_j}{\partial x_i} \frac{\partial e_l}{\partial x_j} + \frac{\overline{\rho u_i' h'}}{\rho} \tag{21.24}$$

Instead of Eq. (21.23), the following equation has been generally used for convenience

$$\frac{\partial \mathcal{D} Q_m}{\partial \tau} + \sum_{i=1}^{d} \frac{\partial C_{mi}}{\partial \xi_i} = \frac{1}{\mathrm{Re}} \sum_{i=2}^{l} \sum_{j=1}^{d} \frac{\partial}{\partial \xi_i} \left(\mathcal{D} \mathcal{V}_{mj} \frac{\partial \xi_i}{\partial x_j}\right) \tag{21.25}$$

On the right-hand side of this equation, the relationships between the transformation metric quantities

$$\sum_{i=1}^{d} \frac{\partial}{\partial \xi_i} \left(\mathcal{D} \frac{\partial \xi_i}{\partial x_j}\right) = 0 \tag{21.26}$$

are not satisfied, but these are used in the derivation of the conservative form of the right-hand side of Eq. (21.21). Depending on the flow conditions, the error introduced by not satisfying Eq. (21.26) may be insignificant.

The approximation referred to as the parabolized Navier–Stokes equations (PNS) omits the time-derivative in Eq. (21.23) or (21.25) and marches the solution in the streamwise direction. For this approximation to be valid, the inviscid outer flow must be supersonic and the upstream influence of the streamwise pressure gradient in the subsonic viscous flow either must be treated approximately or accounted for by introducing some iterative or relaxation procedure, that is by the introduction of *pseudotime*. A special subset of this approximation has been called the *parabolic Navier–Stokes equations*. In these equations, the streamwise pressure-gradient term is deliberately specified in the subsonic region. In hypersonic flows, the viscous-shock layer equations are generally used. These equations do not include viscous and turbulent stress terms in the normal momentum equation.

When a flow field is surrounded by conical boundaries, there is no significant length scale in the conical direction for inviscid flows. This conical flow assumption has been applied locally to viscous flows. The conical Navier–Stokes equations can be determined by using a conical transformation,

$$\alpha = \xi_1, \ \beta = \frac{\xi_2}{\xi_1}, \ \gamma = \frac{\xi_3}{\xi_1} \ \text{and} \ \tau^* = \tau \tag{21.27}$$

together with the assumption of local conical self-similarity in which derivatives with respect to α are neglected. This leads to a set of equations of the form similar to that for the slender-shear layer, but with a source term. Further, the local Reynolds number is determined by the position where the solution is computed. These equations are

$$\frac{\partial \mathcal{D}Q_m}{\partial \tau} + \frac{\partial(F_{m2} - F_{m1})}{\partial \beta} + \frac{\partial(F_{m3} - F_{m1})}{\partial \gamma} + 2F_{m1} = 0 \qquad (21.28)$$

where

$$F_{mi} = C_{mi} - \frac{1}{\text{Re}} V_{mi} \qquad (21.28a)$$

The variables C_{mi} and V_{mi}, that are defined below Eq. (21.21), are appropriately modified for the above coordinate system [see Anderson, Tannehill, and Pletcher (1984)].

Partially parabolized Navier–Stokes equations are usually formulated for an incompressible fluid. Terms in the momentum equations are neglected in order to make these equations parabolic, and the pressure field is determined from an elliptic equation. A discussion of composite equations for an incompressible fluid, as well as solution procedures for these equations, is given by Rubin (1982).

21.2.4 Modeling of Turbulence

Turbulent flows are characterized by a range of scales. Some way to cope with this range of velocity and length scales is required for the successful modeling of turbulence. Most turbulence modeling methods employ only a single time and length scale for describing the various interactions. These methods are called *single scale methods*. Methods that introduce two or more time scales or length scales are referred to as *multiscale methods*. Some methods use only the averaged flow parameters and dimensions to determine the characteristic turbulent time scales and length scales, while other methods use one or two turbulence quantities.

There are two approaches to turbulence modeling—a first-order approach and a second-order approach. In the first-order approach, the Reynolds-stress tensor, a turbulence quantity, is modeled in terms of mean quantities. In the second-order approach, the terms in this tensor are carried along in the computations are dependent variables, having been determined from the Navier–Stokes equations in terms of other turbulence correlations and higher-order tensors which are modeled. In the former approach, one forms the equations for the first-order quantities, such as averaged velocities, and models the second-order quantities that appear in them [see Eqs. (21.12) and (21.15)]. This forms the basis for the models that are conventionally called the *zero-equation* (algebraic), *one-equation*, and *two-equation* models. In the second approach, equations are formed for the first- and second-order quantities ($\bar{u}_i$ and R_{ij} or $\mathcal{R}_{ij}$). The third-order terms representing the transport of R_{ij} or $\mathcal{R}_{ij}$, the pressure-strain redistribution tensor, and the dissipation tensor are modeled requiring at least one length scale or an equation for the rate of dissipation. The equations for the Reynolds stresses may be simplified to yield algebraic-stress relations, still requiring differential equations both for the turbulent kinetic energy and energy dissipation. In practice, the actual form of various basic turbulence models and the manner of applying them often differ in detail from investigator to investigator.

Some first-order turbulence models are based on the *Newtonian* assumption of linearity between the stress and rate-of-strain tensors, and the models are, therefore, *eddy-viscosity models*. Boussinesq's eddy viscosity concept [Boussinesq (1887)] is based on an analogy with the gradient-diffusion mechanism of the kinetic theory of gases. Methods based on this concept are also known as *eddy-diffusivity* or *gradient-transport methods*. Corrsin (1974) has discussed the limitations of gradient-transport methods. These methods are inappropriate when the turbulence energy production is not in balance with the energy dissipation rates, that is, in nonequilibrium (in the turbulence sense) flows.

In eddy-viscosity methods, the kinetic eddy viscosity, ν_t, is assumed to be a scalar and is defined by a Newtonian constitutive equation of the form

$$R_{ij} = \left(\frac{1}{3}\right) v^2 \delta_{ij} - 2(\nu_t \mathbf{S_{ij}} + \nu_t^* \mathbf{S_{ij}^*}) \tag{21.29}$$

$$\mathbf{S_{ij}} = S_{ij} - \langle S_{ij} \rangle; \ \mathbf{S_{ij}^*} = \langle S_{ij} \rangle$$

$$\mathfrak{R}_{ij} = \left(\frac{1}{3}\right) v^2 \delta_{ij} - 2\nu_t \left[S_{ij} - \left(\frac{1}{3}\right) \frac{\partial \bar{u}_k}{\partial x_k} \delta_{ij} \right] \tag{21.30}$$

Here, the Cartesian summation convention is used, and $\langle \ \rangle$ denotes the average over a plane. Equations (21.29) and (21.30) are used to determine the turbulence stresses in Eqs. (21.15) and (21.20), respectively. In Eq. (21.29), $v^2 = R_{ii}$, and the turbulence stresses have been separated into two parts, homogeneous and inhomogeneous turbulence [Schumann (1975)]. (Note that v^2 denotes $v'^2 + u'^2 + w'^2$ which is also twice the turbulent kinetic energy divided by the density.) The v^2 term may be absorbed in the definition of pressure [see Eq. (21.12)]. In Eq. (21.30), $v^2 = \mathfrak{R}_{ij}$ is twice the turbulent kinetic energy. The above relations restrict the Reynolds-stress and the strain-rate tensors to the same principal axes, which is not true in general. As discussed below, it is possible to modify this relation in order to remove this restriction.

The eddy models remove the Reynolds-stress and turbulent heat-flux terms from the equations of motion and approximate their effect by adding some empirically based function, μ_t, to the molecular viscosity, μ. For example, this leads to the alternative approximate form of Eq. (21.22),

$$\sigma_{ij} = (\mu + \mu_t) \sum_{k=1}^{d} \left(\frac{\partial \xi_k}{\partial x_j} \frac{\partial \bar{u}_i}{\partial \xi_k} + \frac{\partial \xi_k}{\partial x_i} \frac{\partial \bar{u}_j}{\partial \xi_k} - \frac{2}{3} \sum_{n=1}^{d} \frac{\partial \xi_n}{\partial x_k} \frac{\partial \bar{u}_k}{\partial \xi_n} \delta_{ij} \right)$$

$$q_i = -\frac{\gamma}{\mathrm{Pr}_l} \left(\mu + \frac{\mathrm{Pr}_l}{\mathrm{Pr}_t} \mu_t \right) \sum_{j=1}^{d} \frac{\partial \xi_j}{\partial x_i} \frac{\partial e_l}{\partial x_j} \tag{21.31}$$

where Pr_l and μ_t represent the turbulent Prandtl number and eddy viscosity, respectively.

Consider below some simple examples of zero- and two-equation models. All of these models are expressed in terms of a turbulence velocity scale $v^{\dagger}$ and a turbu-

---

[dagger]Henceforth the distinction between v [Eq. (21.29)] and $\mathbf{v}$ [Eq. (21.30)] is dropped.

lence length scale l. On dimensional grounds, a combination of these scales determines the value of the kinetic eddy viscosity

$$\nu_t = c_t \nu l \tag{21.32}$$

where c_t is sometimes referred to as the Clauser constant. From this expression, we can determine the dynamic eddy viscosity, $\mu_t = \rho \nu_t$. Algebraic models relate ν_t directly to averaged field quantities without additional partial differential equations. Both one- and two-equation models contain additional partial differential equations for the turbulence scales. One-equation models use a prescribed, empirical formula for the local length-scale, together with a single partial differential equation for the velocity scale, ν. Two-equation models use additional partial differential equations to find both ν and l. Methods using partial differential equations for the turbulence scales are also called *transport-equation methods*.

In addition, the eddy-conductivity concept is used to model the transport of heat owing to the time-averaged product of fluctuating enthalpy and fluctuating velocity. It is assumed that the turbulent heat flux follows a law similar to *Fourier's law*. Although the turbulent Prandtl number varies across the boundary layer, it is commonly considered to be a constant; and it is usually taken to be 0.9 for air. Still more complex modeling of the turbulent heat flux is discussed by Launder (1978).

There are many zero-equation models. The simplest types are constructed using equations that apply uniformly throughout the flow for the turbulence velocity and length scales. Such models are used, for example, in the so-called large eddy simulations (LES). An example based on space averaging* and used by Moin and Kim (1982) in Eq. (21.29) following Schumann (1975) is given by

$$\nu = l(2\mathbf{S}_{ij}\mathbf{S}_{ij})^{(1/2)}$$

$$l = (\Delta_1\Delta_2\Delta_3)^{(1/3)} (1 - e^{(-y^+/A^+)}) \tag{21.33}$$

$$\nu_* = l (2\mathbf{S}_{ij}^*\mathbf{S}_{ij}^*)^{(1/2)}$$

$$l_* = \Delta_3 (1 - e^{(-y^{+2}/A^{+2})}) \tag{21.34}$$

where Δ_i is twice the local grid-spacing in i-direction for $i = 1$ or 3; Δ_2 is the sum of two adjacent, one-half grid-spacings; $y^+ = \nu^* y/\nu$ is a Reynolds number related to the energy containing turbulence with ν^* and y being, respectively, the friction velocity and the x_2-direction; and the value of the *Van Driest damping constant A^+* is taken to be 25. From Eq. (21.32), both ν_t and ν_t^*, which are needed in Eq. (21.29), are computed with $c_t = 0.004225$ and $c_t = 0.065$, respectively, and with the values

---

*This model is formulated with two different space-averaging operators applied to the convective derivatives in Eq. (21.12), $\partial(\overline{u_i u_j})/\partial x_j$. In the horizontal (streamwise and spanwise) directions, the average of the product of averaged velocities is explicitly taken by a Gaussian averaging function, which is a function of only the horizontal coordinates. In the vertical direction x_2 (the direction normal to a solid surface), this outer averaging is implicitly done by a numerical second-order finite-difference scheme. This implicit averaging is the equivalent of using a top-hat averaging function in this direction, when $\partial(\overline{\overline{u_i u_2}})/\partial x_2$ is put in the finite-difference form. But, in the other convective terms, the top-hat averaging is not applied.

of velocity and length scales calculated using Eqs. (21.33) and (21.34). These models are often referred to as *sub-grid-scale models*. Without the exponential damping functions, the form of these eddy-viscosity models is the same as that of a model suggested by Smagorinsky (1963) for atmospheric studies. The length scale, l, without the damping function, is the same as the one used by Deardorff (1970a). The damping functions account for damped turbulence near the wall, as suggested by Van Driest (1956) for extending the proportionality of the length scale to y into the viscous sublayer.

A more elaborate zero-equation turbulence model used in the Reynolds-averaged formulation for two- and three-dimensional flows that are steady is due basically to Cebeci (1971). The turbulent boundary layer is regarded as a composite layer consisting of inner and outer regions. In each region, the distributions of v and l are prescribed by two different empirical expressions. In the inner (log-law) region, l is proportional to y, the distance normal to the wall, and in the outer layer, l is proportional to the shear-layer thickness. In the inner region, the proportionality of l to y is extended to the viscous sublayer with a damping function [Van Driest (1956)] that is modified to take into account the effect of pressure gradient, as was done, for example, by Cebeci and Meier (1979). In the outer (wake) region, the vorticity is used to define the boundary layer thickness, following Baldwin and Lomax (1978), and the Clauser constant c_i is made a function of the local Reynolds number based on the momentum thickness, as suggested by Cebeci (1973). Based on these considerations, the model presented below is a combination of the Cebeci model and the Baldwin–Lomax model.

In the inner layer, $0 \leq y \leq y_c$, the expressions for v and l are

$$(v)_{\text{inner}} = l|\overline{\Omega}_i|$$

$$(l)_{\text{inner}} = \alpha_1 y[1 - e^{(-y^+/A^+)}]$$

$$A^+ = 26(1.0 - 11.8p^+)^{(-1/2)}$$

$$p^+ = -\left(\frac{\mu}{\text{Re}\,\rho^2 u_\tau^3}\frac{\partial\overline{p}}{\partial x}\right) \tag{21.35}$$

with $\alpha_1 = 0.40$ in Eq. (21.35) and $c_i = 1.0$ in Eq. (21.32). In this case, the wall coordinate y^+ can be based on the local value of the density [Cebeci (1971)] or on the value of the density at the wall [Baldwin and Lomax (1978)].

In the outer region, $y > y_c$, the expressions for v and l are

$$(v)_{\text{outer}} = \text{minimum of}\left(L_{\max},\ \frac{U_{\text{diff}}^2}{L_{\max}}\right)$$

$$(l)_{\text{outer}} = y_{\max}C_{BL}\alpha_2 \tag{21.36}$$

If the function $L(y)$ is defined to be

$$L(y) = y|\overline{\Omega}_i|\,[1 - e^{(-y^+/A^+)}] \tag{21.37}$$

the quantity $L_{\max}$ is the maximum value of $L(y)$ that occurs in this equation, and $y_{\max}$ is the value of y at which it occurs. The above exponential term is negligible in the outer part of the shear layer. The quantity U_{diff} is the difference between the absolute velocity at y_{max} and the minimum value of the absolute velocity, the value of C_{BL} is 1.6, and the *Klebanoff intermittency factor*, α_2, is given by

$$\alpha_2 = \left[1 + 5.5 \left(\frac{C_K y}{y_{max}} \right)^6 \right]^{-1} \tag{21.38}$$

where the *Klebanoff constant* $C_K = 0.3$. The Clauser constant, c_t, in Eq. (21.32) is

$$c_t = 0.0168 \frac{(1 + 0.55)}{(1 + \beta)} \tag{21.39}$$

where

$$\beta = 0.55 \, (1 - e^{-0.243 z_1^{1/2} - 0.298 z_1})$$

$$z_1 = \frac{Re_\theta}{425} - 1 \tag{21.39a}$$

The Reynolds number based on the momentum thickness is Re_θ.

The region of validity of the inner and outer scales is determined by y_c, which is defined to be the smallest value of y at which the values of the inner and outer eddy viscosity are the same. The value of α_1 in the inner region and of c_t in the outer region are assumed to be universal constants for $Re_\theta > 5,000$. At lower Reynolds numbers, α_1 and c_t are functions of Reynolds number, as for example, in Eq. (21.39) for c_t [Cebeci (1973)].

The model presented above is based on the absolute value of the vorticity, and for many simple flows it works about as well as a zero-equation model can be expected to. However, in the absence of strain and body forces, it is inappropriate for irrotational shear flow, because it gives an infinite, rather than finite, value of the time-scale. It is also inappropriate for a purely rotating flow because it gives a finite, rather than an infinite, value of the time-scale [Hanjalic (1982)]. It is also clear that the turbulence length scale given above needs to be modified for flows in corners and for other flows bounded by two or more walls. In addition to these limitations, the manner of determining y_{max} in some flows is questionable, and the constants C_{BL} and C_K may be functions of Mach number.

The above models are *mean-field models*; only averaged quantities are used to model turbulence. As an example of a single-scale method utilizing two turbulence quantities, a two-equation model of Wilcox and Rubesin (1980) is presented next. In this model, twice the turbulent kinetic energy ρv^2, which represents indirectly the role of the velocity scale, and the specific energy dissipation $\rho \omega^2$, which in combination with the former gives the length scale, are given by

$$\frac{\partial(\rho v^2)}{\partial t} + \frac{\partial(\rho \bar{u}_j v^2)}{\partial x_j} = -2\rho \Re_{ij} \frac{\partial \bar{u}_i}{\partial x_j} - \beta_1 \rho \omega v^2 + \frac{\partial[(\mu + \beta_2 \mu_t)\partial v^2/\partial x_j]}{\partial x_j} \tag{21.40}$$

$$\frac{\partial(\rho \omega^2)}{\partial t} + \frac{\partial(\rho \bar{u}_j \omega^2)}{\partial x_j} = -\beta_3 l^{-2} \rho \Re_{ij} \frac{\partial \bar{u}_i}{\partial x_j} - \left[\beta_4 + \beta_5 \left(\frac{\partial l}{\partial x_k} \right)^2 \right] \rho \omega^3$$

$$+ \frac{\partial[(\mu + \beta_5 \mu_t)\partial \omega^2/\partial x_j]}{\partial x_j} \tag{21.41}$$

where the length scale is defined by $l = v/\omega$, and the Cartesian summation convention is used. The eddy viscosity is again computed from Eq. (21.32), and the constitutive Eq. (21.30) is used here to provide $\mathcal{R}_{ij}$. The above model requires that in addition to c_t the coefficients β_1, β_2, β_3, β_4, β_5, and β_6 be specified. Wilcox and Rubesin (1980) recommend the following values for these coefficients

$$\beta_1 = 0.09, \ \beta_2 = \beta_5 = 0.5, \ \beta_4 = 0.15, \ \beta_6 = 1/11,$$

$$c_t = [1 - (1 - \beta_6^2)e^{(-\mathrm{Re}_t/2)}]/2 \text{ with } \mathrm{Re}_t = vl/\nu$$

$$\beta_3 = \beta_7 \, [1 - (1 - \beta_6^2)e^{(-\mathrm{Re}_t/4)}]/c_t \text{ where } \beta_7 = 10/9$$

Two-equation models like the one above are the simplest models for which the length scale is not prescribed empirically; the models are, therefore, likely to be less restrictive than the zero- or one-equation models. Examples in which the models are useful are flows with multiple-shock waves, problems with more than a single surface, recirculating flows, and flows with multiple length scales. However, these models are still weak, because the model constants are not universal, and they are found to be applicable to a limited class of problems. Perhaps, their range of applicability can be extended as explained below.

In these examples of eddy-viscosity-type models, the eddy viscosity is assumed to be isotropic. Such models cannot account for any directional influences of turbulence. However, some generalizations to the two-equation models have been developed that do attempt to account for this influence.

Wilcox and Rubesin (1980) have disturbed the alignment of the Reynolds and rate-of-strain tensors by adding

$$-\frac{4}{9} v^2 \left(\frac{S_{im}\overline{\Omega}_{mj} + S_{jm}\overline{\Omega}_{mi}}{\beta_1\omega^2 + 2S_{mn}S_{nm}} \right)$$

to the right-hand side of Eq. (21.30). Here, the vorticity tensor is given by

$$\overline{\Omega}_{ij} = \frac{1}{2} \left(\frac{\partial \overline{u}_i}{\partial x_j} - \frac{\partial \overline{u}_j}{\partial x_i} \right)$$

The Wilcox–Rubesin model is generally inaccurate for boundary layers in adverse pressure gradients. Wilcox (1984) presented a model that uses ω, the rate of dissipation of turbulence per unit energy, instead of ω^2. This model appears to perform satisfactorily in adverse pressure gradients.

Algebraic-stress relations also lead to nonisotropic eddy viscosities. These models are economical compared with the second-order Reynolds-stress models, and the models do maintain some of the universality of the latter. As an example, a model proposed by Rodi (1976) is presented here. The Reynolds stresses for an incompressible fluid are determined from the following algebraic relation

$$R_{ij} = \frac{v^2}{2} \left[\frac{2}{3} \delta_{ij} + \frac{(1 - c_2) \, (P_{ij}/\epsilon - (2/3)\delta_{ij}P/\epsilon)}{c_1 + P/\epsilon - 1} \right]$$

$$P_{ij} = -R_{ik}\frac{\partial \overline{u}_j}{\partial x_k} - R_{jk}\frac{\partial \overline{u}_i}{\partial x_k}; \ P = \frac{1}{2} P_{ii} \qquad (21.42)$$

The rate of dissipation of turbulent kinetic energy ϵ and v^2 are computed from transport equations for these quantities. The constants c_1 and c_2 are determined empirically. The above model is derived by approximating the convective and diffusive terms and introducing assumptions about the pressure-strain and dissipation terms in the Reynolds-stress equations. This model is valid for turbulent flows with negligible effects of molecular viscosity (high Reynolds numbers). Further, the Reynolds stresses are determined iteratively, for they appear on both sides of Eq. (21.42). The reader is referred to Sec. 4.5, Schetz (1993) and Wilcox (1993) for discussions of other turbulence models.

21.2.5 Boundary Conditions

The essential ideas presented below are applicable to both incompressible and compressible fluids, since the physical boundary conditions are basically the same for both types of fluids. The basic equations of motion are valid both in the computational domain and on its boundary. Auxiliary governing equations may be formed from these equations; for example, an equation for the pressure field of an incompressible fluid can be derived by taking the divergence of the momentum equation. When the basic and auxiliary equations are applied at the boundary, the equations must be evaluated subject to the boundary conditions.

Boundary conditions for differential equations, the equations of motion including differential equations for turbulence modeling, are determined by mathematical and physical considerations. The mathematical character of these governing equations dictates the number and types of boundary conditions that determine, along with initial conditions, the well-posedness of a problem, that is, the existence and uniqueness of solutions. Further, this mathematical character is determined by the theory of characteristics. Two kinds of theoretical analyses dealing with boundary conditions are available—one based on the classical energy method [see Elvius and Sundström (1973)] which follows the earlier work of Serrin (1959b), and the other based on the normal mode concept [Kreiss (1970)]. Most of the work regarding boundary conditions has been done for the compressible Eulerian and shallow-water equations. A few studies deal with the compressible Navier–Stokes equations [see Oliger and Sundström (1978) and Gustafsson and Sundström (1978)]. These studies consider both the number and a possible set of admissible forms of the boundary conditions. Such studies serve as a guide rather than as a definitive tool in Navier–Stokes computations. The heuristic discussion of boundary conditions given below is based on both the mathematical character of the equations and on physical considerations, but it is not based on the detailed analytical procedures mentioned above.

The mathematical character of the system represented by the linearized form of Eq. (21.9) or (21.12) is *parabolic–elliptic*. The vorticity equation is parabolic, and the stream-function equation is elliptic. On the other hand, the mathematical character of the system represented by the linearized form of Eq. (21.21) is *incompletely parabolic* [Belov and Yanenko (1971)] or *parabolic–hyperbolic*. Without the time derivative, it is *elliptic–hyperbolic*. The system given by Eq. (21.23) or (21.25) is *incompletely hyperbolic* or *hyperbolic–parabolic*. For example, the composite thin-shear layer system is parabolic in the $(\xi_2 - t)$ plane. The global character of these systems remains the same, even if the local character may be, for instance, purely hyperbolic. Therefore, the boundary conditions are determined by the global character of these systems.

First, consider the boundary conditions for Eq. (21.9), with the vorticity and stream-function equations assumed to be uncoupled. Since the highest spatial derivative in both equations is of second-order in each coordinate direction, two boundary conditions are required in each direction for both variables. These considerations determine the number of boundary conditions. The type of the boundary condition for each variable in any direction is again determined by the highest derivative of the variable in that direction. Following the theory of ordinary differential equations, the boundary condition should be at least one order lower than the highest derivative. This constraint yields boundary conditions that either are Dirichlet or Neumann or of a mixed type.

Next, consider each equation of the system represented by Eq. (21.21) as an equation determining Q_α, separately from the others, assuming the other Q's in this equation to be known quantities. Here, Q_α is a component of vector Q_m. The mass-conservation equation requires one boundary condition in each coordinate direction for Q_1. The second derivative of Q_2 in the Q_2 equation requires two boundary conditions in each coordinate direction. Likewise, two boundary conditions are required for the remaining Q's. This means that at the boundary of the computational domain, conditions specifying $Q_2, \ldots, Q_{d+1}$ are required, and a condition specifying Q_1 is needed on a part of this boundary.

The above heuristic considerations help formulate boundary conditions based on physical considerations. In external flow problems, two kinds of boundaries arise—rigid-wall boundaries and open boundaries. A rigid wall constrains the flow field along the boundary. This physical constraint is relatively easy to formulate and convert into computational boundary conditions. Open boundaries do not provide a material constraint, hence appropriate conditions are not obvious.

A rigid-wall boundary provides velocity and temperature conditions. The behavior of a real gas at ordinary conditions (Knudsen numbers less than 10^{-2}) is accurately described by the no-slip and no-temperature jump conditions. The latter condition corresponds to either the specification of the rigid-wall temperature or the temperature gradient at the wall. These are the only two physical conditions available. (In contrast, for inviscid flows there is only one physical boundary condition, namely, no flow normal to a rigid wall. Please note that, for an inviscid flow past an airfoil, a condition is required at or near the trailing edge of the airfoil, if the inviscid streamline leaving the airfoil is to correspond with that in a viscous flow.) Consider the case of impermeable walls. If the flow is two-dimensional and the fluid is incompressible, for example, the vanishing of surface velocities results in two conditions—the surface is a streamline, and the normal derivative for the stream function is zero. Only the first condition is generally required for the stream-function equation, Eq. (21.9). The second condition is used to determine the vorticity on the surface from the stream-function equation. This condition must also be enforced to solve the stream-function equation, if a numerical method requires the value of the normal derivative for the stream function at the surface as was done by Mehta (1977). Similarly, the first condition must be used to determine the surface vorticity from the stream-function equation, if it contains a first derivative of the stream function along the surface. If the fluid is compressible, the no-slip condition translates into vanishing contravariant velocity components, $\mathfrak{U}_i = 0$ below Eq. (21.21). Further, the temperature condition gives either a Dirichlet or a Neumann condition for the total energy equation.

The mass-conservation equation contains the material derivative of Q_1. Consequently, at a boundary, Q_1 changes if its history is known, otherwise a condition on Q_1 must be specified. This means that if fluid is on the boundary of a computation domain or within this domain, Q_1 is determined by the mass-conservation equation. But, if fluid enters this domain by crossing the boundary, Q_1 must be specified. Therefore, Q_1 cannot be specified on a rigid-wall boundary; it must be calculated from its material derivative. When this recourse leads to numerical difficulties, a new governing equation is formulated by combining the momentum equations to form the normal derivative of pressure

$$\frac{\partial \bar{p}}{\partial n} = (\nabla p) \cdot \left(\frac{\nabla \xi_2}{|\nabla \xi_2|} \right) \tag{21.43}$$

where n represents the direction normal to ξ_2-surface. After expressing pressure in terms of Q's (equation of state) and after using $\mathcal{U}_i = 0$, we have

$$\frac{\partial \bar{p}}{\partial n} \left[\sum_{j=1}^{d} \left(\frac{\partial \xi_2}{\partial x_j} \right)^2 \right]^{1/2} = \bar{\rho} \left(\frac{\partial}{\partial \tau} \frac{\partial \xi_2}{\partial t} + \sum_{j=1}^{d} \bar{u}_j \frac{\partial}{\partial \tau} \frac{\partial \xi_2}{\partial x_j} \right) + \frac{1}{\mathcal{D} \, \mathrm{Re}}$$

$$\cdot \left(\frac{\partial \xi_2}{\partial x_1} \sum_{i=1}^{d} \frac{\partial V_{2i}}{\partial \xi_i} + \frac{\partial \xi_2}{\partial x_2} \sum_{i=1}^{d} \frac{\partial V_{3i}}{\partial \xi_i} + \frac{\partial \xi_2}{\partial x_3} \sum_{i=1}^{3} \frac{\partial V_{4i}}{\partial \xi_i} \right) \tag{21.44}$$

where the first subscript on the vector $\overline{\mathbf{V}}$ indicates the appropriate component of this vector. Since Eq. (21.44) is derived from the momentum equations, and since it replaces the mass-conservation equation, it is not a boundary condition on Q_1. When the right-hand side of this equation is put into a discrete form, the no-slip condition and no-temperature jump condition are enforced. The viscous terms of this equation are zero when the thin-shear-layer approximation is valid. In an orthogonal curvilinear coordinate system, the momentum equation normal to the surface is used to determine Q_1 at the rigid-wall boundary.

The above conditions are also valid for internal flow problems, including those in which simulations of the external-flow problems consider wind-tunnel-wall effects. The latter type of problem is difficult, for there are more surfaces developing shear layers than in either an internal- or external-flow problem. An alternative is not to compute the wind-tunnel-wall boundary layers. In this case, the wind-tunnel walls cannot be considered as open boundaries if the walls distort the flow field around an aerodynamic body from that observed in free flight. An ideal situation is to use measured flow quantities just beyond the wind-tunnel-wall boundary layers as boundary conditions. Probably the next best option is to use only measured pressure values, again just beyond the wall boundary layers, and to consider the boundary formed by the pressure measurement locations as an open boundary. Another approach is to use a slip boundary condition along the walls, provided the walls are contoured such that they coincide with streamlines in free-flight conditions. This is restrictive; for instance, in unsteady flows these free-flight streamlines, at a short distance from the body, are time-dependent, but the wall contours are not likely to be. Instead of these alternatives, free-flight boundary conditions, applied far away from both the wind-tunnel walls and the aerodynamic configuration, may be used when comparing the corresponding computations with data from adaptive-wall wind-tunnels.

The inflow, outflow, and tangent-flow open boundaries require different treatments. Consider first the boundary condition on density. The above discussion dealing with material derivative of Q_1 suggests that Q_1 must be specified on an inflow boundary. On outflow boundaries, Q_1 is determined from the mass-conservation equation, and on a tangent-flow boundary, Eq. (21.44) is used along with the equation of state.

For external-flow problems, idealized or sometimes analytical [see Imai (1951)], boundary conditions are available at infinity or at a great distance from a body. If the inflow boundary is at a short distance, say about 10 times the characteristic length of body, then the influence of the body at that distance is usually negligible, therefore it is possible to use the above conditions as inflow boundary conditions. This leads to the specification of the remaining Q's.

Another prescription for the inflow boundary conditions in external-flow problems is to assume the fluid to be inviscid at this boundary. This leads to specification of boundary conditions that correspond to those for the Euler equations. This prescription is commonly used.

The main difficulty in specifying the outflow boundary conditions across a wake of a body is that the boundary values are part of the solution, hence they are not known *a priori*. However, we do know something about the outflow boundary. There are no physical *boundary layers*, and the flow field is inviscid for all practical purposes. This suggests that the boundary conditions must not introduce any mathematical boundary layer. Similarly, the boundary conditions must not introduce a boundary layer on an open-tangent-flow boundary. Further, on the part of the outflow boundary which cuts the wake region, the flow is rotational. For instance, during passage of vortices or *eddies* through this boundary, the pressure on this boundary varies. The variation depends on the strength of the vortices. In addition, extrapolation of dependent variables* or some combinations of dependent variables along curvilinear coordinates may introduce errors, if one or more of relations between metric coefficients, Eq. (21.26), are not satisfied. This situation is analogous to that between Eqs. (21.23) and (21.25). For these reasons, a possible set of conditions on open-outflow and tangent-flow boundaries are the Euler equations for the Navier–Stokes equations. In other words, the viscous, turbulent and heat-conduction terms are neglected on the outflow and tangent-flow boundaries. This philosophy concerning these boundaries was applied to the incompressible Navier–Stokes equations by Mehta and Lavan (1975) and by Mehta (1977). These conditions satisfy the type of constraint on the boundary conditions as required by the mathematical character of the system represented by Eqs. (21.9), (21.12), or (21.21). The Euler equations at an outflow boundary should be solved as discussed in the next paragraph. When the wind-tunnel flows are simulated with open boundaries, as discussed above, the outflow condition on Q_1 may be replaced by the measured pressure values.

For the system represented by Eqs. (21.23) or (21.25), again heuristic arguments are used for determining the number of boundary conditions. The rigid-wall bound-

---

*An extrapolation of a variable along a curvilinear coordinate is equivalent to equating a derivative of this variable along this coordinate equal to zero, the order of the derivative depending on the extrapolation procedure.

ary conditions discussed above are applicable to the equations of this system. However, open boundaries for these equations require a different treatment. For the coordinate direction in which the viscous stress terms are retained, the above considerations are valid. But, when these terms are absent in a coordinate direction, the system represented by these equations is hyperbolic in that direction. Therefore, the direction of flow of information dictates the boundary conditions. If inflow is supersonic in the *hyperbolic* directions, then all Q's must be specified, otherwise one less specification is required. On outflow boundaries (in these directions) if the flow is supersonic, then nothing can be specified. If it is subsonic, one condition is required. In other words, the local characteristics or eigenvalues determine the number and the admissible forms of boundary conditions based on one-dimensional analysis. For a hyperbolic system, the eigenvalues are real. The number of negative eigenvalues with distinct eigenvectors determines the number of boundary conditions. This number is the same as the number of inward characteristics into the computational domain. In addition, characteristic relations along a characteristic can be extrapolated along this characteristic.

In practice, numerical methods without artificial viscosity require mainly the addition of extraneous diffusion, which is provided by the addition in each of the equations of motion of even-power derivatives of order higher than that of this equation. (This extraneous diffusion is often called *numerical dissipation*.) One point of view is that these terms do not change the mathematical character of the original equation, because they are a part of the truncation errors of a numerical method [Richtmyer and Morton (1967)]. These terms disappear when the computational grid size $\Delta\xi \to 0$. An alternative point of view is to consider these terms as a part of the original equations, since $\Delta\xi$ is never equal to zero. In this case, these terms do not change the character of the original *parabolic* equations. Also, the terms do not change the global character of the original hyperbolic equations, provided they do not introduce any boundary layers at the boundaries. This is achieved by not adding these terms either on the boundaries or next to the boundaries in the direction normal to boundary. This avoids additional boundary conditions for both parabolic and hyperbolic equations. These terms may form interior *boundary layers* such as *captured (smeared) shocks*. In this case, the *additional boundary conditions* for these higher order terms are automatically provided by the appropriate neighboring, interior-flow quantities.

21.3 NUMERICAL ASPECTS OF COMPUTING VISCOUS COMPRESSIBLE FLUID FLOW
Unmeel B. Mehta

21.3.1 Introduction

The numerical aspects of computing the effects of viscosity on fluid flow deal with: 1) devising stable, accurate, and efficient approximating numerical methods for the differential equations and the initial and boundary conditions, 2) generating a grid system, and 3) actually implementing the solution procedure. In this section, some of these aspects are discussed, beginning with a discussion of the concepts of numerical accuracy and mesh Reynolds number. The choice of computational grids based on accuracy requirements in then discussed, and the nonlinear convective phenomena related to shock waves, contact discontinuities, and turbulence are considered. A procedure for developing a class of implicit numerical methods for solving the Navier–Stokes equations for compressible fluids concludes the section. Only finite-difference formulations are considered. The reader is referred to textbooks by Anderson, Tannehill, and Pletcher (1984) and by Hirsch (1988), (1990) and to Chap. 19 for additional information and to a guide by Mehta (1995) for credible computational fluid dynamics simulations.

21.3.2 Numerical Accuracy

In computing the flow of a viscous fluid, the convective and viscous terms appearing in the equations of motion are evaluated by discrete approximations. Although primary interest is in multidimensional flows that are nonperiodic in the spatial dimensions, concepts of spatial, numerical accuracy are illustrated here, for simplicity, in one dimension for a periodic function. This restriction of periodicity facilitates use of *spectral methods* based on Fourier series for analyzing discretization errors. Since spectral methods are much more accurate than the commonly used difference methods, the results of these two methods are compared. In addition, the truncation error of approximating the convective term in a model equation is compared with the magnitude of the viscous term in this equation. Both terms are evaluated using finite-difference methods of the same order. This leads to the concept of *mesh Reynolds number* for viscous computations.

Fourier Series. A discrete, periodic function f with a period λ_{max} can be represented by a discrete Fourier expansion in the x-direction, with k denoting the wave number, as

$$f(x_j) = \sum_{n=-N}^{(N-1)} \tilde{f}(k_n)e^{ik_n x_j} \tag{21.45}$$

where $k_n = \pi/N\Delta x n$ and $N = (J-1)/2$. Here, j and n are the jth grid-point and the nth wave point, respectively, and N and J are the total number of wave numbers and the total number of mesh points, respectively. The value of N is an even integer. A Fourier component is identified by the overtilde. The shortest wavelength and the longest wavelength resolved by a finite-difference mesh are of wavelengths $\lambda_{min} = 2\Delta x$ and $\lambda_{max} = (J-1)\Delta x$, where Δx is the grid-point spacing. The corresponding wave numbers are $k_{max} = \pi/\Delta x$ and $k_{min} = \pi/(N\Delta x)$. Further, the nth Fourier component of the mth-order, discrete derivative is given by

$$\left(\frac{\widetilde{\delta^m f(x_j)}}{\delta x_j^m}\right)_n = i^m \kappa_n \tilde{f}(k_n) \tag{21.46}$$

where $\delta^m/\delta x^m$ is the discrete operator corresponding to $\partial^m/\partial x^m$. The form of this operator determines the value of the modified wave number, κ_n, in the equation. If f is represented by a continuous Fourier series, then

$$\frac{\widetilde{\partial^m f(x)}}{\partial x^m} = i^m k^m \tilde{f}(k) \tag{21.47}$$

In this case, the discrete Fourier components of the mth-order continuous derivative are also given by the above equation, with k^m being discrete, that is, with k_n^m. Some finite-difference forms of the idealized convective and diffusive terms of the equations of motion are considered below.

Convective Terms. The convective terms are of the form $g(\partial f/\partial x)$. First, consider the discrete evaluation of the derivative. If, for example, it is replaced by the second-order, central, finite-difference formula

$$\frac{\delta f}{\delta x} = \frac{f_{(j+1)} - f_{(j-1)}}{2\Delta x}$$

(21.48)

then

$$\left(\frac{\widetilde{\delta f(x_j)}}{\partial x_j}\right)_n = \frac{[e^{ik_n(\Delta x)} - e^{ik_n(-\Delta x)}]}{2\Delta x} \tilde{f}(k_n)$$

(21.49)

The simplification of the above quotient gives the modified wave number as

$$\kappa'_{2n} = \frac{\sin \theta_n}{\Delta x}$$

(21.50)

where $\theta_n = \Delta x k_n$. The prime is used to indicate a modified wave number for first derivatives. The first subscript of the modified wave number denotes the order of the truncation error of this wave number. It is determined by expanding the right-hand side of Eq. (21.50) in terms of a Taylor series, i.e. $\kappa'_{2n} = k_n + O(\Delta x^2)$. This order of truncation error of the modified wave number determines the order of the truncation error of the difference operator. Of course, the latter can also be obtained by a Taylor series analysis of the difference operator. Similarly, the fourth-order, central-difference and Padé schemes [Kopal (1961)] for the first derivative are analyzed. These schemes and the corresponding modified wave numbers are, respectively,

$$\frac{\delta f}{\delta x} = \frac{f_{(j-2)} - 8f_{(j-1)} + 8f_{(j+1)} - f_{(j+2)}}{12\Delta x}$$

$$\kappa'_{4n} = \frac{8 \sin \theta_n - \sin (2\theta_n)}{6\Delta x}$$

(21.51)

and

$$\left[1 + \frac{(\Delta x)^2}{6} (D_+ D_-)\right] \frac{\delta f}{\delta x} = \frac{f_{(j+1)} - f_{(j-1)}}{2\Delta x}$$

$$\kappa'_{4pn} = \frac{3 \sin \theta_n}{\Delta x(2 + \cos \theta_n)}$$

(21.52)

where

$$(D_+ D_-)f = \frac{f_{(j+1)} - 2f_{(j)} + f_{(j-1)}}{\Delta x^2}$$

(21.53)

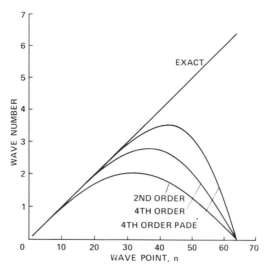

FIGURE 21.19 Comparison of exact and modified wave numbers for the first derivative: $N = 64$ and $\lambda_{max} = 20\pi$.

The exact and modified wave numbers for the above finite-difference formulas are shown in Fig. 21.19; the relative errors are shown in Fig. 21.20. The relative error is defined as $(k_n - \kappa'_n)/k_n$. In Figs. 21.19 and 21.20, results are illustrated for $N = 64$ and $\lambda_{max} = 20\pi$. The following conclusions can be drawn from these two figures. First, it is obvious that higher order methods are more accurate. Second, different methods of the same order of truncation error have different magnitudes of error. Third, as the wave number increases, the error introduced by the finite-difference formulas increases, becoming very large at high wave numbers. Fourth, higher order difference methods require fewer grid-points for the same accuracy. If, for example, a 10% error is acceptable, then the fourth-order Padé is acceptable up to $k = 3.8$, whereas the second-order formula is acceptable only up to $k = 1.6$. Therefore, this Padé formula requires less than half as many grid-points as the second-order formula. Fifth, at small values of the wave number, the slope of the relative error with respect to the wave number corresponds with the order of truncation error of difference scheme.

Next consider the nonlinear convective interaction of $g(\partial f/\partial x)$. Each term of this product may be represented by a Fourier series. This interaction is then the interaction of two waves. In unsteady flows and in relaxation numerical methods, this interaction constantly produces higher wave numbers. For example, the product of the waves $e^{ikx}e^{ilx}$ produces wave numbers $k + l$. Therefore, nonlinear convective interactions are constantly cascading low-order wave numbers to higher order ones. Further, wave numbers higher than k_{max} that can be supported by the grid system, appear as lower wave numbers. This is called the *aliasing error*. A treatment of *unresolved wave numbers* and of the aliasing errors of the nonlinear convective interactions is presented in the next section.

Viscous and Turbulent Diffusive Terms. The diffusive terms in the equation of motion are of the form $\mu_{eff}(\partial^2 f/\partial x^2)$. The coefficient μ_{eff} is the effective viscosity coef-

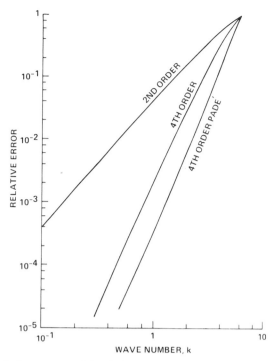

FIGURE 21.20 Relative errors of the finite-difference forms for the first derivative: $N = 64$ and $\lambda_{\max} = 20\pi$.

ficient; it includes both the dynamic viscosity and the eddy viscosity. When the second derivative of the function f is represented by the second-order, central-difference formula, Eq. (21.53),

$$\kappa_{2n}'' = -\frac{2(\cos \theta_n - 1)}{\Delta x^2} \tag{21.54}$$

The double prime is used to indicate a modified wave number for second derivatives. Likewise, the difference schemes and their corresponding κ_n'' for the fourth-order, central-difference and Padé schemes [Kopal (1961)] are, respectively,

$$\frac{\delta^2 f}{\delta x^2} = \frac{-f_{(j-2)} + 16f_{(j-1)} - 30f_{(j)} + 16f_{(j+1)} - f_{(j+2)}}{12\Delta x^2}$$

$$\kappa_{4n}'' = \frac{15 - 16\cos \theta_n + \cos (2\theta_n)}{6\Delta x^2} \tag{21.55}$$

and

$$\left[1 + \frac{(\Delta x)^2}{12}(D_+D_-)\right]\frac{\delta^2 f}{\delta x^2} = (D_+D_-)f$$

$$\kappa_{4pn}'' = \frac{12(1 - \cos \theta_n)}{\Delta x^2(5 + \cos \theta_n)} \tag{21.56}$$

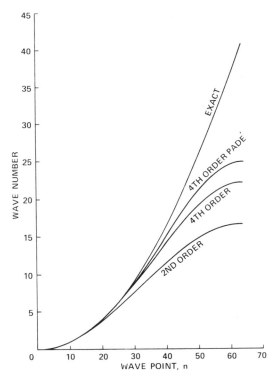

FIGURE 21.21   Comparison of exact and modified wave numbers for the second derivative: $N = 64$ and $\lambda_{\max} = 20\pi$.

The exact and modified wave numbers for the above finite-difference formulas are compared in Fig. 21.21 for $N = 64$ and $\lambda_{\max} = 20\pi$; the relative errors are shown in Fig. 21.22. The relative error is defined as $(k_n^2 - \kappa_n'')/k_n^2$. The conclusions that can be drawn from these figures are the same as those for the first derivative finite-difference formulas. At high wave numbers, the finite-difference formulas are more accurate for second derivatives than for first derivatives (Figs. 21.19 and 21.21 or Figs. 21.20 and 21.22). Therefore, the magnitude of the error is different for first derivatives and second derivatives, if both have the same order of truncation error.

Mesh Reynolds Number.   So far, the convective terms and the viscous and turbulent diffusive terms have been considered separately. Except for Stokes flows, potential flows, and possibly at or in the immediate neighborhood of regions of vorticity production, some components of both of these terms are equally important. Therefore, the important components must be accurately approximated in any computational procedure. Since the errors introduced by the finite-difference form of the convective terms are comparable to these terms at high wave numbers (see Fig. 21.19), the convective errors must not affect the diffusive terms. This is checked by computing the ratio of convective errors to diffusive terms.

Consider the simple model equation representing a balance between convective acceleration and diffusive force

$$\rho u \frac{du}{dx} = \mu_{\text{eff}} \frac{d^2 u}{dx^2} \tag{21.57}$$

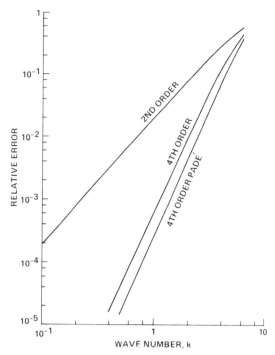

FIGURE 21.22 Relative errors of the finite-difference forms for the second derivative: $N = 64$ and $\lambda_{\max} = 20\pi$.

Assuming that the coefficients in this equation are constant, on substituting the Fourier series in the finite-difference forms of the above derivatives, we obtain for each Fourier component

$$(\rho u)(i\kappa_n')\bar{u}(k_n) = \mu_{\text{eff}}(-1)\kappa_n''\bar{u}(k_n) \qquad (21.58)$$

The Reynolds number is the ratio of the magnitude of the convective term to that of the viscous term. The mesh Reynolds number is then formed by multiplying the absolute value of both sides of Eq. (21.58) by Δx. If the accuracy of a finite-difference method for the first derivative is acceptable up to the maximum value of κ' (Fig. 21.19), corresponding to $n = n_a$, then the maximum mesh Reynolds number for accurate finite-difference of Eq. (21.57) is

$$(\text{Re}_{\Delta x})_{\max} = \frac{\kappa_n''}{\kappa_n'} \Delta x, \qquad n = n_a \qquad (21.59)$$

where $\text{Re}_{\Delta x} \equiv (\rho u / \mu_{\text{(eff)}})_a \Delta x$. If, for example, second-order central-difference formulas [Eqs. (21.50) and (21.54)] are used, then the maximum mesh Reynolds number is 2. For higher order methods, the maximum mesh Reynolds number is larger than this value since n_a is larger for higher order methods (see Fig. 21.19).

In the above derivation of Eq. (21.59), it was assumed that the ratio of the convective term to the diffusive term is unity. However, in practical problems, this ratio is unlikely to be the order of unity because of the presence of the pressure gradient term and the unsteady term in the Navier–Stokes equations.

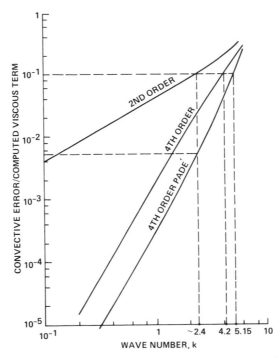

FIGURE 21.23   Ratio of the error in the computed convective term to the computed viscous term for $\mathrm{Re}_{\Delta x} = 1$: $N = 64$ and $\lambda_{\max} = 20\pi$.

Using Eq. (21.58), the absolute value of the ratio of the convective error to the computed diffusive term multiplied by Δx is $\mathrm{Re}_{\Delta x} (k_n - \kappa'_n)/\kappa''_n$. This ratio is plotted in Fig. 21.23 for the above finite-difference methods, with $\mathrm{Re}_{\Delta x} = 1$, $N = 64$, and $\lambda_{\max} = 20\pi$. The curves in this figure for any other $\mathrm{Re}_{\Delta x}$ shift vertically by that factor. If the convective error is 10% of the computed viscous term, the fourth-order Padé scheme extends the wave-number range of the second-order, central-difference scheme by a factor of two. In other words, the former scheme requires $J = 129$ instead of $J = 195$ required for the latter scheme. On the other hand, the former scheme allows $\mathrm{Re}_{\Delta x} \approx 15$ for the above error criteria. The higher order approximations perform better, but they are often difficult to implement for practical problems.

21.3.3   Computational Grids

A computational grid system is a necessary part of any numerical solution method. The primary consideration in selecting a grid system is the requirement for accuracy in the numerical solution. Secondary considerations are the effect of grid choice on computational efficiency of the solution algorithm and the ease of grid generation. Only the accuracy requirements are discussed here (see also Sec. 19.1). The secondary considerations are briefly presented by Mehta and Lomax (1982) and in Sec. 19.1.

Accuracy Requirements. The accuracy requirements of a numerical solution are determined by the particular application one has in mind [Mehta (1990a), (1993)].

If the solution serves the purpose for which it is intended, then the accuracy requirements are satisfied for that particular application. These requirements vary with the particular purpose of the application and frequently tend to be subjective. Unlike accuracy requirements, discretization (truncation) errors of a numerical method are independent of both the purposes of applications and subjectiveness. Therefore, in the discussion that follows, the accuracy constraints are not quantified, and the emphasis is placed on the order of discretization errors.

Simulations of flow regions throughout which the scales of motion are essentially the same in all directions, are probably best carried out by equispaced, Cartesian meshes. In this case, the evaluation of mesh errors on the solution is completely determined by the size of the single space interval. On the other hand, a flow field with a surface along which there is a shear layer is generally computed using a highly *stretched* mesh *normal* to the surface. This mesh is very fine near the surface and is constructed to increase in the direction normal to it. In this case, the errors of the solution are much more intimately tied to the grid structure, and the evaluation of errors is not simple. The situation is again not easy to evaluate when attempts are made to align and *cluster* meshes with shock waves whose positions are not known *a priori*. The relationship between the grid choice and solution errors is very important to the evaluation of viscous flow computations.

It is generally accepted that one of the coordinate families should lie along any surface that is generating a shear layer. This is accomplished most conveniently by using a body-oriented system, which further facilitates application of surface boundary conditions. It is reasonable to expect that the accuracy is best when grid-lines leaving the surface are normal to it, although this does not appear to be crucial. In addition, it is accepted that a nonuniform grid should be used in order to conserve computational resources. This is usually accomplished by clustering points in regions where there are rapid changes of flow variables. As a corollary of the latter process, points are often spread apart in regions where the gradients of flow variables are small.

It also is generally accepted that one of the coordinate families should be made to lie along a shock if this is possible. This is often quite possible for bow shocks which interface with a completely known free-stream flow field. For interior shocks, this is much more difficult, and computations are, generally, done with *shock-capturing techniques* rather than *shock-fitting* ones (see Sec. 20.3). This is primarily a result of the fact that the shock-fitting methods introduce algebraic and data-management complexities. In contrast, the shock-capturing techniques do introduce errors that may modify the shock/boundary layer interaction phenomena by propagating errors into the smooth region of the flow and by influencing the shock strength, location, and thickness. These errors depend on the grid system and on the choice of numerical method. For example, when one of the coordinate families is not aligned with a shock, these techniques tend to thicken the shock-wave region.

The discrete governing equations for flow computations around complicated geometries involve geometrical quantities depending on the numerical method. In finite-difference methods, there are metric coefficients and the Jacobian of topological transformations. In finite-volume methods, we have lengths or surface areas and areas or volumes. The finite-element methods contain shape factors. All these geometrical quantities are obviously grid-dependent, and they appear along with physical quantities in the overall numerical process. Clearly, it is the combination of physical and geometrical quantities that appears in the difference formula that should be accurately resolved in order for the computed flow to be a useful solution of the governing system. This means that geometrical quantities require proper representation just as the physical quantities do.

The generally accepted practice of indicating the order of truncation error of a numerical method does not quantify the truncation errors. The computational accuracy, that is, the level of computational uncertainty is often not easy to determine. A related issue is the modeling accuracy. For example, the effects of turbulence must be modeled, and the models are essentially empirical. These models are usually validated by comparison with experiments and, at times, with analytical results. Experimental results also contain uncertainty. The reader is referred to Mehta (1990b) and Mehta (1993) for a further discussion of the sources of uncertainties in computational fluid dynamics results and in experimental data. A procedure for establishing the credibility of computed results is presented by Mehta (1995). A part of this procedure addresses how to reduce computational uncertainty. For example, confidence in the computed results can be established by repeated grid-refinement studies to demonstrate that these results tend to be independent of numerics.

Methods for Improving Flow Simulation Accuracy. Minimization of discretization errors may be achieved by a proper choice of both the numerical method and the grid system. Usually, there is more freedom in choosing the grid system than in choosing the accuracy of the numerical method. Often the choice of the grid system is determined by *a priori* knowledge about the solution. Most of this knowledge is available in terms of generalities rather than specifics. For example, thin and slender shear layers can be resolved with the help of some stretching function across the known shear layer. But, without the specific information, such as the magnitude and location of gradients in the flow field, the grid system employed can often be wasteful if it is not concentrated satisfactorily in those regions where a better resolution is desirable.

For a better utilization of grid-point resources, there is a growing interest in *solution-adaptive grid systems*. There are currently two basic strategies. The first strategy involves tracking a fluid property, such as density, and inserting additional grid points or regridding so that finely spaced grid points are placed in the region over which that property has steep gradients. The second strategy is to minimize the leading term or terms of the modified equations that determine the order of the truncation error of a numerical method. In a grid method, which allows both nodal amplitudes and nodal positions to move continuously with time, nodes are moved to those regions where the nodes are most needed.

Application of the above techniques to the Navier–Stokes equations is a difficult undertaking. Questions, such as what flow variables to monitor, which truncation errors to minimize, whether all flow variables or truncation errors should be considered simultaneously, and which parts of the flow domain require special checking, need to be resolved. Of course, how to best adapt the grid system is a difficult task. These issues become much more involved when the flow field is unsteady. The obvious payoff of efficient, solution-adaptive grid systems is in terms of efficient use of computer resources. See Sec. 19.7 for more details.

Effect of Grid Refinement on Numerical Stability. The most important side effect of grid refinement on a numerical algorithm is its influence on numerical stability.

Decreasing the grid size decreases the time step of explicit methods, since the time step of these methods is bounded by the size of the space interval. If a single time step is used for advancing the entire solution, an explicit method is generally limited by the smallest space interval in the mesh. This limitation can be very costly if the time step is forced to be very small relative to the time scales of motion that are of interest. In such cases, the algorithm is said to be *mesh stiff*, since the stiffness is caused by the fineness of a space interval in the grid. The most common way to lessen the extent of stiffness is to use implicit, rather than explicit, numerical methods. Almost all codes being used to solve the Navier–Stokes equations have some parts that represent an implicit numerical technique. Even with implicit methods, decreasing the grid size almost always degrades either the stability of these methods or the rate of convergence to a steady state.

21.3.4 Approximations of Unresolved Nonlinear Convective Phenomena

We consider again the frequency content of a Fourier analysis made of the physical variables with reference to either time or space in the interest of further discussing the nonlinear phenomena. The frequency content is associated with a range of scales because of the relationship between the frequency, f, and the wave number, $f = k/(2\pi)$. This association of frequency with scale provides insight into the nature of the nonlinear convective phenomena, just as it is very useful in analyzing numerical accuracy [Mehta and Lomax (1982)].

The nonlinear convective terms of the Navier–Stokes equations for a compressible fluid can support discontinuous solutions referred to as shock waves, and as *contact surfaces*. Across shock waves and contact surfaces, some or all fluid quantities are not, strictly speaking, discontinuous, but continuous because of viscosity. Shock waves and contact surfaces, therefore, have some thickness (see Sec. 21.1.3). Since this thickness is very small, shocks, and contact surfaces are considered *nearly discontinuous*.

The problem of computing turbulence is much more difficult than that of computing flows that contain isolated shock waves and contact discontinuities. Turbulence in an incompressible or compressible fluid is also associated with the nonlinear convective terms. In turbulent flows, the physical variables vary sharply at many places in the flow field, and frequently these variations are almost discontinuous. When turbulence is not modeled, the scales that are to be resolved extend in all three space directions as well as in time. Computational resources limit such calculations to low Reynolds numbers and idealized problems. On the other hand, when turbulence is modeled, it is required only to compute the overall effect of these apparent discontinuities. This is a particularly difficult problem when the flow is complex.

The modulus of the Fourier components, a_k, of a variable having a discontinuity, or an abrupt jump that is nearly discontinuous, is presented in Fig. 21.24. Nearly all of the high wave number components have finite amplitude, and nearly all scales are present. However, any discrete grid system can accurately support only a limited number of low wave number components. The components to the right of the mesh cutoff line are referred to as subgrid components. If their production is permitted, they must alias back into the low wavenumber range causing numerical error. The aliasing error can be severe enough to cause numerical instability. Notice that although viscosity can eliminate components at very high wave numbers, grid reso-

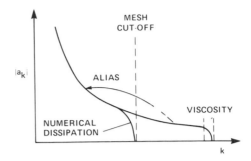

FIGURE 21.24 Spectral analysis of a discontinuous function, effect of viscosity on high-frequency amplitudes, and numerical dissipation of subgrid amplitudes. (From Mehta and Lomax, 1982.)

lution does not permit this. Figure 21.25 summarizes how the physics that cannot be computed is modeled. The concepts that are conveyed in this figure, are discussed below.

Methods of Computing Shocks and Contact Discontinuities. The standard way to cope with the generation of subgrid scales is either to compute the flow properties across the discontinuities separately (shock fitting) or to include in the computing process some form of numerical diffusion* (shock capturing), which removes the subgrid terms before any significant part of them crosses the grid cutoff boundary (see Fig. 21.25). Shock fitting methods are to be preferred when they can be implemented easily. Shock capturing methods are numerical, error control procedures that have nothing to do with any physical viscous or turbulent diffusion or with any physical viscous dissipation. The whole point of a shock capturing method is that the shock forms and moves about in a mesh while some kind of analytic connection is maintained between the flows (the dependent variables) on the two sides of the wavefront. Since shock fitting methods for multidimensional problems can be ex-

*Numerical diffusion is commonly known as numerical dissipation. However, note that smoothing of discontinuities or removal of high frequencies is analogous to viscous diffusion of momentum and vorticity, but not to viscous dissipation that converts mechanical energy into internal energy.

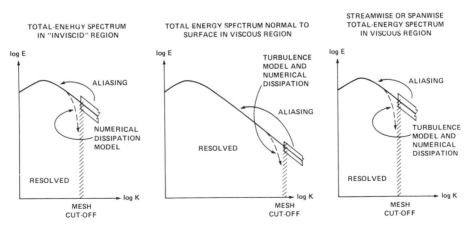

FIGURE 21.25 Modeling of physics that cannot be computed. (From Mehta and Lomax, 1982.)

tremely difficult, only the shock capturing methods are discussed here (see Sec. 20.3 for more discussion).

The numerical dissipation is added in a number of ways. The process can be *hidden* in the differencing scheme. Such is the case for the various Lax–Wendroff (1960) type of schemes where the actual dissipative mechanism, which is provided by the even order derivatives, is uncovered by inspecting the modified partial differential equation, i.e., the differential equation that is actually represented by the discrete equation. A few methods have been developed for constructing nonlinear, second-order schemes. These schemes satisfy the conditions of *total variation diminishing* (TVD) and that of *entropy inequality*. Upwind space differencing schemes also have built-in numerical dissipation, which is again revealed by inspecting the modified partial differential equation. Although a scheme may have inherent dissipation, additional dissipation is sometimes required. Central differencing schemes for a first derivative are well known to be nondissipative, so when these are used in shock-capturing algorithms, higher order dissipation terms are deliberately added. It is interesting to notice that an upwind scheme for first derivatives can be written as the sum of a central difference scheme and a dissipation term.

The shock capturing methods generally have built-in tests that try to isolate the shock location. Quite often they make use of test results to make local adjustments to the differencing scheme to improve its capturing capability. A second consequence of using a shock capturing method is to create the problem of insuring the proper location and strength of the shock as it moves about in the mesh. Lax (1954), (1973) has shown that this can be suitably approximated if the difference equations are locally conservative. The most common way of enforcing this condition is to cast the governing partial differential equations in conservation law form and then to make sure the difference scheme maintains this property. When such a technique is employed, a shock profile, represented, for example, by the pressure distribution, is *smeared* over one or more mesh points, but, for many practical applications, the general position and strength are adequately represented.

Methods for Computing Turbulence Effects. For flow problems in which turbulence is present, the realities of computer resource limits force one to make a severe approximation in formulating the equations of motion even before the numerical methods can be considered, and to accept a severe constraint while computing a turbulent flow field. The approximation is the use of the averaged equations of motion. This eliminates the need to resolve the high frequencies, but it introduces the problem of closure. The constraint is to permit extreme coordinate stretching in only one direction that is normal to shear layers. This permits computation of viscous and turbulent effects in that direction only.

The subgrid scales are constantly being generated by the large-scale structure through the nonlinear wave interactions in the convective terms. Just as in the case of shock capturing, numerical control of the subgrid frequency production is brought about by the addition of dissipation, either through the space derivative approximation or deliberately by additional even order space derivative terms. In either case, the choice is arbitrary, except that it prevents the accumulation of energy in the highest frequencies supported by the mesh (see Fig. 21.25). However, it should not alter the resolved frequencies beyond the error bounds for these frequencies. In particular, it must not change the effect of physical diffusion represented by the viscous and turbulent terms. This selective control is quite often not achieved in computations, and the resolved frequencies are somewhat affected. Because of the lack of adequate grid resolution, numerical control of the subgrid frequencies is essential to the numerical simulation. Although numerical dissipation is not, in conventional terminology, part of the turbulence model, it is in practice. Consequently, the analytical form of the turbulence model is not sufficient to describe its effect on an actual calculation. The numerical turbulence model includes the internal logic controlling the local evaluation of parameters such as the mixing length, the discretization error, and the numerical dissipation. It seems that the only way the effect of numerics can be separated from the analytical turbulence model is by grid refinement studies leading to elimination of numerical dissipation. In the absence of this, the final judgement of the computed results is generally based on a comparison with some experiment or theory.

21.3.5   Alternating Direction Implicit Methods for Navier–Stokes Equations

Finite difference methods for solving the Navier–Stokes equations can be classified as *explicit*, *implicit*, or some *hybrid* combination of the two. These classes of numerical methods have been successfully applied to a variety of flow calculations. The numerical stability bound of a time accurate explicit method limits the time step, and that generally leads to excessively long computation times. The improved stability of an implicit method generally allows the time step limit to be imposed by the time accuracy bound rather than the stability bound. An explicit method very often offers the advantage of ease of implementation, and an implicit method consumes about four times more computation time per time step than an explicit method for two-dimensional problems. The hybrid numerical methods attempt to combine the best features of explicit and implicit methods. Because implicit methods are usually preferred for computing viscous flows, this section is restricted to implicit methods (see Sec. 19.2 for more details).

Often implicit methods for nonlinear problems are based on the alternating direction technique for replacing a multidimensional problem by successive one-dimensional problems and on a noniterative procedure for calculating the solution at each time step. Historically, it appears that Polezhaev (1967) first used an ADI method (without an iterative procedure) to solve the Navier–Stoke equations. Subsequently, the ADI procedure has been applied to partially coupled forms of these equations [Berezin, Kovenja, and Yanenko (1975a), (1975b)] and to their fully coupled form [Briley and McDonald (1973), (1975), (1977), and Beam and Warming (1978)]. Noniterative ADI schemes were developed by Lindemuth and Killeen (1973) for equations of magnetohydrodynamics and by Briley and McDonald (1973) and Beam and Warming (1978) for the Navier–Stokes equations.

A procedure for constructing a class of ADI schemes devised by Mehta is presented below. The steps involved in formulating this class of ADI schemes are as follows. First, in order to devise an implicit scheme, the time derivative is approximated by a Padé formula at a new or unknown time-level, $(n + 1)$. Second, assuming that the effective viscosity and thermal conductivity are not functions of time,*

---

*This assumption is made here in order to keep the description of the steps involved simple. If there are time-dependent equations available that govern effective viscosity and thermal conductivity, then these quantities need to be determined at the $(n + 1)$ time-level.

the spatial derivatives at the $(n + 1)$ level are expressed by a Taylor series expansion in terms of known quantities at the (n) level and of time derivatives of dependent variables at the $(n + 1)$ level. This leads to linearization of the nonlinear terms in the governing equations. The Jacobian and metric coefficients of the spatial coordinates and the time variable are considered to be known, for simplicity, at the $(n + 1)$ level. Third, the left-hand side of the equation resulting from the previous step is approximately factored to form an ADI procedure. The terms that cannot be factored are taken to the right-hand side of this equation. The $(n + 1)$ level time-derivatives in these terms are replaced by corresponding derivatives at the (n) level. Fourth, the continuous space derivatives are replaced by finite differences. Fifth, numerical dissipation is added in the ADI procedure, if it is required. Sixth, physical and analytical boundary conditions are translated into implicit, finite-difference boundary conditions. These conditions are included in the ADI procedure. Seventh, the solution of the equation formed in the previous step determines the time derivative of the dependent variable at the $(n + 1)$ level. Eighth, the time derivative is replaced by a backward finite difference formula, and it is equated to the solution obtained in the previous step. Ninth and last, the solution of the equation in the eighth step determines the dependent variable at the $(n + 1)$ level.

These steps are explained below in detail for the following conservative law form of the Navier–Stokes equations in general curvilinear coordinates:

$$\frac{\partial(\mathcal{D}Q_m)}{\partial t} = \mathcal{R}_m \tag{21.60}$$

with

$$\mathcal{R}_m \equiv -\sum_{i=1}^{d} \frac{\partial C_{mi}}{\partial \xi_i} + \frac{1}{\text{Re}} \sum_{i=1}^{d} \frac{\partial V_{mi}}{\partial \xi_i} \tag{21.60a}$$

where d is the number of spatial dimensions; m is the equation number, varying from 1 to $d + 2$; and Q_m, C_{mi}, and V_{mi} are associated with the Cartesian, conservative dependent variables, convective and pressure terms, and diffusive terms, respectively. The Jacobian of the transformation from the Cartesian coordinates to curvilinear coordinates, ξ_i, is represented by $\mathcal{D}$. The Reynolds number is denoted by Re. The actual forms of Q_m, C_{mi}, and V_{mi} are given in Eq. (21.21a).

Step 1—Let the numerical solution of Eq. (21.60) with appropriate boundary and initial conditions be known at time-level (n). The numerical solution of this equation is required at time level $(n + 1)$ by using an implicit scheme. Therefore, this equation is evaluated at the $(n + 1)$ time level, and the time derivative is replaced by the Padé difference formula containing backward differences

$$\left(\frac{\partial(\mathcal{D}Q_m)}{\partial t}\right)^{(n+1)} = \frac{\left(Q_m \frac{\partial \mathcal{D}}{\partial t}\right)^{(n+1)} + \left(\mathcal{D} \frac{\delta Q_m}{\partial t}\right)^{(n+1)}}{[1 + \Delta t(\phi_3 D_{-1} + \phi_4 D_{-2})]} + O(\Delta t^P)$$

$$D_{-1} f^{(n+1)} = \frac{f^{(n+1)} - f^{(n)}}{\Delta t}, \qquad D_{-2} f^{(n+1)} = \frac{f^{(n+1)} - f^{(n-1)}}{2\Delta t} \tag{21.61}$$

and quantities ϕ_3 and ϕ_4 are constants. The time derivative of the Jacobian $\mathfrak{D}$ is left in the differential form, because it depends on the coordinate transformation which is considered to be a known analytical function of time. The finite difference time derivative in the numerator of Eq. (21.61) is replaced by a backward difference that is not of the Padé type. This backward difference formula is not presented at this point because it is not required in the next few steps. It is defined in the eighth step below along with the values of ϕ_3, ϕ_4, and p.

After substituting the Padé formula, Eq. (21.61), for the time derivative in Eq. (21.60), and after clearing the difference operator in the denominator, we have

$$\left(Q_m \frac{\partial \mathfrak{D}}{\partial t} \right)^{(n+1)} + \left(\mathfrak{D} \frac{\delta Q_m}{\delta t} \right)^{(n+1)} = \phi_5 \mathfrak{R}_m^{(n+1)} - \phi_3 \mathfrak{R}_m^{(n)} - 0.5\phi_4 \mathfrak{R}_m^{(n-1)} \quad (21.62)$$

where $\phi_5 = (1 + \phi_3 + 0.5\phi_4)$. There are unknown quantities on both sides of Eq. (21.62). This makes this equation and the numerical algorithm developed below implicit. This is true even for some other backward difference formula in Eq. (21.61), provided $\phi_5 \neq 0$.

Step 2—First, the left-hand side of Eq. (21.62) is expressed as

$$\left(Q_m \frac{\partial \mathfrak{D}}{\partial t} \right)^{(n+1)} + \left(\mathfrak{D} \frac{\delta Q_m}{\delta t} \right)^{(n+1)}$$

$$= \left(\mathbf{D} \frac{\delta Q_m}{\delta t} \right)^{(n+1)} + Q_m^{(n)} \left(\frac{\partial \mathfrak{D}}{\partial t} \right)^{(n+1)} + O(\Delta t^2) + O[\Delta t^{(q+1)}] \quad (21.63)$$

where the following relations are used

$$Q_m^{(n+1)} = Q_m^{(n)} + \Delta t \left(\frac{\partial Q_m}{\partial t} \right)^{(n)} + O(\Delta t^2) \quad (21.64)$$

$$\Delta t \left(\frac{\partial Q_m}{\partial t} \right)^{(n)} = \Delta t \left(\frac{\partial Q_m}{\partial t} \right)^{(n+1)} + O(\Delta t^2) \quad (21.65)$$

$$\left(\frac{\partial Q_m}{\partial t} \right)^{(n+1)} = \left(\frac{\delta Q_m}{\delta t} \right)^{(n+1)} + O(\Delta t^q) \quad (21.66)$$

$$\mathbf{D}^{(n+1)} \equiv \left(\mathfrak{D} + \Delta t \frac{\partial \mathfrak{D}}{\partial t} \right)^{(n+1)} \quad (21.67)$$

Eq. (21.65) is utilized in order to bring all time derivatives to the same time level. In Eq. (21.66), $q = p$ for $\phi_3 = \phi_4 = 0$, otherwise $q = p - 1$. Next, the terms represented by $\mathfrak{R}_m^{(n+1)}$ on the right-hand side of Eq. (21.62) are expanded in a Taylor series of time t. The expansion of C_{mi}-term is

$$\left(\sum_{i=1}^{d} \frac{\partial C_{mi}}{\partial \xi_i} \right)^{(n+1)} = \sum_{i=1}^{d} \left\{ \left(\frac{\partial C_{mi}}{\partial \xi_i} \right)^{(n)} + \Delta t \frac{\partial}{\partial \xi_i} \left[\sum_{k=1}^{d+2} \left(\frac{\partial C_{mi}}{\partial Q_k} \right)^{(n)} \left(\frac{\delta Q_k}{\delta t} \right)^{(n+1)} \right] \right\}$$

$$+ O(\Delta t^2) + O[\Delta t^{(q+1)}] \quad (21.68)$$

where Eqs. (21.65) and (21.66) are used. The tensor V_{mi} of the viscous term contains first derivatives. These are derivatives of functions that are of the form $\beta = \beta(Q)$.

Symbolically, this is written as $V_{mi} = V_{mi}(\beta_1, \beta_2, \ldots) = V_{mi}(\beta)$. Instead of expanding V_{mi} directly in a Taylor series, it is indirectly expanded by writing each β in a Taylor series, $\beta^{n+1} = \beta^{(n)} + \Delta t \sum_{m=1}^{d+2} (\partial\beta/\partial Q_m)^{(n)}(\delta Q_m/\delta t)^{(n+1)} + O(\Delta t^2) + O[\Delta t^{(q+1)}]$. The viscous term at the $(n+1)$ level is expressed as

$$\left(\sum_{i=1}^{d} \frac{\partial V_{mi}(\beta)}{\partial \xi_i}\right)^{(n+1)} = \sum_{i=1}^{d} \left\{\left[\frac{\partial V_{mi}(\beta)}{\partial \xi_i}\right]^{(n)} + \Delta t\, \frac{\partial V_{mi}(\epsilon^{(n+1)})}{\delta \xi_i}\right\}$$
$$+ O(\Delta t^2) + O[\Delta t^{(q+1)}] \tag{21.69}$$

$$\epsilon = \sum_{m=1}^{d+2} \left(\frac{\partial\beta}{\partial Q_m}\right)^{(n)} \left(\frac{\delta Q_m}{\delta t}\right)$$

with the time level of ϵ depending on the time-level of the time derivative of Q_m in this expression.

Step 3—The second term on the right-hand side of Eq. (21.69) is written as a sum of two terms,

$$\sum_{i=1}^{d} \frac{\partial V_{mi}(\epsilon^{(n+1)})}{\partial \xi_i} = \sum_{i=1}^{d} \left(\frac{\partial S_{mi}(\epsilon^{(n+1)})}{\partial \xi_i} + \frac{\partial X_{mi}(\epsilon^{(n)})}{\partial \xi_i}\right) \tag{21.70}$$

such that the first term on the right-hand side contains no mixed derivatives and the second contains only mixed derivatives. Note that the mixed derivative term is evaluated at (n) instead of $(n+1)$, because it cannot be factored with factors containing only one direction. By using Eqs. (21.63) and (21.68)–(21.70), we can reformulate Eq. (21.62) to

$$\left(\mathbf{D}\frac{\delta Q_m}{\delta t}\right)^{(n+1)} + \Delta t\phi_5 \sum_{i=1}^{d} \left\{\frac{\partial}{\partial \xi_i}\left[\sum_{k=1}^{d+2}\left(\frac{\partial C_{mi}}{\partial Q_k}\right)^{(n)}\left(\frac{\delta Q_k}{\delta t}\right)^{(n+1)}\right] - \frac{\partial S_{mi}(\epsilon^{(n+1)})}{\partial \xi_i}\right\}$$
$$= \Delta t\phi_5 \sum_{i=1}^{d} \frac{\partial X_{mi}(\epsilon^{(n)})}{\partial \xi_i} - Q_m^n \left(\frac{\partial \mathcal{D}}{\partial t}\right)^{(n+1)} + \phi_5 \mathcal{R}_m^{*(n)}$$
$$- \phi_3 \mathcal{R}_m^{(n)} - 0.5\phi_4 \mathcal{R}_m^{(n-1)} + O(\Delta t^2) + O[\Delta t^{(q+1)}] \tag{21.71}$$

where the Jacobian and metric coefficients are evaluated at the $(n+1)$ time-level in $\mathcal{R}_m^{*(n)}$. The approximate factoring of the left-hand side of the above equation leads to the following alternating direction algorithm in *derivative* form

$$\mathbf{D}^{(n+1)}\left(\frac{\delta Q_m}{\delta t}\right)^{I} + \Delta t\phi_5 \left\{\frac{\partial}{\partial \xi_1}\left[\sum_{k=1}^{d+2}\left(\frac{\partial C_{m1}}{\partial Q_k}\right)^{(n)}\left(\frac{\delta Q_k}{\delta t}\right)^{I}\right] - \frac{\partial S_{m1}(\epsilon^{(I)})}{\partial \xi_1}\right\}$$
$$= \text{RHS of Eq. (21.71)} \tag{21.72}$$

$$\mathbf{D}^{(n+1)}\left(\frac{\delta Q_m}{\delta t}\right)^{II} + \Delta t\phi_5 \left\{\frac{\partial}{\partial \xi_2}\left[\sum_{k=1}^{d+2}\left(\frac{\partial C_{m2}}{\partial Q_k}\right)^{(n)}\left(\frac{\delta Q_k}{\delta t}\right)^{II}\right] - \frac{\partial S_{m2}(\epsilon^{(II)})}{\partial \xi_2}\right\}$$
$$= \mathbf{D}^{(n+1)}\left(\frac{\delta Q_m}{\delta t}\right)^{I} \tag{21.73}$$

$$\mathbf{D}^{(n+1)} \left(\frac{\delta Q_m}{\delta t}\right)^{(n+1)} + \Delta t \phi_5 \left\{ \frac{\partial}{\partial \xi_d} \left[\sum_{k=1}^{d+2} \left(\frac{\partial C_{md}}{\partial Q_k}\right)^{(n)} \left(\frac{\delta Q_k}{\delta t}\right)^{(n+1)} \right] - \frac{\partial S_{md}(\epsilon^{(n+1)})}{\partial \xi_d} \right\}$$

$$= \mathbf{D}^{(n+1)} \left(\frac{\delta Q_m}{\delta t}\right)^{(d-I)} \tag{21.74}$$

with a factorization error $O(\Delta t^2)$. When $d = 2$, Eq. (21.73) is not used. The Roman numerals are used to indicate dummy time derivatives.

Step 4—The spatial derivatives appearing in Eqs. (21.72) to (21.74) are approximated by finite difference formulas. The practical limitation on the number of grid points used to approximate these derivatives comes from the bandwidth of the system of equations that must be solved. Generally, the spatial differences that produce tridiagonal systems are used on the left-hand sides of Eqs. (21.72)–(21.74). For example, with three-point central difference approximations, each of these equations requires the solution of a block tridiagonal system of equations with each block having the dimensions $(d + 2) \times (d + 2)$.

The evaluation of the viscous terms in Eqs. (21.72)–(21.74) requires a special treatment, because the effective viscosity is a function of the spatial coordinates. The form of a typical viscous term and, for example, its second-order finite difference form, are as follows

$$\frac{\partial}{\partial \xi}\left(\mu \frac{\partial f}{\partial \xi}\right) = \frac{[\mu_{(j-1)} + \mu_{(j)}] f_{(j-1)} - [\mu_{(j+1)} + 2\mu_{(j)} + \mu_{(j-1)}] \cdot f_{(j)} + [\mu_{(j)} + \mu_{(j+1)}] f_{(j+1)}}{2\Delta\xi^2} + O(\Delta\xi^2)$$

$$\tag{21.75}$$

Step 5—Owing to the lack of sufficient grid resolution in computations of practical problems, aliasing errors are manifested either as oscillations in computed flow quantities or as numerical instability. These symptoms are also caused by poor resolution of the pressure gradient and diffusive terms. These symptoms generally appear when there is a rapid change of some flow quantity in the flow field. As previously discussed, they are controlled by numerical dissipation that is either built in to the numerical scheme or added to the scheme. As an illustration, the fourth-order dissipation, $-\mathcal{D} [\sum_{i=1}^{d} \Delta\xi_i^4(\psi/8d\Delta t) \, \partial^4 Q_{mi}/\partial \xi_i^4]$, term is added to the numerical scheme when central differences are used for the spatial derivatives where ψ is a positive constant of $O(1)$. This dissipative term damps the short wavelength, $(2\Delta\xi_i)$, oscillations. Since it is of higher order than the second-order accurate central differences, it does not disrupt the formal accuracy of the numerical scheme. It is not used at grid points next to the boundaries in the direction away from boundaries. Next to a no-slip boundary, there is no need for numerical dissipation, because physical viscosity should be sufficient. Next to a free boundary, a boundary layer should not be created by numerical dissipation.

The numerical dissipation term can be added to the numerical scheme in a few different ways. First, the above dissipation term is added to the right-hand side of Eq. (21.60). It is then processed through steps (2) through (4). This leads to a pentadiagonal block system instead of a tridiagonal system in each of Eqs. (21.72)–(21.74). If, however, this dissipative term is treated in the same manner as the mixed derivative, viscous terms are treated above, then the tridiagonal block system is maintained. An alternative is to solve the above pentadiagonal system using the procedure of the diagonal algorithm of Pulliam and Chaussee (1981). This alternative lowers the time accuracy. Second, the above dissipation term is added explicitly to the right-hand side of Eq. (21.62) at the (n) time level. In this case, the dissipation constant must not be greater than 1 for numerical stability.

Step 6—A discussion of boundary conditions for the Navier–Stokes equations is given in Sec. 21.2.5. The appropriate boundary conditions that are chosen for the physical problem of interest are written at the $(n + 1)$ time-level. They are expanded in a Taylor series of time, and then they are included in the system formed by Eqs. (21.72)–(21.74). As an illustration, consider the adiabatic condition for internal energy at a solid wall. This condition is

$$\frac{\partial \iota}{\partial \xi_2} = 0 \tag{21.76}$$

where ι is internal energy per unit mass, $\iota = Q_{d+2}/Q_1 - \sum_{i=2}^{d+1} Q_i^2/2Q_1^2$, with $Q_1 = \rho$, $Q_2 = \rho u$, $Q_3 = \rho v$, $Q_4 = \rho w$, and $Q_5 = e$, where e is the total energy. The direction ξ_2 is normal to the wall. The above condition is written implicitly as

$$\frac{\partial}{\partial \xi_2} \left\{ \left(-\frac{Q_{d+2}}{Q_1^2} + \sum_{i-2}^{d+1} \frac{Q_i^2}{Q_1^3} \right)^{(n)} \left(\frac{\delta Q_1}{\delta t} \right)^{(n+1)} - \sum_{i=2}^{d+1} \left[\left(\frac{Q_i}{Q_1^2} \right)^{(n)} \left(\frac{\delta Q_i}{\delta t} \right)^{(n+1)} \right] \right\}$$

$$+ \left(\frac{1}{Q_1} \right)^{(n)} \left(\frac{\delta Q_{d+2}}{\delta t} \right)^{(n+1)} \right\} + O(\Delta t^2) + O[\Delta t^{(q+1)}] = 0 \tag{21.77}$$

This equation is added to the block system represented by Eq. (21.73). The spatial derivative is replaced by a one-sided, first-order, finite difference formula to maintain the tridiagonal form of this system.

Step 7—The noniterative, solution algorithm given by Eqs. (21.72)–(21.74) consists of d one-dimensional sweeps. Each sweep requires the solution of a linear system involving a block system which is generally solved by a block lower-upper decomposition procedure. Except for the last sweep, each sweep determines the dummy value of the time-derivative of Q_m at the $(n + 1)$ time level. The last sweep determines its true value.

Step 8—The finite difference time derivative is replaced by the following finite difference formula

$$\left(\frac{\delta Q_m}{\delta t} \right)^{(n+1)} = [\phi_1 D_{-1} - 2\phi_2 D_{-2}] \, Q_m^{(n+1)} \tag{21.78}$$

As shown in Table 21.1, values of ϕ_1, ϕ_2, ϕ_3, and ϕ_4 in Eqs. (21.61) and (21.78) determine: 1) the numerical scheme devised in the algorithm given by Eqs. (21.72)–(21.74), 2) the order p of time truncation of this scheme, 3) the time level at which $\delta Q_m/\delta t$ is actually evaluated. By itself, $\delta Q_m/\delta t$ is evaluated at the $(n + 1)$ time level, Eq. (21.78); however, in Eq. (21.62), the time level depends on the right-hand side of this equation. This is the time level shown in the table. On the other hand, $\partial Q_m/\partial t$ in Eq. (21.60) is evaluated at the $(n + 1)$ time level along with the right-hand

TABLE 21.1   A FEW IMPLICIT SCHEMES

Scheme	ϕ_1	ϕ_2	ϕ_3	ϕ_4	p	Level
Backward Euler [Laasonen (1949)]	1	0	0	0	1	$(n + 1)$
Three-point backward	2	$\frac{1}{2}$	0	0	2	$(n + 1)$
Crank-Nicholson (1947) type	1	0	$-\frac{1}{2}$	0	2	$(n + \frac{1}{2})$
Lees (1966) type	0	$-\frac{1}{2}$	$-\frac{1}{3}$	$-\frac{2}{3}$	2	(n)

side of this equation owing to Eq. (21.61). In the case of a backward-difference scheme, both in Eqs. (21.60) and (21.62) time derivatives are evaluated at the $(n + 1)$ level.

Step 9—The solution of Eq. (21.78) determines Q_m

$$Q_m^{(n+1)} = \frac{1}{(\phi_1 - \phi_2)} \left[\phi_1 Q_m^{(n)} - \phi_2 Q_m^{(n-1)} + (\Delta t) \left(\frac{\delta Q_m}{\delta t} \right)^{(n+1)} \right] \quad (21.79)$$

From the table and this equation, it can be concluded that the three point backward and the Lees type time differencing schemes are not self-starting.

These nine steps provide a recipe for developing a class of noniterative, fully coupled ADI schemes for the Navier–Stokes equations in derivative form. This recipe is also applicable to the Euler equations for a compressible fluid. We have satisfactorily determined transonic flow past an airfoil of a viscous fluid using the three point backward and the Euler backward schemes with the Euler explicit treatment of the fourth-order numerical dissipation term. Numerical stability analysis is nonexistent for the above class of schemes when applied to the Navier–Stokes equations. Under restrictive conditions, stability analysis is available [Gustafsson (1991)].

21.4   NAVIER–STOKES EQUATIONS FOR INCOMPRESSIBLE FLOW
U. Ghia and K. N. Ghia

Many flows of fluid dynamics interest are dominated by features such as three-dimensionality, unsteadiness, streamwise separation, recirculation, etc. The governing equations for these flows are the complete Navier–Stokes equations. For incompressible flow of a homogeneous fluid, the Navier–Stokes equations can be simplified considerably from their corresponding form for general compressible flow. In vector notation, the governing equations for incompressible isothermal flow in terms of dimensionless variables are

$$\frac{\partial \overline{\mathbf{V}}}{\partial t} + \overline{\mathbf{V}} \cdot \nabla \overline{\mathbf{V}} = -\nabla p + \frac{1}{\text{Re}} \nabla^2 \overline{\mathbf{V}} \quad (21.80)$$

and

$$D \triangleq \nabla \cdot \overline{\mathbf{V}} = 0 \quad (21.81)$$

In Eqs. (21.80) and (21.81), $\overline{V}$ is the velocity vector, p is the pressure, Re is the Reynolds number based on reference quantities, and D is the dilation which is zero for incompressible flow. These are a set of three-dimensional, non-linear, parabolic–elliptic, partial differential equations. Their numerical solution constitutes a formidable task even for flows with simple geometries. During the last 25 years, several numerical methods of solution have been developed, each exploiting certain characteristics of these equations, along with implementing high degrees of sophistication in the numerical analyses. The accuracy and efficiency of the solution is significantly influenced by several factors, e.g., the choice of dependent and independent variables, boundary conditions and the numerical method itself. The discussion here will focus on the various forms of the Navier–Stokes equations rather than on the details of the numerical algorithms available for their solution.

Equations (21.80) and (21.81) are written in terms of the *primitive variables* ($\overline{V}$, p) and are the statement of momentum and mass conservation for a Newtonian fluid. Because the pressure, p, does not appear as the primary dependent or dominant variable (i.e., in a highest derivative term) in Eqs. (21.80) or (21.81), this basic form of the Navier–Stokes equations does not easily lend itself to direct numerical implementation. Several procedures have been devised to address, or circumvent, this difficulty. These consist of either modifying Eq. (21.81) or embodying it in an equation more suitable for the determination of pressure. Alternatively, the total system of Eqs. (21.80) and (21.81) may be recast in forms which are sometimes more convenient to work with. The Navier–Stokes equations, including the primitive-variable form given in Eqs. (21.80) and (21.81), may be represented by one of the three basic formulations:

   i) Primitive Variable System ($\overline{V}$, p)
   ii) Vorticity–Potential System ($\overline{\zeta}$, Φ, $\overline{\Psi}$); also referred to as vector and scalar potential system, this system reduces to the vorticity stream function system for two-dimensional flows
   iii) Vorticity–Velocity System ($\overline{\zeta}$, $\overline{V}$).

Each of these formulations, along with some very fundamental significant information about the numerical methods developed for their solution, will be briefly discussed. The early development of the numerical simulation of incompressible flow using the above systems has been discussed extensively by Roache (1972), and many of the significant contributions have been concisely reviewed by Orszag and Israeli (1974). Hence, many of the earlier studies will not be cited here.

21.4.1. Primitive Variable System ($\overline{V}$, p)

The primitive variable system is widely used and is especially attractive for three-dimensional flow. It uses the physical variables directly and appears to lead to a minimum number of dependent variables for three-dimensional flows. Associated with this system, a staggered grid, rather than a regular grid, is frequently used, since the latter leads to larger truncation error. To account for the appropriate treatment of the velocity–pressure coupling, the Navier–Stokes equations using primitive variables have been solved using three basic forms of these equations: 1) Momentum and Continuity, 2) Momentum and Pressure Poisson, and 3) Momentum and Modified Continuity, also referred to as a form using *artificial compressibility*, and similar concepts.

Momentum and Continuity. This fundamental form consists of Eqs. (21.80) and (21.81). This system contains the pressure, p, as one of the dependent variables; however, the system contains no suitable equation to solve for pressure directly, since Eq. (21.80) is used to yield the velocity vector, $\overline{\mathbf{V}}$, and Eq. (21.81) does not contain p in it. One of the widely used iterative procedures for solving the system of Eqs. (21.80) and (21.81) in an approximate sense is the Semi-Implicit Method for Pressure Linked Equations, abbreviated as SIMPLE. This technique has been described by Patankar and Spalding (1972) as well as by Caretto, Gosman, Patankar, and Spalding (1972). In this method, the momentum equations, Eq. (21.80), are solved using a guessed pressure, p^G, which is updated by a correction pressure, p^C, such that the unknown pressure field, p, is given as

$$p = p^G + p^C \tag{21.82}$$

The pressure correction field, p^C, is obtained by solving the Poisson equation that results from the requirement that the velocity, $\overline{\mathbf{V}}$, satisfy the continuity equation, Eq. (21.81). Since the unknown pressure field is given by Eq. (21.82), this method is classified in group 1 above, instead of in group 2. Humphrey, Taylor, and Whitelaw (1977) used a similar procedure for the solution of system of Eqs. (21.80)–(21.81) to study the laminar flow in a square duct of strong curvature. Raithby and Schneider (1979) have discussed SIMPLE as well as several other techniques for treating the velocity pressure coupling in the system of Eqs. (21.80)–(21.81) for both steady and unsteady flows. Van Doormal and Raithby (1982) have solved the system of Eqs. (21.80) to (21.81) simultaneously on a staggered grid. An Alternating Direction Implicit (ADI)-like method is used, thereby requiring iterations for a multidimensional flow problem. Fluid dynamicists, who have used SIMPLE or related methods, have always treated the convective terms for high Reynolds number flows with first-order accuracy. Roscoe (1976) presented a second-order accurate method for solving Eqs. (21.80)–(21.81) for three-dimensional flows and observed that the method converged only for very low Re flows. Accurate and efficient solution of the system of Eqs. (21.80) and (21.81) for even moderate Re flows still remains a challenging problem. Briley, Buggeln, and McDonald (1982) have solved the corresponding compressible Navier–Stokes equations in a body-fitted orthogonal system. A consistently split, linearized block implicit (LBI) method is developed, and incompressible flow solutions are obtained by setting Mach number $M \approx 0.05$; second-order artificial damping was needed to control the high-frequency oscillations which resulted from the use of central differences for the convective terms.

Momentum and Pressure Poisson. As stated earlier, the unknown pressure field, p, does not appear as the dominant variable in Eq. (21.80). In an alternative form of the governing equations, the continuity equation, Eq. (21.81), is replaced by a Poisson equation for the pressure, p, obtained by forming the divergence of Eq. (21.80), and given as

$$\nabla^2 p = \nabla \cdot \left[\frac{1}{\text{Re}} \nabla^2 \overline{\mathbf{V}} \right] - \nabla[\overline{\mathbf{V}} \cdot \nabla \overline{\mathbf{V}}] - \nabla \cdot \left[\frac{\partial \overline{\mathbf{V}}}{\partial t} \right] \triangleq S_p \tag{21.83a}$$

i.e.,

$$\nabla^2 p = S_p \tag{21.83b}$$

The momentum equation, Eq. (21.80), together with the pressure Poisson equation, Eq. (21.83), form the second system which has also been widely used; the earlier studies have been discussed by Orszag and Israeli (1974). The Poisson equation for pressure in this system can be solved numerically using direct fast solvers. Therefore, this system is especially advantageous for unsteady flows. It should be noted that, in general, Eqs. (21.80) and (21.83) are solved sequentially. However, if only steady-state solutions are desired, it is believed that Eqs. (21.80) and (21.83) should be solved simultaneously.

One of the difficulties associated with this system concerns the boundary conditions for the pressure. Generally, a *Neumann boundary condition* is employed. This boundary condition is given as

$$\bar{\mathbf{n}} \cdot \nabla p = \bar{\mathbf{n}} \cdot \left[\frac{1}{\mathrm{Re}} \nabla^2 \bar{\mathbf{V}} - \bar{\mathbf{V}} \cdot \nabla \bar{\mathbf{V}} - \frac{\partial \bar{\mathbf{V}}}{\partial t} \right] \tag{21.84}$$

where $\bar{\mathbf{n}}$ is the unit normal vector. The divergence theorem requires

$$\int_R \nabla^2 p \, dR = \int_S \bar{\mathbf{n}} \cdot \nabla p \, dS, \tag{21.85}$$

where R and S are the region and surface, respectively, of the domain of interest. Hence, Eqs. (21.83)–(21.84) form a well posed *Neumann–Poisson problem* for pressure only if the Neumann integral constraint given in Eq. (21.85) is satisfied. K. Ghia, Hankey, and Hodge (1979) solved the Navier–Stokes equations in Cartesian coordinates using Eqs. (21.80) and (21.83), and the compatibility condition, Eq. (21.85). K. Ghia, Shin, and U. Ghia (1979) solved the same system of equations using a fourth-order accurate numerical method based on polynomial splines. For higher order accurate solutions of the incompressible Navier–Stokes equations, the reader is referred to the work of Hirsh (1983). Osswald and Ghia (1981) have studied this system rigorously using both the convective and divergence forms for Eq. (21.80). They demonstrated that, once a discretization procedure has been chosen for Eq. (21.80) and the dilation operator D, a preferred discrete formulation exists for the pressure Poisson problem providing the exact relationship between the intermediate pressure field and the advanced algebraic dilation. For a regular grid, the standard five-point star discretization of the Poisson equation, Eq. (21.83), leads to a nonzero dilation field in the steady-state flow solution, whereas for a staggered grid, the steady-state algebraic dilation field is guaranteed to be driven to become arbitrarily small. This is because the algebra satisfied by the continuous partial differential operators and their finite-difference counterparts is identical, through the second-order partial derivatives, for the staggered-grid arrangement. Hence, the staggered grid should be preferred, over the regular grid, for primitive variable solution of the Navier–Stokes equations. Osswald and Ghia (1981) have comprehensively compared the results obtained using the Navier–Stokes equations on regular as well as staggered grids for both the primitive variable formulation and the alternate vorticity–stream function formulation to be discussed later. Bernard and Thompson (1982) solved the two-dimensional, incompressible Navier–Stokes equations, Eqs. (21.80) and (21.83), in curvilinear coordinates. They used an *approximate factorization* (AF) scheme due to Beam and Warming (1977) and Briley and McDonald (1973) to solve the momentum equation, Eq. (21.80), and the *line successive over-relaxation* (LSOR) scheme to solve the Poisson equation, Eq. (21.83), for pressure. Explicit as well as implicit artificial damping was needed to obtain stable solutions at high Reynolds number, and the method was found suitable for $Re < 10^4$.

Momentum and Modified Continuity (Method of Artificial Compressibility and Similar Methods). Yet another formulation using primitive variables is due to Chorin (1967). In this form, the continuity equation, Eq. (21.81), is replaced by the dynamical equation

$$\frac{\partial \overline{\rho}}{\partial t} + \nabla \cdot \overline{\mathbf{V}} = 0 \tag{21.86}$$

where $\overline{\rho}$ is related to the pressure via an artificial equation of state given as

$$p = \overline{\rho}/\delta \tag{21.87}$$

Here, $\overline{\rho}$ is the artificial density and δ the artificial compressibility. Equation (21.80) and Eqs. (21.86) and (21.87) form the new system. The main motivation in suggesting this formulation was that, with the then prevailing methods, the solution of the Poisson equation, Eq. (21.83), was generally very inefficient. The explicit method developed was demonstrated using time independent model problems in two and three space dimensions and a regular grid.

Chorin (1968) also gave an extension of his artificial compressibility method to time dependent problems. In this method, the original system of Eqs. (21.80) and (21.81) is retained, however Eq. (21.81) is satisfied using a pressure iteration scheme, such that

$$|p^{n+1} - p^n| = \alpha D_a \tag{21.88}$$

where the superscript n denotes the time level, α is a relaxation parameter and D_a is the algebraic dilation. Thus, when the dilation approaches zero in the individual cells, the correct pressure solution results. Hodge (1975) used this approach and extended it to solutions of Navier–Stokes equations in generalized coordinates on a regular grid. He successfully computed flow past various airfoils at zero angle of attack. This approach has been shown to be equivalent to that of using the Poisson equation for pressure.

Steger and Kutler (1977) developed an implicit numerical method in which the continuity equation, Eq. (21.81), was replaced by the dynamical equation

$$\frac{\partial p}{\partial t} + \beta \nabla \cdot \overline{\mathbf{V}} = 0 \tag{21.89}$$

which reduces to Eq. (21.86) when $\beta = 1/\delta$, and $\beta \gg 1$. Equation (21.89) is a first-order hyperbolic equation and can be easily integrated using the approximate factorization scheme of Beam and Warming (1977) developed for the integration of hyperbolic, time-dependent equations and valid for obtaining steady-state solutions as $\partial p/\partial t \rightarrow 0$. A fourth-order dissipation term was added to control numerical instability. For large β, the system of Eqs. (21.80) and (21.89) leads to ill-conditioned (*stiff*) coefficient matrices. Kwak, Shanks, Chang, and Chakravarty (1984) further pursued the method of Steger and Kutler (1977) to study the flow in fully three-dimensional geometries using generalized curvilinear coordinates.

21.4.2 Vorticity–Potential System

The Navier–Stokes equations, Eqs. (21.80) and (21.81), in terms of primitive variables, pose some difficulties in satisfying the continuity equation, Eq. (21.81), and in computing the unknown pressure field without an evolution equation for pressure. If this system of Eqs. (21.80) and (21.81) is recast in terms of vorticity and potential, these difficulties can be circumvented. This alternative system of equations to be presented here is due to Aregbesola and Burley (1977) and Richardson and Cornish (1977). The approach aims to satisfy Eq. (21.81) identically by introducing the scalar potential Φ and the vector potential $\overline{\Psi}$ through the relation

$$\overline{V} = \nabla\Phi + \nabla \times \overline{\Psi} \tag{21.90}$$

A vector potential, $\overline{\Psi}$, unique within the gradient of an arbitrary harmonic function, exists such that

$$\nabla \cdot \overline{\Psi} = 0 \tag{21.91}$$

Forming the divergence of Eq. (21.90) leads to

$$\nabla^2\Phi = \nabla \cdot V = 0 \tag{21.92}$$

The equation governing the vector potential, $\overline{\Psi}$, is obtained by forming the curl of Eq. (21.90) and using Eq. (21.91), to give the *vector Poisson equation*

$$\nabla^2\overline{\Psi} \approx -\overline{\zeta} \tag{21.93}$$

where the vorticity, $\overline{\zeta}$, is defined as

$$\overline{\zeta} = \nabla \times \overline{V} \tag{21.94}$$

The evolution equation for the vorticity transport is obtained by forming the curl of momentum equation, Eq. (21.80), as

$$\frac{\partial\overline{\zeta}}{\partial t} + (\overline{V} \cdot \nabla)\overline{\zeta} - (\overline{\zeta} \cdot \nabla)\overline{V} = \frac{1}{\text{Re}} \nabla^2\overline{\zeta} \tag{21.95}$$

For three-dimensional flow, Eqs. (21.92), (21.93), and (21.95) form a complete set defining Φ, $\overline{\Psi}$, and $\overline{\zeta}$. Thus, whereas the primitive variable system, Eqs. (21.80) and (21.81), consisted of only four equations, the vorticity potential system involves seven equations. At each time step, three 3-D Poisson equations need to be solved. Also, the Laplace equation for the scalar potential, Φ, forms a Neumann boundary value problem requiring that a compatibility condition similar to that given in Eq. (21.85) be satisfied. Furthermore, the primitive variables have a direct physical significance. On the other hand, the potentials in the vorticity–potential system do not. However, an important point of significance is that, for problems with no inflow and outflow, $\overline{V} = 0$ at all the boundaries, and Eq. (21.91) has the trivial solution $\Phi = 0$.

This system is particularly useful in two-dimensional flows. In this case, $\overline{\Psi} = (0, 0, \Psi_z)$, so that letting $\psi = \Psi_z$ leads to the conventional definition of stream function ψ in terms of velocity components. In addition, the stream function, ψ, automatically ensures that the velocity field is solenoidal, so that continuity equation, Eq. (21.81), is satisfied, and the scalar potential, Φ, is not needed. Similarly, for the vorticity, $\overline{\zeta} = (0, 0, \zeta_z)$, and hence only one component of vorticity remains, and, for convenience, $\zeta_z = \zeta$. Thus, for two-dimensional flow, Eqs. (21.93) and (21.95) reduce to scalar equations in terms of (ψ, ζ). A considerable development of early numerical methods for the solution of two-dimensional, incompressible Navier–Stokes equations took place using the vorticity–stream function system, and the reader is referred to Roache (1972) for a very comprehensive discussion on this topic. For two-dimensional steady flows, Agarwal (1982) has solved the vorticity and stream function equations using a third-order accurate discretization for the convective terms. For two-dimensional unsteady flows, Mehta and Lavan (1975) successfully developed an implicit–explicit scheme to study external flow past airfoils. On the other hand, for internal flows, Osswald and Ghia (1981) developed a fully implicit method which is very efficient, since it uses the ADI method for the solution of the vorticity–transport equation, with the Block Gaussian Elimination (BGE) method for the solution of the Dirichlet stream function problem. More recently, K. Ghia, Osswald, and U. Ghia (1983) refined this method by ensuring that the grid distribution used honors the physical length scales; they also developed consistent inflow and outflow boundary conditions. Osswald, K. Ghia, and U. Ghia (1983) further explored the efficiency of the method by solving the vorticity-transport equation using a strongly implicit procedure (SIP), while retaining the BGE method for the solution of the stream function, ψ, and provided solutions for persistently unsteady flows inside a doubly infinite backstep channel. Finally, Osswald, K. Ghia, and U. Ghia (1984) have extended their methods to solve the axisymmetric incompressible flow in a pipe with a sudden expansion, whereas McGreehan, K. Ghia, U. Ghia, and Osswald (1984) have used the same method to provide steady and unsteady solutions in a doubly infinite pipe with an orifice in it.

21.4.3 Vorticity–Velocity System

Yet another alternative system to the Navier–Stokes equations, Eqs. (21.80) and (21.81), is the vorticity–velocity system which has been used successfully in both differential as well as integral forms. Here, instead of solving Eqs. (21.92) and (21.93) for the scalar and vector potentials, ϕ and $\overline{\Psi}$, respectively, the Poisson equations for the velocity components are solved directly. These equations are obtained by forming the curl of the vorticity in Eq. (21.94) and substituting the continuity equation, Eq. (21.81), to give the vector equation

$$\nabla^2 \overline{\mathbf{V}} = \nabla \times \overline{\zeta} \qquad (21.96)$$

Dennis, Ingham, and Cook (1979) used this system of Eqs. (21.95) and (21.96) to obtain second-order accurate solutions for three-dimensional flow inside a cubical box. Fasel (1976) used this system to study two-dimensional boundary-layer stability problems.

Instead of solving the vorticity–velocity system $(\overline{\zeta}, \overline{\mathbf{V}})$ using the differential form of Eqs. (21.95) and (21.96), the integral form of Eq. (21.96) or both Eqs. (21.95) and (21.96) can also be utilized to study three-dimensional, incompressible flows. These viscous flow problems are considered to be completely described by their kinetic and kinematic parts. In the first method, the kinetic part consists of the differential form of the vorticity equation, Eq. (21.95), which describes the rate of change of vorticity through the convective and diffusive processes. On the other hand, Eq. (21.96), which describes the kinematics, is recast into an integral representation for the velocity as

$$\overline{\mathbf{V}} = -\frac{1}{A} \left[\int_R \frac{\overline{\zeta}_0 \times (\overline{\mathbf{r}}_0 - \overline{\mathbf{r}})}{|\overline{\mathbf{r}}_0 - \overline{\mathbf{r}}|^d} \, dR_0 \right.$$
$$\left. - \int_S \frac{[(\overline{\mathbf{V}}_0 \cdot \overline{\mathbf{n}}_0) - (\overline{\mathbf{V}}_0 \times \overline{\mathbf{n}}_0) \times] (\overline{\mathbf{r}}_0 - \overline{\mathbf{r}})}{|\overline{\mathbf{r}}_0 - \overline{\mathbf{r}}|^d} \, dS_0 \right] \qquad (21.97)$$

Here, R and S are, again, the region and surface, respectively, of the domain of interest and subscript $_0$ indicates that the variables and integration are in the $\overline{\mathbf{r}}_0$-space. Further, $\overline{\mathbf{n}}_0$ is the outward, unit-normal vector, and the parameters A and d have the values $A = 4\pi$, $d = 3$, for three-dimensional flows, and $A = 2\pi$, $d = 2$, for two-dimensional flows. Wu and Thompson (1973) demonstrated that the integral representation of the kinematic part permits explicit computation of the velocity field and that the computational domain can be confined to the viscous flow region alone. This method was developed for time-dependent problems, with the steady-state solution being obtained as the asymptotic solution in the limit of large time, t. However, Wu and his associates had limited success with this method while attempting to solve complex flow problems.

In order to improve upon their first method, Wu and Wahbah (1976) presented a second method for analyzing steady, viscous flow problems. This method recast into integral form not only the kinematic part, but also the kinetic part and is referred to as the *integral representation method*. Analogous to the velocity field $\overline{\mathbf{V}}$, the vorticity field $\overline{\zeta}$ is given as

$$\overline{\zeta} = -\frac{1}{A} \left[\frac{1}{\nu} \int_R \frac{(\overline{\mathbf{V}}_0 \times \overline{\zeta}_0) \times (\overline{\mathbf{r}}_0 - \overline{\mathbf{r}})}{|\overline{\mathbf{r}}_0 - \overline{\mathbf{r}}|^d} \, dR_0 \right.$$
$$\left. - \int_S \frac{\{(\overline{\mathbf{V}}_0 \cdot \overline{\mathbf{n}}_0) - (\overline{\mathbf{V}}_0 \times \overline{\mathbf{n}}_0) \times\} (\overline{\mathbf{r}}_0 - \overline{\mathbf{r}})}{|\overline{\mathbf{r}}_0 - \overline{\mathbf{r}}|^d} \, dS_0 \right] \qquad (21.98)$$

Equations (21.97) and (21.98) form the new system, and vorticity, $\overline{\zeta}$, is computed iteratively from Eq. (21.98) using the appropriate boundary condition for it on S. In the potential region, the first integrand vanishes, so that the computation of the kinetic part can be limited to the viscous flow region where $\bar{\zeta}$ is nonzero. This method was extended to time dependent viscous flow problems by Wu and Rizk (1978).

21.4.4 Closure

This section has discussed the basic forms of the Navier–Stokes equations for incompressible flows, together with their relative advantages, as well as the difficulties associated with their numerical solution. Methods discussed in Chapter 19 and earlier in the present chapter are applicable for their numerical solution; nevertheless, some pertinent remarks are in order here.

Selection of the form to be employed for a given flow problem should be guided by the conditions of the problem, namely, whether the flow is steady or unsteady and two-dimensional or three-dimensional. The main difficulty with the primitive form is the determination of pressure, while that with systems involving the vorticity is the surface boundary condition for the vorticity. Both of these difficulties have been adequately overcome for a simply connected domain. The former has been discussed in this section; for discussion of the latter, see Davis (1972), K. Ghia *et al.* (1979), and Quartapelle (1981). The formulation introduced by Rubin and Khosla (1982), termed the *composite velocity formulation*, also appears to be promising.

The coordinates and the discretized grid used in a flow problem formulation play a significant role in the accuracy and efficiency of the resulting numerical solutions. With proper coordinates, asymptotic behavior of the flow variables can be appropriately represented, and singularities can be treated more carefully. Use of an appropriate finite-difference grid that honors the length scales of the flow problem permits the physical flow features to be accurately reproduced in the numerical solution [e.g., see Osswald and Ghia (1981)]. With similarity type coordinates, the introduction of similarity-type dependent variables greatly enhances the efficiency of the numerical solution because the weak functional dependence on the streamwise coordinate then allows larger streamwise grid size to be used [see Davis (1972)]. Finally, with the use of proper coordinates, certain coordinate-related approximations may be made in the Navier–Stokes equations, with minimal sacrifice on the physics represented, leading to a hierarchy of approximate mathematical models for flow analysis as presented by Ghia (1984).

As regards numerical methods, implicit schemes are generally preferred over explicit schemes because of their better stability characteristics. However, the choice must now be coordinated with the architecture of the computer used. With vector processors, some previously disfavored schemes can become quite competitive if they are amenable to significant *vectorization* and *parallelization*. The efficiency of iterative schemes can be greatly enhanced by acceleration techniques using the multigrid methodology as shown by U. Ghia, K. Ghia, and Shin (1982) (see also Sec. 19.6).

21.5 NUMERICAL SIMULATION OF TURBULENCE
Saad A. Ragab

Prediction and control of turbulent flows is of great importance in the design of fluid machinery and aero/hydrodynamic vehicles. Computational approaches that are commonly used include the Reynolds-averaged-Navier–Stokes (RANS) equations, direct numerical simulation (DNS), and large eddy simulation (LES). RANS is the most frequently used approach. In this method, all turbulent fluctuations are averaged over a long period of time, and their statistical effects on the mean flow are *modeled*. The turbulence models are usually complex, because they are required to consider all of the turbulent scales including the large scales which may not be the same in different flow fields. On the other hand, DNS represents the other extreme where all of the dynamically significant eddies are computed and none are modeled. In LES, a suitable grid is used that allows the resolution of the large eddies, whereas the statistical effects of the smaller eddies are modeled. Because small scales tend to be isotropic, the turbulence model in LES is anticipated to be simpler than what is required for RANS. Thus, LES may be viewed as a compromise between RANS and DNS.

21.5.1 Direct Numerical Simulation

The three-dimensional, time dependent Navier–Stokes equations have been accepted as a mathematical model of fluid motions including turbulent flows. To predict a turbulent flow, one may use a numerical method to solve these equations for the instantaneous flow variables without the use of a turbulence model. Such a solution is known as a direct numerical simulation (DNS) of the turbulent motion. The mesh and time advancement must resolve all of the dynamically relevant turbulent scales from the largest scales imposed by external effects (such as pipe diameter or airfoil chord) down to the smallest scales which are responsible for the viscous dissipation of turbulent kinetic energy into thermal energy. This resolution requirement puts an upper limit on the Reynolds number that can be successfully simulated on a given computer. DNS has contributed primarily to the understanding of turbulence physics at low Reynolds numbers and continues to provide insight and guidance for the development and assessment of turbulence closure models. For example, data bases from DNS can be used to compute all of the correlations in the Reynolds stress transport equations. The availability of such data, which may be difficult or impossible to determine experimentally, is very valuable in developing and testing a phenomenological turbulence model or a subgrid scale stress model for use with large eddy simulations.

Limits on the achievable Reynolds number by DNS are made clear by estimates of the number of mesh points and CPU time required to resolve all the relevant scales of the turbulence. Chapman (1979) and Reynolds (1989), among others, discussed the resolution requirements at some length. For homogeneous isotropic turbulence, Reynolds estimates the total number of mesh points by $N_{xyz} \approx 1.7 \, \mathrm{Re}_T^{9/4} \approx 0.1 \, \mathrm{Re}_\lambda^{9/2}$, where Re_T is the Reynolds number of the large-scale motion and Re_λ is the one based on the longitudinal Taylor microscale, λ, and the r.m.s. of the velocity fluctuations. The exponent 9/4 shows that a mere tripling of Re_T would require a factor of 12-fold increase in the number of points. For channel flows, Reynolds gives the estimate of $N_{xyz} \approx 4 \times 10^6 \, (\mathrm{Re}_c/3300)^{2.7}$, where Re_c is based on centerline mean velocity and channel half-width. This estimate is guided by the successful simulation by Kim *et al.* (1987) who used 4×10^6 at Re_c of 3300 (note that the formula given by Reynolds assumes that Kim *et al.*, used 2×10^6 points). Similarly, for zero-pressure gradient boundary layers, the estimate is $N_{xyz} \approx 10^7 \, (\mathrm{Re}_x/500,000)^{2.2}$, where Re_x is based on freestream velocity and the distance from the leading edge. Again, this estimate is guided by the simulation of Spalart (1988) at a momentum thickness Reynolds number of 1410 using approximately 10^7 points. The rapid increase in the number of points, and hence in computer memory requirement, with Reynolds number is evident for these basic problems. Another factor that must also be considered is the CPU time needed to complete a DNS. For time-accurate resolution of the small eddies, the time step must resolve the time scale of these eddies. For homogeneous isotropic turbulence, the CPU time is $O(\mathrm{Re}_T^3)$, and for other flows it scales with $N_{xyz}^{4/3}$.

Examples of DNS. The pioneering simulations by Orszag and Patterson (1972) were for isotropic turbulence. Since then, DNS solutions have been obtained for more complex turbulent flows but at low turbulent Reynolds numbers and in simple geometries. Rogers and Moin (1987) simulated homogeneous shear flows, Kim *et al.* (1987) simulated channel flows, and Spalart (1988) simulated spatial boundary layers. The last two flows are presented next. For more applications of DNS, the reader is referred to the review articles by Rogallo and Moin (1984), Schumann and Friedrich (1987), and Reynolds (1989).

Fully Developed Channel Flow: One of the landmarks of DNS in recent years is the fully developed channel flow simulation by Kim *et al.* (1987). A channel flow (a pressure-driven flow between two parallel plates) is of practical interest, yet its geometry is simple enough for numerical treatment. Moreover, the turbulence in this flow is statistically steady and homogeneous in planes parallel to the walls. Thus, periodic boundary conditions can be applied in two directions in any of these planes. This practice is acceptable provided that the computational domain is large enough to accommodate the important large eddies or instabilities as well as allow the turbulence correlations to decay between the periodic boundaries and the interior of the flow. The boundary conditions on the solid walls pose no problem. In the Kim *et al.* simulation, the initial velocity field is obtained from a previous large eddy simulation. In the absence of such a simulation, one may initialize the flow using linear stability eigenfunctions. However, as pointed by Reynolds (1989), the initial conditions are unimportant for statistically steady flows because they eventually establish their own steady-state spectra.

In the Kim *et al.* simulation, the Reynolds number is 3300 based on the centerline velocity, U_c, and the channel half-width, δ, (this corresponds to a Reynolds number of 180 based on the wall shear velocity, u_τ, and δ). A fully spectral method with about 4 million grid points ($192 \times 129 \times 160$ in the streamwise, x, normal to the wall, y-, and spanwise, z-directions) is used. The step sizes range from $\Delta y^+ \approx 0.05$ near the wall to 4.4 at the channel centerline. The streamwise and spanwise steps are $\Delta x^+ \approx 12$ and $\Delta z^+ \approx 7$, respectively. The superscript $+$ indicates the familiar wall units. The presented correlations and spectra confirm that all relevant scales are resolved by this grid. They applied statistical analyses to their data and presented profiles for the mean flow, Reynolds stresses, pressure fluctuations, and several higher-order correlations.

The profile of the mean velocity nondimensionalized by the wall shear velocity is shown in Fig. 21.26 along with the experimental results of Eckelmann (1974) at $\mathrm{Re}_\tau = 142$ ($\mathrm{Re}_c = 2800$). A "correction" to the experimental results is necessary to reconcile differences in the measured profiles on two different occasions in the same facility. The correction amounts to increasing the u_τ of Eckelmann by 6%.

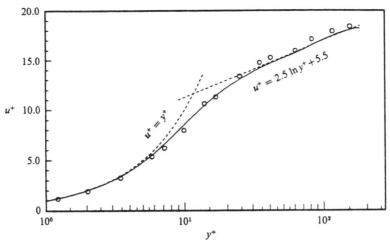

FIGURE 21.26 Mean velocity profiles: ———, upper wall; — – —, lower wall (masked by solid line); ○○○○, "corrected" data of Eckelmann (1974); and - - - -, Law of the Wall. (Reprinted with permission, Kim, J., Moin, P., and Moser, R. D., "Turbulent Statistics in Fully Developed Channel Flow at Low Reynolds Number," *J. Fluid Mech.*, Vol. 177, pp. 133–166, 1987.)

The turbulence intensities normalized by the wall shear velocity are shown in Fig. 21.27, and they are compared with those at $Re_\tau = 194$ from Kreplin and Eckelmann (1979). Kim *et al.* (1987) discuss the discrepancies between the numerical and experimental results, and they basically attribute them to experimental inaccuracies in using the X-hot wire near the wall. The turbulence intensities normalized by the local mean velocity, $\bar{u}$, are shown in Fig. 21.28. The Reynolds stress $\overline{-u'v'}$ is shown in Fig. 21.29 along with the corrected experimental data of Eckelmann. These

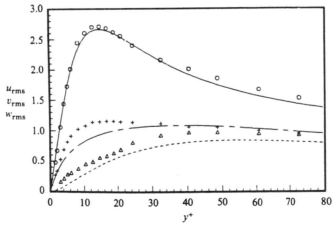

FIGURE 21.27 Root mean square velocity fluctuations in wall coordinates normalized with the wall shear velocity for computations and with "corrected" wall shear velocity for experimental data: ———, u_{rms}; — — — —, v_{rms}; and - - -, w_{rms}. Symbols represent the data from Kreplin and Eckelmann (1979): ○○○○, u_{rms}; △△△△, v_{rms}; and ++++, w_{rms}. (Reprinted with permission, Kim, J., Moin, P., and Moser, R. D., "Turbulent Statistics in Fully Developed Channel Flow at Low Reynolds Number," *J. Fluid Mech.*, Vol. 177, pp. 133–166, 1987.)

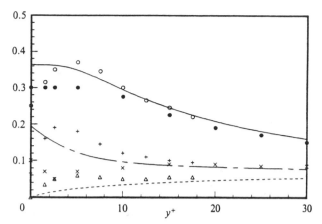

FIGURE 21.28 Turbulence intensities near the wall normalized by the local mean velocity: ——, $u_{rms}/\bar{u}$; – – – –, $v_{rms}/\bar{u}$; and — – —, $w_{rms}/\bar{u}$. From Kreplin and Eckelmann (1979): OOOO, $u_{rms}/\bar{u}$; △△△△, $v_{rms}/\bar{u}$; ++++, $w_{rms}/\bar{u}$. From Hanratty, Chorn, and Hatzia-vramidis (1977): OOOO, $u_{rms}/\bar{u}$; and × × × ×, $w_{rms}/\bar{u}$. (Reprinted with permission, Kim, J., Moin, P., and Moser, R. D., "Turbulent Statistics in Fully Developed Channel Flow at Low Reynolds Number," *J. Fluid Mech.*, Vol. 177, pp. 133–166, 1987.)

quantities are important in developing phenomenological models of wall-bounded shear flows. Moin (1987) points out that the limiting values of 0.37 for $u_{rms}/\bar{u}$ and 0.2 for $w_{rms}/\bar{u}$ at the wall have been confirmed by recent improved measurements by Naqawi and Reynolds (1987).

Kim *et al.* (1987) also used two-point velocity correlations to study the low- and high-speed *streaks* in the near wall region. They found that the computed streak spacing increases with the distance from the wall in agreement with the experimental results of Smith and Metzler (1983). Using the same data base, Moin (1987) presented contours of the instantaneous velocity fluctuations shown here in Fig. 21.30.

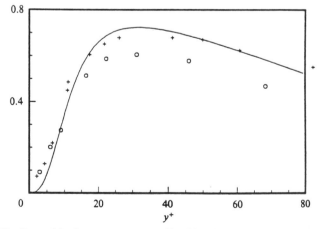

FIGURE 21.29 Reynolds shear stress normalized by the wall shear velocity in wall coordinates, ——, $\overline{u'v'}$; OOOO, data from Eckelmann for $Re_\tau = 142$; ++++, data from Eckelmann for $Re_\tau = 208$; renormalized by "corrected" v^*. (Reprinted with permission, Kim, J., Moin, P., and Moser, R. D., "Turbulent Statistics in Fully Developed Channel Flow at Low Reynolds Number," *J. Fluid Mech.*, Vol. 177, pp. 133–166, 1987.)

(a)

(b)

(c)

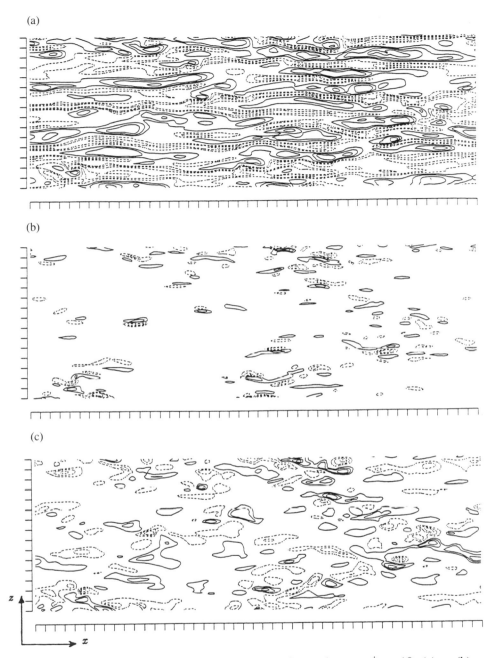

FIGURE 21.30 Contours of instantaneous velocity fluctuations at $y^+ \approx 10$. (a) u, (b) v, and (c) w. Dashed lines indicate negative contours. The streamwise extent of the figures is $2262\nu/v^*$, and the spanwise extent is $1131\nu/v^*$. In each figure, 10 contour levels are drawn spanning the maximum and minimum values of the quantity being plotted in the $y^+ = 10$ plane. (From Moin, 1987.)

He also found that, in the vicinity of the wall, the normal velocity component was highly intermittent and dominated by adjacent regions of fluid moving away from and towards the wall. Single vortices (one legged), with streamwise extent of 100 to 200 wall units, appeared to be the fundamental structures associated with regions of high turbulence production. These vortices sometimes occur in pairs constituting the legs of a horseshoe vortex. Incidentally, similar streaky structures and hairpin vortices have been identified in DNS of homogeneous shear flows by Lee *et al.* (1983) and Rogers and Moin (1987).

Turbulent Boundary Layers: The zero pressure gradient turbulent boundary layer on a flat plate has been studied extensively by experimentalists and theorists. Turbulence statistics have been measured and scaling laws have been proposed and tested. However, direct or large-eddy simulations of this fundamental problem are lagging. Unlike the channel flow, the turbulent boundary layer is spatially developing, and the turbulence is nonhomogeneous in the streamwise direction as well as normal to the plate. Simulation of such a flow requires numerical treatments of turbulence at the inflow/outflow boundaries. Presently, those treatments are approximate or *ad hoc*, and further improvements are needed. In the flat plate case, the inflow boundary may be moved far upstream to include laminar inflow conditions and then trigger transition to turbulence by modifying the wall boundary conditions or by appealing to some receptivity mechanisms [see Rai and Moin (1991)]. However, because of limitations on computing resources, this approach will reduce the streamwise distance over which the boundary layer is fully turbulent leading to a reduction in the achievable turbulent Reynolds number.

To overcome the difficulties with inflow/outflow boundary conditions, Spalart (1988) introduces a coordinate transformation to account for the slow streamwise variations of the boundary layer thickness and the energy level of the turbulence. He also split each of the flow variables into two components; for example, the u velocity is written as $u(x, \eta, z, t) = U(x, \eta) + A(x, t)u_p(x, \eta, z, t)$ where η is the transformed coordinate normal to the wall, U is the mean (over the spanwise direction z and time t), and A is an amplitude function which is proportional to the r.m.s. of the fluctuations. The normalized signal, u_p, has zero mean and its r.m.s. value is independent of streamwise coordinate x. The governing equations for the normalized signals are the Navier–Stokes equations with extra terms that account for the slow variations of growth terms such as U and A. This decomposition makes streamwise periodic boundary conditions seem tolerable, but not fully justified. Yet, the method can provide a good approximation to the local state of a boundary layer that has a slow spatial development.

Spalart used a fully spectral method based on Fourier series in the directions parallel to the plate and an exponential mapping with Jacobi polynomials in the normal, semi-infinite direction. He obtained results at four momentum thickness Reynolds numbers between $Re_\theta = 225$ and 1410. For the case $Re_\theta = 1410$, 1.1×10^7 grid points were used. The step sizes, in wall units, were $\Delta x^+ = 20$ and $\Delta z^+ = 6.7$, approximately. In the y direction the grid was stretched such that 10 points are within 9 wall units. Spalart then applied statistical analyses to the DNS data and presented and discussed different aspects of mean flow characteristics, Reynolds stresses, power spectra, turbulence structures, and the Reynolds stress transport budget. He carefully analyzed the scaling laws of the mean flow and turbulence and discussed the effects of Reynolds number on different terms in the transport equations.

Spalart found that at very low Reynolds number, Re_θ less than 600, the logarithmic layer of the mean velocity profile disappears. He also introduced a new definition of the boundary layer thickness, $\delta = 1.85\,\delta_1$, where δ_1 is the integral with respect to y of the total shear stress normalized by the wall value. The normalized total shear stress when plotted against y/δ shows a good collapse at different Reynolds numbers. Good agreement with the experimental data is obtained for both the mean flow and total shear profiles. However, near the wall, the normal Reynolds stresses do not collapse when normalized with the friction velocity and kinematic viscosity. The vorticity and pressure intensity do not collapse either. Instead, they all show a strong tendency to increase with Reynolds number consistent with the Townsend (1976) and Perry *et al.* (1986) theories. Finally, it is noted that the one-dimensional streamwise energy spectrum shows an *inertial range* that extends over half a decade in wavenumber. This allowed the theoretical estimation of the Kolmogorov constants.

Transitional Boundary Layers: The transition to turbulence remains one of the fundamental fluid dynamics problem yet to be fully understood. The exact mechanism of transition in boundary layers is still a highly debatable issue. During the past two decades, DNS has been extensively used to study the behavior of transitional boundary layers. Historically, temporal simulations have been found successful in realizing experimentally observed features. Even then, the resolution requirements are very severe to satisfactorily predict the late transition stage. Zang *et al.* (1989) showed that coarse grid simulations can give totally erroneous results. The resolution requirements for transitional boundary layers and channel flows are reported in their study.

The first major simulation of transition in boundary layers were performed by Wray and Hussaini (1984). Later Zang and Hussaini (1987) extended the simulation further to include the later stages in transition. These simulation results were found to compare well with the experimental observations. The first detailed simulation of transition in a channel flow is by Gilbert and Kleiser (1990). The reader is referred to Kleiser and Zang (1991) for an extensive review on the numerical simulation of transition in wall-bounded shear flows.

Compressible Flows: Direct numerical simulation of compressible turbulent flows has received a renewed vigor as a result of the current interest in supersonic and hypersonic transport vehicles. Due to the fundamental nature of the problem and the relative ease in implementing the boundary conditions, homogeneous turbulence has been the focus of attention of many researchers in the last decade [Passot and Pouquet (1987), Blaisdell (1990), Erlebacher *et al.* (1990), and Sarkar *et al.* (1991)]. For a recent review of the effects of compressibility on turbulence the reader is referred to Lele (1994), and Rai and Moin (1991) and the references therein.

21.5.2 Large Eddy Simulation

In turbulent flows, the large-scale eddies contain most of the turbulent energy, and they are the primary mechanism of momentum and heat transport between different parts of the flow. These large eddies may vary greatly from one flow field to another. In the RANS approach, it has proven difficult to model, in a universal way, the statistical effects of these eddies on the mean motion. On the other hand, the small-scale eddies tend to be isotropic, hence they are easier to model with some degree of universality. Thus, the objective of LES is to compute the time-dependent, three-dimensional field of the large eddies while modeling the small-scale eddies. In LES, the large-scale eddies are essentially defined by the fineness of the spatial grid used in the simulation. Therefore, it is important that the grid be fine enough to resolve all turbulence production mechanisms (e.g., streak structures near solid walls). Otherwise, a more complex turbulence model will have to be used to account for these mechanisms.

Filtering and the Equations of LES. In LES, the resolved (computed) flow field is smoother than the actual turbulent field being simulated. The resolved field is defined by a smoothing or filtering operation which eliminates small scales that cannot be represented on a specified mesh. Let $u(\bar{\mathbf{x}}, t)$ denote any flow variable in the actual turbulent field. Following Leonard (1974), we define its filtered (large-scale) part $\bar{u}(\bar{\mathbf{x}}, t)$ by

$$\bar{u}(\bar{\mathbf{x}}, t) = \int G(\bar{\mathbf{x}}, \xi) u(\xi, t) \, d^3\xi \tag{21.99}$$

where G is a *filter function*, and the integration extends over the entire flow domain. Note that this is a formal step because $u(\bar{\mathbf{x}}, t)$ is not computed by an LES code. An example of the filter function is the Gaussian filter whose one-dimensional form is given by

$$G(x, \xi) = \sqrt{\frac{6}{\pi} \frac{1}{\Delta}} \exp\left[-6 \frac{(x - \xi)^2}{\Delta^2}\right] \tag{21.100}$$

where Δ is the filter width. The *Gaussian filter* in wavenumber space [the Fourier transform of Eq. (21.100)] is given by

$$g(k) = \exp\left(-k^2\Delta^2/12\right) \tag{21.101}$$

where k is the wavenumber. All wavenumbers are attenuated by the filter except a constant value ($k = 0$) that is returned by the filter intact. The Fourier transform of Eq. (21.99) shows that as Δ is increased, more small scales are practically removed from the spectrum of $u(x, t)$.

A more common filter is the *sharp cut-off filter*, defined by

$$G(x, \xi) = \frac{2 \sin\left[\pi(x - \xi)/\Delta\right]}{\pi(x - \xi)} \tag{21.102}$$

In the wavenumber space this filter has the form

$$g(k) = 0 \text{ if } |k| > \pi/\Delta$$
$$= 1 \text{ otherwise.} \tag{21.103}$$

Hence, it removes all wavenumbers greater than π/Δ and leaves all others unaffected. Both the Gaussian and sharp cut-off filters are usually applied in directions of homogeneity of the turbulent field. For homogeneous turbulence, one may use the sharp cut-off filter in three spatial directions, in which case the filter function is the product of three factors, each of the form given by Eq. (21.102). For fully developed channel flow, the turbulence is assumed to be homogeneous in a plane parallel to the plates (streamwise–spanwise plane). Therefore, the above filters may be used in these two directions. In the transverse direction, the turbulence is non-homogeneous, and a different kind of filter is needed. A *box filter* (also called a *top hat filter*) may be used, which is given by

$$G(x, \xi) = 1/\Delta \text{ if } -\Delta/2 < (x - \xi) < \Delta/2 \qquad (21.104)$$
$$= 0 \text{ otherwise}$$

If the grid is nonuniform, Δ should be a function of x (e.g., Moin and Kim 1982).

The difference between u and $\bar{u}$ is called the *subgrid-scale* (SGS) *part*, u', hence the decomposition

$$u = \bar{u} + u' \qquad (21.105)$$

Filtering Eq. (21.105), we obtain $\bar{u} = \bar{\bar{u}} + \bar{u'}$. If the bar were to indicate the customary Reynolds averaging, we would have set $\bar{\bar{u}} = \bar{u}$ and hence $\bar{u'} = 0$. In the present filtering procedure, a second application of the filter removes further scales from the filtered field $\bar{u}$. Therefore, $\bar{\bar{u}}$ is not equal to $\bar{u}$, and it follows that $\bar{u'}$ is not equal to zero. However, equalities of the respective quantities hold for the sharp cut-off filter [Leonard (1974)].

To obtain the LES equations, the Navier–Stokes and continuity equations, written here for incompressible flows with constant properties (ρ and ν)

$$\frac{\partial u_i}{\partial t} + \frac{\partial u_i u_j}{\partial x_j} = -\frac{1}{\rho}\frac{\partial p}{\partial x_i} + \nu \frac{\partial^2 u_i}{\partial x_j \partial x_j} \qquad (21.106)$$

$$\frac{\partial u_i}{\partial x_i} = 0 \qquad (21.107)$$

are filtered. We assume that filtering and partial differentiation with respect to time and space commute, that is

$$\overline{\frac{\partial u}{\partial t}} = \frac{\partial \bar{u}}{\partial t} \qquad (21.108)$$

$$\overline{\frac{\partial u}{\partial x_i}} = \frac{\partial \bar{u}}{\partial x_i} \qquad (21.109)$$

It can be shown that filter functions of the convolution type $G(x - \xi)$ satisfy the commutative property, if u vanishes on the boundary. That property holds for the filters given by Eqs. (21.100)–(21.102), however it may be violated by a box filter with nonuniform width. The LES equations are then written as

$$\frac{\partial \bar{u}_i}{\partial t} + \frac{\partial \bar{u}_i \bar{u}_j}{\partial x_j} = -\frac{1}{\rho}\frac{\partial \bar{p}}{\partial x_i} + \nu \frac{\partial^2 \bar{u}_i}{\partial x_j \partial x_j} - \frac{\partial \tau_{ij}}{\partial x_j} \qquad (21.110)$$

$$\frac{\partial \bar{u}_i}{\partial x_i} = 0 \qquad (21.111)$$

where

$$\tau_{ij} = \overline{u_i u_j} - \bar{u}_i \bar{u}_j \qquad (21.112)$$

is called the *subgrid-scale* (SGS) *stress tensor*. Equations (21.110) and (21.111) are to be solved for $\bar{u}_i$ and $\bar{p}$. At this level, the term $\overline{u_i u_j}$ is unknown, therefore a turbulence model for τ_{ij} is needed to close the system. Note that LES is a time-dependent, three-dimensional simulation that needs a turbulence model.

Subgrid-Scale Models. The subgrid-scale stress tensor is the mechanism by which energy is drained from the large scales to the subgrid-scale motion, where the energy is eventually dissipated by viscosity. Energy could also be transferred locally (in space and time) from the SGS motion to the large scales; a process known as *backscatter* [Piomelli *et al.* (1991)]. Furthermore, near solid walls, the turbulence structures contain thin streaks of high- and low-speed fluid which scale with the wall units. They play a major role in the dynamics of turbulence and its production and must be resolved or modeled in LES. If turbulence production mechanisms are underresolved as a result of using a coarse grid, the SGS model must also account for those mechanisms [Orszag *et al.* (1993)]. Note also that different filter functions, G, remove different scales from the turbulent field. Thus, one anticipates that the filter function must also influence the formulation of the SGS model. Piomelli *et al.* (1988) found a strong dependence of the structure of the subgrid scales on the type of filter used, hence a consistency between the SGS model and filter should be observed. It is clear that the SGS model is an important ingredient for accurate LES.

The SGS stress tensor, τ_{ij}, may be decomposed into three terms by substituting Eq. (21.105) into Eq. (21.112)

$$\tau_{ij} = L_{ij} + R_{ij} + C_{ij} \qquad (2.113)$$

where $L_{ij} = \overline{\bar{u}_i \bar{u}_j} - \bar{u}_i \bar{u}_j$, $R_{ij} = \overline{u_i' u_j'}$, and $C_{ij} = \overline{\bar{u}_i u_j'} + \overline{u_i' \bar{u}_j}$, are referred to as the *Leonard stress*, *Reynolds stress*, and the *cross stress* terms, respectively. The Leonard stress term needs no modeling, because it can be computed directly from the resolved field once the filter function is specified. However, the term vanishes if the sharp cut-off filter is used. The Reynolds and cross terms must be modeled. In recent simulations, investigators prefer to model τ_{ij} without the splitting shown by Eq. (21.113). This is done to avoid the loss of the Galilean invariance property of the Navier–Stokes equations if the sum $R_{ij} + C_{ij}$ is not properly modeled [Speziale (1985), Germano (1992)]. However, Zang *et al.* (1993) introduced a new splitting that preserves the Galilean invariance property.

Smagorinsky Model: Most of the LES solutions obtained to date used an eddy viscosity model for the SGS stresses. Such a model is given by

$$\tau_{ij} - \tfrac{1}{3}\tau_{kk}\delta_{ij} = -2\nu_T\overline{S}_{ij} \tag{21.114}$$

where δ_{ij} is the Kronecker delta, and

$$\overline{S}_{ij} = \frac{1}{2}\left(\frac{\partial \overline{u}_i}{\partial x_j} + \frac{\partial \overline{u}_j}{\partial x_i}\right) \tag{21.115}$$

is the rate-of-strain tensor of the resolved field. The coefficient of eddy viscosity, ν_T, is given by Smagorinsky (1963)

$$\nu_T = (C_S\Delta)^2\,|\overline{S}| \tag{21.116}$$

where $|\overline{S}| = (2\overline{S}_{ij}\,\overline{S}_{ij})^{1/2}$ and Δ is the filter width, which is usually identified with the grid size or some function of the step sizes in three directions. We note that τ_{kk}, which is twice the SGS turbulence kinetic energy per unit of mass, is not given by the model. For computational purposes, the divergence of the spherical part $\tau_{kk}\delta_{ij}$ can be included with the pressure gradient, and $\overline{p}/\rho$ is replaced by $\overline{p}/\rho + 1/3\tau_{kk}$ in Eq. (21.110). The coefficient C_S is called the *Smagorinsky coefficient*. Unfortunately, C_S is not a universal constant and must be optimized for different flow fields. Lilly (1966) found that, for homogeneous isotropic turbulence with cutoff in the inertial subrange and Δ equal to the grid size, $C_S \approx 0.23$. In his simulation of a channel flow, Deardorff (1970) found that this value caused excessive dissipation of the large scales and that a reduced value of C_S must be used in the presence of mean shear; a value of 0.1 was found adequate. It is noted that the model is always dissipative ($\nu_T > 0$) and therefore cannot account for backscatter. The model also gives nonzero τ_{ij} in laminar flows and thus should not be used in transitional or relaminarizing wall bounded shear layers. A more serious practical limitation of the model is that it does not force the SGS stresses to vanish near solid walls. Thus, it must be modified to reflect the local length and velocity scales of the near wall turbulence. Existing modifications usually introduce *ad hoc* measures such as damping functions.

Dynamic Modeling: Germano *et al.* (1991) introduced a procedure known as *dynamic modeling* for computing the Smagorinsky coefficient. This procedure makes use of the detailed information about the resolved field that becomes available in the course of the simulation. The coefficient $C\,(= C_S^2)$ is computed as the simulation progresses in time instead of specifying it *a priori*, hence the name dynamic modeling. The coefficient, C, provided by this model is actually a function of space and time. The dynamic model overcomes many of the practical shortcomings of the Smagorinsky model. It is worth mentioning here that the scale similarity model introduced earlier by Bardina *et al.* (1983) was also an attempt to use the spectral information of the resolved field to model the SGS stresses. In the dynamic procedure, the large-scale equations are filtered once more using a *test filter* with filter width $\hat{\Delta}$ which is larger than the width of the first filter (called the *grid filter*). By letting $\hat{\overline{f}}$ denote a flow variable which is filtered twice (a bar for the grid filter and a caret for the test filter), one can obtain the equations for $\hat{\overline{u}}$ and $\hat{\overline{p}}$ which are identical in form to Eqs. (21.110) and (21.111) except that τ_{ij} is replaced by T_{ij}, where

$$T_{ij} = \overline{\hat{u}_i u_j} - \hat{\bar{u}}_i \hat{\bar{u}}_j \tag{21.117}$$

Germano (1992) observed that the difference between the yet unknown tensors T_{ij} and τ_{ij} can be computed in terms of the resolved field, $\bar{u}_i$, hence the identity

$$L_{ij} = T_{ij} - \tau_{ij} \tag{21.118}$$

where

$$L_{ij} = \overline{\hat{\bar{u}}_i \bar{u}_j} - \hat{\bar{u}}_i \hat{\bar{u}}_j \tag{21.119}$$

If models are written for T_{ij} and τ_{ij}, the identity in Eq. (21.118) can then be used to determine certain parameters in those models, because L_{ij} is computable independently of the models.

The Smagorinsky model for τ_{ij} written here in the form

$$\tau_{ij} - \tfrac{1}{3}\tau_{kk}\delta_{ij} = -2C\Delta^2|\bar{S}|\bar{S}_{ij} \tag{21.120}$$

may also be used for T_{ij}

$$T_{ij} - \tfrac{1}{3}T_{kk}\delta_{ij} = -2C\hat{\Delta}^2|\hat{\bar{S}}|\hat{\bar{S}}_{ij} \tag{21.121}$$

Substituting these models into the identity in Eq. (21.118), we obtain

$$L_{ij} - \tfrac{1}{3}L_{kk}\delta_{ij} = -2C\Delta^2 M_{ij} \tag{21.122}$$

where

$$M_{ij} = \frac{\hat{\Delta}^2}{\Delta^2}|\hat{\bar{S}}|\hat{\bar{S}}_{ij} - \widehat{|\bar{S}|\bar{S}_{ij}} \tag{21.123}$$

To determine a unique value for C from the five independent equations represented by Eq. (21.122), Germano et al. (1991) contracted it with $\bar{S}_{ij}$ to yield,

$$C\Delta^2 = -\frac{1}{2}\frac{\langle L_{ij}\bar{S}_{ij}\rangle}{\langle M_{ij}\bar{S}_{ij}\rangle} \tag{21.124}$$

where $\langle \ldots \rangle$ indicates averaging over directions or planes of homogeneity. Other forms of averaging such as time averaging have also been used. This averaging is necessary to overcome numerical instabilities [see Akselvoll and Moin (1993) and Lund et al. (1993) for more details about model implementation]. Lilly (1991) used a least squares technique to solve Eq. (21.122) for C, and obtained

$$C\Delta^2 = -\frac{1}{2}\frac{\langle L_{ij}M_{ij}\rangle}{\langle M_{ij}M_{ij}\rangle} \tag{21.125}$$

The value of $C\Delta^2$ from Eq. (21.124) or (21.125) can be substituted into Eq. (21.120) to obtain the final form for τ_{ij}. The only free parameter in the model is the ratio $\hat{\Delta}/\Delta$, for which a value of 2 is found satisfactory by many investigators.

In the derivation of Eq. (21.122), the coefficient C is treated as if it were a constant in the test filtering of τ_{ij} indicated in Eq. (21.118). This mathematical inconsistency has been rectified by Ghosal *et al.* (1992) by a model known as the *dynamic localization model*. Here, the integral over the entire flow domain of the squares of the errors in satisfying the Eq. (21.118) is expressed as a functional of the coefficient C. The necessary condition for minimizing that functional gives a Fredholm integral equation of the second kind to be solved for C. Thus, the error is minimized in a global sense instead of the pointwise least squares minimization suggested by Lilly. An additional constraint that guarantees the positivity of C can also be applied on the variational problem. No averaging over homogeneous directions is needed, therefore the model works for nonhomogeneous turbulence.

Renormalization Group Model: Yakhot and Orszag (1986a,b) formulated the LES equations and SGS model using the renormalization group theory of the forced Navier–Stokes equations. The RNG model has been used in the LES solution of many flow fields including channel flows [Yakhot *et al.* (1989)], backward facing steps [Karniadakis *et al.* (1993)], and transitional boundary layers [Piomelli *et al.* (1990)]. Application of the model to reacting flows is described by Orszag *et al.* (1993).

Subgrid-scale models have been also obtained in the wavenumber space. Derivation of such models can be found in the books by Lesieur (1990) and McComb (1990).

Examples of LES. LES has been applied to many problems in engineering fluid dynamics and heat transfer. Interested readers may consult the volumes edited by Schumann and Friedrich (1988), Ragab and Piomelli (1993), and Galperin and Orszag (1993). In this section only three examples will be presented.

Fully Developed Channel Flow: Deardorff (1970) presented the first LES in engineering fluid dynamics. He considered the fully developed channel flow at infinite Reynolds number, and used the Smagorinsky model to parameterize the SGS stresses. He treated the wall layer (viscous and buffer zones) approximately by using a boundary condition that makes the mean velocity profile consistent with the logarithmic Law of the Wall. With only 6,720 grid points, Deardorff obtained fair comparisons with the experimental measurements. The importance of Deardorff's simulation is that it established the feasibility of LES for high Reynolds number turbulent flows. Since then, the channel flow has been simulated by many authors for different purposes including testing new SGS models, studying Reynolds number effects on wall turbulence, and/or validating a numerical method for LES.

Schumann (1975) developed a finite difference method for LES and applied it to plane and annular channels. The finite difference equations are based on integral conservation equations for each cell. The SGS stresses are obtained as surface mean rather than volume mean values of the fluctuating velocity products. He improved on the Smagorinsky model by solving a partial differential equation for the SGS turbulent kinetic energy from which the velocity scale in the eddy viscosity model is determined. The wall layer was not resolved; instead, an approximate boundary condition based on the Law of the Wall was used. He used up to 65,536 grid volumes and obtained very good agreement with experimental data. Grotzbach (1983) presented further improvements and extensions of Schumann's method to include heat transfer and buoyancy effects in channel flows.

Moin *et al.* (1978) were the first to integrate the LES equations down to the wall in an effort to resolve the wall layer instead of using an approximate boundary condition. However, the grid used was inadequate for resolving the streaky structures near the wall. Moin and Kim (1982) used a fine grid of $64 \times 63 \times 128$ (x, y, z) points and were able to resolve the wall layer at a Reynolds number of 13,800 based on the mean centerline velocity and channel half width. In terms of wall units, the grid steps in the x and z directions were about 63 and 16, respectively. These values are adequate for resolving the near wall large-scale structures whose mean streamwise and spanwise spacings are on the order of 500 and 100 wall units. In this study, the Smagorinsky model with a modified length scale was used. The length scale was exponentially reduced by a Van Driest damping factor as the wall is approached. The computed mean velocity profile, shown here in Fig. 21.31, was in excellent agreement with experimental data and with the logarithmic Law of the Wall. The turbulence statistics and detailed flow structure were also in good agreement with experimental data. The resolvable mean Reynolds shear stress $\langle \bar{u}'' \, \bar{v} \rangle$ and the total Reynolds stress (resolvable plus SGS stress) are shown in Fig. 21.32. The SGS contribution is significant only near the walls. The solid line shows the total shear stress including the viscous stress which is dominant near the wall.

More recently, the channel problem has been simulated by Yakhot *et al.* (1989) using their RNG model, by Germano *et al.* (1991) and Piomelli (1993) using a dynamic model. Both of these models produce zero eddy viscosity in a laminar region and as a solid wall is approached. Therefore, no *ad hoc* damping or intermitency factors are needed with these models. Piomelli's simulations covered a wide range of Reynolds numbers up to 2000 based on the friction velocity and channel

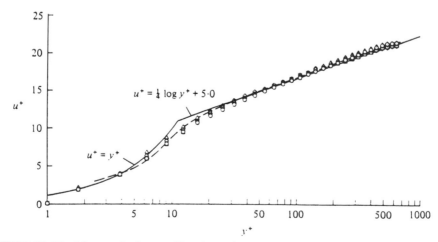

FIGURE 21.31 Mean-velocity profiles from four computed cases and comparison with experimental data: - - - [Hussain and Reynolds (1975)], Re = 13800; △ △ △ △, case 1; ◇◇◇◇, 2; □□□□, 3; and ○○○○, 4. (Reprinted with permission, Moin, J. and Kim, H., "Numerical Investigation of Turbulent Channel Flow," *J. Fluid Mech.*, Vol. 118, pp. 341–377, 1982.)

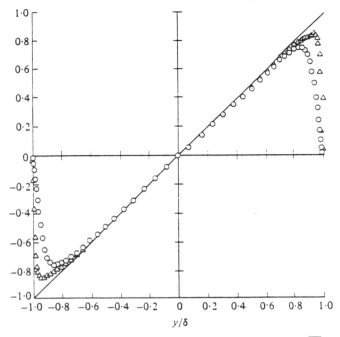

FIGURE 21.32   Resolvable and total turbulence shear stress: ○○○○, $\langle \overline{u''v} \rangle$ and △△△△, $\langle \overline{u''v} \rangle + \tau_{12}$ (Reprinted with permission, Moin, J. and Kim, H., "Numerical Investigation of Turbulent Channel Flow," *J. Fluid Mech.*, Vol. 118, pp. 341–377, 1982.)

half width which corresponds to 47,100 based on centerline velocity. Presently, this is the highest Reynolds number LES solution of a channel flow in which the integration is continued to the wall. The grid consisted of 64 × 81 × 80 points which gave $\Delta x^+ = 244$, $\Delta y^+ = 1.5$ near the wall to 77 near the centerline, and $\Delta z^+ = 40$. The mean velocity profile and turbulent stresses were in good agreement with experimental data, however the wall streaks appeared diffused and underresolved.

Resolving the near wall region with its thin streaky structures continues to pose a challenge to LES at very high Reynolds numbers. To resolve the vortical structures in the viscous layer, grid refinements are necessary not only in the direction normal to the wall but also in a plane parallel to the wall. An interesting discussion and estimates of required grid resolution as well as approaches to modeling the wall layer are given by Chapman (1979) and Moin and Jimenez (1993).

Backward Facing Step: A backward facing step (a sudden enlargement in the flow passage) is a geometrically simple configuration over which the turbulent flow field is rather complex. The complex features include a channel flow, a free shear layer, a recirculation region, a reattachment zone, and a boundary layer. DNS solution of this flow was obtained by Le and Moin (1993), and an LES solution was obtained by Akselvoll and Moin (1993) for the same conditions. The Reynolds number based on the step height and maximum velocity at the inlet was 5100 and the expansion ratio was 1.2. The LES results were obtained using a grid resolution of 192 × 48 × 32 points. The grid resolution in wall units (referred to friction velocity at the inlet of the computational domain) was given by $\Delta x^+ = 39.8$, $\Delta y^+_{min} = 0.8$ and $\Delta z^+ = 16$. The DNS results were obtained on a grid that is a factor of 32 denser than

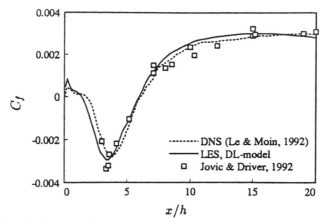

FIGURE 21.33   Coefficient of friction for backward facing step. [Reprinted with permission, Akselvoll, K. and Moin, P., "Application of the Dynamic Localization Model to Large-Eddy Simulation of Turbulent Flow over a Backward Facing Step," *Engineering Applications of LES*, ASME FED., Vol. 162, Ragab, S. and Piomelli, U. (Eds.), 1993.]

the LES grid. Akselvoll and Moin (1993) used a dynamic localization model developed by Ghosal *et al.* (1992). The results of LES were compared with DNS data and the experimental results of Jovic and Driver (1992) which were also obtained for the same geometry and flow conditions. The coefficients of friction and pressure along the lower wall downstream of the step are shown in Figs. 21.33 and 21.34, respectively. The reattachment length in the two simulations was 6.0 (normalized with the step height) which is in the range of 5.9–6.1 reported in the experiments. The mean streamwise and normal velocity profiles at $x/h = 1$ are shown in Fig. 21.35. The curve denoted by "coarse grid DNS" is a simulation on the same grid as the LES grid but with the SGS turned off. It is clear that the SGS model plays a role in the good comparison between DNS and LES shown in the figure. Moin and

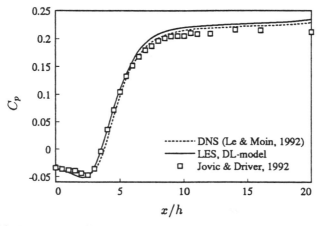

FIGURE 21.34   Pressure coefficient for backward facing step. [Reprinted with permission, Akselvoll, K. and Moin, P., "Application of the Dynamic Localization Model to Large-Eddy Simulation of Turbulent Flow over a Backward Facing Step," *Engineering Applications of LES*, ASME FED., Vol. 162, Ragab, S. and Piomelli, U. (Eds.), 1993.]

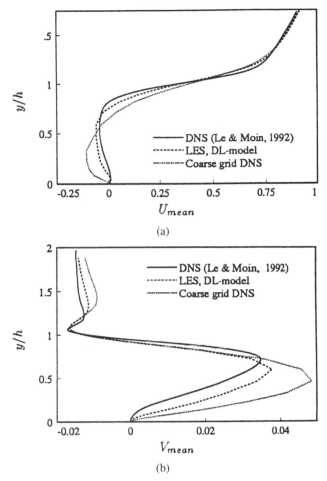

(a)

(b)

FIGURE 21.35 Mean velocity at $x/h = 1$ for a backward facing step: (a) streamwise velocity and (b) normal velocity. [Reprinted with permission, Akselvoll, K. and Moin, P., "Application of the Dynamic Localization Model to Large-Eddy Simulation of Turbulent Flow over a Backward Facing Step," *Engineering Applications of LES*, ASME FED., Vol. 162, Ragab, S. and Piomelli, U. (Eds.), 1993.]

Jimenez (1993) also reported that the turbulent intensities were generally in good agreement with the experimental data.

Karniadakis *et al.* (1993) presented LES results for the backward facing step at transitional Reynolds numbers using the RNG model.

Shear Driven Cavity: The shear driven cavity flow is a flow in a box induced by the motion of one of its walls. The wall moves in its plane at a constant velocity. The simplicity of the boundary conditions (six solid walls) and the availability of experimental data including turbulence statistics make this problem an attractive model for testing numerical methods and SGS models. Zang *et al.* (1993) modified the Germano *et al.* (1991) dynamic model by employing the mixed model of Bardina *et al.* (1983). They obtained LES results for the cavity problem at Reynolds numbers of 3200, 7500, and 10,000 based on the height of the cavity and lid velocity. These values of Re correspond to laminar, transitional, and locally turbulent flow conditions in the cavity. They reported good agreement with experimental measurements for both mean velocity profiles and turbulence intensities.

Jordan (1994) also used the dynamic model with the least-squares formulation of Lilly (1992) and obtained LES results for a cubic cavity. A uniform grid of 101 × 101 × 85 points is used at Re = 10,000. The mean velocity profiles through the cavity centerline at the mid-span plane are compared in Fig. 21.36(a) and (b) with the experimental data of Prasad and Koseff (1989). The root-mean-square (rms) of the horizontal and vertical velocity components are also compared with the experimental data in Fig. 21.36(c) and (d). The results show that the dynamic model performs very well in this transitional-turbulent nonhomogeneous flow.

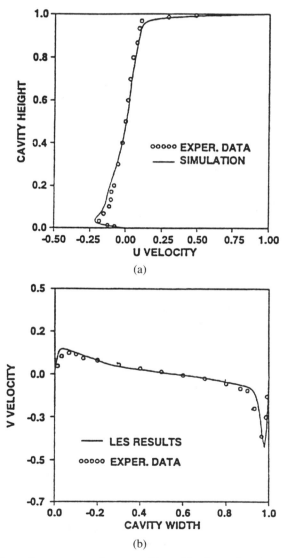

FIGURE 21.36   Centerline mean and rms profiles of horizontal and vertical velocity fluctuations; $u_{\mathrm{rms}} = 10\sqrt{(u')^2}$ for shear driven cavity. (From Jordan, 1994.)

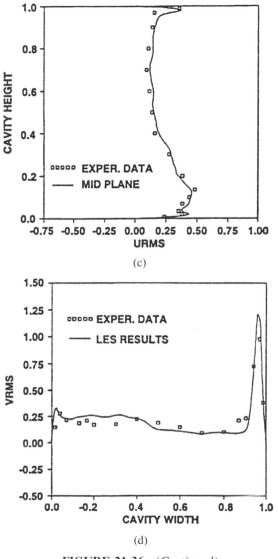

FIGURE 21.36   (*Continued*)

Compressible LES. In compressible flows, the continuity, momentum, and energy equations are supplemented by the equation of state. Erlebacher *et al.* (1992) obtained LES equations for compressible flows in terms of the Favre-filtered (density-weighted) fields. While Erlebacher *et al.* filtered the static energy equation, Ragab and Sheen (1993) filtered the total energy equation. In either case, additional SGS terms such as velocity-temperature, pressure-strain, and/or triple-velocity correlations must be modeled.

Speziale *et al.* (1988) and Erlebacher *et al.* (1992) generalized the scale similarity model of Bardina *et al.* (1983) to compressible flows. Zang *et al.* (1993) used this model in the simulation of isotropic homogeneous turbulence. Ragab and Sheen (1991, 1993) and Ragab *et al.* (1992) used it in the temporal simulation of transitional mixing layers. Moin *et al.* (1991) extended the dynamic model to compressible flows and applied it to homogeneous isotropic turbulence. El-Hady *et al.* (1993) applied the dynamic model to a supersonic transitional boundary layer, and Sreedhar (1994) applied it to a subsonic mixing layer. Sreedhar and Ragab (1994) and Ragab and Sreedhar (1994) also applied this model in the LES of vortices at low Mach numbers.

REFERENCES

Acharya, M. and Reynolds, W. C., "Measurements and Prediction of a Fully Developed Turbulent Channel Flow with Imposed Controlled Oscillations," Report TF-8, Mech. Eng. Dept., Stanford University, 1975.

Adamson, T. C., Jr. and Feo, A., "Interaction Between a Shock Wave and a Turbulent Boundary Layer in Transonic Flow," *SIAM J. Appl. Math.*, Vol. 29, pp. 121–145, 1975.

Adamson, T. C., Jr. and Messiter, A. F., "Analysis of Two-Dimensional Interactions Between Shock Waves and Boundary Layers," *Ann. Rev. Fluid Mech.*, Vol. 12, pp. 103–138, 1980.

Agarwal, R. K., "A Third-Order Accurate Upwind Scheme for Navier–Stokes Solutions at High Reynolds Numbers," *AIAA J.*, Vol. 20, No. 5, pp. 577–584, 1982.

Akselvoll, K. and Moin, P., "Application of the Dynamic Localization Model to Large-Eddy Simulation of Turbulent Flow over a Backward Facing Step," *Engineering Applications of LES*, ASME FED., Vol. 162, Ragab, S. and Piomelli, U. (Eds.), 1993.

Anderson, D. A., Tannehill, T. C., and Pletcher, R. H., *Computational Fluid Mechanics and Heat Transfer*, McGraw Hill–Hemisphere, New York, 1984.

Aregbesola, Y. A. S. and Burley, D. M. "The Vector and Scalar Potential Method for the Numerical Solution of Two- and Three-Dimensional Navier–Stokes Equations," *J. Comp. Physics*, Vol. 24, pp. 398–415, 1977.

Bakewell, H. P. and Lumley, J. L., "Viscous Sublayer and Adjacent Region in Turbulent Pipe Flow," *Phys. Fluids*, Vol. 10, pp. 1880–1889, 1967.

Baldwin, B. S. and Lomax, H., "Thin Layer Approximation and Algebraic Model for Separated Turbulent Flows," AIAA Paper 78-257, 1978.

Bardina, J., Ferziger, J. H., and Reynolds, W. C., "Improved Subgrid-Scale Models Based on Large-Eddy Simulation of Homogeneous, Incompressible, Turbulent Flows," Stanford Report No. Tf-19, 1983.

Batchelor, G. K., *An Introduction to Fluid Dynamics*, Cambridge University Press, 1967.

Beam, R. M. and Warming, R. F., "An Implicit Factored Scheme for the Compressible Navier–Stokes Equations," *AIAA J.*, Vol. 16, pp. 393–402, 1978 (also AIAA 77-645, 1977).

Belov, Y. Y. and Yanenko, N. N., "Influence of Viscosity on the Smoothness of Solutions of Incompletely Parabolic Systems," *Math. Notes, Academy of Sciences, USSR*, Vol. 10, pp. 480–483, 1971.

Berezin, Yu. A., Kovenja, V. M., and Yanenko, N. N., "Numerical Solutions of the Problems of the MHD Flow around the Bodies," *Lecture Notes in Physics*, Vol. 35, Richtmyer, R. D. (Ed.), Springer-Verlag, Berlin, pp. 85–90, 1975a.

Berezin, Yu. A., Kovenja, V. M., and Yanenko, N. N., "Implicit Numerical Method for Blunt-Body Problem in Supersonic Flows," *Computers and Fluids*, Vol. 3, pp. 271–281, 1975b.

Bernard, R. S. and Thompson, J. F., ''Approximate Factorization with an Elliptic-Pressure Solver for Incompressible Flow,'' AIAA 82-0978, 1982.

Bird, R. B., Stewart, W. E., and Lightfoot, E. N., *Transport Phenomena*, John Wiley & Sons, New York, 1960.

Blackwelder, R. F. and Eckelmann, H., ''Streamwise Vortices Associated with the Bursting Phenomenon,'' *J. Fluid Mech.*, Vol. 94, pp. 577–594, 1979.

Blackwelder, R. F. and Haritonidis, J. H., ''Scaling of the Bursting Frequency in Turbulent Boundary Layers,'' *J. Fluid Mech.*, Vol. 132, pp. 87–103, 1983.

Blaisdell, G., ''Numerical Simulation of Compressible Homogeneous Turbulence,'' Ph.D. thesis, Stanford University, Stanford, CA, 1990.

Blasius, H., ''Grenzschichten in Flüssigkeiten mit kleiner Reibung,'' Dissertation, Leipzig, 1907 (also *Z. Math. Phys.*, Vol. 56, pp. 1–37, 1908).

Bonnet, J. P. and Glauser, M. N. (Eds.), *Eddy Structure Identification in Free Turbulent Shear Flows*, Kluwer Academic Publishers, Dordrecht/Boston/London, 1993.

Boussinesq, J., ''Theorie de l'Ecoulement Tourbillant,'' *Mem. Pres. Acad. Sci.*, Vol. 23, p. 46, 1887.

Bradshaw, P., ''Effects of Streamline Curvature on Turbulent Flow,'' AGARD-AG-169, 1973.

Bradshaw, P., ''Prediction of Separation Using Boundary Layer Theory,'' AGARD-LS-94, 1978.

Bradshaw, P., ''Shear Layer Studies—Past, Present, Future,'' Haase, W. (Ed.), *Recent Contributions to Fluid Mechanics*, Springer-Verlag, 1982a.

Bradshaw, P., ''Complex Strain Fields,'' *The 1980–81 AFORSR-HTTM-Stanford Conference on Complex Turbulent Flows: Comparison of Computation and Experiment*, Vol. 2, Taxonomies, Reporters' Summaries, Evaluation and Conclusions, Kline, S. J., Cantwell, B. J., and Lilley, G. M. (Eds.), Stanford University, Stanford, CA, 1982b.

Briley, W. R., Buggeln, R. C., and McDonald, H., ''Computation of Laminar and Turbulent Flow in 90 Degree Square-Duct and Pipe Bends Using the Navier–Stokes Equations,'' SRA Report R82-92099-F, 1982.

Briley, W. R. and McDonald, H., ''An Implicit Numerical Method for Multidimensional, Nonstationary Navier–Stokes Equations,'' United Aircraft Res. Lab. Rep. M911363-6, Hartford, CT, 1973.

Briley, W. R. and McDonald, H., ''Solution of Three-Dimensional Compressible Navier–Stokes Equations by an Implicit Technique,'' *Lecture Notes in Physics*, Vol. 35, pp. 105–110, 1975.

Briley, W. R. and McDonald, H., ''Solution of the Multidimensional Compressible Navier–Stokes Equations by a Generalized Implicit Method,'' *J. Comput. Phys.*, Vol. 24, pp. 372–397, 1977.

Burggraf, O. R., Rizzetta, D., Werle, M. J., and Vasta, V. N., ''Effect of Reynolds Number on Laminar Separation of a Supersonic Stream,'' *AIAA J.*, Vol. 17, pp. 336–343, 1979.

Cantwell, B. J., ''Organized Motion in Turbulent Flow,'' *Ann. Rev. Fluid Mech.*, Vol. 13, pp. 457–515, 1981.

Caretto, L. S., Gosman, A. D., Patankar, S. V., and Spalding, D. B., ''Two Calculation Procedures for Steady, Three-Dimensional Flows with Recirculation,'' *Proc. 3rd Int. Conf. on Num. Meth. in Fluid Dynam.*, Paris, Vol. II, pp. 60–68, 1972.

Catherall, D. and Mangler, K. W., ''The Integration of the Two-Dimensional Laminar Boundary-Layer Equations past the Point of Vanishing Skin Friction,'' *J. Fluid Mech.*, Vol. 26, pp. 163–182, 1966.

Cebeci, T., "Calculation of Compressible Turbulent Boundary Layers with Heat and Mass Transfer," *AIAA J.*, Vol. 9, pp. 1091–1097, 1971.

Cebeci, T. "Kinematic Eddy Viscosity at Low Reynolds Numbers," *AIAA J.*, Vol. 11, No. 1, p. 102, 1973.

Cebeci, T. and Meier, M. U., "Modelling Requirements for the Calculation of the Turbulent Flow Around Airfoils, Wings and Bodies of Revolution," AGARD CP-271, 1979.

Chapman, D. R., "Computational Aerodynamics Development and Outlook," *AIAA J.*, Vol. 17, No. 12, pp. 1293–1313, 1979.

Chorin, A. J., "A Numerical Method for Solving Incompressible Viscous Flow Problems," *J. Comp. Physics*, Vol. 2, pp. 12–26, 1967.

Chorin, A. J., "Numerical Solution of the Navier–Stokes Equations," *Math. of Computation*, Vol. 22, pp. 745–762, 1968.

Coles, D. E. and Barker, S. J., "Some Remarks on a Synthetic Turbulent Boundary Layer," *Turbulent Mixing in Nonreactive and Reactive Flows*, Murthy, S. N. B. (Ed.), pp. 285–292, 1975.

Coles, D. E., "The Turbulent Boundary Layer in a Compressible Fluid," R403-PR, Rand Corp. (also, AD 285-651), 1962.

Coles, D. E., "The Law of the Wake in the Turbulent Boundary Layer," *J. Fluid Mech.*, Vol. 1, pp. 191–226, 1956.

Corrsin, S. and Kistler, A., "The Free Stream Boundaries of Turbulent Flows," NACA TN-3133, 1954 (also NACA TR 1244, 1955).

Corrsin, S., "Limitations of Gradient Transport Models in Random Walks and in Turbulence," *Advances in Geophysics*, Vol. 18A, pp. 25–60, 1974.

Cousteix, J. and Houdeville, R., "Effects of Unsteadiness on Turbulent Boundary Layers," *von Karman Institute for Fluid Dynamics, Lecture Series 1983-03, Turbulent Shear Flows*, 1983.

Crank, J. and Nicholson, P., "A Practical Method for Numerical Integration of Solution of Partial Differential Equations of Heat-Conduction Type," *Proc. Cambridge Philos. Soc.*, Vol. 43, pp. 50, 1947.

d'Alembert, J. le Rond, *Essai d'une Nouvelle Theorie de la Resitance des Fluides*, Paris, 1752.

Davis, R. T., "A Study of Optimal Coordinates in the Solution of the Navier–Stokes Equations," Report AFL 74-12-14, Dept. Aero. Eng., University of Cincinnati, Cincinnati, OH, 1974.

Davis, R. T., "Numerical Solution of the Navier–Stokes Equations for Symmetric Laminar Incompressible Flow Past a Parabola," *J. Fluid Mech.*, Vol. 51, Part 3, pp. 417–433, 1972.

Deardorff, J. W., "A Numerical Study of Three-Dimensional Turbulent Channel Flow at Large Reynolds Numbers," *J. Fluid. Mech.*, Vol. 41, pp. 453–480, 1970a.

Deardorff, J. W., "A Three-Dimensional Numerical Investigation of the Idealized Planetary Boundary Layer," *Geophys. Fluid Dyn.*, Vol. 1, pp. 377–410, 1970b.

Deardorff, J. W., "The Use of Subgrid Transport Equations in a Three Dimensional Model of Atmospheric Turbulence," *J. Fluids Eng.*, Vol. 95, pp. 429–438, 1973.

Dennis, S. C. R., Ingham, D. B., and Cook, R. N., "Finite-Difference Methods for Calculating Steady Incompressible Flows in Three-Dimensions," *J. Comp. Physics*, Vol. 33, pp. 325–339, 1979.

Eckelmann, H., "The Structure of the Viscous Sublayer and the Adjacent Wall Region in a Turbulent Channel Flow," *J. Fluid. Mech.*, Vol. 65, p. 439, 1974.

El-hady, N. M., Zang, T. A., and Piomelli, U., "Dynamic Subgrid-Scale Modeling for High Speed Transitional Boundary Layers," *Engineering Applications of LES*, ASME FED., Vol. 162, Ragab, S. and Piomelli, U. (Eds.), 1993.

Elvius, T. and Sundström, A., "Computationally Efficient Schemes and Boundary Conditions for a Fine-Mesh Barotropic Model Based on the Shallow Water Equations," *Tellus*, Vol. 25, pp. 132–156, 1973.

Emmons, H. W., "The Laminar-Turbulent Transition in a Boundary Layer, Part I," *J. Aeronaut. Sci.*, Vol. 18, pp. 490–498, 1951.

Erlebacher, G., Hussaini, M. Y., Kreiss, H. O., and Sarkar, S., "The Analysis and Simulation of Compressible Turbulence," *Theor. Comp. Fluid Dyn.*, Vol. 2, pp. 73–95, 1990.

Erlebacher, G., Hussaini, M. Y., Speziale, C. G., and Zang, T. A., "Toward the Large Eddy Simulation of Compressible Turbulent Flows," *J. Fluid Mech.*, Vol. 238, pp. 155–185, 1992.

Euler, L., *Principes Généraux du Mouvement des Fluides*, Hist. de l'Acad. de Berlin, 1755.

Falco, R. E., "Coherent Motions in the Outer Region of Turbulent Boundary Layers," *Phys. Fluids*, Vol. 20, No. 10, pp. S124–S132, 1977.

Falco, R. E., "New Results, a Review and Synthesis of the Mechanism of Turbulence Production in Boundary Layers and its Modification," AIAA Paper 83-0377, 1983.

Farren, W. S., "Reaction on a Wing Whose Angle of Incidence is Changing Rapidly, Wind Tunnel Experiments with a Short-Period Recording Balance," Aero. Res. Council R&M 1648, 1935.

Fasel, H., "Investigation of the Stability of Boundary Layers By a Finite-Difference Model of the Navier–Stokes Equations," *J. Fluid Mech.*, Vol. 78, pp. 355–383, 1976.

Favre, A., "Equations des Gaz Turbulents Compressibles," *J. de Mecanique*, Vol. 4, pp. 361–390, 1965.

Ferziger, J. H., Mehta, U. B., and Reynolds, W. C., "Large Eddy Simulation of Homogeneous Isotropic Turbulence," *Proceedings of Symposium on Turbulent Shear Flows*, Pennsylvania State University, University Park, PA, pp. 14.31–14.39, 1977.

Galperin, B. and Orszag, S. A. (Eds.), *Large Eddy Simulation of Complex Engineering and Geophysical Flows*, Cambridge University Press, Cambridge, 1993.

Germano, M., "Turbulence: The Filtering Approach," *J. Fluid Mech.*, Vol. 238, p. 325, 1992.

Germano, M., Piomelli, U., Moin, P., and Cabot, W. H., "A Dynamic Subgrid-Scale Eddy Viscosity Model," *Phys. Fluids A*, Vol. 3, pp. 1760–1765, 1991.

Ghia, K. N., Hankey, W. L., and Hodge, J. K., "Study of Incompressible Navier–Stokes Equations in Primitive Variables Using Implicit Numerical Technique," *AIAA J.*, Vol. 17, No. 3, pp. 298–301, 1979.

Ghia, K. N., Osswald, G. A., and Ghia, U., "A Direct Method for the Solution of Unsteady Two-Dimensional Incompressible Navier–Stokes Equations," *Second Symposium on Numerical and Physical Aspects of Aerodynamic Flows*, Long Beach, CA, pp. I-1–16, 1983.

Ghia, K. N., Shin, C. T., and Ghia, U., "Use of Spline Approximations for Higher-Order Solutions of Navier–Stokes Equations in Primitive Variables," AIAA 79-1467, 1979.

Ghia, U., "Computation of Viscous Internal Flows," *Computation of Internal Flows: Methods and Applications*, Sockol, P. M. and Ghia, K. N. (Eds.), ASME, p. 97, 1984.

Ghia, U., Ghia, K. N., and Shin, C. T., "High-Re Solutions for Incompressible Flow Using the Navier–Stokes Equations and a Multi-Grid Method," *J. Comp. Physics*, Vol. 48, No. 3, pp. 387–411, 1982.

Ghosal, S., Lund, T. S., and Moin, P., "A Local Dynamic Model for LES," *CTR Annual Research Briefs—1992*, pp. 1–22, 1992.

Gilbert, N. and Kleiser, L., "Near-Wall Phenomena in Transition to Turbulence," *Near-Wall Turbulence*, Kline, S. J. and Afgan, N. H. (Eds.), Elsevier, Amsterdam, pp. 7–27, 1990.

Goldstein, S., "On Laminar Boundary Flow near a Point of Separation," *Quart. J. Mech. Appl. Math.*, Vol. 1, pp. 43–69, 1948.

Granville, P. S., "The Determination of the Local Skin Friction and the Thickness of Turbulent Boundary Layers from the Velocity Similarity Laws," R&D Report 1340, David Taylor Model Basin, Hydromechanics Laboratory, Bethesda, MD, 1959.

Grotzbach, G., "Direct and Large Eddy Simulation of Turbulent Channel Flows," *Encyclopedia of Fluid Mechanics*, Vol. 6, 1983.

Gupta, A. K., Laufer, J., and Kaplan, R. E., "Spatial Structure in the Viscous Sublayer," *J. Fluid Mech.*, Vol. 50, pp. 493–512, 1971.

Gustafsson, B. and Sundström, A., "Incompletely Parabolic Problems in Fluid Dynamics," *SIAM J. Appl. Math.*, Vol. 35, pp. 343–357, 1978.

Gustafsson, B., "The Euler and Navier–Stokes Equations, Wellposedness, Stability, and Composite Grids," *Computational Fluid Dynamics*, von Karman Institute for Fluid Dynamics, Lecture Series 1991-01, 1991.

Hadamard, J., "Sur les Tourbillons Produit par les Ondes de Choc, Note III," *Lecons sur la Propagation des Ondes*, Hermann, A. (Ed.), Paris, p. 362, 1903.

Hanjalic, K., "Velocity and Length Scales in Turbulent Flows—A Review of Approaches," *The 1980–81 AFOSR-HTTM-Stanford Conference on Complex Turbulent Flows: Comparison of Computation and Experiment*, Vol. 2, Taxonomies, Reporters' Summaries, Evaluation and Conclusions, Kline, S. J., Cantwell, B. J., and Lilley, G. M. (Eds.), Stanford University, Stanford, CA, 1982.

Hanratty, T. J., Chorn, L. G., and Hatziavramidis, D. T., "Turbulent Fluctuations in the Viscous Wall Region for Newtonian and Drag Reducing Fluids," *Phys. Fluids A*, Vol. 20, p. 112, 1977.

Helmholtz, H. von, "Über Discontinuierliche Flüssigkeitsbewegungen," *Monatsberichte der königl. Akademie der Wissenschaften*, Berlin, Vol. 23, pp. 215–228, 1868 (also *Phil. Mag.*, Vol. 36, Nov. 1868, p. 337).

Herbert, Th. and Morkovin, M. V., "Dialogue on Bridging Some Gaps in Stability and Transition Research," *Proceedings of the IUTAM Symposium on Laminar-Turbulent Transition*, pp. 47–72, 1979.

Hesselberg, Th., "Die Gesetze der Ausgeglichene Atmosharischen Bewegungen," *Beitrage Physik Freien Atmosphare*, Vol. 12, pp. 141–160, 1926.

Hinze, J. O., *Turbulence*, 2nd ed., McGraw-Hill, New York, 1975.

Hirsch, C., *Numerical Computation of Internal and External Flows*, Vol. 1: Fundamentals of Numerical Discretization, John Wiley & Sons, Chichester, England and New York, 1988.

Hirsch, C., *Numerical Computation of Internal and External Flows*, Vol. 2: Computational Methods for Inviscid and Viscous Flow Models, John Wiley & Sons, Chichester, England and New York, 1990.

Hirsh, R. S., "Higher-Order Approximations in Fluid Mechanics: Compact to Spectral," *VKI Lecture Notes*, Brussels, Belgium, 1983.

Hodge, J. K., "Numerical Solution of Incompressible Laminar Flow about Arbitrary Bodies in Body-Fitted Curvilinear Coordinates," Ph.D. dissertation, Mississippi State University, Mississippi State, MS, 1975.

Humphrey, J. A. C., Taylor, A. M. K., and Whitelaw, J. H., "Laminar Flow in a Square Duct of Strong Curvature," *J. Fluid Mech.*, Vol. 83, Part 3, pp. 509–527, 1977.

Hussain, A. K. M. F. and Reynolds, W. C., "Measurement in Fully Developed Turbulent Channel Flows," *J. Fluids Eng.*, Vol. 97, p. 568, 1975.

Hussain, A. K. M. F., "Coherent Structures and Turbulence," *J. Fluid Mech.*, Vol. 173, pp. 303–356, 1986.

Imai, I., "On the Asymptotic Behavior of Viscous Fluid Flow at a Great Distance from a Cylindrical Body, with Special Reference to Filon's Paradox," *Proc. Roy. Soc. (London) Ser. A*, Vol. 208, pp. 487–516, 1951.

Inger, G. R. and Mason, W. H., "Analytical Theory of Transonic Normal Shock-Boundary Layer Interaction," *AIAA J.*, Vol. 14, pp. 1266–1272, 1976.

Inger, G. R., "Nonasymptotic Theory of Unseparated Turbulent Boundary Layer-Shock Wave Interactions with Application to Transonic Flows," *Numerical and Physical Aspects of Aerodynamic Flows*, Cebeci, T. (Ed.), Springer-Verlag, New York, 1982.

Jones, B. M., "The Streamline Aeroplane," *J. Roy. Aero. Soc.*, Vol. 33, pp. 357–385, 1929.

Jones, W. P., "Trends in Unsteady Aerodynamics: 6th Lanchester Memorial Lecture," *J. Roy. Aero. Soc.*, Vol. 67, p. 137, 1963.

Jordan, S., "The Large-Eddy Simulation of Incompressible Flows in Simple and Complex Geometries," Ph.D. dissertation, Virginia Polytechnic Institute and State University, Blacksburg, VA, 1994.

Jovic, S. and Driver, M., NASA Ames Research Center, Moffat Field, CA, unpublished, 1992.

Kaplan, S., "The Role of Coordinate Systems in Boundary Layer Theory," *ZAMP*, Vol. 5, pp. 111–135, 1954.

Karniadakis, G. E., Orszag, S. A., and Yakhot, V., "Renormalization Group Theory Simulation of Transitional and Turbulent Flow over a Backward-Facing Step," *Large Eddy Simulation of Complex Engineering and Geophysical Flows*, Galperin, B. and Orszag, S. A. (Eds.), Cambridge Univ. Press, 1993.

Kelvin, Lord (Thomson, W.), "On Vortex Motion," *Mathematics and Physics Papers*, Vol. 4, p. 49, 1869.

Kim, J., Moin, P., and Moser, R. D., "Turbulent Statistics in Fully Developed Channel Flow at Low Reynolds Number," *J. Fluid Mech.*, Vol. 177, pp. 133–166, 1987.

Kirchhoff, G., "Zur Theorie freir Flüssigkeitsstrahlen," *Crelle J. Math.*, Vol. 70, pp. 289–298, 1869.

Kleiser, K. and Zang, T. A., "Numerical Simulation of Transition in Wall-Bounded Shear Flows," *Ann. Rev. Fluid Mech.*, Vol. 23, pp. 495–537, 1991.

Kline, S. J., Reynolds, W. C., Schraub, F. A., and Runstadler, P. W., "The Structure of Turbulent Boundary Layers," *J. Fluid Mech.*, Vol. 30, pp. 741–773, 1967.

Kopal, Z., *Numerical Analysis*, 2nd ed., John Wiley & Sons, New York, 1961.

Kreiss, H. O., "Initial Boundary Value Problems for Hyperbolic Equations," *Commun. Pure Appl. Math.*, Vol. 23, pp. 277–298, 1970.

Kreplin, H. and Eckelmann, H., "Behavior of the Three Fluctuating Velocity Components in the Wall Region of a Turbulent Channel Flow," *Phys. Fluids A*, Vol. 22, p. 1233, 1979.

Küchmann, D., "Report on the IUTAM Symposium on Concentrated Vortex Motions in Fluids," *J. Fluid Mech.*, Vol. 21, pp. 1–20, 1965.

Kwak, D., Shanks, S. P., Chang, J. L. C., and Chakravarthy, S. R., "An Incompressible

Navier–Stokes Flow Solver in Three-Dimensional Curvilinear Coordinate Systems Using Primitive Variables,'' AIAA 84-0253, 1984.

Laasonen, P., ''Über siene Methode zur Lösung der Wärmeleitungsgleichung,'' *Acta Math.*, Vol. 81, pp. 309–317, 1949.

Lagerstrom, P. A., ''Laminar Flow Theory,'' *Theory of Laminar Flows, Vol. IV. High Speed Aerodynamics and Jet Propulsion*, Moore, F. K. (Ed.), Princeton University Press, Princeton, NJ, 1964.

Launder, B. E., ''Heat and Mass Transport,'' *Turbulence, Topics in Applied Physics*, 2nd ed., Vol. 12, Bradshaw, P. (Ed.), Springer-Verlag, 1978.

Lax, P. D. and Wendroff, B., ''Systems of Conservation Laws,'' *Commun. on Pure Appl. Math.*, Vol. 13, pp. 217–237, 1960.

Lax, P. D., ''Hyperbolic Systems on Conservation Laws and the Mathematical Theory of Shock Waves,'' *Regional Conf. Series in Appl. Math.*, No. 11, SIAM, 1973.

Lax, P. D., ''Weak Solutions of Nonlinear Hyperbolic Equations and Their Numerical Computations,'' *Commun. on Pure Appl. Math.*, Vol. 7, pp. 159–193, 1954.

Le Balleur, J. C., ''Strong Matching Method for Computing Transonic Viscous Flows Including Wakes and Separations—Lifting Airfoils,'' *La Recherche Aerospatiale*, English ed., No. 1981-3, pp. 21–45, 1981.

Le, H. and Moin, P., Stanford Report TF-58, Stanford University, Stanford, CA, 1993.

Lee, M. J., Kim, J., and Moin, P., ''Turbulence Structure at High Shear Rate,'' *Proc. 6th Symp. Turb. Shear Flows*, Toulouse, 1983.

Lees, L. and Reeves, B. L., ''Supersonic Separated and Reattaching Laminar Flows: I. General Theory and Applications to Adiabatic Boundary Layer/Shock Wave Interactions,'' *AIAA J.*, Vol. 2, pp. 1907–1920, 1964.

Lees, M., ''A Linear Three-Level Difference Scheme for Quasilinear Parabolic Equations,'' *Math. Comp.*, Vol. 20, pp. 516–522, 1966.

Lele, S., ''Compressibility Effects on Turbulence,'' *Ann. Rev. Fluid Mech.*, Vol. 26, pp. 211–254, 1994.

Leonard, A., ''Energy Cascade in Large-Eddy Simulations of Turbulent Fluid Flows,'' *Adv. Geophysics*, Vol. 18, pp. 237–248, 1974.

Lesieur, M., *Turbulence in Fluids*, Kluwer Academic Publishers, Amsterdam, 1990.

Lighthill, M. J., ''On Boundary Layers and Upstream Influence: II. Supersonic Flow without Separation,'' *Proc. Roy. Soc. (London) Ser. A.*, Vol. 217, pp. 478–507, 1953.

Lilly, D. K., ''A Proposed Modification of the Germano Subgrid-Scale Closure Method,'' *Phys. Fluids A*, Vol. 4, pp. 633–635, 1992.

Lilly, D. K., ''On the Application of the Eddy Viscosity Concept in the Inertial Sub-Range of Turbulence,'' *NCAR Manuscript*, Vol. 123, 1966.

Lilly, D. K., ''The Representation of Small-Scale Turbulence in Numerical Simulation Experiments,'' *Proceedings of the IBM Scientific Computing Symposium on Environmental Sciences*, pp. 195–210, 1967.

Lindemuth, I. and Killeen, J., ''Alternating Direction Implicit Techniques for Two-Dimensional Magnetohydrodynamic Calculations,'' *J. Comput. Phys.*, Vol. 13, pp. 181–208, 1973.

Lock, R. C. and Firmin, M. C. P., ''Survey of Techniques for Estimating Viscous Effects in External Aerodynamics,'' Royal Aircraft Establishment, Tech. Mem. Aero 1900, 1981.

Lomax, H. and Mehta, U., ''Some Physical and Numerical Aspects of Computing the Effects of Viscosity on Fluid Flow,'' *Computational Methods in Viscous Flows*, Habashi, W. G.

(Ed.), *Recent Advances in Numerical Methods in Fluids*, Vol. 3, Pineridge Press, Swansea, U.K., 1984.

Lugt, H. J., *Vortex Flow in Nature and Technology*, John Wiley & Sons, New York, 1983.

Lumley, J. L., "Coherent Structures in Turbulence," *Transition and Turbulence*, Meyer, R. E. (Ed.), Academic Press, New York, 1981.

Lund, T. S., Ghosal, S., and Moin, P., "Numerical Experiments with HighlyVariable Eddy Viscosity Models," *Engineering Applications of LES*, ASME FED, Vol. 162, Ragab, S. and Piomelli, U. (Eds.), 1993.

Maise, G. and McDonald, H., "Mixing Length and Kinematic Eddy Viscosity in a Compressible Boundary Layer," *AIAA J.*, Vol. 6, pp. 73–80, 1968.

McComb, W. D., *The Physics of Fluid Turbulence*, Oxford Science Publications, Oxford, 1990.

McGreehan, W. F., Ghia, K. N., Ghia, U., and Osswald, G. A., "Analysis of Separated Flow in a Pipe Orifice Using Unsteady Navier–Stokes Equations," *Proc. 9th Int. Conf. on Num. Meth. in Fluid Dynam.*, Saclay, France, 1984.

Mehta, U. B. and Lavan, Z., "Starting Vortex, Separation Bubbles and Stall: A Numerical Study of Laminar Unsteady Flow around an Airfoil," *J. Fluid Mech.*, Vol. 67, Part 2, pp. 227–256, 1975.

Mehta, U. and Lomax, H., "Reynolds Averaged Navier–Stokes Computations of Transonic Flows—The State-of-the-Art," *Transonic Aerodynamics*, Nixon, D. (Ed.), *Progress in Astronautics and Aeronautics*, Vol. 81, pp. 297–375, 1982.

Mehta, U., "Guide to Credible Computational Fluid Dynamics Simulations," AIAA 95-2225, 1995.

Mehta, U., "Aerospace Plane Design Challenge: Credible Computations," *J. Aircraft*, Vol. 30, No. 4, pp. 519–525, 1993 (with errata in Vol. 30, No. 6, p. 1009, 1993).

Mehta, U., "Computational Requirements for Hypersonic Flight Performance Estimates," *J. Spacecraft and Rockets*, Vol. 27, No. 2, pp. 103–112, 1990a.

Mehta, U., "Some Aspects of Uncertainty in Computational Fluid Dynamics," *J. Fluids Eng.*, Vol. 112, pp. 538–543, 1990b.

Mehta, U., "Dynamic Stall of an Oscillating Airfoil," AGARD CP-227, 1977.

Mehta, U., "Physical Aspects of Computing the Flow of a Viscous Fluid," NASA TM-85893, 1984.

Melnik, R. E., "Turbulent Interactions on Airfoils at Transonic Speeds—Recent Developments," *Viscous-Inviscid Interactions*, AGARD-CP-291, 1980.

Messiter, A. F., "Boundary Layer Flow near the Trailing Edge of a Flat Plate," *SIAM J. Appl. Math.*, Vol. 18, pp. 241–257, 1970.

Moin, P. and Jimenez, J., "Large Eddy Simulation of Complex Turbulent Flows," AIAA 93-3099, 1993.

Moin, P. and Kim, J., "Numerical Investigation of Turbulent Channel Flow," *J. Fluid Mech.*, Vol. 118, pp. 341–377, 1982.

Moin, P., "Analysis of Turbulent Data Generated by Numerical Simulation," AIAA 87-00194, 1987.

Moin, P., Reynolds, W. C., and Ferziger, J. H., "Large Eddy Simulation of Incompressible Turbulent Channel Flow," Report TF-12, Dept. Mech. Eng., Stanford University, Stanford, CA, 1978.

Moin, P., Squires, K., Cabot, W., and Lee, S., "A Dynamic Subgrid Scale Model For Compressible Turbulence and Scalar Transport," *Phys. Fluids A*, Vol. 3, No. 11, pp. 2746–2757, 1991.

Monin, A. S. and Yaglom, A. M., *Statistical Fluid Mechanics of Turbulence*, Vol. 1, MIT Press, Cambridge, MA, 1971.

Morkovin, M. V., "Bypass-Transition Research, Issues and Philosophy," *Instabilities and Turbulence in Engineering*, Aphis, D. E., Gatski, T. B., and Hirsh, R. (Eds.), Kluwer Academic Publishers, Dordrecht, pp. 3–30, 1993.

Morkovin, M. V., "Instability, Transition to Turbulence and Predictability," AGARD-AG-236, 1978.

Müller, E.-A., "Theoretische Untersuchungen über die Wechselwirkung zwischen einer einfallenden kleinen Strörung und der Grenzschicht bei schnell strömenden Gasen," dissertation, Universität Göttingen, 1953.

Murlis, J., Tsai, H. M., and Bradshaw, P., "The Structure of Turbulent Boundary Layers at Low Reynolds Numbers," *J. Fluid Mech.*, Vol. 122, pp. 13–56, 1962.

Naqawi, A. and Reynolds, W. C., "Dual Cylindrical Wave Laser-Doppler Method for Measurement of Skin Friction in Fluid Flow," Stanford Report TF-28, Dept. of Mech. Engineering, Stanford University, Stanford, CA, 1987.

Navier, C. M. L. H., "Mémoire sur les Lois du Mouvement des Fluides," *Mém. de l'Acad. des Sciences*, Vol. 6, pp. 389–416, 1826.

Neiland, V. Y., "Towards a Theory of Separation of the Laminar Boundary Layer in a Supersonic Stream," *Izv. Akad. Nauk SSSR Mekh. Zhidk. Gaza*, Vol. 4, pp. 33–35, 1969.

Newton, I., *Philosophiae Naturalis Principa Mathematica*, Book II, London, 1726.

Oliger, J. and Sundström, A., "Theoretical and Practical Aspects of Some Initial Boundary Value Problems in Fluid Dynamics," *SIAM J. Appl. Math.*, Vol. 35, pp. 419–446, 1978.

Orszag, S. A. and Israeli, M., "Numerical Simulation of Viscous Incompressible Flows," *Ann. Reviews in Fluid Mech.*, Vol. 6, pp. 281–318, 1974.

Orszag, S. A. and Patterson, G. S., "Numerical Simulation of Three Dimensional Homogeneous Isotropic Turbulence," *Phys. Rev. Lett.*, Vol. 28, pp. 76–79, 1972.

Orszag, S. A., Staroselsky, I., and Yakhot, V., "Some Basic Challenges for LES Research," *Large Eddy Simulation of Complex Engineering and Geophysical Flows*, Galperin, B. and Orszag, S. A. (Eds.), Cambridge Univ. Press, Cambridge, 1993.

Oseen, C. W., "Ueber die Stokes'sche Formel, und über eine verwandte Aufgabe in der Hydrodynamik," *Ark. Math. Astronom. Fys.*, Vol. 6, No. 29, 1910.

Osswald, G. A. and Ghia, K. N., "Study of Unsteady Incompressible Flow Using Nonuniform Curvilinear Grids, Time Marching and a Direct Method," *Multigrid Methods*, NASA CP-2202, 1981.

Osswald, G. A., Ghia, K. N., and Ghia, U., "Study of Incompressible Separated Flow Using an Implicit Time-Dependent Technique," AIAA 83-1894, 1983.

Osswald, G. A., Ghia, K. N., and Ghia, U., "Unsteady Navier–Stokes Simulation of Internal Separated Flows Over Plane and Axisymmetric Sudden Expansions," AIAA 84-1584, 1984.

Passot, A. and Pouquet, A. "Numerical Simulation of Compressible Homogeneous Flows in the Turbulent Regime" *J. Fluid Mech.*, Vol. 181, pp. 441–466, 1987.

Patankar, S. V. and Spalding, D. B., "A Calculation Procedure for Heat, Mass and Momentum Transfer in Three-Dimensional Parabolic Flows," *Int. J. Heat and Mass Transfer*, Vol. 15, pp. 1787–1806, 1972.

Peake, D. J. and Tobak, M., "Three-Dimensional Interactions and Vortical Flows with Emphasis on High Speeds," AGARD-AG-252, 1980.

Perry, A. E., Henbest, S., and Chong, M. S., "A Theoretical and Experimental Study of Wall Turbulence," *J. Fluid Mech.*, Vol. 165, pp. 163–199, 1986.

Piomelli, U., "High Reynolds Number Calculations Using the Dynamic Subgrid-Scale Stress Model," *Phys. Fluids A*, Vol. 5, No. 6, pp. 1484–1490, 1988.

Piomelli, U., Cabot, W. H., Moin, P., and Lee, S., "Subgrid-Scale Backscatter in Turbulent and Transitional Flows," *Phys. Fluids A*, Vol. 3, pp. 1766–1771, 1991.

Piomelli, U., Moin, P., and Ferziger, J. H., "Model Consistency in Large-Eddy Simulation of Turbulent Channel Flows," *Phys. Fluids A*, Vol. 31, No. 7, pp. 1884–1891, 1988.

Piomelli, U., Zang, T. A., Speziale, C. G., and Lund, T. S., "Application of Renormalization Group Theory to the Large-Eddy Simulation of Transitional Boundary Layer," *Instability and Transition*, Vol. 2, Hussaini, M. Y. and Voigt, R. G. (Eds)., Springer-Verlag, Berlin, pp. 480–496, 1990.

Poisson, S. D., "Mémoire sur les Équations Générales de l'Équilibre et du Mouvement des Corps Solides Élastiques et des Fluides," *J. de l'Ecole Polytechnique*, Paris, Vol. 20, 1831.

Polezhaev, V. L., "Numerical Solution of the System of Two-Dimensional Unsteady Navier–Stokes Equations for a Compressible Gas in a Closed Region," *Fluid Dynamics*, Vol. 2, No. 2, pp. 70–74, 1967.

Prandtl, L., "Über Flüssigkeitsbewegung bei sehr kleiner Reibung," *Verhandlung des drittes Internationalen Mathematiker-Kongresses*, Heidelberg, Teubner, Leipzig, pp. 484–491, 1905 (NACA TM-452, 1928).

Prasad, A. K. and Koseff, J. R., "Reynolds Number and End-Wall Effects on a Lid Driven Cavity Flows," *Phys. Fluids A*, Vol. 1, No. 2, pp. 208–218, 1989.

Preston, J. H., "The Minimum Reynolds Number for a Turbulent Boundary Layer and the Selection of a Transition Device," *J. Fluids Mech.*, Vol. 3, p. 373, 1958.

Pulliam, T. H. and Chaussee, D. S., "A Diagonal Form of an Implicit Approximate-Factorization Algorithm," *J. Comput. Phys.*, Vol. 39, No. 2, p. 347, 1981.

Quartapelle, L., "Vorticity Conditioning in the Computation of Two Dimensional Viscous Flows," *J. Comp. Physics*, Vol. 40, pp. 453–477, 1981.

Ragab, S. A. and Piomelli, U. (Eds.), *Engineering Applications of LES*, ASME FED., Vol. 162, 1993.

Ragab, S. A. and Sheen, S., "Large Eddy Simulation of a Mixing Layer," *Large Eddy Simulation of Complex Engineering and Geophysical Flows*, Galperin, B. and Orszag, S. A. (Eds.), Cambridge University Press, Cambridge, 1993.

Ragab, S. A. and Sheen, S., "Large-Eddy Simulation of a Mixing Layer," AIAA 91-0233, 1991.

Ragab, S. A. and Sreedhar, M. K., "Large-Scale Structure in a Trailing Vortex," AIAA 94-2136, 1994.

Ragab, S. A., Sheen, S., and Sreedhar, M., "An Investigation of Finite-Difference Methods for Large-Eddy Simulation of a Mixing Layer," AIAA-92-0554, 1992.

Rai, M. M. and Moin, P., "Direct Numerical Simulation of Transition and Turbulence in a Spatially Evolving Boundary Layer," AIAA 91-1607-CP, 1991.

Raithby, G. D. and Schneider, G. E. "Numerical Solution of Problems in Incompressible Fluid Flow: Treatment of the Velocity-Pressure Coupling," *Num. Heat Transfer*, Vol. 2, pp. 417–440, 1979.

Ramaprian, B. R. and Tu, S.-W., "An Experimental Study of Oscillatory Pipe Flow at Transitional Reynolds Numbers," *J. Fluid Mech.*, Vol. 100, pp. 513–544, 1980.

Rao, K. N., Narasimha, R., and Narayanan, M. A. B., "Bursting in a Turbulent Boundary Layer," *J. Fluid Mech.*, Vol. 48, pp. 339–352, 1971.

Rayleigh, Lord (Strutt, J. W.), "Notes on Hydrodynamics," *Phil. Mag.*, Series 5, Vol. II, pp. 441–447, 1876.

Rayleigh, Lord (Strutt, J. W.), "On the Question of the Stability of the Flow of Fluids," *Phil. Mag.*, Series 5, Vol. XXXIV, pp. 59–70, 1892.

Reshotko, E., "Boundary Layer Instability, Transition, and Control," AIAA Paper 94-0001, 1994.

Reynolds, O., "An Experimental Investigation of Circumstances Which Determine Whether the Motion of Water Shall be Direct or Sinuous, and of the Law of Resistance in Parallel Channels," *Philos. Trans. Roy. Soc. (London)*, Vol. 174, pp. 935–982, 1883 (also *Scientific Papers*, Cambridge University Press, Cambridge, Vol. 2, pp. 51–105, 1891).

Reynolds, W. C., "Simulation of Turbulent Shear Flows: What Can We Do and What Have We Learned?" *Proceedings of the Seventh Biennial Symposium on Turbulence*, Rolla, MO, 1981.

Reynolds, W. C., "The Potential and Limitations of Direct and Large Eddy Simulations," *Whither Turbulence? Turbulence at the Crossroads*, Lumley, J. L. (Ed.), *Lect. Notes in Physics*, Vol. 357, pp. 313–343, 1989.

Richardson, S. M. and Cornish, A. R. H., "Solution of Three-Dimensional Incompressible Flow Problems," *J. Fluid Mech.*, Vol. 82, Part 2, pp. 309–319, 1977.

Richtmyer, R. D. and Morton, K. W., *Difference Methods for Initial Value Problems*, Interscience Publishers, p. 331, 1967.

Roache, P. J., *Computational Fluid Dynamics*, Hermosa Press, Albuquerque, NM, 1972.

Rodi, W., "A New Algebraic Relation for Calculating the Reynolds Stresses," *Zeitschr. Angewandte Math. Mech.*, Vol. 56, pp. 219–221, 1976.

Rogallo, R. S. and Moin, P., "Numerical Simulation of Turbulent Flows," *Ann. Rev. Fluid Mech.*, Vol. 16, pp. 99–138, 1984.

Rogers, M. M. and Moin, P., "The Structure of the Vorticity Field in Homogeneous Turbulent Flow," *J. Fluid. Mech.*, Vol. 176, pp. 33–66, 1987.

Roscoe, D. F., "The Solution of the Three-Dimensional Navier–Stokes Equations Using a New Finite-Difference Approach," *Int. J. Num. Meth. in Eng.*, Vol. 10, pp. 1299–1308, 1976.

Roshko, A., "The Plane Mixing Layer: Flow Visualization Results and Three Dimensional Effects," *Proceedings of an International Conference on the Role of Coherent Structures in Modelling Turbulence and Mixing*, Madrid, Spain, 1981.

Rubin, S. G. and Khosla, P. K., "A Composite Velocity Procedure for the Incompressible Navier–Stokes Equations," *Lecture Notes in Physics*, Vol. 170, Springer-Verlag, 1982.

Rubin, S., "Incompressible Navier–Stokes and Parabolized Navier–Stokes Solution Procedures and Computational Techniques," *von Karman Institute for Fluid Dynamics, Computational Fluid Dynamics*, Lecture Series No. 1982-04, 1982.

Saint-Venant, B. de, "Note à Joindre au Mémoire sur la Dynamique des Fluides," *Comptes Rendus Acad. Sci. Paris, Ser. A*, Vol. 17, pp. 1240-1242, 1843.

Sarkar, S., Erlebacher, G., and Hussaini, M. Y., "Direct Simulation of Compressible Turbulence in a Shear Flow," *Theor. Comp. Fluid Dyn.*, Vol. 2, pp. 291–305, 1991.

Schetz, J. A., *Boundary Layer Analysis*, Prentice Hall, Englewood Cliffs, NJ, 1993.

Schubauer, G. B. and Klebanoff, P. S., "Contributions on the Mechanics of Boundary Layer Transition," NACA TN-3489, 1955.

Schumann, U. and Friedrich, R., "On the Direct and Large Eddy Simulation of Turbulence," *Advances in Turbulence*, Springer-Verlag, Berlin, 1988.

Schumann, U., "Ein Verfahren zur Direkten Numerischen Simulation Turbulenter Strömungen in Platten- und Ringspaltkanälen und über seine Anwendung zur Untersuchung von Turbulenzmodellen," Ph.D. Thesis, Universitat Karlsruhe, Karlsruhe, Germany, 1973 (also NASA TT F-15,391).

Schumann, U., "Subgrid Scale Model for Finite Difference Simulation of Turbulent Flows in Plane Channels and Annuli," *J. Comp. Phys.*, Vol. 18, pp. 376–404, 1975.

Serrin, J., "Mathematical Principles of Classical Fluid Mechanics," *Handbuch der Physik, Band VIII/1, Strömungsmechanik I*, Springer-Verlag, Berlin, 1959a.

Serrin, J., "On the Uniqueness of Compressible Fluid Motions," *Archives for Rational Mech. and Anal.*, Vol. 3, pp. 271–288, 1959b.

Sherman, F. S., "A Low-Density Wind-Tunnel Study of Shock-Wave Structure and Relaxation Phenomena in Gases," NACA TN-3298, 1955.

Simpson, R. L., Chew, Y. T., and Shivaprasad, B. G., "Measurements of Unsteady Turbulent Boundary Layers with Pressure Gradients," SMU Report WT- 6, Southern Methodist University, Dallas, TX, 1980.

Smagorinsky, J., "General Circulation Experiments with the Primitive Equations. Part I. The Basic Experiment," *Mon. Wea. Rev.*, Vol. 91, pp. 99–164, 1963.

Smith, C. R. and Metzler, S. P., "The Characteristics of Low Speed Streaks in the Near-Wall Region of a Turbulent Boundary Layer," *J. Fluid Mech.*, Vol. 129, p. 27, 1983.

Solonnikov, V. A. and Kazhikhov, A. V., "Existence Theorems for the Equations of Motion of a Compressible Viscous Fluid," *Ann. Rev. Fluid Mech.*, Vol. 13, pp. 79–95, 1981.

Spalart, P. R., "Direct Numerical Simulation of a Turbulent Boundary Layer up to $Re_\theta = 1410$," *J. Fluid Mech.*, Vol. 187, p. 61, 1988.

Speziale, C. G., "Galilean Invariance of Subgrid Scale Stress Models in the Large Eddy Simulation of Turbulence," *J. Fluid Mech.*, Vol. 156, pp. 55–62, 1985.

Speziale, C. G., Erlebacher, G., Zang, T. A., and Hussaini, M. Y., "The Subgrid-Scale Modeling of Compressible Turbulence," *Phys. Fluids A*, Vol. 31, pp. 940–942, 1988.

Sreedhar, M. K. and Ragab, S. A., "Large Eddy Simulation of Longitudinal Stationary Vortices," *Phys. Fluids*, Vol. 6, No. 7, pp. 2501–2515, 1994.

Sreedhar, M. K., "Large Eddy Simulation of Turbulent Vortices and Mixing Layers," Ph.D. dissertation, Virginia Polytechnic Institute and State University, Blacksburg, VA, 1994.

Steger, J. L. and Kutler, P., "Implicit Finite-Difference Procedures for the Computation of Vortex Wakes," *AIAA J.*, Vol. 15, No. 4, pp. 581–590, 1977.

Stewartson, K., "Multistructured Boundary Layers on Flat Plates and Related Bodies," *Adv. in Appl. Mech.*, Vol. 14, 1974.

Stewartson, K., "D'Alembert's Paradox," *SIAM Review*, Vol. 23, No. 3, pp. 308–343, 1981.

Stewartson, K., "On the Flow near the Trailing Edge of a Flat Plate II," *Mathematika*, Vol. 16, pp. 106–121, 1969.

Stewartson, K., "Some Recent Studies in Triple-Deck Theory," *Numerical and Physical Aspects of Aerodynamic Flows*, Cebeci, T. (Ed.), Springer-Verlag, New York, 1982.

Stewartson, K., Smith, F. T., and Kaups, K., "Marginal Separation," *Studies in Applied Mathematics*, Vol. 67, pp. 45–61, 1982.

Stokes, G. G., "On the Theories of Internal Friction of Fluids in Motion," *Trans. Cambridge Philos. Soc.*, Vol. 8, Pt. III, pp. 287–305, 1847.

Stüper, J., "Reduction of Lift of a Wing Due to Its Drag," *Zeitschrift für Flugtechnik und Motorluftschiffahrt*, Vol. 24, No. 16, 1933 (also NACA TM-781, 1935).

Taylor, G. I. and Maccoll, J. W., "The Mechanics of Compressible Fluids," *Aerodynamic Theory—A General Review of Progress*, Vol. 3, Durand, W. F. (Ed.), Dover Publications, New York, 1963.

Telionis, D. P., *Unsteady Viscous Flows*, Springer Series in Computational Physics, Springer-Verlag, 1981.

Temam, R., *Navier–Stokes Equations*, rev. ed., North-Holland Publishing Company, Amsterdam, 1979.

Townsend, A. A., "Organized Eddies in Turbulent Shear Flows," *Eddy Structure Identification in Free Turbulent Shear Flows*, Bonnet, J. P. and Glauser, M. N. (Eds.), Kluwer Academic Publishers, Dordrecht/Boston/London, pp. 5–9, 1993.

Townsend, A. A., *Structure of Turbulent Shear Flows*, Cambridge University Press, Cambridge, 1976.

Townsend, A. A., *The Structure of Turbulent Shear Flow*, 1st ed., Cambridge University Press, Cambridge, 1956.

Truesdell, C. A., *Kinematics of Vorticity*, Indiana University Press, Bloomington, IN, 1954.

Van Doormal, J. P. and Raithby, G. D., "The Simultaneous Solution along Lines of the Continuity and Momentum Equations," *Proc. of IMACS Conf.*, 1982.

Van Driest, E. R., "On Turbulent Flow Near a Wall," *J. Aeronaut. Sci.*, Vol. 23, pp. 1007–1011, 1956.

Van Dyke, M., "Higher-Order Boundary-Layer Theory," *Ann. Rev. Fluid Mech.*, Vol. 1, pp. 265–292, 1965.

Van Dyke, M., *An Album of Fluid Motion*, Parabolic Press, Stanford, CA, 1982.

Van Dyke, M., *Perturbation Methods in Fluids Mechanics*, annotated ed., Parabolic Press, Stanford, CA, 1975.

Vazsonyi, A., "On Rotational Gas Flows," *Q. J. Appl. Math.*, Vol. 3, No. 1, pp. 29–37, 1945.

Webster's New Collegiate Dictionary (s. v. coherent), G. & C. Merriam Company, Springfield, MA, 1976.

Wilcox, D. C. and Rubesin, M. W., "Progress in Turbulence Modeling for Complex Flow Fields Including Effects of Compressibility," NASA TP-1517, 1980.

Wilcox, D. C., "A Complete Model of Turbulence Revisited," AIAA Paper 84-0176, 1984.

Wilcox, D. C., *Turbulence Modeling for CFD*, DCW Industries, Inc., La Cañada, CA, 1993.

Williams, P. C., "A Reverse Flow Computation in the Theory of Self-Induced Separation," *Proc. 4th Int. Conf. on Num. Meth. in Fluid Dynam. Lecture Notes in Physics*, Vol. 35, pp. 445–451, 1975.

Willmarth, W. W., "Structure of Turbulence in Boundary Layers," *Adv. Appl. Mech.*, Vol. 15, pp. 159–254, 1975.

Willmarth, W. W. and Lu, S. S., "Structure of the Reynolds Stress Near the Wall," *J. Fluid Mech.*, Vol. 55, pp. 65–69, 1972.

Wray, A. and Hussaini, M. Y., "Numerical Experiments in Boundary-Layer Stability," *Proc. Roy. Soc. (London) Ser. A*, Vol. 392, pp. 373–389, 1984.

Wu, J. C. and Rizk, Y. M., "Integral Representation Approach for Time-Dependent Viscous Flows," *Lecture Notes in Physics*, Vol. 90, Springer-Verlag, 1978.

Wu, J. C. and Thompson, J. F., "Numerical Solution of Time-Dependent Incompressible Navier–Stokes Equations Using an Integro-Differential Formulation," *Computers and Fluids*, Vol. 1, pp. 197–215, 1973.

Wu, J. C. and Wahbah, M. M., "Numerical Solution of Viscous Flow Equations Using Integral Representations," *Lecture Notes in Physics*, Vol. 59, Springer-Verlag, 1976.

Wygnanski, I. J. and Champagne, F. H., "On Transition in a Pipe, Part I, The Origin of Puffs and Slugs and the Flow in a Turbulent Slug," *J. Fluid Mech.*, Vol. 59, pp. 281–335, 1973.

Wygnanski, I., Sokolov, N., and Freidman, D., "On the Turbulent 'Spot' in a Boundary Layer Undergoing Transition," *J. Fluid Mech.*, Vol. 78, pp. 785–819, 1976.

Yakhot, A. and Orszag, S. A., "Renormalization Group Analysis of Turbulence, Part I. Basic Theory," *J. Sci. Computing*, Vol. 4, pp. 3–51, 1986a.

Yakhot, A. and Orszag, S. A., "Renormalization Group Analysis of Turbulence," *Phys. Rev. Let.*, Vol. 57, pp. 1722–1724, 1986b.

Yakhot, A., Orszag, S. A., Yakhot, V., and Israeli, M., "Renormalization Group Formulation of Large-Eddy Simulation, Part I. Basic Theory," *J. Sci. Computing*, Vol. 4, pp. 139–158, 1989.

Zang, T. A. and Hussaini, M. Y., "Numerical Simulation of Nonlinear Interactions in Channel and Boundary-Layer Transition," *Nonlinear Wave Interactions in Fluids*, Miksad, R. W., Akylas, T. R., and Herbert, T. (Eds.), AMD-87, ASME, New York, 1987.

Zang, T. A., Dahlburg, R. B., and Dahlburg, J. P., "Direct and Large-Eddy Simulation of Three-Dimensional Compressible Navier–Stokes Turbulence," *Phys. Fluids A*, Vol. 4, No. 1, pp. 127–140, 1993.

Zang, T. A., Krist, S. E., and Hussaini, M. Y., "Resolution Requirements for Numerical Simulations of Transition," *J. Sci. Computing*, Vol. 4, pp. 197–217, 1989.

Zang, Y., Street, L. S., and Koseff, J. R., "A Dynamic Mixed Subgrid-Scale Model and its Application to Turbulent Recirculating Flows," *Phys. Fluids A*, Vol. 5, No. 12, pp. 3186–3196, 1993.

22 Computational Methods for Chemically Reacting Flows

PASQUALE CINNELLA
Mississippi State University
Starkville, MS

BERNARD GROSSMAN
Virginia Polytechnic Institute and State University
Blacksburg, VA

CONTENTS

Handbook of Fluid Dynamics and Fluid Machinery, Edited by Joseph A. Schetz and Allen E. Fuhs
ISBN 0-471-12598-9 Copyright © 1996 John Wiley & Sons, Inc.

22.1 INTRODUCTION

The investigation of fluid flows that involve or are dominated by chemical reactions constitutes in part a traditional field of study, with important engineering applications such as combustion and flame propagation problems [Oran and Boris (1987)]. In recent years, relatively *new* or rediscovered subject areas such as hypersonics and hypersonic combustion [Anderson (1989)], biochemical simulations [Szego *et al.* (1993)], and the structure of shock waves [Grossman *et al.* (1992)] have been added to the rolls of flowfields where chemical or even thermochemical nonequilibrium is a major feature. In all cases, the study of the coupling between fluid dynamics and nonequilibrium chemistry is complicated by the fact that many different time and space scales, sometimes orders of magnitude apart, are present in the same application. This induces *stiffness* in the governing equations [Bussing and Murmann (1985)] and sometimes results in stability and/or accuracy problems for the numerical algorithms employed in the simulations. Particularly difficult to tackle are flows in thermochemical nonequilibrium, often referred to as multitemperature flows [Park (1991)], which are important in hypersonic and atmospheric reentry applications. In these cases, the underlying physics is still not well understood and is a subject of ongoing research.

    A great amount of analytical studies and experimental investigations have been performed involving flows in chemical nonequilibrium [Clarke and McChesney (1975)]. Perturbation techniques [Vincenti and Kruger (1986)] were typically employed to overcome the difficulties associated with solving the nonlinear partial differential equations (and sometimes integro-differential equations) that govern these flow regimes. Moreover, existing experimental facilities are usually not capable of producing high speeds and high temperatures at the same time, as required in hypersonic applications [Marvin (1989)]. A potential solution to these difficulties has emerged in recent years with the dramatic improvement of both computational hardware and algorithms for the simulation of fluid flows. Although computational techniques cannot, and probably should not, replace either analytical or experimental approaches, they have emerged as a powerful third prong in the overall effort to understand, reproduce, and improve upon fluid flow features and the devices that exploit them. Unfortunately, flows in nonequilibrium generally require a massive amount of computational resources, especially when compared with simpler physical models such as the ideal gas approach. This results in efforts to simplify the physics, as in the case of the local chemical equilibrium approximation [Cinnella and Cox (1992)], and to improve on the efficiency of numerical algorithms, as in the case of implicit time integration techniques [Walters *et al.* (1992)].

In the following, a survey of computational methods currently available for the simulation of thermochemically reacting flows in presented. Specifically, in Sec. 22.2 the modeling of species and mixture thermodynamic properties is described, including thermodynamic nonequilibrium. In Sec. 22.3, the modeling of chemical activity is detailed, including the special case of local chemical equilibrium. In Sec. 22.4, the governing equations in integral form are introduced for flows in thermochemical nonequilibrium, and their simplification for less *extreme* physical conditions is discussed. Numerical techniques based on the finite-volume approach are reviewed in Sec. 22.5, including flux-split discretizations of the inviscid fluxes and recent developments in the quest for truly multidimensional algorithms. Explicit and implicit integration schemes for advancing the simulations in time are presented in Sec. 22.6. The simulation of radiative heat transfer, the investigation of the structure of shock waves, and some nontraditional biochemical applications are treated in the following three sections. Finally, a small sample of numerical results is introduced, and some conclusions are drawn.

The present chapter is not entirely exhaustive. Some important areas will be left untouched—namely, numerical investigations of laminar and turbulent flame sheets, detonations, ignition phenomena, mixing processes, and turbulent combustion, to name a few. The reader is referred to some of the current literature for details on these topics [Oran and Boris (1989), Dervieux and Larrouturou (1990), and Clavin (1994)].

22.2 THERMODYNAMIC MODELS

A gas mixture whose temperature is high enough to allow for the onset and/or sustenance of chemical reactions will typically deviate from the ideal gas behavior. Internal energy modes which behave nonlinearly with temperature will be activated, and, in extreme cases frequently encountered in hypersonic applications, local thermodynamic nonequilibrium will be established.

22.2.1 Internal Energy

In recent years, a great deal of effort has been spent on physically accurate modeling of *real gas* effects, which is the somewhat misleading terminology used to indicate departures from the low temperature ideal gas laws. Useful reviews of the literature are presented by Gnoffo *et al.* (1989) and Cinnella and Grossman (1991).

A general representation for the species internal energy per unit mass, e_s, suitable for the modeling of problems where thermodynamic nonequilibrium plays a key role, is

$$e_s = \tilde{e}_s(T) + e_{n_s}(T_{n_s}) \tag{22.1}$$

In the above, $\tilde{e}_s(T)$ is the contribution due to translation temperature T. Moreover, $e_{n_s}(T_{n_s})$ is the contribution due to internal modes that are in thermodynamic non-equilibrium, meaning that they are not in equilibrium at the translation temperature T, but may be assumed to satisfy a Boltzmann distribution at a different temperature T_{n_s}.

Equation (22.1) is valid for a very broad range of physical conditions and is susceptible to simplification and reduction to simpler models when the flow regime is less *extreme*. Typically, the contribution in equilibrium at the translational temperature $\tilde{e}_s$ will represent translational and rotational modes, the nonequilibrium term e_{n_s} will coincide with vibrational energy, and the electron/electronic contributions will be either neglected [Candler and MacCormack (1988)], or considered to be in equilibrium at the vibrational temperature T_{n_s} [Park and Yoon (1989)]. Several other possibilities have been proposed in the literature, e.g., see Park (1991) and Cinnella and Grossman (1991).

In the following, a gas mixture composed of N species will be considered, and two major representations of the thermodynamic state of the mixture will be studied. The first one will be denoted *Simplified Nonequilibrium Model*, whereby the first M species will contain a nonequilibrium contribution e_{n_s} [Grossman and Cinnella (1990)]. A simplification of the previous thermodynamic model may be achieved for flows where there is enough time for the internal modes to reach equilibrium at the translational temperature T, typically for flow regimes ranging from low subsonic to low hypersonic. This *Equilibrium Model* will feature only one contribution to the internal energy in Eq. (22.1), namely $\tilde{e}_s$. The usual ideal gas model is a special case of the *Equilibrium Model*, when vibrational and electronic modes are neglected, and translational and fully excited rotational contributions are included in $\tilde{e}_s$.

It is convenient to express $\tilde{e}_s$ in terms of specific heats at constant volume $\tilde{c}_{v_s}$

$$\tilde{e}_s = \int_{T_{\mathrm{ref}}}^{T} \tilde{c}_{v_s}(\tau)\, d\tau + h_{f_s} \tag{22.2}$$

where h_{f_s} is the heat of formation of species s at the reference temperature T_{ref}. The mixture value for the internal energy per unit mass, e, may be written using a standard mass fraction averaged summation, as follows

$$e = \sum_{s=1}^{N} Y_s e_s = \sum_{s=1}^{N} Y_s \tilde{e}_s + \sum_{s=1}^{M} Y_s e_{n_s} \tag{22.3}$$

where the species mass fraction Y_s has been introduced, defined as the ratio of species density ρ_s to mixture density $\rho = \Sigma_s^N \rho_s$. Similarly, the *frozen* specific heats at constant volume, $\tilde{c}_v$, and at constant pressure, $\tilde{c}_p$, may be obtained by means of the mixture rule, as follows

$$\tilde{c}_v = \sum_{s=1}^{N} Y_s \tilde{c}_{v_s}, \qquad \tilde{c}_p = \sum_{s=1}^{N} Y_s \tilde{c}_{p_s} = \sum_{s=1}^{N} Y_s (\tilde{c}_{v_s} + R_s) \tag{22.4}$$

where R_s is the species gas constant. The mixture gas constant R is written as follows

$$R = \sum_{s=1}^{N} Y_s R_s = \tilde{c}_p - \tilde{c}_v \qquad (22.5)$$

A very important subcase of the *Equilibrium Model* is *Local chemical equilibrium*, whereby it is assumed that the kinetic rates in the flowfield are fast enough to bring the chemical reactions that occur in the system virtually instantaneously to their equilibrium values, dictated by the Laws of Mass Action. This assumption has played a significant role in the modeling of real gas effects due to the simplification that it brings in the analysis, as will be shown in subsequent sections. It may be useful to point out that if the flowfield is considered to be in local chemical equilibrium, then it follows from thermodynamic considerations that only two state variables are necessary to describe its condition at a given time for every point in space. Here, density and temperature will be selected as the two fundamental state variables for this subcase [Cinnella and Cox (1992)]. The *equilibrium specific heats* can be obtained as functions of temperature and density derivatives of the mass fractions, as follows

$$c_v = \left(\frac{\partial e}{\partial T}\right)_{\rho} = \tilde{c}_v + \sum_{s=1}^{N} e_s \frac{\partial Y_s}{\partial T}$$

$$c_p = \left(\frac{\partial h}{\partial T}\right)_p = c_v + \frac{R + T \sum_{s=1}^{N} R_s (\partial Y_s / \partial T)}{R + \rho \sum_{s=1}^{N} R_s (\partial Y_s / \partial \rho)} \left[R - \frac{\rho}{T} \sum_{s=1}^{N} e_s \frac{\partial Y_s}{\partial \rho}\right] \qquad (22.6)$$

where the mixture enthalpy $h = e + RT$ has been introduced. Unless otherwise noted, throughout this chapter a partial derivative with respect to the temperature will be taken at constant density, and *vice versa*.

22.2.2 Equations of State

Most flows of practical interest will involve low to moderate pressures and densities. In these instances, *Dalton's Law* will yield the mixture pressure as the sum of species partial pressures

$$p = \sum_{s=1}^{N} \rho_s R_s T = \rho R T \qquad (22.7)$$

where the mixture gas constant R has been defined in Eq. (22.5). This relationship is also known as the *Thermal Equation of State*.

Unlike the ideal gas case, the state relationship of the pressure to the specific internal energy cannot, in general, be written directly, but it occurs implicitly through the translational temperature. A *Caloric Equation of State* such as Eq. (22.3) will relate internal energy, or portions thereof, to temperature. Iterative procedures are necessary for the determination of the temperature due to the inherently nonlinear character of this equation [Cinnella and Grossman (1991)]. Once T is found, the pressure is evaluated from Eq. (22.7).

22.2.3 Speeds of Sound

The speed of sound is a thermodynamic property that plays a key role in most numerical algorithms for the simulation of high speed flows. When the gas mixture is thermochemically active, at least two different speeds of sound may be encountered. The *frozen speed of sound* is defined by freezing chemical composition and thermodynamic nonequilibrium phenomena at the local (instantaneous) values, whereas the *equilibrium speed of sound* assumes that the composition is in local equilibrium and the interal energy contributions have reached equilibrium at the local temperature.

In general, the choice of the speed that enters the description of the propagation of information is dictated by the chemistry model, which will be discussed in Sec. 22.3. When flows that can be studied using either the *Simplified Nonequilibrium Model* or the *Equilibrium Model* are considered, the frozen speed of sound has been found to be the relevant quantity, as long as the gas mixture is not considered to be in local chemical equilibrium [Grossman and Cinnella (1990)]. When the latter assumption is utilized, then the equilibrium speed of sound is the appropriate choice [Cinnella and Cox (1992)]. Formally, the frozen speed of sound for the *Simplified Nonequilibrium Model* can be written as follows

$$a^2 \equiv \left(\frac{\partial p}{\partial \rho} \right)_{s, Y_s, e_{n_s}} = \tilde{\gamma} R T = \tilde{\gamma} \left(\frac{p}{\rho} \right) \tag{22.8}$$

where s is the entropy, and $\tilde{\gamma}$ is the ratio of frozen specific heats $\tilde{\gamma} = \tilde{c}_p / \tilde{c}_v$. The *Equilibrium Model* will feature the same result. However, the nonequilibrium internal energy contributions, e_{n_s}, will formally disappear from the previous expression. The interestingly simple final formula for the frozen speed of sound was described by Clarke and McChesney (1975).

The special subcase of the *Equilibrium Model* that features flows in local chemical equilibrium will have an equilibrium speed of sound defined as

$$a^2 \equiv \left(\frac{\partial p}{\partial \rho} \right)_s = \Gamma R T = \Gamma \left(\frac{p}{\rho} \right) \tag{22.9}$$

where the isentropic index $\Gamma = \Gamma(\rho, T)$ is given by the following

$$\Gamma = \bar{\gamma} + \frac{\rho}{RT} \sum_{s=1}^{N} \frac{\partial Y_s}{\partial \rho} [R_s T - (\bar{\gamma} - 1) e_s]$$

$$\bar{\gamma} = \frac{\partial h / \partial T}{\partial e / \partial T} = 1 + \frac{R + T \sum_{s=1}^{N} R_s (\partial Y_s / \partial T)}{c_v} \tag{22.10}$$

In Eq. (22.10), $\bar{\gamma}$ is the ratio of the partial derivative of enthalpy with respect to temperature at constant density to the partial derivative of the internal energy with respect to temperature, where the latter is the specific heat at constant volume c_v. The use of the isentropic index preserves the simplicity of the formula for the sound speed, but the functional dependence of this index on density and temperature is relatively involved.

There are at least five different γ's that can be defined for a gas in chemical equilibrium: the isentropic index Γ, the ratio of specific heats $\gamma = c_p/c_v$, the ratio of frozen specific heat $\tilde{\gamma} = \tilde{c}_p/\tilde{c}_v$, the ratio of enthalpy and internal energy derivatives $\bar{\gamma}$, and the ratio of enthalpy to internal energy $\gamma^* = h/e = 1 + p/\rho e$. For a frozen flow, the first four of these quantities will reduce to the same value, but for general flow conditions their values will remain different.

22.2.4 Practical Models

The formulation detailed in Secs. 22.2.1 to 22.2.3 is very general, and it allows the utilization of many practical thermodynamic models for the determination of the specific functional form of internal energy, specific heats, and related properties. A few models have emerged as the ones that are preferred by most researchers for numerical simulations of high-speed, nonequilibrium flowfields. More details are given by Cinnella and Grossman (1991).

22.3 CHEMISTRY MODELS

When chemically reacting flows are investigated, a classification of chemical phenomena according to the time available for the completion of reactions is usually employed for order-of-magnitude estimates of the flowfield properties, and this results in the definition of Damköhler numbers [Anderson (1989)]. If reaction times are very large compared with the time scale at which the fluid dynamics is evolving, then reactions have virtually no time to occur, and a *frozen* flow can be assumed. Incidentally, perfect gas results are a particular class of frozen chemistry simulations. The other limiting case occurs when the reaction times are very small compared to the fluid dynamic scales, which results in the reactions having enough time to reach their *local chemical equilibrium* values, given by the *Law of Mass Action*.

The real situation, however, is the general case of *finite-rate chemistry*, whereby in at least portions of the flowfield and/or at some point in the time evolution of the flow, the reaction times are comparable to the fluid dynamic time scales. In this instance, it becomes necessary to simulate the actual kinetic behavior of the chemical system.

22.3.1 Finite-Rate Chemistry

A general simulation of chemical effects can be achieved for a system containing N species where J reactions take place

$$\nu'_{1,j}X_1 + \nu'_{2,j}X_2 + \cdots + \nu'_{N,j}X_N \rightleftharpoons \nu''_{1,j}X_1 + \nu''_{2,j}X_2 + \cdots + \nu''_{N,j}X_N,$$

$$j = 1, \ldots, J \qquad (22.11)$$

where the $\nu'_{s,j}$ and the $\nu''_{s,j}$ are stoichiometric coefficients of species X_s in the jth reaction. Then, the rate of production of the sth species, w_s, may be written as a summation over the reactions [Vincenti and Kruger (1986)]

$$w_s \equiv \frac{d\rho_s}{dt} = \mathfrak{M}_s \sum_{j=1}^{J} (\nu''_{s,j} - \nu'_{s,j}) \left[k_{f,j} \prod_{l=1}^{N} \left(\frac{\rho_l}{\mathfrak{M}_l} \right)^{\nu'_{l,j}} - k_{b,j} \prod_{l=1}^{N} \left(\frac{\rho_l}{\mathfrak{M}_l} \right)^{\nu''_{l,j}} \right],$$

$$s = 1, \ldots, N \tag{22.12}$$

where $\mathfrak{M}_s$ is the species molecular mass. For reaction j, the forward and backward reaction rates, $k_{f,j}$ and $k_{b,j}$, are assumed to be known functions of temperature, and they are related by thermodynamics

$$k_{f,j}(T) = k_{b,j}(T) K_{e,j}(T) \tag{22.13}$$

where $K_{e,j}$ is the equilibrium constant, which is a known function of the thermodynamic state.

The finite-rate chemistry model described remains valid in conjunction with all of the thermodynamic models discussed in the previous section. However, the presence of thermodynamic nonequilibrium in the flowfield is likely to exercise some effect upon the reaction rates, especially when diatomic or polyatomic molecules are involved, whose vibrational modes may be excited. Several attempts to model the interaction between chemical and thermodynamic nonequilibrium have been recorded [Treanor and Marrone (1962), Marrone and Treanor (1963)]. More details are given by Cinnella and Grossman (1991).

The previous chemistry is valid for gaseous systems and involves homogeneous reactions only, thereby excluding solid or liquid surface as well as photochemical reactions. At the present time, wall effects, in general, and wall catalyticity, in particular, are still not entirely understood, although some initial modeling efforts for hypersonic flows have been presented [Bruno (1989)].

22.3.2 Local Chemical Equilibrium

There are many problems that render a finite-rate computation into a nontrivial task. The most important one is the uncertainty in the reaction paths and in the reaction rates. Although the composition of a given mixture in chemical equilibrium as a function of thermodynamic variables can be determined quite accurately by theoretical and experimental means, there is limited knowledge of the actual mechanism with which a given reaction occurs. Moreover, the numerical simulation of flows in chemical nonequilibrium requires the inclusion of N species continuity equations into the algorithms instead of only one global mass conservation equation, as will be discussed in the following. Also, the finite-rate equations become extremely *stiff* when the flow is close to chemical equilibrium [Oran and Boris (1987)]. All of the above reasons render a numerical simulation based upon chemical equilibrium very appealing from a computational standpoint when compared with finite-rate calculations. In addition, for a large number of flows of practical interest, results of considerable accuracy can be obtained using this simplifying assumption.

A close examination of the governing equations, including the combined First and Second Laws of Thermodynamics, indicates that local chemical equilibrium calculations should be performed at constant density and internal energy [Cinnella and Cox (1992)]. The equations to be solved for the determined of the N species densities and the temperature will be: N_e mass conservation equations written for an equivalent number of *elemental species*, $N - N_e$ equations corresponding to a possible formulation of the Laws of Mass Action, and one energy equation, of the type given in Eq. (22.3), specialized for thermodynamic equilibrium [Cinnella and Cox (1992)]. Several methods for the solution of these nonlinear equations have been utilized, and a review may be found in Liu and Vinokur (1989b). Roughly, two different classes of algorithms have been developed, one based on optimization methods, namely minimization of free energy and/or maximization of entropy, and the other based on the direct solution of the Laws of Mass Action, possibly with the introduction of techniques for reducing the number of unknowns, therefore reducing the number of equations and computational time.

Probably the most important gas mixture in engineering is air at standard conditions, which corresponds to 79% nitrogen and 21% oxygen by volume, with traces of other gases. Solutions of the equilibrium composition problem for vast ranges of density and internal energy have been obtained, tabulated, and utilized for the determination of thermodynamic properties of the mixture in terms of curve-fit values [Liu and Vinokur (1989b) and Srinivasan *et al.* (1987)]. The importance of these curve fits should not be underestimated, but unfortunately all the results obtained will be valid only for the specific composition studied. Even slight changes in the base mixture would require new tabulations. Consequently, aside from problems involving the fundamental base air mixture, the simulation of flowfields in local chemical equilibrium for a general gas mixture has to rely on the efficient coupling of chemical composition solvers, based upon the previously mentioned governing equations, with flowfield solvers for the usual form of the Euler or Navier–Stokes equations.

22.3.3 Practical Models

It was previously mentioned that a detailed knowledge of the kinetic behavior of a chemically reacting gas mixture is necessary when finite-rate chemistry investigations are deemed necessary. In recent years, much attention to the kinetic of air and of hydrogen/air mixtures has been registered in the scientific community, spurred by the drive towards hypersonic flight and the necessary propusive tools to achieve it [Anderson (1989)]. A compilation of the most recent data for the kinetics of air may be found in Park (1991), where attention is given to thermodynamic nonequilibrium and its influence upon the reaction rates. Detailed studies of hydrogen/air mixtures have been published recently [Oldenborg *et al.* (1990)]. Moreover, computerized databases are now available, involving several thousand reactions and hundreds of chemical species [Mallard *et al.* (1993)].

When local chemical equilibrium calculations are performed, the actual reaction paths are no longer necessary. Chemical equilibrium is dictated by thermodynamics; the entropy has to be a maximum for a system at constant density and internal energy [Liu and Vinokur (1989b)].

22.4 GOVERNING EQUATIONS

The governing equations for flows in thermochemical nonequilibrium represent the conservation and/or time variation of physical quantities such as mass of a species in the mixture, momentum, nonequilibrium energy contributions, and energy. They are established either for a control volume of arbitrary size (finite or infinitesimal) or for a generic point in the flowfield. The former approach results in an integro-differential form; the latter in a nonlinear, differential system. In the following, the integral form of the equations will be introduced for consistency with the finite-volume-based numerical techniques to be analyzed later. The differential form of the equations can be readily obtained from the integral form and has been extensively discussed in the literature [Grossman and Cinnella (1990) and Lee (1985)].

22.4.1 Thermochemical Nonequilibrium

The governing equations written in integral form are valid both in regions of smooth flows and across discontinuities [Anderson *et al.* (1984)], where they can be shown to reduce to the *Rankine–Hugoniot jump conditions* in the limit of infinitesimal volumes. The control volume employed for the derivation is allowed to move and deform with time, although many important applications take advantage of the simpler case of fixed control volumes. For an arbitrary volume $\mathcal{V}$, closed by a boundary $\mathcal{S}$, the governing equations in integral form read

$$\frac{\partial}{\partial t} \iiint_{\mathcal{V}} \mathbf{Q}\, d\mathcal{V} + \oint_{\mathcal{S}} (\bar{\mathbf{S}} - \bar{\mathbf{S}}_v) \cdot \bar{\mathbf{n}}\, d\mathcal{S} = \iiint_{\mathcal{V}} \mathbf{W}\, d\mathcal{V} \qquad (22.14)$$

where $\mathbf{Q}$ is the vector of conserved variables, $\mathbf{W}$ is the vector of source terms, and $\bar{\mathbf{S}}$ and $\bar{\mathbf{S}}_v$ are the inviscid and viscous flux vectors, respectively. The unit vector $\bar{\mathbf{n}}$ is normal to the infinitesimal area $d\mathcal{S}$ and points outwards. In conjunction with the *Simplified Nonequilibrium Model*, and utilizing a Cartesian frame of reference (x, y, z) whose unit vectors are $\bar{\mathbf{i}}$, $\bar{\mathbf{j}}$, and $\bar{\mathbf{k}}$, respectively, the vectors $\mathbf{Q}$ and $\mathbf{W}$ read

$$\mathbf{Q} = \begin{bmatrix} \rho_1 \\ \rho_2 \\ \vdots \\ \rho_N \\ \rho u \\ \rho v \\ \rho w \\ \rho_1 e_{n_1} \\ \vdots \\ \rho_M e_{n_M} \\ \rho e_0 \end{bmatrix}, \quad \mathbf{W} = \begin{bmatrix} w_1 \\ w_2 \\ \vdots \\ w_N \\ \Sigma_s\, \rho_s g_{s_x} \\ \Sigma_s\, \rho_s g_{s_y} \\ \Sigma_s\, \rho_s g_{s_z} \\ Q_1 \\ \vdots \\ Q_M \\ (\Sigma_s\, \rho_s \bar{\mathbf{g}}_s) \cdot \bar{\mathbf{u}} \end{bmatrix} \qquad (22.15)$$

and the flux vectors $\bar{\mathbf{S}}, \bar{\mathbf{S}}_v$ read

$$
\bar{\mathbf{S}} = \begin{bmatrix} \rho_1(\bar{\mathbf{u}} - \bar{\mathbf{u}}_{\mathcal{V}}) \\ \rho_2(\bar{\mathbf{u}} - \bar{\mathbf{u}}_{\mathcal{V}}) \\ \vdots \\ \rho_N(\bar{\mathbf{u}} - \bar{\mathbf{u}}_{\mathcal{V}}) \\ \rho u(\bar{\mathbf{u}} - \bar{\mathbf{u}}_{\mathcal{V}}) + p\bar{\mathbf{i}} \\ \rho v(\bar{\mathbf{u}} - \bar{\mathbf{u}}_{\mathcal{V}}) + p\bar{\mathbf{j}} \\ \rho w(\bar{\mathbf{u}} - \bar{\mathbf{u}}_{\mathcal{V}}) + p\bar{\mathbf{k}} \\ \rho_1 e_{n_1}(\bar{\mathbf{u}} - \bar{\mathbf{u}}_{\mathcal{V}}) \\ \vdots \\ \rho_M e_{n_M}(\bar{\mathbf{u}} - \bar{\mathbf{u}}_{\mathcal{V}}) \\ \rho e_0(\bar{\mathbf{u}} - \bar{\mathbf{u}}_{\mathcal{V}}) + p\bar{\mathbf{u}} \end{bmatrix}, \quad
\bar{\mathbf{S}}_v = \begin{bmatrix} -\rho_1\bar{\mathbf{V}}_1 \\ -\rho_2\bar{\mathbf{V}}_2 \\ \vdots \\ -\rho_N\bar{\mathbf{V}}_N \\ \tau_{xx}\bar{\mathbf{i}} + \tau_{xy}\bar{\mathbf{j}} + \tau_{xz}\bar{\mathbf{k}} \\ \tau_{yx}\bar{\mathbf{i}} + \tau_{yy}\bar{\mathbf{j}} + \tau_{yz}\bar{\mathbf{k}} \\ \tau_{zx}\bar{\mathbf{i}} + \tau_{zy}\bar{\mathbf{j}} + \tau_{zz}\bar{\mathbf{k}} \\ -\rho_1 e_{n_1}\bar{\mathbf{V}}_1 - \bar{\mathbf{q}}_{n_1} \\ \vdots \\ -\rho_M e_{n_M}\bar{\mathbf{V}}_M - \bar{\mathbf{q}}_{n_M} \\ \bar{\Theta} \end{bmatrix} \qquad (22.16)
$$

where

$$
\bar{\Theta} = (\tau_{xx}u + \tau_{yx}v + \tau_{zx}w)\bar{\mathbf{i}} + (\tau_{xy}u + \tau_{yy}v + \tau_{zy}w)\bar{\mathbf{j}}
$$

$$
+ (\tau_{xz}u + \tau_{yz}v + \tau_{zz}w)\bar{\mathbf{k}} - \bar{\mathbf{q}} - \bar{\mathbf{q}}^R - \sum_{s=1}^{M} \bar{\mathbf{q}}_{n_s} - \sum_{s=1}^{M} \rho_s h_s \bar{\mathbf{V}}_s \quad (22.17)
$$

In the previous equations, the *mass-averaged velocity* $\bar{\mathbf{u}}$ for the mixture has been utilized, whose Cartesian components are (u, v, w), respectively. The mixture velocity $\bar{\mathbf{u}}$ can be defined in terms of species velocities $\bar{\mathbf{u}}_s$ as follows

$$
\rho\bar{\mathbf{u}} = \sum_{s=1}^{N} \rho_s \bar{\mathbf{u}}_s \qquad (22.18)
$$

The difference between species and mass-averaged velocities is a species *diffusion velocity* $\bar{\mathbf{v}}_s$

$$
\bar{\mathbf{v}}_s = \bar{\mathbf{u}}_s - \bar{\mathbf{u}}, \qquad s = 1, \ldots, N \qquad (22.19)
$$

The velocity $\bar{\mathbf{u}}_{\mathcal{V}}$ that appears in the inviscid flux vector is the control surface velocity, which equals zero when the control volume is fixed with time. The first N equations are species continuity equations, relating the time rate of change of species densities ρ_s to convective and diffusive transport and to the creation/destruction of the species due to chemical reactions. The species rate of production w_s has been defined by Eq. (22.12), when finite-rate chemistry models were discussed. Following species continuity are the three components of the momentum equation. The source term $\Sigma_s\rho_s\bar{\mathbf{g}}_s$ represents body forces, e.g., gravity and electric fields, and the viscous fluxes involve the components of the *shear stress tensor*. After momentum,

M nonequilibrium energy equations are written, describing the creation and evolution in time and space of the nonequilibrium components of the internal energy. The viscous fluxes associated with these equations describe the heat transfer due to the transport of nonequilibrium energy by diffusion, $\rho_s e_{ns} \bar{\mathbf{v}}_s$, as well as the conductive heat transfer, $\bar{\mathbf{q}}_{ns}$. The source terms for the nonequilibrium energy equations, Q_s, will be described in more detail below. Finally, the global energy equation describes the time evolution of the total internal energy per unit volume $\rho e_0 = \rho e + \rho(u^2 + v^2 + w^2)/2$. The viscous flux in this case represents viscous dissipation, convective heat transfer due to both equilibrium and nonequilibrium portions of the internal energy, $\bar{\mathbf{q}}$ and $\Sigma_s \bar{\mathbf{q}}_{ns}$ respectively, radiative heat transfer, $\bar{\mathbf{q}}^R$, and diffusive heat transfer, $\Sigma_s \rho_s h_s \bar{\mathbf{v}}_s$, where h_s is the species enthalpy, $h_s = e_s + R_s T$.

In addition to the equations presented, thermal and caloric equations of state will relate pressure and temperatures to equilibrium and nonequilibrium internal energy components, as discussed in Sec. 22.2, and the viscous flux vector entries will be expressed in terms of the entries in the vector of conserved variables.

Simplifications in the physical model employed for the analysis of nonequilibrium flows will bring about corresponding changes in the governing equations. For instance, the Equilibrium Model can be handled simply by dropping all of the nonequilibrium energy equations, along with the nonequilibrium contributions to the global energy equation. The whole process is equivalent to setting M equal to zero. *Inviscid flows* are modeled by dropping the viscous flux vector in conjunction with any of the thermodynamic models discussed.

22.4.2 Local Chemical Equilibrium

The previous form of the governing equations is written for finite-rate chemistry problems. When the important special subcase of the Equilibrium Model that deals with local chemical equilibrium is considered, the vectors that appear in Eq. (22.14) are further simplified. The global continuity equation replaces the N species continuity equations, the nonequilibrium energy contributions disappear, and the final form of $\mathbf{Q}$ and $\mathbf{W}$ reads

$$\mathbf{Q} = \begin{pmatrix} \rho \\ \rho u \\ \rho v \\ \rho w \\ \rho e_0 \end{pmatrix}, \quad \mathbf{W} = \begin{pmatrix} 0 \\ \Sigma_s \rho_s g_{sx} \\ \Sigma_s \rho_s g_{sy} \\ \Sigma_s \rho_s g_{sz} \\ (\Sigma_s \rho_s \bar{\mathbf{g}}_s) \cdot \bar{\mathbf{u}} \end{pmatrix} \tag{22.20}$$

where only body forces are present in the vector of source terms $\mathbf{W}$. Inviscid and viscous fluxes read

$$\bar{\mathbf{S}} = \begin{pmatrix} \rho(\bar{\mathbf{u}} - \bar{\mathbf{u}}_\mathcal{V}) \\ \rho u(\bar{\mathbf{u}} - \bar{\mathbf{u}}_\mathcal{V}) + p\bar{\mathbf{i}} \\ \rho v(\bar{\mathbf{u}} - \bar{\mathbf{u}}_\mathcal{V}) + p\bar{\mathbf{j}} \\ \rho w(\bar{\mathbf{u}} - \bar{\mathbf{u}}_\mathcal{V}) + p\bar{\mathbf{k}} \\ \rho e_0(\bar{\mathbf{u}} - \bar{\mathbf{u}}_\mathcal{V}) + p\bar{\mathbf{u}} \end{pmatrix}, \quad \bar{\mathbf{S}}_v = \begin{pmatrix} 0 \\ \tau_{xx}\bar{\mathbf{i}} + \tau_{xy}\bar{\mathbf{j}} + \tau_{xz}\bar{\mathbf{k}} \\ \tau_{yx}\bar{\mathbf{i}} + \tau_{yy}\bar{\mathbf{j}} + \tau_{yz}\bar{\mathbf{k}} \\ \tau_{zx}\bar{\mathbf{i}} + \tau_{zy}\bar{\mathbf{j}} + \tau_{zz}\bar{\mathbf{k}} \\ \Theta \end{pmatrix} \tag{22.21}$$

where

$$\overline{\Theta} = (\tau_{xx}u + \tau_{yx}v + \tau_{zx}w)\overline{\mathbf{i}} + (\tau_{xy}u + \tau_{yy}v + \tau_{zy}w)\overline{\mathbf{j}}$$
$$+ (\tau_{xz}u + \tau_{yz}v + \tau_{zz}w)\overline{\mathbf{k}} - \overline{\mathbf{q}} - \overline{\mathbf{q}}^R \qquad (22.22)$$

and diffusion has been neglected.

The closure to the mathematical problem of the governing equations for flows in local chemical equilibrium will again be provided by thermal and caloric equations of state, as shown in Sec. 22.2. With temperature and species composition known from the equilibrium solver, the previous equations are tantamount to a general equation of state relating pressure to density and internal energy

$$p = p(\rho, e) \qquad (22.23)$$

When curve-fit properties are employed, an expression such as Eq. (22.23) is readily available [Liu and Vinokur (1989) and Srinivasan et al. (1987)].

After the local equilibrium composition at a given density and internal energy has been determined, the different thermodynamic state variables can be evaluated. Among them, several have a significant practical interest, for example speed of sound a, given by Eq. (22.9), and isentropic index Γ, given by Eq. (22.10), as well as transport properties such as viscosity and thermal conductivity. The evaluation of all of these quantities is made possible by the knowledge of species densities and temperature at equilibrium, but it also requires the values of the partial derivatives of the mass fractions with respect to temperature and density, as shown in Eq. (22.10). In general, these derivatives are determined from the governing equations by means of the *Implicit Function Theorem* [Cinnella and Cox (1992)], due to the fact that Laws of Mass Action and mass constraints are the relationships that implicitly define the behavior of mass fractions with changing temperature and density. For the specific important example of a standard air mixture, tabulations and curve fits are available for a broad range of density and temperature for virtually all of the thermodynamic and transport properties of interest.

22.4.3 Thermodynamic Nonequilibrium

In this section, some basic information is provided as to the current understanding and simulation capabilities of thermodynamic nonequilibrium and its interaction with transport phenomena. More details can be found in the literature that will be referenced.

Two major physical phenomena will contribute to the source terms Q_s in Eq. (22.15), when it is assumed that vibrational nonequilibrium is present. The first one is the change in vibrational energy caused by inelastic collision, that is, creation or destruction of vibrating particles due to chemical reactions [Vincenti and Kruger (1986)]; the second one is due to elastic collisions and is tantamount to energy exchanges between the different internal energy modes. The most important contribution to the latter is the relaxation of nonequilibrium vibrational energy towards the equilibrium levels at the local translational temperature [Lee (1985)]. Additional contributions include energy exchanges among the different vibrating species [Lee (1985) and Candler (1989)].

In reentry aerodynamic applications [Park (1991)], the velocities involved are high enough to warrant the consideration of ionized species and free electrons behind strong shock waves. Moreover, the electronic modes of atoms and molecules can be out of thermodynamic equilibrium. In this instance, it is usually assumed that a Boltzmann distribution of electron/electronic energy levels is maintained at a common temperature T_e, and the relaxation of this distribution to the local equilibrium values is included in the modeling of thermodynamic nonequilibrium [Candler and MacCormack (1989) and Liu and Vinokur (1989a)]. Several assumptions and approximations are usually necessary to simplify the treatment of what would otherwise be a flowfield with two different translational temperatures [Grossman et al. (1992)].

22.4.4   Viscous Fluxes and Transport Properties

In order to complete the description of the governing equations for reacting flows, the stress tensor, the heat-flux vectors, and the diffusion velocities have to be defined and/or related to the conserved variables. Some of the basic results will be examined. For details on their derivation, refer to Bird et al. (1960).

The most common assumptions that are made when dealing with transport properties concern the treatment of the viscous stress tensor. Generally, only *Newtonian fluids* are considered, where there is a linear relationship between stress and rate of deformation. Moreover, *bulk viscosity* effects are neglected, which should be accounted for when rotational nonequilibrium is present [Grossman et al. (1992)], as will be mentioned in Sec. 22.8.

Under these assumptions, the stress tensor components are expressed as

$$\tau_{ij} = \mu \left(\frac{\partial u_i}{\partial x_j} + \frac{\partial u_j}{\partial x_i} \right) - \frac{2}{3} \mu \left(\frac{\partial u_1}{\partial x_1} + \frac{\partial u_2}{\partial x_2} + \frac{\partial u_3}{\partial x_3} \right) \delta_{ij}, \quad i, j = 1, 2, 3 \quad (22.24)$$

where μ is the viscosity coefficient and the symbol δ_{ij} stands for the Kronecker delta. The indices i and j refer to the x, y, and z space directions and velocity components.

For mixtures of gases in thermal equilibrium, the heat flux vector $\bar{\mathbf{q}}$ is modeled by means of the product of thermal conductivity, λ, times the temperature gradient (*Fourier's Law*). The extension of this approach to flows in thermal nonequilibrium is usually done by considering similar contributions for the nonequilibrium temperatures [Candler and MacCormack (1989) and Lee (1985)]. The resulting expressions read

$$\bar{\mathbf{q}} = -\lambda \nabla T, \quad \bar{\mathbf{q}}_{n_s} = -\lambda_{n_s} \nabla T_{n_s}, \quad s = 1, \ldots, M \quad (22.25)$$

This model neglects some supposedly secondary factors like temperature and pressure diffusion, which would greatly complicate the analysis. In addition, the validity of Fourier's Law for nonequilibrium conditions has not been fully assessed to date.

Mass diffusion is a physical phenomenon that arises mostly because of the presence of gradients of mass or mole concentrations in the mixture. Although pressure and temperature gradients as well as the effect of body forces can influence diffusion, for a simple analysis these variables are often neglected. With these simplifications, it is possible to write the Stefan–Maxwell equation [Bruno (1989)]. Solutions to this equation have been attempted in conjunction with numerical simulations of chemically inert flows, as in Baysal *et al.* (1988), where the pressure gradient is also retained. Assuming that the mixture behaves like a binary mixture yields a reduced form of the Stefan–Maxwell equation, the so-called *Fick's Law of Diffusion*

$$\rho_s \overline{\mathbf{v}}_s = -\rho D_s \nabla \left(\frac{\rho_s}{\rho} \right), \qquad s = 1, \dots, N \tag{22.26}$$

where D_s are the diffusion coefficients. Investigations of several transport algorithms have shown that this formula can produce reasonable results at a much lower computational cost [Heimerl and Coffee (1982)].

A theoretical formula for the viscosity coefficient to be used in Eq. (22.24) has been given as a result of the asymptotic analysis that leads to the Navier–Stokes equations [Hirschfelder *et al.* (1954)]. Such a formula is strictly valid for a single monatomic gas, and it relies on an intermolecular potential function being assumed. Results for the Lennard–Jones potential have been tabulated [Anderson (1989)] and extensively used for general gases as well. The corresponding formulas for mixtures of monatomic gases have been formulated [Curtiss and Hirschfelder (1949)], but they are computationally very expensive, due to the appearance of matrices of coefficients.

Curve fit functional expressions have also been proposed, which are based upon experiments on single components. Their advantage is simplicity, when compared to theoretical determinations, and the availability of semiempirical rules for recovering the corresponding mixture values. One of the most widely adopted curve fits is due to Blottner *et al.* (1971). Another very popular approach is the extension of Sutherland's Law for a perfect gas to a generic component [Anderson (1989) and Walters *et al.* (1992)]. Values of the viscosity coefficient for the mixture are usually recovered by means of *Wilke's rule* [Wilke (1950)], which is an extension of a Sutherland-type equation to multicomponent systems, obtained on the basis of kinetic theory and several simplifying assumptions.

Values of the thermal conductivity to be used in Eq. (22.25) are also obtained from an asymptotic analysis for small deviations from equilibrium. That results in expressions that are functions once again of collision integrals, whereby the latter are functions of the chosen intermolecular potential. A more refined theory that takes into account internal degrees of freedom of particles and interactions among different species is quite complex and computationally expensive [Curtiss and Hirschfelder (1949)]. Curve-fit expressions have also been proposed for thermal conductivity, the most accepted one being *Eucken's relation* [Vincenti and Kruger (1986)]. A Sutherland-type formula can also be used, similarly to what has been done for the viscosity coefficient. Results for the mixture thermal conductivity can be obtained again by means of Wilke's rule. In addition, thermal diffusion is universally neglected in the heat flux vector.

The simplest modeling of the diffusion coefficients is obtained by assuming a constant Lewis number, Le. The result can be expressed in terms of the Prandtl number, Pr, as well

$$D_s = D = \frac{\text{Le}}{\rho} \frac{\lambda}{\tilde{c}_p} = \frac{\text{Le}}{\rho} \frac{\mu}{\text{Pr}}, \qquad s = 1, \dots, N \tag{22.27}$$

where only one global diffusion coefficient D is used. Multicomponent effects can be partially taken into account by more complex choices of diffusion coefficients, relating those with binary diffusion coefficients. A complete multicomponent solution has been proposed and utilized [Blottner *et al.* (1971)], but it involves inversion of matrices and evaluations of determinants, resulting in a computationally expensive task.

Theoretical solutions of the Boltzmann equations for mixtures of gases out of thermal equilibrium have not been obtained at the present time. The numerical treatment of transport properties for these cases usually consists of fairly simple assumptions. Diffusion coefficients and viscosity are left unaltered, and thermal conductivity is modified in a straightforward manner, because only the portions of the specific heats that are in thermal equilibrium are included in the equations. The treatment of nonequilibrium thermal conductivity and the determination of λ_{n_s} are dealt with in a simple way, by using an Eucken-type relation [Candler and Mac-Cormack (1988)].

Prediction of transition and modeling of turbulence in the presence of chemical reactions and thermal nonequilibrium are examples of very important problems that are objects of current research. The inclusion of turbulence in the viscous flux vectors is obtained in its simplest form by the addition of turbulent viscosity, thermal conductivity, and diffusion coefficients to their laminar counterparts. The turbulent contributions are then modeled by means of algebraic treatments, foremost among which is the model originated by Baldwin and Lomax (1978), or by introducing and solving additional partial differential equations, for example, for turbulent kinetic energy and dissipation [Morrison (1990)].

In general, the models for turbulence in the presence of thermochemical nonequilibrium have been straightforward extensions of the ones derived for an ideal gas (see Chap. 4). Recent departures from this tendency are the methods based on statistical techniques such as the *Probability Distribution Function* (PDF) techniques [Pope (1985)]. For example, with these methods the effects of random variations in velocity and temperature of a fluid flow on ignition and reaction times and on the heat released are included in the analysis.

22.5   SPACE DISCRETIZATION

In order for a computer simulation to be possible, the governing equations presented in the previous section must be *discretized*, that is, reduced to a set of algebraic equations to be solved. A very convenient discretization approach is the finite-volume technique [Walters and Thomas (1988)], whereby the integral form of the governing equations is solved for the unknown volume averages of conserved variables in some small, but finite, control volume (see Sec. 19.3). The advantage of this method is its use of the integral form of the equations, which allows a correct treatment of discontinuities and also consistently treats general grid topologies. Other discretizations are also possible, for example, spectral techniques [Drummond *et al.* (1986)], which have been successfully applied to flow simulations in a chemically active environment [Eklund *et al.* (1986)], and the more usual finite-difference approach [Anderson *et al.* (1984)] (see Sec. 19.2). Both of these alternative discretizations are based upon the differential form of the equations, and special attention has to be paid to the proper treatment of discontinuities to avoid numerical problems. Utilizing the integral form of the governing equations leads naturally to shock-capturing schemes [Aftosmis and Baron (1989)]. Shock-fitting techniques (see Sec. 20.3) have also been attempted for flows out of chemical equilibrium, but their implementation is at a relatively initial stage for general thermochemical nonequilibrium situations.

22.5.1 Finite-Volume Technique

The finite-volume technique takes advantage of the integral form of the governing equations to achieve a discretized algebraic approximation that can be handled by a digital computer. The physical domain of interest is partitioned into a large set of subdomains, volume-averaged values for the conservative variables **Q** and the source terms **W** are defined for each subdomain (computational cell), and the governing equations described in Eq. (22.14) are written for each cell in terms of the volume averages. The surface integral that involves inviscid and viscous fluxes is partitioned into separate contributions from each of the faces that compose the boundary of the computational cell, and each contribution is written in terms of surface-averaged values of the fluxes. Then, the area-averaged values are related to the unknown volume-averaged values, typically by extrapolation. The resulting set of algebraic equations will be the discretized form of the original problem that is actually solved by means of a digital computer.

In the following, it will be useful to introduce the *generalized inviscid flux vector* $\tilde{\mathbf{S}} = \mathcal{S}\mathbf{S}$, which is the product of the area contribution $\mathcal{S}$ times the area-averaged value of the inviscid flux vector. This vector is essentially the integral formulation counterpart of the inviscid flux vector in generalized curvilinear coordinates employed in the differential formulation [Walters *et al.* (1992)].

22.5.2 Extrapolation Strategies

Relating the area-averaged values of inviscid and viscous fluxes to the volume-averaged values of the conserved variable is the key operation of any finite-volume strategy. A generalized extrapolation procedure to recover values at a face starting from quantities at the center of a volume can be formulated for a generic vector **q**, and this results in two typically different left and right extrapolated vectors, $\mathbf{q}^-$ and $\mathbf{q}^+$, respectively. In order to minimize instabilities and nonmonotonic behavior of the solution, high-order extrapolation formulas will involve *limiters*. A review of commonly used limiters and their properties is given by Sweby (1984).

Two different techniques have emerged in recent years for the treatment of the inviscid flux vector. The first one identifies **q** with the flux vector itself. Consequently, a vector of inviscid fluxes is created from volume-averaged values of the variables in a cell, and, subsequently, face values are created by extrapolation. The second method, called MUSCL extrapolation (*Monotone Upstream-Centered Schemes for Conservation Laws*) [Hirsch (1990)], identified **q** with the vector of conserved variables, **Q**. The extrapolation produces face values for the conserved variables; then the fluxes are constructed from the right and left face values, according to the formulas that will be detailed in the following. It is useful to point out that for many applications, the extrapolation of *primitive* or even *characteristic* variables is preferred to the extrapolation of conserved variables. Numerical experiments have shown that with this procedure more robust codes are obtained for high-speed flows in the case of both perfect gas and reacting flows [Walters *et al.* (1992)]. Specific techniques for the discretization of the inviscid flux vectors will be discussed below.

Recent developments in the arena of extraplation strategies involve subcell reconstruction techniques [Harten (1989)], whereby variations and even discontinuities in the field variables within a computational cell are estimated (or reconstructed) from the volume-averaged data. The tracking of discontinuity within a cell has been shown to have the potential for improving the accuracy of predictions for stiff problems [LeVeque and Yee (1988) and Chang (1989)].

22.5.3 Flux-Split Algorithms

In recent years, flux-split techniques (see Sec. 19.3) have obtained wide acceptance as an accurate means of discretizing the inviscid fluxes. Originally developed for a perfect gas, they have been extended to flows in chemical equilibrium, and to mixtures out of chemical and thermal equilibrium [Cinnella and Grossman (1991)]. They are found to be very accurate and robust when used for transonic, supersonic and hypersonic flows, and they are fully compatible with conservative finite-volume, shock-capturing approaches.

Probably the two most popular schemes in the flux-vector splitting category are the ones due to Steger and Warming (1981) and to Van Leer (1982). The basic premise is to split the inviscid flux vector in one space dimension into two parts, each containing the information that propagates downstream and upstream, respectively. The two parts are constructed using the extrapolation strategies already outlined consistently with the direction of propagation. When the MUSCL approach is used, the result is

$$\tilde{\mathbf{S}} = \tilde{\mathbf{S}}^+ + \tilde{\mathbf{S}}^-, \qquad \tilde{\mathbf{S}}^+ = \tilde{\mathbf{S}}^+(\mathbf{Q}^-), \qquad \tilde{\mathbf{S}}^- = \tilde{\mathbf{S}}^-(\mathbf{Q}^+) \qquad (22.28)$$

where the downstream propagating flux $\tilde{\mathbf{S}}^+$ uses the left extrapolation $\mathbf{Q}^-$ and the upstream propagation flux $\tilde{\mathbf{S}}^-$ uses the right extrapolation $\mathbf{Q}^+$. Alternatively, positive and negative fluxes are extrapolated directly using the negative and positive extrapolation formulas, respectively.

Another very popular scheme, less robust for hypersonic flows, but more accurate, is the flux-difference splitting technique due to Roe (1981). It consists of an approximate Riemann solver, where an arbitrary discontinuity is supposed to exist at the cell surface between the left and the right state, and an approximate solution for this situation is written in terms of waves propagating upstream and downstream. In this instance, the inviscid flux vector is not split, but reconstructed from the upstream and downstream contributions that constitute the left and the right states in the Riemann problem.

Extensions to more space directions are usually made by superimposing pseudo-one-dimensional problems, where the extrapolation formulas to get left and right states keep track of the relevant direction only [Walters and Thomas (1988)]. Much effort has been spent towards the design of truly multidimensional algorithms, which would be at least approximately independent of the grid orientation [Hirsch and

Lacor (1989) and Dadone and Grossman (1992)]. More details are given in Sec. 22.5.8.

22.5.4 Steger–Warming Technique

The flux-vector splitting technique was originally developed by Steger and Warming (1981) for perfect gases. It takes advantage of the homogeneity of the Euler equations to construct those portions of the inviscid fluxes that are associated with the propagation of information in a given direction. The homogeneity property carries over to flows in thermochemical nonequilibrium [Grossman and Cinnella (1990)].

The Jacobian matrices of the inviscid flux $\tilde{\mathbf{D}} = \partial\tilde{\mathbf{S}}/\partial\mathbf{Q}$ can be diagonalized by means of right and left eigenvectors. The diagonal matrix will contain the characteristic speeds of propagation of information and, when split in a nonnegative and a nonpositive contribution according to the sign of those characteristic velocities, will yield a splitting of the Jacobian in two parts, propagating downstream and upstream respectively. Then, the splitting of the inviscid flux into a positive and a negative contribution is only a question of using the homogeneity relation

$$\tilde{\mathbf{S}} = \tilde{\mathbf{D}}Q = \tilde{\mathbf{S}}^+ + \tilde{\mathbf{S}}^-, \quad \tilde{\mathbf{S}}^\pm = \tilde{\mathbf{D}}^\pm Q, \quad \tilde{\mathbf{D}} = \tilde{\mathbf{D}}^+ + \tilde{\mathbf{D}}^- \quad (22.29)$$

with similar results for the other space dimensions. The final result for the positive and negative contributions to the inviscid fluxes can be written as

$$\tilde{\mathbf{S}}^\pm = \frac{\tilde{\gamma} - 1}{\tilde{\gamma}} \lambda_A^\pm \begin{bmatrix} \rho_1 \\ \vdots \\ \rho_N \\ \rho u \\ \rho v \\ \rho w \\ \rho_1 e_{n_1} \\ \vdots \\ \rho_M e_{n_M} \\ \rho h_0 - \rho a^2/(\tilde{\gamma} - 1) \end{bmatrix} + \frac{\lambda_B^\pm}{2\tilde{\gamma}} \begin{bmatrix} \rho_1 \\ \vdots \\ \rho_N \\ \rho(u + n_x a) \\ \rho(v + n_y a) \\ \rho(w + n_z a) \\ \rho_1 e_{n_1} \\ \vdots \\ \rho_M e_{n_M} \\ \rho(h_0 + \tilde{u}^* a) \end{bmatrix} + \frac{\lambda_C^\pm}{2\tilde{\gamma}} \begin{bmatrix} \rho_1 \\ \vdots \\ \rho_N \\ \rho(u - n_x a) \\ \rho(v - n_y a) \\ \rho(w - n_z a) \\ \rho_1 e_{n_1} \\ \vdots \\ \rho_M e_{n_M} \\ \rho(h_0 - \tilde{u}^* a) \end{bmatrix}$$

$$(22.30)$$

where the nonnegative and nonpositive eigenvalues of the Jacobian matrix are given by the following

$$\lambda_A^\pm = \mathbb{S} \frac{\tilde{u} \pm |\tilde{u}|}{2}, \quad \lambda_B^\pm = \mathbb{S} \frac{\tilde{u} + a \pm |\tilde{u} + a|}{2},$$

$$\lambda_C^\pm = \mathbb{S} \frac{\tilde{u} - a \pm |\tilde{u} - a|}{2} \quad (22.31)$$

Here, n_x, n_y, and n_z are the components of the unit vector, $\bar{\mathbf{n}}$, normal to the surface S. Moreover, the relative velocity component normal to the surface $\tilde{u} = (\bar{\mathbf{u}} - \bar{\mathbf{u}}_{\mathcal{V}}) \cdot \bar{\mathbf{n}}$ and the absolute component $\tilde{u}^* = \bar{\mathbf{u}} \cdot \bar{\mathbf{n}}$ have been introduced. It is interesting to point out that simpler forms of the algorithm, including thermodynamic equilibrium and perfect-gas versions, can be recovered by simplifying the equations in a fashion consistent with that discussed in Sec. 22.2.

The derivation of a Steger–Warming-type algorithm for flows in local chemical equilibrium is complicated by the fact that the governing equations are no longer homogeneous of degree one [Liou *et al.* (1988)]. This means that a positive or negative flux cannot be constructed by means of a simple matrix multiplication of the corresponding Jacobian contribution times the vector of conserved variables, as was done in Eq. (22.29) for the general thermochemical nonequilibrium case. An alternative derivation based upon *convective streams* yields a one-parameter family of splittings that is still based upon nonnegative and nonpositive propagation of information, as follows [Vinokur and Montagné (1988)]

$$\tilde{\mathbf{S}}^{\pm} = \frac{\Gamma - 1}{\Gamma} \lambda_A^{\pm} \begin{pmatrix} \rho \\ \rho u \\ \rho v \\ \rho w \\ \rho h_0 - \beta \rho a^2 \end{pmatrix} + \frac{\lambda_B^{\pm}}{2\Gamma} \begin{pmatrix} \rho \\ \rho(u + n_x a) \\ \rho(v + n_y a) \\ \rho(w + n_z a) \\ \rho(h_0 + \tilde{u}^* a) \end{pmatrix}$$

$$+ \frac{\lambda_C^{\pm}}{2\Gamma} \begin{pmatrix} \rho \\ \rho(u - n_x a) \\ \rho(v - n_y a) \\ \rho(w - n_z a) \\ \rho(h_0 - \tilde{u}^* a) \end{pmatrix} \tag{22.32}$$

where the isentropic index Γ plays a central role in the splitting. The nonnegative and nonpositive eigenvalues are given by Eq. (22.31), where the speed of sound is the equilibrium value, defined in Eq. (22.9), and β is equal to

$$\beta = \frac{(\Gamma - \alpha)}{\Gamma(\Gamma - 1)} \tag{22.33}$$

where the parameter α is arbitrary. The value $\alpha = 0$ allows this scheme to reduce nicely to the perfect-gas form and to be formally similar to its nonequilibrium counterpart when the ratio of frozen specific heats $\tilde{\gamma}$ is substituted for the isentropic index Γ.

Other formulations have appeared in the scientific literature. Grossman and Walters (1989) utilize the ratio of enthalpy to internal energy, γ^*, to split the flux and approximate the speed of sound by substituting the same ratio for the isentropic index, yielding $(a^*)^2 = \gamma^* p/\rho$. The final algorithm is formally identical to the one shown in Eq. (22.32) when $\alpha = 0$ and γ^* is substituted for Γ everywhere. Liou *et* *al.* (1988) split only a portion of the fluxes and add one half of the unsplit portion to both positive and negative contributions. Again, their splitting is not based upon the *true* sound speed. It may be useful to point out that no conclusive evidence has yet been presented as to the superiority of one approach over any other.

22.5.5 Van Leer Technique

An alternative flux-vector splitting has been developed for perfect gases by Van Leer (1982). His formulation has continuously differentiable flux contributions and has been shown to result in smoother solutions near sonic points than the Steger–Warming splitting [Van Leer *et al.* (1987)]. The derivation of the algorithm is based upon the splitting of the fluxes by means of low-order polynomials in the components of the normalized velocity parameter $M = \tilde{u}/a$ (often referred to as a Mach number). Then, the mass flux for the vector $\tilde{\mathbf{S}}$, which is $f_m = \rho\tilde{u} = \rho a M$, may be split as $f_m = f_m^+ + f_m^-$, where

$$f_m^\pm = \pm\rho a \left(\frac{\pm M + 1}{2}\right)^2 \tag{22.34}$$

The remaining fluxes may be split to yield

$$\tilde{\mathbf{S}}^\pm = \mathcal{S}f_m^\pm \begin{bmatrix} Y_1 \\ Y_2 \\ \vdots \\ Y_N \\ u + n_x(-\tilde{u} \pm 2a)/\tilde{\gamma} \\ v + n_y(-\tilde{u} \pm 2a)/\tilde{\gamma} \\ w + n_z(-\tilde{u} + 2a)/\tilde{\gamma} \\ Y_1 e_{n_1} \\ \vdots \\ Y_M e_{n_M} \\ f_e^\pm \end{bmatrix} \tag{22.35}$$

where

$$f_e^\pm = h_0 + (\mathbf{u}_\triangledown \cdot \mathbf{n})\frac{(-\tilde{u} \pm 2a)}{\tilde{\gamma}} - m(-\tilde{u} \pm a)^2 \tag{22.36}$$

and m is an arbitrary parameter. The value $m = 1/(\tilde{\gamma} + 1)$ allows this scheme to reduce neatly to the usual perfect-gas version, but numerical experiments do not show any significant effect from the choice of m. The previous formulas are valid in the subsonic range. In the supersonic range, when either $M > 1$ or $M < -1$, the positive (negative) flux becomes the full flux, and the negative (positive) contribution goes to zero. The other space dimensions are treated in a similar fashion.

The derivation of a Van Leer-type algorithm for flows in local chemical equilibrium has been successfully performed by several authors. The final result given by Liou *et al.* (1988) reads

$$\tilde{\mathbf{S}}^{\pm} = \mathbb{S}f_m^{\pm} \begin{pmatrix} 1 \\ u + n_x(-\bar{u} \pm 2a)/\Gamma \\ v + n_y(-\bar{u} \pm 2a)/\Gamma \\ w + n_z(-\bar{u} + 2a)/\Gamma \\ f_e^{\pm} \end{pmatrix} \tag{22.37}$$

where $f_m^{\pm}$ and $f_e^{\pm}$ are given by Eqs. (22.34) and (22.36), respectively, with Γ substituted for $\tilde{\gamma}$ again. The results presented by other authors differ in the choice of the parameter m in Eq. (22.36), although no strong argument has been provided in favor of any specific value. Grossman and Walters (1989) develop a similar splitting, but again the enthalpy to internal energy ratio γ^* has to be substituted for the isentropic index Γ in all of the expressions given, including the definition of an approximate speed of sound.

22.5.6 Roe Technique

The essential features of flux-difference split algorithms involve the solution of local Riemann problems arising from the consideration of discontinuous states at cell interfaces on an initial data line. The scheme developed for perfect gases by Roe (1981) falls into this category and has produced excellent results for both inviscid and viscous flow simulations.

At a cell interface, for a given time, it is possible to define a left state, $(\cdot)_l$, and a right state, $(\cdot)_r$, which correspond to positive and negative extrapolations of cell-volume values, respectively. Then a *jump operator* may be defined

$$[\![(\cdot)]\!] = (\cdot)_r - (\cdot)_l \tag{22.38}$$

The key step in the construction of an approximate Riemann solver [Glaister (1988)], involves determining appropriate averages of eigenvalues, $\hat{\lambda}_i$, right eigenvectors, $\hat{\mathbf{E}}_i$, and wave strengths, $\hat{\alpha}_i$, such that

$$[\![\mathbf{Q}]\!] = \sum_{i=1}^{N+M+4} \hat{\alpha}_i \hat{\mathbf{E}}_i, \qquad [\![\tilde{\mathbf{S}}]\!] = \sum_{i=1}^{N+M+4} \hat{\alpha}_i \hat{\lambda}_i \hat{\mathbf{E}}_i \tag{22.39}$$

for cell interface states which are not necessarily close to each other, so that $[\![\mathbf{Q}]\!]$ is arbitrary.

The solution of the approximate Riemann problem entails determining algebraic averages $\hat{\rho}$, $\hat{\mathbf{u}}$, $\hat{Y}_s$, $\hat{e}_{n_s}$, $\hat{h}_0$, $\hat{\psi}_s$, $\hat{a}$, such that Eq. (22.39) are satisfied. In the above, the definitions

$$\hat{e}_{ns} \equiv \widehat{(Y_s e_{ns})}, \qquad \psi_s = \frac{1}{\tilde{\gamma} - 1} \frac{\partial p}{\partial \rho_s} \tag{22.40}$$

have been used, where the pressure derivatives are taken with respect to the sth species density, and all other conservative variables are kept constant. It is noteworthy that these averages are not unique. As pointed out by Abgrall (1989), the algebraic problem posed in Eq. (22.39) has multiple solutions, and different values have been published in the literature for some of the Roe-type averages [Cinnella and Grossman (1991)]. The major differences between the proposed approaches are in the evaluation of averages of the pressure derivatives, or equivalently of $\hat{\psi}_s$, whereas the same results are obtained for mass fractions and nonequilibrium contributions, if present. No numerical evidence has been presented of the superiority of one scheme over the others.

Following the derivation of Grossman and Cinnella (1990), and using the notation

$$\langle (\cdot) \rangle = \frac{(\cdot)_r + (\cdot)_l}{2} \tag{22.41}$$

for the arithmetic average, the necessary averages are determined to be

$$\hat{\rho} = \langle \sqrt{\rho} \rangle^2 - \frac{1}{4} [\![\sqrt{\rho}]\!]^2 = \sqrt{\rho_r \rho_l}$$

$$\hat{\mathbf{u}} = \frac{\langle \overline{\mathbf{u}} \sqrt{\rho} \rangle}{\langle \sqrt{\rho} \rangle} = \frac{(\overline{\mathbf{u}})_r \sqrt{\rho_r} + (\overline{\mathbf{u}})_l \sqrt{\rho_l}}{\sqrt{\rho_r} + \sqrt{\rho_l}}$$

$$\hat{Y}_s = \frac{\langle Y_s \sqrt{\rho} \rangle}{\langle \sqrt{\rho} \rangle}$$

$$\hat{e}_{ns} = \frac{\langle Y_s e_{ns} \sqrt{\rho} \rangle}{\langle \sqrt{\rho} \rangle}$$

$$\tilde{h}_0 = \frac{\langle h_0 \sqrt{\rho} \rangle}{\langle \sqrt{\rho} \rangle}$$

$$\hat{\psi}_s = \frac{R_s \hat{T}}{\hat{\gamma} - 1} - \hat{e}_s + \frac{\hat{u}_0^2}{2}$$

$$\hat{a}^2 = (\hat{\gamma} - 1) \left[\hat{h}_0 - \frac{\hat{u}_0^2}{2} + \hat{c}_v \hat{T} - \sum_{s=1}^{N} \hat{Y}_s \hat{e}_s - \sum_{s=1}^{M} \hat{e}_{ns} \right] \tag{22.42}$$

where averaged values of temperature, $\hat{T}$, mixture frozen specific heat at constant volume, $\hat{c}_v$, species equilibrium contribution to the internal energy, $\hat{e}_s$, ratio of frozen specific heats, $\hat{\gamma}$, and mixture gas constant, $\hat{R}$, are defined by the following

$$\hat{T} = \frac{\langle T\sqrt{\rho}\rangle}{\langle \sqrt{\rho}\rangle}$$

$$\hat{c}_v = \sum_{s=1}^{N} \hat{Y}_s \bar{c}_{v_s} \qquad \bar{c}_{v_s} = \frac{1}{[T]} \int_{T_l}^{T_r} \tilde{c}_{v_s}\, dT$$

$$\hat{e}_s = (\bar{e}_s) = \frac{\langle \tilde{e}_s \sqrt{\rho}\rangle}{\langle \sqrt{\rho}\rangle}$$

$$\tilde{\gamma} = 1 + \frac{\hat{R}}{\hat{c}_v} \qquad \hat{R} = \sum_{s=1}^{N} \hat{Y}_s R_s = \frac{\langle R\sqrt{\rho}\rangle}{\langle \sqrt{\rho}\rangle} \tag{22.43}$$

In Eq. (22.43), the *integral averages* of the species frozen heats at constant volume $\bar{c}_{v_s}$ have been introduced, which in general will have to be numerically determined with any of the available quadrature formulas. For cases where simple trapezoidal rule is employed, it is easy to recognize that $\bar{c}_{v_s}$ reduces to the *arithmetic average* of right and left values. Moreover, it may be noticed that all of the above definitions reduce to the usual Roe-averages for a perfect gas model.

Rewriting the second of Eq. (22.39) by grouping the repeated eigenvalues, and simplifying the result yields

$$[\tilde{S}] = [\tilde{S}]_A + [\tilde{S}]_B + [\tilde{S}]_C \tag{22.44}$$

The $[\tilde{S}]_A$ term arises from the first $N + M + 2$ terms of the sum in the second of Eq. (22.39), corresponding to the repeated eigenvalue $\lambda_i = \hat{u}$, and may be written as

$$[\tilde{S}]_A = S\left([\rho] - \frac{[p]}{\hat{a}^2}\right)\hat{u}
\begin{bmatrix}
\hat{Y}_1 \\
\hat{Y}_2 \\
\vdots \\
\hat{Y}_N \\
\hat{u} \\
\hat{v} \\
\hat{w} \\
\hat{e}_{n_1} \\
\vdots \\
\hat{e}_{n_M} \\
\hat{h}_0 - \hat{a}^2/(\hat{\gamma}-1)
\end{bmatrix}
+ S\hat{\rho}\hat{u}
\begin{bmatrix}
[Y_1] \\
[Y_2] \\
\vdots \\
[Y_N] \\
[u] - n_x[\tilde{u}] \\
[v] - n_y[\tilde{u}] \\
[w] - n_z[\tilde{u}] \\
[\rho_1 e_{n_1}/\rho] \\
\vdots \\
[\rho_M e_{n_M}/\rho] \\
\Theta
\end{bmatrix}
\tag{22.45}$$

where

$$\Theta = \sum_{s=1}^{M} [Y_s e_{n_s}] - \sum_{s=1}^{N} \hat{\mathcal{V}}_s [Y_s] + (\hat{u}[u] + \hat{v}[v] + \hat{w}[w]) - \hat{u}^*[\tilde{u}^*] \tag{22.46}$$

and $\hat{\hat{u}}$, $\hat{u}*$ are trivially defined starting from the result for $\overline{u}$ as $\hat{\hat{u}} = (\hat{\mathbf{u}} - \mathbf{u}_\nabla) \cdot \overline{\mathbf{n}}$ and $\hat{u}* = \hat{\mathbf{u}} \cdot \overline{\mathbf{n}}$. The $[\![\tilde{\mathbf{S}}]\!]_B$ and $[\![\tilde{\mathbf{S}}]\!]_C$ contributions arise from the last two terms in the summation of the second of Eq. (22.39), corresponding to the eigenvalues $\hat{\hat{u}} + \hat{a}$ and $\hat{\hat{u}} - \hat{a}$, and they are found to be

$$[\![\tilde{\mathbf{S}}]\!]_{B,C} = \mathbb{S} \frac{1}{2\hat{a}^2} ([\![p]\!] \pm \hat{\rho}\hat{a}[\![\hat{u}]\!]) (\hat{\hat{u}} \pm \hat{a}) \begin{bmatrix} \hat{Y}_1 \\ \hat{Y}_2 \\ \vdots \\ \hat{Y}_N \\ \hat{u} \pm n_x\hat{a} \\ \hat{v} \pm n_y\hat{a} \\ \hat{w} \pm n_z\hat{a} \\ \hat{e}_{n_1} \\ \vdots \\ \hat{e}_{n_M} \\ \hat{h}_0 \pm \hat{u}*\hat{a} \end{bmatrix} \tag{22.47}$$

The approximate Riemann solver is implemented by computing the cell face fluxes as a summation over wave speeds [Roe (1981)]

$$\tilde{\mathbf{S}} = \tfrac{1}{2}(\tilde{\mathbf{S}}_r + \tilde{\mathbf{S}}_l) - \tfrac{1}{2} \sum_{i=1}^{N+M+4} \hat{\alpha}_i \, |\hat{\lambda}_i| \, \hat{\mathbf{E}}_i \tag{22.48}$$

which can be written, using the previous results

$$\tilde{\mathbf{S}} = \langle \tilde{\mathbf{S}} \rangle - \tfrac{1}{2}([\![\tilde{\mathbf{S}}]\!]_A + [\![\tilde{\mathbf{S}}]\!]_B + [\![\tilde{\mathbf{S}}]\!]_C) \tag{22.49}$$

with the absolute value of the wave speeds substituted in Eqs. (22.45) and (22.47). When the MUSCL approach is employed, right and left states are evaluated using high-order interpolation formulas applied to primitive or characteristic variables. When the flux interpolation approach is utilized, right and left states correspond to the cell-center values immediately to the right and left of the interface under consideration. Equation (22.49) is only first-order accurate in this instance, and high-order terms may be added [Chakravarthy *et al.* (1985)] to render the scheme second or third-order accurate. These additional terms will involve eigenvalues, right eigenvectors and wave strengths, which have been obtained as part of the derivation [Grossman and Cinnella (1990)].

The derivation of an approximate Riemann solver for flows in local chemical equilibrium is very similar to the one previously outlined for thermochemical nonequilibrium [Cinnella and Grossman (1991)]. Again, the averages that enter the definitions of the flux jumps at the interface are not unique, and the nonuniqueness is reduced to a choice between different formulas for the averages of pressure derivatives. Moreover, even the choice of the two key pressure derivatives to be used in the formulation is different for different authors. Vinokur (1988) employs partial derivatives of pressure with respect to internal energy per unit volume, ρe, and density, whereas Glaister (1988) prefers internal energy per unit mass, e, and density. Grossman and Walters (1989), on the other hand, avoid the explicit use of pressure derivatives by introducing two alternative quantities, isentropic index Γ and enthalpy to internal energy ratio γ^*. Correspondingly, three different expressions for the equilibrium speed of sound are possible

$$a^2 = \left(\frac{\partial p}{\partial \rho}\right)_{\rho e} + h\left(\frac{\partial p}{\partial \rho e}\right)_{\rho} = \left(\frac{\partial p}{\partial \rho}\right)_e + \frac{p}{\rho^2}\left(\frac{\partial p}{\partial e}\right)_{\rho} = \Gamma\frac{p}{\rho} \qquad (22.50)$$

and the definition of the Roe-averaged speed of sound will be affected by the choice made.

Following Cox and Cinnella (1994), the starting point for the construction of the approximate Riemann solver is again the determination of appropriate averages to be used in Eq. (22.39) to determine jumps in conservative variables and interface fluxes.

The solution of the approximate Riemann problem involves determining algebraic averages $\hat{\rho}$, $\hat{\mathbf{u}}$, $\hat{h}_0$, $\hat{\Gamma}$, $\hat{\gamma}$, $\hat{a}$ such that Eq. (22.39) is satisfied. The first three averages are formally unchanged from the previous development, with density being averaged geometrically and velocity and total enthalpy being defined by means of standard Roe-averages. The speed of sound $\hat{a}$ reads

$$\hat{a}^2 = \hat{\Gamma}\hat{R}\hat{T} + (\hat{\gamma} - 1)\left[\hat{h}_0 - \frac{\hat{u}^2}{2} - \sum_{s=1}^{N} \hat{Y}_s \hat{e}_s - \hat{R}\hat{T}\right] \qquad (22.51)$$

where

$$\hat{\Gamma} = \hat{\gamma} + \frac{\hat{\rho}}{\hat{R}\hat{T}} \sum_{s=1}^{N} \overline{\frac{\partial Y_s}{\partial \rho}} [R_s\hat{T} - (\hat{\gamma} - 1)\hat{e}_s]$$

$$\hat{\gamma} = 1 + \frac{\hat{R} + \hat{T} \sum_{s=1}^{N} R_s(\overline{\partial Y_s/\partial T})}{\hat{c}_v + \sum_{s=1}^{N} \hat{e}_s(\overline{\partial Y_s/\partial T})} \qquad (22.52)$$

The mixture gas constant, temperature, species internal energy, and mass fractions are given by Roe-type averages, as in the nonequilibrium case. However, species specific heats at constant volume and mass fraction derivatives are averaged by means of integral averages between left and right state. Vinokur and Montagné (1988) discuss a procedure that would eliminate the discretization error involved in an approximate evaluation for these averages.

The approximate Riemann solver is implemented exactly as discussed for the nonequilibrium case. It may be noticed that when this formulation is used in conjunction with a chemical equilibrium solver, the mass fraction derivatives are readily available. When a curve-fit formulation is used, pressure derivatives have to be obtained by analytical or numerical differentiation or have to be provided alongside of the other relevant pieces of information [Liu and Vinokur (1989b)]. In general, there is no guarantee of continuity and/or smooth behavior of the result obtained by numerical differentiation of the pressure cuves.

22.5.7 Additional Algorithms

The algorithms presented above do not exhaust the category of flux-split techniques. Osher-type formulations exist for perfect gases, and their extension to reacting flows has been attempted [Abgrall and Montagné (1989)]. More recently, new kinds of flux splitting are emerging, whereby the treatment of the pressure gradient plays a central role, and they seem to show promise of improvement over the older schemes. Again, the original development is for perfect gases, although the extension to flows in thermochemical nonequilibrium is relatively straightforward [Bergamini and Cinnella (1993)].

All of the previously mentioned upwind algorithms originate from the Euler equations for gas dynamics. The Euler equations, however, can be obtained as moments of the Boltzmann equation [Vincenti and Kruger (1986)] established in the kinetic theory of gases, provided that the velocity distribution function is Maxwellian. This connection between the Boltzmann equation and the Euler equations has lead to a new class of upwind *kinetic* schemes which are based on the principle that an upwind scheme at the Boltzmann level leads to an upwind scheme at the Euler level. Kinetic-difference schemes for the Euler equations have been developed by Deshpande (1986) and his coworkers. In their approach, the resulting Euler schemes are developed as moments of the discretized Boltzmann scheme with a locally Maxwellian velocity distribution. Splitting the velocity distribution at the Boltzmann level in upwind contributions and taking moments results in a flux-split Euler scheme called *Kinetic Flux Vector Splitting* (KFVS) [Deshpande and Mandal (1987) and Mandal and Deshpande (1989)]. Similar differential-difference schemes with kinetic-based dissipation have been developed in Russia by Elizarova and Chetverushkin (1985), where many applications to high-speed gasdynamic flows have been performed. Recently Weatherill *et al.* (1985) have implemented the KFVS scheme for unstructured grids. Eppard and Grossman (1993a) have extended the KFVS schemes to flows with chemical and thermodynamic nonequilibrium by means of nonequilibrium kinetic theory.

The development of nonequilibrium KFVS is illustrated here for one spatial dimension. The species Boltzmann equation for a gas in local translational equilibrium subject to zero body forces may be written as

$$\frac{\partial}{\partial t}(n_s f_{0_s}) + c_s \frac{\partial}{\partial x}(n_s f_{0_s}) = \sum_r \left\{ \frac{\partial}{\partial t}(n_s f_{0_s}) \right\}_{\mathrm{coll}_{r-s}} \qquad (22.53)$$

where n_s is the species number density, f_{0_s} is the Maxwellian velocity distribution function, and c_s is the component of the particle velocity in the x direction. Moreover, c_s can be written as the sum of the mass-averaged mixture velocity u and the thermal velocity c'_s, as follows: $c_s = u + c'_s$. Note that for chemically reacting flows, the collision integral terms on the right-hand side of Eq. (22.53) must be retained. Next, each species Boltzmann equation is discretized using an upwind procedure

$$\frac{\partial}{\partial t}(n_s f_{0_s})_j + \frac{c_s + |c_s|}{2} \cdot \frac{[n_s f_{0_s}]_j - [n_s f_{0_s}]_{j-1}}{\Delta x} + \frac{c_s - |c_s|}{2} \cdot \frac{[n_s f_{0_s}]_{j+1} - [n_s f_{0_s}]_j}{\Delta x}$$

$$= \sum_r \left\{ \frac{\partial}{\partial t}(n_s f_{0_s})_j \right\}_{coll_{r-s}} \tag{22.54}$$

where the subscript j indicates the discretized space location, and $\Delta x = x_{j+1/2} - x_{j-1/2}$ is the length of the jth cell. Moments of the discretized species Boltzmann scheme are taken by multiplying Eq. (22.54) by the weights Ψ

$$\Psi = \begin{pmatrix} m_s \\ m_s c_s \\ m_s \epsilon_{n_s} \\ m_s \epsilon_{0_s} \end{pmatrix} \tag{22.55}$$

and integrating over the entire velocity space. In the expression for Ψ, m_s is the mass of a particle of species s, and ϵ_{n_s} and ϵ_{0_s} are the particle nonequilibrium and total internal energies, respectively. Summing the species momentum contributions and the species total energy contributions yields an upwind Euler scheme. The components of the split flux vector $\mathbf{F}^{\pm}$ for species continuity, momentum, species nonequilibrium energy, and total energy, respectively, are given in terms of the error function as listed below

$$F_s^{\pm} = \frac{1}{2} \rho_s u \left[1 \pm \mathrm{erf}(\sqrt{\beta_s} u) \right] \pm \frac{\rho_s}{2\sqrt{\beta_s \pi}} e^{-\beta_s u^2}, \qquad s = 1, \ldots, N$$

$$F_{N+1}^{\pm} = \sum_{s=1}^{N} \left\{ \frac{1}{2}(\rho_s u^2 + p_s) \left[1 \pm \mathrm{erf}(\sqrt{\beta_s} u) \right] \pm \frac{\rho_s u}{2\sqrt{\beta_s \pi}} e^{-\beta_s u^2} \right\}$$

$$F_{N+1+s}^{\pm} = \frac{1}{2} \rho_s e_{n_s} u \left[1 \pm \mathrm{erf}(\sqrt{\beta_s} u) \right] \pm \frac{\rho_s e_{n_s}}{2\sqrt{\beta_s \pi}} e^{-\beta_s u^2}, \qquad s = 1, \ldots, M$$

$$F_{N+M+2}^{\pm} = \sum_{s=1}^{N} \left\{ \frac{1}{2} \rho_s u h_{0_s} \left[1 \pm \mathrm{erf}(\sqrt{\beta_s} u) \right] \pm \left(\rho_s h_{0_s} - \frac{1}{2} p_s \right) \frac{1}{2\sqrt{\beta_s \pi}} e^{-\beta_s u^2} \right\}$$

$$\tag{22.56}$$

where $\beta_s = 1/2R_s T$. Extensions of the scheme to two-dimensions in generalized coordinates are given by Eppard and Grossman (1993a).

22.5.8 Truly Multidimensional Algorithms

To obtain accurate numerical simulations for complex flow problems, solvers must be capable of calculating highly resolved shock waves, shear and contact discontinuities, flame fronts, mixing layers, and boundary layers. While there has been considerable progress in these areas over the last decade, some improvements are still necessary. The current generation of Euler and Navier–Stokes solvers are limited by a noticeable degradation in the accuracy of resolution of these fluid-dynamic phenomena when they are oriented obliquely to the computational grid. This problem arises in central-difference codes because of the inability to properly tune the numerical dissipation, and in upwind codes because the current technology is essentially one-dimensional in nature and must be implemented in a dimensionally-split approach for calculations in two and three spatial dimensions. Consequently, excessive numbers of grid points and associated large amounts of computational time and memory are required to accurately resolve the critical flow features. This problem is magnified for calculations involving chemical and thermal nonequilibrium, since additional rate equations must be solved in a coupled fashion, and the species densities and nonequilibrium energies must be stored at each grid point.

One approach towards the development of upwind solvers with multidimensional behavior involves rotated or *multidirectional Riemann solvers* [Van Leer (1993)]. These schemes require the choice of a dominant upwinding direction and a local Riemann solution with left and right states that are functions of the upwinding angle. Efforts along these lines include the work of Davis (1984), Levy *et al.* (1989), and Dadone and Grossman (1992). Solvers of this type have succeeded in calculating shocks oblique to the grid with nearly the same resolution as shocks which are aligned with the grid. However, for three-dimensional flows with complex fluid phenomena, the task of choosing only one pertinent upwinding direction is a formidable one. While these schemes represent an improvement over directionally-split schemes, they are not the final answer and still leave the CFD community in need of truly multidimensional ideas. Other approaches to putting more multidimensional information into Riemann solvers involve the multidimensional flux function approach, advocated among others by Rumsey *et al.* (1991) and Parpia and Michalek (1990).

Recent work on genuinely multidimensional upwind Euler solvers has been centered around the generalization of Roe's one-dimensional scheme. The basic concepts underlying the extension of Roe's scheme for two and three dimensions have been outlined in Struijs *et al.* (1991). Three basic steps are involved: step one requires an eigenvector decomposition of the divergence of the flux vector which will be written (as in Roe's one-dimensional scheme) as the sum of terms of the form (eigenvector) × (wave speed) × (wave strength). This process is not uniquely determined as in one space dimension, and several decompositions have been introduced which *recognize* and select relevant simple wave patterns. Step two is to obtain the discretized counterpart of the previous wave decomposition based on a conservative linearization procedure. Struijs *et al.* (1991) carry out this step for triangular cells in two dimensions and tetrahedra in three dimensions. At this point, the discretized wave structure (orientation and corresponding advection speeds) is known in terms of the unknowns at the cell vertices, and each of these simple waves is governed by the linear scalar advection equation. Step three advances the solution in time in an explicit manner, utilizing multidimensional, fluctuation-splitting schemes for the scalar advection equation, whereby decomposed portions of the flux balance are distributed to the vertices of the computational cell. As a result, the numerical behavior of the system is governed completely by the characteristics of these scalar distribution schemes. Struijs *et al.* (1991) have considered two scalar advection schemes for triangular cells. The first scheme, called the *N-scheme*, is the optimal linear scheme satisfying positivity, because it allows the maximum time step and has the most narrow *stencil*. Unfortunately, it is at most first-order accurate in space. The second scheme is a nonlinear variant of the *N-scheme*, termed the *NN-scheme*, and it is both positive and second-order in space. Positivity is an important characteristic, since it prohibits the occurrence of new extrema and allows these schemes to maintain monotone profiles across discontinuities without the need for limiters. Positivity also imposes stability on the explicit scheme. Preliminary results using these concepts have shown improvements in shock capturing compared to standard solvers. Although these schemes have great potential, more work is needed in the wave modeling stage.

Another attempt at the development of genuinely multidimensional upwind Euler solvers has been based upon kinetic flux-vector splitting described in Sec. 22.5.7 [Eppard and Grossman (1993b)]. In this case, splitting the velocity distribution at the Boltzmann level in an upwind sense is seen to result in a dimensionally-split flux-split scheme (KFVS). The multidimensional linear advection schemes described by Struijs *et al.* (1991) have been applied at the Boltzmann level. The resulting Euler schemes were obtained as moments of the fluctuations in the Maxwellian velocity distribution function. This development is significantly more complicated than standard (dimensionally-split) kinetic schemes, because the Boltzmann discretization depends upon the direction of the molecular velocities, and this must be accounted for in the limits of integration in velocity space. It is important to note that an explicit wave-decomposition model is not required for the development of the multidimensional kinetic flux-vector split (MKFVS) schemes. Encouraging preliminary results have been obtained for perfect gases for the inviscid reflection of an oblique shock wave and for an oblique shear wave. However, these schemes are very complex and inefficient. The MKFVS schemes may give some additional understanding of appropriate multidimensional wave decompositions for the Euler equations.

The multidimensional Euler solvers developed so far consider only frozen mixtures of thermally perfect gases that are in thermal equilibrium. The extension to flows in chemical and thermal nonequilibrium is feasible, but has not been reported yet. However, the benefits realized by the sharp resolution of flowfield structures should be even more pronounced for cases with chemical and thermal nonequilibbrium, because they are very sensitive to the resolution of discontinuities, particularly contact surfaces.

22.5.9 Low-Speed Reacting Flows

The upwind methods examined thus far are well suited for the calculation of high-speed flows (subsonic or higher), but some convergence and/or efficiency problems arise when the Mach number is decreased to low values. Several important applications in combustion and flame propagation involve low-speed flows where compressibility effects are nonnegligible [Bhattacharjee *et al.* (1990)]. The investigation of these flowfields is complicated by the fact that chemical reactions and the ensuing multiple space and time scales are an essential part of the physics.

Two major factors contribute to the inefficiency of standard methods for low Mach numbers: the ill-coupling of pressure and velocity in the continuity and momentum equations and the stiffness of the inviscid fluxes, as measured by the ratio between maximum and minimum eigenvalue of the flux Jacobians (which is proportional to the inverse of the Mach number). *Preconditioning* techniques attempt to improve the performance of high-speed algorithms applied to low-speed compressible (and reactive) problems by rescaling the eigenvalues (characteristic speeds) of the problem and rendering their ratio nonsingular as the Mach number goes to zero [Turkel (1987)].

Peyret and Viviand (1985) present an exhaustive review of preconditioning of the incompressible Navier–Stokes equations. In the low-speed, compressible arena, proposed solutions that alter the steady-state equations have been generally discarded in favor of techniques that alter the time evolution of the problem [Van Leer *et al.* (1991)]. Unsteady problems have been considered by introducing pseudo-time derivatives, which allow the calculations to converge to a time-accurate, unsteady residual [Withington *et al.* (1991)]. Techniques that simplify the governing equations by means of small Mach number expansions have been proposed, but they have the distinct disadvantage of not being accurate for slightly larger Mach numbers. A review of preconditioning techniques for the compressible equations is provided by Turkel (1992). A few preliminary attempts to extend the preconditioning ideas to reacting flow problems have been registered [Godfrey *et al.* (1993)].

A different solution to the low-speed difficulties of flux-split algorithms is represented by the family of SIMPLE and SIMPLER algorithms [Anderson *et al.* (1984)], which have been applied to reactive flow problems [Bhattacharjee *et al.* (1990)], but whose overall accuracy has been repeatedly questioned [Oran and Boris (1987)].

22.5.10 Viscous Fluxes

The viscous vectors are usually discretized using standard second-order accurate central differences. In most applications of engineering interest, the *Thin-Layer* version of the Navier–Stokes equations is actually discretized, whereby the viscous terms in the space directions that are not normal to a solid surface are neglected. On the other hand, when the full Navier–Stokes equations are analyzed, some mixed second-order derivative terms appear in the formulation, and they are usually treated in a diagonal-dominance-preserving fashion [Chakravarthy *et al.* (1985)].

22.6 TIME INTEGRATION

After fluxes and source terms have been discretized and related to the vector of unknown volume-averaged variables, the remaining problem to be solved is how to advance the numerical solution in time, given a known initial state. Some algorithms that were very popular even in recent years use a coupled approach, whereby time integration and space discretization are interconnected. However, the finite-volume technique lends itself to a fully uncoupled procedure, with the time integration segregated from the space discretization procedures.

Two classes of problems usually arise when dealing with time integration. The first class contains the *unsteady* problems, where an accurate time integration is essential for the overall validity of the simulation. Moreover, some unsteady problems could involve moving boundaries and/or moving and deforming control volumes. The second class involves steady problems, where time accuracy is of no concern. A pseudo-transient problem is usually solved in this case, where the solution is advanced in a pseudo-time until convergence to a steady-state is reached. Most algorithms will be able to handle both steady and unsteady problems, provided that the discretized form of the *Geometric Conservation Law* [Thomas and Lombard (1979)] is enforced in the calculation of the time derivative of the control volumes for cases when the control volumes are deforming.

22.6.1 Explicit Schemes

A popular and computationally inexpensive scheme that is second-order accurate in time is the *m*-stage, Jameson-style, explicit Runge–Kutta method [Walters *et al.* (1992)]. Other explicit methods used in CFD applications are discussed in Fletcher (1988). Unfortunately, all explicit schemes have the major drawback of becoming extremely inefficient for *stiff* problems, that is when some of the characteristic times associated with chemical reactions or thermal relaxation become much faster than the fluid dynamics characteristic time. In these conditions, the time step necessary for stability can become even orders of magnitude smaller than the value which would be necessary for an efficient resolution of the overall gas dynamic transient. For this reason, time-implicit schemes are advocated, as they show an increase in efficiency that can be dramatic.

Hybrid schemes have also been investigated [Eklund *et al.* (1986)], whereby the source terms are treated implicitly and the fluxes remain explicit. These techniques show some promise, but they are still the object of current research.

22.6.2 Implicit Schemes

Implicit time integration schemes are very popular especially for steady problems, due to their unconditional stability, the consequent increase in time step per iteration, and the overall efficiency that they make possible. Moreover, for reacting flow problems it has been shown that treating the source terms in an implicit fashion is tantamount to rescaling the governing equations so that all of the time scales in a flow problem become of the same order of magnitude [Bussing and Murman (1985)]. Consequently, implicit algorithms are a viable solution to stiffness problems.

For cases where time accuracy is important, as in unsteady problems, second-order-accurate schemes in time have been developed, such as the implicit two-step Runge–Kutta scheme developed by Iannelli and Baker (1988), or three-time-level schemes [Whitfield *et al.* (1988)]. Examples abound of problems where even first-order algorithms such as *Euler Implicit* are deemed satisfactory, e.g., Taylor and Whitfield (1991). Moreover, for steady problems, the Euler Implicit technique is almost universally accepted as the most efficient approach.

After any one of the implicit techniques is applied to the discretized equations, an extremely large linear problem is obtained, whose unknowns are the volume-averaged values at each control volume and whose left-hand-side comprises the Jacobian matrices of fluxes and source terms. For three-dimensional problems, the storage requirements for an exact solution of this linear problem are too restrictive. Approximate solutions have been devised, namely solutions in a plane in conjunction with relaxation in the third space direction, or some sort of approximate factorization that reduces the original problem to a series of smaller problems [Whitfield *et al.* (1988)].

All of the techniques previously discussed apply to flows in local chemical equilibrium as well as flows in thermochemical nonequilibrium. The difference is the size of the problem due to the increased number of equations that are solved per control volume in the latter case and the fact that in the former case some thermodynamic properties have to be obtained from the composition solver or from curvefit expressions.

In order to reduce the computational requirements for problems involving flows in thermochemical nonequilibrium, loosely coupled or even uncoupled techniques [Oran and Boris (1987)] have been implemented, whereby the species continuity and the nonequilibrium energy equations are solved separately from the gasdynamic equations (global continuity, momentum, and global energy). Larrouturou (1991) proposes a loosely coupled scheme that promises to preserve the positivity of mass fractions, which would be a remarkable improvement over many current techniques, although it is not clear that his results would hold true for complex source terms of the kind presented in Sec. 22.3.

22.7 RADIATIVE HEAT TRANSFER

Radiative heat transfer plays a major role in the thermal loads associated with highspeed flight, as well as many combustion and flame propagation problems. Considering hypersonic applications as an example, Anderson (1969) shows that at the stagnation-point of a blunt body the relative importance of convective heat transfer becomes negligible when compared with the radiative loads for sufficiently high Mach numbers (typically, above 20). For space vehicles like the Galileo Probe, which was designed for Jovian exploration, the heating will be virtually all radiative, because convective heating is hindered due to massive ablation. In combustion applications, the speed of propagation of flames, or even the possibility of flame extinction, can significantly depend on the level of gas-phase radiation.

The mathematical and physical features of fluid flow in the presence of radiative heat transfer were already well established in the 1960's [Vincenti and Kruger (1986)]. The integro-differential equations which govern relativistic flows, and their simplified version valid for usual gasdynamic applications, where the light speed can be considered to be infinite, are presented by Simon (1963). Radiative pressure and radiative energy density are usually insignificant for nonrelativistic applications, and they are neglected by most investigators.

The fact that the governing equations are integro-differential in nature significantly complicates the theoretical analysis as well as the numerical solution procedures. Typically, either a one-dimensional model of the propagation of radiation is employed (the *One-Dimensional Slab model* [Anderson (1989)]), or the equations are reduced to a differential form via perturbation theory. Hartung and Hassan (1992) compared the 1D Slab model to a *Modified Differential Approach* for external aerodynamics applications, and showed that the former is inaccurate if used away from the stagnation point.

Approximate solutions, both analytical and numerical, have been attempted for flow problems involving some form of radiative heat transfer. Zhigulev *et al.* (1963) present analytical results for the supersonic flow over a wedge by using a linearized form of the governing equations and neglecting radiative absorption (*optically thin gas*). The effect of thermal radiation on the internal structure of a shock wave was established by Heaslet and Baldwin (1963), who demonstrated the possibility of discontinuous temperature and velocity profiles even in the presence of strong radiative effects. Other approximate solutions for the unsteady, one-dimensional shock waves have been obtained by Olfe (1964), in connection with the linearized approach, and by Wang (1964), in connection with a self-similar treatment.

A survey of the important effects that influence the analysis of radiating flows is given by Anderson (1969), who shows by comparison that treating a gas as *transparent* or *gray* and using an uncoupled approach can significantly affect the solutions obtained. Other factors to be considered are: the need for including atomic line as well as continuum spectra in the radiative models, the possibility of strong ablative effects for hypersonic applications, and the importance of nonequilibrium chemistry and thermodynamics.

There are two major techniques utilized for the simulation of gasdynamic problems that include radiative heat transfer. Several investigators use an *uncoupled* approach, whereby the thermodynamic state of the flow is determined from a solver without radiative heat transfer, followed by a post-processing phase where the radiative field is evaluated [Hartung and Hassan (1992)]. On the other hand, the *fully coupled* approach is advocated by many [Cinnella and Elbert (1992)], because the decoupling can significantly overestimate both radiative and conductive heat fluxes.

Recent developments in the field include preliminary attempts to perform truly two-dimensional numerical solutions of the radiative heat transfer equations [Elbert and Cinnella (1993)].

22.8   SHOCK STRUCTURE AND THE BURNETT EQUATIONS

Flux-split algorithms have been utilized for flows at high altitudes, where the rarefied atmosphere may be still considered to be a continuum, but where the Navier–Stokes equations cease to be accurate. Recent activity in this so-called *continuum transition regime* has been reported by Lumpkin (1991). Through studies involving the structure of a normal shock wave, the importance of utilizing the Burnett equations in this regime and the importance of modeling rotational nonequilibrium effects has been documented. Some important results on the stabilization of the Burnett equations for numerical calculations have been developed by Zong *et al.* (1991).

The flux-split algorithms with nonequilibrium thermodynamics discussed in Sec. 22.5 have been applied to the shock structure problem in this flow regime in order to examine some aspects of rotational nonequilibrium. In particular, the *bulk-viscosity* model proposed by Goldstein (1972) has been evaluated with the shock structure problem to examine whether it is able to account for rotational relaxation [Grossman *et al.* (1992)].

It is known that rotational nonequilibrium effects are often important in shock structure problems. The rotational relaxation process is often modeled by

$$\frac{\partial e_r}{\partial t} = \frac{e_r^* - e_r}{Z_r \tau_c} \tag{22.57}$$

where e_r is the nonequilibrium and e_r^* is the thermal equilibrium value of rotational energy. The rotational collision number Z_r may be written as

$$Z_r = \frac{Z_r^\infty}{1 + \dfrac{\pi^{3/2}}{2}\left(\dfrac{T_r}{T}\right)^{1/2} + \left(\dfrac{\pi^2}{4} + \pi\right)\left(\dfrac{T_r}{T}\right)} \tag{22.58}$$

where Z_r^∞ is the freestream value, T_r is a characteristic temperature, and τ_c is the mean collision time, given as

$$\tau_c = \frac{\pi\mu}{4p} \tag{22.59}$$

The possibility of including rotational nonequilibrium effects through the bulk viscosity as a possible alternative to rotational relaxation within the shock wave has been considered. Goldstein (1972) developed his bulk viscosity model from the kinetic theory of gases, whereby he arrived at the result for the bulk viscosity associated with rotational nonequilibrium. The coefficient of bulk viscosity may be written as

$$\mu_B = \frac{4}{15} pt_r \tag{22.60}$$

where t_r is approximately the relaxation time for rotation, which may be taken as

$$t_r = Z_r\tau_c \tag{22.61}$$

Preliminary results show that this model is somewhat limited in its capabilities of representing rotational nonequilibrium effects within strong shock waves [Grossman et al. (1992)].

22.9 BIOCHEMICAL APPLICATIONS

Applications involving chemically reacting flows are frequently found in as diverse fields as biomedical research and biochemical processing devices. The flows of air and blood in the human body involve biochemical reactions at bronchial and capillary interfaces. Moreover, the formation and growth of *biofilms* is an important process in many medical and industrial systems, ranging from arterial plaque build-up, which contributes to heart disease, to the corrosion of tubing in water distribution systems, which is frequently linked to bacterial populations at corrosion sites [Characklis and Marshall (1990)]. Similarly, the power industry is plagued by reduced efficiency of heat transfer equipment caused by excessive biofilm accumulation.

Biofilms are defined as the aggregate of microbial cells, extracellular polymer substance (EPS), which binds the cells together, and other particulate matter. A successful attempt to study and model biofilm behavior requires the development of innovative algorithms, because the problem is physically very complicated. In a typical biofilm process simulation, multiple time scales have to be resolved, since biochemical reactions have characteristic times measured in days and fluid mechanic processes change in seconds. Moreover, multiple space scales can exist, because the biofilm thickness is measured in microns and industrial plant components have characteristic lengths measured in meters, or in some cases kilometers. Adding to the complication, multiple phases are typically present in the same problem, for example a bulk liquid phase flowing in contact with a biofilm, which is itself a collection of one liquid and several solid phases. In this context, the interface between biofilm and bulk liquid will move and deform in time, following the biofilm growth and decay processes. In summary, unsteady, reacting, multiphase phenomena with moving boundaries have to be accurately simulated.

A considerable amount of experimental research on the kinetics of biofilm processes has been conducted [Characklis and Marshall (1990)]. The detailed mechanisms of biofilm growth, the specific biochemical reactions that take place at the biofilm interface with the liquid phase, and the effect of different biocides are examples of active research areas. On the other hand, numerical simulations of biofilm processes have not yet reached *maturity*, probably because it is extremely difficult to couple the equations describing all of the biofilm processes to create a simulation model capable of predicting biofilm behavior. Individual biofilm processes are inherently nonlinear, and their response to changes in bulk fluid transport phenomena has not been full documented.

A few models of multispecies biofilm systems have been published. Wanner and Gujer (1986) propose a model that is one-dimensional in space, where a multispecies biofilm is attached to a solid wall and is in contact with a liquid phase through an interface. The biofilm thickness is allowed to change, due to growth, sloughing, or biocide effects, and the transport of substrate elements to the different bacteria in the biofilm is modeled. The *method of lines* [Fletcher (1988)] is applied to reduce the partial differential equations to ordinary differential equations that have to be solved at every time step, and a standard integration package is utilized for that purpose.

The International Association on Water Pollution Research and Control (IAWPRC) has issued a report [Characklis *et al.* (1989)], in which the general governing equations for the kinetics of biofilm systems are derived. Applications to a number of biofilm reactors are discussed, including trickling filters, packed beds and groundwater systems, biological fluidizer bed reactors, and rotating biological contactors. The model is applicable to two and three-dimensional geometries, allows for moving boundaries such as the bulk liquid interface, and includes some modeling of attachment, detachment and sloughing at the interface. Molecular diffusion, turbulent diffusion, and advection are accounted for, and a general model for transformation processes, such as biotic and abiotic reactions, is proposed.

Because the above-mentioned equations are very general, no numerical technique for their solution is proposed or advocated. A finite-volume-based solution procedure has been implemented and at least partially validated [Szego *et al.* (1993)]. The geometric flexibility of finite volumes allows for a simple and straightforward representation of moving interface boundaries, such as the surface film.

22.10 NUMERICAL RESULTS

Extensive numerical results have been obtained in recent years for applications involving chemically reacting flows, and the reader is referred to the literature cited for numerous examples in one, two, or three space dimensions, both steady and unsteady.

A small sample of numerical simulations involving different chemistry and thermodynamic models is presented here. It includes the quasi-one-dimensional flow of air in the divergent portion of a nozzle, the flow of a hydrogen/air mixture in the Space Shuttle Main Engine (SSME) nozzle, a comparison between the KFVS algorithm and those of the Van Leer and Roe types, a study of the structure of a shock wave, and a simulation of biofilm growth at the wall of a circular pipe.

22.10.1 Quasi-One-Dimensional Supersonic Diffuser

The first example involves the quasi-one-dimensional, inviscid, steady simulation of a rapidly expanding supersonic diffuser. The area ratio is $4:1$ over a length of 2 meters, according to the formula [Drummond *et al.* (1986)]

$$ A = \frac{\pi}{16} \left[1 + \sin\left(\frac{\pi x}{2L}\right) \right]^2, \qquad 0 \le x \le L \qquad (22.62) $$

where $L = 2$ m. The calculation was performed on 161 evenly spaced grid points with a third-order upwind-biased spatial discretization and is grid-converged. For fully supersonic calculations, the Steger–Warming and Van Leer flux-vector splitting methods coincide within round-off errors, and the Roe flux-difference splitting technique yields very similar results, indistinguishable within plotting accuracy. The results presented here were obtained using Roe's method. The inlet conditions correspond to local equilibrium at a temperature of $T_{in} = 4000$ K, a density of $\rho_{in} = 8.6756 \times 10^{-2}$ kg/m^3, and a velocity of $V_{in} = 2000$ m/s.

The effects of assuming a frozen chemistry or a local chemical equilibrium for the same problem are shown in Fig. 22.1, where an equilibrium thermodynamic model was used. The air chemistry model involves 11 species and 47 chemical reactions. The temperature predictions for the frozen and equilibrium assumptions are in strong disagreement, with the finite-rate calculations correctly in the middle of the two limiting cases. It can be noticed that some mechanical quantities, such as pressure and velocity, are hardly affected by the differences in the chemical behavior. The mass fractions, however, which are constant for frozen calculations, are greatly affected by the assumption of local chemical equilibrium, as seen in Fig. 22.1b. It is noteworthy that for mass fractions the finite-rate results do not always fall within the bounds set by the two limiting cases, in particular for the NO plots. The reason is found in the large nonlinearities of the model, which specifically shows a maximum for the production of NO in a mass fraction versus temperature diagram. In this example, the assumption of local chemical equilibrium would lead to inaccuracies in the numerical predictions.

22.10.2 Space Shuttle Main Engine Nozzle

Unlike the first, this example yields good comparisons with experimental data when local chemical equilibrium is employed. It involves a simulation of the hydrogen/oxygen flow inside the SSME nozzle with the hydrogen/oxygen flow being represented by six species, namely O_2, H_2, OH, H_2O, O, and H. A second-order accurate space discretization with the Van Leer limiter, in conjunction with a Roe solver,

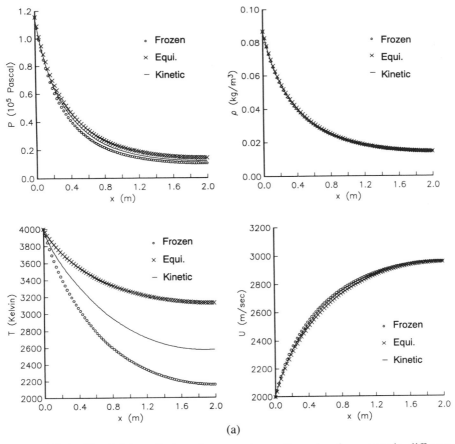

FIGURE 22.1 Flowfield predictions in a quasi-one-dimensional supersonic diffuser. (a) Pressure, density, temperature, and velocity. (b) Mass fractions of N_2, O_2, NO, and O. [From Cinnella (1989).]

was used for the calculation [Cox and Cinnella (1992)]. The geometry of the SSME nozzle and the 88 × 31 grid used for the inviscid computations were obtained by Wang and Chen (1990). Combustion chamber conditions correspond to 100% power at sea level and are: hydrogen/oxygen ratio of 6.0, Mach number $M_c = 0.2$, temperature $T_c = 3639$ K, and pressure $p_c = 20.24 \times 10^6$ Pa. Simulations were made using both the perfect gas model with $\gamma = 1.18$, and the 6-species hydrogen/oxygen model.

The temperature profiles along the centerline and wall of the nozzle are given in Fig. 22.2. The axial distances are referenced to the throat location. Higher temperatures for the reacting solution occur starting at a point just past the throat. The flow in the diverging portion of the nozzle is dominated by recombination reactions, affecting hydrogen and oxygen, and successive combustion of the recombined species. Both sets of reactions are exothermic in nature. The specific impulse of the SSME nozzle was computed to be 460.0 seconds, which is in excellent agreement with the value 460.4 seconds predicted by Wang and Chen (1990) and with experimental data.

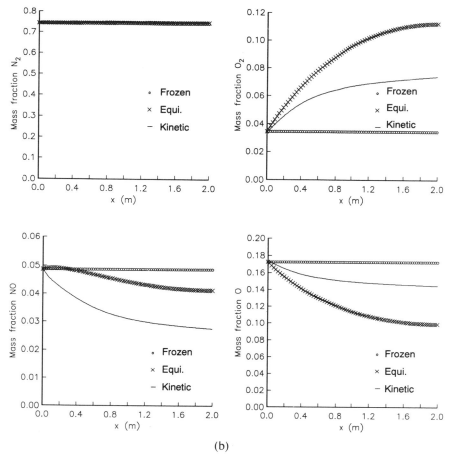

(b)

FIGURE 22.1 (*Continued*)

22.10.3 Kinetic Flux Vector Splitting Performance

Comparisons have been made between the KFVS results and those of the Van Leer flux-splitting and Roe flux-differencing schemes [Eppard and Grossman (1993a)]. The calculations were performed using a two-dimensional/axisymmetric finite-volume Navier–Stokes code for chemical and thermal nonequilibrium. In all cases, MUSCL-type differencing is used in conjunction with either the Van Albeda (smooth) or MIN-MOD limiters.

The example presented here involves the $M_\infty = 5$ inviscid, two-dimensional calculation of reacting air through a supersonic inlet. The inlet walls are ramps composed of a 10-degree compression followed by a 10-degree expansion and are symmetric with respect to the centerline. The air chemistry model consists of 5 species and 17 reactions. The freestream temperature is $T_\infty = 3573$ K, and the initial species densities in kg/m^3 are 0.0651 for N$_2$, 0.00860 for O$_2$, 0.00556 for NO, 1.22×10^{-5} for N, and 0.00900 for O. Figure 22.3(a) and (b) show the pressure and temperature distributions on the lower wall for the KFVS, Roe, and Van Leer schemes on a 101 × 51 grid. The pressure distributions are nearly identical except in the pre-shock region, where all three schemes exhibit a numerical overshoot. The KFVS overshoot

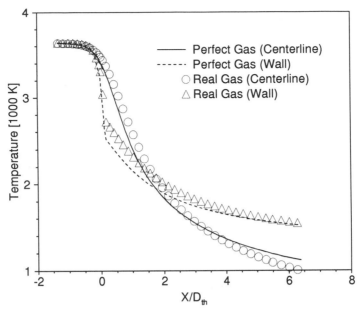

FIGURE 22.2   Temperature versus distance in the SSME nozzle. (Reprinted with permission, Cox, C. F. and Cinnella, P., "General Solution Procedure for Flows in Local Chemical Equilibrium," *AIAA J.*, Vol. 32, No. 3, pp. 519–527, 1994.)

is slightly less than that of the Van Leer scheme and slightly larger than the overshoot produced by the Roe scheme. The temperature distributions are also very similar. The Roe scheme solution exhibits the largest pre-shock oscillation, while again the KFVS and Van Leer schemes produce nearly identical results. The Roe scheme also produces a higher value of temperature past the expansion compared to the two flux-split schemes.

The KFVS scheme has been compared with the flux-vector scheme of Van Leer and the flux-difference splitting scheme of Roe for a series of test cases by Eppard and Grossman (1993a). In all cases, the KFVS scheme compared very closely with the Van Leer scheme, showing excellent shock capturing properties. However, in the quasi-one-dimensional results, the KFVS scheme appears to smear the shock slightly more than the Roe scheme for a first-order calculation. Higher-order calculations seem to agree much better. The Roe scheme solution for the supersonic viscous flow over a cone exhibits much better grid convergence and is much more accurate in the boundary layer than either the KFVS or the Van Leer scheme. The results show that even though the KFVS scheme is a Riemann solver at the kinetic level, its behavior at the Euler level is more similar to the existing flux-vector splitting algorithms that to the flux-difference scheme of Roe.

22.10.4   Structure of a Shock Wave

Shock-structure calculations were performed using an implicit Steger–Warming algorithm, with e_{n_s} corresponding to the nonequilibrium portion of the energy due to rotational relaxation, by Grossman *et al.* (1992). The vibrational energy contribution

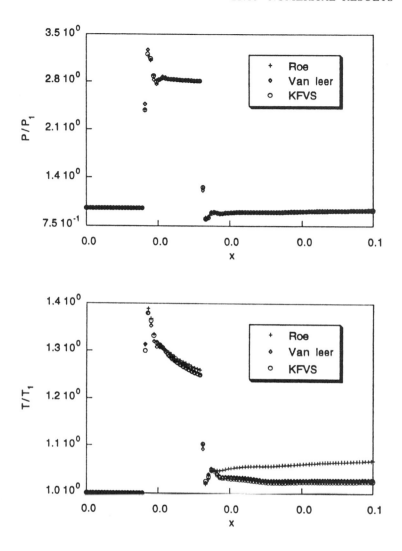

FIGURE 22.3 Flowfield predictions on a 10° ramp with air chemistry. (Reprinted with permission, Eppard, W. M. and Grossman, B., "An Upwind Kinetic Flux-Vector Splitting Method for Flows in Chemical and Thermal Non-Equilibrium," AIAA Paper No. 93-0894, 1993a.)

to the internal energy was neglected for two reasons: only cases where the initial conditions are such that vibrational energy is negligible were considered, and for most shocks the vibrational energy mode requires many collisions to reach equilibrium (on the order of 10^5) and is essentially frozen throughout the shock. The electronic energy contribution to the internal energy was also neglected.

A third-order, upwind-biased extrapolation (in MUSCL form) was used for the inviscid fluxes, while a first-order, upwind extrapolation was used for the inviscid Jacobians appearing in the implicit representation. No limiting was necessary for any of the calculations presented. It is noted that Navier–Stokes equations calculations required a block-tridiagonal inversion at each time step, while Burnett equations calculations required a block-pentadiagonal inversion at each time step.

The one-dimensional solutions are obtained on a grid consisting of 150 equally spaced mesh points. Characteristic boundary conditions were used at the outflow, with the pressure at the exit set at that behind a normal shock. In the calculations presented here, the upstream conditions are denoted by the subscript 1, and the downstream state is denoted by the subscript 2. The distances are referenced to an upstream (hard-sphere) mean-free path $\lambda_1 = 16\mu_1/[5\rho_1(2\pi RT_1)^{0.5}]$. Moreover, for all calculations, the coefficients in the Burnett equations are those corresponding to a hard sphere gas. For the rotational nonequilibrium model in nitrogen the values used for the constants introduced in Sec. 22.8 are: $Z_r^\infty = 18$, and $T_r = 91.5$ K. Density and temperature profiles are presented in a normalized fashion as $(\rho - \rho_2)/(\rho_2 - \rho_1)$ and $(T - T_2)/(T_2 - T_1)$.

Temperature profiles for diatomic nitrogen at Mach 11 are shown in Fig. 22.4a and b. Figure 22.4a shows calculations with the Navier–Stokes equations alone, the Navier–Stokes equations with rotational nonequilibrium, and the Navier–Stokes equations with bulk viscosity, as well as the assumed accurate DSMC result given by Lumpkin (1991). It is seen that the Navier–Stokes calculation alone produces temperature thicknesses that are too small. The calculation with bulk viscosity represents a slight improvement over the Navier–Stokes calculations alone. The calculation with rotational nonequilibrium is able to capture the peak temperature in the shock. However, it can be seen that in all cases the Navier–Stokes terms do not predict large enough temperature thicknesses.

Figure 22.4b shows calculations with the Burnett equations alone, the Burnett equations with rotational nonequilibrium, and the Burnett equations with bulk viscosity, as well as the previous DSMC calculation. It is seen that the Burnett calculation alone also produces temperature thicknesses that are too small. The Burnett calculation with bulk viscosity represents a slight improvement over Burnett alone, and the calculation with rotational nonequilibrium very closely predicts the value and location of the peak temperature in the shock as well as the correct temperature thickness. The Burnett calculation with rotational nonequilibrium most closely matches the DSMC results.

22.10.5 Biofilm Growth in a Circular Pipe

The last example involves the growth of biofilm at the internal surface of a circular pipe carrying water, from Szego et al. (1993). The simple biofilm model consists of one microorganism, *pseudomonas aeruginosa*, and one nutrient, *glucose*. The pipe length is $L = 2$ m, the diameter is $\mathfrak{D} = 4 \times 10^{-3}$ m, and the volume flow rate is $\dot{Q} = 1.4 \times 10^{-6}$ m³/s. One growth reaction is modeled, along with a continuous detachment process that is proportional to the square of the biofilm thickness. An Euler-implicit scheme with a second-order space discretization is employed. The grid size includes 10 points in the biofilm and 21 in the bulk liquid at each axial location. The number of grid points in the axial direction is 20. The results are reasonably grid-converged [Szego et al. (1993)].

The time evolution of the film thickness is plotted in Fig. 22.5. Curve (*a*) shows the thickness at the inlet (actually, in the first column of finite volumes), curve (*b*) is in the middle of the pipe ($x = 1$ m), and curve (*c*) shows the outlet values (the

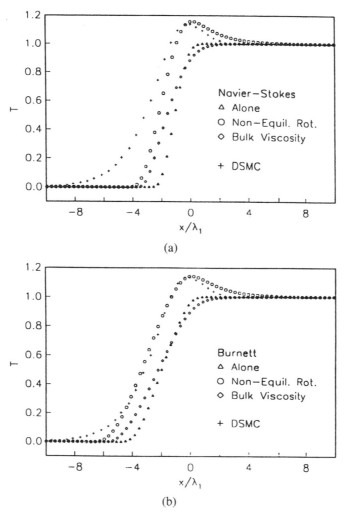

(a)

(b)

FIGURE 22.4 Predictions of the temperature distribution through a normal shock. (a) Navier–Stokes equations calculations. (b) Burnett equations calculations. (Reprinted with permission, Grossman, B., Cinnella, P., and Eppard, W. E., "New Developments Pertaining to Algorithms for Non-Equilibrium Hypersonic Flows," *Comput. Fluid Dyn. J.*, Vol. 1, No. 2, pp. 175–186, 1992.)

last column of finite volumes). It is easy to see that a steady state is reached by the end of the first day. In fact, another run was continued for 6 days of simulated time, and the thickness of the film did not change more than 1%.

Figure 22.6 shows the thickness profiles at different times, namely at 0.1, 0.3, 0.5 and 1 days, along the pipe. In general, the thickness of the film decreases as the distance increases. The reason for this can be found by looking at the steady state concentration profiles of glucose and pseudomonas as a function of the axial distance from the inlet, not shown here. Since the only source of glucose is at the inlet, it gets consumed as the liquid moves in the pipe, thus less and less *food* is available for growth down the pipe.

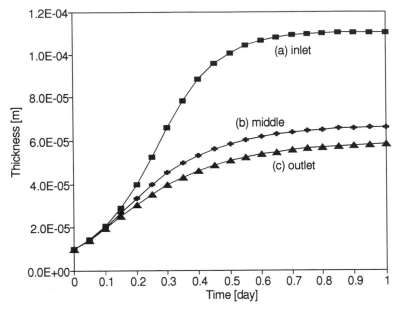

FIGURE 22.5 Biofilm thickness L_F versus time in a circular pipe. (Reproduced with permission, Szego, S., Cinnella, P., and Cunningham, A. B., "Numerical Simulation of Biofilm Processes in Closed Conduits," *J. Comput. Phys.*, Vol. 108, No. 2, pp. 246–263, 1993.)

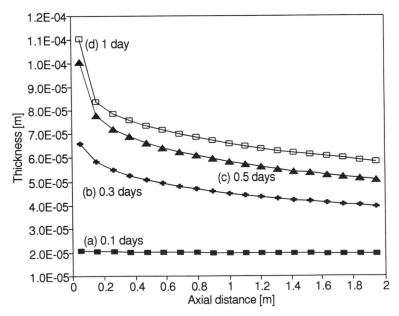

FIGURE 22.6 Biofilm thickness L_F versus axial distance in a circular pipe. (Reproduced with permission, Szego, S., Cinnella, P., and Cunningham, A. B., "Numerical Simulation of Biofilm Processes in Closed Conduits," *J. Comput. Phys.*, Vol. 108, No. 2, pp. 246–263, 1993.)

22.11 CONCLUDING REMARKS

The present chapter has attempted to review some of the current advances in the area of numerical methods for physically challenging flows such as those in thermochemical nonequilibrium. The exposition has been essentially limited to finite-volume techniques and upwind algorithms, although an effort has been made throughout the text to mention other approaches.

Future work in this field is very promising due to the availability of bigger and faster computers and the advances in physical modeling and numerical algorithms. Parallel computers are likely to play a major role in making numerical simulations of reacting flows affordable. A better understanding of turbulent combustion and nonequilibrium aerothermodynamics is necessary, and some progress is being made in this field. Last, but not least, the *open* field of biochemical and biomedical applications involving one- and multiphase fluids, chemical reactions, and interface exchange phenomena is ripe for more systematic advances in the state-of-the-art simulation capabilities [see Board (1993)].

REFERENCES

Abgrall, R., "Extension of Roe's Approximate Riemann Solver to Equilibrium and Non-equilibrium Flows," *Note on Numerical Fluid Mechanics*, Wesserling, P. (Ed.), Vol. 29, pp. 1–10, Vieweg, Braunschweig, FRG, 1989.

Abgrall, R. and Montagné, J. L., "Generalization of Osher's Riemann Solver to Mixture of Perfect Gas and Real Gas," *Rech. Aerosp.*, No. 1989-4, pp. 1–13, 1989.

Aftosmis, M. J. and Baron, J. R., "Adaptive Grid Embedding in Nonequilibrium Hypersonic Flow," AIAA Paper No. 89-1652, 1989.

Anderson, J. D., Jr., "An Engineering Survey of Radiating Shock Layers," *AIAA J.*, Vol. 7, No. 9, pp. 1665–1675, 1969.

Anderson, J. D., Jr., *Hypersonic and High Temperature Gas Dynamics*, McGraw-Hill, New York, 1989.

Anderson, D. A., Tannehill, J. C., and Pletcher, R. H., *Computational Fluid Mechanics and Heat Transfer*, Hemisphere, New York, 1984.

Baldwin, B. S. and Lomax, H., "Thin Layer Approximation and Algebraic Model for Separated Turbulent Flows," AIAA Paper No. 78-257, 1978.

Baysal, O., Engelund, W. C., and Tatum, K. E., "Navier–Stokes Calculations of Scramjet-Afterbody Flowfields," *Symposium on Advances and Applications in Computational Fluid Dynamics*, 1988 Winter Annual Meeting of ASME, Chicago IL, 1988.

Bergamini, L. and Cinnella, P., "A Comparison of "New" and "Old" Flux-Splitting Schemes for the Euler Equations," AIAA Paper No. 93-0876, 1993.

Bhattacharjee, S., Altenkirch, R. A., Srikantaiah, N., and Vedhanayagam, M., "A Theoretical Description of Flame Spreading over Solid Combustibles in a Quiescent Environment at Zero Gravity," *Combust. Sci. and Tech.*, Vol. 69, pp. 1–15, 1990.

Bird, R. B., Stewart, W. E., and Lightfoot, E. N., *Transport Phenomena*, John Wiley & Sons, New York, 1960.

Blottner, F. G., Johnson, M., and Ellis, M., "Chemically Reacting Viscous Flow Program for Multi-Component Gas Mixtures," Report No. SC-RR-70-754, Sandia Laboratories, 1971.

Board, J. A., Jr., "Grand Challenges in Biomedical Computing," *High-Speed Performance in Biomedical Research*, Pilkington, T. *et al.* (Eds.), CRC Press, Boca Raton, FL, 1993.

Bruno, C., "Real Gas Effects," *Hypersonics*, Bertin, J. J., Glowinski, R., and Periaux, J. (Eds.), Vol. 1, pp. 303–354, Birkhäuser, Boston, MA, 1989.

Bussing, T. R. A. and Murman, E. M., "A Finite-Volume Method for the Calculation of Compressible Chemically Reacting Flows," AIAA Paper No. 85-0331, 1985.

Candler, G., "Translation-Vibration-Dissociation Coupling in Nonequilibrium Hypersonic Flows," AIAA Paper No. 89-1739, 1989.

Candler, G. V. and MacCormack, R. W., "The Computation of Hypersonic Ionized Flows in Chemical and Thermal Nonequilibrium," AIAA Paper No. 88-0511, 1988.

Chakravarthy, S. R., Szema, K.-Y., Goldberg, U. C., Gorski, J. J., and Osher, S., "Application of a New Class of High Accuracy TVD Schemes to the Navier–Stokes Equations," AIAA Paper No. 85-0165, 1985.

Chang, S.-H., "On the Application of Subcell Resolution to Conservation Laws with Stiff Source Terms," NASA TM No. 102384, 1989.

Characklis, W. G. and Marshall, K. C. (Eds.), *Biofilms*, John Wiley & Sons, New York, 1990.

Characklis, W. G., Bouwer, E., Gujer, W., Hermanowicz, S., Wanner, O., Watanabe, Y., and Wilderer, P., "Modelling of Biofilm Systems," IAWPRC Scientific and Technical Report, 1989.

Cinnella, P., "Flux-Split Algorithms for Flows with Nonequilibrium Chemistry and Thermodynamics," Ph.D. Dissertation, Virginia Polytechnic Institute and State University, Blacksburg, VA, 1989.

Cinnella, P. and Cox, C. F., "Robust Algorithms for the Thermo-Chemical Properties of Real Gases," *Comput. Fluid Dyn. J.*, Vol. 1, No. 2, pp. 143–154, 1992.

Cinnella, P. and Elbert, G. J., "Two-Dimensional Radiative Heat Transfer Calculations for Flows in Thermo-Chemical Non-Equilibrium," AIAA Paper No. 92-0121, 1992.

Cinnella, P. and Grossman, B., "Flux-Split Algorithms for Hypersonic Flows," *Computational Methods in Hypersonic Aerodynamics*, Murthy, T. K. S. (Ed.), pp. 153–202, Computational Mechanics Publications, Southampton, UK, 1991.

Clarke, J. F. and McChesney, M., *The Dynamics of Real Gases*, Butterworth & Co. Ltd., London, 1964 (second edition under the title *The Dynamics of Relaxing Gases*, 1975).

Clavin, P., "Premixed Combustion and Gasdynamics," *Ann. Rev. Fluid Mech.*, Vol. 26, pp. 321–352, 1994.

Cox, C. F. and Cinnella, P., "General Solution Procedure for Flows in Local Chemical Equilibrium," *AIAA J.*, Vol. 32, No. 3, pp. 519–527, 1994.

Curtiss, C. F. and Hirschfelder, J. O., "Transport Properties of Multicomponent Gas Mixtures," *J. Chem. Phys.*, Vol. 17, No. 6, 1949.

Dadone, A. and Grossman, B., "A Rotated Upwind Scheme for the Euler Equations," *AIAA J.*, Vol. 30, No. 10, pp. 2219–2226, 1992.

Davis, S. F., "A Rotationally-Biased Upwind Difference Scheme for the Euler Equations," *J. Comput. Phys.*, Vol. 56, pp. 65–92, 1984.

Dervieux, A. and Larrouturou, B. (Eds.), *Numerical Combustion*, Lecture Notes in Physics, Springer-Verlag, New York, 1990.

Deshpande, S. M., "On the Maxwellian Distribution, Symmetric Form, and Entropy Conservation for the Euler Equations," NASA Technical Publication No. P-2583, 1986.

Deshpande, S. M. and Mandal, J. C., "Kinetic Flux Vector Splitting (KFVS) for the Euler Equations," Fluid Mechanics Report No. 87-FM-2, Dept. of Aerospace Eng., Indian Institute of Science, Bangalore, India, 1987.

Drummond, J. P., Hussaini, M. J., and Zang, T. A., "Spectral Methods for Modeling Supersonic Chemically Reacting Flowfields," *AIAA J.*, Vol. 24, No. 9, pp. 1461–1467, 1986.

Eklund, D. R., Drummond, J. P., and Hassan, H. A., "The Efficient Calculation of Chemically Reacting Flow," AIAA Paper No. 86-0563, 1986.

Elbert, G. J. and Cinnella, P., "Axisymmetric Radiative Heat Transfer Calculations for Flows in Chemical Non-Equilibrium," AIAA Paper No. 93-0139, 1993.

Elizarova, T. G. and Chetverushkin, B. N., "Kinetic Algorithms for Calculating Gas Dynamic Flows," *U.S.S.R. Comput. Maths. Math. Phys.*, Vol. 25, No. 5, pp. 164–169, 1985.

Eppard, W. M. and Grossman, B., "An Upwind Kinetic Flux-Vector Splitting Method for Flows in Chemical and Thermal Non-Equilibrium," AIAA Paper No. 93-0894, 1993a.

Eppard, W. M. and Grossman, B., "A Multi-Dimensional Kinetic-Based Upwind Solver for the Euler Equations," AIAA Paper No. 93-3303-CP, 1993b.

Fletcher, C. A. J., *Computational Techniques for Fluid Dynamics*, Vol. 1, Springer-Verlag, New York, 1988.

Glaister, P., "An Approximate Linearised Riemann Solver for the Three-Dimensional Euler Equations for Real Gases Using Operator Splitting," *J. Comput. Phys.*, Vol. 77, pp. 361–383, 1988.

Gnoffo, P. A., Gupta, R. N., and Shinn, J. L., "Conservation Equations and Physical Models for Hypersonic Air Flows in Thermal and Chemical Nonequilibrium," NASA Technical Paper No. 2867, 1989.

Godfrey, A. G., Walters, R. W., and Van Leer, B., "Preconditioning for the Navier–Stokes Equations with Finite-Rate Chemistry," AIAA Paper No. 93-0535, 1993.

Goldstein, S., "The Navier–Stokes Equations and the Bulk Viscosity of Simple Gases," *J. Math. Phys. Sci.*, Vol. 6, pp. 225–261, 1972.

Grossman, B. and Cinnella, P. "Flux-Split Algorithms for Flows with Nonequilibrium Chemistry and Vibrational Relaxation," *J. Comput. Phys.*, Vol. 88, No. 1, pp. 131–168, 1990.

Grossman, B. and Walters, R. W., "Flux-Split Algorithms for the Multi-Dimensional Euler Equations with Real Gases," *Comput. and Fluids*, Vol. 17, No. 1, pp. 99–112, 1989.

Grossman, B., Cinnella, P., and Eppard, W. M., "New Development Pertaining to Algorithms for Non-equilibrium Hypersonic Flows," *Comput. Fluid Dyn. J.*, Vol. 1, No. 2, pp. 175–186, 1992.

Harten, A., "ENO Schemes with Subcell Resolution," *J. Comput. Phys.*, Vol. 83, pp. 148–184, 1989.

Hartung, L. C. and Hassan, H. A., "Radiation Transport around Axisymmetric Blunt Body Vehicles Using a Modified Differential Approach," AIAA Paper No. 92-0119, 1992.

Heaslet, M. A. and Baldwin, B. S., "Predictions of the Structure of Radiation-Resisted Shock Waves," *Phys. Fluids*, Vol. 6, No. 6, pp. 781–791, 1963.

Heimerl, J. M. and Coffee, T. P., "Results of a Study of Several Transport Algorithms for Premixed, Laminar Flame Propagation," *Numerical Methods in Laminar Flame Propagation*, Peters, N. and Warnatz, J. (Eds.), Vieweg & Sohn, Braunschweig, 1982.

Hirsch, C., *Numerical Computation of Internal and External Flows*, Vol. 2, John Wiley & Sons, Chichester, UK, 1990.

Hirsch, C. and Lacor, C., "Upwind Algorithms Based on a Diagonalization of the Multi-dimensional Euler Equations," AIAA Paper No. 89-1958, 1989.

Hirschfelder, J. O., Curtiss, C. F., and Bird, R. B., *Molecular Theory of Gases and Liquids*, John Wiley & Sons, New York, 1954.

Iannelli, G. S. and Baker, A. J., "A Stiffly-Stable Implicit Runge-Kutta Algorithm for CFD Applications," AIAA Paper No. 88-0416, 1988.

Larrouturou, B., "How to Preserve the Mass Fraction Positivity when Computing Compressible Multi-Component Flows," *J. Comput. Phys.*, Vol. 95, pp. 59–84, 1991.

Lee, J.-H., "Basic Governing Equations for the Flight Regimes of Aeroassisted Orbital Transfer Vehicles," *Prog. Astronaut. Aeronaut.*, Vol. 96, pp. 3–53, 1985.

LeVeque, R. J. and Yee, H. C., "A Study of Numerical Methods for Hyperbolic Conservation Laws with Stiff Source Terms," NASA TM No. 100075, 1988.

Levy, D. W., Powell, K. G., and Van Leer, B., "An Implementation of a Grid-Independent Upwind Scheme for the Euler Equations," AIAA Paper No. 89-1931-CP, 1989.

Liou, M.-S., Van Leer, B., and Shuen, J.-S., "Splitting of Inviscid Fluxes for Real Gases," NASA TM No. 100856, 1988.

Liu, Y. and Vinokur, M., "Upwind Algorithms for General Thermo-Chemical Nonequilibrium Flows," AIAA Paper No. 89-0201, 1989a.

Liu, Y. and Vinokur, M., "Equilibrium Gas Flow Computations. I. Accurate and Efficient Calculations of Equilibrium Gas Properties," AIAA Paper No. 89-1736, 1989b.

Lumpkin, F. E., III, "Accuracy of the Burnett Equations for Hypersonic Real Gas Flows," AIAA Paper No. 91-0771, 1991.

Mallard, W. G., Westley, F., Herron, J. T., Hampson, F. R., and Frizzell, D. H., *NIST Chemical Kinetics Database: Version 5.0*, National Institute of Standards and Technology, Gaithersburg, MD, 1993.

Mandal, J. L. and Deshpande, S. M., "Higher Order Accurate Kinetic Flux Vector Splitting Method for Euler Equations," *Notes on Numerical Fluid Mechanics*, Vol. 24, pp. 384–392, Vieweg, Braunschweig, 1989.

Marrone, P. V. and Treanor, C. E., "Chemical Relaxation with Preferential Dissociation from Excited Vibrational Levels," *Phys. Fluids*, Vol. 6, No. 9, pp. 1215–1221, 1963.

Marvin, J. G., "CFD Validation for NASP," Paper No. 5, 7th National Aero-Space Plane Symposium, 1989.

Morrison, J., "Flux Difference Split Scheme for Turbulent Transport Equations," AIAA Paper No. 90-5251, 1990.

Oldenborg, R., Chinitz, W., Friedman, M., Jaffe, R., Jachimowski, C., Rabinowitz, M., and Schott, G., "Hypersonic Combustion Kinetics," NASA TM No. 1107, 1990.

Olfe, D. B., "The Influence of Radiant-Energy Transfer on One-Dimensional Shock Wave Propagation," *Supersonic Flow, Chemical Processes and Radiative Transfer*, Olfe, D. B. and Zakkay, V. (Eds.), Macmillan, New York, 1964.

Oran, E. S. and Boris, J. P., *Numerical Simulation of Reactive Flow*, Elsevier, Amsterdam, 1987.

Park, C., *Nonequilibrium Hypersonic Aerothermodynamics*, John Wiley & Sons, New York, 1991.

Park, C. and Yoon, S., "A Fully-Coupled Implicit Method for Thermo-Chemical Nonequilibrium Air at Sub-Orbital Flight Speeds," AIAA Paper No. 89-1974, 1989.

Parpia, I. and Michalek, D., "A Shock Capturing Method for Multidimensional Flow," AIAA Paper No. 90-3016-CP, 1990.

Peyret, R. and Viviand, H., "Pseudo-Unsteady Methods for Inviscid or Viscous Flow Computation," *Recent Advances in the Aerospace Sciences*, Casci, C. and Bruno, C. (Eds.), Plenum, New York, 1985.

Pope, S. B., "PDF Methods for Turbulent Reactive Flows," *Prog. Energy Comb. Sci.*, Vol. 11, pp. 119–192, 1985.

Roe, P. L., "Approximate Riemann Solvers, Parameter Vectors, and Difference Schemes," *J. Comput. Phys.*, Vol. 43, pp. 357–372, 1981.

Rumsey, C. L., Van Leer, B., and Roe, P. L., "Effect of a Multi-Dimensional Flux Function on the Monotonicity of Euler and Navier–Stokes Computations," AIAA Paper No. 91-1530-CP, 1991.

Simon, R., "The Conservation Equations of a Classical Plasma in the Presence of Radiation," *J. Quant. Spectrosc. Radiat. Transfer*, Vol. 3, pp. 1–14, 1963.

Srinivasan, S., Tannehill, J. C., and Weilmuenster, K. J., "Simplified Curve Fits for the Thermodynamic Properties of Equilibrium Air," NASA Research Publication No. 1181, 1987.

Steger, J. L. and Warming, R. F., "Flux Vector Splitting of the Inviscid Gasdynamic Equations with Applications to Finite-Difference Methods," *J. Comput. Phys.*, Vol. 40, pp. 263–293, 1981.

Struijs, R., Deconinck, H., De Palma, P., Roe, P. L., and Powell, K. G., "Progress on Multidimensional Upwind Euler Solvers for Unstructured Grids," AIAA Paper No. 91-1550-CP, 1991.

Sweby, P. K., "High Resolution Schemes Using Flux Limiters for Hyperbolic Conservation Laws," *SIAM J. Numer. Anal.*, Vol. 21, No. 5, pp. 995–1011, 1984.

Szego, S., Cinnella, P., and Cunningham, A. B., "Numerical Simulation of Biofilm Processes in Closed Conduits," *J. Comput. Phys.*, Vol. 108, No. 2, pp. 246–263, 1993.

Taylor, L. K. and Whitfield, D. L., "Unsteady Three-Dimensional Incompressible Euler and Navier–Stokes Solver for Stationary and Dynamic Grids," AIAA Paper No. 91-1650, July 1991.

Thomas, P. D. and Lombard, C. K., "Geometric Conservation Law and Its Application to Flow Computations on Moving Grids," *AIAA J.*, Vol. 17, No. 10, pp. 1030–1037, 1979.

Treanor, C. E. and Marrone, P. V., "Effect of Dissociation on the Rate of Vibrational Relaxation," *Phys. Fluids*, Vol. 5, No. 9, pp. 1022–1026, 1962.

Turkel, E., "Preconditioned Methods for Solving the Incompressible and Low Speed Compressible Equations," *J. Comput. Phys.*, Vol. 72, pp. 277–298, 1987.

Turkel, E., "Review of Preconditioning Methods for Fluid Dynamics," ICASE Report No. 92-47, NASA Langley Research Center, Langley VA, 1992.

Van Leer, B., "Flux-Vector Splitting for the Euler Equations," *Lecture Notes in Physics*, Vol. 170, pp. 507–512, Springer-Verlag, New York, 1982.

Van Leer, B., "Progress in Multidimensional Upwinding," *Lecture Notes in Physics*, Vol. 414, pp. 1–26, Springer-Verlag, New York, 1993.

Van Leer, B., Thomas, J. L., Roe, P. L., and Newsome, R. W., "A Comparison of Numerical Flux Formulas for the Euler and Navier–Stokes Equations," AIAA Paper No. 87-1104-CP, 1987.

Van Leer, B., Lee, W.-T., and Roe, P. L., "Characteristic Time-Stepping or Local Preconditioning of the Euler Equations," AIAA Paper No. 91-1552-CP, 1991.

Vincenti, W. G. and Kruger, C. H., Jr., *Introduction to Physical Gas Dynamics*, Krieger, Malabar, FL, 1986.

Vinokur, M., "Flux Jacobian Matrices and Generalized Roe Average for an Equilibrium Real Gas," NASA CR No. 117512, 1988.

Vinokur, M. and Montagné, J.-L., "Generalized Flux-Vector Splitting for an Equilibrium Real Gas," NASA CR No. 117513, 1988.

Walters, R. W. and Thomas, J. L., "Advances in Upwind Relaxation Methods," *State-of-the-Art Surveys on Computational Mechanics*, Noor, A. K. (Ed.), ASME, New York, 1988.

Walters, R. W., Cinnella, P., Slack, D. C., and Halt, D., "Characteristic-Based Algorithms for Flows in Thermo-Chemical Nonequilibrium," *AIAA J.*, Vol. 30, No. 5, pp. 1304–1313, 1992.

Wang, K. C., "The 'Piston Problem' with Thermal Radiation," *J. Fluid Mech.*, Vol. 20, Part 3, pp. 447–455, 1964.

Wang, T. and Chen, Y., "A Unified Navier–Stokes Flowfield and Performance Analysis of Liquid Rocket Engines," AIAA Paper No. 90-2494, 1990.

Wanner, O. and Gujer, W., "A Multispecies Biofilm Model," *Biotechnol. Bioeng.*, Vol. 28, pp. 314–328, 1986.

Weatherill, N. P., Mathur, J. S., Natakusumah, D. K., and Marchant, M. J., "An Upwind Kinetic Flux Vector Splitting Method on General Mesh Topologies," *Proceedings of the 4th International Symposium on Computational Fluid Dynamics*, Davis, CA, pp. 1192–1197, 1991.

Whitfield, D. L., Janus, J. M., and Simpson, L. B., "Implicit Finite Volume High Resolution Wave-Split Scheme for Solving the Unsteady Three-Dimensional Euler and Navier–Stokes Equations on Stationary or Dynamic Grids," EIRS Report No. MSSU-EIRS-ASE-88-2, Mississippi State University, 1988.

Wilke, C. R., "A Viscosity Equation for Gas Mixtures," *J. Chem. Phys.*, Vol. 18, No. 4, 1950.

Withington, J. P., Shuen, J.-S., and Yang, V., "A Time-Accurate, Implicit Method for Chemically Reacting Flows at All Mach Numbers," AIAA Paper No. 91-0581, 1991.

Zhigulev, V. N., Romishevskii, Y. A., and Vertushkin, V. K., "Role of Radiation in Modern Gasdynamics," *AIAA J.*, Vol. 1, No. 6, pp. 1473–1485, 1963.

Zong, X., MacCormack, R. W., and Chapman, D. R., "Stabilization of the Burnett Equations and Application to High-Altitude Hypersonic Flows," AIAA Paper No. 91-0070, 1991.

Index